Painting in Stone

Painting in Stone

Architecture and the Poetics of Marble from Antiquity to the Enlightenment

Fabio Barry

Yale University Press New Haven and London

yalebooks.com/art

Designed by Leslie Fitch and Tina Henderson
Printed in China by 1010 Printing International Limited

Library of Congress Control Number: 2019940767
ISBN 978-0-300-24816-6 (hardcover); 978-0-300-24817-3 (paperback)

A catalogue record for this book is available from the British Library.

This paper meets the requirements of ANSI/NISO Z39.48-1992 (Permanence of Paper).

10 9 8 7 6 5 4 3 2 1

Frontispiece: Stephen Cox, *Dreadnought: Problems of History—The Search for the Hidden Stone* (1990-ongoing). "Brecciated" porphyry block found at Mons Porphyrites (Gebel Dokhan, Egypt), as displayed at Chatsworth House, 2015. Dimensions: 106 1/4 x 35 1/2 x 35 1/2 in. (270 x 90 x 80 cm). Vertical (chisel?) marks are Roman workings, possibly for a standing statue. Cox has drilled out the greyer xenoliths and repaired a fracture with concealed pins and a gilded epoxy resin seam (in both cases to evoke ancient micro-mining of gold ore).

Endleaves: Detail of bookmatched Proconnesian panels (eroded), west exterior wall of the Treasury of San Marco, Venice. Merged 3D render of color and relief information by Factum Arte, Madrid (Image: Factum Foundation).

For Peter Carl

Contents

Acknowledgments

The ideas in this book germinated in a quickly drafted article, when Joseph Imorde gave me ten days to plug a gap in a themed issue (on architecture and color) of the now-defunct periodical *Daidalos*. The subject of colored marbles was just as quickly forgotten, only to be revived as a dissertation topic at Columbia University on the urging of the late Hilary Ballon. In those early days I was only interested in Early Modern architecture, but fortunately Thomas Dale opened my eyes to the whole field of Byzantine art, David Rosand introduced me to Venice, Richard Brilliant revived my interest in ancient Rome, Anthony Grafton introduced me to sympathetic magic and much besides, and I began dimly to perceive the scope of the subject. My doctoral advisor, Joseph Connors, enthusiastically encouraged me to pursue the topic and proved to be both a model of sage restraint and an enlivening spark; in the final stages he was also a merciless yet humorous enforcer and castigator of my then embroidered style. Above all, I took to heart his words, "You must keep growing," and it is in this spirit that the present text came to its ultimate form.

It would have been hard to develop my ideas without the two happy years that I spent at the National Gallery of Art in Washington, D.C.: first as David E. Finley Fellow at the Center for Advanced Studies in the Visual Arts under the aegis of Elizabeth Cropper, and then the illuminating experience of working as an intern in the Department of Sculpture and Decorative Arts for Nicholas Penny, who wore his erudition lightly and made me look hard. I must also thank Faya Causey, Caroline Elam, Alison Luchs, and Debra Pincus for all their support and suggestions—Caroline Elam also for saving me from a few howlers and vetting my Latin translations. In Rome, which was a recurrent base for research, I had the good fortune to profit from the friendship and erudition of Robert Coates-Stephens, who read early drafts and never forgot to mention, e-mail, and even text the most recondite primary sources or notices of archaeological remains that he thought might interest me. Without his generosity this book would not have been so rich in detail and would probably have been finished some time earlier. In the process I rediscovered my love for Greek and Latin and rekindled my youthful flirtation with archaeology. My original intention had been to study marble symbolism over only two or three centuries (the fifteenth through eighteenth), but as I looked for evidence and slowly grasped the millennial continuity of the themes that took possession of me, it grew and grew and grew. While the final product was wide-ranging, it also wandered widely, so I set about publishing articles and book chapters on particular issues that not only required their own mental space, but that also helped recalibrate my trajectory each time I returned to the greater body of writing. Only fragments of that original dissertation are now preserved in this book. A complete rewriting of the text, which eventually entailed halving it in length while adding three new chapters at the same time, took place at Stanford University, where I reveled in the exceptional resources of the library, the encouragement of my colleagues (especially Nancy Troy and Richard Meyer), and the wider frontiers that art historical writing seemed to offer. It was the financial resources provided by the university that enabled the subvention necessary to publish the book, for which I must thank Debra Satz, Dean of the School of

Humanities and Sciences. A week spent on a restoration course at Yale University, and a "lightbulb moment" when I painted a small patch of fresco for the first time in my life, resulted in the entire argument of Chapter 4. Special thanks go to Caroline Cima, who accompanied me on long trips to find and photograph monuments in Italy, Greece, and Turkey, always with unbridled enthusiasm, and whose observations and queries frequently alerted me to some question I had overlooked or that required greater thought. In the last stages of rewriting, I enjoyed the high spirits and learned discussion of Paul Gwynne, whose help with some particularly difficult translations amounted to an advanced Latin course, from which I hope I will never recover.

I have never conducted research that has met with such interest and friendly support as this book. I hope I will be forgiven by those whose names I have forgotten in the following list, but for references, generosity, and goodwill let me thank: Felix Ackerman, Al Acres, Gwen Ajello, Sam Barnish, Sergio Bettini, Elisabeth Bolman, Dorothy Bosomworth, Spike Bucklow, John Bushell, Deniz Calis, Marietta Cambareri, Robin Cormack, Lucos Cozza†, Maarten Delbeke, Rodney Fitzsimons, Beate Fricke, Milette Gaifman, Sharon Gerstel, Michael Greenhalgh, Chiara Guarnieri, Jeffrey Hamburger, James Harper, Ingo Herklotz, Michael Hill, Salima Ikram, Joseph Imorde, Philippa Jackson, Nate Jones, Duncan Keenan-Jones, Meg Koster, Christopher Krebs, Ann Kuttner, Bob La France, Anne Laidlaw, Didier Laroche, Benedict Leca, Manfred Leithe-Jasper, Estelle and Stuart Lingo, Gitte Loenstrup, Julian Luxford, Ruth Macrides†, Henry Maguire, Flavia Marcello, Vassileios Marinis, the inimitable Robin Middleton, John Mitchell, Jennifer Montagu, Andrew Morrogh, Alexander Nagel, Gülru Necipoglu, Robert Nelson, Alona Nitzan-Shiftan, Giorgio Ortolani, Irina Oryshkevich, Steven Ostrow, Alessandro Pierattini, Sheryl Reiss, Eva Renzulli, Michael Roberts, Maddalena Scimemi, Cinzia Sicca Bursill-Hall, R. R. R. Smith, Gavin Stamp†, Paul Tegmeyer, Marjorie S. Venit, Rick Vinograd, Alicia Walker, Nicholas Warner, Tristan Weddigen, Claudia Wedepohl, Richard Wittman, Vitale Zanchettin, and Kaspar Zollikofer. In addition, I must thank Katherine Boller at Yale University Press for accepting the project and for providing sustained encouragement and exemplary professionalism, as well as the rest of the Yale team, including Raychel Rapazza, Heidi Downey, Laura Hensley, Tina Henderson, MaryEllen Oliver, and Mary Mayer.

As Pelé Cox has instructed me that under no circumstance should I express any gratitude for the enormous help she gave me in choosing and assembling the final selection of illustrations, as well as many other matters necessary to get this book to press, I will simply thank her for having a poetic soul and marrying me into the bargain. Let me also remember my daughters, Vanessa and Connie, who long suffered, not always silently, my repeated desires to slip into another medieval church or clamber up another Sicilian hill, often in melting heat, to take "just one more photo," but readily assented when they knew the day would end at a poolside.

I reserve my final thanks for Peter Carl. Peter shaped my architectural thinking as an undergraduate and graduate at the Department of Architecture in the University of Cambridge. It would be better to say that he opened my eyes to what architecture is and how we can usefully speak about it, rooting in me a conviction that the architecture and arts of all eras and cultures assemble in a humane continuum, and therefore that none should be overlooked in their pertinence to the present. Over the intervening years he has continued to be a long-distance companion and interlocutor in my research. Peter read various drafts of this book, sometimes just stray paragraphs, and responded in voluminous e-mails packed with dense commentary and his own insights. They would always lead me down some new path or to a higher ground from which to assess the problem at hand. As even these thanks seem paltry, it is right and fitting to dedicate the book to him as well.

Introduction

This book narrates a history of premodern architecture through its lithic materials, mostly colored marbles, but also white ones, and sometimes hard stones and gems, as well as synthetic substitutes for all of them. It is a biography of a material, written from the viewpoint of a medium, as a model for the history of the imagination. It identifies three salient characteristics that sustained a "lithic imagination" over almost five millennia: marbles as images of substance according to a premodern concept of matter and premodern notions of geology; the perceived indwelling of astral light in telluric stones; and the longue durée that colored marbles constituted a form of natural (hence, divine) painting thanks to their vivacious veining, rainbow palette, and chance images.

The examples range from the Atlantic to the Euphrates, lap the shores of the Mediterranean, and occupy an arc of time from the fourth millennium BC to the early eighteenth century. This book does not presume to be comprehensive, but hopes to be representative; the architecture and artworks under examination are selective to the ends of the narrative, and a great many have been sacrificed to the needs of brevity. However, this book *does* set building within the gamut of the visual arts, as architecture has never existed as a thing apart from them.

The themes investigated were not the conceptual monopolies of Europe or the Mediterranean Basin. For example, the aspiration to carve buildings from a single piece of stone (discussed in Chapter 3) became hard fact all the way from the Hindu Kailasa Temple at Ellora in India (eighth century) to the rock-hewn churches of Lalibela in Ethiopia (thirteenth century). This communion with geology, but also the tight-fitting masonry of Greco-Roman architecture (also in Chapter 3), invites further and accurate comparison with the culture of Inca masonry (imaginatively analyzed in Carolyn Dean's *A Culture of Stone: Inka Perspectives on Rock* [2010]). And one can quickly find global resonances for many other themes that fill this book. For example, the marble table screens that once sat on Chinese scholars' desks or adorned their study walls, inspiring their waking dreams and often inscribed with short poems inviting scrutiny of their markings, belong to the same vein as the whole Western tradition of in-seeing stones discussed throughout this book. However, as much as I would have liked to engage such phenomena, and more, four-five millennia is already a handful for any book, and they will await future study.

But there is another reason they must be excluded from the present book. The geography of the narrative presented here is circumscribed by the historic evolution of marble architecture, which is primarily a Western phenomenon. For most of the period under review, marble architecture simply didn't exist elsewhere. We should take special notice of the incomprehension of the Chinese ambassadors who were confronted with marble buildings for the first time in late antiquity, a sight so alien to them that they could only conclude in astonishment that they must have been made from glass (see Chapter 5). Once we enter the Early Modern period, marble architecture ceases to be a European monopoly. Nonetheless, I barely discuss Islamic architecture, except to mention the affinity between Byzantine and Umayyad revetments, even though intricate marble inlays became as popular in mosques and palaces from Cairo to Istanbul to Delhi as they were anywhere in Western Europe. Cultural exchange also continued, and one must mention one salient episode, the pavilion for public appearances (*jharokhā*) of Shah Jahan, Mughal emperor of India. Around 1648 Shah Jahan had its back wall clad with 318 *pietra dura* panels of birds and flowers, some imported from Florence, that local craftsmen then skillfully imitated, all topped by a central scene of Orpheus playing the fiddle; all of which syncretically assimilated the ruler to Solomon, Aflāṭūn (Plato), and the Qur'ānic David, all of whom exercised musical dominion over the animal world (see Ebba Koch's *Shah Jahan and Orpheus: The Pietre Dure Decoration and the Program of the Throne in the Hall of Public Audiences at the Red Fort of Delhi* [1988]). However, Shah Jahan (and the Taj Mahal) do not feature in this book for the same reasons that I barely examine the Opificio delle pietre dure in Florence nor delve into the *intarsie* of Naples and Palermo. However exciting and delicate and mind-bogglingly skillful these spectacles of artistry are, they are neither geologically allusive nor project the sort of innate painting that this book investigates.

There have been numerous studies of marbles before now, and the publications continue to amass. Among the pioneering works, those by Raniero Gnoli and Gabriele Borghini range widely among the decorative marbles quarried by the Romans, the latter more a historico-petrological guide, the former with an anthology of quotations collected from classical and medieval authors.[1] The bulk of marble literature still largely concerns the Roman period and focuses on specialist issues: quarry identification, stone extraction, the economics of distribution, the fixing and typologies of revetments, and so on.[2] Studies of colored marbles for earlier or later periods are comparatively scanty and tend to pursue "niche" issues: in the medieval era, for example, the uses of spolia or the typologies of tessellated pavements. However, it would be unfair to say that there have not been some strident exceptions. Georges Didi-Huberman's *Fra Angelico: Dissemblance et Figuration* (1990) is particularly memorable for its pioneering and thought-provoking interrogation of the painter's depiction of marble. More recently, a string of monographs and articles on the painting-on-stone genre of late sixteenth- and early seventeenth-century Italy and Europe have also added much to our historical knowledge. Other essays that break the mold of archaeological analysis in favor of more speculative interpretation are engaged in the text and notes of this book.

When it comes to the material's representational role in architectural history, it has long been a default position that costly marbles served the agendas of conspicuous consumption and social prestige. A patronal impulse toward ostentation is often undeniable. However, such interpretations stem from a literary tradition, originating in antiquity and surviving through the medieval era, that attacked marbles as the trappings of wanton luxury, and so moral laxity; the engines were social conservatism in the first instance, notions of Christian humility later on, and puritanical distrust of material display after that—whether the hard-liners were religious or architects. Another argument, that the wherewithal to procure marbles demonstrated political power, has become virtually a mantra. Four centuries saw Imperial Rome import multicolored marbles from every quarter, from the furthest frontiers, and so voluminous was this influx that after the demise of the Western empire (476 AD), the *caput orbis* herself became the quarry of Europe for another millennium. So it is that we are often told that the Roman use of exotic marbles in public buildings represented the reach of empire and, therefore, that periodic revivals of polychrome display amount to *imitatio imperii.*

This book seeks to avoid these well-worn ideas and occasionally debates them, not because they are unfounded but because they are too reductive. The upper echelons participated in a broad range of auto-representation that cannot be generalized into the anodyne concept of "power," which is a legacy of Marxist critique even if the authors are frequently unaware of it.

The usual doctrines of display or political coding also iron out the distinctions and continuities between one culture or epoch and another, and they totally ignore the role of architects and artists. Patrons paid for their pleasures, but they generally left artistic decisions at the discretion of their practitioners, for whom marble was neither only a precious commodity nor some raw material to be consumed. Just as paintings are not simply pigments with relative dollar values, marbles were media capable of summoning a range of imaginative associations and responses. Indeed, more poets praised wall revetment as a true art form than the moralists who reviled it as a costly accessory.

In recent years, as materials and media have begun to take center stage in scholarly investigations of artistic creativity, a more sympathetic attitude has developed toward historical marble uses. This is not the place for a review of all the recent and extremely healthy developments in materiality as a field of study. Space simply does not permit it. As far as marbles go, an often-iconological approach has broached color symbolism and, while this is sometimes useful for particular situations and artifacts, hardly any colored marbles, just like colors themselves, can be said to hold settled significations. One of the rare exceptions is Egyptian porphyry (properly a granite, not a marble), whose identification with imperial purple remained constant over centuries. Even then, the saturated red of porphyry and the chain of association with the "blood" (in truth, glandular secretions) of the mollusk that produced the dye meant that it seemed blood-like too; the understanding of the elemental composition of all matter, not just stones, gave it a particularly fiery identity as well. This book accordingly explores colored marbles in terms of the metaphysical commonalities that valorized them: namely, marbles as premodern images of earthly matter; the indwelling of light; and colored marbles as a form of natural or divine painting.

Preceding all other considerations are premodern notions of the formation of stones and marbles. Today, geologists group stones into the categories of igneous, metamorphic, and sedimentary. Greek science and its pre-Enlightenment posterity taught that marbles were solidifications of the earth's "exhalations," moist vapors or dry gases that were frozen or fired solid by earthly humors, which together accounted for their variously earthy, watery, fiery, or airy appearance. Some stones, gems in particular, seemed to have "trapped" light, which "surfaced" when they were buffed up to a high polish. Uniting these perceptions was the belief that all matter consisted in a cocktail of the four elements (water, earth, fire, air), which were in theory utterly inter-transmutable—alchemy—because they were permutations of "primary matter" under the influence of the four principles (hot, dry, cold, moist). In medieval science these permutations could be encouraged by astral influences, a belief that pointed in the direction of sympathetic magic and the influential consonances between stars and sublunary materials (although the theory largely applied to gems and hard stones, and to marbles only by analogy). Alongside this embryonic geology persevered the idea of Mother Earth, thence "living rock," meaning a rock that was "born" or "engendered," which could even regenerate itself once it had been "pruned" in the quarries. Marble was organic. It was either a vegetable thing, which grew and flowered, or a stony flesh through which veins of life-giving water coursed visibly.

In the dialogue between artist and material, therefore, the animate marble always spoke first. A more distributive agency was at work in creating artworks, whether buildings or paintings, than the imposition of the artist's will on some passive material. Mind did not yet dominate matter. The material might be an author, the artist a respondent, the process a courtship. This conception held many implications for marble use in architectural decoration, where modernist doctrine tacitly perceives "ornament" to be a subtractable addendum to "structure." Instead, a central goal of this book is to trace the materiality of buildings as metamorphic architecture. Marble had more evident liminal qualities than any other material because it presented an internal image of chthonic generation but, with polishing, it acquired a surface slick of astral light. It seemed both a rock and a gem, both of this earth and heavenly, so an ideal medium to bridge between immanence and transcendence. These qualities recommended the use of marbles in temples and churches that wished to be heavens on earth, also in the palaces of rulers who considered themselves (semi-)divine and aspired to haunt heavenly abodes, and even in an object that one could hold in the hand.

Marble was an art form too, of its own nature. Marbles that were colored and had stirring veining evoked painting in both aspect and—because of the stone's mobile medium—process. Marbles especially resembled paintings when they seemed to figure images of men, animals, trees, and the like, painted by the hidden hand of Nature, either for her own delight or as a divine proxy. The symmetries released by the technique of book-matching also seemed to disclose images to the receptive observer, as if by accident or perhaps divine design. From antiquity to the seventeenth century, slabs of such "natural painting," book-matched or not, were erected in counterpoint to meticulous inlays of marble marquetry (*opus sectile*), inviting a comparison between the skill of man and that of nature, and both profiting by the comparison. When painters eventually began painting directly onto marble slabs in the sixteenth century, one of the main attractions, as more than one observer remarked, was that because the painter must adapt his composition to the existing veining, then "the skill of the artist played with the art of nature."

As the marbling of church interiors was revived in the sixteenth century, and became widespread in the seventeenth, emboldened artists, particularly Gian Lorenzo Bernini, exploited colored marbles to enable an intrinsically "painted" architecture. Thanks to its dual nature, both painting and stone, this new architecture spanned between the facts and fictions of sculpture and painting to compose a seamless intermediality that guaranteed the autonomy of the church or chapel as a parallel world. Moreover, in works such as the S. Andrea al Quirinale, and then the German and Eastern European churches that emulated its effects in *Stuckmarmor* (faux marble), the triumvirate of divine light, gassy geology, and natural painting blended to represent and embody an otherworldly cloud architecture as elusive as the building was tangible.

With the gradual erosion of the four-element theory, the advent of the periodic table, and the birth of geology and geochemistry as analytical sciences, the poetic imaginings that fill this book were sacrificed and lost to folklore. To visualize premodern attitudes toward marbles there are few modern scholarly models. A precocious example, from the 1930s and 1940s, is Gaston Bachelard (1884–1962), a philosopher of science, who coined the idea of a "material imagination" and divided perceptions of substance, as though it were some unitary matter, under the traditional rubric of the four elements. Bachelard proposed a cognition of matter (which he likened to a *reverie*, or "waking dream") that preceded intellectual apprehension, in which "direct images of matter" took precedence over form, facture, or function.[3] In terms of artistic production, Bachelard's "material imagination" seeks out the more immediate media(tions) between making and perceiving, and this approach is particularly suggestive for pre-Enlightenment culture, before the distancing of elemental evocations and poetic analogies by the scientific emphasis on empirical properties. We have now lost much of the intimate contact with materials that existed before the era of mechanization and mass production, whose standardization has distanced, if not erased, the natural origins of the material, while the resulting view of the manufact as a consumer article has also removed it from the realm of myth to the domain of the market and product.

Just as significant as the elemental model that Bachelard re-proposed for the perception of materials is the fact that his literary sources are almost exclusively snatches of poetry, one following the other in a stream of association that rambles through the mind. It is no coincidence that the bulk of the texts adduced in this book are also poetic, whether epigrams, sonnets, encomia, panegyrics, epic poetry, or ekphrases. In fact, at the conclusion of writing, this book has become as much a history of poetry on marble as of the marble medium in its own right, and at this stage, I realize, a conversation back across the intervening centuries with those poets. Prose chronicles, building records, inscriptions, correspondence, and notarial documents do appear, whenever they are necessary to historical fact. Likewise, the unguarded or expectant comments of travelers and pilgrims testify to local perceptions or traveling tales, and the commentaries of theologians or art theorists provide us with the ideological bedrock for religious representation and artistic debate. But it is in the realm of poetry that ekphrases pose a particular mode of reenacting the artifact verbally and answer the challenge of constant dialogue with the art object. Poetry, more generally speaking, is the closest textual parallel to artistic intention in the visual arts. It is the hard currency of the imagination that echoes the poetics of artistic creation across media, unburdened by the contingencies of history. The essence of poetry is not only its compression and circumvention of prose in the employ of intuition, that emotional impact that prose struggles to rival, but also its seemingly spontaneous utterance; for poetry is ποίησις, a making.

1 A Medium Foretold

Material Synthesis and Heavenly Stones in the Ancient Near East

At the dawn of writing, before history was yet recorded, kings and priests in the world's earliest cities raised the first buildings in colored stones. At Uruk in Mesopotamia, around 3300 to 3100 BC, the so-called *Steinstifttempel* ("stone-cone temple") was constructed within the Eanna precinct ("house of heavens"), the earthly residence of the heavenly queen, Inanna, who had long ago wrested it from the sky god Anu. The temple walls were cast, layer upon layer, from a cement that set like stone, and this was clad in a thick stucco into which were pressed many thousands of miniature cones in red or black limestone, and white gypsum, which altogether triangulated surface patterns of chevrons and lozenges (fig. 1.1). The same technique was applied at a temple in nearby Eridu, where the heads of the gypsum cones were even covered with copper plating, but all other buildings in the region were made from mud brick alone.[1]

No contemporary texts record the motivations for these displays—in fact, literature had yet to be invented. Uruk and Eridu were ancient even to the Sumerians, who revered them as the first cities built after the Great Flood; the *Sumerian King List* (late third millennium BC) tells us that, even earlier, Eridu had been the spot where human "kingship descended from heaven." The temples were the stuff of legend too, as transpires from the epic *Enmerkar and the Lord of Aratta,* written more than a thousand years after the Uruk temple, but harkening back to "days of yore, when destinies were determined." This epic relates how Enmerkar, king of "Unug-Kulaba" (Uruk), had been "chosen by Inanna in her holy heart from the bright mountain," and beseeched the goddess to compel Aratta, a far-off competitor state probably in northeastern Iran, to provide the precious materials to build her temple in Uruk (the Eanna sanctuary):

> [Let them cut for my sake] polished lapis lazuli from
> its block;
> [Let them work for my sake] the translucent smooth
> lapis lazuli;
> [Let them] build [for my sake] the holy mountain in
> Unug [Uruk]!
> A temple [descended] from heaven—your place
> of worship,
> The shrine Eana—let [Aratta] build that!
> [. . .]
> Let the people of Aratta

> Bring down for me the stones of their hills and mountains
> And build for me the great shrine, erect for me the great abode!
> For my sake, let them make illustrious the great abode, the abode of the gods!
> Let thus my power become evident in Kulab.

For the temple at Eridu, he entreats her:

> Let the Abzu flourish like the holy mountain,
> Let Eridug scintillate like the hill ranges,
> For my sake, let them make shrine Abzu illustrious like silver in the lode![2]

To visit Uruk and Eridu today is to enter a dustbowl; but the cities were originally founded in wetlands, and, before the course of the Euphrates shifted, its waters lapped the walls of Uruk and traversed the city as canals. To Sumerians, the marshes of Uruk and Eridu suggested the primeval watery abyss of the creation myths, from which the mound of the earth first emerged, and they were far detached in space and mind from the mountains where the limestone was quarried. The mountains were the giants on the horizon, the peaks where the gods might alight, and the ends of the earth on which the vault of the sky rested.

It is often assumed that the pointillist patterns of the temple walls of Uruk and Eridu imitate woven fabrics or hanging reed mats, but this hypothesis brings us little closer to understanding their significance. The pyramid patterns, which easily devolve into zigzags, may stand for mountains, like the clustered peaks on which Inanna stands in the famous Adda Seal (mid-second millennium BC). Indeed, in a later mural (c. 1750/1700 BC), in the Temple of Ištar (the Akkadian name for Inanna), within the Palace of Zimri-Lim at Mari (Abu Kamal, Syria), we see the moon god Sin receiving a libation from the king while seated on a veritable rainbow of hillocks, which closely resemble the polychrome walls at Uruk (fig. 1.2).[3] Perhaps the various patterns are also geometrical notations for

FIG. 1.1. Stone-cone façade, High Gate, Eanna precinct, Uruk, c. 3300–3100 BC. Vorderasiatisches Museum, Berlin.

FIG. 1.2. *Sin Receives a Libation from the King Whilst Seated in the Mountains*, c. 1750/1700 BC. Watercolor after mural. Temple of Ištar, Palace of Zimri-Lim at Mari (Parrot 1958b).

stone veining, yet the overall ambition was to create a flickering polychromy that made nuggets out of its bricks, and made the wall "shine like the silver in the lode."

This fastidious masonry did prosper, reaching southern Syria and even as far afield as the Nile delta, but in every case the cones were now made from clay and their bases painted black, red, and white, to simulate stone.[4] While there were older stone buildings elsewhere (for example, the megalithic temples of Malta, 3600–2500 BC), and the Egyptians would begin to build colossal ashlar pyramids and temples from the mid-third millennium, remarkably there would be no more polychrome stone buildings for over a thousand years, until the "Treasury of Atreus" at Mycenae (c. 1250 BC; see fig. 2.3). Yet monarchs, priests, and builders would dream of hewing whole temples from a single stone, or piling them up from sparkling gems, stones that were of the earth but captured astral radiance, or had seemingly fallen from heaven itself.

IMPROVING ON NATURE IN THE ANCIENT NEAR EAST AND EGYPT

Until such time as buildings and statues would come to be made from colored marbles, artifacts in synthetic substances predestined their discovery. Reproducing the saturated hues and radiance of precious gems and hard stones was long the goal of parallel technologies in Mesopotamia, Syro-Palestine, Egypt, and the Aegean, and this praxis was the origin of alchemy.

These synthetic technologies can be traced to fifth-millennium BC Egypt, where artificial gems were first created by adding colored glazes to steatite and quartzite. However, the fact that steatite (soapstone) was easy to carve also meant that it was too soft to take a high polish, so this deficiency was remedied by firing it in a kiln as well. The quartz available, on the other hand, was very hard but colorless, so green- and blue-tinted glazes were needed for it to achieve the appearance of turquoise, lapis lazuli, or green feldspar.[5]

It became possible to produce more sophisticated "gems" only once glass had been invented, which happened somewhere in Syria or Mesopotamia, sometime around 2500 BC. A very early text (Ur III period, c. 2000 BC) recounts that stoneworkers (*zadim*) had made "gems" that resembled *dusû*-stone (perhaps quartz or topaz) and *saggilmut*-stone (amethyst). Glass can only be the material in question because the verb used to describe their manufacture was "*sur*" ("to liquefy" or "sprinkle").[6] The Sumerians continued to dye quartzite with honey and wine to simulate more precious amber, brown agates, and sardonyx, but, in time, they also came to replicate lapis lazuli, obsidian, and alabaster in glass.[7]

The Sumerians regarded these vitreous creations not as counterfeits but syntheses. As a chorus of scholars has pointed out, although we now differentiate between "stone" and "glass," Akkadian texts from the second half of the second millennium describe one as "lapis from the mountain" (*uqnū šadī*) and the other as "lapis from the kiln" (*uqnū kūri*). Marco Beretta has reasoned, "This clear distinction between the natural and the artificial . . . recognizes the technical capability to create a perfect imitation of a natural object. . . . Glass was not seen as an entirely artificial product but as a natural stone."[8] However, for the Sumerians this would evidently have been a distinction without a difference. We might better grasp their mindset by comparing the distinction between "lapis from the mountain" and "lapis from the kiln" to that which we make between wildflowers and nursery-grown ones, where nature is cultivated into varieties that still belong to the same species. Stones were anyway not distinguished according to any probing geological taxonomy, but rather they were associated prima facie, on the basis of their appearance.[9]

Conceiving gems as simply a harder form of glass also becomes readily more understandable when one considers that a *naturally formed*, volcanic glass—obsidian—had been traded in ingots, blades, and jewelry throughout the Near East and Europe for more than a thousand years already (since at least the fifth millennium BC) (fig. 1.3).[10] As obsidian was mined in the depths of Mother Earth, when glassmakers built kilns to create the same kind of heat that they knew could liquefy "stone" (though glass is actually non-mineral because it does not crystallize), they may have conceived of these kilns as man-made wombs. Glass recipes copied in the seventh century BC actually specify that human embryos be built into the furnace walls, and the kiln-womb pairing certainly existed in the minds of Sumerian brickmakers.[11]

And Sumerian artisans were turning clay into "stone" as well. Clay occupied the lowest position in the material hierarchy, but it was also the base matter for life. According to the Babylonian creation myth, the *Enûma Eliš* (early second millennium BC), it was from clay that all humans and all things had been made, and this raw good could be transformed into temples as well. Once the Sumerians had fired clay vessels to acquire the superficial hardness of stone, first in pits and (from c. 3500 BC onward) in kilns, they began (about 1500 BC) to dip them in alkaline glazes as well, which made the pots and bowls impermeable but also gave them the color and sheen of rare stones.[12] Real stone vessels had already been carved for a millennium and a half, and had circulated across the Mediterranean for five centuries or more.

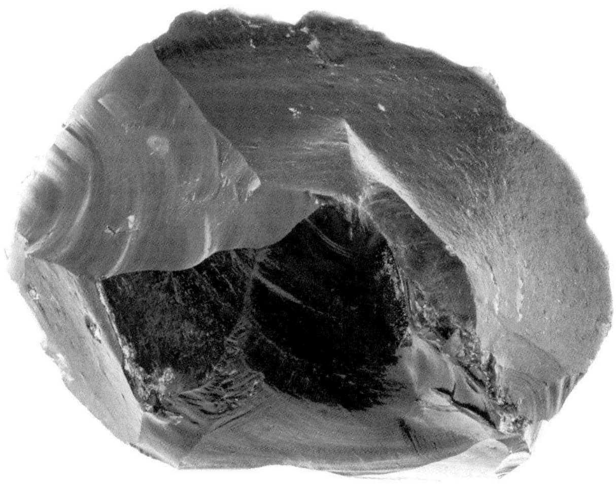

FIG. 1.3. Obsidian.

Glass was used in Mesopotamia from about 2500 BC, but it would take a thousand years for it to become known in Egypt and, when it did, it was worth its weight in gold. Indeed, the pharaoh Thutmosis III (r. 1479–1425 BC) was so excited by the ingots of colored glass sent to him as diplomatic gifts from Syro-Palestine (1458/1438 BC) that he christened them "Menkheperre-lapis-lazuli" and "Menkheperre-turquoise" (or "-malachite"). "Menkheperre" was the pharaoh's theophoric name, meaning "enduring in the apparition of Ra," so not only was this a signal honor, but the moniker also underscored the glass's miraculous luminosity. These ingots must have seemed like the biggest gems ever known, all the more miraculous because they had been manufactured. The Egyptians too spoke about glasses in the inherited terms of "stones of casting" (*jnr n wḏḥ*) or "stone of the kind that flows" (*ꜥ3t wḏḥt*).

Among the earliest artifacts to survive from the first wave of these imports to Egypt were "marbled" glass vessels that caused as much of a stir as glass alone, and which were prized enough to be represented in the murals of the tomb of Thutmosis III's vizier, Rekhmire (fig. 1.4).[13] One such vessel was included in the dowry of one of the foreign wives of Thutmosis III and survives intact, a goblet "marbled" in brick red, camel, light and dark turquoise, and white, which was probably made at Nuzi (Yorghan Tepe, Iraq) (fig. 1.5).[14] It is an unthinking convention to call this goblet "marbled"—unthinking because the artisans of Nuzi had never seen mottled marbles. None were yet quarried. They may have had agates in mind; or they may have meant to improve upon nature by creating a new mineral from their own imaginations.

In this respect, it seems to have gone completely unnoticed that the epic *Enmerkar and the Lord of Aratta* (c. 2112/1763 BC) had already celebrated such a technological coup several centuries earlier. King Enmerkar of Uruk, gifted with divine insight by the artificer god Enki, accepts the daunting challenge of creating a regal scepter of no known substance: he pours a gluey concoction into a hollow reed, which he cools and heats and cools again, splits open after five or ten years to reveal a hardened substance, and finally lacquers with oil. The outcome is something that is "neither wood nor metal nor stone"—in other words, not of this earth, but out of this world, and divinely inspired. Such a substance was fitting for Sumerian scepters that were believed to be living shoots handed down from heaven, conferring divine assent on earthly kings, and so holy that they were sometimes

FIG. 1.4. "Marbled" glass vessels, wall painting, c. 1400 BC. Tomb of Rekhmire (TT100), Transverse Hall, South Wing, West Wall, Register 4, Thebes (Sheikh Abd el-Qurna), Egypt (de Garis Davies 1935, pl. ix).

elevated to the status of divinities in their own right, even with sacrifices made to them.[15]

Enmerkar's scepter must count as the first synthetic manufact described in world literature, unless we count the creation myths themselves. The technique for making the actual Nuzi goblet was just as pioneering, a composite of glass and "faience" (i.e., a paste of finely powdered quartz fused together with small amounts of alkali and/or lime through partial heating). The goblet maker sandwiched doughy wads of differing faience mixes, which he rolled together and sliced into varicolored and swirling slabs. Glass, either nuggets or powdered, was also thrown into the mix to make translucent eyes in this storm of veining, and the metamorphosis into "stone" was completed by marvering and flame-polishing.

Now the Egyptians did have their own home-grown stones, which they exported throughout the eastern Mediterranean, especially a translucent "Egyptian alabaster" (which is actually a calcite-travertine, but I will continue to use the traditional term) that they extracted at Hatnub (*ḥwt-nbw*, "mansion of gold"), and from which they began cutting vases and platters in the late third millennium BC (fig. 1.6).[16] These vessels were so prized that those who could not afford them sometimes had clay copies made as grave goods.[17] However, the Egyptians had also begun (before 3000 BC) to produce synthetic stones in a superb faience of their own, which they glazed to achieve the highest luster. They called faience "*ṯḥnt*," meaning "that which is brilliant." In the pyramid

FIG. 1.5. Drinking cup, c. 1479–1425 BC. Probably made at Nuzi (Yorghan Tepe, Iraq). Thebes (Wadi Gabbanat el-Qurud, Wadi D), Tomb of the Three Foreign Wives of Thutmosis III. Glassy faience, gold, height: 4 in. (10.2 cm); diam.: 2¾ in. (7 cm). Metropolitan Museum of Art, New York.

FIG. 1.6. Alabastron (perfume vase), 6th–5th century BC. Cypriot, reflecting early Egyptian influence. Calcite alabaster, height: 4⅞ in. (12.4 cm). Metropolitan Museum of Art, New York.

FIG. 1.7. Wall tiles from the "Blue Chambers" (subterranean funerary apartment), step pyramid of King Djoser, c. 2630–2611 BC. Saqqara. Faience tiles in limestone matrix, individual tile, 2⅖ × 1⁵⁄₁₆ in. (6.1 × 3.4 cm). Metropolitan Museum of Modern Art, New York.

complex of the pharaoh Neferefre (mid-25th century BC) at Abusir, "some of the faience tiles bear added decoration in gold leaf. . . . Much later texts pair faience with gold, as though they were complementary lunar and solar aspects of light . . . and this may be what was meant at Abusir."[18] Indeed, the substance entered the kiln a dry and dull paste and emerged, alchemically transformed, glossy and a saturated hue; earth had become ether. The epithet "ṯḥnt" also came to be bestowed upon deities, and eventually upon the semi-divine pharaohs, and faience therefore embodied divine luminosity. More rarely, the Egyptians called faience "*ḫsbḏ*," the same word they used for lapis lazuli, again pausing to distinguish between "true/real lapis lazuli," *ḫsbḏ m3* ʿ(.*t*), and "man-made lapis lazuli," *ḫsbḏ iry.t*. In actual fact, works in faience encompass a range of hues between blue and green, recalling not only lapis lazuli from Afghanistan, but also the turquoise from the Sinai that the Egyptians considered to hold magical properties. In Egyptian "to be green" meant "to grow," or "flourish," and the evergreen bloom of faience metaphorized life, rebirth, and immortality.[19]

Faience was used to make all manner of small votive items and cult objects, but was also used as architectural cladding; the earliest and most spectacular use is in the "Blue Chambers" hewn from the bedrock deep below the step pyramid of the Third dynasty King Djoser (c. 2630–2611 BC) at Saqqara (figs. 1.7, 1.8).[20] About thirty-six thousand greenish-blue plaquettes were sewn into the limestone walls with copper wire to simulate reed-mat hangings, even down to the way the revetments are "rolled up" above the fictive doorways. The whole ensemble therefore recalled the reed huts and primeval architecture of the Nile delta, which, in turn, evoked the rushy waters around the primordial mound of the Egyptian cosmogony; it could also summon up the "Field of Reeds," or Egyptian paradise. All in all, the faience reeds are wettened either by primal or eschatological waters, and in either case they have acquired an otherworldly gleam.

These "Blue Chambers" are the terminus in an itinerary of implied petrification that runs through the whole funerary complex of Djoser. The pyramid was surrounded by limestone ramparts (when all towns and palaces were still mud brick); the monumental columns (the earliest in the history of architecture) of its entry passage imitated bundled reeds and supported a ceiling that simulated logs or bales of rushes (fig. 1.9); and the step pyramid itself was revolutionary not only in its form but also because it was the first Egyptian tomb built in cut stone rather than mud

FIG. 1.8. "Blue Chambers," step pyramid of Djoser, c. 2630–2611 BC. Saqqara (Lauer 1936).

FIG. 1.9. Entrance corridor, pyramid complex of King Djoser, c. 2630–2611 BC. Saqqara (reconstruction: Lauer 1936).

brick, and again on a scale that beggared belief. It was a shock to anyone's imagination, effectively to make a mountain from scratch on the desert sands. The particular temporal rationale behind this ghost palace in the desert was that it was to be an eternal venue for the Sed festival, which the pharaohs normally celebrated after thirty years on the throne. Djoser had forgone this festival when he died after only nineteen years in power (2667–2648 BC), so this lost moment of a repeating cycle was ossified in an abode that must be as lasting as the eternal time loop.

ENGINEERING THE HEAVENS FROM SUMERIA TO BABYLONIA

Any marble architecture still lay far in the future, and the earliest conceptions of a truly monumental building in multicolored stones belong to a cosmic architecture with absolutely no corollary in earthly construction. From time immemorial, the idea that some supreme deity dwelled in a cave or stone house in the sky, or even that the whole sky was mineral, has been a widespread belief throughout East Asia, India, Australia, Nigeria, and even among the Navajo and indigenous peoples of California.[21] In Sumeria too some hymns show that the sky was conceived as bowl-shaped, or mention a "zenith," which in itself implies a vault.[22] What could such a vault be made of? What material could account for its varying appearance from day to night? Since the Sumerian word for "tin" is "*an-an*" ("metal of heaven"), they perhaps imagined the daytime sky as beaten from this or some other burnished ore.[23] Because some hymns instead describe the dawn as "Utu [the sun god] who comes from the lapis sky" (*[d]utu an-za-gìn-ta è-a*), the Sumerians must have conceived of the starry firmament as hewn from this saturated blue stone, with its twinkling calcite and pyrite specks.[24] What made the equation all the more compelling was that lapis had to travel from the Sar-e-Sang valley in the Badakhshan mountains of northeast Afghanistan to reach Mesopotamia, an almost unimaginable distance of about 2,200 miles, from what seemed the ends of the earth that grazed the sky. To hold lapis was therefore to grasp heaven in the hand; no wonder they sought to synthesize it in glass and bottle its luminosity.

The Babylonian creation myth, the *Enûma Eliš* (early second millennium BC), says nothing about the material of the heavens, merely mentioning that Marduk, the creator of the world, "made strong . . . the floor of the celestial regions."[25] However, at some later point, apparently after the Kassite period (c. 1595–1155 BC), the heavens came to be conceived much more elaborately. Indeed there were three heavens now, each with a dome of diaphanous hard stone:

> The Upper Heavens are sardonyx [*luludānītu*]. They belong to [the sky god] Anu. He settled the 300 Igigi gods inside.
>
> The Middle Heavens are amethyst [*saggilmut*] of the Igigi. Bel [i.e., Marduk] sat on the high dais inside, in the lapis lazuli sanctuary. He lit a lamp of amber [*elmēšu*].
>
> The Lower Heavens are pale blue chalcedony [*ašpû*]. They belong to the stars. He drew the constellations of the gods on them.[26]

All the stones mentioned can be reliably identified as sardonyx, amethyst, and chalcedony. All were types one might find in a treasury or a temple, but none were building stones, nor came in dimensions bigger than a man's fist (fig. 1.10).[27] Nonetheless, the hard-stone heaven of the Kassite cosmology was imaginable as a construction because glassmakers had been manufacturing semiprecious "stones" (i.e., glass) for centuries. Indeed, one could countenance the idea of the creator and ur-artificer Marduk himself using the same techniques on an industrial scale to cast the heavens—as the Psalms (19:1) will say of the Jewish God: "and the firmament sheweth His handywork." Moreover, all the stones in the Kassite cosmology are themselves diaphanous, from which one deduces that the changing appearance of the sky from sunrise to sunset was imagined as the effect of the lamp of the sun shining through a series of colored lenses. A passage written by the Roman philosopher Seneca many centuries later (56/62 AD) unexpectedly mirrors the Kassite vision:

> If a house were gifted to you, which resplended with marble and gleamed with a ceiling of gold or spread with colors, you would hardly call the present mediocre. Yet, he [i.e., god] has built you a huge mansion . . . a house in which you see, not flimsy veneers thinner than the very blade with which they were sawn, but virgin masses of the most precious stone, whole masses of a substance with such a variety of markings that the tiniest fragments of it fill you with wonder, and a ceiling gleaming in one way by night, and in another by day. (*Ben.* 4.6.2)

We can also probably give more credence than is usual to the account that the Parthian king Vardanes (40/45 AD)

UPPER HEAVENS
sardonyx (*luludānītu*)

Sanctuary of Marduk
lapis lazuli (*uqnû*) amber (*elmēšu*)

MIDDLE HEAVENS
amethyst (*saggilmut*)

LOWER HEAVENS
chalcedony (*ašpû*)

FIG. 1.10. Kassite cosmology.

gave judgment in Babylon, below a dome clad in lapis lazuli in which gilded images of the gods glittered and twinkled, and exactly the same arrangement in the throne room of the Sasanian king Ḵosrow II (r. 590–628) at Ctesiphon, whose dome even revolved around the king. It has also been suggested that, had they been able to procure enough lapis, the Medici would have reproduced the scheme in their gargantuan burial chapel at S. Lorenzo in seventeenth-century Florence (see fig. 12.4).[28]

STAIRWAY TO HEAVEN: FROM ZIGGURATS TO THE HEAVENLY JERUSALEM

Indeed, the celestial gem architecture of Mesopotamian scripture represented a symbolic ideal that would become the object of simulation in practice, from the development of new architectural media such as glazed bricks to the much later exploitation of colored marble. In reality, the only structural stones known to the Mesopotamians were limestone and basalt, and even these were used very sparingly. However, mud was abundant in the "fertile crescent," and once this was baked, first under the sun, then in the kiln, it could make claim to being the first form of artificial stone. As soon as the early builders had learned to veneer mud-brick buildings with kiln-fired bricks, and then glaze them as well, they also acquired a radiant alterity.

This aspiration in Mesopotamia can be traced back even before the Steinstifttempel with which this chapter opened, to the second half of the fourth millennium BC. The "White Temple" (carbon-dated to c. 3517/3358 BC), again in Uruk and thought to have been dedicated to the sky god Anu, was covered in gypsum plaster that reflected sunlight like a mirror. It also sat upon a battered platform—a proto-ziggurat—seventy feet high, thickly coated with mud plaster, and traversed by a monumental stairway paved in white limestone slabs (fig. 1.11).[29] From the great plain of Sumer it had all the qualities of a mirage, a concrescence of light, a heaven brought down to an earth that rose up to meet it. It seemed a cosmogony frozen in the moment of birth, a suddenly crystal stratigraphy of mediation between earth and ether. The aspirations of its builders are summed up by the temple hymn (c. 2600 BC), among the world's oldest surviving literature, that likens the Temple of the Mother Goddess at Keš to the moon against the sky, its cosmic dimensions filling the world, a "platform of the land . . . growing as high as the hills, embracing the heavens."[30]

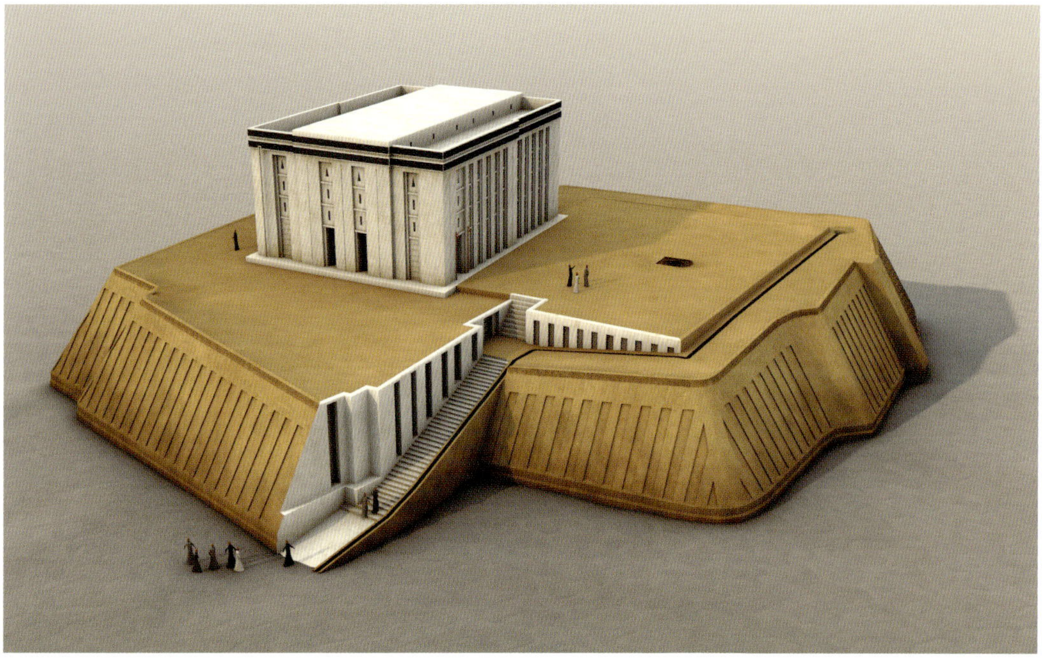

FIG. 1.11. White Temple, c. 3517/3358 BC. Uruk (reconstruction: Artefacts, Berlin).

Such gleaming plaster, or maybe just whitewash, must have become common enough for the high priestess Enheduanna (the world's first author recorded by name) to write, around 2300 BC, of temple "brickwork glistening and pure, its burnished clay placed firmly (on the earth)."[31] This same Enheduanna can be seen offering a libation before just such a shining tower, on a (heavily restored) disk of translucent white calcite (fig. 1.12) that, by and of itself, almost certainly represented the full moon, Nanna, the god to whom Enheduanna ministered as high priestess (*En'hedu'anna*).[32]

FIG. 1.12. Disk of Enheduanna, c. 2300 BC. Ur, Iraq. Translucent white calcite, diam.: 9¹¹⁄₁₃ in. (25 cm); depth: 2¾ in. (7 cm). University of Pennsylvania Museum of Archaeology and Anthropology.

The chief god of the Sumerian pantheon was Enlil, and his principal temple was the Ekur ("mountain house") at Nippur, originally built c. 2500 BC. A hymn to Enlil lauding this ziggurat highlights the metaphorical meaning of its materials, which were perhaps painted:

> The shrine of plenty, the Ekur, the "lapis lazuli" house,
> he raised up out of the dust,
> Planted it in a pure place like a (high) rising mountain,
> [. . .]
> Its brickwork of red metal, its foundations of
> lapis-lazuli.[33]

Radiance was a cardinal aesthetic in Mesopotamia, a prerequisite of temple and palace architecture for denoting the aura of gods or monarchs, and hymns and inscriptions recurrently acclaim the effects that must be translated into materials. One of the earliest is a dedication inscription (c. 2125 BC) telling us that Gudea, ruler of Lagaš (r. c. 2144–2124 BC), has "decorated [the temple to Niĝirsu] with the radiance of heaven."[34] At Ur, the mausolea of the Third dynasty kings, especially those of Shulgi (r. c. 2029–1982 BC) and his son Amar-Sin (r. c. 1981–1973 BC), give us some idea of how temple cult rooms might have been; the tombs had gilded walls, with shining mosaic and cut-out work of gold, carnelian, and agate.[35] The Middle Assyrian king Tiglath-Pileser I (r. c. 1115/1076 BC) also speaks of building ziggurats to the gods Anu and Adad, with one "interior . . . decorated like the interior of heaven. I decorated its walls as splendidly as the brilliance of rising stars."[36]

Such brilliance might be produced not only by shining, shimmering, or twinkling surfaces but also iridescent ones. The rainbow patterns of the Steinstifttempel at Uruk forecast the genuine iridescence of the twin columns (figs. 1.13a, 1.13b) that flanked the entrance to the Temple of Ninhursag at Tell al-'Ubaid, which was built or restored by King A'annepada (c. 2400–2250 BC).[37] These columns were inlaid with tesserae of mother-of-pearl, pink limestone, and shale, and glued with bitumen to palm-trunk cores. The pattern traced the imbrications of the palm bark beneath, almost as though the tree had petrified and crystallized for evermore. These jeweled columns immediately recall a passage in the later *Gilgāmeš* epic (written c. 2100 BC), where the eponymous hero—king of Uruk—reaches the Garden of the Sun in the farthest east, a proto-paradise where the "trees of the gods" and all their fruits are gems:

> A carnelian (tree) was in fruit,
> hung with bunches of grapes, lovely to look upon.
> A lapis lazuli (tree) bore foliage,
> in full fruit and gorgeous to gaze on.
>
> [. . .] cypress [. . .]
> [. . .] cedar [. . .]
> its leaf-stems were of *pappardilû*-stone [banded agate or onyx] and . . .
> Sea coral . . . *sāsu*-stone,
> instead of thorn and briars [there grew] an.za.gul.me stone.
> He touched a carob, [it was] *abašmu*-stone,
> *šubû* stone [agate?], and haematite [or obsidian].[38]

FIG. 1.13A. Column, Temple of Ninhursag, c. 2400–2250 BC. Tell al-'Ubaid. Wood, shale, mother-of-pearl, limestone, height: 23¼ in. (59 cm); diam.: 12³⁄₁₆ in. (31 cm). University of Pennsylvania Museum of Archaeology and Anthropology.

FIG. 1.13B. Column, Temple of Ninhursag (excavation photograph: Hall and Woolley 1927).

In the land of everlasting life, the petrified fruits are without season, continuously ripe, for time is suspended. The words that described them were also eternal, perhaps handed down by the gods, but to write them the scribe had to press a stylus into wet clay that then had to be dried or baked, to become like stone, or else crumble to dust.

One must await the Middle Assyrian period (1392–936 BC) for the invention of a glazed brickwork that will make both temples and palaces truly resplend. It was a short step from glazing pots to glazing bricks. Temples were now clad with blue- and silver-glazed bricks and, as glass was considered a kind of semiprecious stone, these bricks effectively assimilated the materials of the super-celestial order, so stirringly described in the post-Kassite stratification. The ascending stages of the temple or ziggurat changed in color and mineral allusion.[39] Thus, a first millennium BC hymn praising the Ezida (Temple of Nabû) in Borsippa itemizes,

> its bricks of deep red stone [*lamassu*],
> its cornice of reddish gold [*ṣāriru*], choice gold [*liqtu*]
> its foundation is of alabaster [*gišnugallu*],
> pure glass [*zakakātu*] is the robe [i.e., cladding] of Ezida![40]

However, the most extensively preserved remains of glazed brickwork are far later and come from Babylon, where Nebuchadnezzar II (r. c. 604–562 BC) had both the court preceding his throne room and the famous Ištar Gate (c. 575 BC) clad with them (fig. 1.14). The inscription on this gate is careful to announce, "I laid the foundation of the gates down to the ground water level and had them built out of pure blue stone [*uqnû*]."[41] Nebuchadnezzar also restored the ziggurats at Babylon and Borsippa, the E-temen-anki and E-ur-me-imin-anki, respectively, erecting atop both of them a "chamber of repose [for Marduk] resplendently of pure tiles [of] lapis lazuli." In other inscriptions Nebuchadnezzar also states, "I completed them [i.e., the ziggurats], making them [gleam] bright as the [sun]" and "bright as day."[42]

Nebuchadnezzar likened the whole city of Babylon, lying on the Euphrates and surrounded by moats, to the primordial mound that had arisen from the primal waters at the beginning of time; within the precinct of the

FIG. 1.14. Ištar Gate, c. 575 BC. Babylon. Blue-, turquoise-green-, and yellow-glazed bricks. Reconstruction (1928–30), Vorderasiatisches Museum, Berlin.

E-temen-anki there was even a platform called "the pure hill," and a well-known map depicted the city as the center of the world. For all these reasons, Babylon was the linchpin of heaven, earth, and abyss, and its spiritual center was the E-temen-anki, whose Sumerian name translates as the "House-Foundation-Platform of Heaven and Underworld" (likewise, that of the E-ur-me-imin-anki at Borsippa signified the "House that Controls the Seven *Me*'s [divisions] of Heaven and Underworld"). This Babylonian ziggurat was said to be "dug out of the depths," meaning that it was rooted in the deep—what they called *apsû*, the origin of our word "abyss"—the bottomless, watery pool from which the mound of the world had arisen at the beginning of time. Indeed, when the workmen dug the foundations of the E-temen-anki they would have struck the high water table and must have thought they had uncovered the face of creation (even today the exposed foundations are flooded). The bricks of this man-made mountain were also made of clay, and because this was the primal substance making and consecrating the first brick (the "foundation stone") of any temple was of the utmost ritual importance.[43] Conversely, foundation deposits of precious stones show that temples also had their grounding in the sky, because these materials had celestial associations, origins even. This may be what is meant when the aforementioned hymn on the Ekur at Nippur says that the ziggurat's "foundations" were of lapis lazuli.

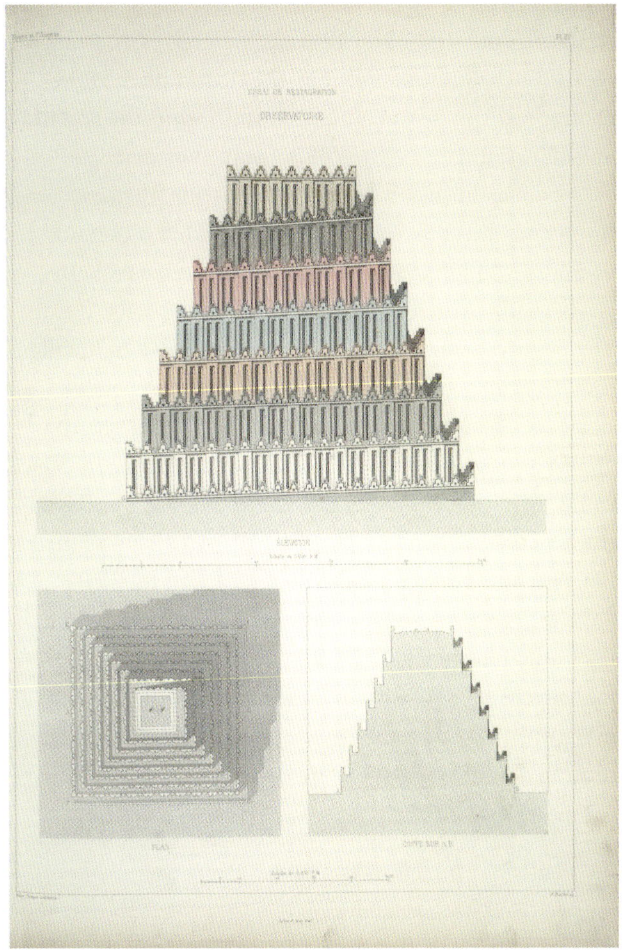

FIG. 1.15. Ziggurat of Sargon II, c. 717–705 BC. Khorsabad (Place 1867–70).

The ziggurat was, after all, not only the stairway to heaven for the king and his priests, but also the ladder by which the tutelary god might descend to the city and be among men. The liminal materiality of the temple as a bridge between worlds already appears in the *Dream of Gudea* of Lagaš (c. 2125 BC), wherein the god Ninġirsu commands Gudea to build him a temple, the goddess Nisaba directs it to be aligned with propitious stars, and the god Nindub draws the plan on a tablet of lapis lazuli, and all the while Gudea has before him the "holy brick mold" and "fated brick"—that is, the first brick laid, which was a metonym of the whole temple and consecrated at its molding.[44]

As temples became more lofty, their materials became more elaborate. The ziggurats of the Sumerian period (third millennium BC) and early Assyrian epoch (mid-second millennium) had been typically divided into three or four stages. However, in late Assyrian (911–612 BC) and neo-Babylonian times (626–539 BC), these structures became more towering affairs with seven stages—for example, those that Sargon II built at Khorsabad (c. 717–705 BC) and Nebuchadnezzar II at Borsippa (c. 604/562 BC), or that which Nabonidus reconstructed at Ur (c. 556–539 BC). There is some evidence that the individual coloring of the various stages now relayed a celestial stratification according to the seven known planets, and that this was achieved with painted plaster on the lower stages and glazed bricks on the final ones (fig. 1.15).[45] In other words, ziggurats had graduated from being a "platform of the land . . . growing as high as the hills, embracing the heavens" to telescoping towers that signified ascent through the planetary cosmos. Fittingly, the summit of the ziggurat was probably an eyrie for astronomy and divination. Moreover, the stage-by-stage progression of the building from earth to highest heaven—which rehearsed the emergence of substance from chaos—was mirrored by its own material transubstantiation: the once-wet clay had been baked to become dry stone; after glazing, this stone became glass *qua* precious stone; and when this precious stone was blue-glazed, it transformed into lapis lazuli, which was the stuff of the heavens themselves. Marduk's shrine, at the summit and amid the clouds, pierced the chalcedony sky, and in his lapis throne room, the god lit an amber lamp (the sun) to illumine the whole world.

Nebuchadnezzar's favorite commission, the Ištar Gate, brought all this down to the ground in what is as much a vision as a building, whose "lapis" bricks tell us that this is a heavenly structure, not an earthly one (see fig. 1.14). The Ištar Gate divided the chaos of apsû from the cosmos that Marduk had constructed, and it was the portal to Babylon, "the gate of the gods" (*Bab-ili*). In daylight, the glazed blue wall shone like the stone of a seal and, as Marc van de Mieroop has remarked, called to mind a passage in the *Erra* epic (c. 800–612 BC) where the god Ishum likens Babylon to "a gem-stone seal on the neck of the sky."[46]

The gemmed heavens of the Kassite cosmology, and the ramparts and ziggurats of Babylon, had a long afterlife in the Jewish and Christian imaginations. While a captive in Babylon in 593 BC, the Jewish priest Ezekiel ben-Buzi records that he was struck by a divine vision:

> The likeness of the firmament . . . which was as the color of a terrible crystal. . . .
>
> And stretched out over their heads was the likeness of a throne, as the appearance of lapis lazuli. . . .
>
> And I saw as the color of amber [חַשְׁמַל *chashmal*], as the appearance of fire round about within it.[47]

The words themselves make plain the transmission: "chashmal" appears nowhere else in the Old Testament and

FIG. 1.16. Heavenly Jerusalem, S. Maria Maggiore, c. 431 AD. Wall mosaic. Rome.

is clearly a loan from the Akkadian "elmēšu," and that calls to mind the amber sun lamp of the Kassite cosmology. Another text, Exodus, also composed in the sixth or fifth century BC, is even more telling; we are told that when Moses and the elders saw god, there was "under His feet, as it were, a paved work of lapis lazuli [סַפִּיר *cappiyr*] and, as it were, the substance [עֶצֶם *'etsem*] of heaven in its clarity." The crucial clue to understanding that this vision is again tangibly based on a real building, the giant ziggurat of the E-temen-anki, is in the original Hebrew: the noun normally translated as "paved" (לִבְנָה *libnah*) also means "brick" and is derived from the verb "לָבַן *laban*," "to make bricks."[48] It comes as a shock now to realize that the all too familiar but evanescent image of eternal heaven as fashioned from some supersized gem in fact has its feet firmly planted in the bricks and mortar of historical Babylon.

The later *Book of Enoch* (c. 300 BC) also remembers ziggurats when it describes a whole mountain range of precious stones underlying God's throne, "but the middle one reached to heaven like the throne of God, of alabaster, and the summit of the throne was of lapis lazuli."[49] The imagery of jeweled walls such as those lining the Processional Way in Babylon instead appears in Isaiah (this section also written during the Babylonian Captivity, 593/571 BC) and Tobit (c. 225/175 BC), not to mention Aramaic texts such as the *Description of the New Jerusalem* (second/first century BC). The final installment, so to speak, is the book of Revelation (21:18–21), written still centuries later (c. 96 AD), whose material eulogy of the ultimate city in the sky, the "Heavenly Jerusalem" (fig. 1.16), marries the walls of Babylon with the Kassite cosmology, when it itemizes the gems that will make up her ramparts:

> And the wall was of jasper, and the city of pure gold, as clear as glass.
>
> (19) And the foundations of the wall of the city were decorated with every kind of precious stone. The first foundation was jasper; the second, sapphire; the third, agate; the fourth, an emerald;
>
> (20) the fifth, onyx; the sixth, ruby; the seventh, chrysolite; the eighth, beryl; the ninth, topaz; the tenth, turquoise; the eleventh, jacinth; and the twelfth, an amethyst.
>
> (21) And the twelve gates were twelve pearls, each gate made of a single pearl. The great street of the city was gold, as pure as transparent glass.[50]

2 A Medium Fulfilled

The Emergence of the Marble Temple

Only in Egypt had a monumental ashlar architecture emerged in the third millennium BC, thanks to her rich holdings of sandstone, limestone, travertine ("alabaster"), and granite. The result was the colossal temples, obelisks, and pyramids that still stand in abundance. There was no such abundance of stone in Mesopotamia, and no such architecture until the sixth century BC.

MINOAN AND MYCENAEAN MASONRY

In the Greek world stone architecture made only a creeping advance. Ashlar basements appear at Knossos (c. 1900–c. 1600 BC) and other Minoan sites, but the superstructures were half-timber packed with clay and rubble and plastered over. It is only in the Mycenaean era (c. 1600–c. 1100 BC) that we see a more substantial use of stone, in the massive limestone boulders and conglomerate ashlars used to build the walls of Mycenae and Tiryns. At Mycenae, conglomerate stone was used to face the *dromoi* (entrance passages) to important tholos tombs, the bastions framing the Lion Gate, as well as palace lintels, thresholds, and pillar bases. Conglomerate was not found on-site at Mycenae, and most scholars tend to stress that the capacity to quarry and move it signified the power of the state, or visually linked the monuments across the landscape, and that its centuries-long exploitation made visible the continuum between the state and its origins.[1]

But why pick conglomerate at all, when the local limestone was so plentiful and much easier to work? Conglomerate contains pebbles of various sizes (clasts) cemented together in a finer-grained matrix (fig. 2.1)—pebbles that make the stone difficult to work and liable to crack. At the same time, they also gave the stone that prized mottling that had already been imitated in earlier Minoan fresco (see the faux-stone border in fig. 4.2).[2] Moreover, as the cement is yellowish and the pebbles are calcite, when the blocks were first hewn and set, they would have shone and sparkled in the sunshine, as befitted Homer's "Mycenae rich in gold" (*Il.* 7.180).

Mycenae also saw the greatest strides toward architectural polychromy in colored stones, at the Lion Gate and "Treasury of Atreus" (both thirteenth century BC). Huge conglomerate blocks adorned the Lion Gate and flanking bastions, while the remainder of the fortifications were built of smaller blocks cut from the

FIG. 2.1. Conglomerate blocks, Dromos, Treasury of Atreus, c. 1250 BC. Mycenae.

FIG. 2.2. Lion Gate, 1250 BC. Limestone bedrock, conglomerate, and limestone blocks. Mycenae.

dark limestone bedrock (fig. 2.2). On the Lion Gate not only does the eponymous relief in white limestone contrast with the conglomerate all around it, but the lions' heads, now missing, were of yet another contrasting stone (possibly dark steatite), or even gilt bronze.[3]

However, this sensitivity to color is trifling when compared with the façade of the so-called "Treasury of Atreus," which is altogether such a bolt from the blue that it is surprising so little attention has been paid to it (figs. 2.3a–d). This was the first façade built in colored stones since Uruk, the first in marbles, and it had the earliest stone columns in Greek architecture (there will be no others for seven hundred years). Its red and gray-green marbles were again not local but brought all the way from Kyprianon in Laconia.[4] Two stories of green half-columns bracketed the façade, a giant order finely carved with chevrons, supporting a dwarf one carved with spirals. Over the lintel of the colossal portal (*stomion*) ran a green frieze of disks (representing beam-ends); above this a frieze of running spirals; above this yet another red frieze (like an archaic metope and triglyph); and above all this, and filling the now-triangular void, ran bands of red stone with running spirals (like a wine-dark sea), some of their "eyes" probably inlaid with glass paste for added sparkle.[5] Bronze clamps or dowels fixed all these ornaments onto the stomion, itself an ashlar wall of tawny-speckled-white conglomerate. The long-lost doors may have been sheathed in bronze, maybe gilded, maybe with ivory and other precious inlays. Another monumental tomb, the slightly later "Tomb of Clytemnestra," was decorated in a similar way, this time with fluted half-columns of gypsum (possibly from Crete), a frieze of bluish limestone, and yet more greenish-gray marble; its threshold was certainly sheathed in bronze and, therefore, probably the doors as well.[6]

FIG. 2.3A. Façade, Treasury of Atreus, c. 1250 BC. Mycenae. (reconstruction: Higgins et al. 1968).

FIG. 2.3B–D. Colored limestone fragments from frieze and tympanum. British Museum, London.

Such sumptuous façades perhaps prompted the travel writer Pausanias (writing c. 150 AD) to assert the sepulchral mounds were "treasuries" of Atreus and his children at Mycenae (*Desc.* 2.16.6).[7] Yet a tomb of such opulence can only have been built for the *wanax* ("king") of Mycenae—it is sometimes also called the "Tomb of Agamemnon"—and its polychrome-stone panoply may actually be a hardier version of the frescoed ornaments that once adorned the façade of the palace *megaron* (throne room).[8] If so, the idea of a monumental stone structure for the dead, in dialogue with an ephemeral one for the living, has a rather Egyptian tinge to it. A scene painted on the near contemporary Hagia Triada Sarcophagus (c. 1370–1320 BC), from Crete but under Mycenaean influence, perhaps offers an eschatological view of such a monumental portal, with the deceased in his winding sheet standing before it (fig. 2.4).[9]

FIG. 2.4. Hagia Triada Sarcophagus, white figure before monumental portal, c. 1370–1320 BC. Archaeological Museum, Heraklion.

BUILDING A RELIGION: THE PETRIFACTION MYTH AND STONE TEMPLES

On neither Minoan Crete nor the Mycenaean mainland were there any buildings that we could properly call temples. There were hallowed caves and shaded sanctuaries on mountains and in woods, but religion within the urban settlements lay in the hands of a ruler, who could orchestrate cultic feasts for the community and host them in his palace. The same held true in Mycenaean-era Greece, and probably explains the monumental dimension of the hearths that dominated the center of the megaron (see figs. 4.12, 4.13).

It is only toward the end of the eighth century BC, as ruling elites seemingly lost their grip on communal religion, and the civic sodality that we now know as the polis began to emerge, that the first urban temples appear.[10] These new buildings borrowed so many of their features from noble dwellings that it is not always possible to distinguish between their remains. During the sixth century, one of these features, the peristyle, an orbital ring of wooden posts supporting the eaves, metamorphosed into the orbital stone colonnade (*peripteros*) that would become a hallmark of the classical Greek temple. The sixth century also saw an increasing shift toward ashlar as the primary material of temples, but the transition was prolonged. Temples were largely built in mixed materials (stone, brick, and timber), and the mix differed greatly from region to region, and even within them. Even so, stone temples were now built at Corinth, Athens, Aegina, Corfu, Kalydon, and in the Argolid; and from the early sixth century, monumental limestone temples began to be built in the Greek colonies of Sicily and southern Italy as well (see figs. 2.7, 2.17).[11]

A century or two later, this shift in material technology, ultimately to buildings completely of stone (even the roofs), became an issue that Greek architects felt they needed to theorize. The explanation that persisted for centuries was that passed on by the Roman engineer and architect Vitruvius in his *De Architectura* (c. 27/19 BC). Vitruvius traces the genealogy of all construction back to the earliest men, who, he says, interlaced the branches of trees into their first shelters, and thereafter learned to build log cabins once they began to establish towns (*De Arch.* 2.1.1–7). Later stone architecture was, he continues, modeled on these timber cabins, and builders simply maintained the tectonic forms even when they switched materials, such that the stone ornaments of the (Doric or Ionic) temple were fossils of the prior carpentry (*De Arch.* 4.2.2–6). By implication, this meant that the stone column replicated the wooden post, although Vitruvius only actually discusses the entablature, stating that Doric triglyphs represent the plaques affixed to beam ends, or Ionic dentil rafters.[12] The justification he gives for these retentions is "that a distribution would not be correct in a copy which could not exist in the prototype. For the perfection of all works depends on their fitness to answer the end proposed, and on principles resulting from a consideration of Nature herself, and they approved those only which, by strict analogy, were borne out by the appearance of utility."[13]

Vitruvius's "petrifaction" theory has proven so abiding that it has long been thought that the Temple of Hera at Olympia (begun c. 610/590 BC) was originally constructed with timber columns that were then substituted piecemeal by the surviving stone ones, while other scholars contend that the *peristasis* (peripteral colonnade) was stone from the start.[14] Whatever the truth in the matter, Greek temples were built in mixed materials for at least a century before they turned uniformly to stone, and capital types continued to evolve long after they began to be carved in stone, so there was no "moment" of translation from one material to another, no archetype "set in stone."

Vitruvius compiled his account from treatises of either the Classical or Hellenistic periods, when the petrifaction theory must have become established. Indeed, the Doric order that Vitruvius describes seems to be the slim Hellenistic version of the third and second centuries BC, not one of its stouter precursors (*De Arch.* 4.3.4; 4.3.9). The very fact that he quotes competing theories (*De Arch.* 4.2.4, on whether metopes represented the panels between the beam-ends or instead mezzanine windows) also reveals that the origins of the ornament were debated, and suggests that the temple was held to be an icon of construction. It is easy to see how such ideas might have arisen when stone temples were replacing older houses of the gods in mixed materials or built alongside them, and when Greek homes were still built in adobe and timber, soon to become stone themselves. In fact, some buildings, such as the aforementioned Temple of Hera at Olympia, will have appeared to be "in transition," for, besides the columns and stylobate, only the lower courses of the *naos* (cult room) were made from stone, while their upper parts were in mud brick, and the entablature was all timber.

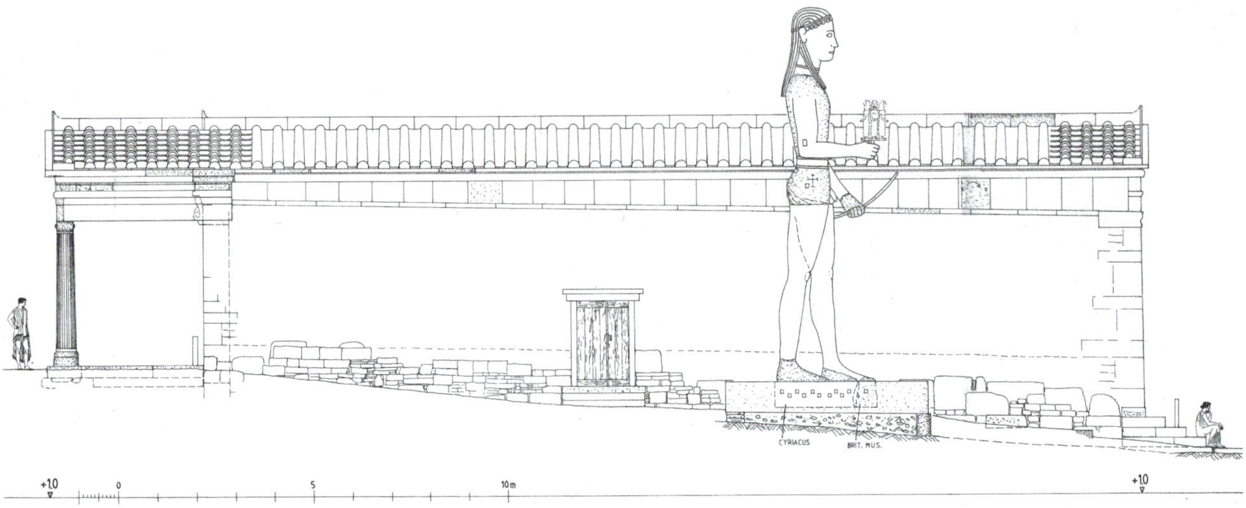

FIG. 2.5. The "Oikos" and Colossus of the Naxians, c. 590–580 BC and c. 600/590 BC, respectively. Delos (reconstruction: Gruben 1997).

Yet conspicuously absent from the Vitruvian account and, presumably, the Greek sources on which he drew, is any reference to foreign influences in the adoption of stone technology. Apparently, the Greeks preferred a chauvinistic view of their architecture as purely indigenous (invented by first Dorian, then Ionian, and finally Corinthian Greeks) and rationalized it as the result of a progressive *techne,* "resulting from a consideration of Nature herself."

Yet, even if Greek temples were as homegrown as Vitruvius's sources pretend, it would be extraordinary if they did not receive some impetus from the limestone and sandstone temples of Egypt, as the Greeks had no contact with any other stone architecture to speak of.[15] Plato famously commends the wisdom of Egypt, and Herodotos praises its stone architecture. After a Greek settlement was set up at Naukratis in the Nile delta (by at least 620 BC), Egyptian artifacts began to flow along Greek trade routes, eventually finding their way into the homes and workshops of Greece. Conversely, Greek travelers were exposed once more to the wonders of the Egyptian architecture and sculpture that had been removed from them since the Bronze Age.[16] In the new stone temples that rose on mainland Greece, the similarity of the nascent Doric order to Egyptian columns (most famously at Deir el-Bahri) has been pointed out several times. More significantly, the "bundled-reed" columns and "log" ceilings of the entrance pylon to the funerary complex of Djoser at Saqqara (see fig. 1.9), not to mention the Blue Chambers there (see figs. 1.7, 1.8), witness a far older history to the petrifaction myth as well as a metaphysical dimension expunged from the later recountings. The skeuomorphic allusion to an earlier wooden technology was plain to see, and probably served two purposes: the first was to signify the antiquity of the cult by preserving the image of its aboriginal housing; the second was to display the transformation of the ephemeral into the eternal, as befitted a house of a god rather than a man. Petrifaction introduced a moment of ceaseless stillness into the tide of human history.

Sculpture certainly played its part in transmitting a lithic imagination from Egypt to Greece. The striding male figure is the oldest type of Egyptian standing effigy, and there has been consensus (since Diodorus Siculus, 60/30 BC) that it dictated the posture and proportions of the Greek *kouros,* a statue type that emerged about 600 BC. Almost as soon as the kouros appeared, an Egyptian sense of scale appeared with it. One of the earliest kouroi is also one of the largest, a colossal Apollo, about thirty feet

FIG. 2.6. Torso fragments, Colossus of the Naxians, c. 600/590 BC. Delos.

high, erected by the Naxians circa 600/590 BC on the holy island of Delos, the mythical birthplace of Apollo (figs. 2.5, 2.6). It is impossible to consider this colossus without thinking of those of the pharaohs (e.g., Senusret III, Ramesses II, and so on) that the Egyptians had erected before their own temples, and which Herodotos later praises.[17]

In the open precinct of the Greek temple, sacrifice was offered at an external altar, while the temple itself was oriented to one or two mountains on the horizon, because they were the hilltop homes of the local gods, and the temple's petrifaction made it a material surrogate.[18] Martin Heidegger's famous and bewitching description of the Temple of Hera II (c. 470–460 BC) at Paestum (fig. 2.7) is the most powerful description of the Greek temple's densely material and elemental presence:

> Standing there, the building rests on the rocky ground. This resting of the work draws up out of the rock the mystery of that rock's clumsy yet spontaneous support. . . . The luster and gleam of the stone, though itself apparently glowing only by the grace of the sun, yet first brings to light the light of the day, the breadth of the sky, the darkness of the night. The temple's firm towering makes visible the invisible space of air. . . . The Greeks early called this emerging and rising in itself and in all things *phusis.* It clears and illuminates, also, that on which and in which man bases his dwelling. We call this ground the earth. What this word says is not to be associated with the idea of a mass of matter deposited somewhere, or with the merely astronomical idea of a planet. Earth is that whence the arising brings back and shelters everything that arises without violation.[19]

From about 580 BC, the new Greek temples were raised on a *krepidoma,* a stepped platform typically consisting of three stages, often incrementally decreasing in height (see fig. 2.17). Because these stages were normally too tall to climb with any ease, and intermediate steps, separate stairs, or ramps had to be added, the krepidoma belonged to a world of representation, not ergonomics. Its closest affinity was to the stepped altar, which seems to have originated in Egypt and made its way to Ionia in the early sixth century, then to mainland Greece.[20] Like a ziggurat in miniature, the krepidoma raised up the earth to meet the sky, and its platform imagined the plateau. The stones still belonged to the earth, but with this uplifting gesture the temple metamorphosed into a supernatural object, a quasi-Olympus. To complete the transformation, the stones must transform themselves, and the sleek luminosity of marble rematerialize earthy substance as solid light.

FIG. 2.7. Temple of Hera II, c. 470–460 BC. Paestum.

CHRYSELEPHANTINE STATUES AND AKROLITHS

The image of classical buildings as glistening bodies of white marble is now so widespread that it requires a concentrated act of historical recollection to realize that such buildings only became familiar from the fifth century BC, commonplace only under the Roman empire, and ubiquitous in the architectural imagination only from the sixteenth century onward.

Their whiteness requires explaining too. Nowadays white is automatically equated with purity in religious contexts but, in antiquity, white primarily meant light. The Greek word *"leukos,"* which is so often translated merely as "white," signifies radiance. It is no coincidence that the Latin word meaning "to shine" is *"lucere,"* or that all the Romance and Germanic derivatives (e.g., *"luce," "luz," "licht," "light,"* even *"lumière"*) find their Indo-European roots (*leuq-/*louq-) in words that unite brightness with whiteness. To Homer, white epitomized solarity: "as *leukos* as the sun," he says.[21] It was this radiant "whiteness as brightness" that dictated the materials of white temples and white statues, which can trace their genealogy back to first cycladic sculpture, and then the chryselephantine (ivory-and-gold) effigies of archaic and classical Greece.

Miniature chryselephantine sculpture was produced in the Aegean as early as the fifteenth century BC, and

FIG. 2.8. Alan LeQuire, Replica of the Athena Parthenos (original c. 438 BC), 1982–90, 2002. Gypsum cement and ground fiberglass over a steel and aluminum armature, acrylic paint and gilding. Nashville Parthenon. The original base possibly gilt bronze figures on an Eleusinian limestone background (compare fig. 3.16).

life-size statuary was introduced to mainland Greece by the mid-sixth century BC.[22] The undisputed paragons of this material vocabulary were the colossal effigies of Athena Parthenos (c. 438 BC) and Zeus Olympios (c. 432 BC), both by Pheidias. Both statues have been destroyed, but we may vividly imagine the former from its modern replica in Nashville (fig. 2.8). These were the largest chryselephantine statues ever made, reaching the temple rafters, and it was said that if the enthroned Zeus stood up, he would knock off the roof (figs. 2.9a, 2.9b). The flesh of these gods was crafted from great tusks of ivory, and their robes were gilded plates tricked out with colored glassy pastes. The importance of all these materials resided not so much in their preciousness, but their rarefaction and light quotients: gold, ivory, and other shimmering materials seemed the heavenly stuff that could absorb the divine essence into the statue. In the eyes of the faithful, they brought a piece of heaven to earth as much as those aniconic effigies that were said to have fallen from the skies themselves. As Lucian says, "Those who enter the temple [of Zeus at Olympia] see not ivory from India nor gold from Thrace but the real son of Kronos and Rhea translated to earth by Pheidias."[23]

The dismembered scraps of this rich material tradition are sadly piffling, but a well-preserved ivory face from the cult statue of a goddess does offer us a glimpse of their spectral presence (fig. 2.10).[24] One might assume here that ivory represented an attribute of femininity, untanned skin—Penelope's skin was "whiter than new-sawn ivory" (Hom. *Od.* 18.190–96)—but it could hardly recommend itself to male subjects, especially not the terrifying Zeus, king of the gods. When such brilliant media composed the bloodless effigies of divinities, they exceeded the mimesis of human flesh alone.

When the gods revealed their true natures, their flesh was dazzling.[25] One has only to consider the story of Pelops, and the Osiris myth in which it originated, to understand that lustrous ivory could summon the supernatural tissue of those unaging gods who feasted on nectar and pure ambrosia—food that was unavailable to humans, and whose original meaning was "uncontaminated by blood."[26] According to myth, Tantalus had butchered his

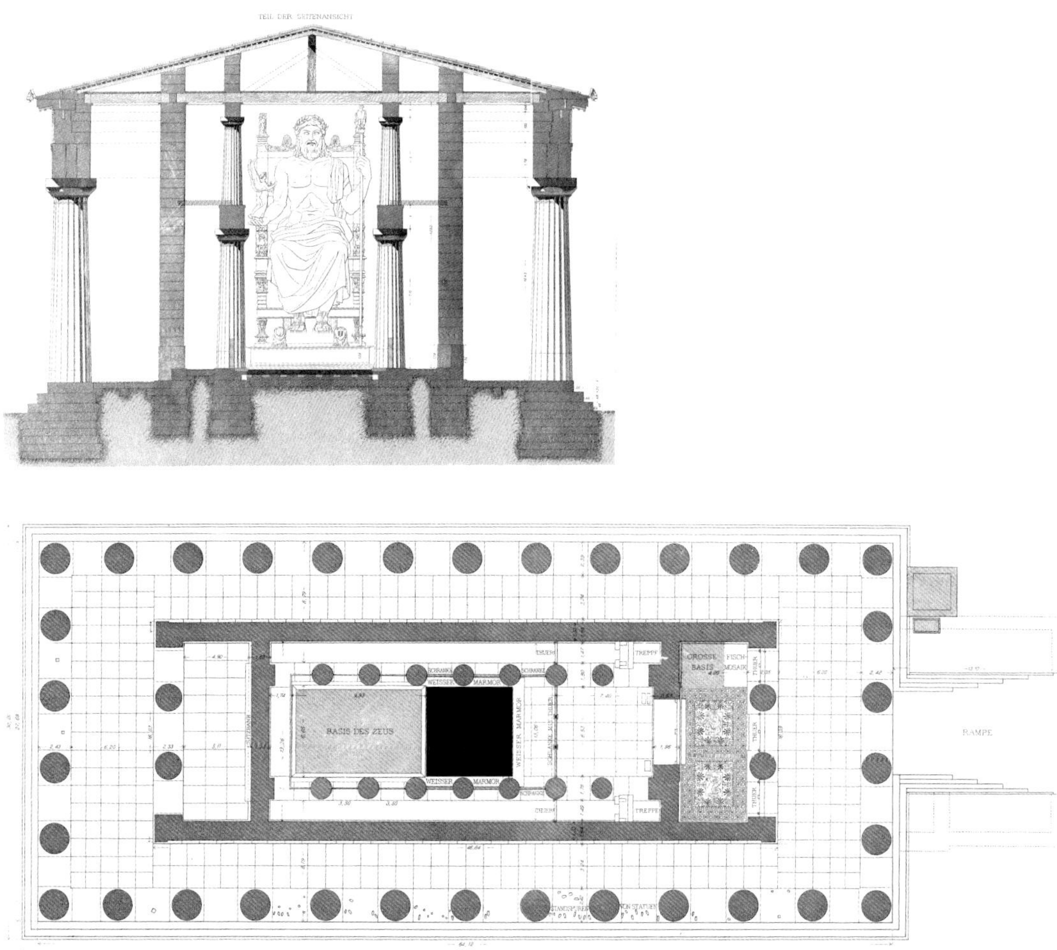

FIG. 2.9A–B.
Temple of Zeus, c. 470–456 BC. Naos remodeled by Pheidias c. 432 BC. Olympia (section and plan with the pool of oil marked in black: Curtius et al. 1890).

FIG. 2.10. Euphranor?, Face of a goddess, c. 350 BC. Ivory, life size. Museo Nazionale Archeologico di Palazzo Massimo, Rome.

son Pelops and served him up to the gods as a meal to test them, though only grief-stricken Demeter absentmindedly ate part of his shoulder. Thereafter, when the outraged gods recomposed Pelops's body and restored him to life, they substituted the missing flesh with a shoulder in ivory.[27] Pliny even says that when he visited the sanctuary of Edlis, in Greece, he saw this shoulder blade displayed there as a relic. The irony—and one that demonstrates the symmetry between rarefied materials and immortal flesh—is that what he saw was probably a fragment of an earlier chryselephantine statue.[28] The double irony is that ivory, the elephant's tusk, was itself once a living tissue.

The visionary materials of chryselephantine sculpture came to be translated into stone in the genre of akrolithic statuary (what Kenneth Lapatin has dubbed "pseudo-chryselephantine"). In akroliths (from "*akrolithos,*" "with extremities of stone"), gilded robes or real drapery concealed a scaffold of wood, to which were fixed the heads and limbs of shimmering white marbles, especially Parian, which substituted for the former ivory.[29] A particular, regional variant was to make the whole statue from stone, with limestone for the body and reserving the finest marble only for the flesh (Clemente Marconi has called this technique "pseudo-akrolithic"). The most splendid example is the Sicilian "Goddess of Morgantina," probably Persephone (c. 420–410 BC; fig. 2.11), but even on the metopes of the Temple of Hera at Selinous, Sicily (c. 460–450 BC), inserts of Parian marble pick out the flesh of "white-armed Hera" (Hom. *Il.* 15.78) within slabs of the local, calcareous limestone (fig. 2.12).[30]

BUILDING IN LIGHT: WHITE MARBLE TEMPLES

The first stone temples that the mainland Greeks built in the sixth century were made of *pōros* (or *poros*), a term so vague that it includes travertine, limestone, and sandstone—in other words, stones from the nearest quarries to hand.[31] These sedimentary rocks were compact enough to be dressed and honed, but too holed and grainy to take a polish. Marbles are metamorphic rocks, limestones whose calcite has recrystallized under intense heat and pressure.

FIG. 2.11. The Morgantina Goddess (Persephone?), c. 420–410 BC. Parian marble and local wackestone, height: 90 in. (220 cm). Museo Archeologico, Aidone.

FIG. 2.12. Metope, Temple A (Hera?), c. 460–450 BC. Selinous, Sicily. Parian marble inserts in calcareous stone (Cusa). Museo Archeologico A. Salinas, Palermo.

FIG. 2.13. Temple of Demeter (?), c. 530 BC. Sangri, Naxos.

As they are denser and harder, they take a sheer polish that reflects light with stunning efficiency, and because they are crystalline, light can penetrate and, if they are thin enough, pass through them. In other words, marbles radiate and shine, and their light seems not alien to substance but embodied within it. Indeed, the root of the noun "*marmaros*" ("marble") was the verb "*marmairein*" (meaning "to shine," "glisten," "sparkle," or "flash").[32]

The marble temple actually begins as a local idiom, and at the same time that stone temples were beginning to appear on the Greek mainland and in its western colonies. The invention occurs on Naxos and Paros, because these Cycladic islands offered brilliant white marbles in plenty.[33] Naxian marbles have the biggest calcite crystals, but they are also micaceous, meaning that a dusting of glitter seems to overlie the inherent glow. Naxos was also renowned for its emery (in Latin, "*naxium*"), the tough abrasive that made shaping and polishing the marble first possible, and that led to the first systematic sculpture in marble here around 3000 BC, the world-renowned "Cycladic Sculpture," and then the first monumental Greek sculpture, the kouros (and perhaps the first *kore*), from about 600 BC.[34]

At the end of the seventh century, the Naxians began boating marble thirty miles away to the island of Delos, which was the mythical birthplace of Apollo and therefore one of the most important sanctuaries in the ancient world. Here they built first a terrace of lion statues (much like an avenue of Egyptian sphinxes) and then a massive kouros, also from Naxian marble, whose attributes identified him as Apollo (c. 600/590 BC). This stood about thirty feet high—that is five to six times life-size—weighed twenty-five tons, and was a monolith (see figs. 2.5, 2.6).[35]

The Apollo of the Naxians was the largest standing statue in the Greek world, the largest artifact in marble, and, in fact, the largest monolithic structure of any kind. It pointed the way to entire buildings made in this medium and at this superhuman scale. It dwarfed the building next to it, the so-called "Oikos of the Naxians," which has been variously interpreted as a treasury, banqueting hall (*hestiatorion*), or the first monumental temple to Apollo on the island.[36] In effect, there is no reason it should not be all three. Within a few years (590–580 BC) this "Oikos" was rebuilt and, in an amazing innovation accredited to the Naxian sculptor Byzes, roofed over with marble tiles sawn so thin that they were translucent and illuminated the room within.[37] All gods revealed themselves in light, but Apollo's particular epithet was "*phoibos*" (meaning "bright," "pure," or "radiant"), and the roof was fashioned from the statuary marble that figured the colossus of the god right next to it. One can scarcely imagine the creeping radiance in the cult room as the dawn approached, and the next thing one expected (as Kallimachos says in his own later hymn to Apollo) would be to hear Apollo's beautiful foot kicking at the door.

In the same era that the Naxians rebuilt the "Oikos" on Delos, they began buildings entirely of white marble on their own island. The first was a temple at Yria (c. 580 BC), whose main (east) façade and roof were made from it. It was probably dedicated to Dionysos, the most popular god on Naxos (another colossus of the god, even bigger than that on Delos, still lies in the quarry at Apollonas, and was possibly intended for a gargantuan temple at Palatia).[38] At the slightly later Temple of Demeter at Sangri (c. 530 BC), every single block was of Naxian marble, including its roof tiles and even its rafters and purlins (fig. 2.13). This temple must have seemed a concrescence of light itself, and within it a chryselephantine cult statue materialized from the numinous aura, like a glistening bather rising from a tub.[39]

Naxos exported kouroi to Attica throughout the late seventh and early sixth century BC, but nearby Paros boasted an even better statuary marble, which gradually ousted Naxian from the market.[40] However, both islands exported marbles not only to make sculpture but also to make tiles for temple roofs. A roof tile has even been found in Athens inscribed "By" [-zes?], and Parian tiles have been retrieved, for example, from the fifth-century BC "Temple C" (dedicated to Athena?) at Metapontum, in Calabria (Magna Graecia), and Temple A (Hera?) at Selinous, in Sicily (fig. 2.14).[41] These roofs made the temples flash like lightning or sparkle like stars. They brought the sky to earth, as much as the krepidoma raised the earth to meet it. Just how many of these roofs also worked as internal diffusers to illuminate the cult room is a controversial question.[42] Yet it would seem perverse that later architects adopted the Naxian invention of the marble roof only to discard its most wondrous characteristic, and there is a strong case to make that such ceilings were highlights of the Parthenon (447–438 BC) and great Temple of Zeus at Olympia (c. 470–456 BC).

Both temples housed the shimmering chryselephantine cult statues of Athena Parthenos (c. 438 BC) and Zeus Olympios (c. 432 BC), respectively (see figs. 2.8, 2.9a). Both temples were also roofed in marble tiles: the Parthenon of Pentelic marble, and the Temple of Zeus of Parian.[43] The architects of the Lincoln Memorial (1914–22) in Washington, DC, whose statue is clearly indebted to the Olympian Zeus, and the Parthenon replica at Nashville (1920–31, interior 1925–31) unwittingly replicated the arrangement when they installed translucent laylights.[44] In the Lincoln Memorial, the ceiling panes are white Alabama marble saturated in beeswax and paraffin for an alabaster look, and they come alive at night when they are electrically backlit (fig. 2.15).[45]

It has been calculated for the Temple of Zeus at Olympia that each of the thousand Parian roof tiles filtered a light equivalent to a twenty-watt bulb, resulting in a luminance of about twenty thousand watts and a ground illuminance of at least two hundred lux (the

FIG. 2.14. Roof tiles, Temple A (Hera?), c. 460–450 BC. Selinous, Sicily. Parian marble. Museo Archeologico A. Salinas, Palermo.

modern minimum for a lobby).[46] One would love to have explicit contemporary texts describing such luminous interiors, but the only testimony is Pliny the Elder's description of the much later Temple of Fortune in Rome that the emperor Nero rebuilt (64/68 AD). This was fashioned from a Cappadocian stone "as hard as marble, brilliant and translucent, even in those parts that were streaked with yellow veining, for which reason it is called *phengitēs* ["splendor-stone"] . . . by which means, even when the doors were shut, it gleamed like the day . . . as if the light was enclosed within, rather than transmitted" from without (*HN*. 36.46.163). This stone has not been identified, but the etymology is clear: "*phengitēs*" derives from "*phengos*" (meaning "light" or "splendor").

This is not the end of the story, however. In both the Parthenon and Temple of Zeus at Olympia, Pheidias also excavated shallow reflecting pools before the statues, dissolving the floor into a mirror. The Parthenon pool contained water, and that in Olympia a film of olive oil over dark Eleusinian limestone.[47] The result in both cases was a mist of light that filtered from above and reflected from below, a topsy-turvy world with a liquid floor and a marble sky. We can glimpse the effect in a modern installation by the sculptor Richard Wilson, who filled a top-lit gallery

FIG. 2.15. Henry Bacon and Daniel Chester French, Lincoln Memorial, 1914–22. Ceiling panes of white Alabama marble saturated in beeswax and paraffin. Washington, DC.

FIG. 2.16. Richard Wilson, *20:50*, 2018. Cor-Ten steel and recycled engine oil. Hayward Gallery, London.

with recycled engine oil to create a space "where the internal volume is greater than its physical boundaries," a volume pleated by a horizon of a hair's breadth (fig. 2.16).[48]

In Olympia and on the Akropolis, colossal chryselephantine cult images seemed to mediate between the liquid earth and the marble sky. In Greek, ceiling and heaven are both *ouranos*, like Latin *caelum*, and it is no wonder that the Roman poet Propertius will later (c. 23 BC) describe the temple at Olympia as "the mansion of Elean Jove, that counterfeits heaven" (*El.* 3.2.20: *Iovis Elei caelum imitata domus*). The Parthenon may have basked in a more golden light. Its Pentelic roof transmitted a yellowish light (unlike the white of Naxian or the orange of Parian), the massive naos doors were gold and ivory, and the whole exterior seems to have been glazed to give it an ivory tint (an ocher patina remains).[49] And, of course, the naos was oriented to face the rising sun.

POROS AND PAROS

Marble roof tiles weighed a few pounds apiece and made nice ballast, but quarrying a whole marble temple, and then shipping it block by block, would have cost the earth. Builders must instead rely on the best poros available, and because poros is porous, add a coating of impermeable lime plaster to defend the pores from water and frost.[50] That is the practical explanation, but supernatural radiance had been the goal of brilliant white plaster since the "White Temple" at Uruk (see fig. 1.11), and the shimmering stucco skins on Greek temples did not so much counterfeit marble as denature the stone altogether. It is all very well to talk about simply imitating marble, but in places such as Sicily, there were no marble quarries, and the only time the locals saw white marble was on imported statues of gods, or the akroliths in which white marble portrayed divine flesh (see figs. 2.11, 2.12). White marbles were as alien to these climes as blocks of ice, and their crystalline structure must have seemed just a rung below rock crystal, which the ancients held to be simply a form of super-frozen ice. Thus, when the masons of Selinous concealed their honey-veined limestone under a coat of white plaster, they conjured up not only a marble temple but almost an ice palace (fig. 2.17).[51] Hellenistic Greeks may have carried this transubstantial cosmetic as far afield as Pakistan.[52]

By about 540 BC, Parian marble seems to have definitively displaced Naxian as the stone of choice for statuary.[53] While Naxian marble had a blue-gray shimmer, Parian had a golden glimmer. Its low index of refraction allows light to penetrate several millimeters into the stone before it scatters out, which gives Parian a yellowish cast when placed next to other white marbles. Pindar compared it with refined gold (*Nem.* 4.81–83; c. 473 BC), while it instead reminded the Hellenistic poet Theokritos of the color of teeth (*Id.* 6.38; third or second century BC). Like ivory it was translucent, which brought it close to human flesh and made it an ideal material for akroliths. In classical poetry Parian became a favored metonym for the porcelain complexion, the alabaster ankle, or the ivory neck. Poets similarly speak of firm and glowing breasts from Sumeria until late antiquity.[54] Horace will tell us, "The sparkle of stunning Glycera, purer than Parian marble, sets me on fire"; and, in Petronius, Encolpius claims that the flesh of his adolescent lover "puts Parian marble in the shade."[55] A Byzantine intellectual, Michael Psellos (eleventh century), is surely describing Parian marble when he says of a statue of Eros, "It gives the impression of frozen milk or snow before it is heaped up and turns to ice" (*Orat. min.* 34). A fire-scorched head of a goddess, from Italica in southern Spain, best makes us appreciate by its very damage the corpuscles of light that seem to be sealed within Parian's macrocrystalline grain (fig. 2.18).

The accelerating use of Parian marble throughout the Greek mainland brought enormous wealth to the small island. Between the mid-sixth and late fifth centuries, several Greek cities elected Parian to build their treasuries at the sanctuary of Apollo at Delphi, where they sparkled like diamonds sprinkled on the calcareous tuff slopes of Mt. Parnassos.[56] It was the diminutive scale of these treasuries that made an entire building of marble even feasible. Every block had to be conveyed the 240 miles from Paros to Delphi, by land and sea, along with the masons who knew how to work them. Every block was a treasure in its own right, in buildings that were both votives to the divinity in the form of miniature temples, and caskets of offerings made from other precious materials. The touchstone of Parian's value is the fact that the Athenians

FIG. 2.17. Temple of A (Hera?), c. 460–450 BC. Stylobate and stuccoed columns. Selinous, Sicily.

FIG. 2.18. Head of a goddess, 1st or 2nd century AD from Italica. Parian marble, life size. Archeological Museum, Seville.

FIG. 2.19. Treasury of the Athenians, c. 489 BC. Parian marble. Delphi.

still used the marble for their own treasury, c. 489 BC (fig. 2.19), at a time when Paros was a hostile state, and when Athens was building its most important temple, to Athena Polias, from its own local and equally exquisite Pentelic marble.[57]

Within thirty years of the earliest treasuries in Parian marble, the Athenian Alkmaionid family made a material bid for conspicuous piety on the Panhellenic stage by rebuilding the Temple of Phoibos Apollo (514/507 BC) at the very heart of Delphi (fig. 2.20a). The original contract had specified that the temple was to be built from the local poros, but the Alkmaionids poured money into making the east façade entirely of Parian.[58] Many gods revealed themselves in light, but Apollo seemed the very incarnation of radiance, and this identity was so strong that from the fifth century he was increasingly identified with the separate sun god Helios. It was the dawn sun that brought the Parian façade to life, awakening its innate glow. By no coincidence, the tympanum of the east façade featured a frontal epiphany of Phoibos Apollo, again in Parian marble, arriving at Delphi from Delos by a circuitous route and bringing his radiance with him. As the contemporary (pseudo-Homeric) *Hymn to Pythian Apollo* recounts: "Flashes of fire flew from him thick, and their brightness reached to heaven. He entered into his shrine between priceless tripods and there made a flame to flare up, showing forth the splendor of his shafts, so that their radiance filled all of Krisa [Delphi], and the wives and the well-girdled daughters of the Krisaeans raised a cry at that outburst of Phoibos."[59]

The new temple had quite an impact on the Greek imagination. Within twenty-five years, the poet Pindar remarked, "The story is doing the rounds in all cities, of how the citizens of Erechtheus [i.e., Athenians, the Alkmaionids], Apollo, made your temple in divine Pytho a marvel to behold" (*Pyth.* 7.9–12; 486 BC). Even after all the gleaming buildings of the Athenian Akropolis had been built, Euripides will write, "Not only in sacred Athens are there halls of the gods with beautiful columns . . . but also in the house of Loxias [Apollo], Leto's son, there is the lovely-eyed light of twin faces" (*Ion* 184–89; 414/412 BC); these enigmatic words may evoke the soul-searching eyes of Apollo and his twin, Artemis (in the pediment?), or the brilliant gable-end façades of the temple.

An earthquake devastated the temple in 373 BC, and when it was reconstructed in 330 BC, stucco was lathered over poros to counterfeit the lost marbles so that the façade remained almost as much an apparition as before. Within years (c. 325/300 BC) the Rhodians had raised a towering pedestal for a gilt-bronze statue group of the sun god Helios, driving his solar chariot straight at the temple pediment, the horses rearing to suggest the rising sun, thereby giving shape and substance to a relationship that had until now been manifested only by light (fig. 2.20b).[60]

FIG. 2.20A. Late archaic Temple of Apollo, 514/507 BC. Delphi (Courby 1927).

FIG. 2.20B. Rhodian monument with quadriga of Helios, c. 325/300 BC. Delphi (Didier Laroche).

The Athenians built their Delphian treasury from Parian, but back home they were building the Temple of Athena Parthenos (begun c. 490 BC) from the exquisite white marble of Mt. Pentelikon, only eleven miles from the city. This was to be the first all-marble building on a monumental scale anywhere in Greece, and provisioning it now opened Athenian eyes to the vast stocks (a whole mountain, not just a seam) at their disposal. When it came to rebuilding the Acropolis after the Persian sack of 480 BC, they set about building every temple from it, not just that of Athena. (And even though the marble was local, the extraordinary cost still saw Pericles effectively loot the treasury of the Delian League to pay for it.) Henceforth, Pentelic marble became the premier white stone for architecture, alongside to some degree Hymettan, and the Acropolis became the paragon of a shining citadel for the gods.[61] Xenophon (c. 354 BC) says that Pentelic became coveted by Greeks and barbarians alike (*Por.* 1.4), regardless of their political sympathies. Thus, at Samothrace, just as at Delphi before it, a monumental pavilion erected by the Macedonian kings Philip III Arrhidaios and Alexander IV (323–317 BC) was mainly built from the local Thasian marble, which was brilliant white if rather coarse-grained, but the façade was instead made from shimmering Pentelic.[62]

As white marble became a fact of premier building, city-states began to prospect their own supplies. It is a sign of how widespread the diffusion of brilliant white marbles had become by the end of the Classical period that in 323 BC, the year Alexander the Great died, the town council of Ephesos in Asia Minor was publicly debating exactly which kind they should import to rebuild the recently burned Temple of Artemis—whether from Paros, Prokonesos, Herakleia-under-Latmos, or Thasos—until a shepherd (Pixodaros) providentially discovered local quarries that removed any need (Vitr. *De Arch.* 10.2.15).

WHITE TEMPLES AND WHITE ROME

On the Italian peninsula, however, there were no marble buildings anywhere, because there were no marbles in Magna Graecia, and the now-famous quarries at Carrara (ancient Luni) would not be opened until the mid-first century BC.[63] Etruscan and Italic temples were largely timber and brick affairs on tufa podiums, and those wholly built from calcareous stones or tufa only began to appear in the third century BC. Following Greek precedent, a skin of brilliant stucco was applied to conceal the shabby stones and make the temples gleam. This immaculate disguise is textually attested in Rome itself by 179 BC, when the censors M. Aemilius Lepidus and M. Fulvius Nobilior "embellished the temple of Jupiter Capitolinus and the [votive] columns around it with white stucco," and later building inscriptions (at Assisi and Brescia) continue to applaud such *opus albarium* or *marmoratum.*[64] These stucco coatings covered a multitude of sins: mixed materials, construction phases, fire damage, or they just updated a temple's ornament; whatever its troubled past, a coat of stucco ensured that the shrine always looked new.[65] Conversely, when Cicero indicted Verres for corruption, he charged that he had literally "white-washed" the columns (*columnas dealbari*) of the Temple of Castor and Pollux in the Forum Romanum, rather than having newly quarried ones delivered (*Verr.* 2.5.145, 2.5.147).

The first abortive steps by the Romans to exchange these stucco fictions for marble facts did not come until 173 BC. In this year the censor Q. Fulvius Flaccus became so enamored of the Pentelic and Parian roof tiles on the Temple of Hera (c. 475–450 BC) at Cape Lacinium, near Kroton, in Calabria, that he stole half of them to reuse on his own Temple of Fortuna Equestris in Rome. It was not to be. The scandalized Senate cried sacrilege, using the term "*spolia*" for the first time in the sense that we now know it, and forced the tiles' return.[66]

The definitive step was taken after the conquest of Macedon in 146 BC, when the returning victor, Q. Caecilius Metellus Macedonicus, dedicated Rome's first marble temple to Jupiter Stator. The architect of this trophy, Hermodoros of Salamis, came from Greece, as no doubt did the marble and the masons to carve it.[67] The foreign stones substantiated Rome's new hegemony in her power literally to move mountains. Temples to Mars and Neptune followed suit, both in marble from Mt. Pentelikon.[68] Although these temples are long lost, another just as Hellenic and also Pentelic, the Temple of Hercules Victor, still stands to vouch for the pristine and prismatic edifices we have lost (fig. 2.21). According to one scholar this temple was another trophy, built c. 142 BC by L. Mummius Achaicus, the conqueror of Achaea and despoiler of Corinth; according to another it was commissioned, c. 100/90 BC, by a merchant who had grown rich on the trade in Aegean olive oil.[69]

For the next two centuries virtually all temples in Rome continued to be built from white marbles, whether Greek or Italian. After visiting the Greek theater at Mytilene in 62 BC, Pompey the Great decided to build

FIG. 2.21. Temple of Hercules Victor, c. 142 BC or c. 100/90 BC. Pentelic marble, later repairs in Luna (Carrara) marble. Rome.

FIG. 2.22. Temple of Apollo Sosianus, 34/29 BC?. Rome.

Rome's first permanent theater (completed 55 BC), incorporating a temple to Venus Victrix, which was presumably in Pentelic marble. Pompey's competitor, Julius Caesar, made white marble a more ubiquitous fact of the Roman landscape when he (almost certainly) exploited the quarries at Luni (modern Carrara) on numerous projects he instigated for Rome's urban regeneration: the first independent Forum Iulium with its own temple to Venus Genetrix, the new Curia, and the Saepta Iulia.[70] Nevertheless, and despite the expense, Pentelic would remain a marble of choice long after Carrara had become a busy quarry.

Caesar was assassinated in 44 BC, a brief civil war ensued, and then a prolonged power struggle between Caesar's lieutenant, Mark Anthony, and his nephew, Octavian. In 31 BC the young Octavian finally disposed of Mark Anthony at the battle of Actium, changed his name to Augustus (27 BC), and in stealthy steps substituted the troubled republic with an imperial regime. Within three decades the empire had come to span lands between the English Channel and Parthian frontier, and from the Rhine to the Nile.

FIG. 2.23. Cella, Capitolium, late 1st century BC. Fresco imitations of marble revetment. Brescia.

The variety of marbles that Augustus's engineers prospected within these stabilized frontiers enlarged dramatically, yet Augustus was resolute in building only white temples (fig. 2.22). Such "color blindness" is all the more conspicuous if one remembers that the cult rooms, and often the precinct porticoes, were lavishly adorned with the gamut of marbles now arriving from Europe, Asia, and Africa.[71] These effects can now best be imagined from their imitation in fresco, even before the materials themselves became widely available (fig. 2.23). A temple here or there may have diverged from the trend, but only with Hadrian (r. 117–138 AD) was the color bar truly broken and do we regularly witness temples with columns of fiery granites, purple-streaked Phrygian marble, or sea-green Carystian, such as the Pantheon (118–124 AD), the Temple of the Divine Trajan (119/128 AD), and the Temple of Antoninus and Faustina (begun 141 AD). Nonetheless, in the mid-second century the rhetorician Aelius Aristeides would still exalt a Rome so crammed with glistening white buildings that it seemed an Alpine landscape of snow-clad slopes and gorges.[72] A silvered glass orb, from Mérida, bearing a view of the civic center of Pozzuoli outlined in white, best conveys this ancient imaginary of the resplendently marbled metropolis (fig. 2.24).[73] The question remains, however, why were the temples white, and why did they remain white for so long?

FIG. 2.24. Silvered glass ampule with view of Pozzuoli, 2nd century AD?. Diam.: 4⅛ in. (10.6 cm). National Museum of Roman Art, Mérida.

A WHITER SHADE OF PALE: THE VIA LABICANA AUGUSTUS AND OTHER AKROLITHS

At this point it is important to qualify that the Romans perceived that there were at root two types of whiteness: *albus* and *candidus.* In the words of the fourth-century literary detective Servius: "It is one thing to be shining [candidus], that is to be bathed in a certain brilliant light, and another to be white [albus], which amounts to approaching pallor [*pallor*]."[74] "Pallor," incidentally, does not mean "paleness" so much as the absence of color: Vergil, for example, only uses "*pallidus*" to describe individuals who are white with fear—that is, drained of color—or blanched. The opposite of *pallidus* was *rubeus,* which signified the capacity to blush.[75] In material terms, the acme of "candidus" was an Indian glass prized above all others because it was as refractive and transparent as rock crystal (Pliny. *HN.* 36.66; Isid. *Etym.* 16.16). Clearly, the Latin "candidus" matches the Greek "leukos," and indicates "white" only insofar as this means the zenith of visible light; there was a dark analogue to "albus-candidus" in the distinction that Latin also made between "*niger*" and "*ater,*" both words for "black." "Niger" was to "ater" as "ebony" is to "charcoal."[76]

The famous statue of Augustus from the Via Labicana demonstrates how the distinction between "albus" and "candidus" translated into sculptural practice, and thence how such sensibilities clarify the material preoccupations of white temples as well (fig. 2.25). Augustus is portrayed as the supreme priest of all pagan cults, the Pontifex Maximus, pouring a libation and head covered in deference to the god.[77] Though totally white, the statue is made from two different marbles, Italian Luna for the toga and Greek Parian for all Augustus's visible flesh.

It is obvious that these material distinctions adopt the religious material vocabulary of that pseudo-akrolithic cult statuary that had translated chryselephantine into stone, where Parian marble substituted for ivory. The taste for pseudo-akroliths had continued unabated into the Roman era, whether statues of gods or divinized emperors, which came in extra-large to colossal dimensions. Unhappily there is no catalogue of them, but the fragmentary akrolith of Domitian (c. 96 AD) in the Temple of the Flavian Emperors at Ephesus (fig. 2.26) provides an appropriate example, not just for its impressive scale but also because the court poets Martial and Statius lovingly catalogued the sanctity of this emperor's flesh.[78] Fittingly, the most ingenious statement that such statues may be less effigies than epiphanies comes from the same

FIG. 2.25. Via Labicana Augustus, 1st century AD. Pseudo-akrolith, Parian and Luna marbles, height: 81½ in. (207 cm). Museo Nazionale Archeologico di Palazzo Massimo, Rome.

reign, Martial's address to the *statue* of Julia, Domitian's deified niece:

> Julia, who would not think that you were shaped by
> Phidias's chisel
> or were not the work of Athena's art?
> The beaming Parian answers me with a not-mute image
> [OR *the beaming Parian matches a speaking likeness*]
> and living beauty shines in your serene face.[79]

The Via Labicana Augustus must be one of the seminal examples of this material tradition in imperial portraiture. Combinations of fine and finer white marbles had already figured in the aforementioned "Goddess of Morgantina" (see fig. 2.11), as well as Hellenistic examples such as the famous "Antium Girl" (250/230 BC), which, significantly, most commentators now consider to represent a priestess.[80] A verse once inscribed on a long-lost statue by another famed sculptor also highlighted its subtle play on shades of white: "Praxiteles formed Danae and the Nymphs / in Parian marble, but me, Pan, of Pentelic stone."[81] However, for a sculptor to suggest that Augustus had effeminate, lily-white skin would have been a hanging offense, and the material juxtapositions of his effigy instead announce that the emperor's flesh is so radiant that he is *brighter than white.*

Just how bright becomes clear from the toga that envelops him. These woolen robes were themselves whitened with fuller's earth, chalk, and on occasion even marble dust. They weighed a ton and were no longer worn for comfort but only to accommodate tradition. In fact, Augustus endorsed and enforced their use. The most famous of these brilliant togas, the *toga candida* worn by those seeking office, was positively loaded with chalk to present the bearer as a spotless "candidate" for the future burdens of office (Isid. *Etym.* 19.24.3; Pers. *Satir.* 5.177). Likewise, Augustus's priestly robes would have been of the whitest cloth so that his purity and divine office shone about him.

The Romans never stopped being aware of the potential associations of the brilliant toga. Even in Christ's Transfiguration, his seamless robes are described as "shining, excessively white [candidus], like snow, whiter than any fuller on earth could make them" (Mark 9:3). Fra Angelico's white-on-white version of the event (fig. 2.27) is one of the best representations of this transcendent moment when, as he would have known, Christ discards

FIG. 2.26. Colossal akrolith of Domitian, Temple of the Flavian Emperors, c. 96 AD. Ephesus. White marble (from Aphrodisias?). Archaeological Museum, Ephesus.

FIG. 2.27. Fra Angelico, *Transfiguration*, 1440–42. Fresco. Cell 6, Convento di S. Marco, Florence.

his human nature and reverts from *lumen de lumine* (the "light from light," the incarnate light) to *lux,* God's essential nature; that supernal and uncreated light which is so blinding it is unintelligible. "Lux" is to "lumen" as "candidus" is to "albus."

In the Via Labicana Augustus the toga gleams but the flesh glows, thanks to the translucency of Parian marble. Moreover, this is not just Parian marble, but the most refined quality called *"lychnites"* (lamp-like; from *"lychnos,"* meaning "lamp"; cognate with "leukos," meaning "white" or "bright"). Lychnites unlikely earned its name because (according to Varro) it was quarried in shafts by lamplight, but rather because this word signified a particular brilliance.[82] The two marbles of the Via Labicana Augustus both gleam, but differentially, and the effect was as intentional as Dan Flavin's juxtaposition of daylight and cool white fluorescent tubes (fig. 2.28). The two marbles of the Via Labicana Augustus distinguish between a relative and an absolute white, between albus and candidus, between pallor and what we might call an immaterial or transubstantial white. However, "candidus" has an ethical meaning as well, one that is preserved in English. To be candid is to be transparent in one's meaning; to speak candidly is to tell the unvarnished truth. So the candid marble of Augustus's body also manifests his moral probity. He is high-minded and *pectore candidus*—pure of heart. Lychnites, *purus* and *clarus,* was perfect for the job.[83]

But Augustus is also god-like. It is immaterial whether the Via Labicana Augustus was carved before or after his death, as scholars debate. Posthumously, Parian would forecast the luminosity of Divus Augustus, destined like his divine father (Julius Caesar) and all later apotheosized emperors to become a star in the firmament.[84] In life, Augustus's bright flesh might project his ensoulment of destiny, the inner glow of his guiding spirit, the *genius.* And Augustus's divinity had begun in life: he was both equated with Jupiter and had ushered in a new golden age under the aegis of his patron, the sun god Apollo. Augustus encouraged the belief that he was the son of Apollo, and Horace hails him as the "beautiful sun" whose "countenance like spring shines upon the people" (*Carm.* 4.2; 4.5).

As soon as he assumed the reins of the state, not only did he dedicate a temple to Apollo on the Palatine (36–28 BC), built from the best Luna marble, with ivory doors and gilded statuary, but he also rebuilt his own house around it.[85] Propertius (c. 28 BC) celebrated its material program, recalling literary topoi reaching back to Homer:

> Then in the middle rose the temple of dazzling marble,
> Dearer to Phoebus even than his Ortygian home:
> Upon the pediment of this stood the chariot of the Sun,
> And the doors were noble works in Libyan ivory.

Reciprocally, when Ovid later (c. 7 AD) described the Palace of the Sun, it was clearly with Augustus's temple in mind:

> The Palace of the Sun was raised high on towering columns,
> And shimmered with the glint of gold and fiery bronze;
> Its high pediment was covered with shining ivory,
> While its twin doors shone bright with silver.
> The workmanship was even more superb than the materials.[86]

Around this incandescent shrine stretched a ring of columns in yellow Numidian marble like the spokes of Helios's halo, and the temple shone all the brighter for this frame of gold-colored stones.[87]

Under Augustus, as Propertius and Ovid intone, all Rome had suddenly become golden (*nunc aurea Roma est*), meaning not only actual gilding but the brilliant light from her new marble temples.[88] Indeed, the Romans were

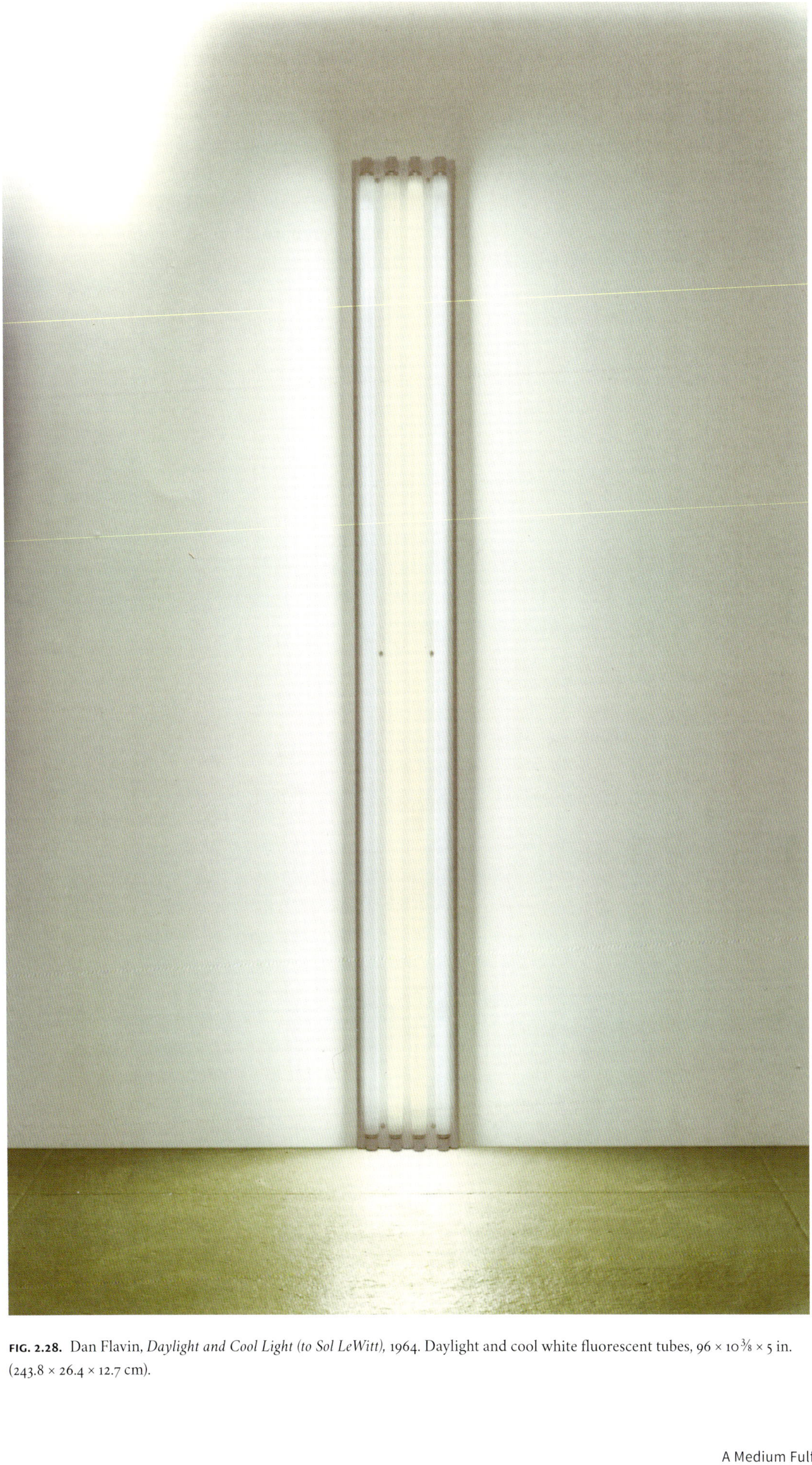

FIG. 2.28. Dan Flavin, *Daylight and Cool Light (to Sol LeWitt)*, 1964. Daylight and cool white fluorescent tubes, 96 × 10⅜ × 5 in. (243.8 × 26.4 × 12.7 cm).

avid polishers: the Pentelic shafts of the Capitolium were supposedly ruined once they reached Rome after 86 BC, because the local masons polished them so much that they lost their correct proportions (Plut. *Popl.* 15.4). The already-high polish of both architectural and statuary marbles could achieve maximum reflectivity with the further addition of oils or wax: Hellenistic inscriptions on Delos record that temples there were waxed and oiled, like wrestlers, and in the Imperial period architectural and statuary marbles received a film of wax (a technique called "*ganosis*").[89] The results were sometimes so reflective that pilgrims were actually advised to shield their eyes from the glare (Pliny. *HN.* 36.4.32). When the poet Rutilius Namatianus strolled through Rome in 417 AD—more than a decade after the temples had been officially closed (395 AD)—he still felt that this superabundant luster meant that the gods could not be far away: "And the glittering temples dazzle the wandering gaze: I could well believe that the gods themselves dwelt like this." Around 800 AD, the temples of Verona still seemed to bask in the same light.[90]

It was not only the sun from which snowy stones borrowed their light, but also his sister, the moon. From at least the days of Enheduanna, the moon was not made of green cheese, but a great, glowing orb of stone pouring its own light from the nighttime sky (see fig. 1.12). The philosopher Anaxagoras was actually impeached (c. 450 BC) by the Athenians for blasphemy when he instead proposed that the moon shone with light reflected from the "red-hot stone" of the sun. Pliny was probably enchanted by the moonlight when he patriotically claimed that Luna (Carrara) marble shone even brighter (*candidior*) than Parian, and sixteenth-century observers would enthuse that it glowed like the moon reflected in the bottom of a well.[91] Strabo (*Geog.* 5.2.5) says that the Greeks called this town "Selene" ("Moon"), so Rutilius Namatianus was repeating an old pun when he capitalized on the similarity of the place-name "Luni" to the word for moon, "*Luna.*" Describing the glistening quarry walls, he writes of the "sister [*Luna*] who draws her radiance from the Sun [and] is the bestower of the city's name [*Luni*]" (*De red.* 2.63–64).

FIG. 2.29. Temple of Venus and Rome, detail of *Presentation of Christ in the Temple,* triumphal arch mosaic, S. Maria Maggiore, 432–440 AD (restored 1929–31).

Rutilius continues, "The native crag shines painted with soft sheen, and its rocks surpass smiling lilies. The soil is rich with marbles, which, sumptuous in the light of their color, challenge the virgin snow" (*De red.* 2.65–68). Highlights danced on the polished temples like the glints on snow-clad peaks, and this brilliance indexed the glacial purity and supernatural shimmer of the slopes, the traditional domain of the Olympians and the apex of virgin Nature. The Macedonian poet Poseidippos (c. 250 BC) imagines "the ridges of snow-clad Parnassus or Olympus" before calling Apollo's temple at Delphi "the snow-white house of Parian [marble]."[92] And Vergil evokes the same environment when he speaks of "the snowy threshold of [the temple of] brilliant Phoebus," and calls Paros "snowy" (*Aen.* 8.720; 3.126). "*Niteo*" ("to shine") is also cognate with "*nix*" ("snow"), and Ovid celebrates the future emperor Tiberius's reconstruction (7–10 AD) of the Temple of Concordia as the "snow-white shrine . . . of the shining goddess" (*Fast.* 1.637). But Josephus is most explicit when he describes the great temple that Herod had rebuilt in Jerusalem (20–19 BC): "Persons straining to look at [it] were compelled to avert their eyes, as from solar rays. To approaching strangers it appeared from a distance like a snow-clad mountain; for all that was not overlaid with gold was of purest white."[93] The irony is that no marble seems to have been used in the construction of this temple, and Herod only used the local, pinkish limestone, so that these effects, just like those at Selinous, were achieved with a coating of fine white stucco, like a summer snowfall.[94]

Even Christian mosaicists acknowledged the polished whiteness of Rome's temples by inventing a diagonalizing, *changeant* convention to express their shimmer, as when they depicted the Temple of Venus and Rome on the triumphal arch of S. Maria Maggiore in the early fifth century AD (fig. 2.29).[95] The same arch has the earliest representation of the heavenly edifice that will sweep away the pagan temples, the Heavenly Jerusalem that must descend at the end of time and whose walls were built from twelve gems the size of boulders (Revelation 21:18, 21–25). The juxtaposition of pagan and apocalyptic seems to foretell the words of Fulgentius of Ruspa (468–533) visiting the city in the early sixth century: "Oh brothers, how beautiful must be the heavenly Jerusalem, if earthly Rome can shine with such brilliance!"[96]

POLYCHROMY

Of course, even white temples were not exclusively white, and long-standing debate over the extent of their polychromy has reignited in recent decades. The Neoclassical mental regime of all-white monuments cutting glacial silhouettes against azure skies has long been exiled to the camp of "chromophobia." A host of recent publications intent on resurrecting the faded or invisible polychromy once applied to ancient sculpture has opened our eyes wide to the phenomenon, but in some cases the reconstructions are submerged in paint of the most saturated hue, leaving no patch of white anywhere.[97]

Moreover, it is a gross simplification to be confronted with the stark choice between all-white objects or those with the chromatic subtlety of a traffic light. Polychromy was far more partial and semantically selective than either camp will allow. In architecture, the friezes, cornices, eaves, coffers, and details of the capitals were painted, brightly too, but the temples remained predominantly white. A particular problem regarding architecture is the tacit assumption that the polychromy remained at the same unchanging extent and intensity over the entire ancient period. James Summitt has instead postulated, for example, that there was a growing appetite for monochromy in architecture during the Hellenistic period, in favor of the play of light and shadow.[98] Whatever the case, white had an intrinsic value that was anything but neutral: its luminosity. It was not a blank sheet to be "colored in."

The same holds for sculpture. The Via Labicana Augustus would have been painted only sparingly so the white still reigned supreme and the stone shone its own light. The broad purple hem of the *toga praetextata* (a privilege of the pontifex) was apparently painted on the statue, as traces were still visible in 1911. Otherwise, only quite delicate touches of paint would have been applied to the head. Augustus's complexion was "between dark and fair," not pancake; his hair was "flaxen," not ginger (Suet. *Aug.* 79: "*colorem inter aquilum candidumque . . . capillum subflavum*"); his eyes must be painted not to mask the marble but exploit its luminosity, for Augustus "had clear and bright eyes, in which he liked to have it thought that there was a kind of divine power such that he rejoiced if anyone at whom he looked promptly averted their gaze as if dazzled by the rays of the sun" (Suet. *Aug.* 79).

If one imagines a less opaque and more modulated use of pigment on the statue, the surface would have bloomed in the same way that contemporary cameos did. Cameos are especially apposite comparanda for the original polychromy of such statues because, notwithstanding the chromatic restrictions of the material, the technique required stripping back the stone's colored layers to allow their vestigial juxtaposition with the bright-white core. The "Blacas Cameo" (fig. 2.30), carved from a three-layered sardonyx, demonstrates just such a contemporary sensitivity toward embodied coloration that is as delicate as the petal of a flower. This is to observe just the surface; when the cameo was held up to the light the background would glow, materializing the halo of the emperor's genius or divinity. Augustus is transfigured, he is eternally young, he has become candidus, he shines like the sun, and to ancient eyes his divinity would have been blindingly obvious.

FIG. 2.30. Blacas Cameo (Augustus), c. 27 BC–14 AD. Sardonyx cameo, 5$\frac{1}{16}$ × 3$\frac{2}{3}$ in. (12.8 × 9.3 cm). British Museum, London.

3 Ancient Geology, Living Rock, and *Ex Uno Lapide*

As temples in marble, then other civic buildings as well, became common in fifth-century Greece, her natural philosophers came to reflect on minerals, and the first scientific literature on geology was born.

Only a century earlier, Thales of Miletus (c. 624–c. 546 BC) had explained the world in terms of a hylozoistic monism that made water the origin and basis of all life and matter (Arist. *Metaph.* 983 b21f, b6 8–11, 17–21; *Cael.* 294A). He also seems to have believed, based on observation of alluvial deposits, that water could turn into earth. Much later sources claim that, long before Thales, the Egyptians had conceived of matter as elementally quadripartite, and the idea is mentioned in Greek texts from Homer onward. However, it is Empedokles (c. 490–c. 430 BC) who is normally given the credit for arguing that every substance was some combination of earth, fire, air, and water.[1] This elemental tetrad became the canon after Plato (428/427 or 424/423–348/347 BC) and his pupil Aristotle (384–322 BC) adopted it in the fourth century BC, and remained basic to the understanding of matter until it was displaced by quantitative chemistry in the eighteenth century and definitively superseded by the periodic table (invented in 1869).

The ancient elements did not exist independently as discrete substances in a pure state (a clump of earth, a naked flame, and so on), but all worldly things were compounds of them. The four elements signified four modes of being: earth the most solid, water the most liquid, air the most gaseous, and fire the consuming mode. As every mundane substance was a composite that had an affinity with one predominant element, an antipathy to another, and was neutral with respect to the remaining two, its behavior was perceived in terms of the element that dominated its mixture. Each substance sought to find its proper place in the universe according to its relative gravity. Thus, more earthy compounds hungered to deposit in the terrestrial bowels, more airy ones yearned to depart for the heavens. Moreover, under the actions of heat or cold, one substance could transform into another. These metamorphoses of matter were ongoing, every substance in an open-ended state of transformation that was the result of an efficient cause or awaited one.[2] Any material, stones included, was therefore some compound of elements, or in a stage of transformation between them, and in the process of finding its proper place in the world.

Pondering the four-element theory would lead Plato and Aristotle to theorize the generation of stones and metals. Thus Plato, writing c. 360 BC, explained that when earth is strained through water, the water transforms into air and departs the ground to find its proper place in the cosmos (i.e., evaporation); and that the pre-existing air displaced by this updraft presses down on the earth, compressing it into a stony mass (*Ti.* 8.60b–c6). This opaque sketch of geological formation was the departure point for Aristotle and his successor, Theophrastos (c. 371–287 BC), to elaborate their own ideas. Aristotle's *Meteorology* (336/323 BC) was the first attempt to formulate a chemical theory of matter, while Theophrastos's *On Stones* (almost certainly 315/314 BC) was the earliest extant scientific work devoted to minerals and would not be superseded until the *De Re Metallica* of Georgius Agricola (Georg Bauer) in 1556.[3]

Aristotle explained that the heat of the sun caused the earth to emit two kinds of "exhalation" (*anathumiasis*), one a moist vapor, the other an inflammable gas. When these exhalations could not escape the earth, they formed two substances below its surface: "things that can be mined" (*ta metalleuomena*) and "things that can be quarried" (*ta orukta*), or, roughly speaking, metals and stones. His ensuing reasoning follows a general principle that "all things harden either by heat which dries dampness, or by cold which squeezes it out." Thus, the moist exhalation condensed into a liquid that itself solidified into metals, apparently when it came into contact with freezing rock (for which reason metals are often extracted from ores); the dry exhalation, "the cause of the fire that is generated within the earth," instead combusted earthy matter to produce a powder whose color was determined by its degree of burning, and which then compacted by freezing into stone. While the moist exhalation was the actual matter of which metals were formed, the dry exhalation was the efficient cause of stones, only supplying the heat that formed them (according with the idea that "fire" is a consuming force rather than a substance).[4] While metals were essentially watery for Aristotle, stones were to some degree as well, because the "dry exhalation" was the *efficient* not *material* cause of their generation.

Theophrastos's own work ignored Aristotle's theory of procreative "exhalations" and instead argued that stones acquired their "smoothness, solidity, luster and transparency" because earth transformed into "a pure and uniform matter" through "filtration" (*diēthēsis*) and "conflux" (*surroē*). In other words, particles of earth suspended in rainwater percolated down through fissures in the bedrock, which trapped the coarser increments, and deposited the filtrated particles in subterranean cavities where they settled into microgranular layers (today we would call this "sedimentary stone"). Thereafter, when these layers were subjected to heat or cold, they petrified into one mass (today we would call this "metamorphic stone").[5]

Theophrastos, like Aristotle, did not believe that stone was composed exclusively from earth, only that earth was the predominant element, because stone must contain some water to cohere at all (the "water" in question being not simply the potable liquid, but signifying an elemental fluid and matter in its state of liquidity). Thus, Theophrastos talks of a "stone half-*smaragdus* half-*iaspis*" (an emerald-jasper) "as though the transformation from water were not yet complete," and a fragment of Aristotle mentions that not only was air transparent, "but also water and certain things composed of it, such as glass, crystal and certain bright stones" (presumably gems).[6]

Some of Aristotle's and Theophrastos's formulations derived, it seems, from conversation with artisans and other practitioners. Theophrastos, for example, discusses "artificial" stones and uses the term "*autophuēs*" ("self-grown") to characterize natural azurite in contrast to its synthetic imitation (*De Lap.* 55)—not far from Sumerian ideas. Likewise, when Aristotle talks about rock as a condensed powder whose color is determined by the degree of burning, it sounds as though he has conceptually reverse-engineered the production of mineral pigments, as ochers were routinely heated to darken and intensify their hues. More generally, both Aristotle and Theophrastos could not help but be informed by the natural world around them, and both must have been particularly alive to the lithic environment. Aristotle came from Stageira in Macedonia, possibly the only town in Greece built of granite, while Theophrastos was born at Eresos on Lesbos, a village that lies amid an extensive petrified forest of sequoias and palm trees. (For Theophrastos the proposition that a Doric column was conceptually a petrified tree trunk must have been easy to accept.) Aristotle's doctrine of the "dry exhalation" may well be influenced by either volcanic landscapes or observing marsh gas (methane), but the ideas both men formulated on stone formation are most likely indebted to the "karst topography" that predominates the Greek mainland and many of its islands.

The Aegean is abundant in limestone and marble, stones that easily dissolve in and crystallize from water. When rainwater accumulates carbon dioxide from the atmosphere, an acid is produced that enlarges fissures in the bedrock to form sinkholes and caves (what geologists call a "karst topography"); when the concentration of carbon dioxide drops, calcite crystallizes from the water to produce stalactites and stalagmites within these cavities. Through the same drainage system, rain or melted snow will sink down into the rocks and reemerge as spring water on rock faces or within grottos.

Such grottos had become cult sanctuaries in the wilds of the Peloponnese and on the Greek islands from at least the Neolithic period. In these barely accessible and gloomy recesses water came into the world, healed the faithful, and birthed new rock formations (fig. 3.1).[7] These emanations became magnets of worship because they suggested that stone grew in the earth, like a child in the womb. From at least the eighth century BC, the formation of minerals was in fact described with the Greek verb "*gignesthai*" and the Latin one "*nasci*" ("to be born"), or "*phuein*" and "*vivere*" ("to grow," "to live"), or "*gignere*" and "*procreare*" ("to beget," "to engender"). For the same reason, the Romans called the bedrock "*saxum vivum*" ("living rock").[8] This expression survives in English, and the terms "quicklime" (burned lime that has not yet been slaked with water, or calcium oxide, "*calx viva*" in Latin) and "quicksilver" (mercury) preserve the same sense of vitality, as in premodern English "quick" meant "alive." Though some stones were mortal (Lucretius speaks of mountains aging until they decay and crumble away; *Rer. nat.* 1.326–27), others were so vital that they could regenerate. It is symptomatic that a legal term applied from Cicero to Theodosius to describe reused stones (what would later be called "spolia") was "*lapis redivivus*" or "*saxum redivivum*" ("stone that lives again").[9]

The idea of the engendering earth as a colossal and constantly pregnant body can be traced to Egyptian and Assyro-Babylonian stone lore and is likely prehistoric. The Egyptian word for "mine shaft" (*bi̯*), for example, also means "uterus." Mining minerals from "Mother Earth" was therefore an obstetric operation, and it has

FIG. 3.1. Cave of Archedemos (Nympholyptos Cave), c. 480/440 BC. Near Vari, Attica, Greece. Shrine of Apollo, Pan, and the Nymphs, discontinuously occupied from the 6th century BC. to the 6th century AD.

been suggested that when prehistoric copper mines were backfilled after extracting a seam, the "surgeons" were attempting to heal the wound postpartum.[10] In the first century AD, Pliny claims of Spanish lead mines that "it is a wonder that only do these quarries, once they have been abandoned, return to life more fruitful. The air seems to do this by pouring in through the opened vents and filling the place, just as miscarriages make some women more fertile." Here Pliny imagines insemination by a propagating air rather than water, not so different from the idea of a god that can breathe life into clay.[11]

Marbles reproduced as much as other minerals. The philosopher Papirius Fabianus (c. 35 BC–c. 30 AD) recounts that "among the many other marvels of Italy itself, is that marble actually grows in the quarries. . . . The quarrymen, moreover, assert that the scars on the mountain sides fill up of their own accord." The myth took a long time to die and was repeated by the monk Restoro d'Arezzo in 1282, the jurist-poet Scipione Capece in 1546, and even the Neoclassical theorist Francesco Milizia in 1785. This particular tradition of self-generating marble is at least of Hellenistic origin, because when the geographer Strabo mentions (c. 24 AD) that the iron mines on Elba, stone quarries on Rhodes, and salt mines of India replenish themselves after lying fallow, he adds the marble quarries on Paros to the list. Strabo cites the Alexandrian historian Kleitarchos (mid- to late fourth century or mid- to late third century BC) as his source for the salt mines.[12] It may also well be that Theophrastos had already discussed marble growth in his lost work, *On Petrifactions* (*Peri tōn apolithoumenōn*).

On Bronze Age Paros, sculptors made do with the marble fragments they could scavenge off the ground where the Skarpathiotis torrent had exposed them along its banks. However, when larger blocks became needed in the sixth century, tunnels were burrowed into the bedrock

FIG. 3.2A. Relief of Adamas, Cave of the Nymphs, c. 350–325 BC. Malathri, Paros (Reinach 1909–12).

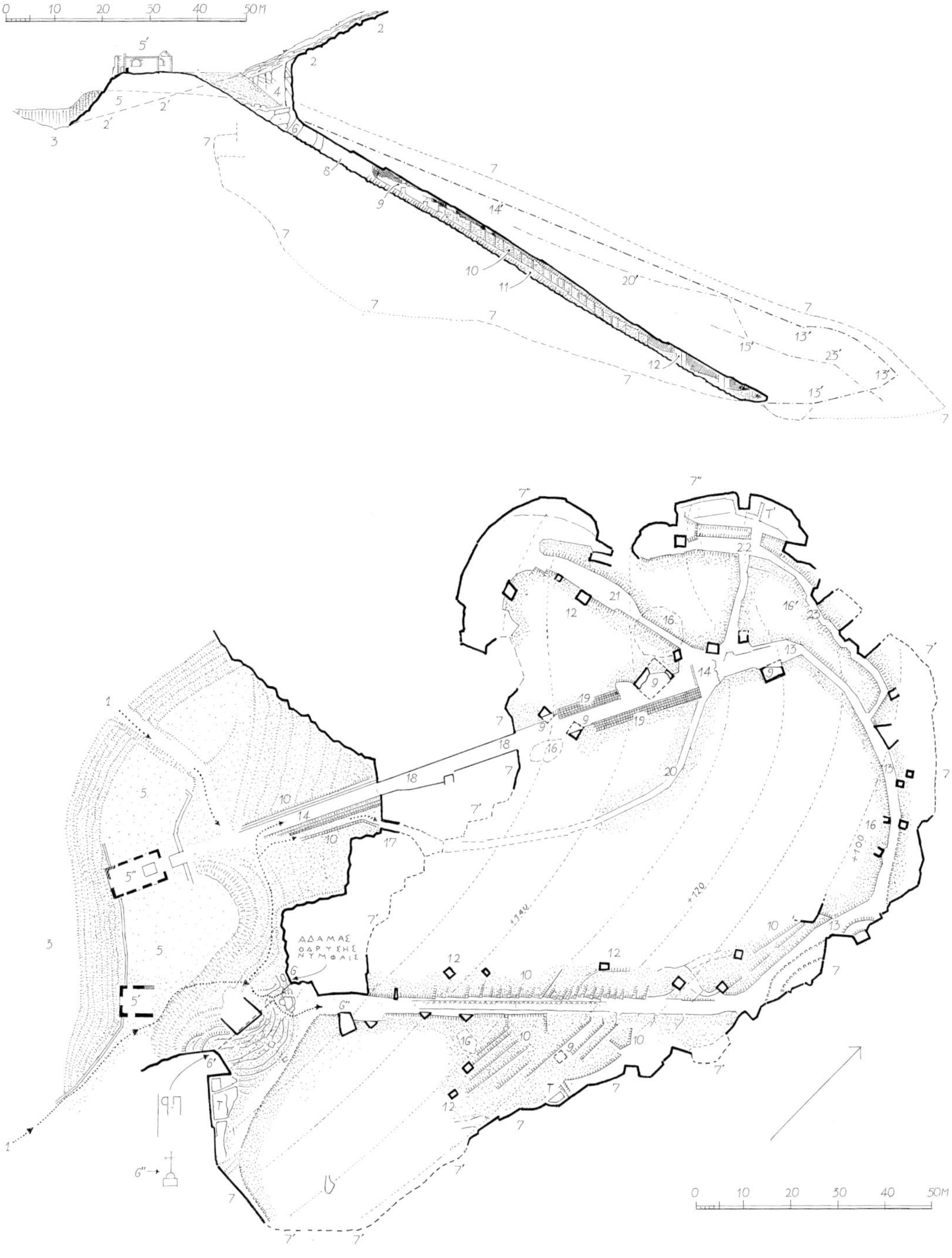

FIG. 3.2B–C. Shaft quarries. Malathri, Paros (plan and section: Manolis Korres).

to chase the best seam (Parian *Lychnites*), ten to thirteen feet thick, thereby avoiding the graphite-stained strata above and below. The most important of these tunnels (near Malathri) sank to a depth of six hundred and fifty feet at thirty-five degrees (figs. 3.2b–c), and at its bottom stalactites began to drip through the porous rock where generations of quarrymen could witness their sluggish growth. This marble mine, also known as the "Cave of the Nymphs," became intimately associated with the tradition of the sacred and life-giving grotto, some time before the so-called "Adamas Relief" was carved at its entrance (c. 350–325 BC). This depicts Pan and the Muses (who are water spirits), and later tradition insisted that the relief had been naturally formed and only "finished off" by sculptors (see fig. 3.2a).[13]

EAST MEETS WEST: ALEXANDRIA (C. 300–C. 250 BC)

The point of contact between the older material technologies of Egypt and the ancient Near East, and the marble reality and natural philosophy of classical Greece, was Alexandria, the city founded by Alexander the Great (331 BC) during his stunning conquests from Greece to India. His successors, the Ptolemies, made Egypt their kingdom (323–30 BC), and Alexandria soon became the intellectual capital of the Hellenistic world. Here new lithic media developed from the contact between the two cultures. Macedonian pebble mosaic developed into tessera mosaic and micromosaic; cameo was invented, a cross between a veined marble and an intaglio gem, from which images could be coaxed by judiciously stripping away the laminations (see fig. 2.30); and hard stone cups (see fig. 6.15), especially Myrrhine ware, became treasured luxuries.[14]

Ptolemy I Soter (r. 323–283 BC) or Ptolemy II Philadelphus (r. 283–247 BC) founded the Great Library to gather all knowledge in one place, and Ptolemy III Euergetes (r. 246–222 BC) requested texts be brought from every known country for translation into Greek. The Mouseion (or Musaeum) that adjoined the library was built in the same period (300–250 BC), becoming its think tank in both the humanities and sciences, all under the protection of the Muses who were the continual founts for their inspiration and mingling.[15] Library and Mouseion lay within the compound of the royal palace, and although no descriptions survive of either, one literary nugget (c. 250 BC) describes a long-lost fountain house somewhere on the grounds:

> Gladly accept this gift of King Ptolemy,
> who added to his residence a rich work
> in stone; having unleashed the limpid (water-)drop
> of *Paros,*
> he fashioned it into a hemisphere [i.e., basin]: the
> *Lychnites* plinth
> supports the frieze with columns in the Ionic style
> and within the hollow molding the speckled
> *Syenite* glistens
> at their bases; such is the foundation of the column.
> The *stone of Hymettus* gushes out the draught of
> spring-water,
> Moistened and receiving the waters that pour over the
> natural rocks.
> He [had] your images softly modelled in rich *Lygdian*
> *marble,*
> and in the midst he placed [Queen] Arsinoe,
> [so she may be] a companion to the Nymphs the whole
> year round.
> Come in due order to the fountain, oh nymphs of
> the spring.[16]

Because this fountain was dedicated to the nymphs, which the Muses were themselves, it is quite possible that it stood near the Mouseion. As the Mouseion also possessed a garden containing the most exotic flora and fauna, as well as a grand exedra, our fountain may have stood in one of these places.

The idea of a Mouseion, as both a cultural institution and a built setting where scholars could gather, traced its origin back to the precinct for the Muses that Plato had made within the gymnasium he established in the grove of Academe, as well as the annex with its own Mouseion that Theophrastos added to it.[17] Theophrastos, as we know, had written *On Stones.* The four stones on the Alexandrine fountain house that the poem identifies span the Mediterranean from Athens (Hymettan) to Aswan (Syenite) by way of the island of Paros (Lychnites, "Lygdian marble"), and this geographical range matches the travels of the poet to whom the text is normally attributed, Poseidippos (c. 310–c. 240 BC). Poseidippos had been born in the Macedonian capital of Pella but spent most of his career in Alexandria.

He was also the author of the earliest surviving Greek lapidary, the *Lithika,* a long collection of poems that mainly describe engraved gems, and that was only rediscovered in 2001.[18] It seems that Poseidippos made a specialty of writing epigrams on stones, producing countless numbers of them. This was a real novelty, as there do not appear to be any Greek lapidaries before him at all, let alone in verse. Lapidaries, lists of precious and semiprecious stones that catalogued their amuletic or therapeutic properties, had, however, long existed. The Ebers Papyrus (c. 1500 BC), an Egyptian medicinal text, contains many cures that use minerals, rocks, and fossils. The earliest-known lapidary proper is Assyrian, the *Abnu-šikinšu* (probably from the reign of Tiglath-Pileser, 1114–1076 BC), but two much later ones are also recorded from the Near East, one by a certain Sotakos (c. 320–270 BC), and the other attributed to "Zoroaster" (c. 500/300 BC), a fictional character who embodies Persian or Near Eastern learning. Such books were likely among those collected and translated into Greek at the Mouseion, with the upshot that the stone lore of Egypt and the ancient Near East must have fused here with Greek scientific tradition. The tradition continued in the works of other Babylonians, such as Zachalias, who "in the volumes which he dedicates to King Mithridates, attributes man's destiny to the influence of precious stones," and Soudines, who was apparently astronomer-astrologer to Attalos I of Pergamon (r. 241–197 BC).[19]

While several scholars have stressed the metapoetic qualities of Poseidippos's *Lithica*, others have realized that he was also well acquainted with Theophrastos's *On Stones* or at least the same mineralogical terminology.[20] Theophrastos, for his part, certainly knew something of Egyptian building stones, as he mentions basalt (or diorite, or diorite-gneiss), two types of Egyptian alabaster, and even specifies a particular limestone or travertine (poros) that he says was used for cornices or coursing.[21] Poseidippos could well have read Theophrastos at the Mouseion, where the poets and the scientists worked side by side. One of Theophrastos's pupils (Demetrios of Phaleron) may have been the chief librarian of the Great Library, and in any case a copy of *On Stones* was almost certainly deposited there.[22]

Returning to the poem on the nymphaeum attributed to Poseidippos, one line proves crucial in its interpretation, where the poet seems to refer to Parian marble as watery: "having released the limpid (water-)drop of Paros / [Ptolemy] fashioned it into a hemisphere." Editors of the poem have found the choice of words counterintuitive, wondering how any marble could be a "water-drop" (*stagōn*) that had been "unleashed" (*ekpodizein*).[23] However, the conceit becomes more acceptable when one remembers the widespread belief that marbles grew in the quarries, including those on Paros, and "water-drop" may well allude to the shape of the basin as well.

The continuing belief that marbles were inherently water-based, however earthy, may explain the curious observation of a rabbi visiting Rome in the first century AD, who remarked that "marble columns were covered with tapestries so that they might not crack during the heat and not congeal during the cold."[24] It may also elucidate the nuances in the description by Kallistratos (late third or late fourth century AD) of a statue of Narcissus, the youth who became enamored of his own reflection in water: "Whereas the marble was in every part trying to change the real boy so as to match the one in the water, the spring was struggling to match the skilful effects of art in the marble, reproducing in an incorporeal medium the likeness of the corporeal model and enveloping the reflection which came from the statue with the substance of water as though it were the substance of flesh."[25]

THE ROMANS AND LIVING ROCK (FIRST CENTURY BC–FIRST CENTURY AD)

The nymphaeum described by Poseidippos may have incorporated rocks in the manner of an artificial grotto. Indeed, the Ptolemies built numerous faux-grottoes, and thereby transplanted the primordial landscape and sacred springs of Greek deities to Egyptian soil.[26] The sanctuary of Pan (the Paneion) at Alexandria, for example, was a "conical eminence, man-made, and resembling a rocky mound" (Strab. *Geog.* 17.1.10).

Artificial grottos became a staple ingredient of Italian architecture from the late Hellenistic period as well, often semicircular in plan and with a half-dome, to summarize the idea of a cave. But in these incarnations (perhaps because southern Italians were more familiar with a volcanic landscape than a karst one) tufa and pumice were used to feign the calcareous stalactites. As Pliny says (*HN.* 36.154), pumice is "the name given to the eroded rocks in the buildings they call *musaea*, where they hang down to represent by artifice the image of a cave." Pliny also calls this stone "spray of the sea," whether from its foamy texture or from observing the surface detritus of subaqueous eruptions. In fact, in the Italian versions, the original spring-fed grotto also became more associated with the sea as seashells were applied alongside the pumice. Occasionally, these translations of grotto into architecture give us a glimpse of the more creative geology of the Roman mind—for example, in the nymphaeum of the Via degli Annibaldi in Rome, dating from circa 50 BC and presumably part of some patrician residence (fig. 3.3). The walls were encrusted with pumice; with periwinkle, cockle, mussel, and oyster shells; with white marble chips; and with shards of streaky turquoise, dark blue, yellow, and white glass, and Egyptian blue. However, in among them are various shapes in terra-cotta representing bucrania, cornucopiae, cuirasses, shields, and so on.[27] All this was originally viewed through a film of water trickling down from a lead pipe concealed in the string course, suggesting a submarine lair and vivifying the effects of the collage. While the shells belong to this marine world, and the marbles and glass suggest its glints and sparkle, the

FIG. 3.3. Nymphaeum, Via degli Annibaldi, c. 50 BC. Rome.

figured stones instead insinuate the fossils of artful nature, just as Hellenistic artists had imagined them (see fig. 4.17) and medieval ones would continue to do (see fig. 6.25). Theophrastos's *On Petrifactions* must also have discussed fossils, after a fashion, as Pliny says that he discussed the "bone-stones" that grew in the earth (*HN.* 36.29.134: *lapides osseos*).

The artificial grotto would enjoy a long afterlife across Europe, and when Early Modern observers regarded them as crucibles of generation and laboratories of natural art, they took their cue from Roman poets.[28] Over centuries this poetry sustained a ready appreciation for the natural architecture of grottos, which they counterposed to man-made construction. Thus, circa 8 AD, Ovid (*Met.* 3.158–60) commends the freshwater grotto of Gargaphia, sacred to Diana,

> Worked by no art: Nature, with her own skill, had imitated art;
> For with living pumice and light tufa
> She had drafted a natural-born arch.

Likewise, circa 470, Sidonius Apollinaris (*Carm.* 22.224–26) praises another tufa grotto (at the villa of Pontius Leontius near Bordeaux) because

> It needs no embellishment, for Nature has given it beauty.
> It pleases that no fiction has pleased here, no pomp through art whatsoever,
> No hammer with resounding blow shall dress the stones,
> Nor will marbles supplement the eroded tufa.

The Romans also put to use the sand made of compacted volcanic ash that they found in Campania to make their own artificial stone, concrete. At Puteoli (Pozzuoli), in the caldera of the Phlegraean Fields, they excavated this volcanic sand (*puteolanus pulvis,* or pozzolana) and mixed it with lime and water to make a mortar with a compressive strength higher than that of modern concrete. Because pozzolana has microstructures of intertwining fibrous minerals, and because the lime-pozzolana compound has "hydraulic" properties (it can set even underwater), it also cured rapidly. The inspiration to invent this compound grew from the geological characteristics of the terrain itself. The caldera is the lid of a hydrothermal system: the deposits of pozzolana sit over a meta-carbonate basement, and when carbon dioxide is released through this, it pyroprocesses the carbonate to form lime, which bonds with the pozzolana to form a natural concrete.[29] When Seneca mentions this pumice in his *Natural Questions* (c. 62/64 AD) he comments, "If Pozzuoli ash *touches* water, it becomes stone" (3.21: *Puteolanus pulvis si aquam attigit, saxum est*). It is far from clear that he recognized this as a chemical reaction rather than a metamorphosis permitted by the theory of the four elements, for the comment comes in the context of explaining how a certain river ("of the Ciconians") will petrify objects that it touches.

The engineering stone that most effectively communicated ideas of watery generation and petrifaction was, however, travertine. "Travertine" owes its name to (*lapis*) *tiburtinus,* the "stone of Tivoli," where it was and still is extensively quarried, scarcely twenty miles from Rome. Travertine is a sedimentary stone formed by the precipitation of carbonate minerals from aqueous solution. When the water was particularly chalky, ancient maintenance engineers observed firsthand how this stone grew (Frontin. *Aq.* 122), because it clogged up the aqueduct conduits (*cuniculi*), like the furring in a kettle or a boiler. The faster the flow, the greater the deposit, and the rectangular section of the cuniculi also meant an asymmetrical distribution of velocity and therefore greater sedimentation.[30] At Hierapolis (Pamukkale, Turkey), the entire population participated in the process. As Vitruvius noted (*De Arch.* 8.3.10), their city was built out of, sat upon, and arose from a huge calcareous mass; and lime-charged water still streams through its ruined streets in search of long-vanished bathhouses, depositing lime as it goes (fig. 3.4).

Travertine was most expressively used on two imposing monuments built during the reigns of Claudius and Vespasian, the Porta Maggiore (52 AD) and the Claudium (not Claudianum, 69–early 70s AD), both of which are fantastically rusticated, and both of which had water flow through them.[31] The Porta Maggiore is part of an aqueduct, a type of structure that was always celebrated as a feat of engineering and the alteration of nature (fig. 3.5). Roman poets loved the "vagrant water that thrives on piles and is conducted suspended on innumerable arches," the "hanging streams on airy arches, where not even the rainbow could raise its rainy waters," in fact, "you would say that mountains had grown as far as the stars." As late as the 1250s the Nicaean emperor Theodore II Lascaris (r. 1254–58) exalted an aqueduct in Pergamon as "a river bridged by tall arches which . . . you would not think to be composite, but rather to have grown up naturally as a single block of stone."[32]

The Aqua Claudia and the Anio Vetus conveyed their waters from springs miles distant to their outlets in the city's spigots, fountains, and baths. The Porta Maggiore rose at the point where these conduits crossed the fork of the Via Labicana and the Via Praenestina, transforming brick into travertine to make a huge triumphal arch that marked the threshold between city and country. The rustication easily presents an image of "living rock" in the columns that sprout up like palm trees, while the pedestrian passage between the two great arches resembles more a cuniculus than an urbane underpass. A little earlier, the antiquarian Verrius Flaccus (c. 55 BC–20 AD) had felt it necessary to differentiate that "there are two kinds of rocks, one of which is the natural rock which juts into the sea . . . the other kind shaped by hand" (Fest. *Apogr.* 226: *Petrarum genera sunt duo, quorum alterum naturale saxum prominens in mare . . . alterum manufactum*). It seems that not only the waters, but the grotto also, have been conveyed to the city. The rude blocks, which are the ultimate deposits of the distant springs, compose a "dry" nymphaeum, the name Romans gave to the dramatic and theatrical terminal fountains that occasionally commemorated the distant epiphany of water in miraculous grottos. Only at the attic level are the stones drafted, and epigraphy incised, where the architect seems to have completed what nature had begun.

The Claudium also had expressively rusticated blocks. This was a massive podium for the Temple of the Divine

FIG. 3.4. Travertine deposits. Hierapolis (Pamukkale), Turkey.

FIG. 3.5. Porta Maggiore, 52 AD. Rome.

Claudius that extended and regularized the Caelian Hill. It is as though the topsoil had been removed and the engineers had drafted the travertine bedrock into a building of a single stone (substituting the real geology, which is tufa). The tallest (eastern) façade was a good 650 feet long, and to this was added a monumental nymphaeum, down whose serried niches streamed rivulets from a branch of the Aqua Claudia (the Arcus Neroniani), with the result that the man-made hill did become a nymphaeum for the emperor's eponymous aqueduct.

The rustication of the stone, the "unfinished" look, expressed the mutability of travertine, its seeming ability to "grow," dissolve, and recalcify. Growth and ruin mirrored each other, one a birth from nature, the other a return to it. The tribute to the ruined Theater of Pompey that Cassiodorus wrote much later (507/511 AD) might apply to the Porta Maggiore just as well: "It would be conceivably easier for mountains to fall than for that solidity to tremble: For that very mass is so entirely formed from vast blocks that, but for the added craftsmanship, it too might be thought the work of nature. . . . Those arched vaults, with their overhanging stonework, are so beautifully shaped with invisible jointing that you would suppose them the caverns of a lofty mountain, rather than anything man-made." The Romantic poet Percy Bysshe Shelley registers the same reaction to the wreck of the Colosseum: "The Coliseum is unlike any work of human hands I ever saw before. It is of enormous height and circuit, and the arches built of massy stones are piled on one another, and jut into the blue air, shattered into the forms of overhanging rocks. It has been changed by time into the image of an amphitheatre of rocky hills."[33]

All these geological "facts" of ancient natural science have unforeseen implications for sculptural practice as well.

It is a commonplace that Roman sculptors used particular stones in a straightforwardly mimetic way: that Egyptian schist imitated bronze (a resemblance mentioned by Pliny. *HN.* 36.11.58), that *bigio morato* depicted the black skin of Africans, or that herms of Dionysus were made from yellow Numidian marble (*giallo antico*) to mimic the saffron robes of his devotees.[34] The material pun occasionally extended to names as well: Pescennius Niger (anti-emperor to Septimius Severus in 193–194 AD) was presented with a statue by the Egyptian city of Thebes "carved in Theban marble" (basalt) bearing verses that reported, "Black is the surname he bears, and black have we made you, so that form and mineral agree" (*S.H.A.* 12.4–6).

However, ancient geological understanding adds nuance to the meaning of these stones beyond their surface color. Pausanias (*Desc.* 8.24.12), for example, relates, "All statues of River Gods are made from white marble, except for the Egyptian Nile. It is customary to make effigies of the Nile from black stone, because it descends to the sea through Ethiopia." The Egyptians had long called their country the "black land" (*Kemet*), from the fertile alluvium the Nile deposited on the banks and fields, and it is for this reason that the earliest statues of the life-giving Isis were made in black stone. In Roman eyes, then, and perhaps Egyptian ones too, these stones were made by the same process of deposition that they saw in

FIG. 3.6. Fountainhead with a bacchant, c. 117/138 AD. Taenarian marble (rosso antico). Museo Nazionale Romano, Palazzo Altemps, Rome.

a travertine, or the hot sun of Egypt could be imagined to have cooked the stone almost to the point of charring, as it supposedly had the natives.[35] A similar idea may explain why Taenarian marble (*rosso antico*) was the marble of choice for sculpture with Bacchic subjects. This wine-red stone might be compared either to the lees and dregs or to *merum*, pure wine before it was mixed with water. Such a notion may be consummated in the Hadrianic fountainhead of a bacchant with grapes in her hair, which could once even have gushed wine, although this hypothesis is as luxurious as the object (fig. 3.6). However, Rolf Schneider has ingeniously hypothesized that Taenarian marble may have been selected for a statue of a fluvial deity—a *unicum*—found at Sessa Aurunca, because the quality of the local water was comparable to wine.[36] Perhaps the same held for the fountain.

MOTHER EARTH, THE LIVING ROCK, AND VEINS

As there is no modern study of geological theory and mineralogy between Theophrastos (315/314 BC) and Agricola (1556), one is compelled to collect diverse fragments to extrapolate any picture of Roman beliefs. Agricola would count twenty-seven Greek authors on stones that the encyclopedist Pliny the Elder cites in his *Natural History* (77/79 AD), but it has been said, "The information . . . obtained from these vanished authors would indicate indeed that the contributions they made to sciences of the mineral kingdom were of such trivial importance that their loss is one to which we may reconcile ourselves, even with thanks."[37] Aristotle and Theophrastos's theories certainly remained current in the Roman world and were reprised by the Stoic philosopher Poseidonios (c. 135–c. 51 BC) and the philosopher-poet Lucretius (90s–50s BC), who passingly describes the formation of earth and heavens from exhalations (Lucr. *Rer. nat.* 5.460–65). Ovid (43 BC–17 AD) also alludes to petrifaction theories here and there (Ov. *Met.* 3.397–99; 9.226–29) and observations on geology were gathered by more obscure figures such as the aforementioned Papirius Fabianus, the retired senator Gaius Licinius Mucianus (30s?–c. 72/78 AD), and others only whose names we know.

However, one strand of thought that does emerge clearly is the enduring conception of the great geological body of Mother Earth. This conception prospered thanks in part to Stoic vitalism. The Stoic school of philosophy, beginning with its founder, Zeno of Citium (333–264 BC), had personified and virtually deified Nature as a creative power, and the ideas of the Stoics were resumed by Seneca, Pliny, and others.[38] Pliny asserts, "The power of Nature is that which we call God" (*HN.* 2.5.27), and he contrasts the unitary *divina natura* with the *imbecillitas humana* of pantheism (*HN.* 2.14–27). The Stoics linked nature especially with *pneuma*, the "breath" that Anaximenes of Miletus (c. 585–c. 528 BC) had propounded as the basis of life, and that inspired Aristotle's own doctrine of twin procreative vapors, as though, in fact, he thought the earth were breathing.

The Stoics certainly did believe the earth was a living body (Sen. *QNat.* 2; 60/65 AD), and Pliny attributes to Mother Earth veins, arteries, and entrails. He mentions the earth's "vitals" (*HN.* 33.1.2: *viscera*), calls Italy "pregnant" (*HN.* 37.77.202: *gravida*), and remarks that when the earth is mined, she "gapes" and "trembles" (*HN.* 33.1.1: *dehiscere . . . intremescere*). Pliny also opens his Book Thirty-Six on stones and marbles with the following indictment: "Nature made mountains for herself to serve as a sort of structure for pressing together the viscera of the earth, and at the same time to subdue the assaults of rivers, and to break the tides, so as to curb the least becalmed elements with her hardest matter; we cut these [mountains] up and haul them away for no other cause than pleasure. . . . Now these selfsame mountains are cut into a thousand kinds of marble."[39]

Even if stone was earth that had solidified into a rocky mass of bones and cartilage, veins of living water still coursed through its flesh, and antique authors distinguished this "living water" (*aqua viva*) from rainwater or "heavenly moisture" (Sen. *QNat.* 3.7.3). Seneca would develop, or reprise, an elaborate analogy between veins and arteries and the earth's inner channels for water and air (*QNat.* 3.15.1), but the idea had pre-circulated enough for Vitruvius, decades earlier, to liken the different qualities of waters to the various juices in the human body: "Because even in the human body, which consists in part of the earth, there are many kinds of humors, such as blood, milk, sweat, urine, and tears; wherefore, if such diversity of flavors is found in only a small portion of the earth, it should not be surprising if countless varieties of juices are found in the bulk of the earth, through whose veins the force of water runs, and becomes saturated with them before arriving at the outlets of the springs" (*De Arch.* 8.3.26).

Ovid's verses in the *Metamorphoses* (published 8 AD) capitalize on this terrestrial physiology to explain the very

FIG. 3.7. Left: *Marsyas*, 3rd–4th century AD. White marble, 25¾ × 7¹³⁄₁₆ × 6⅞ in. (65.4 × 19.9 × 17.5 cm). Right: *Marsyas*, 14/68 AD. Phrygian marble (pavonazzetto), 21⅛ × 7½ × 6⅜ in. (53.6 × 19 × 16.2 cm). Worcester Art Museum, MA. Sarah C. Garver Fund. 1971.89.

origins of the human race, and he draws on the experience of sculpture production as a metaphor (compare figs. 4.32 and 4.33). After the Great Flood with which the gods had exterminated mankind for its wickedness, the sole survivors, Deucalion and Pyrrha, sowed a new race of men from the rocks they tossed over their shoulders as they descended from their mountain refuge. Where they fell to earth, the stones

> began to discard their hardness and rigidity
> And after a while grow soft and, having softened,
> take on form.
> Soon, wherever they grew and a riper nature
> possessed them,
> Some semblance of man could be seen, though not
> yet plain,
> As though only begun from marble, not yet exact,
> And most resembling roughed-out statues.
> Nonetheless, any part of them that was earthy and
> Moist with juice, converted to corporeal use;
> What was solid and inflexible, turned into bones;
> What was lately a vein, continued under the same name.
> (*Met.* 1.400–410)

The "flesh and blood" of stone in fact gives a physiological depth to several statues of the satyr Marsyas carved from Phrygian marble, because this white stone, varicose with violet veining, embodied the flesh of the unfortunate satyr exposed after his flaying by a vindictive Apollo (fig. 3.7). Marsyas was Phrygian, a devotee of the Earth Mother Cybele, and his demise came about at Apamea-Kelainai (Dinar, western Turkey), about a hundred miles from the quarries at Docimium (İscehisar) from which the marble came. The blood from Marsyas's body drained into the earth there and gave birth to the river that would take his name. One poet decided to celebrate the achievement of carving Marsyas statues in Phrygian marble, stressing that art was necessary to complete nature in distinguishing tree from body, especially because both were fashioned from the same stone. At the same time, he implicitly compares the sculptor who cuts away the stone to reveal the veined effigy to the executioner who flayed the satyr to expose his bloody innards:

> From the airy branch hangs Marsyas defeated.
> The natural redness shows his chest is taut.
> A skilled hand has filed the stone into different limbs,
> And through this skill shines forth the truth of tree and man.[40]

A medieval Byzantine poem will echo the same conceit to imagine the rent flesh in marble statues of the companions of Odysseus mangled and devoured by Cyclops.[41]

"Pavonazzetto" is the name that Phrygian marble has been accorded since at least the sixteenth century, and in modern Italian *"paonazzo"* can mean simply purple, or it can indicate a flushed or bloodshot complexion. In the first century AD, the poet Statius had this image in his mind when he invoked the figure of Attis to explain the marble's distinctive veining, which "Attis himself had bloodied with shining splatter in Synnas's hollow Phrygian cave."[42] According to myth, Cybele had struck Attis mad for his infidelity to her, and his lunacy climaxed in a tantrum of self-castration. From his blood sprang the violet. Because Cybele was worshipped at Pessinous (modern Ballıhisar), also in central Phrygia, in the shape of a black meteorite—that became well known to all Romans after it was transferred to their capital (204 BC)—it was apposite that Attis should leave his bloody mark on a white one. As Attis was not just a shepherd but also a fertility god with his own cult, we may perhaps understand his blood as inseminating the goddess's earthly body. Moreover, the priests of Attis's cult (the *Galli*) not only emasculated themselves in emulation of their hero, but every spring spattered his white altars with offerings of their own blood. However gruesome, this ritual was a well-known sight.

The veins that coursed through the earth and fed the stones recalled both the blood in animals and the sap in plants, because both fluids were species of the elemental, life-giving water. As long as rocks remained attached and plants rooted, they were "living" and enjoyed the nourishment that circulated from vein to vein. Thus, a line in Ovid that would otherwise seem a paradox makes perfect sense: Ovid calls heartless Anaxarete "harder than (Noric heat-)tempered steel, and the rock that is kept alive by its root" (*Met.* 14.712–13: *durior et ferro, quod noricus excoquit ignis / et saxo quod adhuc vivum radice tenetur*).

Ovid also supplied the most famous version of the story of Daphne, the naiad (freshwater nymph) whose father, the river god Ladon (or Pineios), had transformed her into a laurel tree to prevent her rape by the god Apollo. "A heavy numbness seized her limbs, her soft breasts were girdled by thin bark, her locks sprouted into leaves, her arms into branches, and her feet so swift a moment ago now held fast with slow-moving roots" (Ov. *Met.* 1.549–52).

Another poem, by the same author as the elegiac couplets on the Marsyas statue, capitalizes on this vegetal metaphor by praising an effigy of Daphne:

> An expert hand took care that a suitable color appears
> Within the leaves and limbs sculpted.
> Sculpture and painterliness conjoined produce a marvelous beauty
> When variegated marble reveals two statues.[43]

No statue of the mutating Daphne survives from antiquity. Perhaps the poet is describing a relief. Or perhaps his statue was simply a product of his imagination, and then, if so, imagined from the raw materials and sculptural possibilities. Yet we in turn might imagine its lost appearance from a contemporary Syrian mosaic, another technique that is both stone and painting (fig. 3.8).

As stones were born and animate, they could enjoy an occasional sexuality themselves. It was the "attraction" between stones, their mutual magnetism, that animated lodestone (Pliny *HN.* 2.96.211). And the poet Nonnos (*Dion.* 2.493–94; c. 470 AD) tells us that when male and female flints met, sparks flew. The Romans therefore called flint *"lapis vivus."* Likewise, sard and onyx even mated to produce the race of sardonyx, the premier material of cameos (see fig. 2.30). Thus, the idea of a hybrid stone may underlie those statues that portray hybrid creatures—for example, a second-century *Triton*, a merman whose torso is white but coiling bifurcated fish tails are purple, not only to express the difference between skin and scales but also because he was the hybrid offspring of Poseidon and the Nereid Amphitrite (fig. 3.9).[44] The same may hold for

FIG. 3.8. *Apollo and Daphne,* late 3rd century AD. Mosaic pavement, 6 ft. 2 11/16 in. × 19 ft. 5 1/16 in. (190 × 592 cm); figured panel: 42 1/8 × 41 5/16 in. (107 × 105 cm). Princeton University Art Museum. Gift of the Committee for the Excavation of Antioch to Princeton University, y1965-219.

FIG. 3.9. *Triton,* 2nd century AD. Found north of the Hippodrome, Istanbul. Phrygian marble. Archaeological Museum, Istanbul.

a *Europa and the Bull* found at Aphrodisias, where the former is white and the latter purple, because the subject is the unnatural coupling between the chaste Europa and the aroused bull (Zeus).[45] In other words, the mythical outcome was already embodied in the genealogy of the block used to represent it.

FIERY AND AIRY STONES

The belief in an animate and generative earth, and the arteries that fed it, might explain the composition of veined, earthy-looking stones, but not paler and more translucent ones, and certainly not gems. Such stones should be more watery and airy, as Aristotle and Theophrastos had implied, and Seneca presumably had them in mind when he commented, "In the earth also there are several kinds of moisture . . . which change from liquid to stone" (*QNat.* 3.15.2–3: *in terra quoque sunt umoris genera complura . . . quae in lapidem ex liquore vertuntur*). The baseline for understanding these less earthy and more watery stones was rock crystal, which was widely believed to be super-frozen ice (in Greek, "*krustallos*" means both "ice" and "crystal"). It was therefore best retrieved from mountain slopes, held to possess refrigerating properties, and highly favored for drinking vessels. These enduring beliefs meant that some rock-crystal hanging lamps in Islam and Byzantium were shaped like beakers and engraved with swimming fish, and it may also explain why desert Muslims would later toss rock-crystal talismans into dry wells.[46]

As Strabo (*Geog.* 2.3.4) says of some Indian gems, "They are found by digging, being solidified from a liquid state, just as our crystals are." It is still current usage in all Romance and Teutonic languages to speak of "the water" of a gem when assessing its luminosity and transparency, and the finest jewels are said to be "of the first water." Marbles were not gems, of course, and they occupied a lower rung on the ladder of creation, but they still shared the same crystalline essence as gems and crystals by virtue of their watery base, just with more earthy matter mixed in. One word, "*metallum,*" in fact covered all the categories. Although this is frequently mistranslated as "metal," the Greek verb "*metalleuein*" meant only "to mine," the Latin derivative "*metallum*" meant "quarry," and the plural "*metalla*" came to indicate the broad category of "minerals."[47] In the long epoch before quantitative science, spectators often equated the surface appearance of minerals with their secret substance, and this appearance either suggested a moment in their evolution or the particular cocktail of their base elements. Color and grain told whether various minerals were truly fiery, gassy, earthy, or aqueous.

The heat of the sun not only caused the warm exhalation but also determined the color of the earth's fruits. When Diodorus Siculus writes of Arabia (*Bibl. Hist.* 2.52.2–9; 60/30 BC), he says that even rock crystal, whose generation almost everyone else thought to be cryogenic, was "pure water which has been hardened, not by the action of cold, but by the influence of a divine fire." He continues that it was the sun's heat that hardened subterranean water into stones and made them resplend; while some stones gained their color from other agents (e.g., emeralds and beryls from sulfur), rubies and garnets drew it directly from the action of the sun. Pliny also explained that fire-veined agate was the result of extreme heat. Other explanations were given for the fiery appearance of some stones. Aristotle's later commentators had considerable difficulty understanding his doctrine of the "dry exhalation," and thought that rocks must be composed from the dry exhalation when it caught fire, which was therefore their *material* (not *efficient*) cause.[48] A stone like porphyry might therefore be a product of combustion that materialized flame, just as it looked to some.

Other stones, such as Egyptian alabaster (see figs. 1.6, 5.8, 5.9) and the mysterious Phengite seemed at once more airy, and this could also be explained using the Aristotelian model. According to Aristotle (*Mete.* 3.6.378 a 18ff.), clouds are generally moist exhalations, but dry ones produce dark thunderclouds that create lightning and condense to form stones (by which he means meteorites). Pliny follows this lead when he explains, "It is certain the earth exhales a damp mist and at other times a smoky one due to vapor, and that clouds are formed out of moisture rising to a height or air condensed into moisture" (*HN.* 2.42). As marble was itself a solidification of the earth's exhalations, then it followed that cloudy alabaster *was* what it seemed. Pliny, in fact, makes this observation about "specular stone" (see fig. 6.9), just before he describes the mysterious Phengite Temple of Fortune: "It is manifestly apparent that this humor, like rock crystal, has been frozen by some soul of the earth and hardened into stone" (*HN.* 36.45.161).

EX UNO LAPIDE, ANATHYROSIS, MAGNIFICENTIA, AND THE *RES PUBLICA*

The integrity of an animate earth finds reflection in the term that was occasionally applied to congratulate sculpture that seemed *ex uno lapide* (Greek, *ex henos lithou*), literally "of one stone"—but, as we shall see, meaning "all of a piece." The phrase first appears inscribed on the plinth of the massive kouros that the Naxians had erected on Delos (c. 600/590 BC), which asserts, "Of the Same Stone Am I, Statue and Base" (see figs. 2.5, 2.6). Scholars have found the words a conundrum, because this plinth is actually a separate piece of stone from the statue, and it is not even settled that they were cut from the same marble.[49] It has therefore been claimed that the lines mean "cut from the same block," "from the same quarry," "of the same quality," "of the same marble," "of the same marble and cut from the same block," or "each block is separately a monolith," all of which seem too prosaic and anecdotal to go to the trouble of versifying on a colossus.[50] Philippe Bruneau instead argued that "*lithos*" is a semantic double, indicating not only a single stone, but also a single identity; in other words, the statue is not only objectively of one piece but materially of one kind. Moreover, the verses speak in the first person singular, presumably Apollo, whether present or by proxy. Patricia Butz has accordingly expanded on Bruneau's aperçu, of doubleness as sameness, to argue that the inscribed plinth echoes Apollo's voice as much as the statue represents his superhuman stature. The archaic Greeks did in fact name and identify some stones with particular gods, some of which they even attributed oracular agency. While the colossus may therefore stand in for Apollo, his poetic voice may also resound from the verse on its base, and one medium fulfill the other.[51] Certainly, very recent investigation has shown that archaic Naxian temples, including one dedicated to Apollo, took great care to incorporate a significant outcrop of bedrock in their walls, which suggests that the local earth was an object of veneration. Moreover, one of the Naxian quarries (Apollonas) was actually sacred to Apollo.[52] The Greeks also traced the practice of cult statuary to *xoana,* aniconic idols of wood or stone, objets trouvés that had been found on earth or fallen from the sky, and that were pregnant with potential form but not yet representational artifacts.[53] The huge colossus that still remains prone and roughed out in the quarry of Apollonas pictures this arrested evolution from bedrock to effigy.

A later echo of this idea of the mantic rock is the oral tradition recorded by Pausanias, when describing Damophon of Messene's colossal cult statues of Despoine and Demeter enthroned (early second century BC) in the temple of the former at Lykosoura, in Arcadia. Pausanias remarks that the effigies along with their double throne and footstool were hewn from a single block (which is plainly not the case from the remains), that "no part was fastened on with iron or mortar," and, most intriguingly, that the sanctuary keepers "say this stone was not imported, but they say that, after a vision in a dream, they discovered it by digging the earth within the precinct." In reality, the statue group is carved from faraway Pentelic marble.[54]

The same scholarly uncertainty affects the most famous example, the statue of Laocoön (probably between 27 BC and 68 AD), which Pliny also praised as ex uno lapide.[55] This has proven very vexing for some scholars, who have concluded that he must have been describing another statue, because that now exhibited in the Vatican is actually assembled from seven interconnecting pieces. However, if we read a later poem by Claudian, "On a Marble Chariot" (c. 395/404), we realize that such acclaim cannot be understood as merely literal description:

> Who has wrought countless appearances from a single marble?
> The chariot merges into the charioteer, and paired reins keep
> The teamed horses in equal check: they that form has released,
> A kindred matter tethers seamlessly.
> The man becomes the chariot again: the team is led by the axle,
> And each fashions itself from the other. What utter mastery!
> A single rock binds all these members, the stony mass
> that suffers the chisel, guided by art, transforms into varied limbs.[56]

The chariot group is a metaphor, expressing the reciprocity between ordering art and unruly matter, and the lively forms emerge from what is "kindred" to both, the mass of marble. Ex uno lapide emerges as an ideal, a criterion of conceptual integrity rather than material fact.

The same ideal arguably existed in architecture, though no body of theoretical writing now survives from antiquity to attest to it. If Herodotos is anything to go by, fifth-century Greeks certainly appreciated monolithic feats of building, for he marvels how Egyptian obelisks, temple roofs, and even small shrines were carved from a single block (Hdt. *Hist.* 2.111; 2.155; 2.175). A building hewn from the landscape in a single block, rather than constructed stone upon stone, aspired to be an extraction from nature that assimilated the monument to a rock formation and, in the case of temples, the sacred mountain. The dedication inscription of the Lateran obelisk (357 AD), for example, calls it "a not inconsiderable chunk of mountain . . . hewn from the red quarry, that has leapt up and knocks at heaven's door" (*CIL.* 6.1163: *haut exiguam partem montis . . . rursus rufis avulsa metallis emicuit pulsatque polos*).

Indeed, the act of quarrying must itself have raised the comparison, for once the blocks had been cut from the quarry (from medieval Latin "*quareria,*" from classical

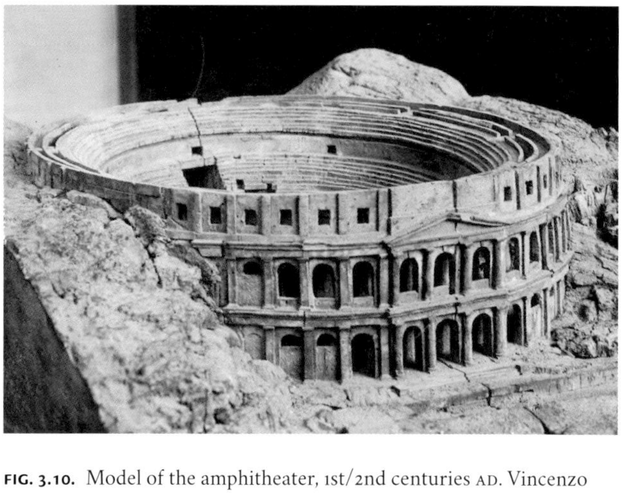

FIG. 3.10. Model of the amphitheater, 1st/2nd centuries AD. Vincenzo Crespi, 1888. Cagliari, Sardinia.

FIG. 3.11. Penelope Painter, *Gigas and Athena*, c. 480/440 BC. Skyphos, height: 7⅝ in. (19.9 cm); diam.: 9⅙ in. (23.1 cm); width: 13¼ in. (33.7 cm). Musée du Louvre, Paris G372.

Latin "*quadrum*," or "square"), what was left was a stepped and architectonic negative. Moreover, the roughed-out block or column drum remained a shaping of the "living rock" up until the very moment it was levered away to become a building block. As Adrian Stokes put it, "A Greek temple is an ideal quarry reconstructed on the hill."[57] Some select buildings actually were ex uno lapide and later became the stuff of legend in medieval mirabilia—those Greek theaters carved from rocky slopes and their followers (e.g., Petra in Jordan, first century AD), and some Roman amphitheaters such as that at Sutri in Italy (late first century BC). There is perhaps no better example than the amphitheater at Cagliari, the lion's share of which was excavated from the gully between two mountains, leaving only a fraction of the façade to be constructed, and gently protruding as though it had been birthed from the mountains (fig. 3.10).[58]

However, all such buildings were exceptional. In Greece the middle ground between geology and building had been the "cyclopean masonry" of archaic Mycenae and Tiryns. The term was coined by antique observers (Paus. *Desc.* 2.16.5; 2.25.8) who imagined that only giants were capable of the brute dexterity necessary to coax irregular blocks into polygonal unison, and Theophrastos singled out Tiryns as the origin (Pliny. *HN.* 7.80.195). A fifth-century skyphos seems to depict the feat, with Athena directing a giant (Gigas), who carries a large rock (or rocks) to build the originally cyclopean walls of the Acropolis (fig. 3.11). Even when the Acropolis walls were completely rebuilt, stretches of the old wall were still preserved, and a window made into one bastion (of Athena Nike) to glimpse the earlier circuit, because it was a relic of heroic or mythic origins. The technique also prospered in archaic Italy, and in some Latial sanctuaries, such as the second-century BC Temple of Juno Moneta at Segni, where massive polygonal blocks were used to make temple podia many centuries after the technique's heyday, presumably to make manifest the antiquity of the foundation.

FIG. 3.12. Wailing Wall, 20–19 BC. Jerusalem.

However, in practice the gold standard of ex uno lapide was unfeasible in architecture, which is the art of assembly, and the ideal was instead communicated, paradoxically, by valorizing the joins between the individual stones themselves. In Greek terminology, that which Vitruvius transmitted to Roman tradition, the term used to denote the perfect jointing necessary was "*harmonia*," cognate with "*harmoge*," meaning a "joint," whether in timber or stone, thence "harmonia" denoted the integrity of both the body and the body politic. "Harmonia" could therefore apply to both the working unison of different organs of government and the organic joining of human bones.[59]

The traveling semantics of harmonia were allegorized in the legend of Amphion, whose lyre enchanted the walls of Thebes into building themselves, their cohesion engineered by the harmony of his song. This Hellenistic myth may still be remembered in a sixth-century poem commemorating the rebuilding by the Vandal king Thrasamund (r. 496–523) of the Alianas Baths near Carthage, which he says "exhibited stones that run to be easily assembled." Analogously, it was held that the akropolis of Megara, where there was a sanctuary of Apollo, had been built by the god, and one stone maintained musical qualities because he had rested his lyre upon it.[60]

In Greek monumental masonry this concern with fine fitting had given rise to the technique known as *anathyrosis*. To achieve the perfect fit, masons would hollow out the centers of the contacting surfaces between blocks and only finish their edges, thereby minimizing the labor by avoiding the task of leveling the whole surface. The outer faces of these blocks showed the inverse: the middle of the block was left proud (as a boss), and only a peripheral channel was chiseled down to ensure the block's alignment with its neighbors. The remaining bosses were meant to be chiseled off later to leave only a network of superfine coursing lines, unless the wall was considered utilitarian, such as the famous "Wailing Wall" in Jerusalem that was the retaining wall of Herod's Temple (fig. 3.12).[61] There was absolutely no need to leave the bosses on the facing walls of premier monuments, unless they enjoyed some emblematic value. There was even less reason to simulate the technique on veneers (see figs. 2.21, 3.13); still less when that veneer was stucco and concealed real ashlar in perfect coursing, as on the Mausoleum of Lucius Munatius Plancus at Gaeta (c. 15 BC).[62] Moreover, the broad drafted margins around the joints were considered so important that they were miniaturized on depictions of temples on coins. Vitruvius even refers to the technique when he writes, "The projecting edges surrounding the vertical joints and beds will produce a more *graphic* effect to the eyes" (*De Arch.* 4.4.4: *circum coagmenta et cubilia eminentes expressiones graphicoteran efficient in aspectu delectationem*). Vitruvius transliterated "*graphicoteran*" from some Hellenistic treatise (it is the comparative of the Greek "*graphikos*"), and it derives from the verb "*graphein*," which can mean "write," "draw," or "paint." By a stroke of linguistic luck, "graphikos" can be happily translated with the English pun of "drafted."

The result was that the blocks concorded not only in harmonia but also in *symmetria*, a term that signifies not symmetry but commensurability. As Vitruvius again says, "Symmetria is the convening in agreement of the members of a work itself, from the different parts to the appearance of the entire figure, in correspondence with a calculated unit" (Vitr. *De Arch.* 1.2.4: *symmetria est ex ipsius operis membris conveniens consensus ex partibusque separatis ad universae figurae speciem ratae partis responsus*). It was the perfect fit between the separate blocks that allowed them to join in unanimity (*consensus*), and the phrases "*consensus lapidum*" and "*coniunctio lapidum*" were occasionally inscribed on monuments and repeated as late as the ninth century.[63] A nice example can be excerpted from a very long inscription on the Mausoleum of the Flavii (c. 150 AD) in Tunisia:

> Here on earth there remains this impressive tomb
> in its eternal newness, thus the shining stones
> perfectly cohere, and as the steps rise from their root
> they have grown finer, so that every corner has
> been traced
> as if with the stuff of malleable wax.[64]

By stressing the joins, anathyrosis professed the solidity of the wall by implying a depth to its blocks, an image of solidity that was indispensable in conferring a moral depth as well. Seneca both applauds the decent and manly construction of Scipio's "villa built of drafted stone" and also rants that "under marble and gold dwells slavery."[65] Men of substance were not superficial.

The fine joining of the stones also proved that no alien bond was essential to the wall's stability, that mortar was superfluous (though in reality cramps and ties were hidden). It was the honesty of the seam that proved the integrity of the construct, and the overall network became the very picture of solidity. Vitruvius applauds the bonding in depth of Greek walls because it "attains a perfection that will endure to eternity" (*De Arch.* 2.8.5). The obvious paradox—that one could see all the joins—only resolved itself if one regarded them as the seams of union rather than the borders of segregation, and we might better apply the term "fractionizing" rather than "subdivision," because it better preserves this sense of overall cohesion. The same paradox will be encapsulated in a medieval adage about the Vatican obelisk, "If it is a single stone, tell me by what art it has been raised? If many stones, tell me where are the joins?"[66]

The ensemble of keenly drafted stone with mortarless joints of a hair's breadth presented an image of steadfastness that matched the cohesion of the building blocks of the state itself. Frequently quoted are Augustus's deathbed words that he had found a Rome of brick but left a city of marble (Suet. *Aug.* 28.3), less so Dio Cassius's interpretation (c. 210/222 AD) that "he did not thereby refer literally to the appearance of its buildings, but rather to the strength of the Empire" (Dio Cass. *Hist. Rom.* 56.30.3–4). This metaphor held the same significance even before Augustus, because late Republican paintings and reliefs summarize the foundations of cities with the rebus of ashlar walls carefully joined under the noses of the tutelary deities that safeguarded the state (fig. 3.14).[67] Vitruvius associated this construction ideal with the culmination of stable government and society, as his own narrative takes us from the aboriginal cave and woods to the masonry city. Indeed, the allegory became a reality in city walls such as those of Pompei, which were made from local volcanic rubble but stuccoed to project a brilliant image of *opus quadratum* (squared ashlar) far afield.

This idea became so common that, when the rhetorician Aelius Aristeides much later eulogizes Hadrian's massive temple at Cyzicus (166 AD), it is natural for him to compare well-jointed masonry to the regimen of the well-ordered city. The temple is a colossal mass, whose individual blocks and ornaments (citizens) are in harmony

FIG. 3.13. Isodomic cladding, Luna (Carrara) marble cella wall, Temple of Venus Genetrix, 98/117 AD. Rome.

FIG. 3.14. *Foundation of a City,* Basilica Aemila, 1st century BC–1st century AD. Frieze. Rome.

with the whole (the state).[68] The chain of hierarchy that unified the body public into a seamless state, pictured in this masonry composite, may be further understood from a passage from Seneca. Seneca distinguishes between continuous bodies (man), composite bodies (ships, houses), and collective ones (the army, the people, the senate), but he concludes, "We believe nothing is good, which is composed from things that are distinct" (*Ep.* 102.7). His densest formulation pronounces:

> Some things are continuous, others composite. A composite is [defined by] the contact between two bodies joined to each other. Continuity is the uninterrupted joining of parts to each other. Unity is continuity without the composite feature [. . . and] when I speak of the one [. . .] I am referring not to a number, but to the characteristic of a body that is cohesive by its own oneness, without any external help. (*QNat.* 2.2.2, 2.2.4)[69]

While the dominion of empire was manifested in the power to command even Nature herself, the ulterior ambition was for buildings to appear almost ready-formed by nature in all its seamless perfection and detail, with the architect or artist consequently acting only as a sort of midwife to their birth. This idea was so rooted that there was potentially even a silent equivalent to the harmonia of Amphion. For the Old Testament records that Solomon's Temple was "built of stone made ready before it was brought thither, so that there was neither hammer nor axe nor any tool of iron heard in the house while it was in building" (1 Kings 6:7). These words seem to record a precocious episode in off-site prefabrication occasioned by the necessity that not even the sounds of labor violate the sacrality of the sanctuary. However, the same passage can also signify that the alienation of labor was the first step to erasing all the indices of facture. Josephus, in fact, goes on to specify: "With great skill the whole construction of the Temple was born from stones cut fine and laid together so neatly and smoothly that to the beholder there appeared no sign of the use of mallets or other work-tools, but all the material seemed to have fitted itself together naturally without the use of these things, so that their fitting together seemed to have come about of itself rather than through the force of tools" (*AJ.* 8.69).

Associated with the "you can't see the joins" topos in pronouncing the miraculous birth of the building is another that has been equally debased by over-repetition, namely, "It didn't take very long either." When the hand is removed from the wholesome artifact's making, the process of its manufacture is elided, and with it the dimensions of history or human time in favor of the undivided moment of perfect birth. These considerations subtend the remarks of Plutarch (c. 45–125 AD) on the Parthenon, five hundred years after its inauguration (432 BC), who marveled not only at its skilled precision, but also that the whole building was "created in a short time for all time. . . . Its beauty was even then and at once antique; but in the freshness of its vigor it is, even to the present day, recent and newly wrought. Such is the bloom of perpetual newness, as it were . . . which makes [it] ever look untouched by time, as though the unfaltering breath of an ageless spirit had been infused into [it]."

The Parthenon was unblunted by the ages, as immortal as the Olympians themselves, and as "unaging and imperishable" as the artifacts of the divine blacksmith Hephaistos.[70] It had come into being just as Athena had irrupted into the world fully formed and fully armed from the head of Zeus, and it seems no coincidence that the German expressionist architect Bruno Taut (1880–1938) would compare the Parthenon with the Japanese Ise shrine, which has been reconstructed in facsimile with virgin old-growth timber every twenty years since about 500 AD, making it at once ancient and brand new. As Taut remarks on the Parthenon, "Perhaps it was created and dropped down from Heaven."[71] In Roman architecture, the desire for such "perpetual newness" (as the Mausoleum of the Flavii says) even led inscriptions to claim that buildings were newly built when they had received only the most modest of restorations.[72]

The ideals of ex uno lapide still steep the description of Rome's monuments made in the early sixth century by Cassiodorus (480–c. 575 AD), rhetorician and administrator for the Ostrogothic kings of Ravenna (505/540 AD). Cassiodorus's praise is all the more remarkable because it occurs in what is ostensibly a bureaucratic document, the *Formula Addressed to the Prefect of the City on the Appointment of the Architect.* Before concluding that all Rome is a miracle, and having lauded Rome's statuary as "another population in stone" procreated by nature, he asks: "What should we say about the lofty columns as slender as reeds? The buildings' towering shafts look as though they have been bundled together from tall canes, and hollowed out by flutings in such uniformity that you would rather believe they had been poured from a vessel; you would think that what you see polished from the hardest minerals was actually made from wax; you would say that the joins of the marbles were generative veins; that wherever the eyes are fooled, the [marbles] deserve to be praised as having grown miraculously."[73] Cassiodorus still sustains the criteria of harmonia and ex uno lapide, for he speaks of these marble piles as though Nature herself had cast them whole in the blink of an eye. But he also alludes to the peak of an imagined process of growth and compares the joins of the man-made assembly to the ligatures of nature. In so doing, he relies on long-standing beliefs about the generation of the stones.

AUGUSTUS AND COLORED MARBLES: GEOLOGY AND GEOPOLITICS

So far we have only heard about white temples, although they were never, as is now well known, completely white. All Greek temples from their earliest incarnations were complemented by painted polychromy, most of it in the superstructure: the entablature, acroteria, tympana, and often the capitals. However, it was on that first monumental complex of all-white-marble architecture, the fifth-century Athenian Acropolis, that architecture in differently colored stones made its classical appearance, in the Propylaia and Erechtheion.

In the Propylaia (437–432 BC), the monumental entrance gateway to the Acropolis, designed by Mnesikles, orthostates in dark-gray limestone, from Eleusis in Attica, line the walls of the internal columned passage hall (fig. 3.15). As their height makes up the difference between the floor level within the hall and that of the Acropolis beyond, the dado is like a tide mark, and to experience the liminality of the propylaion is as if to wade through shallows. The same dark stone resurfaces externally, as the stereobate of the propylaion wings, where it protrudes like a geological substratum and optically discounts the relative proportions of main façade to side façades.[74]

The Erechtheion (c. 421–406 BC) was dedicated to Erichthonios, a legendary early king of Athens also called "Erechtheus" in some versions. A legend, which the Athenians cherished, related that he was conceived when Athena thwarted her rape by Hephaistos, whose impetuous semen she had to wipe from her thigh with a scrap of wool, which she threw to the ground, where it inseminated the Earth Mother, Gaia. This legend was important to the Athenians, who called themselves the "*Erechtheidae*" ("descendants of Erichtheus"), and who were proud to think themselves *autochthonous* ("born from the earth itself"). There could also be no better foundational allegory for the Erechtheion, and stone architecture in general, than the birth from Mother Earth of a child of the artificer-god, by the prophylactic midwifery of the goddess of wisdom. The Erechtheion had a frieze (made 410/406 BC) and tympanum of dark Eleusinian stone onto which separately carved figures in Parian marble were attached with metal dowels (fig. 3.16). The figures were to glow phantomatically against the dark background and (as cameo had yet to be invented) the frieze was likely derived from either ivory-on-ebony friezes on furnishings, or the gilt-bronze appliqués with which Pheidias

FIG. 3.15. Mnesikles, Propylaia, Akropolis, 437–432 BC. Pentelic marble and Eleusinian limestone. Athens.

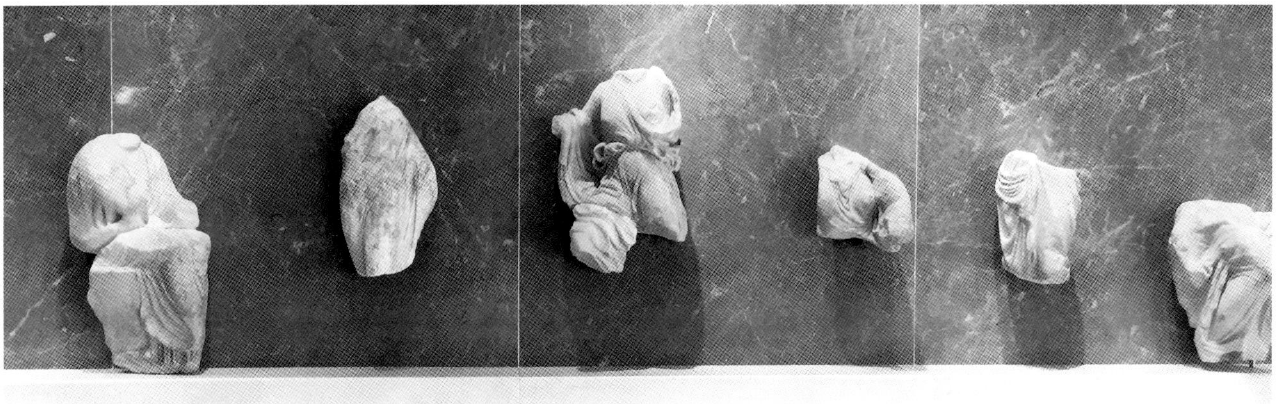

FIG. 3.16. Tympanum sculpture, Erechtheion, Akropolis, 408/407 BC. Parian marble and Eleusinian limestone (restored). Akropolis Museum, Athens.

and his followers decorated cult statue bases, the best examples being the pedestal of Zeus (c. 432 BC) at Olympia and that of Athena (c. 438 BC) in the Parthenon itself (see fig. 2.8). Besides all this, the Erechtheion had another unusual touch: colored glass beads were inserted into the carved guilloche of the Ionic capitals, to bejewel them and make them twinkle (figs. 3.17a, 3.17b).[75] The Erechtheion is normally attributed to Mnesikles, but not on any firm grounds. It seems entirely possible that its inventive stone polychromy should instead be attributed to Iktinos, the architect of the Parthenon, because an equally subtle and bold building that may be nearly contemporary with the Erechtheion was attributed to him already in antiquity.

This is the Temple of Apollo Epikourios at Bassai, a hundred and fifty miles away in Arcadia, which dates to somewhere between 450 and 425 BC (some argue c. 429–427 BC). It is more famous for its Ionic naos and as the debut of the Corinthian order, but its innate polychromy is just as revolutionary. The temple sits on a windswept limestone plateau, 3,700 feet above sea level, on the southern slope of Mt. Kotilion, and three or four miles from the ancient city of Phigaleia (but only as the crow flies). It was a huge challenge to convey marbles here, by pack animals up steep trails, so Iktinos used limestone from the site itself for the bulk of the temple (fig. 3.18.a–d). This was variable in hue, and he reveled in its coloring, of three different types: "grey, very light in the full sun, tending to lavender in the half shadow and purple-blue in the deep shadows; the very frequent patches of crimson-pink discoloration [a reaction of fossils in the stone to water and wind] now add to the lavender hue." Inside the naos, the shafts of this tinted limestone rose to within a few inches of the white marble capitals. Externally the temple was all limestone, the architrave of color-contrasting types, and above it floated the customary roof of glistening white marble tiles and floral akroteria.[76] Today, after centuries of battering by the elements, these stones are gray and fissured all over. Yet Pausanias asserts "of all the temples in the Peloponnese this may be placed first (along with the one at Tegea) for the beauty of the stone and symmetry of its proportions" (*Desc.* 8.41.8). It is astonishing, given all the debate on temple polychromy that has ensued since the early nineteenth century, that this innately colored building has not received the attention that its exceptionalism demands.

Moreover, the idea that white marbles were watery generations might just explain one spectacularly obsessive detail of this temple. The guttae (conical ornaments suspended from the mutules in the Doric cornice) were not carved in relief from the limestone architrave, but were separate plugs of white marble and, since there were 1,404 of them, the labor involved was mind-boggling (see fig. 3.18d).[77] Guttae are normally interpreted as skeuomorphs of the hardwood trunnels that were driven through the mutules (triglyph soffits) to clamp the rafters onto the beams of the putative timber prototypes of the Greek temple. According to the same logic, the conical shape of the gutta served as a water-repellant, a "drip detail" that stopped the capillary infiltration of rainwater. In Latin "gutta" does, in fact, mean "droplet," and, although the corresponding Greek term is unknown, Vitruvius presumably translated "*stagōn*" ("water-drop"). At Bassai, these Parian guttae would have glistened like dewdrops against the blue-gray architrave, and Iktinos may even have been inspired by embryonic stalactites like those we still see today in the bowels of the Parian quarries.

The innately lithic polychromy of the temple at Bassai remained, however, an anomaly until the advent of an

FIG. 3.17A. Ionic capital, Erechtheion, 421–406 BC. Pentelic marble. British Museum, London.

FIG. 3.17B. Thomas Leverton Donaldson, *Colored Glass Beads Set into Guilloché Molding on Erechtheion Capitals*, 1820. Watercolor and graphite. British Museum, London, inv. 1857,1212.10.

empire of resources, the Roman empire, that made possible global provisioning, the combination of far-flung stones at the capital city, and then the dissemination of a new polychrome model back to the provinces. The taste was acquired already under Julius Caesar, who had the *rostra* in the Forum (45–44 BC) clad with panels of Chian marble (*porta santa*) divided by strips of Lucullan (*africano*), which still survive in patches.[78] But it was Augustus who definitively accelerated the use of marbles in public settings and who initiated an imperial procurement system that left little leeway for entrepreneurs; a century would pass before the emperors loosened their grip on supplies by moving the quarry system from on command to on demand.[79]

Four centuries saw Imperial Rome become the entrepôt for a Mediterranean-wide marble trade that she

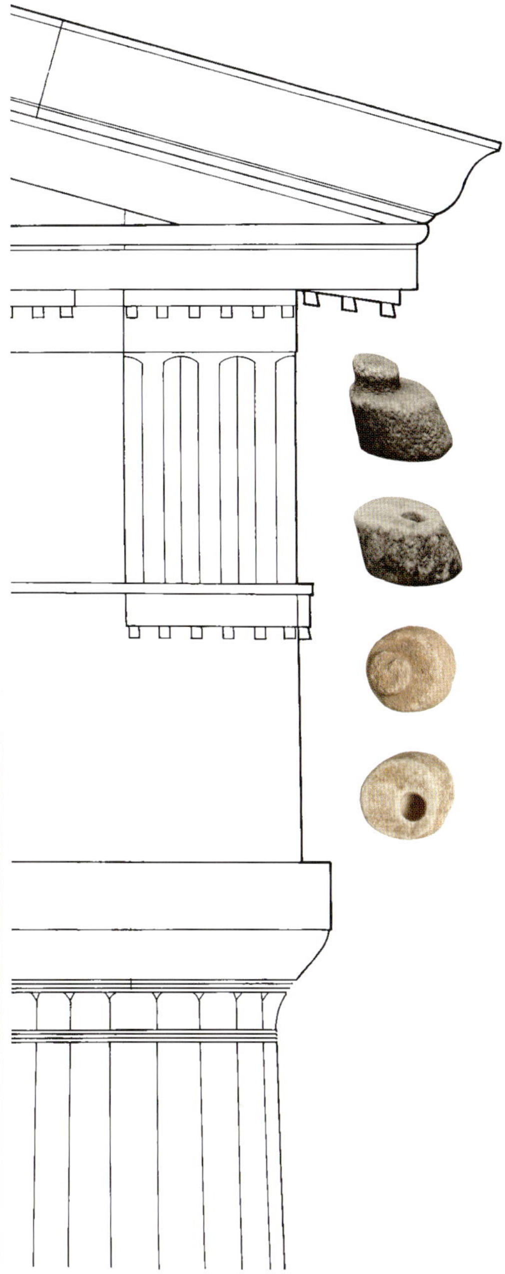

FIG. 3.18A–D. Iktinos, Temple of Apollo, 450/425 BC. Bassai, Greece. Left, from top: a) reassembled architrave, local limestone; b) partially restored capital; c) temple from the north (1966); above, d) elevation of architrave, and guttae in Parian marble.

pioneered. Multicolored marbles flooded in from every quarter, from the furthest frontiers, and on a scale so unprecedented it would not be matched again until the mechanized era. So voluminous, in fact, was this influx that after the demise of the Western empire (476 AD), the caput orbis herself became the quarry of Europe for another millennium. Augustus's engineers prospected a palette so diverse that the geographer Strabo would exclaim, as early as circa 7 BC, that white marble had begun to lose any value (Strab. *Geog.* 9.5.16; 12.8.14). Indeed, although snowy-white marble assumed a dominant key in Augustan poetry, it would cease to do so after his reign.

In fact, even if the temples Augustus built or completed were still predominantly white, their interiors were lavishly colored with the gamut of marbles now arriving from Greece, Asia, and Africa. Many, such as those of Apollo Sosianus (34/29 BC?), Concordia (10 AD), Mars Ultor (42–2 BC), and possibly also Agrippa's original Pantheon (27–25 BC), were not just shrines but accessible museums, virtually national galleries of looted art. As the poet Manilius evocatively wrote (c. 10–20 AD) of the booty deposited by the first emperor, "The gifts of Augustus glitter in the consecrated temples, their golden gleam vying with the flames of the Sun, and the fires of gems flash light from the shadow."[80] These treasuries deserved brilliant halls both to meet the new demands of display and to honor the gods who had brought Rome such good fortune. Although the remains of revetment are, as ever, scanty, we can get a lively feeling for their appearance from earlier fictive representations such as the late Republican sanctuary at Brescia (89/75 BC; see fig. 2.23).[81]

On the public stage, there was no sating the growing appetite for ever more diverse marbles. Apart from temples, the earliest monument lavishly to employ colored marbles was the Basilica Aemilia, again in the Roman Forum, on whose second story (rebuilt 14 BC) colossal statues of Phrygians assumed poses of support, either balletically standing or bearing their loads kneeling (fig. 3.19).[82] These petrified captives were hewn from Phrygian and Numidian marble, and Rolf Schneider has exhaustively argued, and innumerable scholars have since reiterated, that their foreign presence materially embodied the recent subjugation of the states from which the stones originated, Anatolia and North Africa, respectively. Such material iconography, the argument continues, also underlay Augustus's use of the same marbles in his own Forum, where a message of vengeance would have reached fever pitch when the public glimpsed the standards recaptured from Parthia and rededicated within the temple cella of Mars Ultor ("Mars the Avenger"). The exotic "Phrygians" that decorated it therefore doubled for exotic Parthians, commemorating Augustus's putative subjugation of Parthia in 20 BC.

This stone embodiment of conquest is attested long before Augustus, in fact in North Africa. In 152 BC King Massinissa (r. 202–148 BC), a Roman ally, annexed territories from the Carthaginians, discovering Numidian marble in the process and opening the quarries. Almost as soon as he died, his son and heir, Micipsa (r. 148–118 BC), commemorated Massinissa and his feats by erecting a double-story funerary altar to him, decorated with reliefs of military trophies, entirely from Numidian marble, and on the highest peak of the same quarries (Djebel Chemtou).[83] Ironically, when King Bocchus I of Mauretania erected a victory monument on the Capitoline in 91 BC, portraying himself handing over Micipsa's successor, Jugurtha, to Sulla (in 106 BC), its base was now made from gray limestone slabs brought over from Numidia. And it has been argued that the twenty-foot memorial column erected to Caesar in the Forum Romanum (Suet. *Iul.* 85) was fashioned from Numidian marble to commemorate his formation of the new province of Africa Nova.[84] In the same light, the Greek marble of Metellus's temple to Jupiter Stator (146 BC) may itself be considered victory spoil as well; Ovid punningly describes the surrounding Porticus Octaviae, built by Augustus circa 27 BC, as "a work rich in foreign marble" or "rich with marble coating" (*Ars Am.* 1.70: *externo marmore dives opus*).

A similar agenda of triumphalism is generally attributed to the marbles of the Forum of Trajan (106–113 AD) adjoining the Forum of Augustus, where the parade of colossal captives in porphyry, Phrygian, and Numidian marbles would now celebrate the emperor's conquest of Dacia (106 AD).[85] The more that such triumphal marbles were deployed, the more the unimaginable extent of Rome's new dominion was visibly summarized into a mineral map before the eyes, the narrative normally goes. These arguments are persuasive for the Augustan period, but if marble's geopolitical symbolism persisted during the High Empire, by which time the shock of the new materials had long worn off, its expression must have become grammatical. In actual fact, no writer from this period talks about marbles in these terms. The only text that does is much later, desperately nostalgic, and oddly

FIG. 3.19. Kneeling Parthian, after 14 BC. Phrygian marble (black marbles are Renaissance restorations). National Archaeological Museum, Naples.

unquoted. Sidonius Apollinaris (c. 430–479/480) describes the throne of Roma:

> The throne rises from stones that were shorn
> Away from the ruddy mountain in Ethiopia,
> Where an inborn purple scorches the cliffs
> Inflamed by the nearness of the sun.
> Here Synnadian marble joins with it, there Numidian
> stone is added,
> Which counterfeits ancient ivory; next a row of the
> grassy marble
> From Laconian rock flourishes radiantly.
> Such that when she sat herself amidst the throne,
> The whole earth flocked to her all at once.[86]

Before becoming bishop of Clermont-Ferrand, Sidonius was Praefectus Urbis (470 AD), and his audience hall was just off the Forum Romanum, so he may have had in mind the cult statue of Roma herself that dominated the apse within the nearby Temple of Venus and Rome (built c. 125–145; rebuilt after 307 AD). And the marbles Sidonius itemizes came from Aswan, Numidia, Greece, and Asia, which vaguely correspond to equidistant quarters, and therefore the idea that "the whole earth flocked to her."

Nonetheless, this triumphalist view of colored marbles is all too vulnerable to semantic determinism—the attribution of too exclusive a range of meaning—and begs greater nuancing. Indeed, before porphyry became king, Phrygian had held pride of place in Roman marble exhibitionism.[87] Shot through with purple, Phrygian was one of the costliest marbles in circulation and, therefore, the target for special abuse from the first to fourth centuries. Tibullus complains of domestic columns, Horace adds it to the catalogue of purple horrors that closes his ode to simplicity, Juvenal itemizes it when he ridicules the house of the millionaire Licinus, Ovid's Dido finds craggy caves preferable to it, and Ausonius exalts sandy shores over any Phrygian pavement.[88]

Again, although the Dacians in Trajan's Forum were all carved from frontier stones, these same minerals were luxury goods that intimated luxury fabrics: embroidered robes, cloth of gold, and purple-dyed tunics (Phrygian, Numidian, and porphyry). Not coincidentally, in late antique Latin, the word regularly used for marble panels, "*crusta*," could also indicate a fabric that was richly woven, embroidered, and jeweled. Porphyry was particularly exalted because it resembled imperial purple, a fabric prized because of the expense involved in gathering its dye from the secretions of thousands of mollusks. It is little coincidence that the extensive quarrying of porphyry from the Hadrianic era onward coincided with the most intensive exploitation of purple dye. Ever after, any emperor, pope, or cardinal raised to the purple would exploit the stone for this association.[89] For similar reasons, although real porphyry was beyond the reach of the city of Meninx in Tunisia—an epicenter of purple-dye production—purplish limestones were used in its lieu to adorn the barbarians of her forum basilica (Trajanic-Antonine), while the others were of golden-hued limestone.[90] This kind of subterfuge seems to be hinted at in a late second- or third-century epigram inscribed on a relief from Sagalossos (Anatolia), which concludes, "Even if art through skill wishes to show that this is Phrygian stone, it deceives: for the stone has grown in this land."[91]

At the Flavian triumph in Jerusalem of 70 AD, not only the emperor but also his entourage wore purple, and the captives themselves were also richly adorned (Joseph. *BJ.* 7.123–38). Trajan's triumph in 107 AD may have adopted the same wardrobe, and Thomas Morton has therefore argued that the parade of captive statues in the Fora would have represented a permanent parade. It follows that the aristocracy of stones in Trajan's Forum could have stood for the *purpurati* of foreign lands, not the soldiery or commons, with the overall implication that Trajan (or Augustus for that matter) was a prince over princes.

Even if such variegated display did map the empire, it mapped the fertile lands she encompassed and, for the same reasons, a later empire would mine the wealth of her dominion for museum-display in her capital. When the Museum of Practical Geology was founded in London in 1851, Prince Albert remarked that it was thanks to the "immense mineral riches granted by the bounty of Providence to our isles, and their numerous colonial dependencies." Likewise, when Richard Owen addressed the British Association for the Advancement of Science in 1858, he declared, "Our colonies include parts of the earth where the forms of plants and animals [and minerals] are the most strange. . . . Naturalists consequently visit England anticipating to find in her capital and in her National Museum the richest and most varied materials for their comparison and deductions."[92] Bringing marbles to Rome from the provinces was much like bringing wheat from Egypt, lions from Africa, or any of the numerous wines whose origins Horace so lovingly lists. The *Pax*

Romana ensured that the same counters of wealth and dominion traversed the empire from one boundary to another, as marbles found from Crete to Britain demonstrate, establishing an endless commonwealth dealt in the same coin wherever one went.[93] Eventually, by the reign of Trajan's successor, Hadrian, Roman material vocabulary had departed so far from its Greek origins that this philhellenic emperor even exported massive amounts of Phrygian marble to Athens (Paus. *Desc.* 1.18.9).

Furthermore, if marble display meant "the physical embodiment of empire and the capacity of the imperial regime literally to move mountains," these efforts were as much a victory over nature as over barbarians, and some writers employ the same idioms whether discussing quarrying or war.[94] For emperor and soldier, nature could be tamed and vanquished like a new province, but for philosophers and naturalists, this exploitation represented defilement and mutilation.[95] Pliny feared that extensive quarrying equaled vivisection, undermining the mountains that clamped together a seething and brewing earth. While Ovid lightheartedly quipped that the mountains "were shrinking" thanks to marble quarrying, Pliny's sour riposte was that, since marbles could supposedly regenerate, "there is reason to hope that there will always be marble sufficient to satisfy luxury demands."[96] Ultimately, the power "to tame" (*domare*) the mountains signified an overwhelming force so primeval that, long after recycling building materials had become habit, some citizens still felt it necessary to specify that they had beautified their cities with stone "cut from the mountains, and not removed from ruined monuments."[97] In the same light, it becomes easier to understand why fourth- and fifth-century emperors would eventually mobilize against spoliation.[98] Spoliation meant that the empire was visibly ailing, its body sickly, its provinces splintering, and the stones in its frontier walls loosening under the pressure of the marauding hordes. In Rome itself, by the late fourth century, spoliation was even capable of provoking public riots.[99]

If the panoply of marbles available in the first century had testified to the empire's growth, by the fifth, increasing shortages made visible its decline. When the Byzantine historian Zosimus (early sixth century) describes the fire that destroyed the Senate House in Constantinople in 404 AD, he emphasizes that it was "a beautiful and lavish building adorned with . . . marble in many colors no longer quarried." Since this writer's militantly pagan *New History* aimed at exposing how Christianized Romans had lost their empire in less time than it had taken to build it, the loss of the stones implied the loss of the provinces.[100] With marble starvation, the body of empire had visibly begun to consume itself.

4 Painting in Stone

From Knossos to Rome, from Fresco to Marble

Marbles were not simply tokens of empire. They were prized for the natural artistry of their markings. Indeed, by the late first century AD, when the homes of Rome's super-rich had become encrusted in colored marbles, Pliny the Elder was moved to lament (c. 70/77 AD) that fresco painting had "been completely ousted by marbles. . . . No longer are simple panels pleasing, nor the expanses of mountains spread about the bedchamber; *we have even begun to paint in stone*" (*HN.* 35.1.2–3). Pliny's obituary of painting was premature, but he was also woefully anachronistic in accusing his peers of "painting in stone." Such a medium had been invented in the form of *buon fresco* by Minoan painters around the turn of the second millennium BC on the island of Crete.

THE MINOAN INVENTION OF FRESCO (C. 2000–C. 1450 BC)

The fresco technique entails covering a masonry wall with a rough coat of lime and sand render (*arriccio*), onto which one or more layers of much finer lime plaster are troweled and smoothed, the final layer (*intonaco*) spread over an area as large as one can paint in a single day (*giornata*). The pigments must be diluted in limewater (calcium hydroxide) before they are daubed onto the plaster, into which they sink, whereafter pigment and plaster dry and bond into an insoluble amalgam. The fine residue of pigment particles remaining on the surface dries into a protective crystalline mesh known as the "lime crust," which can be buffed and burnished after the plaster has completely dried. A Gallo-Roman relief (c. 75–125 AD) shows painters and plasterers working side by side, executing every step in the process simultaneously (fig. 4.1).[1]

FIG. 4.1. Drawing of Gallo-Roman relief, with painters and plasterers working side by side, c. 75–125 AD. Musées de Sens.

FIG. 4.2. Bull-leaping fresco, Palace of Knossos, 1700/1450 BC. Heraklion. Fresco, 30 13⁄16 × 41 1⁄8 in. (78.2 × 104.5 cm). Archaeological Museum, Heraklion.

The most famous Minoan frescoes, in the palace at Knossos, include scenes such as the famous rodeo of fearless toreadors somersaulting over charging bulls (fig. 4.2). Yet this painting dates somewhere between 1700 BC and the destruction of the palace about 1450 BC—in other words, several centuries after the invention of the new medium. The earliest datable fragments of fresco (c. 1900–1750/1700 BC) were instead decorated with wavy patterns that scholars have either deemed "abstract," and thence "decorative," or concluded represent colored stones, and are therefore "mimetic" (fig. 4.3).[2] Because Minoan figurative wall painting arrives so much later, its eventual emergence was once taken to witness the graduation of mural painting from "craft" to "art," from an artisanal way of ornamenting a wall to the birth of an artistic consciousness.[3]

Where fresco recognizably doubles for stone, it is also assumed that this simulation obviated the difficulty of quarrying the real thing. However, the indigenous Cretan gypsum that was used to clad floors, stairs, benches, and dados was available in plenty and easy to cut and smooth.[4] Moreover, since the early Bronze Age (c. 2500 BC) and under Egyptian influence, the Minoans had devoted great skill and endless hours to drilling and abrading vessels, vases, and beakers—some enormous—from all manner of much harder stones: marble, breccias, basalt, calcite, diorite, obsidian, schist, serpentine, and others.[5] It also has to be said that the stones Minoan painters imaged in fresco hardly resemble the soapy-looking Cretan gypsum, even the *balatino* variety with swirling reddish veins, and have more affinities with the luxury vessels the Minoans imported from Egypt and the conglomerates and stones with deeply colored veins that they imported from mainland Greece.[6]

Instead there is an argument to be made that the fresco technique was devised precisely in order to synthesize colored stone. Minoan ceramicists had been cloning stone in clay for some centuries before fresco appeared, beginning with the vessels now known as "Vasiliki ware." Around 2600 to 2200 BC, clay cups and beakers were

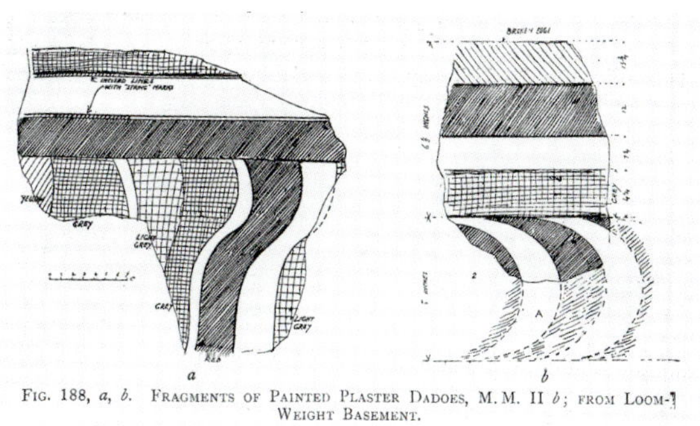

FIG. 4.3. Fresco fragments, "Loom-Weight Deposit," Palace of Knossos, c. 1900–1750/1700 BC. Crete (Evans and Evans 1921).

brushed and speckled with amorphous patterns of blacks, reds, browns, and oranges (on clay of reddish yellow to grayish color); they were burnished before firing, and their luster was then augmented by the vitrification of the slip in the heat of the kiln.[7] These vessels immediately recall the veining of genuine stone ones, and in some cases the similitude is uncanny (fig. 4.4). Vigorous variants, splashed or stippled presumably to imitate the mottle of stones, have been dubbed "Spatter ware" (c. 1900 BC) and even the "Jackson Pollock Style" (c. 1600–1450 BC).[8] In the next generation of Minoan pottery, the so-called "Kamares ware" (c. 2000–1700 BC) that was coeval with the invention and development of fresco, Minoan painters sometimes corrected the unruly veining of the stones they aped into pleasing geometrical conventions (see fig. 4.4c), which would be repeated down the centuries as shorthand for marble veining. The relationship between stone, ceramic, and fresco also became reflexive. Thus, fragments of wall plaster (c. 1900–1800 BC) from Knossos reproduce an unusual class of Kamares pottery, sometimes called "Print ware," in which a pattern of black curling lines was painted onto a white background. Elsewhere, a love of exuberant veining for its own sake resulted in frescoed borders that have been called "splash painting."[9]

All these concerted simulations suggest that the development of fresco was neither accidental nor utilitarian. It may have been the very experience of working in clay, painting it, seeing it harden and the surface vitrify, that inspired fresco painters. Firing these vessels in the kiln transformed the wet and malleable clay into a dry and hard fossil of its former, more vital state. Fresco was another technological step, devised to substitute stone and impersonate its most engaging characteristics. It belonged again to that Bronze Age culture of material synthesis (brick, glass, faience, and so on) that allowed artisans not only to imitate nature but "improve" upon it, in this case at its painterly best.

Meanwhile, their Egyptian colleagues were completely unaffected by the Minoan discovery of fresco and continued simply to paint tempera (pigment in a binding agent of glue, honey, milk, and commonly egg) straight onto walls covered with dried gypsum plaster (our "plaster of Paris"). The resulting painting was an emulsion, a vulnerable skin that may chip and flake over the years. By contrast, in fresco the coloration is as inherent and petrified as that which the Minoans cherished in colored stones. And all the raw materials of painting were products of the

FIG. 4.4A–C. a) Vasiliki ware, 2600–2200 BC; b) stone vessel, c. 2600–2200 BC; c) Kamares ware, c. 2000–1700 BC. Archaeological Museum, Heraklion.

FIG. 4.5. Room 2, Delta House, c. 1700–1650 BC. From Akrotiri, Santorini. Fresco. Archeological Museum, Athens.

earth about them. Paint did not come in tubes nor plaster in sacks. The pigments were colored earths—sometimes even the blues—rocks and clays that were naturally tinted by the mineral oxides they contained, predominantly iron but also manganese.[10] Painters prospected the land for these earths, digging boulders out of the hillsides, crushing, washing, grinding, and drying them, then rewashing them to increase the intensity of their color, or burning them so that yellows turned to reds.

Plaster was also the salt of the earth. The lime needed to make it had once been limestone (calcium carbonate, $CaCO_3$) that had been calcined—that is, baked—in an oven or on a huge bonfire until it decomposed into lime (calcium oxide, CaO). It was then pulverized and slaked with water in a violent exothermic reaction that transformed it into the quicklime [calcium hydroxide, $Ca(OH)_2$] for mortar and plaster. The transaction with nature was brutal and elemental, a battle of earth, fire, and water, with the remnants consumed by the air. The finished fresco itself was another case of "just add water," for no other binder was necessary for the pigments to fasten except the fabric of the plaster itself. Fresco was an amalgam.

To sum up, in fresco the pigment, and therefore the image too, is *in* the wall not *on* it, and so earthy are all its constituents that we can say that fresco represents the earliest example of "painting in stone." Even twentieth-century fresco painters will reminisce that one had to paint quickly "before the paint transformed to stone."[11] Fresco was therefore a means of synthesizing nature rather than simply coloring plaster, materializing stone by reconstituting it from its basic elements. The materiality of fresco gave it the essential potential to create a secondary reality, by substantiating something that would otherwise remain only depicted. As Seth Estrin has pointed out, later concepts of illusionism are extraneous to Minoan wall painting because the means of artistic description are embodied ("not *on* the wall but *in* it"), and there is accordingly no picture plane and no "window" onto a virtual reality apart from the viewer.[12] While Minoan frescoes will become mimetic insofar as they describe birds, trees, bulls, flowers, people, and so on, the image is always immanent in the media.

A nice example reconciles all these aspects and reveals how the materiality of the wall rewards its pictorial representation. On the island of Thera (Santorini) in the Cyclades, Minoan or Minoan-trained painters painted a domestic room with the earliest surviving "pure landscape" in Western art (c. 1700/1650 BC; fig. 4.5). Its subject is the natural world, a fecundating image of spring or summer, with flowering lilies on the mountain slopes and the chirruping courtship of swallows overhead. There is no perspective, of course, nor any progressive diminution of the components to intimate spatial depth. By the same token, there is no background. The scene is frameless. Instead, the earth presses on the wall and swallows up the room at its most pregnant situation, on the ground floor.

Created in the space of years between a violent earthquake and the volcanic eruption that would eventually bury the town, the painters could not have been more aware of the chthonic force of nature in action. They striped the mountains they painted with bands of greenish-ocher, terra-cotta red, and bluish-gray paint, which could respectively stand for clay, pumice, and ash, even the ultimately fertile lava flows that the floral eruptions intimate. Altogether, the fresco reveals a geological image of the paint itself, the pigments that the earth offers up to the prospecting painter, and the basic conditions of nature.[13] The most spectacular examples of such bounty are the dunes known as the "Seven Earths" in the Mauritius—which result from the decomposition of volcanic rock first into clay, and then soil tinted by iron and aluminum—or the heavily mineralized "Rainbow Mountains" in China and Peru. The stratigraphy of Thera is no less remarkable even if the palette is hardly so kaleidoscopic. All in all, the Thera fresco truly embodies the subject as much as it depicts it, and in a more distributive agency than maker-made image. It is thus prophetic of Pliny's "expanses of mountains spread about the bedchamber" and, because fresco is a stony medium, it may lay claim to being the first fully marbled room as well.

Minoan materiality shines through another fresco, this time at Knossos, where undulating lines separate and surround naturalistic partridges (fig. 4.6). These lines have been interpreted as rivers or just decorative patterns, but the earlier Vasiliki ware suggests that they actually represent gypsum veining, and a consciousness of the painted image as consubstantial with its support. This concern with "presencing" the image also meant that much Minoan painting was rarely flat. The surfaces heave and ho, rarely in unison with the contours of the painted figuration. It would be presumptuous to categorize the whole as a "painted relief" because—as in those prehistoric cave paintings where the artists (or shamans) capitalized on the natural contours of the rock by smearing round and over them with colored earths—there is no reason to believe that such distinct (and a posteriori) categorizations existed in Minoan minds. The encounter was a duality. In the dialogue between maker and material, the material spoke first, the lithic matter (whether fresco or rock faces) provoking the reagent artist, who in turn saw the material as not simply passive, but a manifestation of the "living rock." Nonetheless, from a modern viewpoint, the art form was predestined to give birth to molded images, and the trajectory is completed when whole figures eventually emerge from the wall as crisp reliefs, as in the famous *Charging Bull* from about 1600 BC (fig. 4.7).

From an early date, fresco spread also to the floors. At first it was simply the mortar between real slabs that was colored, but because fresco was hard enough to tread

FIG. 4.6. Scene with partridges, House of the Frescoes, c. 1600/c. 1500 BC. Knossos (Evans and Evans 1921).

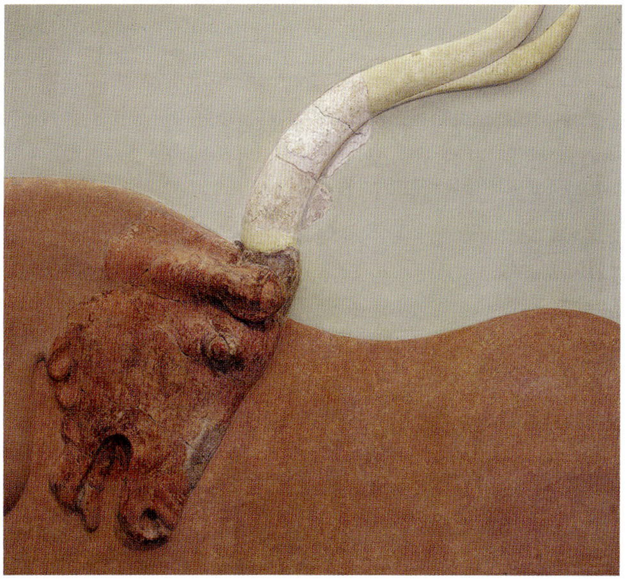

FIG. 4.7. Bull-relief fresco, North Stoa, Palace of Minos, c. 1600 BC. Knossos, Crete. Archaeological Museum, Heraklion.

underfoot, eventually smooth floors were painted all over, even outdoors, and over eighty are known from palaces and mansions on Bronze Age Crete.[14] A stunning example comes from, of all places, eastern Syria. "Of all places" because, while it was believed for most of the twentieth century that the art of mural painting must have been brought west from Sumer or Assyria, and fresco technique with it, in the last twenty years archaeologists have instead argued that itinerant Minoan painters took their skills and material imaginary east, and their works have been found at sites in Syria, Turkey, Israel, and Egypt. In the Palace of Zimri-Lim at Mari (Abu Kamal, Syria) it was almost certainly Minoan painters who frescoed (c. 1750/1700 BC) the upper face of a dais in the *papahum* (large hall that was the focus of political and administrative duties)—perhaps for a king to sit on—in reds, yellows, and bluish-green ochers (fig. 4.8).[15] The dais materialized another imaginary stone, probably inspired by agates or the other veiny stones that the king craved from abroad, as we know from letters written back to him by his overseas agents.[16] In fact, the material envisioned by the painters was so alien to the lithic environment at Mari, a mud-brick city built on the silty banks of the Euphrates and surrounded by desert, that it might as well have been a meteorite. At Mari, the illusion of assembly, right down to the "error" of a plaquette that interrupts the border, redoubles the claim of fresco to be "the real thing" and so estrange still further the alterity of the materials from their setting. At the Palace of Yarim-Lim at Alalakh (Hatay province, Turkey), it was also probably Minoan painters who painted the orthostates (a dado of upright slabs) to imitate veined stones (c. 1750/1700 BC).[17]

Evidence that imported stone vessels may have influenced this taste for veining comes from another room

FIG. 4.8. Podium, Hall 64, Palace of Zimri-Lim at Mari, c. 1750/1700 BC. Abu Kamal, Syria (photograph: c. 1935).

in another house on the island of Thera (c. 1700/1650 BC; figs. 4.9, 4.10). Not only did the painters create dados of faux-"alabaster" all around the room—and no real revetments have so far been found on the island—but they also painted imported Egyptian vases in the jambs of a deep window niche where the sun would have illumined them the best and, presumably, conjured up the translucency for which these vessels must have been prized.[18] Egyptian alabaster was a solar, heavenly material, a fundamental reason it became a luxury item in Egypt and abroad, and its greatest quarries were in Upper Egypt at Hatnub, whose name means "Mansion of Gold." The remaining decorations in the room represent the *ikrion* (the oxhide bridge of a ship), even allowing the possibility that this house belonged to a maritime trader, one who might have brought Egyptian and Minoan artifacts and raw materials to Thera.

Meanwhile, in Egypt itself, Minoan painters simulated the very "alabaster" that sparked their imaginations in the first place, on the walls of the palace at Avaris (Tell el-Dab'a), in the Nile delta, dating to the reigns of Hatshepsut/Thutmosis III (1479–1458 BC) and Amenhotep II (1425–1398 BC). Elsewhere they decorated the walls of the late Bronze Age palace at Qaṭna (Tell el-Mishrife, Syria) (fig. 4.11). At a Canaanite palace in Tel Kabri (Galilee, Israel;

FIG. 4.9. Room 4, West House, c. 1700–1650 BC. Akrotiri (redrawn from Televantou 1994).

FIG. 4.10. Egyptian alabaster vase in window jamb, Room 4, West House. Akrotiri.

FRAGMENT DE PEINTURE MURALE TROUVÉ DANS LE PALAIS
A L'EST DE LA SALLE DES DEUX BASES. EPOQUE MITANNIENNE
Les deux clichés se font suite (repère A). A droite, coupe de l'enduit.

FIG. 4.11. Fragments of wall plaster with painted veining, Palace of Qaṭna, late Bronze Age. Tell el-Mishrife, Syria (du Buisson 1935).

now dated to between 1742 and 1640 BC), Minoan or Cycladic painters fashioned the whole floor of the throne room as a chessboard of facsimile gypsum or travertine; and other examples are still coming to light.[19]

The taste for such pavements reached its acme in Mycenaean Greece in the grand palaces of Homeric fame, Mycenae and Pylos, as well as those at Tiryns and Gla, all of which date from around 1300 BC (figs. 4.12, 4.13).[20] When these floors were first uncovered in the late nineteenth century and at the beginning of the twentieth, the patterned grids were thought to represent textiles, but it is now long understood that the variegated lines and dots are conventions for stone veining and mottling (like the Kamares ware mentioned earlier; see fig. 4.4c). Even so, the floors may combine evocations of stone and textile side by side and, together with marine animals often figured, compose an assemblage that completely supersedes any possible reality. All frescoes were always highly polished, like the best-quality stones—that is gypsum, Egyptian alabaster, and marble—accentuating their affinity through luminosity, but also subtracting some of their earthbound nature. All in all, the viewer now inhabited a total pictorial

FIG. 4.12. Piet de Jong, Reconstruction of the interior of the Megaron of the Palace of Nestor at Pylos. Ephoreia of Olympia.

FIG. 4.13. Piet de Jong, Floor plan of the Megaron of the Palace of Nestor. Archaeological Museum, Chora.

FIG. 4.14. King's Megaron or Hall of the Double Axes, Palace of Knossos, 18th–15th century BC (reconstructed by Piet de Jong and Sir Arthur Evans, 1928). Crete, Greece.

environment, which elevated time and place onto an oneiric plane, a parallel world composed from hyperreal materials. Piet de Jong's imaginative reconstruction does some justice to the original appearance of the Megaron at Pylos (see fig. 4.12), the type of hall in which Telemachus blurts out, "This is the way the court of Olympian Zeus must be within, such untold riches lie here: I am overawed at the sight" (Hom. *Od.* 4.74–75). Seeking parallels more familiar to the modern world, we might view this environment through the lens of "magical realism" in twentieth-century Latin American literature, or as the avatar spectator in the virtual worlds of the contemporary internet. In the megaron it was only the pavement that gave any solid footing in mundane reality, and the shimmering surface made even that footing seem precarious. Medieval Arabic stories of the queen of Sheba being fearful to tread on a floor so polished that it resembled water may be reminiscences of this tradition.

Traces of the origins of Minoan fresco in the synthesis of colored stone still linger in the bull-leaping fresco, in the border of overlapping and striated scales that never receives much comment because it lies at the margins of the figural scene (see fig. 4.2). These scales may imitate the conglomerate stone that was so popular for making vessels on Crete (and the later monumental masonry of Mycenae) (see fig. 2.1).[21] Closer inspection suggests that this border belonged to a continuous frieze around the room, one that materialized what lay beneath the plaster—for the walls themselves were built from a cage of wooden posts and beams, filled with rubble packed in clay. It is a retrospective projection of modernist expectations to think of this X-ray transparency in terms of "structural honesty," and it is instead tempting to think of the frieze as a petrification of the wood within the wall. Indeed, when the occluded framework emerges from the wall, as it does in the "Hall of the Double Axes," it is in the shape of hefty beams and posts (fig. 4.14). The distinctive triple banding of Minoan murals also reflected the introduction of orthostates, the upright slabs of stone that lined the base of the wall inside and out.[22] Functionally, orthostates were "kick zones" that also prevented walls from wicking water. Symbolically, they communicated with the generative earth, miniatures of the mountain ranges that filled the room in Thera (see fig. 4.5), stones that arose into the light.

THE "MASONRY STYLE" IN THE CLASSICAL AND HELLENISTIC WORLD (C. 400–14 BC)

Stone and fresco dados continued to line the palaces of Mycenaean Greece, but the role of colored plaster in materializing revetments and masonry did not halt there. Across the eastern Mediterranean in the Classical period, and then from Egypt to Rome in the Hellenistic period, a means of sheathing walls with "marbleized" stucco developed, only to be supplanted by real marble revetment in the first century AD (as Pliny bemoaned). This fictive masonry in stucco with drafted margins has been variously termed the "Incrustation Style," "Structural Style," "Architectural Relief Style," and most recently the "Masonry Style."[23]

The Masonry Style originated in Greece in the first half of the fifth century BC, in Athens, whence it was transmitted to northern Greece, first in public buildings and then domestic settings.[24] Fragments survive in buildings from Athens, Delos, Pella, Pergamum, and many other Greek cities besides, but the earliest examples are in the Agora of Athens and the private houses of Olynthus.[25] The best-preserved remains are in the Hellenistic mansions of Delos, where even whole rooms stand intact (fig. 4.15). These are difficult to date and usually assigned, on no conclusive evidence, to the island's second period of Athenian rule (169–88 BC).[26] However, the most impressive example to survive comes from the Macedonian capital of Pella, being a lofty wall section reassembled from fragments found in a banqueting room in the late third century BC (?), "House of the Wall Plasters" (figs. 4.16, 4.17).[27] The structural walls in stone rose only to two or three feet, and the rest was in baked brick, but all of it camouflaged by this refined architecture in stucco. Again, the illusion is painted but not pictorial, for the architecture still arises as relief, and only that remnant that is colored blue suggests that there could be a sky beyond, something outside of the wall.

In time, the color schemes and molded relief of these Hellenistic walls spread to the Italian peninsula, where they were reincarnated in the Roman idiom now known as the "First Style." The practice made its entrance into Italy through the Greek colonies of southern Italy, from whom it was adopted by the Etruscans in central and northern

FIG. 4.15. Room, House of Hermes, 2nd century BC. Delos.

FIG. 4.16. Wall from the "House of the Wall Plasters," late 3rd century BC. Pella. Archaeological Museum, Pella.

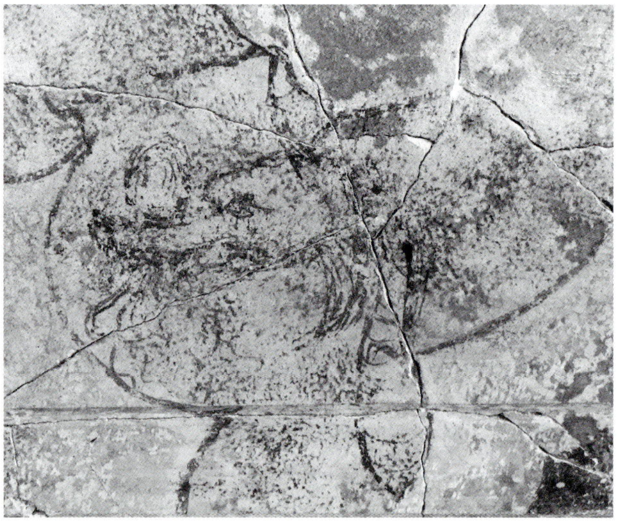

FIG. 4.17. Detail of *Dolphin Capturing an Octopus,* wall from the "House of the Wall Plasters." Fresco painted in imitation of breccia. Pella.

Italy.[28] The back chamber of the Tomb of the Saties (François Tomb), Vulci (c. 325/300 BC), is a precocious example in the "Zone Style" (where only paint and scored lines imitate masonry, not relief as in the "Masonry Style"), and, because the plinths for sarcophagi suggest couches, it seems that polychrome masonry was already imitated in Etruscan dining rooms of the fourth century BC (fig. 4.18).[29] Two centuries later, more dazzling remains come from the loggia of a house at Populonia (c. 125 BC), in a now-Romanized Etruria (fig. 4.19). The technique had not changed much since Knossos. The mud-brick walls (*opus craticium*) were given an adobe coating, then covered with coarse plaster of lime and siliceous aggregate, and topped off with a finishing coat, which was extremely hard and brilliant white, thanks to the sparry calcite crushed and mixed in with the lime. From this gleaming matrix the

FIG. 4.18. Rear chamber, Tomb of the Saties (François Tomb), c. 325/300 BC. Vulci.

painters conjured the illusion of at least twenty different "alabasters" and colored marbles that could hardly have been available on the open market, and which they either painted from memory, fantasy, or samples (the earliest marble floors in Italy are littered with offcuts).[30]

However, the Masonry Style, or rather the First Style, became far more common in Italy after the Roman defeat of Macedonia at the battle of Pydna (168 BC), the transformation of Delos into a free port (166 BC), and the conquest of Corinth (146 BC), when Greek painters were enslaved or immigrated to Italy of their own volition.[31] Examples of the First Style appear up and down the Italian peninsula, in both public buildings and homes, but the best-preserved examples began to clad house walls in Pompei and Herculaneum by at least 150 BC (fig. 4.20).[32]

The individual blocks of these walls also had drafted margins to indicate their genealogy in solid ashlar rather than superficial revetment (that is, they imitated anathyrosis). Scholars have therefore long argued that the Masonry Style began in Greece with the imitation of public buildings, and that its dissemination represented a progressive ennoblement of the Hellenic house. In other words, the knock-on effect of increased ashlar construction in civic architecture was that it raised the stakes for the dignity of the patrician home, and scholars have concluded that these domestic walls were intended to associate the private (or semiprivate) space of the house with the public arenas of political life.[33] In fact, Vitruvius remarks (*De Arch.* 6.5.2) that there are "common spaces [in houses], into which even uninvited people may enter in their own right, like vestibules, atriums, peristyles," and that therefore "the houses of advocates and orators ought to be still more handsome and spacious [than those of bankers and taxmen] for receiving delegations," while for the houses "of nobles, who in holding offices and magistracies, must attend to duties for their fellow citizens, regal and lofty vestibules must be provided," not to mention "libraries, picture galleries, and basilicas, finished with a magnificence similar to that of public works; because in their houses, public council affairs as well as private judgments and arbitrations are more often determined." At Pompei, not only is the city's basilica (89/80 BC) decorated with First Style stuccoes (which were also once highly colored), but even the towers of the city walls.[34] Faux-masonry sometimes covered domestic façades as well, such as the House of the Ceii at Herculaneum, and some of these must also have been highly colored.[35] Faux-masonry

FIG. 4.19. Loggia, house at Populonia, c. 125 BC (Cavari and Donati 2015).

FIG. 4.20. *Fauces* (vestibule), First Style, House of the Samnites, late 2nd century BC. Herculaneum.

could therefore offer the most urban(e) face of the domus, a domestic rampart in the camouflage of the public realm, and a particularly strident example is the townhouse of Julius Polybius (early first century AD), where channeled faux-masonry papers a *vestibulum* that has swelled to become almost an internal piazza.[36] In circa 91 BC the progressive Tribune Marcus Livius Drusus had even rebuffed an architect who offered to build him a house free from the public gaze, with the words, "You must, as much as your skills allow, design my house such that, whatever I do, anyone may perceive it" (Vell. Pat. *Hist.* 2.14.3).

Furthermore, First Style stuccoes do not line every part of the Roman house, but tend to be restricted to the public areas, the vestibule alone, or the atrium as well. The atrium was not only the face of the household (in fact, quite literally, as here hung the portraits of the ancestors) but also the site of the lararium, the altar of the household gods. It was the heart(h) of the home and a shrine.

Borrowing the formula from the public sphere helped to imitate the public, but also sacralize the domestic space. The desire to intimate the proper gravitas, or lay claim to *prisci mores* (ancient custom), may also be the reason that the First Style was occasionally produced in the same locales in the Augustan and Julian-Claudian eras, by which time it was long out of fashion.[37]

At the same time, however, we must also recognize that in Republican Italy, the First Style was common to houses, basilicas, and temples. Whether it is true that the house was ennobled by association, before all else the stuccoes were a self-referential sign, an architecture about architecture—that is, it was they that made the place a "building," block upon block. Moreover, a relative ranking for First Style masonry only became possible after the more elaborate illusions of the Second, Third, and Fourth Styles begin to infiltrate the home, leading to any possible differentiation.

Two hundred years after the arrival of the First Style in Italy, Vitruvius himself (writing c. 27/19 BC) will explain this species of wall decoration by saying: "The ancients who first used polished surfaces [*expolitiones*] originally imitated the varieties and juxtapositions of marble veneers [*crustae marmoreae*], and thereafter the various combinations of cornices and ochrish bosses."[38] This seems straightforward, that first came colored marble cladding, and then its imitation in stucco. However, scholars have debated when these particular "ancients" were active, and whether Vitruvius actually meant "marbleized coatings" rather than marble veneers.[39] Indeed, there is not a scrap of evidence for the colored stone prototypes Vitruvius postulated to justify his a posteriori rationalization. Before the medieval advent of spolia construction, there are no walls assembled from heterogeneous blocks of varicolored marbles. Not one example comes to mind, and surely in the centuries of archaeological excavation some relic, however paltry, would have been found. It seems no coincidence that the sole text to mention walls of this sort is Plato's *Critias* (c. 360 BC), which describes a never-land, Atlantis: "And the stone they quarried beneath the island . . . some of it being white, some black, some red . . . and some of the buildings they framed of one simple color, in others they wove a pattern of many colors by mingling the stones for the sake of decoration so as to confer upon the buildings a natural charm" (116a–b).

Nor, in fact, is there much evidence for colored marble revetment, besides ancient mention of blue-gray, Proconnesian orthostates in the Palace of Mausolos at Halikarnassos (377/353 BC) and some dark-gray orthostates found in a palace at Pergamon (197–159 BC). This is not surprising because the vast majority of Greek houses, even palaces, had rubble footings and sometimes a low plinth or orthostate in stone, but the rest was mud brick. Bedding and fixing an orthostate was much easier than cladding an entire wall many feet high, which demands a counteracting mass greater than mud brick to bear the cumulative load of slab after slab. Even the fictions of colored marble in Masonry and First Style stuccoes do not really make structural sense if one examines them. Blocks "fold" around internal corners, and margins are nonsensically a different color from the face of the block in which they are supposedly drafted.[40] Most tellingly, orthostates (i.e., cladding) appear below blocks with drafted masonry (i.e., ashlar), a fiction that could only be consistent with an imaginary prototypical wall whose base was stone and its upper part stuccoed, as one would never build a fair-faced ashlar wall over a clad one.

When Second Style painters at Pompei and Herculaneum let their imaginations loose on the inherited schemata, they departed still further from any constructional credibility, multiplying the coursing into a series of plinths, demi-plinths, orthostates, and various faux-ashlar courses separated by any number of stringcourses and fictive reliefs. Moreover, the only Hellenistic text to describe a "Masonry Style"—a contractual memorandum (c. 255 BC) from the Alexandrian painter Theophilos for the House of Diotimos at Philadelphia, in the Fayum—does not mention any specific stones to be imitated but only variegation and patterning, specifying "to paint the cornice with a purple border, and the upper part of the wall variegated [*poikilos*], the lower band the color of vetch-seed," and "the orthostate with swirling veining [*phleboperimetrion*]."[41] It remains to be said that all this stucco can never have been just an economy, that is, a cheaper substitute for real marble cladding. The vermilion paint on the wall at Pella (see fig. 4.16), for example, was made from cinnabar, one of the most expensive raw goods in the Mediterranean such that Pliny castigated it as showy (*color floridus*).[42]

In summation, all the make-believe of the Masonry and First Styles was an image not of *what was* but of *what could be,* and even though these painted reliefs could almost pass themselves off as plausible construction, they had no true referents in reality. They are "simulacra," copies without an original. This modality recalls Aristotle's comment on mimesis, that any "maker of likenesses must

FIG. 4.21. Cubiculum 16, Villa of the Mysteries, mid-1st century BC. Pompei.

necessarily in all instances represent things in one or other of three aspects, *either as they were or are, or as they are said or thought to be or to have been, or as they ought to be*" (Arist. *Poet.* 1460b.25: ἤ γὰρ οἷα ἦν ἢ ἔστιν, ἢ οἷά φασιν καὶ δοκεῖ, ἤ οἷα εἶναι δεῖ). The representation of the wall is an idealization (even heroization) of its material possibilities, lifting the prose of construction into a creative dimension where it becomes a thing of poetry. Looking for real-world models is futile because the painted and molded fiction of the architecture aims at the kind of wonder that was unavailable in actual construction.

By Vitruvius's day (20s BC) Mycenaean palaces had long fallen into ruin and the Minoan ones sunk into total oblivion under their own debris. By now the idea that mural painting was mimetic was taken for granted and, consequently, also the idea that First Style representations must have had referents in real-world prototypes—that is, marble walls. An eon had passed since the Aegean invention of fresco, and the origin of the fresco medium as a substantial reification rather than an optical simulation was long forgotten. By Vitruvius's day, the "First Style" had also been supplanted by the "Second Style" (c. 80–c. 14 BC), wherein the wall hosted perspectival illusions of a theatral architecture that framed yet more fantasy architecture in receding vistas (fig. 4.21). The clerestories of attached pilasters or half-columns that adorned Hellenistic buildings had come to earth as fictive screens to illusions that were all sealed beneath its surface of a sheer and relief-less wall (see figs. 4.16, 4.19).[43] At the same time, perspective has dissembled the First Style wall into layered space in which everything becomes an object.

The difference between the original Minoan creations and these later frescoes is effectively the difference between a civilization that melds nature and its representation into a material (perhaps animistic) continuum, and another that was now influenced by the science of things and optics, and used to the representations of theater. Nonetheless, while the Second Style tends to be regarded as an evolutionary step toward illusionism and the emancipation of the viewers and their roving eyes, it remains hermetically self-referential. The proliferation of columns in the frescoes witnesses the growing use of marble ones in the home, so to portray them is to materialize the house even more. The illusion of depth seems almost a by-product of the increasing complexity of the representation, whose overwhelmingly tectonic vocabulary is still guaranteed by the stoniness of the medium. The Second

FIG. 4.22. Cubiculum B, Villa della Farnesina, c. 28–19 BC. Museo Nazionale Archeologico di Palazzo Massimo, Rome.

Style is also reflexive, knowingly referring back to the "style" that historically preceded it and that continued to coexist with it: the porticoes depicted in the Second Style rooms of the Villa of the Mysteries at Pompei (see fig. 4.21) are "built" in the First Style.

In the very years in which Vitruvius wrote, the Second Style was itself giving way to the "Third Style." Indeed, the culmination of the progressive estrangement from substance and growing devotion to objecthood is a Third Style fresco such as that in the cubiculum of the Villa della Farnesina, Rome, from the mid-first century BC (fig. 4.22). The relatively shallow recession of the virtual wall acts to make it the stage for a variety of represented paintings: classical Greek panels in tempera or encaustic, with or without folding shutters to protect them, fresco representing fresco painting, and even small-scale sculpture. The elision of space has led to a deficit in substance and the wall has become a *pinacotheca,* an art gallery, an art world.[44]

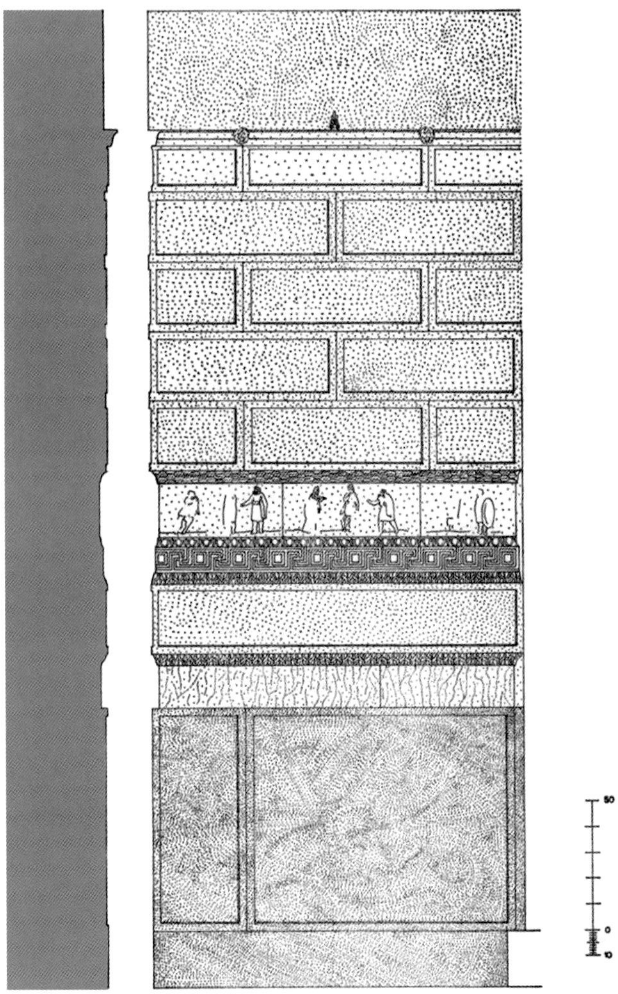

FIG. 4.23. Division of a Masonry Style wall, House of the Comedians, Room N, 2nd century BC. Delos.

"WE HAVE BEGUN TO PAINT IN STONE"

It is within the context of the First Style, and its ambition to show not only *what is* but *what could be,* that the absent marble realized all its painterly potential when it began to birth "chance" images. The artisans who executed First Style frescoes did not stop at coloring the wall as fictive stones, but also painted those images that were said to arise spontaneously within them.

Images in stone are mentioned as early as the Skeptic philosopher Karneades (214/213–129/128 BC), whose lost work Cicero cites when he scorns the portentous character that gullible dupes award the chance artifacts of nature.[45] Karneades had mentioned a head of Pan that quarrymen had discovered in a block on Chios, and as this island would yield one of the most popular streaked marbles (*Marmor Chium,* now known as "porta santa"), the prodigious image must have been perceived in the stone's veining. When Pliny repeats the topos, he transfers the image to snowy Paros, where James "Athenian" Stuart (1713–1788) and several travelers after him claimed to recognize the image in a relief carved at the mouth of a quarry (see fig. 3.2a).[46] In another passage, Pliny also mentions images perceivable within hard stones, especially an agate owned by King Pyrrhus (319/318–272 BC) in which one saw Apollo holding a lyre and accompanied by the nine Muses, an image "not made by art, but by nature of her own accord," whose "markings spread in such a way that even the individual Muses had their appropriate emblems allotted to them" (*HN.* 37.1.5).

Cameo-cutters must have been prompted by such natural wonders when they sought to "extract" latent images by selectively peeling away the laminations of tricolored sardonyx (white through ocher, to a wine-rust brown) or bicolored agate (see fig. 2.30). Some muralists in turn mimicked cameo. Throughout the Hellenistic world painters leavened the "Masonry Style" with figural friezes (at eye level, the better to examine them), but on Delos they painted such friezes to resemble cameo, creating a fiction consonant with the lithic illusions of the wall around it (figs. 4.23, 4.24).[47] Cameo provided the palette for these "monochromata," which were painted *a secco,* layer upon layer over the dried plaster, as actual laminations.[48]

Ancient painters also evoked the potential images discovered in veining when they frescoed faux marble. The earliest surviving example is buried in the imitation breccia of the late third-century BC wall at Pella, and figures a dolphin eating an octopus (see fig. 4.17). Perhaps the painter knew fossiliferous stones that included seashells and other

FIG. 4.24. *Hypnos and Ariadne,* 2nd century BC. Archaeological Museum of Delos, B 17654.

marine creatures, and synthesizing stone in fresco meant that he could create his own fossils as well, however impossible they were in nature. Similar fancies are the human head that emerges from veining in the House of the Gilded Cupids at Pompei (fig. 4.25) and the phantom vine leaf in a villa at Boscoreale (c. 40–30 BC; fig. 4.26). The painters stippled the latter image onto a dado as though it were emerging from the minute grain of granite, one of the stones least likely to give rise to natural images.[49]

Roman poets also associated marble with painting thanks to its brushy veining and rainbow color. Martial wrote of "marble-painted Numidia" and Statius of "marbles shining brightly with painted vein."[50] Other poets delighted in stones that evidenced the brush of nature, where "nature artfully designs figures, when she fashions the veins of the stones into images, skilfully bathes their whiteness with rosy vigor, and paints the fitting members with scattered specks."[51] Even the Latin word for painting,

FIG. 4.25. *Human Head,* before 79 AD. Faux-marble dado, lararium. House of the Gilded Cupids, Pompei.

FIG. 4.26. Fresco with vine-leaf detail, Cubiculum F, Villa of P. Fannius Synistor, c. 40–30 BC. Boscoreale. Metropolitan Museum of Art, New York.

"pictura," does not mean applied images alone—that is, "pictures"—but also the very medium of these images, and even the idea of variegation (*varietas*).[52]

The Latin notion of pictura is long preceded by a shift in late classical Greek painting away from closed outlines and graduated shading to a love of painterliness for its own sake. Pliny singles out the speed (*celeritas*) of the fourth-century artist Nicomachos of Thebes ("no-one was faster in that art than he") and his pupil, Philoxenos of Eretria, who, "imitating the speed of his master, invented a more cursory method of painting than those that already existed"—and using only four colors (white, yellow, red, black). Presumably Nicomachos and Philoxenos painted *alla prima* with a loaded brush, and the frozen spontaneity of the painting imparted the vitality of imaginative creation. For Pliny also admires those "paintings left unfinished" at painters' deaths, which "were admired more than those they finished, because in them are seen the artists' remaining lines and very thoughts"—in other words, all the evident signs of their creativity.[53] It takes little to see how a painter's cursive strokes might resemble marble veining and paint blots marble mottling, and vice versa, how the stone seemed an "action painting" itself. When Pliny condemns Protogenes (also fourth century) for ruining his paintings by overlaboring them, he singles out the painter's delineation of its spittle in a rabid dog: what "displeased [Protogenes] was artifice itself, which he could not diminish, and seemed excessive, and far removed from truth," which he only overcame after throwing a sponge at it in rage, "and thus did chance represent Nature in a painting" (*HN.* 35.36.102–3). Seneca also says that Roman eyes would "not tolerate marble unless it were mottled . . . a table unless it were marked by a vein" (*Ira.* 3.35.5). The taste for ever more varied veining was fed by the latest discoveries in the quarries, where a new seam could deliver unsuspected characteristics—for example, veins that "curled over like waves so as to form curls."[54]

Of course, fresco painters were already predisposed to see marble as a painterly medium, as they saw their paintings petrify before their eyes. And once they had cleaned their brushes, the murals were buffed up to as high a polish as their marble- or alabaster-dust impregnated stucco and pigments would allow.[55] The association between painting and "painting in stone" was also sealed in the realm of mosaic pavements, where the micro-mosaic centerpieces (*emblemata*) were made as separate easel pieces and often copies of well-known paintings with which they vied.[56] In some rare cases, geometrical emblemata even migrated to painted walls, as for instance in the House of the Griffins on the Palatine (fig. 4.27).[57] In short, the idea that mosaic or revetment was a form of painting was by no means remote, and the artisans who worked between these media must have been highly aware of the expressive potential of the ensemble, a *varietas* of modalities within a diversity of lithic media.

An excellent example is the exedra in the House of the Faun in Pompei (c. 125/100 BC), executed for a fabulously wealthy patron, and maybe even by Alexandrian artists. Every available wall surface was covered in First Style stuccowork (begun about 110 BC), though most of it has been bleached and obliterated since the house's discovery in 1831. This stuccowork also covered the sidewalls of the

FIG. 4.27. House of the Griffins, Room 2, 100 BC?. Rome.

FIG. 4.28. *Symposium of Centaurs,* atrium exedra, House of the Faun. Pompei. Watercolor by Giuseppe Marsigli, October 23, 1831. Naples Archaeological Museum, inv. ADS 405.

exedra set between the first and second peristyles, which housed the massive "Alexander Mosaic."[58] Figures painted in the swirls of faux-veining, such as a bird in one block, adjoined fictive reliefs, such as the *Symposium of Centaurs,* wherein the painted veining dipped and rose to follow the contours of the "carved" surface (fig. 4.28).[59] The latter fiction closely recalls the figural friezes on Delos painted to resemble cameo (see fig. 4.24). The beauty of fresco was that it allowed artists to create such "virtual collages," but eventually, real mural collages were also made, for Vitruvius remarks, "Some people cut out slabs [of fresco] from old walls and use them as panels, and these same stuccoes, divided up into panels and mirrors, have around them projecting frames."[60] Overall, the collage walls of the House of the Faun created a "painting in stone" to compete with the famous "Alexander Mosaic" on the floor, which, as a skillful copy after an earlier Greek mural, just possibly a fresco by the aforementioned Philoxenos of Eretria, is a mosaic that manifestly vies with painting in its meticulous facture.

"PAINTING IN STONE": MARBLE REVETMENTS

There is some evidence that the same discourse of intermediality and meta-painting occasionally informed real marble revetments, even in public commissions. In the Forum of Augustus (inaugurated 2 BC), the porticoes were of the choicest marbles or *poikiloi lithoi* ("colored stones") to resume and petrify the Greek *stoa poikile* ("painted stoa"). In the Stoa of Attalos I (c. 241–223 BC) at Delphi, large painted panels had even been set into recesses so they were flush with the wall, to appear integral with it.[61] At the Forum of Augustus, in the public cult room (*Aula del Colosso*) that headed one of the porticoes, the material dialogue between paintings and "painting in stone" became more acute. Here stood the Parian colossus of Augustus against a sheer wall of Luna marble painted from floor to ceiling in simulation of a vibrantly blue curtain embroidered in vermilion and gold.[62] Instead, on the sidewalls of the chamber the pilasters, friezes, and dado in marbles (Phrygian and Numidian) naturally "paint" the wall with their own hues and veining. Casements in these sidewalls may have displayed panel paintings by Apelles, the "Prince of Painters" (fig. 4.29). If this hypothesis is correct, the missing panels resembled the way that easel pieces (*pinakes emblētoi*) were embedded in domestic interiors, and the shuttered *parerga* that illusionistically "hang" in Third Style murals (see fig. 4.22). These selfsame murals mimicked the hanging of artworks in public

FIG. 4.29. Side wall, Aula del Colosso, Forum of Augustus. Rome.

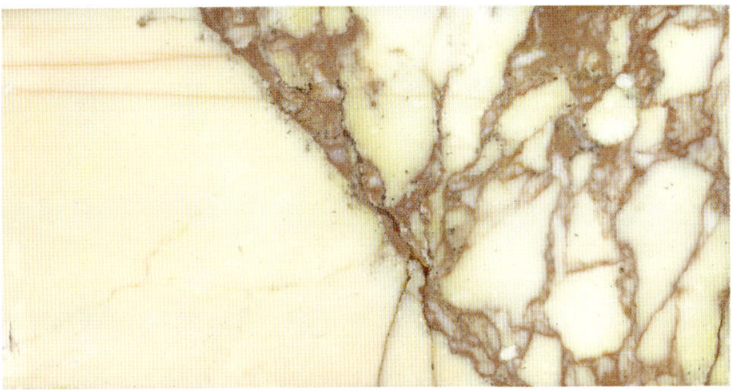

FIG. 4.30. Numidian marble (giallo antico), Corsi Collection, no. 3. Museum of Natural History, Oxford University.

settings, especially public galleries (*pinacothecae*) and the temple precincts in which they were votive offerings.[63]

Pliny's lament, that "we have begun to paint in stone," was written c. 70/77 AD, in the Flavian period. Only in the Neronian era (54–68 AD) had marble decorations begun to catch up with what had long been imagined in paint, and spread up the walls, where they sometimes joined mosaic. Complex inlays (*opus interrasile*) were also devised, slabs of dark slate or rosso antico inlaid with pale marbles, often heated with a hot iron to shade them, as well as glass paste and mother-of-pearl, depicting figures, flora, and often Dionysian scenes.[64] Pliny singled these out for abuse, when he wrote that painting "has now been completely ousted by marbles . . . and not only to the point that whole walls are covered, but *even by intaglio marbles and slabs with wriggling contours that represent objects and animals.* No longer are simple panels pleasing, nor the expanses of mountains spread about the bedchamber; we have even begun to paint in stone."[65]

However, the meaning of the last phrase has always been controversial. All interpretations hinge upon what version of Pliny's text one has at hand. One manuscript tradition gives the ablative (*coepimus et* lapide *pingere*), another the accusative (*coepimus et* lapidem *pingere*). The difference is substantial. Either Pliny spoke metaphorically, "We have begun to paint *in* (or *with*) stone," or quite literally, "We have begun to paint (*on*) the stone."[66] However, it was natural for Pliny to use this metaphor. As a good Stoic he deified and personified Nature. Elsewhere, he characterizes her as an artist, *Natura Pictrix,* because she painted the seasons with flowers, speckled the shells of crustaceans, dappled the hides of wildlife, and (although it is not one of his examples) mottled and streaked the stones that flowered in the bowels of the earth (Pliny. *HN.* 7.3; 9.52; 21.1).

The case ending does become crucial when one moves on to the next passage, where Pliny (writing c. 70/77 AD) remarks a recent technical innovation: "This [painting in stone] was devised in the reign of Claudius [41–54 AD], but in that of Nero [54–68 AD] by inserting markings that were not present in the slabs, they could give variety to uniformity so that Numidian marble might be ovate lines and Synnadic [Phrygian] marble be divided by purple, just as luxury would have liked them to have been by nature. These are our standby when the mountains fail us."[67]

These words have baffled interpreters since the Renaissance, unsure whether Pliny was describing inlay or some way of staining marbles.[68] However, it is clear from the outset that Pliny meant that these manipulations were necessary only when the desired markings were not naturally present in the marbles available. Numidian marble (*giallo antico*) comes in varieties ranging from blocks that are fairly uniform to breccias that are richly mottled and dappled (fig. 4.30). If the "breccia effect" were required but the available samples were thinly veined, then one solution would be to mechanically insert oval inlays into the slabs to mimic the effect, and thereby make nature rise to the artistry expected of it. In theory, to do so was only to simulate the fabric of a real breccia itself, a rock in which mineral fragments "float" as foreign bodies cemented together by their fine-grained matrix. Yet late antique mosaic imitations of this marble (fig. 4.31) show us exactly what Pliny had in mind when he mentioned ovals, especially if one considers his choice of words: "So that Numidian marble might be ovate [*ut ovatus esset Numidicus*]." "*Ovatus*" literally means "egglike"; the mosaic patterns resemble halved eggs, and there is even a Roman recipe for boiled eggs in a fish paste and pine-nut sauce that may have looked little different from our

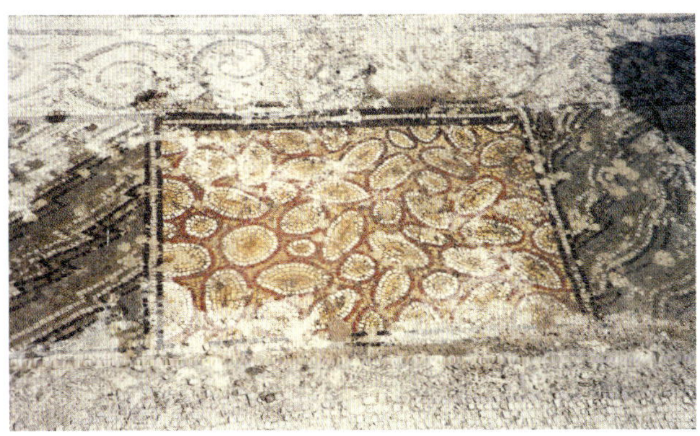

FIG. 4.31. Numidian marble breccia imitated in mosaic, 3rd century AD. Bulla Regia, Tunisia.

pseudobreccia. While such yolk patterns, or "bull's-eyes," are never found in Numidian breccia, they do appear in Egyptian travertine ("alabaster"). The ancient mosaicists were therefore borrowing characteristics from one marble and applying them to another, combining ingredients from different stones to "cook" completely new ones.[69]

Likewise, Synnadic marble (Phrygian marble or *pavonazzetto*) was famously shot through with purple veining such that, as one early church father observed, "Thanks to the chance scattering of the purple dye in the whiteness of the marble, the Phrygian stone becomes a luxury for the most voracious eyes, because she paints the multifarious and manifold dispersion of the color within the white."[70] Monolithic busts of the stuff show that the same block could also be streaked at one end and absolutely immaculate at the other (figs. 4.32, 4.33). It was therefore entirely possible to have blocks of Phrygian marble that were top quality but scanty in the colored markings for which it was prized.[71]

Again Pliny can only be describing a mason's technique that survived well into the nineteenth century. This was the practice of inserting veined shards into otherwise plain slabs to relieve their monotony, but making them seem all of a piece by cutting them with undulating contours, as if jigsawed, so that the joins would be hidden in the veins of the stone.[72] The most extreme example is a column in Ostia, which must have been a special order, an unexciting shaft of *Africano* that was veneered all over with the more heavily veined *bigio africanato* (fig. 4.34).[73]

Once again, the idea that one could improve upon nature did not go unnoticed in the larger art world. The relationship between such "reconstituted" marbles and painted revetment became still more reflexive. While the masons pieced together shards to create veinings not available in nature, painters copied their efforts in fresco, painting marbles that could not possibly exist, but which Natura Pictrix might one day make herself (fig. 4.35).[74]

FIG. 4.32. Bust of an unknown woman ("Poppaea"), c. 225–50 AD. Phrygian marble monolith, pedestal integral with bust, height (including base): 26 in. (66 cm). Capitoline Museum, Rome.

FIG. 4.33. Roughed-out bust, 2nd/3rd centuries AD. Phrygian marble. Archaeological Museum, Afyonkarahisar, Turkey.

FIG. 4.34. Column, Africano shaft veneered with tessellations of bigio africanato, 2nd/3rd centuries AD. Ostia.

FIG. 4.35. Faux-marble dado imitating Numidian marble with artificially assembled veining, Oecus (h), House of D. Octavius Quartio. Pompei.

5 Homes Fit for Heroes

Luxury and Light from the *Per a'a* to the *Domus*

EGYPT AND MESOPOTAMIA

Revetments in colored stones do not adorn palaces in Egypt before the Ptolemaic period (305–30 BC). This may seem surprising in a country where so much labor and tonnage had gone into gargantuan stone pyramids and temples from the third millennium BC onward. Yet the Egyptians built their palaces almost exclusively of mud brick, used limestone sparingly, and reserved the great ashlars in granite and limestone for temples and tombs, what they called the "works of eternity" because gods are eternal and the afterlife lasts an eternity.

On the other hand, they adopted shimmering revetments in gold and faience to transfer supernatural radiance from the temple to the throne room. The "Blue Chambers" below the step pyramid of Djoser at Saqqara (c. 2630–2611 BC) are the only well-preserved example of such interiors (see figs. 1.7, 1.8), but there is no way of knowing whether the same decorations adorned the king's long-lost palace at nearby Memphis. Faience plaques and inlays certainly continued to be used in temples and palaces on door jambs, furniture, and miniature statue shrines (*naiskoi*), the most impressive survivals being remnants of the ceiling from the Temple of Amun at Malkata (c. 1390–1352 BC), where the network of faience and gold whorls evoke the starry firmament (fig. 5.1).[1]

The Egyptian word for "faience," "*ṯḥnt*," or "that which is brilliant," bestowed as an epithet on many gods, including Hathor, Thoth, and Horus, whose appearances were likened to bursts of celestial light, was eventually applied to the pharaohs from the Eighteenth dynasty (1549/1550–1292 BC) on.[2] The Palace of Seti I (1294–1279 BC) at Pi-Ramesses (Qantir) in the Nile delta also displayed faience cladding, of which one portal now in the Louvre is a tantalizing fragment of these lost interiors.[3]

In Egypt, sovereign, palace, and deity were intertwined from predynastic times (Naqada II period, c. 3400 BC) by the pictogram known as the "Horus name" (or "serekh name"). This was the earliest crest used by rulers, and it represented a deity (most often Horus) perched on a rectangle portraying a palace façade (*serekh*) with a courtyard behind, within which the ruler's name was figured in more hieroglyphs (fig. 5.2). As the king was identified as the semi-divine Horus, the original sun god in the heavens, whose role he assumed in the world, his palace was inevitably sacred, something like a temple.

FIG. 5.1. Reconstructed ceiling panel, Temple of Amun, c. 1390–1352 BC. Malkata. Relief plaster, faience, and gold leaf. Metropolitan Museum of Art, New York.

After death, the king would become the Osiris-king, and his palace therefore anticipated a celestial afterlife, in which gold and faience forecast more heavenly mansions. The stone mortuary temple at the king's burial place was the more permanent incarnation of his palace. Conversely, any temple was the palace of a god, and the cult performance that took place within it mirrored the approach to the sovereign within his audience hall; and this audience hall, in turn, was the analogue to the offering chamber in his mortuary temple. As the *Tale of Sinuhe* (c. 1950 BC) tells us, those who were successfully granted a royal audience were ushered through a sequence of vast courtyards and pillared halls until they reached the throne room, where they might find the pharaoh, on a great throne "in the portal" or "within a canopy" (*wm.t n.t ḏʿm*) of "electrum" (a naturally occurring gold-silver alloy).[4]

Another description, from the poem "Instructions of Amenemhat," thought to have been written by the Middle Kingdom pharaoh Senusret I (c. 1991–1962 BC), might be a description of such a palace interior, except that it was "equipped for everlasting life" and is therefore more likely his mortuary temple:

> I made my house adorned with gold,
> its ceilings in lapis lazuli,
> its walls in silver, its floors in hard stone,
> its doors in copper,
> its bolts of bronze,
> made for eternity, equipped for everlasting life.[5]

Nonetheless, the Papyrus Lansing, written a century or so later (1878/1839 BC), does bear testament to a transfer of values from temple to palace by describing a domestic decor that must have been influenced by the royal household. The papyrus tells us that the mansion of the scribe of the army and overseer of the cattle of Amun, Nebmarenakht (called "Raia" for short), was avowedly "constructed like a work of eternity," and that "one does not tire of looking at it. One is gay at its door and drunk in its halls. Handsome doorposts of limestone, carved and chiseled; beautiful doors, freshly carved," and there were "walls inlaid with lapis lazuli [i.e., faience]."[6]

The material brilliance of the Egyptian palace was the embodiment of the semi-divine king. Egyptian kings were so closely connected with their royal residences that by the reign of Amenhotep IV (later called Akhenaten; c. 1352–1347 BC) one of the names for the palace, the *pr-ʿꜣ* (or *per a'a*, "great house"), which had perhaps indicated the innermost apartments, came to signify the ruler himself, the "pharaoh."[7] Eventually palace-temple complexes were built as well, and many of the royal mortuary temples of the New Kingdom on the West Bank at Luxor have small

FIG. 5.2. Funerary stele of King Djet, showing the "Horus name" or "Serekh name," c. 3000 BC. Limestone. Abydos. Musée du Louvre, Paris.

palaces attached, and notably some of these are stone constructions for the first time. This epoch, the New Kingdom, may be said to be the high-water mark for solar kingship. The pharaoh's accession to the throne and mounting of his chariot were likened to the sun's rising; his public manifestation at the "Window of Appearance" to the sun's appearance on the eastern horizon. In the fifth year of his reign, Amenhotep IV ("Amun is satisfied") changed his name to Akhenaten ("of great use to the Aten") and substituted Egyptian polytheistic religion with the henotheistic cult of the Aten (the sun disk), proclaiming himself its only true son. Akhenaten (r. 1353/1351–1336/1334 BC) built a new city in the desert for the new theocracy, Akhetaten ("Horizon of the Aten," now Tell El Amarna). Scarcely ten miles away were the alabaster quarries of Hatnub ("Mansion of Gold"), and the capacity of these stones to bear and radiate light may have been a consideration in this location of Akhenaten's new solar capital. Much of his great palace was certainly constructed of stone, although it would be dismantled when the city was abandoned and Akhenaten's memory obliviated after his death. Even so, Akhenaten's palace was the exception that proved the rule.

In Sumeria and Assyria too, a cult of supernatural brilliance dictated the agenda for choosing palace materials. From the late third millennium BC, ruling elites throughout Mesopotamia began to speak of their frontiers as ranging from sunrise to sunset, and of themselves as the sun (gods) of those lands. The Babylonian king Hammurabi (r. 1792–1750 BC) combined both these allusions in the prologue to his famous law code, styling himself "the mighty king, solar disk of the city of Babylon, who spreads light over the lands of Sumer and Akkad." This trend continued under the Hittites, whose empire occupied Anatolia between the seventeenth and twelfth centuries BC, and whose kings also assimilated solar deities.[8] The brilliance of their throne rooms expressed this identity, and loyal subjects and foreigners alike were exposed there to the "terror of radiance" (Akkadian, *puluḫti melammi*) effected by the king's supernatural powers.[9]

For the same reasons, glazed bricks would eventually enter the material vocabulary of Assyrian palaces, as well as stones of varying brilliance. The palace of the Assyrian king Tiglath-Pileser I (r. c. 1115/1076 BC) at Nineveh was

FIG. 5.3. "Assyrians Moving a Winged Bull on a Sledge" (moving the lamassu from the quarry of Balatai, 700/681 BC). From Layard 1849–53, vol. 2, pl. 13.

built, an inscription recorded, "with bricks glazed (the color of) obsidian, lapislazuli, *pappardilû*-stone [agate or onyx] and *parūtu*-alabaster." When Ashurnasirpal II (r. 883–859 BC) completed his palace in 879 BC at the imperial capital, Kalhu (modern Nimrud), he too decorated it with glazed bricks. More significantly, he initiated a defining feature of neo-Assyrian palaces by lining the principal rooms with enormous gypsum bas-reliefs (orthostates) showing the king, divine genies, royal hunts, and warfare, their outlines picked out in paint but the stone background mostly left clean to shine its own light (fig. 5.3). When Tiglath-Pileser III (r. 745–727 BC) expanded the palace, he richly decorated with ivory and exotic woods, silver and gold; he set up statues of lions and *lamassu* (apotropaic statues of hybridic deities, having a human head, the body of an ox or a lion, and bird wings) at the entrances and "laid threshold slabs of gypsum and parūtu-alabaster beneath them and thus brightened the exits"; he also set up a "sun chamber" (*atman šašši*) that was "inlaid with precious stones."[10]

This gypsum (often called "Mosul marble") had some iridescence because it was mottled with fossilized shells, but some panels were particularly translucent, blossoming with light (fig. 5.4). Slabs of basalt and white limestone also decorated the lower parts of important rooms in Assur and Kar-Tukulti-Ninurta. Both Sennacherib (r. 705–681 BC) and his son Esarhaddon (r. 681–669 BC) also decorated their palaces with bricks glazed the color of obsidian and lapis-lazuli, but they also increased the

FIG. 5.4. Relief of eagle-headed demon, Palace of Ashurnasirpal II, 883–859 BC. Kalhu (Nimrud), Assyria (Iraq). Calcite ("Mosul marble"), 15⅜ × 12⅝ in. (39 × 32 cm). Fitzwilliam Museum, Cambridge, ANE.1.1908.

range of stones, and thereby the meanings attributed to them. Sennacherib, who razed Babylon in 691/689 BC, boasted that he was "the one who replaces brickwork . . . with mountain limestone, which none of the kings before me had done in Aššur," and that he had despoiled the limestone from conquered settlements to underpin his palace walls at Nineveh, while he brought "precious basalt [*kašurru*] from a distant land" to the same ends and through the "power of the scepter" divinely entrusted. The gypsum for his palace was instead quarried twenty odd miles upstream, at Balatai (modern Eski Mosul). He attributed his discovery of alabaster to a divine revelation, and identified the laborers as "inhabitants of hidden mountain regions, conquest of my hand." The stones either revealed themselves (*pendû*-stone, *turminabandû*-breccia) or were revealed by the gods (parūtu-limestone, *pīlu peşû*). The task of conveying them, especially the massive lamassu "made out of a single block of stone" and forty or fifty tons apiece, which made up the jambs of the most important portals, "their bodies radiant like a bright day," were recorded on the orthostates of the palace walls (see fig. 5.3). While gypsums, limestones, and alabaster recur in the building inscriptions, those of Sennacherib and his son Esarhaddon mention for the first time breccia, basalt, and dolomite, as well as red and brown stones, some of which were regarded in the same breath as gems and accorded apotropaic properties.[11]

EAST MEETS WEST: THE MARBLE HOME FROM MYCENAE TO AIGAI

The first domestic revetment in varicolored stone, creamy and streaked gypsum, is found in the Minoan palaces of Crete. The Hall of the Double Axes at Knossos (see fig. 4.14), for example, was sheathed with large gypsum slabs, ranging in height from eighteen inches to almost six and a half feet. The best-preserved remains are in the comparatively late villa at Agia Triada, where the gypsum pavements and wall slabs must have made the rooms radiate. Small fragments at Knossos also suggest that where horizontal timber beams were externally exposed, they too were sometimes clad with stone veneers to make decorative stringcourses.[12] The luminosity of the palace interiors was also as important to the Minoans as it was to the Egyptians and cultures of the ancient Near East, and several Minoan palaces seem to have been planned with rooms that offered a theatral view of dawn throughout the solar year, and whose alignments allowed its rays to strike their penetralia. The most impressive single example is the throne room at Knossos, a place that slept in the penumbra for most of the day, but whose four wall embrasures were apparently aligned to capture the rising sun at the solstice and equinox.[13]

Revetments in Mycenaean palaces did not measure up to their Minoan precursors. A colored stone capital (or base?) was recovered from the megaron porch at Mycenae, but nothing else.[14] At Pylos, a stone dado course (about 10 inches high) cut from a very hard limestone still stands in the porch of the megaron.[15] Throughout the palace in Tiryns (c. 1400–1200 BC) the floors were made of pure lime or of lime mixed with small pebbles to make them glitter, but far more intriguing are the fragments of an elaborate frieze that extended across the façade wall of the megaron. This was made from alabaster or gypsum (from Egypt or Crete?), sculpted with an elaborate pattern of palmettes and rosettes, and inlaid with blue glass paste, an ensemble that testifies to the continuity of a material iconography across the eastern Mediterranean. Moreover, faience objects have been found at Mycenae, and an inscription of Amenhotep III (1390–1351 BC) in the Temple of Kom-el-Hetan even records a voyage of the sovereigns all the way to Mycenae (*Mukanu*) where their faience *cartigli* were eventually found.[16]

Knowledge of Near Eastern palatial architecture certainly reached the Greek world along with the epic of *Gilgāmeš*, and the effulgent palace of the mythical king Alkinoos that Homer describes (*Od.* 7.81–102) seems derived from the Assyrian palaces of Ashurnasirpal II and his successors.[17]

> Meanwhile Odysseus
> was walking towards the renowned home of Alkinoos; and his heart
> was greatly stirred, as he stood there, before he came to the brazen threshold.
> For it was like the radiance of the sun and the moon
> Throughout the lofty house of great-hearted Alkinoos.
> For brazen walls stretched on either side
> From the threshold to its innermost recess, and around them was a band of blue.
> And golden doors closed off the well-built house
> And silver columns stood upon the brazen threshold
> With silver lintel upon them, while the handle was of gold.
> On either side were gold and silver dogs,
> Which Hephaistos made with knowing skill
> To guard the home of great-hearted Alkinoos,
> Being deathless and ageless for all time.
> And inside were thrones leaning against the wall, everywhere
> From threshold to its innermost recess continuously, on which cloth
> Fine and well sewn was placed, the work of women.
> On them, the Phaiakian leaders were accustomed to sit
> As they ate and drank, for they feasted continuously.
> And golden youths stood on well built plinths,
> Holding blazing torches in both hands
> Providing light for the banqueters throughout the house by night.

This house shines like the sun and moon, in silver and gold, yellow and white metals; the walls are also "brazen" (bronze banding?); and the blue band (7.87: *thringkos*

FIG. 5.5. Apadana of Darius I, Susa, Persia. 520–486 BC. Faux-marble bricks in brown, cream, and yellow siliceous pastes. Now in Musée du Louvre, Paris.

kuanoio) running around them sounds just like the frieze at Tiryns.

The cultural exchange was still alive centuries later under the Achaemenid dynasty of Persia (c. 550–330 BC), whose kings employed workmen from throughout their empire and beyond their borders, and recorded the feat in palace inscriptions. Greek masons almost certainly worked at the palace at Pasargadea.[18] At Persepolis and Susa in Persia, gray limestone was now used for plinths, portals, and columns, some carved by Ionian Greeks, but the rest was glazed brickwork fashioned by Babylonian craftsmen. The most famous remains are the reliefs of the immortals and other standing warriors. However, in the Palace of Darius (c. 520/486 BC) at Susa, in among them, on dados or floors, were bricks made from swirling siliceous pastes of yellow, cream, and brown mixed in the mold, then glazed and fired (fig. 5.5).[19] They may represent Egyptian alabaster, which had long been imported, but they also look back all the way to the goblet from Nuzi (see fig. 1.5).

Both Vitruvius and Pliny vouch that the first example of colored marble revetment was in the Palace of King Mausolos (377–353 BC) at Halikarnassos (Bodrum, southwest Turkey). These slabs (from the island of Prokonnesos, in the Sea of Marmara) were orthostates, probably not much higher than a man, and Vitruvius (who may never have seen the palace) seemed even more impressed by the fact that it "has walls built of brick that are exceptionally firm to this day, with stuccowork so polished that it seems to have the translucency of glass."[20] Even this passing reference suggests that these were Masonry Style adornments, and if any fragments will ever come to light, they may resemble either the aforementioned bricks from Susa or even the reliefs from the Achaemenid palaces. It would not be strange for Mausolos's palace to be a melting pot of Western and Eastern modes. He was a Carian, of Anatolian ethnicity—though, intriguingly, Herodotos, who was from Halicarnassus himself, says Carians were descended from Minoans—and a Satrap (provincial governor) of the Achaemenid empire. At the same time, he pursued an agenda of Hellenization fifty years before Alexander the Great's arrival in Asia Minor, and his palace was probably designed and executed by the same Ionian Greek architects (Satyros and Pythios of Priene) and craftsmen who executed his eponymous Maussolleion (Mausoleum). Mausolos's access to marble was so limitless that this tomb was a real behemoth of limestone and marble, involving more precious stone, and more statues, than all the buildings on the Athenian acropolis put together. Nonetheless, even within the very heart of the Mausoleum, the walls of Mausolos's burial chamber were still dressed in an architecture of burnished plaster.[21]

Meanwhile, in Athens, marble columns had already begun infiltrating the patrician home, and with sufficient speed for them to become a target of moralizing arrows, no doubt because they disturbed an atavistic sensibility to the column as a sacred object in itself, one of the totems that made up a temple. Thus, in Aristophanes's *Clouds* (423/417 BC), a scathing lampoon of Sokrates, the protagonist, Strepsiades, scorns his spendthrift son by telling him to hurry over to the house of the tycoon Megakles and see if he can "live on columns alone." The pseudo-Platonic dialogue *Eryxias* (after 337/318 BC) also disparages men who reject wisdom and yearn instead for cypress wood and Pentelic marbles like those in the House of Poulytion, which was apparently the last word in luxury.[22] No such houses have been unearthed, but we get some idea of the intrusion from the towering marble columns in the second-century BC houses on Delos (fig. 5.6). Because

FIG. 5.6. Atrium, House of Dionysos, 2nd century BC. Delos.

the Greek *oikos,* hitherto made of adobe and timber, was now in the process of being monumentalized by the new materials, and posts exchanged for columns, the experience of this transformation may also have contributed to the myth received by Vitruvius of wooden huts that transformed into marble temples.[23]

In the Greek-speaking world, the culmination of this domestic monumental appropriation from the sacred and public world was undoubtedly the palace at Aigai (Vergina-Palatitsia), built by Philip II of Macedon (r. 360–336 BC), probably in the second half of his reign. At least twenty quarries of the local poros were exhausted to raise this complex, an exalted landmark on a terrace above the city. It was "the architectural manifesto of a new era," possibly designed by Pythios after finishing his work on the Mausoleum at Halikarnassos. It was entered through a grand propylon flanked by double-story public *stoai* in which the king perhaps gave judgment, and devoid of the introverted privacy of the usual Greek house. The profusion of Doric and Ionic columns on all scales borrowed their forms from the latest fashions in temple design, and the monumental and sacred peripteros was now turned inside out to make the first courtyard peristyles in Greek architecture. At its heart was not a cult statue (although there was a shrine to Herakles), but rather the anointed vessel of divine blessing, Philip II of Macedon, the "Zeus-born/ descended" king.[24]

Over the next decades colored marbles may have made some inroad into the Athenian home, for the playwright Menander (c. 342/341–c. 290 BC), who was also a pupil of Theophrastos, remarks on "varicolored mottling and the whole pomp of marbles" (Pliny. *HN.* 36.8.44). He likely meant house fittings because Pliny calls Menander "a most assiduous observer of *luxury,*" a private vice, not a public virtue. Not much later, the poet Phoenix of Kolophon (an Ionian city), in a poem written c. 290/270 BC and addressed to "Poseidippos" (possibly the selfsame author of the *Lithika*), outdid all his predecessors by deploring those men who would even build "houses made of emerald stone [Spartan green porphyry], if they could somehow manage such a thing," and would "have porches with four columns, worth many Talents."[25]

There is now very little material evidence for the marble appetite these men of letters reviled. Philip II's palace at Aigai only had marble thresholds, but the rest of the fabric was plastered poros. In "Palace V" (attributed to Eumenes II, 197–159 BC) at Pergamon, dark stone orthostates were actually found in two important rooms, though again the walls above them would have been plastered in the usual way.[26]

ALEXANDRIA

Real strides toward a polychrome marble architecture came in Ptolemaic Egypt. The successors of Alexander, the Ptolemies (323–30 BC), continued to build the capital he had founded at Alexandria. Over the next three centuries each successive sovereign, each Greek pharaoh, added to a palace complex so sprawling that it eventually consumed a quarter, maybe even a third, of the entire city (Strab. *Geog.* 17.1.8–9). We do know from the rich nymphaeum that Poseidippos describes (see above, pp. 54–55) that choice marbles were imported from Greece and combined with Egyptian stones, suggesting that the early Ptolemies tapped a material tradition as old as the pyramids and fused it with their newcome Greek identity.

The palace must have been an exultant affair. Unfortunately, Strabo only mentions its varicolored suites (*poikilas diaitas*), but we do get a zestier flavor from the description made by Kallixeinos of Rhodes of the throne room that Ptolemy Philopator (r. 221–204 BC) built on a vast floating pleasure barge (the *Thalamegus*). This barge sported colonnades and pavilions in rare woods, and a propylaion in ivory—perhaps standing in for Parian marble—as well as real stone columns imported "from India" (whatever these were).[27] Kallixeinos also specifies that some walls were striped, for the Egyptians "variegate the walls with white and black plaques, and sometimes also with the stone called *alabaster*" (Ath. *Deip.* 5.206c). Faience plaques were still in use, because sixty of them were unearthed in Alexandria's royal quarter in the 1930s, but the Hellenistic craftsmen may have begun substituting them with squares of basalt, just possibly obsidian.[28] As it happens, several Alexandrian tombs represent this sort of scheme, and one example is very revealing of the mingling of Egyptian and Greek material culture.[29] At first, Ptolemaic tombs had been frescoed in the Greek "Zone Style," with a course of orthostates painted to imitate alabaster, a stringcourse above, and friezes simulating isodomic blocks; however, when the early second-century tomb (Anfushy Tomb II, Room 1) was modernized in the first century BC, these Greek decorations were painted over to simulate the Egyptian "Chessboard Style" (fig. 5.7).[30]

Some of the palace streets and courts may just have been cobbled with red porphyry (from Egypt) and green

FIG. 5.7. Anfushy Tomb II, Room 1. Wall in "Zone Style" with isodomic masonry (c. 2nd century BC), overpainted with faux-chessboard revetment and alabaster string courses (c. 1st century BC). Alexandria.

porphyry (from Sparta), for when this technique was introduced into the Palace of the Caesars in Rome in the early third century AD it went under the name of "*opus alexandrinum*."[31] A pale echo of the lost splendors comes from the Septuagint, the translation of the Old Testament from Hebrew into Greek that our earliest sources relate was made in Alexandria under Ptolemy II (r. 283–246 BC). When the Greco-Jewish translators sought to make sense of the much earlier and very generic Hebrew words used to describe stones, they looked around at their own environment for inspiration. Thus, a courtyard or reception hall in an early fifth-century BC description (Esther 1:6) of the Palace of Ahasuerus (either Xerxes or Artaxerxes II) in Shushan (Susa) is said to be "paved with Spartan green porphyry, alabaster, and Parian marble."[32] To Seneca, *alexandrina marmora* ("Alexandrine marbles") were a byword for material opulence, meaning the quintessentially Egyptian stones such as sunny Egyptian alabaster and fiery porphyry that are simulated with overwhelming frequency in Pompeian frescoes and throughout Italy (see fig. 4.27).[33]

Whatever the case, the throne room of the last pharaoh, Cleopatra (r. 51–48, 47–30 BC), which was the last Ptolemaic addition to the palace, must have far exceeded the older buildings, if we can trust the word of the Roman poet Lucan (61/65 AD):

> The place itself was the very picture of a temple,
> One that a more corrupt age can scarcely build;
> Both the coffered roofs flaunted wealth,
> And solid gold smothered the rafters.
> And the shining palace was encrusted not in the choicest,
> marble veneers;
> But the agates were stout freestanding blocks, and the
> porphyry too;
> And onyx spread throughout the whole hall to trample;
> Ebony from Meroe did not cover the massive door-posts,
> But took the place of their humble oak,
> Not for appearance but to support the palace.
> Ivory clothed the halls and Indian tortoise-shell lay upon
> the doors,
> Dyed by hand and thickly dappled with emeralds.
> Gems glittered on the couches, and the cups tawny
> with jasper
> Weighed down the tables; the varicolored coverlets
> gleamed,
> Most of them long boiled again and again in Tyrian dye,
> One part shining with imbricated gold, another blazing
> with scarlet,
> As is the Pharian way of mingling warp and wheft.
> (*Phars.* 10.5.113–26)

FIG. 5.8. Floor from the Horti Lamiani, c. 40 AD. Egyptian alabaster and other marbles. Palazzo dei Conservatori, Rome.

Lucan's outraged description might be passed off as fanciful hyperbole, but some of the details are attested elsewhere. Ivory and tortoiseshell doors and their jambs, for example, appear in several Campanian frescoes. We also gain some idea of Cleopatra's onyx floor from other Ptolemaic pavements in Alexandria, or the more exciting example in the Roman palace in the Horti Lamiani (c. 40 AD; fig. 5.8).[34] Yet these are thin veneers, and as Lucan is at pains to insist that Cleopatra's throne room was hewn from solid blocks, it perhaps appropriated the material vocabulary of "works to eternity" that spanned from the Old Kingdom to Alexander the Great. In the impressive Valley Temple of Khafre at Ghiza (c. 2520–2494 BC), an "alabaster" floor undergirded granite monoliths of pink Aswan granite. This temple, on the banks of the Nile, was where the deceased pharaoh's journey to becoming a god began; here his cadaver was washed, eviscerated, and embalmed, and the "opening of the mouth" ceremony enacted, to ensure that the deceased was able to take nourishment in the afterlife. This temple was also a throne room of sorts because it was the setting for multiple basalt statues of the enthroned Khafre. Today, the floor has been bleached by the sun to uniform pallor, but it was once vividly streaked with golden veins, and striding across it must have seemed like treading on sun-filled cloud.[35] In Alexandria, five massive monoliths of the same stone assemble a surreal portal to a now-vanished tumulus, quite probably the antechamber to the tomb of Alexander the Great himself, like ingots of solid ether, the crystalline quintessence above the sky in which the stars burned and where the deceased hero would find apotheosis (fig. 5.9).[36]

LUXUS

Egypt was annexed as a province of Rome in 31 BC after Octavian (later Augustus) defeated the combined forces of Cleopatra and Mark Anthony at the battle of Actium and then captured Alexandria.

The marble monopoly that Augustus and his successors gradually imposed ensured that no private contenders could outshine them in the city's public commissions. Augustus's drive to clad all Rome with imported marbles also reclaimed for public magnificence a material that had become morally tarnished by displays of private opulence, and his initiative appeased the guardians of Roman morals by sanitizing a corrupting luxury. With the provinces newly acquired during the late Republic had come an influx of costly and exquisite paintings, bronzes, fabrics, and finally marbles, which had awakened Rome's elite to foreign opulence. The Romans soon absorbed and metabolized these influences into their own art forms, but later writers were appalled that the clear streams of Roman modesty had been polluted by these Eastern vices. Stolid, rural values combined with fierce, self-sacrificial militarism were sacred cows of the Roman self-image. They cherished the tale of Cincinnatus returning to his plow and the grimy humility enshrined in Scipio's dismal home at Liternum. These were the chords that Seneca struck (c. 64 AD) when he contrasted the latter with the modern vice for marbles: "One seems poor and mean if one's walls are not resplendent with large and costly mirrors, unless Alexandrine marbles are not set off by panels of Numidian marble, if their borders are not faced over on all sides with difficult patterns, arranged in many colors like paintings; if our vaults are not buried in glass; if our swimming pools are not lined with Thasian marble, once a rare and wonderful sight in any temple" (*Ep.* 86.6). For writers of his ilk, "magnificence" and "luxury" were antonyms, the first denoting public service and decorum, the second personal ambition and social transgression. "The Roman people disapproves of private luxury, but admires public magnificence," Cicero had bravely asserted in the face of all evidence to the contrary (*Mur.* 76).

The magnificent homes with marble floors and gilded ceilings that Seneca condemned had begun to sneak onto the Roman scene in the early second century BC when Cato (234–149 BC) inveighed against houses with ivory and citrus-wood trimmings and "Punic pavements" (probably mortar floors with marble fragments in them, *opus signinum*).[37] Next would come mosaic floors.[38] The diffusion of such luxuries was gradual, but the literary tracts of later moralizers such as Pliny and Seneca, who railed pitilessly

FIG. 5.9. "Tomb of Alexander," c. 323 BC?. Alexandria.

against marble luxury with varying degrees of vim and wit, settled on a list of bogeymen responsible for ushering in the excesses. According to these, the consul Marcus Aemilius Lepidus raised the bar in 78 BC by installing thresholds of golden-yellow Numidian marble in his townhouse. A modest concession to luxury, perhaps, but in the eyes of conservative zealots it seemed as though Lepidus would not hesitate to trample gold underfoot (Pliny. *HN.* 36.8.49: esp. *in massa ac vilissimo liminum usu*). This was just the thin end of the wedge, and four years later (74 BC) the consul L. Licinius Lucullus was lambasted as "Xerxes in a toga" for importing so much marble for his domestic consumption that a foreign stone for the first time soiled the name of a Roman citizen: Marmor Luculleum (Pliny. *HN.* 36.8.49–50; Vell. Pat. *Hist.* 2.33.4).[39] Thereafter private luxury proliferated in such leaps and bounds that within thirty years Lucullus's house could not even enter the top hundred in Rome's most beautiful homes (Pliny. *HN.* 36.24.109–10). In circa 48/45 BC, Mamurra became the first person in Rome to veneer walls with marble as well (Cornelius Nepo cited in Pliny. *HN.* 36.7.48). As Mamurra had been Julius Caesar's commander of military engineers (*praefectus fabrum*) in Gaul, and there was not a column in his house that was not Carystian or Luna marble, he may well have been Caesar's quarry supervisor at Luni (Carrara) as well. This taste must have quickly gathered pace, because Caesar almost immediately devised new revenues by instituting a tax (the *columnarium*) on every column that was imported into Rome (Cic. *Att.* 13.6.1; March 45 BC).[40]

Mamurra's personal corruption was notorious enough to enter the verse of Catullus (*Carm.* 27, 54), and Pliny thought this made the opulence of his home even more of an outrage (Pliny. *HN.* 36.7.48). It was, however, the aedile Marcus Aemilius Scaurus who became everybody's whipping post for luxury and whose construction "insanity," Pliny exaggerated, had exceeded that of the emperor (*HN.* 36.6). Pliny's grievance was that, in 58 BC, Scaurus had installed four thirty-eight-foot-high, monolithic columns of Lucullan marble in the atrium of his own home. As Scaurus had been prosecuted for embezzling funds from the province of Sardinia, his domestic magnificence seemed to mirror his inner corruption and, as Cicero says, now only marble columns could satisfy his mindless vice (*Scaur.* 2.45).

Growing opulence not only suggested a disregard for cherished Republican virtues, but broadcast the capacity for largesse, thereby political wherewithal, and ultimately an open gate to tyranny. By the mid-first century AD, the taste for marble had become such a mania that Seneca ranted that the world was upside down, with Rome's citizens consumed by marble frippery while the *urbs* itself teetered on the brink of ruin (Sen. *Ira.* 3.35.5). Seneca balked at merely discussing the marble luxuries that were now appearing merely in the baths of freedmen. Luxurious marble had become a medium for subversion whereby freedmen acted like patricians, patricians like despots, and despots like gods. While the faux-marbles of the First Style had, by contrast, been a "grave and healthy splendor, a pleasure without luxury" (the villa of Manilius Vopiscus, Stat. *Silv.* 1.3.92–93: *gravis sanusque nitor luxuque carentes / deliciae*), the real thing now warranted serious charges. What had once been a transcendent medium had become a capital commodity.

Seneca, Pliny, and Lucan wrote decades and decades after the introduction of the adornments they vilified, so it must have been obvious to those "in the know" that they were striving to lock the stable of a horse that had long since bolted. Moreover, Lucan's ambiguous attack on the safe target of Cleopatra's palace echoes the diatribes of his uncle, Seneca, against the excesses of more contemporary Romans. Both authors were almost certainly displacing the blame. Since both Seneca and Lucan, uncle and nephew, were compelled to commit suicide by Nero, and since Nero's Domus Aurea was an infamous money pot of luxuries, it is questionable whether their veiled target was not, in fact, the emperor himself.[41] Nero expropriated swathes of the city, virtually deified himself, and fostered a palatial opulence that smacked of Satraps. Augustus had been far subtler in ostensibly shunning private opulence, and Suetonius approvingly writes that only peperino piers could be found in his house on the Palatine, not even a marble pavement (Suet. *Aug.* 72.1). In truth, the same house adjoined the precinct of Apollo Palatinus, and in this sense the god's solar marbles could even count as the public façade of Augustus's private chambers. Such pious sleight of hand assured splendor by association and declared Augustus *primus inter pares.*

Conservative indignation over domestic marbles was not simply snobbery, then, for many commentators protested that stone's sacral character was so abiding that to squander it on one's own home was almost a sacrilege. Thus, Augustus's court poet, Horace, decries modern luxury and even cites (maybe invents) ancient legislation proscribing the use of newly cut stone outside of temples and public buildings (*Carm.* 2.15.13–20). And Seneca noted that while once white marble alone had been a spectacle in any temple, his contemporaries had now debased it to lining their swimming pools. Likewise, Pliny's rants against the marble luxuries of Lepidus, Lucullus, Scaurus, and Mamurra only make sense if he is castigating the privatization of the divine, since cut marble had long been used for public building. If ivory, gold, gems (and now marbles) were to decorate the home, he asked a century after such luxuries had become commonplace, "*Quid omnino diis reliquimus?*"—"What, then, if anything, do we leave to the gods?" (*HN.* 36.2.6).

For Pliny, the transgression, that migration of supernatural splendor into worldly luxury, was epitomized by the fact that when Scaurus had dragged his thirty-eight-footers along the Via Sacra to his private home on the lower slopes of the Palatine, they had filed by the terracotta pediments of Rome's earliest temples.[42] Probably no less harrowing was the fact that Scaurus's domestic columns were smuggled into Rome under the pretense of erecting them in a public theater. When Pliny, a few lines later, castigates the orator Licinius Crassus for installing ten Hymettan marble columns in his house (95 BC), it is because his was "an epoch in which public buildings still lacked a single marble column."[43] (Intriguingly, when attacked for his columns, Crassus confronted his accuser with the fact that his own house was worth half its value without its nettle trees: Val. Max. *Facta.* 9.4.) Similarly, in the *Satyricon* (c. 54/66 AD), the ex-slave Trimalchio can claim that he has finally "arrived" because his house boasts a couple of marble colonnades, so that it "used to be a shack but now it is a temple" (Petr. *Sat.* 77.4). The column was continuing its journey away from constructional necessity toward extravagant surplus, becoming, as Seneca censured, something "that supports nothing, built for decoration, merely in order to spend money" (Sen. *Ep.* 86.7).[44] It is a sign of the times that even Hero of Alexandria (10–70 AD), who taught at the Mouseion, began figuring calculations for column shafts into his mathematical exercises (*Stereom.* 49). With all this in mind, it seems obvious that Julius Caesar's earlier column tax, the *columnarium* (48/45 BC), a sumptuary law issued just over a decade after Scaurus's home improvements, was fiscal opportunism disguised as a morality tax.

So began a debate about conspicuous consumption that would linger on for centuries and would rescue neither temples, nor eventually churches, from the charge of materially intrinsic evil. Two centuries after the rot had set in, the historian Velleius Paterculus (c. 19 BC–c. 31 AD) will still conclude some fulsome praise of conservative Tiberius, by singling out old Metellus as "the first man to build a temple of marble in Rome and consequently the pioneer of this type of magnificence—or, one might say—this type of luxury" (1.11.5. Cf. Pliny. *HN.* 36.4.42–43). Marble luxury would long remain a means of indicting an individual or an age. When Ammianus Marcellinus tells us, in the fourth century, that the Roman soldiery "even sought out marbled houses," we know that their moral degeneration presages imminent catastrophe (Amm. Marc. *Res Gest.* 22.4.6: *quaerebantur et aedes marmoreae*).

Whatever the ethical qualms of Roman conservatives, they were right about the sheer expense of importing and working marble. In his day, Cicero says that just one column in a private house cost twenty or forty thousand (depending on source) sesterces (*Verr.* 2.1.147). And it has been calculated that by the High Empire, twenty-seven thousand workers, the equivalent of four or five legions, were employed in quarrying throughout the empire. Moreover, although the quarries were often so inhospitable that only slaves or convicts crewed them, their subsistence wages still cost the exchequer somewhere between three and four million sesterces per annum.[45] And this figure does not include the costs of urban administration or all sorts of extras, such as custom-built cargo ships, nor the retail end of the market: once marbles were purchased and delivered, Diocletian's *Edict of Maximum Prices* (301 AD) demonstrates just how ruinously expensive they were to work and lay.[46] While contractors grew fat on this incessant trade, private fortunes were supposedly ruined in the race to deck domestic halls with ever more costly marbles.[47]

The caveats of the Roman conservatives had fallen on deaf ears. The scale and decor of the *domus* had always been commensurate with the citizen's public dignity, and it was unlikely that they would now be denied the honor of marble, for "reputation and following hung on a man's opulence, housing, and trappings."[48] At least the floors of any decent, late Republican house would have been marbled, as the numerous remains at Pompei, Herculaneum, Rome, and elsewhere demonstrate.[49] Julius Caesar himself was so infatuated with "*tessellata et sectilia pavimenta*" that he even carried them around with him on campaign, like oriental rugs (Suet. *Iul.* 46).

The fact that conservative Roman voices are so partisan about the evils of luxury has also somewhat poisoned the well for later interpreters, who regularly fall back on the default doctrine that precious marbles were simply displays of wealth, prestige, and power. It can be taken for granted they did fulfill this role, and we have a record of the crazed Caligula scuttling around his residence in the Horti Lamiani demanding that all the fittings be made "more costly" (*polutelesteras*), including windowpanes of transparent stone (Philo. *Leg.* 351–67, esp. 358–59, 364). But even this is a biased report, by the leader of an embassy of Alexandrian Jews in 40 AD, whom Caligula was berating for not worshipping him as a god. Moreover, patrons knew well enough to leave matters of taste and execution in the hands of the artist. Thus, when Pliny the Younger (61–c. 113 AD)—nephew of Pliny the Elder—asked the architect Mustius to reconstruct the Temple of Ceres on his estate, he wrote: "I consider that it will be an act of piety and munificence, when rebuilding this temple on the noblest scale, if I add to it a spacious portico; the first for the service of the Goddess, the other for the use of the people. I beg you therefore to purchase for me four marble pillars, of whatever kind you shall think proper; and also a quality marble for laying the floor and encrusting the walls."[50]

LIVING LIKE GODS IN THE PALACE OF LIGHT (FIRST–SIXTH CENTURIES AD)

Whatever their dollar value, however precious the stone, however pleasing the patterning, once marbles had been set edge-to-edge to form sheer surfaces, a laborious process of polishing ensured that they would seem of one piece and reflect like giant mirrors. The slabs were so well joined, so polished, that one could not get a thumbnail between them and the fingers glided over the surfaces without a hitch. The expression for such precision, *ad unguem* ("to the fingernail"), continued into the Renaissance.[51] Already at the beginning of the second century BC, long before the advent of marble, the comedian Plautus describes a desirable house as "mirror-bright, pure radiance" (*Most.* 5.642; c. 193/184 BC: *speculoclaras, candorem merum*). And still in the late fourth century AD, the senator Symmacus says of the marbles in his father's house, they were "so polished that despite the composition of many pieces they give the illusion of being one sole piece."[52] The house of the lord and master must really shine, and when a VIP was expected, the marbles received as much spit and polish as the family silver (Juv. *Sat.* 14.59–63. Cf. Plaut. *Pseud.* 161).

FIG. 5.10. Cubiculum, villa, 2nd century AD. Silin, Libya.

At home, this hall of mirrors projected the household onto another plane. The most exalted patrons from the emperor down were happy to become temporary gods by basking in the shimmering reflections that suggested they haunted solar abodes. The master of the house was the light of his home. So palpable was this notion that when Annia Regilla, wife of the billionaire magnate-cum-rhetorician Herodes Atticus and the very "light of his home," expired prematurely around 160 AD, her husband "altered the appearance of his house in her honor by making the paintings and decorations of the rooms black by means of hangings, dyes, and Lesbian marble, which is a gloomy and dark stone" (fig. 5.10).[53] Herodes's decision to wreathe his house in permanent mourning was all the more poignant because he had habitually embellished Greek cities with "immortal monuments" in brilliant, Pentelic marble.[54]

The most opulent homes looked up to divine palaces, like that of Cupid described by Apuleius (c. 124–c. 170 AD) in the *Golden Ass:* "A royal palace, constructed not by human hands but divine arts. For you will know from the very moment you enter that you are looking at the brilliant and delightful abode of some God . . . and all the walls, which were constructed from bars of solid gold, glow with their own brilliance, so that the house creates its own daylight even without the sun's permission: thus the rooms, thus the porticoes, thus even the doors themselves blaze."[55]

The man to beat in this game, however, was not Cupid but the sun himself. The motif of the ruler's domain as a Palace of the Sun reached back to the Egyptians, Assyrians, Hittites, Minoans, and others. From at least Homer's description of the palaces of Alkinoos and Menelaos

(Hom. *Od.* 4.43–46; 7.81–133), poets imagined palaces clad in fiery bronze and glowing ivories, flashing with amber and flaming with jewels. When Telemachus bursts into the Palace of Menelaos, it is exactly this sort of display that moves him to exclaim that surely the Palace of Zeus on Olympus must be this way (Hom. *Od.* 4.71–75). Each palace must be *thauma idisthein,* "a wonder to behold," and the corollary was that seeing was believing.

It was not just the palaces of men. Lucan imagined the alabaster abode of Cleopatra as the "very image of a temple" in which the queen reigned as a goddess, and Statius describes the palace of a real goddess, Venus, in comparable terms:

> A lofty mansion lays bare the shining shrine. . . .
> The dwelling is worthy of a goddess, and is no less pure than the shining stars.
> Here is Libyan and Phrygian stone, here the hard rocks of Laconia are verdant,
> Here are winding alabaster and the veins the color of the deep sea,
> And the rocks glisten such that the dyer of the Oebalian and Tyrian cauldron is often red with envy.
> Pediments hang in the air fixed on countless columns,
> The beams glitter united by Dalmatian ore
> And the light seems to come from within, for
> The cold that sinks from the ancient woods shuts out the rays of the sun,
> And transparent springs live in marble. (*Silv.* 1.2.145, 147–55)

The idea that the house shines with its own light becomes a topos of the literature on palaces, baths, audience halls, and, later, churches. It is a credit to the splendor of the materials, but in the case of the palace we may also take it as the halo of the ruler, and this notion becomes ever more rote from the Roman period on.

Ernst Kantorowicz methodically traced the idea from the Roman and Byzantine emperors through to Louis XIV (the "Sun King"), and even Napoleon, that all these sovereigns were happy to be considered the rising suns that illuminated their cities and kingdoms. The Roman emperor was lauded as the rising sun, the *Oriens Augusti,* because this title clinched the moment when sunbeams flood the world and rid it of darkness. It was also said that in the emperor's power this moment became everlasting, signifying his "rising in timeless perpetuity," and therefore implied an *imperium sine umbris,* an "empire on which the sun does not set." These same emperors expected posthumous divinization, so it was fitting that in this world too they live like gods, in houses that competed with that of Helios-Sol himself and on a scale so huge that they were visible in their entirety only to his supernal eye.[56]

Marbles were instrumental in achieving the desired effect, transforming the interior into a mirage of light that seemed to emanate from the ruler himself. This widespread perception was stoked by the epideictic oratory and poems churned out by court literati, who often borrowed their plaudits from earlier descriptions of temples. Perhaps no emperors took up the gauntlet with greater gusto than Nero (r. 54–68 AD) and Domitian (r. 81–96 AD). Nero's lust for light meant that his Domus Aurea (64–68 AD), the "Golden House," was entirely south-facing, and so flooded with sunlight from dawn until dusk.[57] The pinnacle of this luminosity was the Temple of Fortune that he rebuilt and incorporated into the complex, which "gleamed like the day . . . as if the light was enclosed within, rather than transmitted" (Pliny. *HN.* 36.46.163).

Nor was Nero content with sheathing the entire edifice in marble, for it was famously studded with gems.[58] Although the emperor's palace was stripped bare after his fall, a vivid impression of its effects comes from near-contemporary Campanian murals, showing columns of gilt bronze with entwined tendrils that bud into gems (fig. 5.11). Actual fragments of such decors have come to light, the most spectacular being those from the palace in the Horti Lamiani that Caligula refurbished in 40 AD: "A gallery two hundred and seventy-six feet long, the pavement of which was inlaid with the rarest and costliest specimens of alabastrine-agate, while the ceiling was supported by twenty-four columns of giallo antico resting on gilt bases."[59] On the walls, gems, quartzite, rock crystal, and agates accompanied foliate decorations in gilt copper that (probably) made up friezes over the revetment. Similarly, the few surviving vaults of the Domus Transitoria, Nero's house before the fire of 64 AD, have borders once studded with "gems" of vitreous paste.[60] All Nero's brilliant encrustations identified him as Sol-Helios, the new sun that would heal Rome after the disastrous fire of 64 AD, and the Domus Aurea as his proper abode, a gilded house and a sun palace.[61]

The equally demented Domitian was the author of the equally vast Domus Flavia on the Palatine (81–92 AD). Although Suetonius says that Domitian forestalled assassins by lining his palace colonnades with Phengites (because he could "see in its brilliant surface the reflection

FIG. 5.11. Triclinium, "Villa of Poppaea," c. 60–40 BC. Oplontis.

of all that went on behind his back"), this scurrilous anecdote marginalizes the material's solar aspirations. Others lamented that Domitian thought himself a new Midas, "that everything became gold and stone at [his] touch."[62] Summing up the entire palace (built 81–96 AD), Martial announces that here "the day sees nothing brighter in the whole world . . . sated with the hidden light of the rising sun. . . . [It is] a house equal to heaven" (*Ep.* 8.36.4, 9–10, 11). He means all parts of the palace, but especially the Aula Regia and Triclinium that lay at its heart, overwhelmingly grandiose constructions premeditated to rival the scale of temples precisely because they presented Domitian as a living god. Domitian was the first and last emperor to make his official title Dominus et Deus ("Lord and God"), and his Triclinium (the so-called "Dining Room of Jove") was the first to have an apse, a feature that formerly housed cult statues in temples.[63] It is in this Triclinium that Statius (writing 93/94 AD) imagines that he must be "reclining with Jupiter amongst the stars":

> An edifice august, immense, and magnificent not with a hundred columns,
> But enough to support heaven and the gods above
> Were Atlas to be eased of his burden.
> The neighboring temple of the Thunderer is agape at it,
> And the gods rejoice that you are installed in a seat equal to their own
> —you need not hasten then to ascend to the vast heavens;
> So broad is the spread of the building, such the extent of its halls,
> Broader than any plain, embracing the Ether within its shelter,
> Yet still lesser than its Lord;
> He fills the house and gladdens it with his mighty life-force.

There rival the Libyan mount [Numidian marble] and gleaming
Stone of Troy [Phrygian], and much Syenite [Aswan granite],
Chian [Portasanta] too, and the rock that vies with the gray-green sea [Carystian];
And Luna [Carrara] enough laid down to carry the columns.
It stretches beyond sight: your exhausted gaze scarcely grasps the summit
And you would think it the gilded ceiling of the sky.[64]

Four centuries later, the rhetoric of the emperor's sun palace remained intact. The Homeric Palace of Menelaos remained the gold standard in the ruler-god's imagination. In the second century AD, Lucian is not surprised that "Homer's island boy was astounded at the house of Menelaus and to compare its ivory and gold to the beautiful things in heaven because he had never seen anything so beautiful on earth," and in the fourth century Themistius (384 AD) says that the houses of Constantinople were so richly marbled that they would put Menelaus to shame, who had only used gold and silver.[65]

In the sixth century, an anonymous epigram lauds "the bright house" of the emperor Justinian (r. 527–565) in the same glowing terms, as "a marvel for the sun to view at its rising," because "never before when he mounted his celestial path did he see such beauty on earth."[66] Corippus, the court poet to Justinian's successor, Justin II (r. 565–578), went into considerably more detail when he penned the panegyric, *In Praise of the Younger Justin* (566/567 AD). In one passage, the poet offers this ekphrasis of the emperor's throne room in the recently completed Sophiae Palace:

There is a hall within the upper part of the palace
That shines with its own light as though it were open to the clear sky
And so gleams with the intense luster of glassy minerals,
That, if one may say so, it has no need of the golden sun
And it ought really to be named the "Abode of the Sun"
So pleasing is the sight of the place and still more wondrous in appearance.

One commentator has taken this passage to mean "a kind of solarium, with walls and perhaps roof of glass."[67] Instead, the poet means a chamber sheathed in glass mosaic (*vitrei . . . metalli*) and abounding in such gleaming reflections that the room *seemed* to illuminate itself. It was a throne room worthy of the sun, and the emperor Justin is that sun. This interpretation is more than confirmed by the permutations on the theme that pervade the rest of the poem: Justin acceded to the throne at dawn, and light filled the palace at his acclamation; his father, Justinian, was "the light of the city and the world"; Justin's royal limbs emit light; the "imperial palace with its officials is like Olympus. Everything is as bright . . . shining with light"; Justinian overshadows the empire as the sun eclipses the stars; the great consistory (throne room) "gleams with a mosaic sun"; a "glorious light shone from the inner chambers and filled the entire Consistory" before "the emperor came forth"; the Avar ambassadors shuddered on entering the throne room because "they believed the Roman Palace was another heaven"; "the sun shone out from a clear sky . . . another clear light [i.e., Justin] had increased this brilliant light [i.e., the sun] . . . ; New Rome [Constantinople] shone and the lovely palace vied with heaven itself"; even Justin's throne is "proud with gold and jewels, having its own light without the sun"; and when we see Justin himself, "like the holy light, his appearance surpassed all gems and gold, his angelic eyes equaling the stars in heaven . . . he cast terror into the people by his appearance, and his eyes flashed: as when the golden sun . . . illuminates the whole world with the glory of his radiance."[68] Justin's appearances seem divine epiphanies, and the palace might be easily mistaken for a church. Indeed similar imagery fills the contemporary ekphrases of Hagia Sophia by Paul the Silentiary (563) or Prokopios (c. 554/560), and Corippus could not be more explicit in making the parallel: "There are two wondrous things imitating brilliant heaven, both founded with the guidance of God, the venerable temple and the glittering building of the new Sophianae Palace. One is the hall of the emperor and the other of God" (*Laud. Just.* 4.285–88).

At the other end of the Mediterranean, even the Vandal dynasty that Justinian annihilated (533–534 AD) had subscribed to the same imagery. Their compliant court poet, the aptly named Luxorius, enthused (523/530 AD) that the "audience chamber of the King [Hilderic] at Anclae [Carthage] gleams, wondrously made by art, labor, skill, riches, wealth. From here the sun itself captures the rays that it could give it. You would believe that another day is [born] in the marbles." Luxorius deemed the light in Hilderic's palace so intense, in fact, that it seemed to melt the floor: "The unclouded pavement seems to be thickly spread snow. When your feet stand upon it, you would think they could sink into it."[69]

FIG. 5.12. Nilotic scene, c. 360/375 AD. Glass opus sectile. Iseum, Kenchreai.

None of these revetted interiors survive, but the shock effect of all this splendor can still be appreciated through the eyes of the Chinese diplomats who infiltrated the empire's eastern borders in late antiquity. Used to jade and other hard stones, but certainly no marble architecture, when these stupefied newcomers came face to face with marbled palaces, they could only conclude that their "kingposts" were of coral, their walls and "eaves" (i.e., cornices) of "opaque glass," their pillars of bluestones, their pedestals of crystal, their floors even of gold.[70] It is ironic that these Asian interlopers mistook marble for glass, for marble revetment achieved the acme of radiance precisely when it was replicated in colored glass. The most extensive example of such vitrine revetments is the suite of panels retrieved from the Iseum at Kenchreai (c. 360/375 AD), where intricately inlaid Nilotic scenes and figures of philosophers are framed by intarsiated pilasters and sit over high dados, whose glass is mixed to imitate marble veining (fig. 5.12).[71] But the practice must have been reasonably widespread, for fragments have been retrieved from the Villa of Lucius Verus, just outside Rome (fig. 5.13), while wall panels (of glass, bone, and marble) were reused in a mid-fifth-century reflecting pool in the villa at Faragola in Puglia, and other remains were found at the Villa di

FIG. 5.13. Villa of Lucius Verus, 4th century AD?. Faux-marble, glass revetment. Museo Nazionale Archeologico di Palazzo Massimo, Rome.

FIG. 5.14. Triclinium, House of the Relief of Telephus, 62/79 AD. Carystian, Numidian, Phrygian, and Chian marble. Herculaneum.

San Vincenzino at Cecina (villa of the Praefectus Urbis in 415 AD, Caecina Decius Aginatius Albinus).[72] In some cases, glass opus sectile panels of swimming fish transformed Triclinium walls into aquaria with portholes.[73] Still others have been found and lost, and yet more are known from ancient authors.[74]

THE HALL OF THE BUILDING OUTSIDE THE PORTA MARINA, OSTIA

Precious little has survived to vindicate the poetic rhapsodies that greeted marble revetments. Hopeful expectations are disappointed at Pompei, where only seventeen houses show any vestiges of incrustations, and these were robbed by the survivors or scavengers once the ash and pumice had cooled.[75] The only well-preserved example (so far) at Herculaneum is the Triclinium in the "House of the Relief of Telephus" at Herculaneum, where the skirting and dado were marbled somewhere between 62 and 79 AD (fig. 5.14).[76] Staggeringly little marbling otherwise survived the depredations that accompanied Early Modern—and modern—excavations. The extraordinary alabaster furnishings of the baths at Allianoi (Paşa Ilıca), Turkey, disappeared after they were photographed in 1912 (fig. 5.15), and the whole complex was in any case flooded by a new dam in 2010.[77] In Rome, a line engraving and a monochrome photograph are apparently all that now remain of a revetted bath hall (mid-third century) that was uncovered in 1865–66.[78] Second-century revetments were discovered still stockpiled in a patrician villa on the Janiculum in 1999, and some partial revetments (c. 300/350 AD) were found in situ in a domus under the Palazzo Valentino in 2005. There are more extensive schemes, saved by a landslide, only now being reconstructed in the Hanging Houses at Ephesus (second and third century AD), but yet to be completed.[79] However, this is a very short list, and the ranked sockets for the iron cramps that once affixed the revetments have left cautious archaeologists with little option other than to reconstruct spiritless grids of insipidly anonymous slabs.[80] We depend on the domestic imitations in paint (some of which, as I have tried to show, are loving fantasies) or the few, and

FIG. 5.15. Baths, Allianoi (Paşa Ilıca), early 2nd century AD?. Revetments in Egyptian alabaster. Turkey (Conze 1912).

FIG. 5.16. Aula, Domus outside the Porta Marina, c. 394 AD. Ostia. Museo dell'Alto Medioevo, Rome.

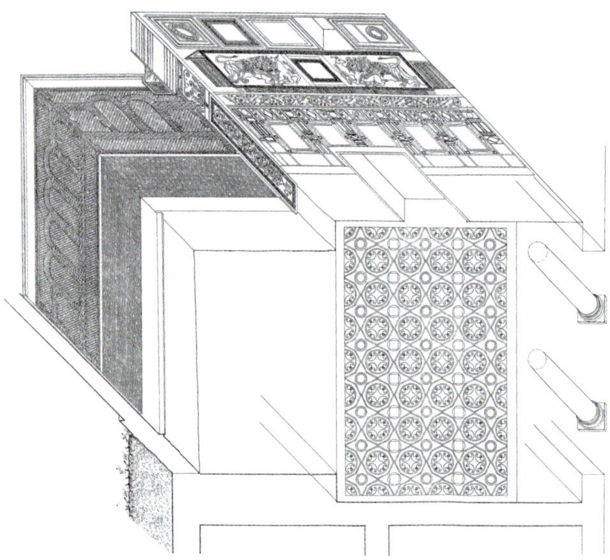

FIG. 5.17. Aula, Domus outside the Porta Marina, c. 394 AD. Ostia (axonometric: Becatti 1969).

rather fragmentary, revetted cubicula in catacombs that must be material reflections of the late domus interior.[81]

However, a stroke of misfortune has ensured that one grand interior has survived sufficiently intact to provide a spellbinding vision of what has been lost, and it is, in fact, the most complete example of a revetted interior to survive from antiquity. This is the late fourth-century hall from a building outside the Porta Marina, Ostia (figs. 5.16–5.18). Giovanni Becatti, who excavated the room, considered the complex in which it stood to be the seat of some college, and this particular chamber a Christian cult room. Nowadays most scholars would instead accept that the remains actually belong to a deluxe domus.[82] At the risk of pedantry, because its distended plan hugged the ancient coastline, we might even call it a suburban maritime villa.

Around the year 394 AD, after masons had finished revetting most of its walls, the new hall caved in, for unknown reasons.[83] Fortunately, because the room collapsed like a house of cards, there were relatively few gaps in the

FIG. 5.18. *Lion Overpowering an Antelope,* Domus outside the Porta Marina. Opus sectile wall panel. Ostia. Museo dell'Alto Medioevo, Rome.

decorations when the various layers were peeled away during excavation (1959–66). Below all the detritus, an opus sectile floor was also found, ready to lay, and it too was largely intact, although only sufficient to fill about a third of the floor area, suggesting that some other pattern would have been laid elsewhere (say, in the exedra) or even that a mosaic scene (*emblema*) was intended to punctuate it.[84]

The hall itself is oriented north–south and divided into two corps, each richly marbled from head to toe (see fig. 5.18).[85] The main body was a tall chamber, facing the Mediterranean through a loggia of paired columns, opposite which lay a narrower and lower alcove, or exedra. The hall proper was revetted in five registers, increasing in scale as they advanced toward the roof, presumably to ensure that the upper panels maintained their legibility. The lower wall adopted the paneled revetment that was a staple in many public halls, although enlivened by small emblemata and pelta motifs, and with vertical, geometrically subdivided panels substituting for pilasters. On the centerline of one long wall a panel portrayed a blessing figure that Becatti identified as Christ, but that might just as well be a charismatic philosopher.[86] Above all this ran a richly detailed floral frieze with a geometric border, all in minute opus sectile. Further up, huge panels of lions and tigers overpowering deer and gazelles were divided by simpler paneling and bookended by spiral colonnettes bearing broken pediments. Finally, the uppermost register was occupied by counter-framed paneling, some enclosing elaborately bordered roundels (*rotae*). Throughout the revetments small amounts of glass were inserted to extend the palette, feign translucent marbles (serpentine, white-veined red), as well as make buds, gems, and eyes glisten, scintillate, and sparkle.[87]

The decorative sources of this part of the hall were drawn from an expansive pattern book of mural repertoires built up over centuries. They recombine architectural motifs from Second and Third Style mural painting (i.e., the *scenarum frontes* of triple colonnettes bookending the lateral lion panels) with megalography (the large panels of lions and tigers assaulting gazelles and antelopes), imitations of panel painting (i.e., the *pinax* of Christ or a philosopher), and perhaps even tapestry (the inhabited-scroll *antependia* pilasters on the *antae* of the exedra). Several of these motifs reappear in the only building whose decorations can now be compared to the Ostia room, the audience hall of the Palace of the Bassii (the "Basilica of Junius Bassus," 331 AD), in Rome. Although destroyed in the seventeenth century, its revetment (c. 13,000 square feet of it) is recorded in several detailed drawings, and individual panels also survive (fig. 5.19).[88]

The other corps of the Ostia hall, the exedra, is quite a different animal. Although its upper walls are exquisitely and meticulously marbled, all the artistry has gone into reproducing reticulate tufa-work divided by brick bands and quoining brickwork (*opus mixtum* or *opus compositum*), right down to the mortar joints. It is as though the origins of Roman mural decoration in the First Style, which Vitruvius had explained "imitated the variety and juxtapositions of marble facing," had come full circle. Like the illusionist paraphernalia of a Mannerist room, in fact, the Ostia room condenses onto its mural surfaces almost every artifice used to decorate a wall during the previous seven centuries. The retrospection is purposeful. Even the marbles had been quarried three hundred years earlier.[89] In this sense, the collection of effects is the built equivalent of the late antique *cento* ("patchwork"), a genre of verse composed entirely from fragments of earlier poets, which had originated in the second century but had its greatest diffusion in the fourth and fifth centuries.[90] The cento is, indeed, the very definition of intertextuality. Michael Roberts has influentially linked this genre to a "jeweled style" in late antique poetry and suggested a similar aesthetic in the visual arts, wherein a preference for episode over whole resulted from the revival of the original meaning of "*varietas*" (*poikilia*)—that is, "colorfulness."[91] The willful contrast between hall and exedra, the former

FIG. 5.19. Giuliano da Sangallo, c. 1480 (Barb. lat. 4424, fol. 33v). Revetted wall of the Basilica of Junius Bassus, 331 AD. Rome. The vertical panels on the right are the bronze doors from the Curia Senatus.

evidencing several arts, the latter an "unadorned" and aniconic alcove, makes it also tempting to enlist rhetorical conceits and see one half expressing a "decorated mode" (*modus ornatus*) and the other a "humble mode" (*modus humilis*). One might even cite Augustine's definition of the beauty of speech as *contraria contrariis opposita:* "The beauty of the course of this world is built up by a kind of rhetoric, not of words but of things, which employs this contrast of opposites" (*De civ. D.* 11.18).

But such love of contrast does not entirely explain why the exedra wall reproduces, at enormous expense, such comparatively humble brickwork (albeit an *opus* that is itself *mixtum,* so to speak). Whoever the patron of this stately home, he was so influential that his residence now blocked the Porta Marina, and with it access to the seashore, thereby effectively turning the town's decumanus into his front drive. The sheer cost of the marbles used throughout the hall, and the access to the imperial depots necessary to acquire them, also confirms that the patron could only have belonged to the highest echelons. Our room, thanks to its scale and splendor, was most likely the *aula,* or audience room, for this grandee, and may have doubled up as a dining hall.[92]

As the empire progressively declined from the fourth century on, public functions were absorbed into the houses of the super-rich, and while many rich houses were increasingly abandoned, those that remained became still richer.[93] It would have been the stage for the *salutatio,* or dawn greeting to the aristocrat from friends and clients, a practice that had not ceased with the Republic but persisted until the eclipse of empire. Thus, when Ammianus Marcellinus castigated the nobility for the same old grievances, only about thirty years before the Ostia Aula was built, he singled out the salutatio sycophants who "greet every word uttered by the great man with various expressions of hypocritical applause, like the parasite in comedy who inflates the pride of the boastful soldier by attributing to him heroic exploits in sieges and in battles against overwhelming odds." Ammianus even seems to have a room like the one at Ostia in mind when he writes that these parasites especially "admire the rows of columns hanging on the towering façade, and the walls gleaming with the remarkable colors of precious stones, and they exalt these noblemen above mortals" (*Res Gest.* 28.4.12).

The quotation of opus mixtum in the exedra was a doubly self-conscious gesture. *Opus reticulatum* had survived in the masons' repertoire at Ostia until at least the fourth century, whereas elsewhere it had become obsolete a hundred and fifty years earlier.[94] When fair-faced, as represented here, it was an external revetment, and one can find windows and doors in brick walls blocked in exactly the same way, with opus reticulatum, just a few blocks up the decumanus.[95] Not only that, but the checkerboard reticulation simulated can be found on tombs at the other end of the same street.

The fiction that the exedra was an external space was also corroborated by its mosaic ceiling, where gold vine tendrils once spread across an azure sky. In this sense, the faux-brick wall seems to pull the image of public life from the city streets into the interior of the domus, at the site of the salutatio, in a manner analogous to First Style stuccowork. But the vine-tendril ceiling also suggests that this exedra was used for dining, because of the vine's obvious Bacchic associations; in fact, a whole floor of a second-century dining hall at St.-Romain-en-Gal (Vienne) was occupied by the same design, with Lycurgus battling the vine at its center.[96]

As for the pseudo-brick of the exedra walls, such humility came at an exorbitant price. Indeed, if precious materials only signal social status by conspicuous consumption, as some scholars automatically assume, then making bricks out of marble was brazen to the point of decadence. A more sympathetic answer again resides in the manifestation of luminosity, and Roman yearning for a masonry that seemed to shine through its very pores is well attested. Pliny, for example, tells us that an architect made the walls of a temple to Jupiter at Cyzicus glow by lining the ashlar joints of the cella with "small gold tubes . . . such that very fine filaments of light shine through the interstices" (*HN.* 36.21.98). Moreover, the imitation of brickwork in marble was itself unique neither to the Ostia room nor its epoch. Patches of *opus spicatum* (herringbone-brick paving) made from pink pieces of Numidian marble have come to light in the first-century Villa of Domitian at Sabaudia and the (probably) second-century villa at Villamagna, while a "terra-cotta" tiled floor in the same materials has been found in Rome itself.[97]

Is it just possible that this fiction of gleaming bricks recollects those long-ago palaces in glazed bricks at Babylon, Ecbatana, and Susa? The walls of Babylon remained a world wonder, erroneously attributed to Queen Semiramis, who had herself become a legend. Martial mentions them when (90s AD) he writes that Tucca has preferred costly marbles to build his baths

(*Ep.* 9.75), and Apuleius (c. 124–c. 170 AD) repeats a Hellenistic source marveling at the flashing walls of the palaces at Susa and Ecbatana.[98] Cassiodorus (c. 536 AD) eulogizes the Ravenna palace of the Ostrogoth queen Amalasuntha with words that could just as easily have addressed the lord of this home in Ostia:

> You have also made, Lady, a palace that makes You manifest to the still unacquainted, when they grasp the greatness of the resident from the enormity of her abode. The marble revetment gleams the same color as gems, sprinkled gold shines on the plating, the gift of mosaic delineates the circling vaults of stones; and the whole thing is adorned with colored stones, wherever encaustic painting is declined. It is recounted that Queen Semiramis built the circuit of Babylon's walls from bitumen mingled with sulfur, but—is it not a wonder!—that this circuit did not yield to steel, that city which could not defend herself from war? The house of Cyrus, King of the Persians, is said to have been built with varied stones bonded with gold, and at Susa [. . . *illegible* . . .].[99]

All in all, the patron of the maritime villa outside the Porta Marina probably wished to sustain the usual conceit that his hall was envied by the sun. Although he was neither god nor emperor, he wished to be "exalted above mortals" (as Ammianus protested) and needed a home fit for a hero. In the eastern empire we even hear of two aristocrats who converted temples into luxury residences under Constantius (r. 337–361), and another who built himself a villa out of one, only to earn the censure of the emperor Julian (r. 361–363) when paganism was briefly restored.[100] More superhuman but not quite blasphemous magnificence may instead be the gist of Nero's much earlier quip about the Domus Aurea: "Finally I begin to live in a house worthy of a man" (Suet. *Ner.* 31.2). The menagerie of massive lions and tigers overwhelming gazelles and antelopes that populate the upper walls certainly implies some power discourse, for they wear harnesses as though they were hunting hounds, and their harnesses are jeweled too, so these are the "hounds" of a king or god (see fig. 5.18). The master has tamed the prowling beast and harnessed its power, and these ferocious felines might next pounce on the parasitic *clientes* milling below.[101]

The Ostia patron may also have been a devotee of the charismatic philosopher portrayed on his wall, whose inner virtue burns so brightly that his head is surrounded by a halo. In a south-facing hall flooded with light reflected off the Mediterranean, the marbles would have rippled and reflected upon the lord of the house to appoint him all the more (il)lustr(i)ous. The Christian poet Ennodius makes the same parallel when describing a bishop's house in Milan (c. 503–513), saying that "two things have established the spirit of this home for eternity: that it shine either by marbles or the virtue of its master" (*Carm.* 2.10.1–2). It was as commonplace in antiquity as it is today to speak of people who "shine," are "brilliant" in intellect, or "dazzle" with their speech. The Manichee Secundinus even flatters St. Augustine that his polished eloquence shines brighter than the marbles in the Palace of the Anicii in Rome.[102] By extraordinary coincidence, it was these very marbles that Theoderic would despoil, between 507 and 512 AD, probably for his own palace in Ravenna, so that they might become "the joyful witnesses of our government, the sparkling embodiment of our reign, the herald's call of sovereign power. We show these things with admiration to the ambassadors who can easily identify the sovereign with his residence."[103]

At Ostia, much like Pylos (see figs. 4.12, 4.13), we are in the territory of magic realism. Either this hall was so outlandish that even the bricks were made of marble, or its inner light was so strong that fired clay leaked a burning light. Significantly, such intarsiated brickwork would reappear in later palaces of light, the Christian basilicas—for example, in the arcade spandrels of Santa Sabina in Rome (422–432 AD; fig. 5.20) and the central, apse vignette in the Euphrasius Basilica at Poreč (530/550 AD).[104]

As important as luminosity per se was the shifting surface achieved by deploying tesserae of varying color. Today we think of luminosity predominantly as the intensity of a light source, although we still speak of some colors as "bright." Text after text from antiquity onward applauds the brightness of individual colors—purple, for example, but also multicolored walls or paintings, as though the juxtaposition of color alone were enough to make them sparkle. We may think of this as the ancient equivalent of fluorescence. Much like the walls of Uruk (see fig. 1.1), the Ostia reticulations are confected from nuggets of porphyry, serpentine, Numidian, and Phrygian marble, and the wall seems threaded with gems, so that the net surface will become as iridescent as the rainbow or a peacock splay.[105] Indeed, a famous passage in Ovid compared Arachne's weavings with the rainbow, saying, "Although a thousand different colors shine in it, the eye cannot detect

FIG. 5.20. Nave arcade, S. Sabina, 422–432 AD. Opus sectile spandrels. Rome.

the change from each one to the next. . . . There too they weave in poliant threads of gold" (*Met.* 6.65–66, 68).

Changeant surfaces such as the walls at Ostia are extolled by the poet Nonnos (c. 470 AD), who describes a "king's courts . . . flashing from afar with its varicolored kosmos of stone" and a floor that "shone with the many branching patterns of multicolored mineral mosaics" (*Dion.* 18.62–63, 83–84). Likewise the anonymous *Digenis Akritis* epic (tenth/eleventh century) will remark of the hero's imaginary castle, "the craftsman so beautified his work that you might think that what you saw was woven out of the precious stones' bright and multi-formed appearance."[106] At Ostia the metaphor, rather than mimesis, of the woven wall explains the broad band of small checker stones underlying the faux-opus mixtum (see fig. 5.16). Becatti himself argued that this band resembled a checkered textile wall hanging (*parapetasma*). In so far as the pattern seems equally inspired by weaving and mosaic, it creates a zone in which technique and allusion blend. A fragment of an earlier floor makes the metaphor more literally apparent in its "weaving" of "three-ply" tesserae (fig. 5.21).[107] All in all, the cumulative effect in the Ostia aula is not variety alone, but a variety of varieties, the acme of the aesthetic of varietas or poikilia.

FIG. 5.21. Fragment of floor mosaic. Museo Nazionale Archeologico delle Terme, Rome.

Exactly the years in which this Ostia room was decorated saw the last gasp of paganism, the desperate revolt of Nicomachus Flavianus, consul and praetorian prefect of Italy, on behalf of the usurper Eugenius (r. 393–394 AD).[108] Within a few more years (410), the Goths would be at the gates and Ostia depopulating and headed into ruin. But this artistic *summa* saw little discontinuity with the Hellenistic conventions and Alexandrine culture to which it ultimately laid claim. Its meticulous artfulness was as much a testament to the enduring heritage that the patron embraced as the materials were to his personal wealth. In the mind of the patron all these trappings put the domus squarely in the literary tradition of the house of light, which reached back to Egypt and was constantly rejuvenated in the extravagant descriptions of fairy-tale palaces in lands where the sun rose.

Let us conclude with a passage from the Latin version (340/361 AD) by Julius Valerius, of the Greek *Alexander Romance* (third or second century BC) by Pseudo-Kallisthenes. This description of the mythical palace in the Sudan that Queen Candaces shows off to Alexander the Great is close in time and spirit to the Ostia interior, and the nearest we now have to the desires of the original patrons to live out their imaginations:

> The palace was adorned with truly stupendous workmanship and arrayed with such minerals that an observer would consider its richness, splendor, and the loftiness of the enterprise all on a par; nor was there any base thing inside unequal to the rest, and everywhere it shone brilliantly with a particularly burning light. All things gleamed everywhere, covered and adorned with silk drapes of purple color, woven in with gold and gems. Whether you marveled at the size of the regal gems, were bewildered by the colors, or overwhelmed by their sparkles, your opinion was not a figment of some delirium. Amid all the gold there was purple cloth, gems and much ivory, but it was in the ivory that you appreciated the greater artistry. There was plenty of onyx and columns made of gems, all of which were taller than our trees, and the highest of them were also inlaid with the best ebony; polishing gave yet others the look and color of the blue sky, yet many of them were purple, and another stone feigned ebony.
>
> And there the scores of effigies in both varied marbles and minerals of every conceivable sort were beyond number, and their profusion vanquished the heart of any inquisitive person. Among all these things one could see that the most outstanding of the queen's possessions was a temple carved from a single gem, including its columns hewn from the same material. [. . .]
>
> And Candaces led Alexander with her through her hall and all the secret chambers and recesses of the palace, and on departing them, showed him certain buildings made of one stone, which, because its brilliance is so fiery (they even call it "lychnites") seemed to beholders to be made from the same stuff as the rising sun. In the same quarters he also saw a triclinium of another type of stone, which, since it includes fiery spots or embers just like meteorites [?], seemed to be really flaming, and made you wonder if the seven heavenly bodies were running hither and thither and the stars performing their celestial dance in all eagerness and under a cloudless sky. And the revolving roof was just like an entire structure made of precious marble.[109]

6 Medieval Substitutions in the Western Church

In 312 AD, after a short struggle for supremacy, the emperor Constantine vanquished his rival Maxentius at the Pons Milvius, just outside Rome, and made himself master of the West. He later credited the victory to a winning vision from the Christian God on the eve of battle, and the following year (313) came the Edict of Milan, which declared religious tolerance of Christianity throughout the empire. Christians could now assemble and worship publicly. As the old temples reeked of paganism and had not been designed to hold crowds, Christians instead took secular basilicas as their model, huge public rooms that had served as courtrooms and audience halls for grand functions, "covered piazzas" divided up by files of columns into a central nave with one or two side aisles. The access point to these secular basilicas had tended to be mid-axis on one of their long sides, but the new Christian ones were all entered at the gable end, such that their interiors induced tunnel vision toward an all-important apse. Apses had been built in only a few Roman temples, to host the divine cult statue, and in the new churches they supported a half-dome, which maintained the celestial connotations of a full one. By the early fifth century, this vault was adorned with shimmering and figurative mosaics, some presenting an epiphany of Christ as a ruler (hence judge and ultimate authority). Eventually, the apse became the location of the high altar where Mass was celebrated and the Eucharistic sacrifice performed. While nowadays these Early Christian churches often seem like unadorned barns, their adornments have fallen away or been stripped and they were once covered with murals, mosaics, and especially marbles. Marble's radiance, color, and transparency allowed the new sacral architecture to assimilate flowery meadows, starry skies, jeweled tapestries, and particularly heavenly gems. Brilliantly polished marble was especially suited to the evocation of light (the first substance of creation and God's true nature), and paramount in the presentation of divine light were such church windows as were themselves mineral filters. From the outset marble fulfilled a vital role throughout the church interior as a gleaming reflection of the heavens.

THE REMAINS

In Western Europe actual remains are, however, paltry, surviving mostly in Rome and, even then, the most extensive examples are now only known from drawings. No marbled interior survives intact from the period

FIG. 6.1. Narthex, Baptistery, St. John in Lateran, 432–440 AD. Opus sectile revetment. Rome.

between the fourth and sixth centuries, and many are relatively recent casualties. The Mausoleum of S. Costanza (337/351), for example, retained its intricate cladding and magnificent dome mosaic until 1620, when they succumbed to the vandalism of the fanatical Cardinale Veralli.[1] Although a Christian foundation, S. Costanza was built as a mausoleum, not a church, so the earliest surviving ecclesiastical revetment is instead represented by the nave spandrels in S. Sabina, Rome (422–432; see fig. 5.20).[2] A large patch of revetment also survives in the narthex of the Lateran Baptistery, Rome (432–440; fig. 6.1), but the extensive cladding of the baptistery proper also went missing in the seventeenth century (fig. 6.2).[3] The adjoining Oratory of the Holy Cross (461–468) was clad all over on an even more monumental scale, but this structure too fell victim to late sixteenth-century demolitions and is only known from drawings and engravings (fig. 6.3).[4] The city's great basilicas, not to mention a host of other smaller foundations, must have been revetted all over, but nothing survives except the odd apse (e.g., S. Agnese fuori le Mura, 625–638). Old descriptions record traces of marbling in St. John in Lateran (c. 313) and S. Maria Maggiore (432–440), but for Old St. Peter's (326–329) the only documentation is for its columns and pavement.[5] In other buildings, such as S. Maria Antiqua (fourth/sixth century) or the Baptistery of S. Clemente (late fourth century), both in Rome, the revetments have not survived, but at least their backing mortar is sufficiently preserved that it retains the imprint of the long-lost slabs.[6] In still other cases, such as S. Crisogono (314–335), Rome, we at least get a picture of the missing revetments from the surviving frescoes that were meant to imitate them.[7] Besides all this, numerous floors survive in part or total, beginning with the Constantinian remains at St. John in Lateran.[8]

Outside Rome, even less remains. In Milan, just a few opus sectile panels (fourth century) survive from the apse of S. Ambrogio, and fragments (late fifth/early sixth century AD) from the cenotaph of Julius in the Basilica of S. Giulio on the island of S. Giulio d'Orta (near Varallo, Piemonte).[9] At Cimitile thin strips of paving and apse revetting are all that testify to the panoply to which Paulinus of Nola dedicated so many resources and poems in the early fifth century.[10] The most intact remains seem to be those in the Orthodox (or Neonian) Baptistery (c. 458), Ravenna, but these were much restored in the 1890s.[11]

FIG. 6.2. Nicolas Beatrizet, *Lateran Baptistery*. Section showing original revetment removed in the 17th century (Antoine Lafréry, *Speculum Romanae Magnificentiae*, 1550–73).

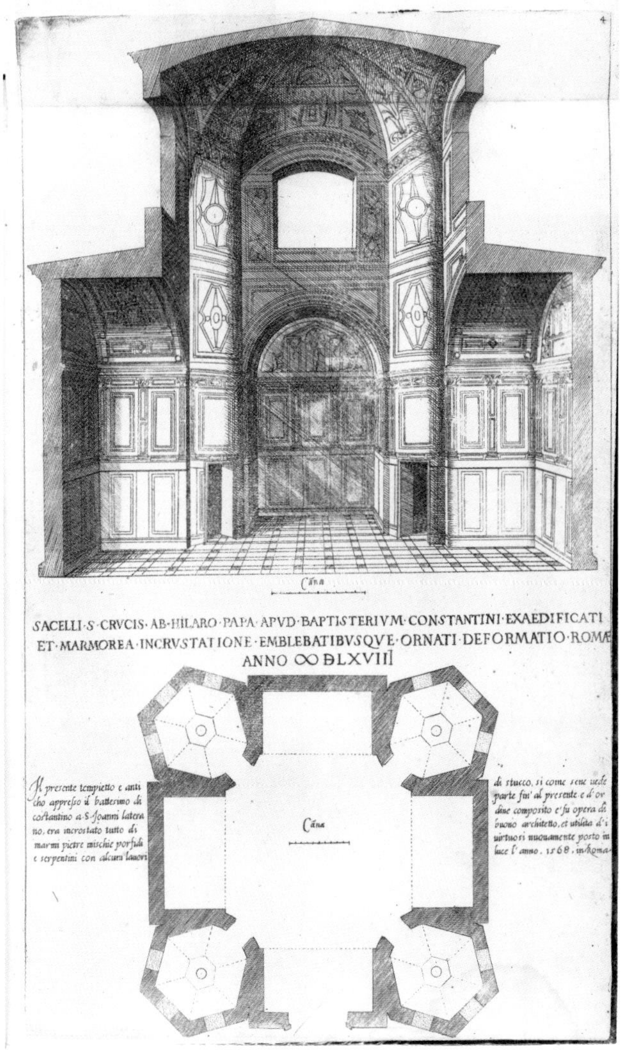

FIG. 6.3. Antonio Labacco, *Oratory* of the Holy Cross. Sectional perspective. Labacco, *Libro . . . appartenente a l'architettura nel qual si figurano alcune notabili antiquita di roma*, 1568.

THE HOUSE OF THE RISING SUN

From the mid-third century AD, a vastly diminished imperial workforce meant that freshly cut marbles were in ever-shorter supply, especially in the West. In 363 Emperor Julian went so far as to surrender the imperial monopoly on marble quarrying to entrepreneurs, expressly so that "countless veins of glittering stones might come to light."[12] This liberality proved futile; the quarries were closing, and the spoliation of marbles from demolished or decaying buildings had already become a fact of life. If anything, the practice accelerated.

The more the luster of empire tarnished, the more the jeweled style of late antique poetry salvaged motifs from a vanishing world to conjure up edifices that were mirages of flashing gems and marbles.[13] The old topoi continued to haunt late antique poetry and supply nostalgic and eloquent literati such as Prudentius (348–c. 410), Sidonius Apollinaris (c. 430–479/480), and Venantius Fortunatus (c. 535–c. 605) with raw material for conceptualizing the new breed of construction, the Christian basilica. Every church was now a "Palace of Dawn," and Christian epithalamia still drew on Statius's description of Venus's marbled palace to show that marriages were truly made in heaven:

> Afar shine and glitter the goddess' many-colored halls,
> Green-gleaming by reason of the encircling grove.
> Vulcan built this too of precious stones and gold,
> wedding their costliness to art.
> Columns cut from rock of hyacinth support emerald
> beams;
> The walls are beryl, the high-builded thresholds of
> polished jasper,
> the floor of agate trodden under foot.[14]

Late antique dreams of heaven, such as Prudentius's rhapsodies, easily collated images from such models and in turn fed Arabic poetry, or both recycled earlier common sources.[15] Indeed, little distance separates Nonnos's neo-Homeric description of the mythical Palace of Lyaios (c. 470 AD) from contemporary eulogies of churches:

> While Botrys was yet arranging the feast for Lyaios, the king of magnificent bounty displayed to Bacchus the artist's hand in the stonework of his hall, from which poured a shining brightness of many colors and shapes like the sun and his reflecting moon. The walls were white with solid silver. There was the lychnite, which takes its name from light, turning its glistening gleams

in the faces of men. The place was also decorated with the glowing ruby stone, and showed wine-colored amethyst set beside sapphire. The pale agate threw off its burnt sheen, and the snakestone sparkled in speckled shapes of scales; the Assyrian emerald discharged its greeny flash. Stretched over a regiment of pillars along the hall the gilded timbers of the roof showed a reddish glow in their opulent roofs. The floor shone with the intricate patterns of a tessellated pavement of minerals; and the huge door with a baulk of wood delicately carved looked like ivory freshly cut.[16]

Concurrently, as client-patron symbiosis withered, and with it the need for vast spaces of self-representation, mansions became monasteries and their defunct halls churches. The change in use did, however, ensure a material continuity between the new Christian assembly halls and their secular predecessors, which included administrative buildings.[17] Thus, the senate house (*Curia*) became the Church of S. Adriano, and the audience hall of the domus of the Praefectus Urbis became the nave of SS. Cosma e Damiano. Both had elaborate marble floors and revetments that survived into the Renaissance.[18] So did the civil basilica built (331) by Junius Bassus, which was rechristened (568/583) S. Andrea in Catabarbara (see fig. 5.19). Early churches also assumed the royal title of "basilica" (*basilika*, meaning "kingly"), and both church and palace were indiscriminately termed "aula" or "palatium."[19] Moreover, in panegyric and theology, Christ now acceded to the old imperial title of the "Rising Sun," "Who had risen again so as never to repeat His setting," and who had "rent the darkness of the nether world to shine forth with the light of resurrection."[20] It naturally followed that his feast day fell on Sun-day and that the new churches should face east, toward the sunrise, wherever possible.

Since there is such scant evidence for natural illumination in most classical temples, it is normally assumed that they were lamp-lit, although, as we have seen, several Greek temples, and probably some Roman ones as well, may have been illumined by translucent marble roofs. Others had windows. A building such as the Pantheon was a very special case. The rotating spotlight of sunlight cast by its oculus made the interior a cosmic clock, "the admission of light at the top symboliz[ing] the fact that the sun, by sending in its light from the highest part of the heavens, illumines the whole world and that at its rising all things become visible."[21] Wherever pagan temples were inaccessible to both light and the general public, such hermeticism reflected the fact that—in strictly religious terms—they were embassies for the divine presence and the god's private house.[22]

The cult structures of the new Christian religion were quite the opposite, places of assembly for communion with God and public fellowship in Christ. In the course of time, the sacrifice came to be made within the church to summon God, not outside the temple to placate and supplicate gods. But long before the Eucharistic rite became a diurnal occasion in the church, a communion by light was achieved when the church was filled with those divine illuminations that were the subject of countless dedicatory inscriptions (*tituli*).[23]

The much-quoted but virtually untranslatable titulus in the episcopal chapel of St. Andrew, Ravenna (c. 495 AD), is typical:

> Either the light is born here or, made captive, it here reigns freely.
> Either it is the light of before, whence came the modern glory of heaven,
> Or else deprived [of light] the building engenders the gleaming day
> And the trapped radiance glows though heaven is secluded.
> See how the marbles are rejuvenated by the serene rays,
> And all stones struck in the starry purple vault.
> These gifts resplend thanks to their author, [Bishop] Peter.
> To him belongs honor, to him belongs merit for composing small things
> In such a way that they overcome the large by compacting the intervals.
> It is no mean thing for Christ to possess a well-jointed house.

"God is Light and in Him is no darkness at all" (I John 1:5), and tituli as well as *encomia*, consecration speeches, and theological texts all proclaimed that the beams flooding the church interiors exorcized the site with their cleansing light. In these divine throne rooms of the cosmic *basileus*, the surplus of light reminded the devotee that "the happiness of the Lord King on His arrival wipes away from the world, like the twilight of the sun, the trembling fear of gaping darkness."[24] Another, later titulus, in S. Agnese (625–638), Rome, spells this out:

The golden painting rises forth from cut minerals
And the day at once embraces it only to be itself
imprisoned;
You might believe this was the dawn silently advancing
from snowy springs
Through the lowering clouds, moistening the fields
with dew
Or such light as the rainbow brings among the stars
And the purple peacock itself, shining in color.
He who had the power to bring an end both to the night
and to the light
Banished chaos hence from the tombs of the martyrs.[25]

It hardly needs stressing that the sun's rising stood for the Resurrection, and the impenetrable night for death.[26] This catharsis made shadows flee, and the nocturnal supplement of oil lamps turned night into day, or conjured starry skies in place of local suns. At the threshold, the outside world faded to gray and any human calendar was eliminated to make way for the epiphany of the telos. In fact, the consecration ceremony programmed every church to become an unveiling of heaven ("apocalypse"), when the full glory of God would be exposed rather than known only through a glass darkly.[27] Moreover, as bright as the physical light and its reflections were, they could only advertise the superior light of faith and anticipate its ultimate consecration at the Resurrection, when the new, spiritual sun would blot out the old, physical one.[28] Then the eternal church—the Heavenly Jerusalem—would descend, empty save for light because it "had no need of sun, neither of the moon, to shine in it: for the glory of God did lighten it, and the Lamb is the light thereof" (Revelation 21:23). In fact, all surviving texts stress that it is always dawn or high noon in the early church. Only much later, from about the ninth century onward, would awe-inspiring gloom become the consequence of, or motivation for, the growing use of stained glass.[29]

The light of the early church was, like Creation, a generative light; like Pentecost, a revelational light; and, like Ascension, an eschatological light. In the eyes of Paulinus of Nola (c. 354–431), this brilliant plasma reverberated around the interior, renewing and regenerating itself ("*novat et novatur*," he says), succumbing to temporary entrapment in matter and assuring the faithful that the numen might be resident.[30] When Avitus (c. 470–523), archbishop of Vienne, consecrated the Church of St. Peter at Tarentaise (Savoy), he even called it "a prison of light."[31] Such light foretold the final reward, union in the blaze of glory that was the Godhead, and the earliest surviving panegyric of a church, Eusebius's exegesis of the church at Tyre (*Hist. Eccl.* 10.4.46ff; c. 317 AD), explains its internal splendor as matching that of Christ's resurrected body, or the garments of salvation (see fig. 2.27).

The baptistery was the location for the Christian's birth into the real life, a prelude to the life eternal; the day of baptism was therefore the *dies natalis,* and so baptisteries must be glittering and sun-filled interiors forecasting a bright future. It almost goes without saying that as much as the Christian basilica dressed itself in the garb and plaudits of its secular precursor, baptisteries borrowed the praises that had been previously poured out on ancient bath buildings, and even, to some degree, the allusions to them as geo-systems; that water came forth from rock, or that the room was a fission chamber for the mingling of earthly and heavenly elements. Late antique epigrams on baths could just as well apply to baptisteries, that "blazing lights here congregate with clear springs, and Phoebus and wave mingle to make a new day. Indeed, so much light enters our baths that you can see naked people blush for themselves." The faithful were clothed in light, and in this baptismal light they rediscovered themselves purified and reborn in Christ.[32]

In baptisteries or vaulted churches, especially in the East, it was the gold-glass mosaics that trapped the sun and diverted its light to the faithful.[33] The first-century philosopher Philo of Alexandria had long before made gold emblematic of divine diffusion throughout the world, owing to the metal's fusible qualities (*Quis haer.* 44; *Leg. alleg.* 20). Marble revetments were bafflers, absorbers, and reflectors of this liquid light. When Avitus (c. 450–c. 518/526), bishop of Poitiers from 490, speaks of the new churches, he writes that the faithful must "polish with praise the glory of the marbles, from which only jealousy of their size denies the title of jewels," and laud "the daylight, somehow collected within by man's labor, [so that the church] is alive with the light of so many gleaming minerals."[34] Coffered ceilings, as Paulinus zealously writes of his own church near Nola (401–404), also allowed it to be "arrayed in a new garment purified from old age.... Behold, you see how great a brilliance emanates from it, as if the church were regenerated, while the ceiling imparts an effect of shimmering waves to its decorations in relief."[35] Sometimes even the tie beams in the open roof were gilded too (Prudent. *Perist.* 11.219–20). Glowing tendrils of light reached down in the shape of pendant lamps, which twinkled like stars to fend off the night and once more remind devotees of the true light to which they must aspire.[36]

PARADISE REGAINED: THE JEWELED GARDEN AND THE JEWELED CITY

There were two biblical archetypes for making a church heaven on earth, one rural, the other urban. The first was Ezekiel's Eden (Ezekiel 28:13, 14), a garden carpeted with nine precious stones that flashed with fire, an imaginary that reached back to the grove of the gods in the *Gilgāmeš* epic. The second, St. John's City of God (Revelation 21:18–21), was a divine stronghold whose walls were encrusted with twelve gems that, as we also know, looked back to Babylon. From the early fifth century on, this divine city recurrently made cameo appearances in the mosaics of church apses and triumphal arches, encouraging a comparison between her jeweled walls and the revetted interiors (see fig. 1.16).[37] By the late Middle Ages this apocalyptic gem symbolism had also infiltrated consecration ceremonies[38] and spread to church bell towers, now studded with porphyry and serpentine slabs as well as brightly colored ceramic plates.[39] In the minds of the orators, churchmen, and patrons who constantly extolled the twin archetypes, city and meadow fused into the idea of a walled garden ("*paradeisos*" in Greek), and it is this atemporal synthesis of Eden redeemed and paradise regained—the basis of the eschaton—that we see, uniquely represented, on the eleventh-century vault of S. Pietro al Monte, Civate (fig. 6.4).[40]

Where funds permitted, real gems did stud the furniture of Early Christian churches and, because pearly gates were said to swing on the heavenly hinges,[41] the Byzantine Church of S. Euphrasius at Poreč (530/550), for example, radiates with inset oystershells the size of dinner plates (fig. 6.5).[42] But thanks to the scriptural prototypes, it is also abundantly clear that early commentators were apt to describe marbles in the same language and regard them either as jewels or earthly reflections of the upper, jeweled heaven. The most stirring is Prudentius's *Psychomachia* (c. 400), which portrays the ideal church as hewn out of gems the size of boulders, from whose "clear depths the light from On High pours living, breathing colors" and columns of rock crystal (834–37, 851–72). Their fabulous scale comes into thrilling focus through images of whirring cranes that haul them to their mountainous summit.

Mosaic representations of church columns show that jeweled surfaces were desired and even accomplished in actual construction. The columns of the templon

FIG. 6.4. *Christ in the Heavenly Jerusalem-Paradise*, S. Pietro al Monte, 11th–12th century. Vault fresco. Civate.

FIG. 6.5. Apse, S. Euphrasius, 530/550 AD. Poreč, Croatia.

(sanctuary screen) in Hagia Sophia were studded with gems, while those in several sixth-century churches in Constantinople, most notably Hagios Polyeuktos (524–527), mimicked the display with humbler substitutes: that is, white marble shafts intarsiated with lozenges of colored glass, marble, and hard stones just like mounted gems (fig. 6.6).[43] Maybe Ennodius (474–521), bishop of Pavia, had something similar in mind when he wrote, of the Baptistery of Laurentius in Milan, "Just as you say that the fleeces of China are dyed with purple, / so the sojourning light adorns inlaid gems / whenever art makes the snowy stone blush more beautiful."[44] These incised patterns were themselves then imitated in mosaic and in ceramic tiles made to clad columns, columnettes, pilasters, and wall panels, and they also recur in manuscript illuminations.[45]

Marbles were also directly compared to gems, especially the most prized varieties that had become the legal preserve of the imperial family. Mark the Deacon likened (after 420) the Carystian columns in the cathedral at Gaza to emeralds, and in Gaul it was Avitus (513/516) again who claimed that only the size of church columns prevented the onlooker from calling them jewels.[46] For centuries after, the popular imagination maintained that many churches, such as the Basilica Ursiana in Ravenna, were entirely jeweled, when they were of course merely clad with marble.[47]

FIG. 6.6. Templon column, Hagios Polyeuktos, 524–527 AD. Constantinople. White marble filled with amethyst and colored paste. Archeological Museum, Istanbul.

Since John's Heavenly Jerusalem in Revelation was a building, it was an obvious symbolic template for church decoration east and west, but the meadow metaphor was equally influential, and city and meadow were fused in the walled garden that was potentially every church. Ekphrasists frequently compared the walls and paving of the church, sometimes the whole interior, to a springtime meadow.[48] Many historians have considered this simply a rhetorical convention. Yet, while it is obvious that paradise was a garden, and that any flowery metaphor therefore suited the church as heaven on earth, there were other nuances to the idea of architecture as gardens in stone that encouraged its dissemination.

The springtime garden signified creativity in nature and the arts alike. This pairing is exemplified in Lucian's flowery speech on an opulent civic hall (mid-/late second century AD), a formula that medieval ekphrasists adopted almost verbatim when discussing marbles: "The frescoes on the walls, the beauty of their colors, and the vividness, exactitude, and truth of each detail might well be compared with the face of spring and with a flowery field, except that those things fade and wither and change and cast their beauty, while this spring is eternal, field unfading, bloom undying. Naught but the eye touches it and culls the sweetness of what it sees."[49]

Spring was a theme appropriated from descriptions of the fecundating power of Venus, to become an integral part of the Christian celebration of Easter and the Annunciation.[50] In the Sumerian *Gilgameš* epic, the "trees of the gods" are perpetually fruiting because they bear forth hard stones and gems. In the Christian paradise, as Avitus says, "Autumn with its fruits and Spring with its flowers fill the whole year." The open-ended cosmic spring of the church interior now signified the perpetual advent of the Lord. Ezekiel's garden had been filled with gems, and marbles were easily compared to them; spring buds ("bud" is "*gemma*" in Latin) were the gems of nature; and to complete the circle, the ekphrases that exalted the gem-encrusted meadows of church interiors were themselves commended as jeweled pastures (*prata gemmea*) of verse.[51] The gemmy meadow was therefore the rhetorical image of fecund variety in any artistic pursuit.

As the generative light spilled down the walls of the church, it left images of procreation everywhere, especially floral ones (see fig. 6.1), and these in turn crept up the walls on their way to the supernal light. This metaphysical photosynthesis stretched as high as the ceiling mosaics, and Prudentius, again, speaks of a church that the patron had "spanned with curving arches of brilliant multicolored glass, like meadows bright with springtime flowers" (*Perist.* 12.53–54, on S. Paolo fuori le Mura). The topos of the springtime meadow was an all-encompassing metaphor that overruled material particularities to envisage an abundant palette, sunny relucency, and the season of nature's regeneration. Inscriptions even suggest that the walls "bloom" (*virent*) under the divine lamp.

Finally, though gems, and the materials that resembled them, conformed both to the letter of scripture and the jeweled style of late antique letters, they also pursued an anagogical program. All glittering materials would bedazzle spectators, who "transferring that which is material to that which is immaterial" would find themselves "dwelling, as it were, in some strange region of the universe which neither exists entirely in the slime of the earth, nor entirely in the purity of Heaven; and by the Grace of God, be transported from the inferior to the higher world in an anagogical manner."[52] These phrases of Suger, written in the mid-twelfth century, are now famous and widely cited, but the abbot was simply repeating much older metaphysical commonplaces. Six centuries earlier, Hypatius, bishop of Ephesus (531–538) and one of Emperor Justinian's leading religious advisors, had explained a little less rapturously: "We allow even material adornment in the sanctuaries . . . because we permit each order of the faithful to be guided and led up to the Divine Being in a manner appropriate to it, [and] because we think that people are guided even by these [materials] towards intelligible beauty and from the abundant light in the sanctuaries to the intelligible [i.e., understood with the intellect alone] and immaterial light."[53]

WATERY LIGHT AND LIQUID GEMS

One of the essential reciprocities in the medieval church interior, therefore, was between the seeping rays of the sun that spilled downward, nurturing and arousing the light potentialities of lustrous materials and, conversely, their photosynthetic growth toward the light. But what goes up must come down. The Greek word for baptism was "*Photisma,*" which can mean both "illumination" and "enlightenment," and Ennodius seems to make great play on these affinities in his ode "On the Font of the Baptistery of St. Stephen, and the Water Which Came by Means of the Columns":

> Behold, cloudless it rains under a roof in a serene
> shower,
> And the pure face of heaven supplies the waters.
> Flowing streams run from the sacred marbles,
> And a rock again produces moisture.
> For the dry upper-storey pours forth liquid springs,
> And water from on high comes to the reborn.
> The sacred water flows through the recesses of heaven,
> summoned by the ministry of the priest Eustorgius.[54]

In some eighth-century Lombard baptisteries, the idea of bringing the holy baptism to earth from heaven was accomplished by actually turning the piers into hollow downspouts for rainwater.[55] Whether Ennodius's epigram refers to such hydraulic contrivances or more poetically alludes to the shimmer and dewy properties of marbles, he does equate light with water, a metaphor of course highly suitable for a baptistery. It is also noteworthy that Ennodius seems to draw on Martial's description of the baths of Etruscus:

> Nowhere is the cloudless [vault] so empty of sparkle;
> light itself lasts longer there, and from no place does the
> day recede more tardily . . .
> plunge in the natural stream of the Virgin Marcia,
> which glitters so bright and clear
> that you will not notice any ripples there,
> and believe that the Lygdian [Parian] marble shines
> empty. (Mart. *Ep.* 6.42.8–10, 17–21)

But the perception of light as in some way liquid has other implications for the use of shimmering stones on the church interior. Here marbles approached the upper light because of their capacity to scintillate, reflecting the Aristotelian and Theophrastian geology that all stones must retain some measure of water for their particles to cohere. The example par excellence was rock crystal, and other stones that approached the lucidity of gems were also supposed to be of watery origin. Such metachemistry supplied physical rather than poetic reasons for believing in marble's heavenliness, and the material science that underpinned the whole construct would remain seriously unchallenged until the seventeenth century. As Erkinger Schwarzenberg has suggested, the key to understanding premodern material hierarchies is to consider the physical nature attributed to rock crystal.[56] Crystal was widely believed to be super-frozen ice that could resist the highest temperatures of all earthly furnaces. Crystal must therefore have been formed through fierce freezing beyond normal, terrestrial conditions, which had taken place during the primordial cold of Creation. It followed that the products of this divine permafrost were fossils of the original light, and therefore preserved traces of Creation and of the performing power of the Logos. In other words, divine light was frozen into their composition.

As for marbles, they might be lower on the ladder of Creation, but they still shared the same crystalline essence as gems and crystals by virtue of their watery base, just with more earthy matter mixed in. They were all still called *"metalla,"* in fact. Any minerals that could receive a polish shared a quintessential luminosity, their relative diaphaneity originating in a frozen, aqueous medium. Thus, light was inherently a dormant physical property with an innate bond to the heavenly spheres. Marble could recover this original light in a surface slick when it was polished, much as any stone becomes lustrous after it has been immersed in water. As counterintuitive as it now seems, marbles were both solids and manifestations of a liquid light. Already Poseidippos (c. 225/200 BC) celebrates "a stone that if wettened, [looks as though] its entire mass is surrounded by light, a marvel of illusion." Observers were, in fact, entranced by the way "a ray of sun flashing upon a wall and trembling with the movement of the moisture which the beam has taken up in mid-air [is] checked by the hard surface [and sets] up a strange quivering."[57] To sum up, the common resemblance of shimmer to wetness, the "wet look" that mosaics and marbles alike could achieve, pointed beyond the surface to a substratum of physical affinities in an original nature.

GLASS MADE OF STONE AND WINDOWS OF DIVINE LIGHT

As dedicatory inscriptions seem to imply, marbles, like gems, lay dormant until solicited by the generative light from above, and even then, the light only indwelled the stones for moments. In some cases, however, particularly luminous stones were believed to shine actively of their own accord.[58] Suger attributes this property exclusively to "carbuncle," but many gems and semiprecious stones with a high refractive index (like cats' eyes) were said to function as the night-lights of temples, churches, and mosques.[59] As regards marble, Theophrastos had remarked that the earliest reference to a self-illuminating stone, Herodotos's description of an emerald column that lit up the Temple of Herakles at Tyre, actually referred to a monolith of Spartan green porphyry or malachite.[60] In time, the divine luminescence perceived within stones, including windowpanes of stone, would also come to fuel travelers' tales of marble slabs as ever-burning lamps.

Thus, travelers fêted the gleaming marble or alabaster panes, probably transenna panels, in Hagia Sophia and Hosios Loukas until the eighteenth century,[61] and a medieval Russian pilgrim had perhaps a similar phenomenon in mind when he described the lamps in Hagia Sophia as existing "between the walls."[62] Certainly the most attested and elaborate case of these glowing marbles is provided by the twin slabs in the medieval apse of the Christianized Parthenon, which were famed from Turkey to Iceland. Tales originally circulated about an inextinguishable lamp that burned in the temple, but with time this legend came to be transferred to the building's marbles.[63] When an Italian traveler visited the church in 1394/95, he remarked, "In a chink of light in the wall of the church a light shines, as if from a lighted fire that never goes out. It is believed that the body of a saint is enclosed in the wall just there."[64] A succession of expectant and credulous European travelers would repeat the observation, comparing the gleam to a glowing coal, to a carbuncle, or to a twinkling star. Moreover, Turkish guides or latter-day Athenians assured these Europeans that the reddish stone was a perpetual lamp or the frozen paschal flame, that everybody from St. Paul to Mahomet had ignited it, and that it shone by day and night.[65]

For the divine light to bring the walls alive, the church had to be fenestrated.[66] Where churches were freestanding, such as S. Apollinare in Classe (c. 534–539 AD) and S. Apollinare Nuovo in Ravenna (c. 500 AD), or S. Giorgio in Velabro (847–855 AD) and S. Paolo fuori le Mura in Rome (324 AD), large windows punctuated both nave and aisles. In this respect they followed the audience halls on which they were modeled, themselves converted to worship, such as the basilica at Trier (c. 305/312 AD) and the Basilica of Junius Bassus in Rome (see fig. 5.19). In all cases the high position of the windows was necessary to avoid mundane distractions—that is, the outside world. Moreover, the windowpanes of the early basilicas were translucent rather than transparent, and even in those apertures where there might have been such clear glass as was available, the only glimpse afforded the faithful was "a view of the sky, which will not divert the minds of the celebrant or supplicant from divine matters" (Alberti, *De re aed.* lib. VII, cap. 12, f. 127).

But windows were far from being passive tools of illumination and, as the entrances of divine light, they became autonomous presences. In the early eighth century, the Venerable Bede tried to convey the symbolic agenda of church windows when he explained (c. 729–731) why the windows in Solomon's Temple had been funnel-shaped (i.e., with their jambs splaying out on the interior): "The windows of the temple are the holy teachers and all those spiritual people in the Church to whom, when in divine ecstasy, it is granted more specially than others to see the hidden mysteries of heaven. And when they reveal publicly to the faithful what they have seen in private, they fill the inner recesses of the temple as windows do with the sunlight they let in."[67]

Such symbolism was old by the time Bede repostulated it. It crops up in Byzantine sources such as the dedication hymn for the cathedral at Edessa (c. 543–554), which states, "One light shines forth also in its sanctuary by three open windows, and announces to us the mystery of the Trinity, of the Father and the Son and the Holy Spirit. And the light of its three sides abides in many windows. It portrays the apostles, Our Lord, the prophets, martyrs and confessors."[68] If we turn to the mosaics within S. Apollinare Nuovo in Ravenna (556–569), the windows alternate and are paired with book-wielding saints, the so-called "holy teachers" (fig. 6.7). Likewise in the nave tympana of Hagia Sophia, the ninth-century mosaics of prophets and the luminaries of the Eastern church also assumed a parity with the windows next to which they stand, overshadow them, or are even ensconced in blind arcades that imply fictive windows. Furthermore, the pyramid of windows and saints, whose apex is a trinity of windows, becomes a flow diagram of dissemination from on high, like the Pentecost. The cumulative message was that however many the windows, the light always derived from the same source, and this testified at one and the same time to the singularity of the Lord and the multitude of his agents.[69]

At S. Apollinare Nuovo, each didactic saint is circumscribed by an implicit niche, whose vault is an umbrella-shaped *velarium* (in the next register up) that rhymes with the shape of the adjacent window (see fig. 6.7). Lucretius had written stirringly of the rainbow effects that resulted when sunlight passed through real amphitheater velaria, "painting the people below," and it must have been a ready metaphor for a multicolored, heavenly light in the early centuries of official Christianity, when observers were still familiar with the spectacle.[70]

Like the velarium, the window was a filter, the agent for commuting daylight into divine light and converting the church interior into a heaven. In theological terms the difference between this original light, which was God's

FIG. 6.7. Clerestory, S. Apollinare Nuovo, 556–569 AD. Mosaic. Ravenna.

FIG. 6.8. Mausoleum of Galla Placidia, c. 454. Ravenna. Looking south: clerestory, Apostles below the velarium, and doves flanking a fountain of paradise; lunette, St. Lawrence rushing to Martyrdom on the gridiron.

essence, and its emanation in visible or created forms eventually came to be codified as the divide between "increate light" (*lux*) and "created light" (*lumen*).[71] Because the church was a catoptric and cathartic vessel, the windowpane was a necessary diaphragm between the two lights, without which the ordinary mortal would have been overwhelmed and consumed by the supercharged splendor of his visage. Moreover, just as Christ himself was the *lumen de lumine* (the reflected "light from light," the incarnation of lux in matter, flesh), the saints were the savior's own receptors and reflectors, his own windows on the world.

Because the true brilliance of the Godhead, surpassing all understanding, would blind human perception once revealed, the author of Revelation chose translucent stones to epitomize the supreme deity's brightness in recognizable terms, "like a precious stone, even like a jasper stone, clear as crystal" (Revelation 21:11). The force of the simile was twofold. While crystal expressed the super-rarefied clarity of God's light, the color of jasper embodied its substance and difference from daylight. In the church this alterity was desirable to re-create created light (lumen) as increate (lux), and the tint of diaphanous stones or colored glass made daylight metamorphose more decisively than any simple transparency.

Windows were not absences but figures, not voids but mediators. In Eusebius's words, the windows of the ideal church "shed light with apertures towards the light" (*Hist. Eccl.* 10.4.64). They were antennae, both emitters of divine effluence into the human realm and portholes into the heavenly one. In fact, some etymologists even derived the Latin word "*fenestra*" from the Greek "*phoster*" ("star"). When divine light (lux) did make its entrance into the world, windows could make this emanation, a part of God, temporarily corporeal. In the clerestories of the so-called "Mausoleum of Galla Placidia" in Ravenna (c. 454; fig. 6.8), and the S. Zeno Chapel (c. 817–824) at S. Prassede in Rome, the saints flanking the windows actually turn inward to face the illumination, as though Christ himself had stepped over the lintel. In the S. Zeno Chapel, the light streams across the chapel vault from the revelational window to a mosaic of the *etimasia,* or the prepared throne that remains empty until the Final Coming.[72] At Galla Placidia, the apostles flanking the clerestory windows stand in perspectival rooms (created by the wall plane "folding back" to make platforms), which have their own velarium-shell vaults. This perspectivity guarantees that the window can only read as a body, and something very much like the *Shekinah* of God that guided the Israelites across the Sinai to the Land of Milk and Honey (Exodus 13:21: "And the Lord went before them by day in a pillar of cloud to lead them along the way, and by night in a pillar of fire to give them light"). The promise of the window as presence and pathway, to God, to paradise, is confirmed by the St. Lawrence mosaic in the lunette below. Here, the instrument of his martyrdom, a wildly flaming grate hinges on the window just as the doors swing open from the bookcase. The implication is that fire pales in comparison to lux. As the inscription on the triumphal arch of his basilica in Rome, S. Lorenzo fuori le Mura (570–590 AD), announces,

> MARTYRIUM FLAMMIS OLIM LEVVITA SVBISTI
> IVRE TVIS TEMPLIS LVX VENERANDA REDIT.
>
> Oh Levite, once you endured martyrdom in flames.
> By right, in your temples, the holy light returns.

The substance of windowpanes was therefore of the utmost importance, and from antiquity to the Middle Ages, rough glass, mica, and "specular stone" were all used. Thin slabs of translucent "specular stone," a transparent form of gypsum also called "selenite" (or "moonstone"), have been found at Pompei and Ostia (fig. 6.9); we hear of Caligula ordering windows of "diaphanous stone" like "white glass" in the Horti Lamiani; and there were busy mines of the stuff throughout Italy and Spain.[73] In late antiquity this mineral became so associated with windows that it lent them its name (*speculae*), and thence to the

FIG. 6.9. Specular stone window panes, 1st century AD. Pompei.

tower-top rooms for star-gazing or "speculation."[74] Pliny (*HN.* 35.36.97) compared the dim light that shone through them to a softening glaze that Apelles had used on his paintings, and Lactantius (c. 240–320 AD) used specular panes to explain human cognition: the mind peeps out of the skull through the eyes, "as though through windows covered with pellucid glass or specular-stone."[75]

Specular stone was certainly used in churches. The mineralogist Agricola mentions panes still extant in German ones in the 1530s and 1540s, and the *Liber Pontificalis* (chronicles of the early popes) records windows "of gypsum-stone" (*ex metallo gypsino*) being installed in the most important Roman basilicas in the ninth century.[76] A few have come to light, especially those at S. Sabina, which may date back to the fifth century (fig. 6.10). Here the specular panes were cast into a grille of gypsum that, ex uno lapide, together with all the revetments and mosaics, made the walls once glow from tip to toe as a "prison of light."[77] Sidonius Apollinaris, describing the Church of St. Justus at Lyons, must allude to an interior like this when he comments, "Marble, adorned by varied flickers, spreads over the ceiling, floor, windows" (*Epist.* 2.10.11–12: *distinctum vario nitore marmor / percurrit cameram solum fenestras*). A late fourth- or early fifth-century riddle about specular stone could well have been written about the windows of S. Sabina:

> Specular Stone.
> I am looked clear through and do not prevent vision,
> Transmitting the wandering glance within my members;
> Nor does the cold pass through me, yet the sun shines out within me.[78]

In the hands of theologians, these same properties also made it a substantial metaphor of the Conception and Incarnation. As a fifth- or sixth-century sermon explained: "The ray of the sun penetrates the specular stone; it both pierces its solidity with imperceptible subtlety and what exists outside seems to exist within. It does not scatter when it enters, nor does it injure when it exits; because both in its entry and exit the specular stone remains steadfast and intact. The ray of the sun did not therefore break the specular stone; could the entrance and exit of the ray have harmed the truthful integrity of the Virgin?" Likewise, Eugenius I (c. 600–657), archbishop of Toledo writes, "Just as entering Phoebus knows not to destroy the specular stone, so did the Virgin Mary remain [intact] when engendering Christ."[79]

Dichroic glass was also used for windows, though very rarely. Such glass is opaque and jade-colored in reflected light, but shifts to amber in transmitted light, like the celebrated Lycurgus Cup now in the British Museum. Not only did it miraculously commute the daylight, but since dichroic effects are achieved by suspending minute quantities of gold and silver in glass, it must have seemed to follow Revelation almost to the letter. The walls of the Heavenly Jerusalem were said to be of a gold so pure that they would be completely transparent to the lux within them (Revelation 21:18, 21–25). In reality, producing glass that aped this higher reality was costly, and in at least one case prone to theft: the glass in the church at Yzeures (Auvergne) was so golden-hued that a cat burglar made off with the panes in order to extract, unsuccessfully, the gold in his crucible.[80]

FIG. 6.10. Integrally cast gypsum panes and glazing bars, S. Sabina, possibly 5th century AD. Rome.

Observers from at least Alberti's day onward firmly believed that "the windows that provide the temple with light would have . . . thin slabs of translucent alabaster or a lattice of bronze or marble . . . filled . . . with specular stone," and Vasari confidently held that the ancients glazed their baths and other rooms with agates, alabasters, and veined marbles.[81] However, the only antique text to mention anything of the sort describes the Temple of Fortune built from the mysterious Phengites, which Pliny says made the temple "gleam like the day" within "as though the light was enclosed within, rather than transmitted" (Pliny. *HN.* 36.46.163). Christian alchemical writers used the word "Phengites" to describe a gem whose radiance was so impressive that they compared it with the brilliance of God.[82] When they read Pliny's description, it is easy to see how they could have thought that a temple might actually have been hewn from a gem. Pliny says that Phengite came from Cappadocia, but in the next breath he mentions another translucent stone used for windows in Arabia, and might well be referring to the alabaster panes of Yemen that have now become famous. The *quibla* bay of the Umayyad mosque in Sana'a, Yemen, is still capped by lanterns of translucent slabs that probably date to the caliphate of al-Walid (c. 705–715; fig. 6.11).[83] Now much blackened, these luminous hollows must once have encapsulated the words of the *Qur'an* (24.35), "His light is like a niche [*mishkat*] within which is a lamp, the lamp enclosed in glass, the glass as if it were a pearly star. . . . Light upon Light, God guides whom He will to His light."[84]

Most observers would have found that onyx or translucent calcite-alabaster reproduced the effects that ancient texts had so stirringly described, and it is hard to imagine that medieval patrons in the West avoided using slabs where they were available. But hard evidence is completely lacking. The earliest extant, and securely documented, examples known to me are the splendid alabaster windows in the nave of the duomo at Orvieto, which began to be installed in 1292 (fig. 6.12).[85] The large slabs lighting the apse of S. Miniato, Florence, could conceivably date to some point in the church's construction, between 1018 and 1207, but they are first mentioned in 1510.[86] Travelers mention now-missing examples

FIG. 6.11. Lanterns capped by alabaster slabs, quibla bay, Great Mosque, c. 705–715 AD. Sana'a, Yemen.

FIG. 6.12. Alabaster window (alabastro di Volterra), Duomo, 1292. Orvieto.

in Turkey, and Patrik Reutersward also suggested that alabaster panes can be inferred from the faux-alabaster window splays in the early twelfth-century abbey church at Gurk, Carinthia, or the late twelfth-century Church of St. George at Kurbinovo (near Lake Prespa, modern Macedonia).[87]

But this is paltry evidence, and calcite-alabaster windows can never have been used with the regularity that we have come to assume unquestioningly, and as Sir Lawrence Alma-Tadema emblematized them in a 1907 painting of a very virginal *Orant* (fig. 6.13). The image maybe prospered in modern England thanks to the reception of a particularly visionary passage in Isaiah (54:11–12) that includes the key phrase וְשַׂמְתִּי כַּדְכֹד שִׁמְשֹׁתַיִךְ (*w'sam'Tiy Kad'khod shim'shotayikh*), which both the Septuagint and the Vulgate translate as "I will make your bastions of jasper" (καὶ θήσω τὰς ἐπάλξεις σου ἴασπιν, *et ponam iaspidem propugnacula tua*). The King James Bible actually seems closer to the Hebrew when it styles the phrase "and I will make thy windows of agates," because "Kad'khod" indicates something red-hot, and "shim'shotayikh" is the feminine plural of "*shimsha*," which is cognate with "*she'mesh*" ("Sun").

Whatever their understanding of scripture, it goes unrecognized that, besides Orvieto and S. Miniato, virtually all the alabaster and onyx windows now visible in medieval churches are nineteenth- or twentieth-century reconstructions. The best-known examples in Rome, the nave windows of S. Paolo fuori le Mura, are materially authentic inasmuch as they are a genuine Egyptian alabaster, probably Cairene, but the stone was a diplomatic gift to the papacy from King Fouad I of Egypt, after an accidental explosion at the Forte Portuense blew the old windows out in 1891. Likewise, the most frequently cited examples, the panes in the Mausoleum of Galla Placidia (c. 454) at Ravenna, were not installed until 1911 and were offcuts from the Tomb of Umberto I in the Pantheon (see fig. 6.8).[88]

FIG. 6.13. Sir Lawrence Alma-Tadema, *Orant*, 1907. Oil on wood. Private collection.

FIG. 6.14. Marbled window glass, refectory, S. Vincenzo al Volturno, 9th century AD. Molise.

The early medieval *desire* to use veined and diaphanous stones is, however, attested by their simulation in colored glass (fig. 6.14). In antiquity, marble was certainly imitated in glass (see fig. 5.13), and fragments of glass windowpanes with conspicuous fictive veining, dating from the sixth to ninth centuries, have been found at sites as dispersed as Jarrow (Northumberland), Müstair (Switzerland), Liège, Farfa (Lazio), and S. Vincenzo al Volturno (Molise).[89] These panes come in all stripes, but red and green are preponderant. Some scholars who have observed this "veining" have considered it a botched attempt to produce colored glass.[90] Were this true, of course, one would feel compelled to inquire why glassmakers, from one end of Europe to another, installed these "failures"—which so consistently resemble marbling—in such prominent locations.

It can only have been the desire for gemlike windows that underwrote the fenestration in the refectory of the Benedictine abbey at S. Vincenzo al Volturno (eighth century), where red-and-green-streaked (agate-type?) panes adorned a room liberally painted in imitation of

FIG. 6.15. Chalice of Abbot Suger of St. Denis, sardonyx cup, 1st or 2nd century BC; mounting, gilding, and ornamentation, c. 1137–40 AD. Height: 7¼ in. (18.4 cm); overall (diam. at base): 4⅝ in. (11.7 cm); overall (diam. at top): 4⅞ in. (12.4 cm). National Gallery of Art, Washington, DC. Widener Collection.

marble. Pride of place was given to a roundel of translucent, veined, royal-blue glass situated near the pulpit, itself painted with disks and marbled veins, where the material concert arguably evoked the idea that the reader pronounced the Word of God from something like his sapphire throne.[91]

The difficulty we might now experience in matching most of these simulations to any particular materials is that the exact appearance of the original minerals (Phengite, jasper, alabaster, and the other miscellaneous stones and jewels mentioned in the Bible) were unknown to the Middle Ages, and their identities always confused. When medieval craftsmen did look to material remains, it is more than likely that they took inspiration from the antique agate, sardonyx, and myrrhine cups that were desired for chalices in the most prestigious church treasuries (fig. 6.15). Pliny's description of myrrhine cups and their physical constitution is, in fact, as stirring as any torrent of praise for marbles: "It is thought to be a humor that has thickened underground through heat. . . . There is a splendor to them, but without intensity, more truly a glimmer than a splendor. But their value lies in the variety of colors, the veins swirling around themselves in purple and brilliant white, or a third color mixed from both, the purple burning or the milk-white reddening, as though by the transit of the color. There are some who praise their edges most, and those ones where the clustering of colors resembles those visible in the innermost rainbow" (*HN.* 37.8.21–22). Myrrhine vases were replicated in glass in antiquity, and the practice was eventually revived in the "chalcedony glass" of mid-Quattrocento Venetian glassblowers.[92] The aforementioned Lycurgus Cup itself almost certainly imitates the effects of vessels carved from huge, single agates such as the famous "Rubens Cup" (c. 400 AD), now in the Walters Art Museum, Baltimore. Moreover, medieval observers could not always tell the difference between the original and its simulation or, like the Sumerians, they saw the affinities over the distinctions. The most egregious case is the huge dish venerated in S. Lorenzo, Genoa, as the "Sacro Catino," which was renowned as "a single emerald." It is in reality green glass.[93]

In any case, the overriding desire for using these glasses, as with selenite and specular stone, was to create an entire wall with a seemingly innate radiance, a prison of light for the new day. The eventual descendent was the Gothic interior, in which the all-too-earthly masonry is gradually eradicated (as much as statics would allow) to make way for vast expanses of stained glass that made the entire surface into a wall of glowing gems.[94] The choicest example is surely the Sainte-Chapelle in Paris (fig. 6.16), the archetype for which can only be the Heavenly Jerusalem, but whose particular case begs a more historical comparison. Louis IX built this palace chapel to house the relic collection of the Byzantine emperors, the most important Christian relics of all, those of Christ's Passion. These had once been kept in the Pharos Chapel of the Imperial Palace, in Constantinople, and astonished Latins had themselves marveled that it "was so rich and noble that . . . there was no column that was not of jasper or porphyry or some other rich precious stone," and this was the jewel box that Louis sought to re-create in his own Sainte-Chapelle.[95]

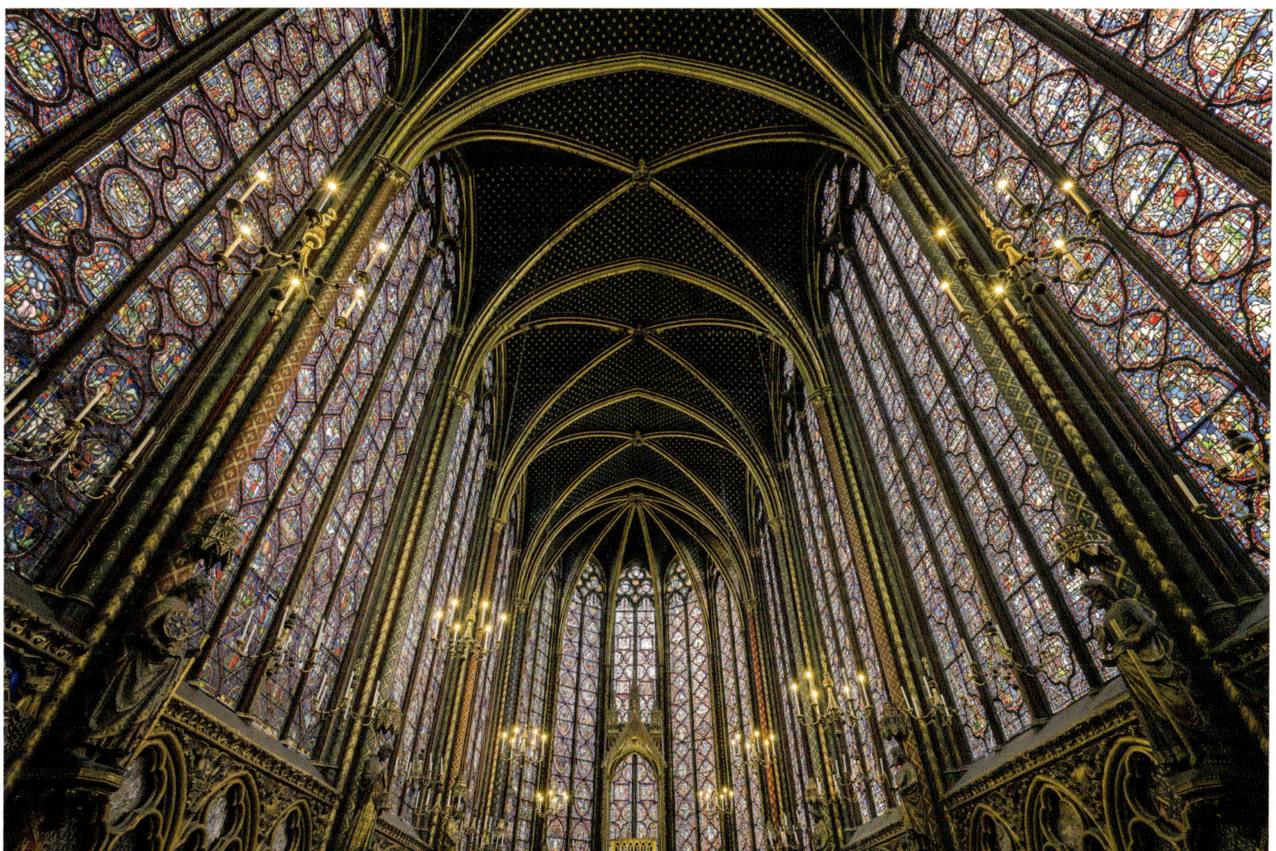

FIG. 6.16. Interior, Sainte-Chapelle, 1238–48. Paris.

VANITAS

Colored marble revetments continued to enjoy a glorious afterlife in Byzantium until the fall of Constantinople in 1453, even though the city and empire had to cannibalize existing buildings to live up to the past expectations that medieval ekphrasists make clear still persisted. The architectural paradigms of the Greek East also remained relatively ossified, when compared to the emergence of the Romanesque, Gothic, and finally the reinterpretation of classical architecture in Western Europe. In the West, the only points of comparison for the material opulence of Byzantium are the rich floors and church furniture, from ambos to candelabra, that the Cosmati workshops turned out for the wealthiest patrons in Italy, and Cosmati floors also owed their sophistication to the pavements that were developed in Byzantium in the Middle Ages. Marble revetment died out or became a scrappy hotchpotch thrown together from whatever was available. Finding a suite of marble columns that were the same color or even size became virtually impossible, even in Rome. The glaring exception to this rule, and the most elaborate marble spolia construction in Europe, outside and in, is S. Marco in Venice (twelfth and thirteenth centuries) (see figs. 8.5, 9.2), a beautifully blended salmagundi of slabs torn from buildings all over the Mediterranean, with columns everywhere, seeming "to have been made by angels rather than men," and vying with any Constantinopolitan church to create "almost another Byzantium."[96]

Otherwise the story of Western marbling until the fifteenth century is not particularly engrossing. In fact, it is rather depressing. The despoilment of the once-shimmering edifices stripped them to the bare bones, and offered a painfully obvious picture of the passing of worldly glories. From the moment that colored marbles had begun to enjoy any popularity, they had anyway numbered among all those worldly chimaeras that distracted humanity from its inevitable demise. As Horace cautions:

> Neither ivory nor gilded
> Coffers gleam in my home
> Nor do beams of Hymettan marble
> Rest on columns quarried in farthest Africa
> [. . .]
> You order marbles to be sawn,
> Though you are at the point of death,
> Oblivious of the tomb, you build a palace.[97]

In late antiquity, it became all the easier to say such things, when the empire was on the point of ruin, the city collapsing about one's ears, and speculators going from house to house to size up columns for reuse, or consign statues to the lime kiln. It is in this context that a late antique poem, "On Riches and the Dishonest Soul," resumes the old charge that rich furnishings were symptoms of corruption:

> The fact that your lofty home has a thousand solid columns,
> The fact that your portal gleams with marble jambs
> The fact that your highest ceilings glitter with gilded coffers
> The fact that precious slabs cover the lowly floor
> The fact that rich ornament covers your halls all around,
> Does this really raise your spirits, oh happy one?
> Although gems smother everything throughout the whole house,
> It is vile, and nothing is more vile than its lord.[98]

Moreover, not only did this obsession with luxury distract one from one's end, but marble could actually hasten it along. Agathias's quite malicious account of the untimely end of an imperial chamberlain in the Constantinopolitan earthquake of 558 AD confirmed the moral, even mortal, perils of marble luxury:

> Anatolius was sleeping at the time in his customary bedchamber. The apartment was adorned with a variety of marble plaques attached to the wall, of the kind that are lavishly and ostentatiously displayed by those who are inordinately fond of such superfluous and unnecessary bric-à-brac. One of these plaques, which was fastened to the wall next to the bed, was shaken loose from its fittings and wrenched off by the violence of the tremors. It came down with all its weight on his head, and smashed his skull. He had barely enough time to utter a deep and muffled groan of pain and then sank back on his bed. Death had overtaken him.[99]

Reading between the lines, the underlying irony becomes obvious: Anatolius's bedroom has become his sepulcher. This symmetry was easy enough to apprehend. Mausolea had reached the grandest proportions and received the most lavish decorations two or three centuries earlier, and other homes for the patrician dead, such as the third- and fourth-century sepulchral cubicula in Italian catacombs, were often indistinguishable from those for the living. In effect, Agathias echoes Augustine's much earlier critique of the elite preoccupation with building marbled tombs: that ultimately they were just as futile as the mansions that wealthy men erected for themselves in life. Neither could live up to the rewards of eternity. "I have a marbled home," Augustine says, "which I will leave, and I do not dwell upon my eternal home where I will be always" (*Enarratio in Ps. 48; PL* 36, col. 554: *habeo marmoratam domum quam relicturus sum, et non cogito mihi aeternam domum ubi semper ero*).

Even Christian emperors had to come to terms with the vanity of marble for their own tombs, "the only things left of those Roman Emperors of old who boasted of glorious deeds . . . their coats of stone, those cold and last garments."[100] Early emperors such as Nero and Hadrian had been buried in porphyry caskets, and emperors from Constantine onward again sought to steep themselves in the purple for eternity with enormous sarcophagi of the same stone.[101] But within a century and a half of the relocation of the imperial capital to Constantinople, they could no longer count on even this cold comfort. Weakening borders and dwindling resources soon meant that outlying quarries such as those of Mons Porphyrites were luxuries the empire could no longer afford. Convicts were still sent there as late as 308 AD (Euseb. *MP.* 8.1), but the last porphyry sarcophagus for a Byzantine emperor, Marcian, was cut in 457.

On the other hand, stockpiles of porphyry must still have been available, though not perhaps the monolithic blocks suitable to an emperor's sarcophagus, and one wonders whether the decision to avoid the material did not also represent some sort of postmortem humility. From classical times, the *servus publicus* had stood in the chariot of every *imperator,* and then every emperor, who triumphed through Rome, whispering in his ear, "Look behind you, remember that you are only a man" (Tert. *Apol.* 33.4: *respice post te, hominem te memento*). In Byzantium the practice persisted in the public processions of the emperor and his entire court from the palace to Hagia Sophia. In his left hand he held a jeweled cross but his right gripped a gilt wood capsule (the *akakia*) containing a bag of tomb earth, while every two paces a court functionary goaded him, "Remember death."[102] This moral message carried over into Byzantine investitures, where the memento mori would take the form of marble samples. Two sources detail the salient ceremony, beginning with Leontios of Neapolis (641–642), who relates that while St. John the Almoner pondered his own mortality, he recalled:

> When an emperor is crowned, first of all the members of the guild of tomb-builders have access to the royal presence while the whole Senate and the army are in attendance; directly after the crowning the builders of the imperial tomb come in and bring with them four or five small pieces of marbles of different colors and say to him: "Of which mineral does Your Majesty desire his tomb to be made?" thus suggesting to him that, as a corruptible mortal who soon passes away, he should take thought for his own soul, and govern his kingdom righteously. The blessed man [St. John the Almoner] imitated this truly praiseworthy custom and gave orders for a tomb to be built for him in the place where the previous patriarchs were buried, but to leave it unfinished until his death so that on some great feast-day when the clergy were present, the zealous Christians, as they are called, should come in and say to him, "Your tomb, master, is still unfinished. Allow us, we pray, to finish it because you do not know at what hour the thief will come."[103]

Leontios's account happens to coincide with the first coronation of an emperor, Constans II, in the ambo of Hagia Sophia (641), but the practice may date back to the late sixth century, perhaps even earlier.[104] A second source, Ignatius of Smolensk, suggests either that the practice persisted for several hundred years, or that it was revived for the investiture of Manuel II by the patriarch Antonius IV, which he witnessed in 1392.[105] In fact, by word of mouth or Latin translation, this rite of passage seems also to have made its way West. In 1354 the Florentine Dominican Jacopo Passavanti also recounts the ceremony in sermons, and it came to be incorporated, however briefly, in the papal investiture as well.[106] The Welsh priest and chronicler Adam of Usk records that at the 1404 investiture of Innocent VII ("just like that of the Emperor"), after the burning of the tallow, which itself symbolized the *Gloria Mundi,* Roman *marmisti* offered stones "worked with every artifice," adding the pregnant question, "And from what stone will your tomb be made?"[107] The pope himself received these ministrations seated in a massive porphyry throne, the infamous *sedia stercoraria.*

Throughout the medieval period, marble still embodied a dual valency. While many of the Church Fathers vented against its useless vanity, it remained nonetheless a prestige symbol par excellence and a material of heavenly character. The Christian ethos was, in fact, continually torn between a desire to exploit the eschatological and honorific character of gold, gems, and precious marbles, and a fundamental humility that strove against terrestrial wealth and mundane ostentation. In the eternal city, and every other decaying metropolis of antiquity, the two conditions were all the more obvious. In Rome's churches and cathedrals, lustrous marbles might seem supernatural markers, but in the city at large their veins dissembled with weathering, and the abandoned and despoiled fora, sanctuaries, and circuses that had become marshes and pastures were littered with blocks scorched and cracked in the heat of the city's numerous sacks. Such decay and spoliation made the ruins monumental vanities, empty tokens of the fall of earthly splendor. In some eyes, the living rock was now just dead stone, and numinous stones had become barren idols.[108] The city was cannibalizing itself and, by the end of the eleventh century, the wrecking crews and lime-burners had done such a good job that there was even a shortage of good-quality, monumental columns.[109]

On the one hand, then, it was right and fitting to use such rich materials to outfit the House of the Lord; on the other, they smacked of worldly corruption. This disparity of signification in turn broached a larger dilemma: whether God was best honored with all the riches that ecclesiastical pomp could muster, or whether man was even capable of fabricating the House of God; whether, in fact, only a spiritual edifice might properly honor a deity defined by spirit.[110] Christian writers in the latter camp took up Seneca's charge that an inner, spiritual edifice excelled any material counterpart, and Cyprian, Augustine, Caesar of Arles, and others encouraged the faithful to disregard the splendor of marble walls and gilded ceilings and instead embellish their souls as temples of God illumined and painted by virtue.

The dilemma between external and internal architectures, overt splendor and internal brilliance, itself depended on a larger ambiguity as to the nature of beauty. It was of two kinds: a sensuous quality of a material object that seduced the physical eye; and the intangible, imperceptible, but not entirely insensible moral beauty that was the ornament of the soul and the promise of higher glory. As Ellert Dahl has argued, the whole dilemma was summed up by contrasting interpretations of the Psalmic verse, "Lord I have loved the beauty of Thy House" (Psalm 25:8: *Domine, dilexi decorem domus tuae*). This division of interpretation became so important in the medieval period that, Dahl concludes, it amounted to "a fundamental cleavage in the attitude towards church-building in the twelfth and thirteenth centuries in Western Europe."[111]

MARBLE SIMULATION AND FAUX–BOOK-MATCHING

One solution to the dilemma was simulation. Simulating marble in paint sustained its image in absentia but, as we shall see, simulation of white marble masonry in particular allowed patrons a loophole in the debate over worldly vis-à-vis heavenly magnificence, straddling both sensible beauty and moral intangibility.

Despite the dearth of freshly quarried marbles, and the decreasing supply of spoliated ones, desire for the material remained strong across Europe, and patrons and artists wishing to exploit it were driven back on either spoliation or simulation. Simulation was de rigueur in territories as outlying as Northumbria, Georgia, or Cappadocia, with no history of imperial patronage, and where it remained impossible to obtain anything but fragments of real marbles. Simulation had always been achieved by patterning walls with colorful squiggles and spatter because marble veining was just as important as its color, but veining was to become an equally international style of ornament when geometrically abstracted.

FIG. 6.17. Nave, Durham Cathedral, 1093–1133. Durham.

Strident zigzags, chevrons, spirals, and reticulate patterns decorate the nave columns in a series of late eleventh- and early twelfth-century churches in England—Durham Cathedral, 1093 to 1133, being the most famous (fig. 6.17)—and many others from Scandinavia to Georgia. Spiral patterns such as those at Durham are generally associated with the spiral barley-leaf columns of the chancel screen in Old St. Peter's erected under Constantine, and the earliest example of this idiom is perhaps the crypt of St. Peter in Utrecht (consecrated 1048). It also seems certain that these and the other geometrical patterns quoted the decoration of the varied columns that decorated the aedicule of Christ's tomb with the Church of the Holy Sepulcher in Jerusalem.[112] However, these columns themselves can date from no earlier than the eleventh century, as in 1009 the Fatimid caliph Al-Hakim bi-Amr Allah had completely leveled the tomb, and it was rebuilt at the behest of the Byzantine emperor Constantine Monomachos only in 1042 to 1048. Moreover, as early as the eighth century, when the Al-Aqsa Mosque on the Haram al-Sharif (Temple Mount) in Jerusalem was renovated, the columns of the prayer hall were plastered and painted both with marbling and diaper patterns.[113] In short, the tradition must be considerably older.

"Lightning-bolt" patterns, nested chevrons, diapers, and so on were actually long-standing, antique, pictorial conventions for representing marble veining.[114] One finds "lightning-bolt" veining conventions on the fourth-century mosaics throughout North Africa and elsewhere in the empire.[115] The so-called "zebra-stripe" pattern may be found adorning the walls of houses and baths up and down Italy, and beyond (fig. 6.18), where it obviously mimics book-matching.[116] Chevron patterns were another convention of great longevity. Rainbow chevrons decorate the columns of Cappadocian rock-cut churches, such as the eleventh-century Karanlik Kilise ("Dark Church") near Göreme, another area where true marble was unavailable.[117] When these patterns reappeared on elaborate, Cosmati church furniture, the patterns were frequently inlaid with glass and marble tesserae, as if they too were glittering veins.

In the medieval world, manuscripts as much as masons were probably the primary channels for the convention's diffusion, especially the faux-marble architecture that framed canon tables in late antique Gospel books and the medieval ones that copied them. Page after page of the Lorsch Gospels (c. 810), for example, show these marbled

FIG. 6.18. "Zebra-stripe" frescoes, "Villa of Poppaea," c. 60–40 BC. Oplontis.

arcades and make up a hefty virtual church in themselves (fig. 6.19).[118] Such illuminations even became the models for Qur'an frontispieces, but the canon tables themselves also, ironically, fell victim to spoliation when they were reused as flyleafs or integrated into later codices. After the fall of Constantinople in 1204, there was vastly increased opportunity for such influences. The looting of the city brought not only the booty of relics and actual marbles, but also resulted in a diaspora of manuscripts to Italy and France, including several with miniatures that showed, or purported to show, marble-clad Byzantine churches.[119] Such sources would explain both chevron and spiral patterning, but the reticulate patterns may harken back to the "jeweled" columns that had ennobled so many Constantinopolitan churches (see fig. 6.6).

There were still more direct forms of transmission. The earliest columns to bear such motifs in England are from Lanfranc's first cathedral at Canterbury (c. 1071–78), and both he and his architect, Gundulf, were of Italian

FIG. 6.19. Columns with human heads and zoomorphic images in their veining. Detail, canon table, Lorsch Gospels, c. 810.

(Lombard) origin.[120] Likewise the bishop who commissioned Durham Cathedral, William of St. Calais, had just returned from exile (1088–91) at Rouen, and there is every likelihood that the architect of Durham was in his retinue. While all these transmissions are possible, and probable, we must also recognize that such notations for stone veining are one of the greatest longue durées in the history of ornament. They stretch back to Minoan "Kamares ware" (see fig. 4.4c) and the repertoire zigzags and parallel wavy lines that decorated the fresco floors of Mycenaean palaces (see figs. 4.12, 4.13). We might in fact say that they stretch all the way back to the very walls of Uruk (see fig. 1.1). The same patterns are of such longevity that they reappear in the Venetian *Hypnerotomachia Poliphili* (1499), again probably by way of Byzantine manuscript illustration, most notably in the images of a garden containing the tomb of Adonis, where the precinct enclosure mimics a transenna of marble.[121] In Byzantine church decoration, such geometricized veining also survived until at least the late eighteenth century, perhaps until the present day.

The choice of marble, a material scarcely available in the British Isles, but one whose image persisted in the bewildering descriptions of classical texts, still re-evoked matter that was intrinsically sacral because it recalled the Heavenly Jerusalem fashioned from glittering gems. At Durham, these patterns vary from column to column, as they do in the other contemporary and comparable English churches, though one assumes that by the time such patterns were cut in relief, they had lost any memory of their origins (see fig. 6.17). The particular variation of the columns in these English churches may also, as Eric C. Fernie extensively argued, have served to make them liturgical landmarks. At Durham the spirally grooved columns flanked the important Jesus Altar, and in Rome columns of prized marbles had long emphasized the site of the nave altar, or the division between the lay and clerical parts of the nave.[122] Whatever the case, the longer that zigzags and their like remained in use as a notation for polychrome marble, the longer they continued to figure the medium's miraculous image-making veining. Graphic evidence of this comes again from the canon tables of the Lorsch Gospels (c. 810), in whose columns of abstracted veining human faces are clearly perceptible (see fig. 6.19).[123] It is equally possible that the rigorous geometries of Durham's columns were the inevitable product of simulating the symmetry that had become so distinctive of "book-matched" marble revetment from the sixth century on.

This was visibly the case at the monastic Church of St.-Savin-sur-Gartempe (1060–1115) in the Poitou (fig. 6.20), the contemporary St.-Hilaire-le-Grand and

FIG. 6.20. Nave, St.-Savin-sur-Gartempe, 1060–1115. Fresco and paint on plaster, heavily restored in the 19th century. St.-Savin-sur-Gartempe, Poitou.

Notre-Dame-la-Grande in Poitiers itself, and several others where the nave columns are unmistakably painted in direct imitation of book-matched marbling (even though book-matched columns never apparently existed).[124] These churches must be considered the "missing links" between the sort of patterning one finds at Romanesque Durham and its ancient Mediterranean forebears.

ROMANESQUE FAUX-MASONRY AND ITS SIGNIFICANCE

St.-Savin's columns are painted in a warm and earthy palette of pinks and rust reds, but the walls behind them are masked by a layer of fine, white plaster, on which evenly spaced mortar joints have been carefully ruled to give the illusion of finely drafted masonry (see fig. 6.20). Such decorations once coated scores of Romanesque and Gothic churches, not just in the Poitou but throughout Europe and, as at St.-Savin, the faux-coursing frequently covered a real structure of finely drafted masonry. In fact, it is disconcerting to find in some instances (such as St. Alban's in Hertfordshire) that the faux-masonry is visually much inferior to the real thing, which is finely finished and well drafted, while the painted coursing meanders drunkenly over the surface.[125] Nonetheless, whatever the practice's technical shortcomings, it only seems explicable if the artisans' goal was to substitute all-too-earthly stones with the immaterial image of a more heavenly masonry. Although Paul had called the false priest a "whited wall" (Acts 23:3), in the Romanesque churches, representation would now triumph over matter.

More precisely, the origins of the phenomenon arguably lie in the Greek mistranslation of several Old Testament passages, subsequently perpetuated in Jerome's Vulgate, where the generic Hebrew "*Shayish*," which means only a whitish stone or just possibly gypsum, is rendered as "Parian."[126] From now on everyone took this as bible. Of course, no French or English monk had ever laid eyes on a block of Parian marble, nor would they have known it if they had. But they also had it on the word of Josephus that the Temple had been a brilliant white, even though he was actually describing Herod's building, not Solomon's. Moreover, many monastery libraries conserved copies of Pliny, or at least Isidore's *Etymologies* (c. 600/625), and these underscored Parian marble's glimmering and dazzling whiteness. The monks were happy to apply the name in precisely this qualitative sense, and consequently began to term any high-grade, white stone as "Parian." The glaring exception that proves the rule is Abbot Richard of Ware (d. 1283), who supposedly imported slabs of Thasian marble (from the North Aegean island of Thasos) to Westminster Abbey in 1259, but the terminological specificity seems more an optimistic literary allusion than any petrological determination.[127]

A church that was built, or appeared to be built, from "Parian" marble was desirable because it signified the eternal church, of which the Jewish Temple had been the archetype, and the brilliant body of Christ, the new incarnation. Prudentius had foreseen a new Rome devoid of paganism when "at last the marbles will shine because they are cleansed of all blood."[128] More importantly, the medieval commentators who succeeded him spared no opportunity to compare gleaming Parian to the spotless purity of the elect, thereby emphasizing how Solomon's Temple reflected the community of heaven in material terms. The Venerable Bede, in his treatise *On the Temple* (c. 729–731), was the first to assert, "It is obvious to everyone that the white marble of which the house of the Lord was built denotes the pure manner of acting of the elect as well as their conscience purified of all blemish of corruption." Bede also mentions that Parian "is born" (*gignere*), and the long-standing tale that it generated itself on mountain slopes, almost like snow, can only have added to its immaculate appeal. Bede's formula was disseminated by a number of other influential theologians, who often repeated it verbatim.[129] Hrabanus Maurus, writing about 844, only paraphrases him when he writes, "Parian marble is of a brilliant hue, and signifies the spotlessness of the elect. For which reason one reads that the temple walls in the Temple of Solomon were built from this very stone: which signifies that the holy church is constructed out of the spotless life of the saints and most excellent abode." Next, Raoul Glaber's *Historiae* (written 1046/50) used the white garments of Christ's Transfiguration (see fig. 2.27) as a metaphor for those with which the savior clothed Glaber's age. In 1003, he says, "It was possible to see almost everywhere in the world, but particularly in Italy and Gaul, the reconstruction of church buildings. . . . One might have said that the world itself stirred to shake off its old garment in order to cover itself everywhere with a white cloak of churches."[130]

Bede and Maurus had openly didactic aims, seeking in particular to educate the monastic communities that would one day commission or actually decorate buildings like St.-Savin. For this reason, they wished it to be obvious that God infused every block and cruet in the place of worship. As Bede exhorts us in a dedication homily, we should note carefully "the marvelous workmanship that went into the construction of the Lord's earthly house . . . so that these details, when spiritually understood, will arouse our minds to more ardent love of the heavenly dwelling place."[131] At St.-Savin the salient detail is the precision of the painted masonry, whose evenly spaced mortar joints simulate the ashlar-type that Vitruvius had called *"opus isodomum"* (*De Arch.* 2.8.6). Some monasteries certainly owned copies of Vitruvius, and many learned men knew the work either in toto or through excerpts.[132] Moreover, more than one abbot saw himself as a successor to Augustus and explicitly bragged that he too had found a city of brick (or wood) and left one of marble.[133] Yet the depiction of finely drafted masonry at St.-Savin, and churches like it, is arguably less a generic reclamation of antiquity than an extension of the much-traveled metaphor of the elect as the living stones of the eternal church, an aspiration that was voiced during every dedication.[134] In the hands of other theologians the metaphor could be surprisingly literal, since the saints were living rocks, honed by adversity, well-squared by tribulation and polished by virtue. The wall of the "living church" is therefore formed of perfectly squared and close-fitting blocks.[135]

In the church building, this picture of assembly could allegorize the sanctuary interior into a likeness of the universal church of living stones, whose unity was epitomized by their fine fit. The visual strategy was little removed from that of Augustan Rome, but now the solidity of masonry did not speak of the integrity of the imperial state but the totality of a super-populated heaven teeming with the white-clad elect, as in Paschal I's mosaic in the apse of S. Maria in Domnica (817–824; fig. 6.21). In other words, the proof of solidity was a jointing so fine that the masonry would appear to be ex uno lapide. These qualities were applauded in the Eastern Church. The Patriarch Photius (864) describes the façade of the Constantinopolitan Pharos Church in these very terms: for its "slabs of white marble, gleaming bright and cheerful, occupy the whole façade, and by their evenness and smoothness and close fitting they conceal the setting of one to another and the juncture of their edges, so that they suggest to the beholder's imagination the continuousness

FIG. 6.21. *Paschal I clasps the foot of the enthroned Virgin with Christ Child, surrounded by the angelic hosts.* Apse, S. Maria in Domnica, 817–824. Mosaic. Rome.

of a single stone with, as it were, straight lines ruled on it."[136] Nearer in time, William of Malmesbury makes comparable comment on the fabric of the new Salisbury Cathedral (1107–39), that it is "so befittingly composed from an ordering and joining of stones that it presses upon the gaze, and the whole wall feigns a single stone."[137]

Furthermore, at St.-Savin, the fictive blocks of the masonry are shown divided by double, concentric lines rather than single ones. These might be taken to represent thick mortar joints, but if the blocks are "living stones" then the convention represents the drafted margins of anathyrosis. This idea of valorizing the joint to evoke all the virtues of monumental masonry was understood from one end of Europe to the other, for it arguably underlies the "*cloisonné* masonry" of middle and late Byzantine churches in central and southern Greece (tenth to fifteenth centuries). The masonry envelopes of these churches evoke monumental masonry by outlining squared plaques of limestone with red brick courses and dividers. All the surviving diminutive churches of Athens, a city where the model of antique, monumental precision-masonry was particularly overbearing, are characterized by this technique. The most relentless example is the Panagia Parigoritissa at Arta (c. 1290), and the earliest the Theotokos Church at Hosios Loukas (946/955), where the brick interstices gradually cajole the rough blocks into a harmonious coexistence (fig. 6.22).[138]

FIG. 6.22. Theotokos Church, 946/955. Hosios Loukas, near Steiri, Greece.

By the same token, the insubstantiality of the faux-masonry in the Romanesque interior was ultimately in keeping with the precepts of parallelism on which it depended. It should be a pale reflection, because the church interior was not heaven itself, but merely a likeness, one of many "heavens on earth." As one particularly expressive twelfth-century chronicler puts it:

> I have actually considered it superfluous to say anything about material buildings, since every day they are renewed, the old ones destroyed and the new ones built, and, God willing, continually changed for the better. On the other hand, one must know about spiritual buildings because this house is founded on a firm bedrock, which is Our Lord Jesus Christ, which is raised on seven columns and rises through the clouds, exceeds the stars, and stretches right to the throne of the Supreme Judge, where the King of Kings wields the scepter, manages the reins of the world, and steadily regulates its speeding course. In it, great stones, living stones, precious stones are hewn, squared and polished, from which the heavenly Jerusalem is built so as to ensure the participation of the citizens in its own self.[139]

This passage is heir to a series of Early Christian texts, which posit an immaterial heaven in order to anathematize the material splendors coveted by this world. In the eyes of some observers and the hands of some artists, heaven could reside in the material world only through metaphor or allegory. For others, the ultimately insensible heaven could not be figured in any material at all. Thus Prudentius, the poet who penned so many evocative descriptions of gemmed edifices, can also compose the following thoughts, in which rhetorical topoi are inverted to constitute virtually an anti-ekphrasis.[140] In this text (c. 402/403) Prudentius argues for a temple not of marble, but of the heart (*c. Symm.* 2.244–55):

> Come then, oh mortal, build a temple to me alone
> And worship me as the sole God. I seek no quarried
> stones,
> Neither the rock that Paros nor the Punic cliff cuts,
> Nor those that green Lacedaemon or veined Synna
> possess;

> Let no man consecrate natural red stone to me.
> I love a temple of the mind, not one of marble.
> In it the golden foundations of faith stand true,
> The edifice soars shining with holiness snow-white,
> Righteousness coats its lofty roof, and within
> Life-giving purity dyes the floor with blushing flowers
> Of modesty scattered about, and keeps the courts.
> This is the house that befits me, the beauteous abode
> Worthy of its eternal and heavenly guest.

The purity of the heavenly masonry is wonderfully allegorized in a still more visionary passage from a very Early Christian tract, *The Shepherd of Hermas.* This Greek text dates back to the late first or mid-second century, but was soon translated into Latin, became canonical, and enjoyed wide diffusion. In the parable, virgins build a heavenly tower (the ideal church), gathering

> from all the mountains stones of varied colors, hewn by men. The maidens carried them through the portal and delivered them for the building of the tower. And whenever these stones of varied colors were set in the building, they all became identically white, shedding their former colors. But some stones brought by the men for building did not become radiant but remained such as they had been before they were set: for they had not been handled by the maidens or passed through the portal. Such stones had to be removed and set aside that the maidens might handle them: for if they are not brought through the portal by the maidens' hands, they cannot change color.[141]

Not only does all color bleed away as the blocks coalesce into the ideal church, but the author concludes, "They adhered so closely one with another that their joining could not possibly be detected and the building of the tower appeared as if it were built of one stone."[142]

SUBSTITUTE STONES

All this said, polychromy had not disappeared, nor the desire for colored stones. In continental Europe there were still plenty of ruins from which marbles could be salvaged for skirtings or floors, and a motley array of columns had been recycled to build churches in the early Middle Ages. However, the stocks were constantly diminishing, and a more coherent architectural style, the Romanesque, replaced the fetish for monolithic columns with masonry piers, banishing the optical havoc of ad hoc spolia assemblies that was all too redolent of worldly vicissitudes. The materiality of these new interiors could be "upgraded" with paint and, in some very rare cases, such as St.-Savin-sur-Gartempe (see fig. 6.20), marble columns were simulated. The practice may have been more widespread than today's physical remains suggest, for recipes certainly circulated, one of them surviving in the collection of "Pseudo-Heraclius," a twelfth- or thirteenth-century tome that is variously said to have been compiled in northern France, England, or eastern Italy.[143]

In the British Isles, Roman buildings had never enjoyed the material opulence of their cousins in the other European provinces, so scavenging marbles was not a possibility anyway. When Roger, prior of Durham (1137–49), decided to pave the church with marble, "because he had heard that foreign churches in overseas countries were resplendent in the beauty of such work," he even went to the lengths of convening continent-bound pilgrims to procure him individual slabs. One poor soul, Harpin, a knight of Thornley, trudged all the way to Rome and back, returning with a slab in the lap of his saddle, which the friars of the church greeted with such devotion that they wore it down with their rubbing fingers.[144] This is characteristic of the new marble trade of the Middle Ages, the transit of fragments, of which portable altars are a conspicuous phenomenon (fig. 6.23). These were porphyry or serpentine slabs, more rarely veined marbles and *breccias* (see fig. 7.18), richly mounted in gold and silver, and often with an accompanying inscription.[145] These slabs were marvels of nature, relics and indices of holy places (one inscription makes sure to remark, "This stone is from Rome," which is as much an idea as a place). They were precious in themselves, like red gold, but they were all the more precious because they connoted the imperial purple; thence also Christ's robe at the Mocking; and by appearance alone, but also the fallacy that the Tyrian dye had been the blood of mollusks (instead of their glandular secretions), Christ's precious blood as well (some inscriptions mention "*crus,*" meaning "gore"); and because it was a stone anyway, then Christ the cornerstone; or the Stone of Unction on which his body had been laid, or his sepulcher; and so on, and so on. All of these allusions made porphyry tablets intense worlds of tropological density when functioning as platforms for the Eucharistic sacrifice.

And this medium-message could and did travel. Tiny fragments have been discovered as far afield as Armagh, and in 1071, for example, Abbot Baldwin of Bury St.

FIG. 6.23. Anglo-Saxon portable altar, 11th century. Porphyry and gold, 10½ × 5⅜ in. (26.6 × 13.7 cm). Musée National du Moyen Âge, Hôtel de Cluny, Paris.

Edmunds had received just such a porphyry altar from Pope Alexander III while in Rome.[146] The slab that Harpin of Thornley brought back from Rome may well have been a portable altar as well.

One slab, of course, would not fulfill the aspirations for a complete heaven on earth, except in microcosm, so the next prior of Durham (1153–95), Hugh de Puiset, who had been consecrated in Rome (1153), "decorated the whole building with marble from afar," and adorned the resting place of the Venerable Bede with precious stones, importing the "columns and bases of marble from beyond the sea."[147] The columns in question still exist in the Galilee, and are made from Purbeck marble, quarried on the isle of that name (in reality a promontory), near Corfe in Dorset, at the other end of the country.[148] It was "beyond the sea" only in the sense that the stone had to be shipped to Newcastle before traveling overland to Durham.

Purbeck marble is actually not a marble at all, but rather a limestone capable of taking a high polish, which ranges from blue-gray to reddish-brown, green, and deep purple. Purbeck's range of hues hinted that it could be the indefinable "Jasper" mentioned throughout the Bible (and which has nothing to do with modern definitions of the mineral), but the purple kind was very likely regarded as a substitute for imperial porphyry. On other occasions (e.g., Worcester Cathedral) the stone was blackened to resemble black marble.[149] Various other limestones were also exploited in England as "marbles," and finding substitutes for the ancient marbles no longer available was widespread across Europe.[150] In the Rus, where antique marbles had been sporadically imported from Byzantium for use in floors, local materials or ceramics were enlisted after the eleventh century to imitate them.[151] Masons also prospected various types of red slates and sandstones.[152] In Hungary too, from at least the twelfth century, red limestone was quarried, probably at the behest of King Béla III (r. 1172–96), who had been raised in the Byzantine court, where he would have known porphyry well.[153]

In England, Purbeck *qua* porphyry came to serve for the tombs of kings, knights, and bishops, and eventually churches too.[154] The earliest use of major church piers of Purbeck is in the Temple Church in London (built c. 1160, consecrated 1185), which was a conscious recollection of the great rotunda of the Holy Sepulcher in Jerusalem, as well as the richly marbled Dome of the Rock, now misidentified as Solomon's Temple (fig. 6.24). In the late twelfth century, Purbeck was also being used to make columns at Canterbury that apparently copied classical proportions as well.[155] Between roughly 1250 and 1350 it came to be used in almost all the cathedrals of southern England and Wales, including Exeter, Ely, Norwich, Chichester, Salisbury, Lincoln, Llandaff, Southwark, Canterbury, and Westminster Abbey.

What only added to Purbeck's appeal was its fossiliferous nature (it contains the snail *Vivipara carinifera*), which made it a potent image of Nature the creator and God's presence in the world. The perception of marble as a form of painting and an image-maker also remained in full force, as can be seen in canon tables of the Lorsch Gospels (see fig. 6.19) and no doubt in church murals, such as the little-known *Christ between Six Saints* (1255) that survives in SS. Giovanni e Paolo, Rome (fig. 6.25). The painter took some care in imitating a breccia for the arcade columns shielding the protagonists, with a barely perceptible spiraling of the shaft, within which human profiles, lizards, crabs, and fish appear.[156] The promise of Purbeck, that it could be both heavenly gem and natural painting, is explicit in Henri d'Avranches's description, written about 1235, of the Purbeck columns in the choir of Lincoln Cathedral (built between 1192 and 1210):

> When this stone is examined it may leave minds
> Uncertain whether it be jasper or marble; but if
> Jasper, a faint jasper; if marble, a noble marble.
> In that place are columnettes, which have so girded
> the columns
> That they seem to perform some ring-dance there.
> Their outer surface, more polished than a fresh-grown
> fingernail,
> Opposes the bright stars to the reflected rays of sight,
> For chance [*or* nature] has painted there so many varied
> forms that,
> If art toiled to copy a similar picture,
> However sustained the endeavor, it would barely
> imitate nature
> [*or* it would be as futile as plowing sand].[157]

Henri d'Avranches was an itinerant priest and no mean wordsmith. When he turns to the rest of the cathedral, he resumes all the topoi that reach back to Bede, and Augustus long before him, with words that could also have been written about St.-Savin-sur-Gartempe (see fig. 6.20). The cathedral has its feet firmly planted but it "jostles the

FIG. 6.24. Rotunda, Temple Church, consecrated 1185. London.

clouds," and its vault spreads out like the wings of a bird. As to the walls:

> The sticky lime glues together the white stones,
> all of which the mason's hand has cut to the utmost
> precision
> And although the wall is constructed from a pile of them,
> it is as though it disdains this, and feigns a continuum
> of contiguous parts; it seems to exist not by art
> but by nature; not a thing unified, but a single entity.

After his stirring itinerary of the whole cathedral, Henri d'Avranches itemizes the allegorical meaning of every component of the church, the stones and "marbles" included, and concludes with a grand synopsis (*Consummatio Totius Allegoriae*) that states, predictably, "Thus do the senseless stones conceal the mysteries of the living rocks" (*sic insensibiles lapides mysteria claudunt / vivorum lapidum*).[158]

This is not quite the end of the story in England. In 1268 Westminster Abbey received a magnificent Cosmati floor below its crossing tower (fig. 6.26), almost twenty-five foot square, with a wealth of classical marbles inlaid

FIG. 6.25. *Christ between Six Saints,* 1255. Fresco. SS. Giovanni e Paolo, Rome.

FIG. 6.26. Odoricus, Cosmati floor, Westminster Abbey, 1268. Egyptian alabaster, porphyry, Spartan green porphyry, and yellowish limestone with red, turquoise, cobalt blue, and bluish white glass in a Purbeck marble matrix, 24 ft. 10 in. (7.58 m) square. London.

FIG. 6.27. Chancel, cathedral, Ávila, Spain (begun 1172).

in the most intricate patterns by a workshop that traveled especially from Rome, and an equally elaborate inscription illuminating its meaning: "If the reader prudently considers all that is set down he will find here the end of the *primum mobile*," and, around the central rota, "The sphere shows the archetype, this globe shows the macrocosm."[159] In other words, the central rota figures the archetype (the idea of the world in God's mind *before* its realization in matter), and the marginal rotae the elemental macrocosm (the sensible world formed from the four elements). But this floor was a royal commission, from Henry III, and it remained absolutely unique in the British Isles.

Looking for churches elsewhere in Europe that were built from colored stones is no easy task. We have mentioned S. Marco in Venice, but let us conclude with a building far less famous but no less exceptional, the cathedral at Ávila in central Spain (begun 1172). The body of the church is made from Caleño granite, quarried at La Colilla, just outside Ávila, the same stone used for the walls that still encircle the town, an ocherish stone that comes alive in sunlight, but lies dormant in the shadows of the church interior. As is normal still in Spanish cathedrals, any view of the chancel is blocked off by a canons' choir that is a freestanding building in itself and occupies the head of the nave. It comes as a considerable surprise to skirt this obstacle, venture into the sanctuary, and be met with the vision of walls and vaults built from another stone that is bloodshot with blotches and splashes ranging from rust-red to violet (fig. 6.27). The previously stolid walls now seem to spark and flicker before the eyes. This is *piedra sangrante* ("bleeding stone"), a granite like no other, laced with iron oxyhydroxides that give it this seeping mottle.[160] Externally, this same chancel and its apse created a massive bulge in the city walls (fig. 6.28) that made the church into a rampart, just like the Heavenly Jerusalem in Revelation (and also Isaiah 54:11–12, "I will make your bastions of jasper"), and placed it under the protection of the almighty. That this spattered stone made up the internal walls of the chancel is no coincidence. Here was the site of the Eucharistic sacrifice, where bread became flesh, and where the transformation of wine was as miraculous as getting blood from a stone.

FIG. 6.28. Model of cathedral and town wall. Ávila, Spain.

7 Hagia Sophia and Byzantium

By the fifth century, marbles adorned the churches of the fractured Western world only where they could be scavenged, but in the Eastern empire a few quarries continued to function, as did an imperial treasury ready to subsidize parades of marble long after the Western empire had fallen. Churches built in Byzantium, or by Byzantine workmen beyond its borders, have fared far better than the church incrustations of Western Europe, yet, even so, the remains are still just a fraction of the original output.[1]

THE REMAINS (324–1453)

From the moment that Constantine transferred the imperial capital to the site of Byzantium in 324 and renamed it Constantinople, marbles were lavished on public buildings, palaces, and churches with no less zeal than they had been in old Rome. Of the palaces and civic buildings, next to nothing has survived the successive depredations of the Latins, Ottomans, and even the Byzantines themselves. Of the churches, only about thirty have come down to us from an estimated five hundred. Of these, three retain largely intact marble revetment: Hagia Sophia (532–537), the Theotokos Kyriotissa Akataleptos (Kalenderhane Camii, c. 1200), and the Chora (Kariye Camii, c. 1316–21). Columns and floors, but no revetment, can be found in the Church of St. John Stoudios (454/463), the Pantokrator Church (Zeyrek Camii, 1118–36), and the Parekklesion of the Theotokos Pammakaristos (Fethiye Camii, c. 1310–14).

Outside the capital, one may count the intact apse of the Euphrasius Basilica at Poreč in Istria (530/550; see fig. 6.5);[2] S. Vitale in Ravenna (532/533–548), whose cladding has been largely lost but can be restored with some confidence;[3] the apse of St. Catherine's below Mt. Sinai (548/565); the extensive revetments of Hagios Demetrios in Thessaloniki (c. 500/c. 620), largely destroyed in a 1917 conflagration but known from survey drawings and old photographs;[4] the near-intact ones in the Nea Moni on Chios (c. 1042–55)[5] and in Hosios Loukas, near Athens (1011/48);[6] the meager remains in the Church of the Dormition at Daphni (1081/1111);[7] and the opus sectile revetment in the presbytery of the Theotokos Chrysokephalos in Trebizond (1214/35).[8]

In addition, a series of buildings survive that are not strictly Byzantine, but lay within the empire's orbit and employ the same material vocabulary. These include

S. Marco in Venice (c. 1084–c. 1260), which emulated Constantinopolitan churches and was possibly revetted by Byzantine craftsmen;[9] Monreale Cathedral (1172–16) and the Cappella Palatina (c. 1140–43), in and around Palermo, which also followed Byzantine-Arabic models;[10] and the most important surviving Ummayad constructions, the Dome of the Rock (692–702) in Jerusalem and the Great Mosque at Damascus (709–715).[11] The marbled reconstruction of the most sacred of Muslim constructions, the Ka'bah in Mecca (693), by the same Ummayad caliph who commissioned the mosque at Damascus, Al-Walid ibn Abd al-Malik (668–715), has not survived but evidently followed suit.[12] Indeed Byzantine marbles, and possibly marble-workers, may have traveled as far afield as Baghdad and Kiev.[13] Finally, a great number of marbled floors have survived, too numerous to list, while still others are known from old drawings and photographs.

EKPHRASES AND OTHER TEXTS

The most important and most influential of all Eastern churches was the cathedral of Byzantium, Hagia Sophia (532–537; fig. 7.1).[14] This edifice stretched contemporary technology beyond all known limits, setting an unreachable bar for all later Byzantine churches, and was still the gold standard when Bramante set about reconstructing St. Peter's in 1506, when Sinan raised the Süleymaniye Mosque in 1550 to 1557, and when Borromini built his own temple of wisdom, S. Ivo alla Sapienza, in 1643 to 1660.[15]

Of all Byzantine buildings, Hagia Sophia is the sole survivor that can now be measured against descriptions in a spectrum of contemporary texts. Besides the usual chronicles, these divide into ekphrases, exegetic hymnody, folkloric descriptions, and travelers' accounts. Since these eyewitness sources are vital to understanding contemporary perceptions of Hagia Sophia, the whole miscellany requires a brief introduction.

The literary compositions known as ekphrases survive in quantity from the sixth to fifteenth centuries and cast a raking light on the aesthetic and symbolic criteria of all artistic production, including architecture, and with frequent emphasis on marble work. Three survive for Hagia Sophia: those of Prokopios (c. 554/560), Paul the Silentiary (563), and Michael the Deacon (c. 1140/50). Until recent years, scholarship was prone to marginalize such ekphrases as insufferably long-winded and obscure. In large part, this myopia was based on three misconceptions: that ekphrasis was a precious and self-indulgent academic genre; that its excessive literary wrapping stifled any would-be objectivity; and that any departure from pure description was gratuitous because ekphrases exclusively treated artworks. Instead, we now know that such compositions stemmed from a broader species of rhetorical exposition, itself a subsection of panegyric, within which architectural ekphrasis belonged to the specialized evocation of places (*topoi*).[16] Periphrasis and periegesis were endemic, and empiricist objectification was antithetical because ekphrasis—literally "a telling in full," but "vocalizing" would be a more immediate equivalent—intended "a speech [about persons, places, times, events, and later plants, animals, and festivals] which leads one around, bringing the subject matter vividly before the eyes."[17] In fact, as Ruth Webb importantly clarified, the modern "distinction between narration (unfolding in time) and description (the problematic verbal representation of an object, constituting a 'narrative pause')" was in any case irrelevant to periegetic descriptions whose success depended chiefly on their degree of *enargeia* ("vividness").[18]

Not only that, but ekphrasis was a technique of evocation that aimed at *interpreting*, not *describing*, what the audience saw, compensating for the limitations of vision by emotively addressing the *subject* of representation. In the words of Nikolaos Mesarites (1198/1203), which are much quoted, the goal was "to look at things with the eyes of sense and to understand them with the eyes of the spirit . . . to understand ultimate things and to enter the secret places."[19] Indeed, some ekphrasists render the verb "to see" with the ambiguous "*noein*," a verb that denotes both physical and intellectual perception.[20]

Furthermore, ekphrases enjoyed *ut pictura poesis*, a performative function parallel to the artwork. Ever since Statius had sung for his supper by praising the marble baths of his host between courses, ekphrases were declaimed within the actual buildings to audiences who had every opportunity to compare the words with the facts.[21] Thus, Paul the Silentiary's ekphrasis of Hagia Sophia in 563 was pronounced from one of the church galleries to the assembled court and clergy with a grandstand view of the interior.[22] Moreover, not only were ekphrases like his actually intoned at a church's consecration, but some were then even inscribed on its walls, so that the building spoke for itself and its words returned to life each time a visitor recited them.[23] At least one such verse inscription, engraved in the atrium of

Hagios Polyeuktos (524–527), a church whose Solomonic pretensions Hagia Sophia sought to surpass and supplant, introduced the marble manuscript of the church interior (see below). Ekphrases were, therefore, anything but textual adjuncts after the constructional fact, and enjoyed a more reflexive relationship. Indeed, architectural ekphrases as well as Western tituli (dedicatory inscriptions) also enjoyed independent circulation as models for composition.[24]

Two other sorts of texts supply parameters for the symbolic understanding of the building and its materials, especially at Hagia Sophia: exegetic descriptions and consecration hymns. The former begin as early as Eusebius's panegyric on the church at Tyre (c. 317), which is carefully crafted to underline parallels with Solomon's Temple.[25] The latter, consecration hymns, are particularly important because, like ekphrases, they cannot be dismissed as ex post facto "poetic fantasies" either. They were specifically devised for the consecration—the moment when the supreme religious function of the church was expounded, when a building stopped being just a building and became a sort of time-space portal—and were sung to the court and populace, who often signaled their assent in choreographed responses. The most densely allusive such hymn is that for the dedication of the cathedral at Edessa, commemorating its Justinianic reconstruction under Bishop Amazon (or Amidonius) in c. 543 to 554.[26] This hymn has particular bearing on Hagia Sophia because, despite its Syrian origins, the church in question was almost contemporary, built from imperial funds, domed, marbled, mosaicked, and probably enjoyed the same dedication—to Holy Wisdom.[27] To some extent this hymn also fills the vacuum left by the loss of the original consecration hymn for Hagia Sophia (537), although this loss is more than compensated by the survival of the elaborate *kontakion* (canticle, or homiletic anthem) for the second inauguration, on Christmas Day 562, publicly delivered a few days before Paul the Silentiary's panegyric, and celebrating the reconstruction of the dome after its partial collapse in 558.[28]

In the final categories, popular description and folklore, a plethora of pilgrims' accounts survive. One Turkish history even purports to record the stunned reaction of Mehmet the Conqueror when he entered Hagia Sophia for the first time on the day of Constantinople's fall.[29] But most important is the ninth-century *Diegesis* (or *Narratio*) of Hagia Sophia, supposedly a chronicle of its construction, which is accurate in several respects but woven with legendary events and peppered with interpretations that, however fantastic to modern ears, record the popular theology the building absorbed, transmitted, or instigated.[30]

HAGIA SOPHIA: WHAT'S IN A NAME?

The emperor Justinian built Hagia Sophia as a cathedral for a rejuvenated empire and as a national unity monument. Abroad, his generals had recovered swathes of the Western empire lost to the barbarians a century earlier, and at home he had brutally suppressed the Nika revolt (January 532) that had almost dethroned him and burned the old church to the ground. The new building that rose from its ashes became the mascot for a rejuvenated empire under one law and one religion, headed by this "Great Church" as its new Capitolium.[31] We can sense better the all-embracing nature of its massive construction if we remember that, almost two centuries earlier, Ammianus Marcellinus had marveled that the Pantheon seemed to encompass an "entire quarter," within a city that itself seemed the "temple of the whole world."[32]

However, for the new building to become the heart and soul of empire it must tempt down the almighty to reside in it and live amongst men. "But will God indeed dwell on the earth?" Solomon himself had asked, on completing the Great Temple in Jerusalem; "Behold, the heaven and the heaven of heavens cannot contain Thee; how much less this house that I have builded?" (1 Kings 8:27; 2 Chronicles 2:6). St. Paul echoes the sentiment in the New Testament when he specifies, "The God who made the world and everything in it, He who is Lord of heaven and earth, does not live in shrines made by human hands" (Acts 17:24; cf. Acts 7:48). The Edessene Hymn, sung at the dedication of the cathedral there in c. 543 to 554, also implies that for any temple to house the uncircumscribable divinity is a paradox equal to the Incarnation itself.[33] A true church must be soul architecture.

Skill, not just scale, were needed to make a container for the uncontainable. The most authoritative precedent was the home that God had made for himself, the cosmos, for he himself had said, "Heaven is my throne, and the earth is my footstool" (Isaiah 66:1–2). The historical next precedent had been set by God himself when he became incarnate in Christ, also an image not made by human hands, and one that had come into being within the Virgin, who was herself a vessel for divinity. The actual buildings

FIG. 7.1. Nave, view toward apse, Hagia Sophia, 532–537 AD. Istanbul.

FIG. 7.2. Olafur Eliasson, *The Weather Project*, 2003. Monofrequency lights, projection foil, haze machines, mirror foil, aluminum, and scaffolding, 87 ft. 7 3⁄16 in. × 73 ft. 1 15⁄16 in. × 509 ft. 10 1⁄8 in. (26.7 × 22.3 × 155.4 m). Turbine Hall, Tate Modern, London.

that had been hotels for the almighty were Moses's tabernacle in the desert (Exodus 25–31; 35–40), and Solomon's Temple in Jerusalem (1 Kings 5:15–18:66; 2 Chronicles 1–7:10), and although these had been manufactured, their design was divinely mediated. The sum of these archetypes predicted that if a church were contrived with the right materials and sufficient sleight of hand, God would always feel at home there.

Furthermore, to achieve such a vision of the heavens, in analogy to the Incarnation, the credibility of that vision also depended on evading blatant evidence of earthly substance or artifice (or, in rhetorical terms, reductio ad materiam and reductio ad artificium). Hagia Sophia must seem a temple made by natural or supernatural forces, the materials must fit together in harmony as if by themselves, the earthly designers must seem not to create but only to dispose, and the proof of their success should be a product of literally unspeakable perfection. Alongside these conceits thrived the usual topoi: that the church was a paradise and its walls and floor meadows of gems, or that it was as much a microcosm as the cosmos was God's vastly expanded temple.[34] And while any church was typically a "heaven on earth," a new Temple of Solomon, Mosaic Tabernacle, microcosm, or all of the above, Hagia Sophia and those churches that shared its dedication had particular claim to represent "Wisdom building Herself a house."

HAGIA SOPHIA AS THE WORD MADE FLESH

The key to understanding Hagia Sophia in fact lies in its name. For Hagia Sophia was dedicated neither to a saint, the Virgin, nor the person of the savior. It was dedicated to Wisdom as Christ, a concept incarnated in a person.[35] The earliest writers call it "the Great Church," "Wisdom," "Holy Wisdom," or "the Temple of Wisdom," but later sources expanded this to "the Great Temenos of the Wisdom of the Logos of God" (*to mega temenos tēs tou Theou Logou Sophias*). This more exegetic title married St. Paul's definition of two wisdoms: the original Wisdom that had shaped and infused the universe, and the Incarnation of Wisdom in the world through Christ, the Word made flesh (*logos sesarkomenos*).[36] The upshot was that any church dedicated to Wisdom automatically enjoyed a range of reference spanning from the cosmos to Christ's own body, and broached the contingent themes of Creation, Incarnation, and Christ's two natures.

Indeed, the *kontakion*, the anthem sung at Hagia Sophia's second inauguration (562), explains that only because Christ once dwelled in the flesh will God now consent to indwell the Temple of Hagia Sophia: "Having once resided in flesh the Word consents, by the operation of the Spirit, to reside in temples built by hand, assuring His presence by mystical rites; and He who cannot be contained, nor even approached, by the whole universe lives by grace among mortals. And not only does the Heavenly One share a roof with those on earth, but he welcomes them as table[-fellows] and entertains them with the banquet of His Flesh, which is set before the faithful by Christ" (strophe 4).

Furthermore, because Christ had spoken of his body as a temple (John 2:19, 21), Paul later insisted that his risen body had actually substituted the Old Temple as "a greater and more perfect tabernacle, not made with hands, that is to say not of this building" (Hebrews 9:11; see also Mark 14:58; 2 Corinthians 5:1; Colossians 2.11). Conversely, the indwelt or ensouled temple (the "*empsychos naos*," as the exegete and historian Eusebius calls a church at Tyre) could be compared to "the spiritual edifice in our souls." Indeed, Plato had described the body as a "vessel for the soul" (*to tēs psuchēs . . . kutos*), and one Christian poet, Ennodius, had praised another, Avitus, as a man "in whom [human] wisdom enclosed herself as if in a shining abode."[37]

Further to this, because Proverbs (9:1) introduced the idea of "Wisdom having built herself a house," this architectural image could be taken to reconcile the structure of the cosmos, through the Incarnation, with Jesus's persona as a "temple." For this very reason, the same verse was declaimed at both Marian feasts and at the consecration ceremonies (*enkainia*), in which the supreme divinity was invited to inhabit his earthly vessel. The bow of this internal loop was tied by the fact that the book of Proverbs had been written by Solomon, the builder of the Ur-Temple and himself a byword for Wisdom. All of this meant that, in the religious imagination, the minute one built a church and dedicated it to Wisdom, a wealth of symbolic factors became effortless expectations, overlapping, fusing, and defying segregation. It is this same polysemy that is reflected in the ekphrastic texts, which—not so differently from Sumerian hymns—come to itemize the architecture (and acknowledge the materiality) of Hagia Sophia.

Moreover, the exact nature of Christ's Incarnation was on everybody's minds when Hagia Sophia was being designed in 531/532. The best part of a century had passed since the Council of Chalcedon (451) had rejected

Nestorius's segregation of Christ's human nature from his divinity (the Logos), and instead embraced Cyril of Alexandria's counter-position that Christ's two natures existed in fusion. But the council's ruling failed to pacify any parties loath to accept that the Trinity could ever have suffered in human flesh. Only once Hagia Sophia was finished did the maverick theologian Leontios Byzantios cut this Gordian knot—and found today's Orthodox theology—in his *Three Books against the Nestorians and Eutychians* (540/543).[38] His solution is too abstruse to detail here, but suffice it to say that he traced "all things in being" (*panta ta onta*) to a unifying substantiality (*ousia*), whereby the relations between *natures* and *persons* could result both in the union of diverse persons in the same nature (i.e., the Trinity), as well as the union of diverse natures in one person (Christ, the Word made flesh). Christ as the Word made flesh therefore unified divine and human natures within himself (*enypostatic* union) and mediated between God and humanity (*hypostatic* union), to become the ultimate *medium* between divine and human, universal and particular, metaphysical and substantial. Through this absolute mediation, he derived his full soteriological power to unite mankind with the Logos.

Leontius's difficult metaphysics is of particular interest because, in 532, just before the old cathedral went up in smoke and the new one rose from its ashes, he joined the team of Orthodox bishops who convened at the adjoining imperial palace to debate Christ's nature with their Monophysite counterparts. Although the meeting proved abortive, it is undeniable that Christ's exact nature was more than normally topical when it came to designing Hagia Sophia—the Holy Wisdom of the Word made flesh—a few months later.

LIGHT AND SURFACE

The most apparent expression of Christ the medium in Hagia Sophia was the light that infused and saturated the interior. Not only was Holy Wisdom the lamp of truth but the first of God's creations, brighter even than the first light. It was the light that had inseminated the Logos in the Virgin (Ecclesiastes 1:3; Luke 15:8; Proverbs 8:22–31; Wisdom 7:24–25), an event to which the ekphrasist Michael the Deacon presumably alludes when he characterizes Hagia Sophia as the "life-giving Queen" or claims that it "could be pregnant with many thousands of bodies."[39] Likewise, Venantius Fortunatus, who arrived in Gaul from Ravenna in 565, with the city's Byzantine monuments fresh in his mind, would pronounce that a church in Bordeaux shone "like the womb of the Virgin."[40] Everyone knew that a luminous, God-filled cloud had filled Solomon's Temple at its consecration (1 Kings 8:10–13; 2 Chronicles 5:13–14), and this image remained so powerful in the Byzantine mind that, on the eve of Constantinople's eventual fall in 1453, the populace would lament that the almighty had forsaken them when they saw a thick fog shroud the city only to drift out to the Sea of Marmara and there disperse.

Wisdom infused the universe and was of a piece with the incarnated Christ. So, if we can think of Hagia Sophia as an elusively divine body, we must remember that the body in question belongs to the larger Creation, and that its flesh takes on as many varieties as the products of Creation.

Hagia Sophia was conceived as a concrescence of light. The only way that this Wisdom could be apprehended within the building sentiently, and to "spiritual eyes," was through the emanation and manifestation of light.[41] All the ponderous apparatus of engineering was pushed to the exterior, where immense masses aggregate to buttress the dome. Within, by contrast, is a delicate vessel of light that cascades from a dirigible dome. The court poet Paul the Silentiary said that looking into it was like gazing at the midday sun (*Descr. S. Sophiae.* 671–72), and the historian Prokopios that "nothing so beautiful could possibly be supported on piers *raised by human hands*, but must be suspended from heaven on a golden chain" (*Aed.* 1.1.34, 46), recalling the notion that, unlike any human structure, the Creation had begun on high and the cosmos built from the top down. Although there was no end of columns in the galleries, there was no attempt to align them with those below, and as Paul says, the designer "has not hesitated to fix their bases over empty air" (*Descr. S. Sophiae.* 394; see fig. 7.4). Furthermore, as Anthony Kaldellis has pointed out, Prokopios's "golden chain" is a quotation from Homer (*Il.* 8.19) dear to neo-Platonic metaphysical exegesis because it was believed to allude to the Platonic chain of being that connected all levels of reality, from the heavens to the earth and sea.[42]

Prokopios also resorts to a well-worn topos when he adds, "Its interior is not illuminated from without by the sun, but the radiance comes into being within it, such an abundance of light bathes this shrine." To see through his eyes we must think of something like Olafur Eliasson's *The Weather Project* (2003), in the Tate Modern's Turbine

Hall (fig. 7.2). Eliasson installed a mirrored ceiling and a half-circle glowing lamp at the end, which, through reflection, constituted a full sun that neither rises nor sets, within an environment in which the viewer was immersed thanks to tiny vapor jets spraying the hall with a fine mist. In Hagia Sophia the luminosity was instead sustained and multiplied by the varying reflectivity of every single surface.

Inside Hagia Sophia the great canopy of light comes to ground as a tetrapylon inscribed within an aisled box. Its original appearance is best appreciated from Henri Prost's worm's-eye view from the early 1900s, itself possibly inspired by recently republished ekphrases (fig. 7.3).[43] On its way earthward, the massive dome buds into clustered hemidomes over the trefoil exedra at either end of the nave, all covered in millions of gold-backed tesserae. The mosaics offered both a superficial transparency (glass) and an ulterior reflectivity (gold leaf), and this paradoxical non-surface in a non-color made the divine essence visible but denied it any substance, except the light-charged air, a constituted radiance.[44]

Upon the building's completion in 537 AD, all mosaics in Hagia Sophia were aniconic, and the expansive vaults were studded everywhere only with talismanic crosses. Neither figures nor narratives were meant to trivialize the primacy of an otherworldly vision. There was no Pantokrator in the dome. The only figures that may have existed in the original building were the cherubim in the pendentives (although those we see now date from much later), because cherubim were the only figurations that God had ordained for the Ark of the Covenant, preserved in the Holy of Holies in Solomon's Temple. These cherubim overshadowed the Ark's lid, the so-called Mercy Seat, from which the almighty would deliver his commands to the Israelites. The cherubim of Hagia Sophia are also those that appear *under* the throne of God in the vision of Ezekiel, which describes the spiritual temple that is only dimly foreshadowed by the material one. These two worlds synchronized and meshed during the divine liturgy, when the priests of Hagia Sophia concelebrated in unison with the entire heavenly host. The cherubim that adorn the great central baldachin of Hagia Sophia

FIG. 7.3. Henri Prost, Worm's-eye view of Hagia Sophia, 1905/11. Académie d'Architecture, Paris. III6 HP.DES.10/9.I.

therefore both signify the church as the place of direct communion and the earth as God's throne.

In Hagia Sophia, the original dome was also twenty feet lower than the present one, with the result that it continued the curvature of the pendentives in one seamless arc.[45] The same negation of structure can be observed in the way that the huge tympanum windows were set flush with the interior wall plane, allowing no perceptible reveals; that many of the window transennae (solid panels) were of translucent marble; and that the vaults of the upper galleries were covered with dark-blue tesserae with few gold tesserae, to make the arcade in front a tracery. Moreover, the marble paneling coats the nave walls with no regard to structural expression, like a patchwork quilt. Prokopios adds that the marbles shone more brightly than the vaults, again effacing terrestrial substance in favor of heavenly shimmer, and, as Michael the Deacon writes, "What manner of stonework is this that has fastened around the building, striving with its variegated coloring and smoothness against the gold shining because of its smoothness and, because of its diversified bloom having something that surpasses even gold, which is of one color?"[46]

The same applied to the only exterior that everyday visitors could actually see. Like the Pantheon in Rome, the bulk of Hagia Sophia was largely occluded by adjoining buildings (the Great Palace, secretariat, library, and other satellites), and at close quarters, in the large but shallow forecourt, it was therefore not the dome but the west façade that loomed large. While the atrium walls extended the varicolored paneling of the nave,[47] the upper register of the west façade was clad with Proconnesian panels, some of which still survive in situ.[48] These polished, white panels streaked with gray could again only dissolve the bulk of construction and melt into the lowering sky.

SECOND NATURE: THE GROVE, THE MEADOW, AND VEGETAL MARBLES

According to Paul the Silentiary and Prokopios, the counterpart to the descent of light was a nave dominated by the worldly metaphor of landscape. The surface area of marbles used to clad Hagia Sophia runs to many thousands of square feet.[49] About ten varieties were used on the walls (if we include Egyptian alabaster and porphyry), but the diversity of their veining increased the apparent variety. This variety not only represented the material tribute from the empire to the emperor—a theme more explicit in ekphrases of other churches—but mapped the macrocosm with stones from its four quarters and the ends of the earth.[50]

Paul the Silentiary not only eulogizes the marbles' geographical origin but conceives of the church as a *Weltlandschaft,* both paradise and pasture. Indeed, the odd coupling of paradisiacal microcosm and the Word made flesh are announced at the very threshold to Hagia Sophia, at the "Royal Door," where the Justinianic inscription conflates two passages from St. John (10:7 and 9) to describe the cathedral as both Christ the gate and the pasture of the blessed: "Our Lord hath said I am the door of the sheep; by Me if any man enter in, he shall go in and out, and find pasture."[51]

Once inside, again according to Paul the Silentiary, we enter a "sacred grove [*alsos*] . . . high-crested and blooming," as though "with bright flowers."[52] Constantine Rhodius will also describe "boundless forests of many-colored marbles" within the Apostoleion (931/944).[53] As Pliny had written, "Once upon a time trees were temples of the deities. . . . Country places even now dedicate a tree of exceptional height to a god; and we do not pay greater worship to images shining with gold and ivory than to the forests and to the very silences that they contain" (*HN.* 12.2.3). The god Dionysos had built himself a round temple, on the mythical Mt. Nysa (India), by interlacing the branches of laurels with vine and ivy tendrils (Philostr. *VA.* 2.8). Apollo's first temple at Delphi also had been built the same way with laurel branches from the Vale of Tempe in Thessaly (Paus. *Desc.* 10.5.9). The nave columns of Hagia Sophia are made from green Thessalian marble (fig. 7.4).

In the Vitruvian architectural myth, columns easily doubled for trunks, and the basket capitals of the nave at Hagia Sophia bud and bloom as acanthus, and then sprout intarsiated "arcs laden with fruit, baskets and leaves, and . . . birds perched on boughs; the twining vine with shoots like golden ringlets winds its curving path and weaves a spiral chain of clusters" (Paul. *Descr. S. Sophiae.* 348–54). The Thessalian column shafts in the nave only augment the organic affinities, and magnolia trees would immediately come to mind were they indigenous to Europe. However, flowering laurel would have been a common sight. It hardly matters. Such horticultural comparisons were favored by classical and late antique poets for whom serpentine slabs, for example, conjured images of boxwood, turf, and rue (fig. 7.5).[54]

Vegetal associations would easily have sprung to mind, as such symmetries between art and nature had a long

FIG. 7.4. Nave arcades, Hagia Sophia, 532–537 AD. Thessalian marble columns, opus sectile spandrels. Istanbul.

FIG. 7.5. Lacaedamonian marble, also known as Spartan green porphyry or serpentine.

FIG. 7.6. Large Palaestra, 1st century AD, Pompei.

prehistory before Hagia Sophia. Pompeian interiors had juxtaposed painted hedges, trees, and whole landscapes with their real counterparts, and classical villas and public gardens had inflected landscape by aligning garden trees with portico columns. The ranks of plane trees that spanned gymnasium courtyards, in file with the encircling columns, provided an ideal venue for the discussions of *ars* and *natura* that took place as the interlocutors strode between them (fig. 7.6).[55] Ennodius had sung that on house floors "grass-green rocks simulating verdant pasture may entice the eyes," and in a few cases (for example, Hadrian's villa at Tivoli) floors were paved in Carystian slabs, like petrified lawns.[56] Villa walls extended into the landscape in the shape of hardy perennials and topiary evergreens that seemed as immune to seasonal change as marble: "a lovely box-tree with clipped foliage, the likeness of Spartan marble, spreads out its trimmed surface; not even Parthian arrows can pierce it," writes the poet Flavius Merobaudes.[57] The sap-colored veining of Carystian marble could mimic palm bark, as on a palm-tree column once in Constantinople's Forum Tauri and now in the courtyard of the Istanbul archeological museum. In medieval Florence another column of this marble became the ideal simulacrum for a miraculous elm, with which the coffin of St. Zenobius had collided, causing the startled tree to bloom even in the dead of winter.[58]

Agriculture and horticulture were regarded as "creating almost a second nature within nature's world" (Cic. *Nat. D.* 2.40.152: *in rerum natura quasi alteram naturam efficere*) and a lengthy ekphrasis, for example, by the Greek poet Longus (second century AD) of a garden (*paradeisos*) pursues this theme. His perfect garden must include the greatest variety of plants possible in order to flower and fruit for as many months as possible and thereby produce the gamut of fertile nature and an undying garden. Longus dwells on the interplay between skill (*techne*) and nature/growth (*physis*), or earth (*ge*), and not just man's design (between "borders produced by the earth and others by skill"), but even between nursery plants and those that grow wild.[59] This is not a world away from the Sumerian understanding of "lapis from the mountain" and "lapis from the kiln."

The reciprocity between architectonic nature and organic architecture (the "contest between art and nature," *certamen artis et naturae;* Pliny. *HN.* 21.4) also underlies Pliny the Younger's famous description (c. 100 AD) of his Tuscan villa.[60] It sat among "florid and gemmed meadows," providing a view "less real, than some painted landscape." Around the villa there were "plane-trees covered with ivy . . . twining around the trunk and branches, spreading from tree to tree and connecting them together," all of which was echoed by a vine that "trained upon four small pillars of Carystian marble" (*Ep.* 5.6.11, 13, 32, 36). In a late antique domus at Ostia, in fact, Carystian transformed a whole exedra, possibly the *stibadium* (dining room), into a garden room (fig. 7.7). The wall's largest fields are "crazy-paved" with Carystian slabs, so irregular that they have been mistaken for modern restoration.[61] The Carystian offcuts are collaged together like a bed of fronds, their strident veining used like giant hatching, creating expressionistic juxtapositions that figure a garden or even a grotto.[62] The whole room is in any case an exedra off the domus's central garden, onto which it looks through a (vanished) loggia of paired columns and, beside them, intarsiated door leaves that emphatically point the room beyond its perimeter. It literally held a mirror up to nature or, to borrow a phrase from Pliny

FIG. 7.7. Aula, late antique Domus ("College of the Augustales"), 4th century?. Revetment in Carystian marble. Ostia.

the Younger's description of a marbled chamber in his own villa, "reflects as great an ornament . . . as it borrows from it" (*Ep.* 5.6.38). A modern descendant of this room can be found in the perimeter wall of Mies van der Rohe's Barcelona Pavilion (1928–29), which elides building and nature, crystallizing an architectural seam between the water from which it draws its polish and the firs from which it absorbs its color (see fig. 13.4).

The greedy engine of Christian hermeneutic swallowed up all these allusions, to imagine entire churches as a sacred clearing in a flowering grove. Here is Sidonius Apollinaris (c. 430–479/80) on St. Justus, Lyons:

> Within the light trembles and moved thus
> is the sun to the gilded coffers
> That it strays to the yellow metal, to share the
> same color.
> Marble, adorned by varied flickers
> Spreads over the ceiling, floor, windows
> And below variegated designs,
> A flourishing and grassy crust
> Swirls its sapphirine jewels over the leek-green glass . . .
> And across the central pasture a stony copse spreads
> In widely spaced columns. (*Epist.* 2.10.8–15, 20–21)

Paul the Silentiary also reads into the marbles of Hagia Sophia a whole world complete with heavenly vault, a floor like a sea, flower-banked streams, and walls with "wheatfields and sheltering woods, playful flocks of sheep and gnarled olive trees, spreading vines." Such allusions did not halt at superficial similitude because, as we know, the passionate perception of flora sustained a spiritual significance, not only as a paradisiacal meadow but also as an image of the artistry of nature. Anyone knows that cut flowers will continue to bloom for days, and marbles offered an eternal flowering, as if by divine command and with time out of mind. Tessellated Cosmati floors, east and west (see figs. 6.26, 8.5), were "adorned with pieces of marble colored like flowers, truly like a spring meadow except for the fact that flowers wither and die, and this is a meadow that will never wither but will last forever, preserving in itself an eternal spring" (Philagathos of Cerami, 1143/46).[63] Because heaven was "covered and scented with beds of rich roses," in the Spanish church of a virgin martyr, "the luster of shining marble, foreign and native, lights up the nurturing hall, the worshipful soil tends her remains and holy ashes in its bosom . . . and sawn stones so variegate the ground that you would think it a rose-colored meadow reddening with varied blooms" (Prudentius).[64]

In Byzantium the same ideas flourished with the same vigor. One epigram, really a mini-ekphrasis, was inscribed on the narthex entrance of Hagia Sophia's rival, Hagios Polyeuktos (524–527). This text preannounced that within the church:

> The facing walls in innumerable paths
> Are clothed in marvelous mineral veins of color, like
> flowery meadows
> Which Nature made to flower in the depth of the rock,
> And hid their glory, keeping them for the House of God,
> To be the gift of Juliana, so that she might produce a
> work of wonder,
> Following in her toil the stainless dictates of her heart.[65]

Elsewhere the same poet speaks of the "deep-rooted [*bathurrizos*] foundations, springing up from below." Indeed, flowering and blossoming stones transcended metaphor by recalling that marble was organic and rooted, a *living rock,* "that is kept alive by its root," as Ovid had said (*Met.* 14.713). Michael the Deacon describes the columns in Hagia Sophia as "greenish in color, as though they had grown out of the ground just there"; Nikolaos Mesarites makes the same comment (1198/1203) on the Apostoleion; and Michael concludes that the marbles of Hagia Sophia "in many respects convict the flowers of being easily withered, since it is also cut from the mines of the earth, but preserves its flowery dye even after severance from its own root." Perhaps Themistios (*Orat.* 13.179a) also had Thessalian marble in mind when he praises Emperor Gratian (376 AD), saying, "If he had raised for us columns of Phrygian or Pentelic or Egyptian marbles, we would do well to praise him; but since he has erected for us these living and animate columns that rest on the earth but rise so high, should we not praise him even more?"[66] These eternally flowering stones outlawed the sterility of man-made construction to create a divine work, a living, blossoming church of incarnate stones, the *empsychos naos.* This is what Prokopios means when he explains the marble meadows of Hagia Sophia, saying, "Whenever anyone enters this church to pray, he understands that it is not by any human power or skill, but by the influence of God, that this work has been so finely turned. And so his mind is lifted up toward God and exalted, feeling that He cannot be far away, but must especially love to dwell in this place that He has chosen."

The church interior should vaunt the infinite detail of a real landscape, not only because it was a microcosm, but also because the metaphorical "meadow" meant fecundity in nature and the arts alike. Thus, when Gregory of Nyssa writes about the Martyrium of St. Theodore at Euchaita, he says, "The painter has spread out the blooms of his art . . . and adorned the church like a beautiful meadow."[67]

Viewers were to lose themselves in the meadow of art and understand that the only principle that could transcend its variety of varieties was the mind of God. In other words, the infinite but singular God can only be represented by the greatest diversity, which simultaneously best expresses non-multiplicity. It is also no coincidence that in Greek the word for "ornament" (*kosmos*) also signifies "universe," a pairing that induces the mind to move from contemplation of one to absorption in the other.

BOOK-MATCHED REVETMENT AND INNATE ORDER

Even so, all these sprawling meadows must be accommodated to architecture, to produce "a second nature." The mechanism to do so was the technique now called "book-matching," whereby the quarried block was sawn parallel to its surface, and the unfolded panels set edge to edge like the facing leaves of an opened book (figs. 7.8–10; see figs. 7.12, 7.14). The block could be unfolded once to produce bilateral symmetry ("twinning"), or twice to produce quadrilateral symmetry ("quartering," i.e., mirror-like in two directions). Multiple pairs of single unfoldings resulted in a continuous bellows or accordion pattern.

Book-matching became a trademark of Byzantine architecture, a technique that workshops exported along with the raw materials to both the Latin West and Islamic East.[68] Where there were neither blocks nor slabs

FIG. 7.8. Scheme for book-matching marble slabs. Slicing of block into veneers and consequent mirroring (twinning and quartering).

FIG. 7.9. Nave revetment, Hagia Sophia. Book-matched and individual slabs divided by bead-and-reel moldings.

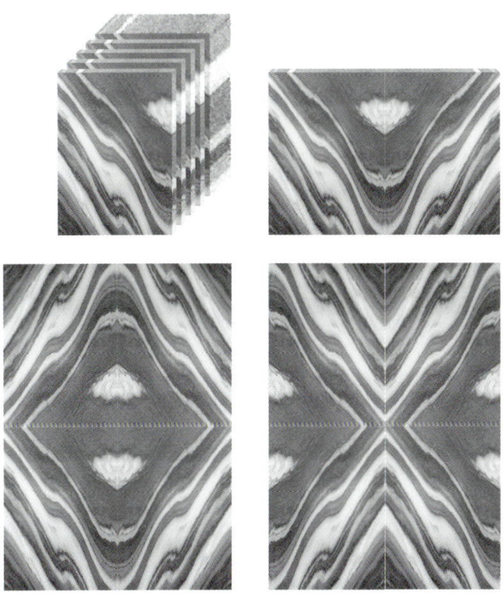

FIG. 7.10. North Gallery, Hagia Sophia. "Accordion"-style book-matching.

available to recycle, enterprising masons resorted to sawing columns lengthwise to achieve the effect, a practice still exploited in 1920s Jaffa, as we can see from perhaps the only photograph ever taken of spoliation in action (fig. 7.11).[69] The narrower slabs resulted in the more cardiographic patterns that reanimate the naos revetment of the Chora Church (c. 1316–21; fig. 7.12).[70]

FIG. 7.11. Men sawing column of Proconnesian marble lengthwise. Jaffa (Scholten 1929; 1:fig. 93).

Book-matching was not, however, a Byzantine innovation. It is simulated in Pompeian dados and throughout the empire well into the fifth century.[71] Real examples survive in the second-century villa at Silin in Libya (see fig. 5.10) and at Ephesus.[72] This technique may originally have been learned from carpentry, when splitting planks revealed an instant symmetry, and opus sectile was probably equally indebted to wood marquetry. Indeed, items of furniture survive from Pompei and Herculaneum that mirror marble floors.[73] In Hagia Sophia, book-matching is used in two opposing but complementary manners. In the nave, twinned slabs form isolated panels framed by beaded

FIG. 7.12. Naos revetment and paving, Chora Church, 1316–21. Istanbul.

moldings (see fig. 7.9), while in the aisles, galleries, and (originally) west façade, twinned panels wrap around all the wall surfaces without break, accordion-style (see fig. 7.10).

By a simple cut, a tectonic order automatically regulated the veining from within. Marbles with pronounced veining were vital to the effect, and great care was taken that the veins always ran true. Ekphrasists missed few opportunities to commend the tight fit (*harmonia*) of the marbles and the image of unicity that this sponsored, and the criterion of ex uno lapide (*ex henos lithou*) continued to thrive in Byzantium. Eusebius (*Vit. Const.* 3.36) says of Constantine's Church of the Holy Sepulcher in Jerusalem (326–335) that its "exterior walls, adorned by the dressing of the stones, one to another, provided a spectacle of surpassing beauty, no whit inferior to the appearance of marble." Gregory Nazianzenus also commended "the squared and dove-tailed stonework" on the church his father had built at Nazianzos (374 AD), and a ninth-century building in the Great Palace at Constantinople would even acquire the name "*Mousikos,*" "on account of the precise joining of its marbles."[74]

The invisible seams and mirror-imaging of book-matching instilled axes of visual repose and an innate ordering of lozenges, X-forms, and V-forms. When the mirroring of the slabs extended in series, like an accordion, the sequential unfoldings promised an exponentially unlimited extensibility. Like any repeating pattern, wherever it is cut, it seems complete. The aim was to suppress substance and distance any arduous history of manufacture. As Nikolaos Mesarites says, "The craftsmen in finishing the stone brought it to such thinness that the wall seems to be covered with many-colored woven cloths. . . . Indeed the stone bursts forth into such a shimmer that its glistening surface vanquishes any flower."[75] It was as if the wall had simply unfolded (or blossomed), automatically reproducing itself again and again, rather than built block upon block. This was especially true of "quartering," which presented an instant commensurability of parts. In the mid-fifteenth century Leon Battista Alberti, possibly with the nave revetments of S. Marco, Venice, in mind, recognized the merits of these schemes, when he wrote: "Even the minutest elements must be so arranged in their level, alignment, number and appearance, such that the right exactly matches the left, top matches bottom, adjacent matches adjacent, and equal matches equal . . . so they appear twinned. The ancients attached such value to this balancing of equal parts that they even tried to match their marble panels exactly. . . . Veining must join with veining, color with color, and so on, so that they each enhance one another."[76]

Theodore Metochites's tortuous description of the church he commissioned, the Chora (c. 1316–21; see fig. 7.12), makes the same allusions, saying that he had adorned it,

> with splendid marble stones,
> In the most beautiful colors that can be seen everywhere,
> Each and every one with each other, each one throughout the work
> Beautifully begotten; those underneath on the ground,
> Those that stand upright on each side, hither and thither,
> Reaching up high on high most seemly: a wonder to behold,
> a pleasurable mingling, how each thing is fitted together:
> With all measured joining they are reconciled,
> Fitted together in both alternating and interweaving fashion,
> These with these, and those with those, and nothing that is not.[77]

BOOK-MATCHING AND PAINTING IN STONE

In the galleries of Hagia Sophia, book-matching functions as a sort of limitless background, dissolving the perimeter. Viewing the Proconnesian backdrop beyond the Thessalian columns is like glimpsing the sky through trees (see fig. 7.10). But, in the nave, architecture and painting are interwoven into a giant and counterchanged patchwork (see fig. 7.9). Ekphrasists praised such joining because, as Paul the Silentiary says, "The joining [harmonia] of the finely cut marbles resembles the art of painting. For you may see the veins of the quarter-cut and eighth-cut stones meeting so as to form ornament: connected in this way, the stones imitate the glories of painting." Paul exults in the palette of Thessalian marble "that in one stone various beauties mingle," and Prokopios exults in the marbles, "which Nature like some painter varies, with most contrasting colors." The orator Chorikios even exhorts painters to look to marble veining for inspiration because they "are so variegated by their natural colors that they do not fall short of human painting."[78]

Not only poets praised this conjunction of stones as rivaling painting, but also prospective patrons. In 508/509 AD, when Theoderic the Great, the Ostrogoth king who had spent his youth in Constantinople, wanted to decorate a new building in Ravenna, he knew what to ask for. As he wrote to Agapitus, the prefect of Rome: "We depute Your Lordship to spare no effort in sending us the most expert marble-workers of the City, so they may join together those things that have been choicely divided, and that, when they are connected and their veins sport together, they may admirably feign a natural appearance. Let what conquers nature come from art: let the variegated surface of the marbles be woven into the loveliest variety of coloration [*picturarum varietate*]; because that which has been devised for beauty is always of value."[79]

Theoderic's missive speaks of painting, true, but not paintings. His "*picturarum varietas*" is often translated "variety of paintings," although "pictura" indicates not only images applied to surfaces but also the medium of those images, so that "picturarum varietas" (like its Greek equivalent, "*graphike poikilia*") may equally infer "variety of coloration."

It has become rote to assume that the praise of "painting" that ekphrases accord marble revetments can only refer to the images that were latent in book-matched symmetry, in the manner of a Rorschach inkblot (*klecksogram*). From this assumption, scholars such as John Onians have developed sophisticated theories of a late antique way of seeing, in which the responsibility of perceiving fully defined figuration was "transferred" from the artist to viewer. John Mitchell has even catalogued what he considers to be columns with anthropomorphic veining, arguing from the "visual evidence" of the marbles themselves, without recognizing, it seems, that his perception of the patterning is entirely subjective and personal; he even concludes that the image-bearing blocks must have been "discovered and revealed by the agency of percipient quarry-masters." In short, while it is true that modern eyes have often relegated the importance of the medium to the sidelines, there has been a more recent blindness to an unacknowledged bias that these columns and revetments must have held a coherently mimetic significance, and one that was figurally imagistic.[80]

However, Byzantine observers of the early centuries hardly ever identified figural images in revetment. When Paul the Silentiary talks of sheep he does not mean that he sees them in the stone, but in his mind's eye, where they graze in marble pastures. This is not to say that medieval viewers were incapable of detecting such images. When the Granadan scholar Muḥammad ibn Aḥmad Ibn Jubayr (1145–1217) admired the ka'ba at Mecca in 1183, he remarked on the "marble slabs of pure and unstained whiteness in which Great and Glorious God had fashioned, at its first creation, remarkable designs, inclining to blue, of trees and branches, and another beside it with the same designs exactly, as if they were parts (of the same stone); and if one were placed over the other, each design would correspond with its opposite. Beyond a peradventure each slab is the half of the other, and when the cut was made they divided to make these designs and each was placed beside its sister."[81] However, Ibn Jubayr perceived a vegetal image in the branching patterns of veining. He might just as well have seen rivers. Absent are the heads and animals that were painted in faux-marbles (see figs. 4.17, 4.25. 4.26, 6.19, 6.25), foreign inclusions that might be imagined from the clasts of a breccia, or which were real fossils in a conglomerate, rather than the laminations of a veined marble, whose serendipitous veins could suggest halo-like contours instead (see fig. 10.4).

Byzantine ekphrasists also preferred to dwell on the variegation that made Nature herself a painter and stall any train of thought that led to praise of skill in painting, the capacity to portray emotions or the like, things that might point back to human artistry. In sixth-century accounts, emergent figuration is always arrested before it becomes recognizable. A very simple statistic suffices here. In the entire corpus of Byzantine poetic descriptions of revetment, there is only one line, by Paul the Silentiary, that has been taken as a reference to perceiving human figures in marbles, but a glance at the Greek shows that he only says "living things." By contrast, an anonymous two-line epigram, of uncertain date but at least late antique, describes a piece of black-and-white veined marble (fig. 7.13): "You see how great the beauty of the stone / Lies in the disorderly order of the veining."[82]

One should not take the statistic too seriously, but on textual grounds, there is already double the evidence for an appreciation of veining per se than for the idea it was intended to be perceived as compositions of humans

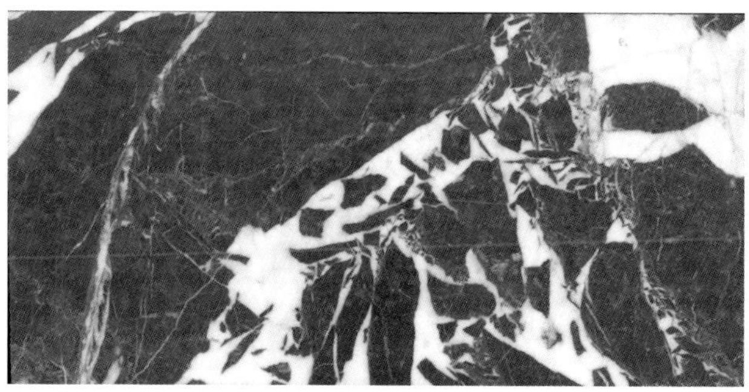

FIG. 7.13. Marble from Aquitaine (*bianco e nero antico*).

and animals. Paul the Silentiary also draws attention to the veining "applied as ornament" (609–610), and in one instance "the brilliant marbles with a thousand sinuous veins . . . Iassos, red-on-white" (629–632). There were images in the stones, but they need not be figural, and the language of in-seeing in Byzantine ekphrasis most often assimilates architecture to landscape. The quarries are mentioned so frequently that the cliff-like walls of the church seem reconstituted mountains. The stones were vital and elemental. They represented rock and earth, sky and sea. They imaged their own elemental substanceness, not ulterior images. As a result, we can see the loving reception of marble images as a genuine aesthetics of abstraction, but only insofar as no figural images were prized that were alien to its matter.

Only from the ninth century onward do humans, animals, even entire figural compositions become the default mechanisms of perception in book-matching.[83] Superstition played its part in popularizing this perception, but it is no coincidence that its ascent coincided with the final demise of Byzantine iconoclasm. In the debates that wracked the legitimacy of images, the *acheiropoieton* had occupied pivotal importance in disproving the Mosaic prohibition against graven images. It could not pertain to those images of Christ that had been made-without-human-hands: the Mandylion, the Keramion, the images of Camuliana, Melitene and Memphis, and, in the West, the Veil of Veronica and the Shroud of Turin.[84] All these portraits had imprinted themselves into their fabrics without human agency and, in the case of the

FIG. 7.14. Book-matched pier revetment, S. Vitale, c. 532/548. Ravenna.

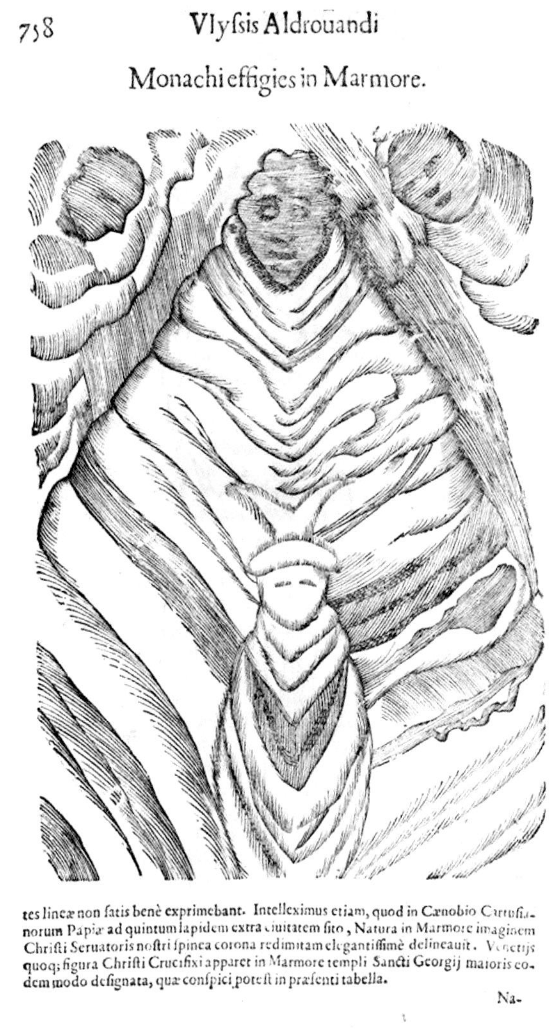

758 Vlyſsis Aldrouandi

Monachi effigies in Marmore.

tes lineæ non ſatis benè exprimebant. Intelleximus etiam, quod in Cænobio Cartuſianorum Papiæ ad quintum lapidem extra ciuitatem ſito, Natura in Marmore imaginem Chriſti Seruatoris noſtri ſpinea corona redimitam elegantiſſimè delineauit. Venetijs quoq; figura Chriſti Crucifixi apparet in Marmore templi Sancti Georgij maioris eodem modo deſignata, quæ conſpici poteſt in præſenti tabella.

Na-

FIG. 7.15. Image of monk and chorister in book-matched revetment, S. Vitale. Ravenna (Bernia et al. 1648, pl. 758).

Keramion, a tile had even reproduced the mirror-image of the Mandylion that it covered. The images in marbles were equally innate, and in the case of the Keramion and even the Shroud, there were immediate affinities between book-matching and the reproductive symmetries of the relics.[85]

The triumph of the iconodules made it licit to see divine portraits everywhere. Moreover, all the high medieval responses to marbles as containing miraculous images of saints, and so on, derive from a different sort of response, pilgrim literature, that was ready and willing to identify real presences in sacred sites and make relics of their individual stones. Today it is quite hard to find the images that pilgrims perceived, but if one wishes to look over medieval shoulders, there is no better example than that at S. Vitale, Ravenna. There, the image-bearing, book-matched slabs remain in situ (fig. 7.14) and may easily be compared with Ulisse Aldrovandi's engraving and the descriptions of Renaissance and Baroque visitors, who generally perceived a monk and an altar boy celebrating the Mass (fig. 7.15).[86] The Λ- and V-shapes typically produced by book-matching spontaneously suggested the monk's cowl or hands joined in prayer, and such images, especially one of a monk at S. Marco in Venice (who stood "with hands apparently in prayer"), inspired Giovanni Battista Cima da Conegliano to superimpose the hands of praying donors over a book-matched dado in his Brera *Virgin and Child with Saints* (c. 1486–88; fig. 7.16). The wit underlying this instance is that Cima has overlain the steepled fingers of the donors over lithic ones not made by human hands.

Divine artistry was ongoing, and heaven-sent images were liable to appear at any moment to seal the sanctity of the place with a true presence. Indeed, stories of the Virgin or saints appearing in columns and revetment traversed the Mediterranean with greater ease than the materials themselves, miracles waiting for a place to happen. In the eyes of pilgrims, Hagia Sophia was soon full of them, including an image of John the Baptist and the whole cast of the Crucifixion.[87] They repeated the stories when they returned home to others departing for the metropolis who were now forearmed with expectation. A good example is supplied by the *Cantigas de Santa Maria*, an anthology of poetry written in the 1270s for Alfonso X, king of Castile and Leon. The *Cantigas* illustrate images of the Virgin and Child that suddenly appeared in church piers thanks to the light irradiated by the divine presence (fig. 7.17):

> Rightfully can God reveal the likeness of Himself or of His Mother in His Creations, for He formed them. . . . Therefore if He causes images to appear on stones, no one should be amazed at this, nor likewise on plants, for He causes them to grow and gives them many colors to appear beautiful to us. Therefore it happened in Constantinople, as I learned, that the good Emperor Don Manuel ordered a very noble church built there. And, as I heard, he had blocks of marble brought there from far away and sawn in the middle to make great tablets to place around the altar of the Holy Virgin, Mother of Our Lord. While they were sawing one of them, they saw Her image inside, painted in colors, just as God had painted it . . . holding her son, Who took flesh from Her, in Her arms. When the emperor heard of this, he mounted his horse at once, and when he saw the image he worshipped it and had it placed in the main entrance. And there it sits today, and all hold it in great reverence.[88]

Such stories were urban legends, for the emperor mentioned in the above *cantiga*, Manuel I Komnenos (1143–1180), is renowned for not building a single new church in Constantinople.[89]

ART AND NATURE: HUMAN AND DIVINE PAINTING

The "glory of painting" in the sixth century infers not paintings as pictures, but pictures as painterly. To put this another way, painting is not an object but an activity, it is a verb not a noun, an action rather than an outcome. Nature was a painter, as God's proxy, and brushy veining a miracle of divine artistry, because it was made "without hands," and any images that arose from it were incarnated in substance and disclosed by the providence of book-matching. This is the miracle to which the Edessene Hymn refers when it celebrates the cathedral's cladding explicitly as "like an image not made by human hands," an acheiropoieton.[90] In the early sixth century, before the term *acheiropoietos* became restricted to miraculous icons, it referred to natural objects made by God rather than by man, the human body, heavenly as opposed to earthly objects, and spiritual rather than material things.[91] It could be translated as "natural," were this word free of its modern nuances. Any building that was to reveal itself as divine would inevitably have to bear the signs of its nature. And marble was a painting that never faded for gardens that never withered. Moreover, the idea that marble was a

FIG. 7.16. Giovann Battista Cima da Conegliano, *Virgin and Child Enthroned with St. Sebastian, St. John the Baptist, St. Mary Magdalen, St. Roch and Members of the Confraternity of San Giovanni Evangelista,* c. 1486–88. Brera, Milan.

form of natural painting would only become more immediate as a metaphor for, or sign of, the Incarnation when Byzantine theologians came to compare the permeation of divine light in human flesh to the paint trapped in the weft of canvas or linen.

There are independent reasons to believe that the ideators of Hagia Sophia wished to avoid any figural contamination in its decorations. Upon completion in 537, the cathedral's mosaics were aniconic: the figural imagery that we now see (or know from old descriptions) would not be added until the ninth century, and the expansive vaults were instead tattooed only with ubiquitous crosses and possibly the cherubim of the crossing (see fig. 7.3).[92] Even these crosses were possibly signs to consecrate the structure, their serial repetition multiplying their power. The magical power of the cross resulted in their use as caprices on both city walls and also the external ones of Hagia Sophia, and so a fourth-century floor mosaic announces that "not by big stones, not by firm iron and tawny bronze, not even by steel are these buildings girded, but, by the much-invoked signs of Christ."[93]

The omission of grand figural mosaics, which seems so extraordinary in a building of such artistic ambition, is normally explained by reference to the Monophysite sympathies of Empress Theodora, Justinian's spouse.[94] But not all Monophysite churches were aniconic, and icons were certainly allowed in Hagia Sophia, while plain gold mosaics had also adorned the much earlier apses of Constantine's Roman basilicas, St. Peter's and St. John in Lateran.[95] At Hagia Sophia, not only was the substantiality of light to predominate, but also no figuration must compromise the implications of the materials and thence the autonomous metaphoricity of the whole construct.

Not only book-matched marbling, but opus sectile and mosaic were considered species of painting in stone, and it is only in the latter arts that writers explicitly compare the craft with human painting. Chorikios extols not only the "painting of pigments, but also the type which imitates

FIG. 7.17. *Cantigas de Santa Maria*, 1270s (Escorial Ms T.I.1, fol. 44r).

it in tesserae"; and Photios, in the ninth century, goes so far as to dismiss all the illusionist feats of antique painting as child's play when compared with the fitting of an opus sectile so snug that it rivals atomic bonding.[96] From the third through fifth centuries, Tunisian workshops had been so enthralled by the natural painting of book-matching that they had imitated it in mosaic, effectively creating a second art within art (see fig. 4.31).[97] In other churches and at later times, a unity of lithic arts seems to have thrived from this subtending theme of pictura, or poikilia graphike. Thus, in the naos of the Chora Church (c. 1316–21) large mosaic panels intrude into the revetment-like overscaled icons, inviting a comparison between painting in stone and stone as painting, a continuity that was all the more legible because Byzantine painting is oblivious to perspectival depth and, therefore, more integral with its surface (see fig. 7.12).[98]

In Hagia Sophia, the spandrels of the nave arcades are sheathed with elaborate opus sectile (see fig. 7.4), budding fronds intertwining around porphyry disks like giant cherries, below which flourish the delicate, basket-weave capitals. Altogether, the conjunction of book-matching and opus sectile created a unifying dialectic of art and nature by friendly rivalry, in which nature represents and is represented. The demonstrable artistry of man (opus sectile) was the measure of nature's handiwork (book-matching) and, vice-versa, natural art was the proof of man's artistry (techne).[99] This reciprocity was sustained by the differences in facture. Both were selected, sawn, and set, but unfolding book-matching suggested a natural and apparently artless evolution, while pieces of inlay must be skillfully shaped and configured. Likewise, nature exhibited her variegation within single slabs, while multifarious art assembled varied fragments into new wholes. Ennodius (473/474–521) reproduced these sentiments in dense and difficult verses, which are scarcely translatable because the allusions of many words have no counterpart in English. Ennodius was a poet, but also bishop of Pavia (from c. 515), so his poem may have been inspired by works he commissioned at the Basilica of St. Michael, or those he saw in Rome or Milan. It is unclear whether he had simple slabs or opus sectile in mind, as the concepts could apply to either, which may well have been his intention:

> On a Patchwork in Marbles.
> When the innards of stones have been assembled by the law of fusion,
> They (re)create nature with the artificers' skill.
> The marbles hardened into a solid whole from broken members,
> And the spirit is gathered into a crust from the parts.
> He who could compose a single appearance from various fragments,
> Has bent the stone to his will.[100]

Moreover, the combined effects of natural art and artful nature must overwhelm the senses, a blizzard of stimuli leaving viewers purblind in the miasma, and persuading them that the single principle that transcended such diversity was the mind of God himself. As Nicolas of Cusa would later put it, "Infinity, which cannot be multiplied, is best unfolded by reception in a variety of things, for greater diversity best expresses nonmultiplicity." It bears repeating that in Greek (*kosmos*; and Latin, *mundus*) the word for "ornament" also signifies "order" and "universe," a pairing that assists the mind in moving from one to the other.[101] Ekphrasists loved to stress the disorienting effects of this anagram of the divine mind: "With contracted brows" spectators departed "still unable to understand the skilful craftsmanship" and always "overwhelmed by the bewildering sight." Indeed, as Photios says of the Church of Our Lady of the Pharos (864), the observer, "through his whirling about in all directions and being constantly astir, [by that] which he is forced to experience on all sides, imagines that his personal condition is transferred to the subject."[102]

The culmination was the greatest diversity within the greatest cohesion, in a marriage between skill and matter. Thus, when Chorikios praises the revetment of Hagios Sergios at Gaza (before 536 AD), he crams into one sentence the most loaded concept-words in the theoretical lexicon. The aisle walls, he writes, "On the one hand, are covered with marble slabs that by one skill [*techne*] hold together in unique cohesion [*harmonia*], while by the nature [*physis*] of the material [*hyle*], they are equally adorned [*ke-kosmenenai*] with painterly diversity [*graphike poikilia*]."[103]

HORIZONS

In the New Testament, all the attributes of Wisdom were reconciled into the figure of Christ-the-Word (1 John 1:9–10, 18). The Word exists *ab aeterno* with God, participates in the creation of the world, is the light among men that shines in the darkness, assumes the flesh of man, and as the first and last manifestation offers the bread and wine, the sacrifice of salvation. John's genealogy offers a narrative for Hagia Sophia, from the descent of light in the dome to the eventual culmination of all liturgy at the eucharistic consecration in the sanctuary. In fact, the building synthesizes a processional, basilical space with a cosmological, domical one. The stratification of the Great Church also calls to mind the universe as described by the Alexandrian cosmographer Cosmas Indikopleustes (writing c. 550), who describes the sky as a "dome" (which he is at continual pains to point out was not made with hands), but also talks of the "arena-like space below."[104]

This ground-scape is activated by the divine liturgy, the event that, according to Byzantine theology, made two worlds synchronize and mesh.[105] The earthly clergy concelebrated in unison with the heavenly hosts, such that the faithful could both imagine themselves in heaven and apprehend the church as a heaven on earth, in which God dwelled and walked.[106] Russian ambassadors who entered the church in 987 were, by their own account, converted on the spot both by the vision and the conviction that the angels themselves were descending from the gold mosaics to celebrate the liturgy.[107] The two worlds intersected at the Transubstantiation, when God became once more incarnate in the world, and at the communion, when ingesting the eucharistic wine and bread made the recipients—however briefly—Christ-like, walking temples themselves, and, of course, vessels of divine light.[108]

Communion took place in the nave, which became an arena for liturgical spectacle (see figs. 7.1, 7.3). The patriarch and emperor would march its entire west–east length when they made their grand entrances from the narthex on important feast days. The sanctuary responded by projecting a gangway (*solea*), like a seaside pier, all the way to an elevated lectern (*ambo*) that stood almost under the center of the dome.[109] The public looked on from the surrounding aisles and galleries. The twin poles of this bifocal amphitheater were therefore the ambo and the sanctuary, a binary that reflected the essential division of the divine liturgy into the Liturgy of the Word and the Liturgy of the Eucharist. Step one took place at the ambo, in the most luminous part of the building, where the gospels were read and the Word pronounced with three readings before an opulent offertory processed to the sanctuary, where the eucharistic sacrifice was consummated.[110] Here the columns in the corner exedrae are porphyry, imperializing the sanctuary (as they did in St. Stephen at Gaza), but also bearing the stain of royal blood.[111]

In short, while the domed nave was the place of the Word, the sanctuary was the place of the flesh, where Christ's two natures finally, fleetingly, and recurrently mingled. If we consider the morphology, light, materials, and liturgy all together, we begin to apprehend that the idea of Christ's Incarnation, the Word made flesh, is enacted across the whole volume of the church and infiltrates its materiality, all the way from dome to altar. The light of Wisdom, which portends Incarnation, descends on the church, from its dematerialized upper reaches to infuse its lower regions, and through the enacted liturgy and with the encompassing participation of the enlightened and inspired faithful, it materializes in the eucharistic gifts within the sanctuary.[112]

In fact, Hagia Sophia was arguably invented to meet this new challenge in architectural representation. Just how dramatic its rupture was with the past becomes obvious if one recalls that the Theodosian building it replaced was a five-aisle basilica, quite probably with transepts, and that the contemporary Apostoleion (c. 536–550), another huge and star imperial commission, was built on a cruciform plan apparently by the same architects who designed Hagia Sophia, Anthemios and Isidoros. In other words, the design of Hagia Sophia was not a timely exercise in architecture but conceived to fulfill a specific metaphysical agenda. However, there are echoes of symbolism in the later Middle Ages when Nikolaos Mesarites (1198/1203) says of the Apostoleion's crossing that it "binds the other four [arms of the church] to itself . . . and stands there as a kind of mediator and a reconciler of those which formerly were separated from each other, in this, I believe, imitating the mediator between God and man, who is portrayed in the midst of it [in the dome], Christ, truly the square-cut stone."[113]

This much the Apostoleion may have shared with the conception of Hagia Sophia, but the figure subtending Justinian's Hagia Sophia was to be anything but the symbol par excellence of the Passion, the cross. Hagia Sophia was to embody Holy Wisdom as the Word made flesh. Justinian's reconstruction may not have been the first cathedral to eschew a basilical, cross-, or even T-shaped plan, and its structural solutions have traceable precedents, but it was not the culmination of some Darwinian evolution, wherein the historical progression of structural innovation was an end in itself. Rather, it was, sui generis, the solution to an unthinkable ambition.[114] Hagia Sophia is neither purely a centralizing nor longitudinal structure, but fuses both into a unified volume. Functionally, it synthesizes a processional and basilical space with a cosmological and domical one. It is as though a basilica had mated with the Pantheon.[115]

WITHIN THE SANCTUARY: TRANSFORMATIONAL MATERIALS

Within the sanctuary, the platform for Transubstantiation, the altar top, was an alchemist's delight. It was made in 562 to replace one destroyed by falling vaulting in 558. Before the Crusaders melted it down in 1204, we are told it was an amazing alloy of "gold, silver, various precious stones, pearls and mother of pearl, copper, electrum, lead, iron, tin, glass, and every other mineral," pulverized, smelted in a furnace, and then cast in a mold. Anyone who saw it specified that it was not encrusted or cloisonné, and all agreed: "Who can behold the appearance of the altar-table without being amazed? Who indeed can comprehend it as it changes color and brilliance, sometimes appearing to be gold, in other places silver, in another gleaming with sapphire—in a word, reflecting seventy-two hues according to the nature of the stones, the pearls and all the metals?"[116] Another traveler (1136/43) even thought that the altar was "cast from brilliant emerald."[117] How such a thing could have been made, without burning or cracking the precious stones, is a mystery, but it does seem likely that the alloy matrix was at least heat-treated or lustered for iridescence. What this looked like is also equally mystifying, but a rare portable altar may be a pale reflection (fig. 7.18). The metallurgy of the wondrous slab in Hagia Sophia recalled the Mosaic Tabernacle and Solomon's Temple, whose artisans had been divinely inspired to work all metals and set stones (Exodus 31:3–5; 2 Chronicles 2:7). Moreover, the long dedicatory inscription on the altar also began by invoking the eucharistic gifts ("We your subjects bear to you, Christ, those your things that belong to you"), and the transmutation of stones and ores in the altar top paralleled the Transubstantiation of the sacramental act. While the floor of Westminster Abbey was a divine diagram of the dissolution of primary matter into the four elements (see fig. 6.26), the Altar of Hagia Sophia was a reconstitution of that matter which had been divided at Creation, in an act of tabletop alchemy, like casting a slab from stardust.[118]

The vessels for the eucharistic gifts were probably no less transformational. No description survives of those used in Hagia Sophia, and many must have been gold and bejeweled, but others would have been of hardstones, such as the later imperial paraphernalia that survive in the treasury of S. Marco, Venice.[119] Translucent onyx, both cloudy and fleshy, was ideal for the patens that received the consecrated host, his body (fig. 7.19). As Paul the Silentiary says of onyx, it had "a rosy bloom mingled with pallor, or the fair brightness of human fingernails."[120] Agate chalices (fig. 7.20) were even more implicative, for their color and translucency suggested a substance somewhere between solid wine and frozen fire. By a figure of speech, the material cup was often used as if it were synonymous with the precious blood itself, for St. Paul had written, "The cup of blessing which we bless, is it not the communion of the blood of Christ?" (1 Corinthians 10:16). Porphyry vases were venerated here and there as the vessels of the water that the savior had miraculously transformed into wine at the marriage of Cana.[121] The chain of association was facile: the fine flecks in this burgundy stone seemed almost to bubble, and when ruby wine filled crystal goblets, Muslim poets imagined that the wine was one with its vessel.[122] Likewise, the blended veins of agate or sardonyx vessels, as exemplified in

FIG. 7.18. Itinerant or Sicilian master?, Portable altar, second half of the 12th century. Hertfordshire puddingstone, wood, brass, copper, silver, enamels, 15⅜ × 7½ in. (39 × 19 cm). Cathedral Treasury, Agrigento.

FIG. 7.19. Onyx paten, 10th/11th century. Treasury of S. Marco, Venice.

Suger's famous chalice (see fig. 6.15), might recall the tempering of wine with hot water, a practice that continues in the Orthodox rite until this day.[123] In the Catholic rite, (tepid) water is also added to the wine, expressly to symbolize the Incarnation, the water representing Christ's humanity, and the wine Christ the vine. Moreover, during the sacramental sacrifice the wine in the chalice transubstantiated into Christ's blood, and since the dark liquid harbored the divine light, it was fitting that translucent agate glowed like a coal, or seemed as fiery as well-stirred wine.[124] Those who drank from the "mystic cup" imbibed "a stream of repentance" that flowed directly from the Holy Spirit, and the chalice miniaturized the surrounding church as a vessel of brilliant wisdom.[125]

In the sixth and seventh centuries, pilgrims to Jerusalem were led to believe an agate chalice venerated at the Holy Sepulchre was Christ's own, as was another in Valencia made from the same stone.[126] However, the best evidence for the impact of hard-stone vessels on the imagination comes from a later myth that seems an elaborate allegory of communion. In Wolfram von Eschenbach's *Parzifal* (c. 1197–1209/15), the Holy Grail is a clear "garnet hyacinth," which the apostles drain at the Last Supper and use to catch Christ's blood at the crucifixion (book 5). In later versions of the legend, the mineral eventually superseded the receptacle in importance, and a red powder (or panacea) was ground from the philosopher's stone and then mixed with alcohol to create the elixir of eternal life.

Inscriptions further emphasized the material latencies in Byzantine vessels. During communion, the elite consumed the host off onyx patens inscribed "Take, eat, this is my body" (see fig. 7.19) and drank wine from agate or sardonyx chalices that commanded, "Drink all of this, this is my blood" (see fig. 7.20).[127] As these beckoning phrases were quoted directly from the eucharistic consecration, they are not labels but invocations. Amulets often require such adjuncts, and lapidaries not only specify the miraculous properties of individual stones but the inscriptions necessary to activate them, agate and sardonyx included. In the West, by the thirteenth century, lapidaries are quite explicit on this point, even beginning by comparing the power of gems to the act of Transubstantiation.

DIVINE ARTIFICERS

As Prokopios says, visitors to Hagia Sophia must understand immediately that it had been fashioned "not by human power or skill, but by the influence of God." The designers chosen, Anthemios of Tralles (c. 474–before 558) and Isidoros of Miletus (442–537), were the ideal men for the job. They were not architects, but *mechanopoiei*

FIG. 7.20. Agate chalice, 10th/11th century. Treasury of S. Marco, Venice.

or *mechanikoi* (*mechanici*). The science of mechanics was traditionally divided into theory and practice. Theory included geometry, arithmetic, and physics; practice meant works in metals, architecture, carpentry, and painting. Only a man who had fulfilled all the requirements of the curriculum, and few could, was granted the title of "*mechanikos*," far superior in rank to an *architekton* or *oikodomos* (architect/builder). "Mechanics" were almost oracles, because their command of *theoria* enabled them to "prepare in advance designs of future constructions" (Procop. *Aed.* 1.1. 24), and their feats of engineering, control of nature, and ability to simulate its effects often suggested that they were more magicians than engineers.[128] As Cassiodorus had put it (c. 507 AD): "All the disciplines, the whole endeavor of the wise, seek to know the power of nature so far as they can. Only the art of mechanics [*mechanisma*] tries to imitate it by contraries, and in some things, it is proper to say so, even seeks to surpass it. . . . The *mechanicus*, it is proper to say, is almost a partner of nature, unlocking her secrets, changing what she reveals, playing with wonders, and making such exquisite counterfeits that what we take for truth is certainly artificial" (*Var.* 1.45.10–11).

Anthemios and Isidoros in particular blended armchair physics with practical engineering. They wrote treatises on conical sections, designed parabolic burning mirrors, built bridges, and dammed rivers. Anthemios even devised earthquake and lightning simulators and was, as the contemporary historian Agathias (c. 530–582/594) says, "one of those who apply geometrical speculation to material objects and make models or imitations of the natural world."[129] It must have been this seemingly occult knowledge and paranormal insight that appealed to Justinian when he sought to raise a cathedral that was out of this world, a construction that would be not simply an accumulation of technical feats but an object that concretized supernal geometries and controlled light to result in something, like Christ, "not of this building." As even the kontakion says, the church surpassed "the whole of mankind's knowledge of technology" (strophe 5), and the unwanted proof of this ambition was that the dome would collapse no less than three times (in 558, 989, and 1346).

From a biblical perspective, because Anthemios and Isidoros worked as a team, they replayed the roles of Bezalel and Aholiab, the ur-artificers who had built the tabernacle to God's own blueprints (Exodus 31:3, 6). Probably for the same reason, the Edessene Hymn stresses the paired architects, Asaph and Addai, in its dedication. Justinian, for his part, became a new Moses and a new Solomon in masterminding the enterprise and interpreting the will of the *Sapiens Architectus*, that is, the almighty himself.[130] Thus, Prokopios writes that when Hagia Sophia's vaults threatened to collapse during construction, Justinian found the remedy "impelled by I know not what, but, I imagine by God."

Anthemios and Isidoros were, above all else, geometers and heirs to an unbroken succession of Neoplatonic teaching on geometry and its relation to philosophy that stretched, via Alexandria, all the way back to Proklos (412–485 AD), and beyond him, Plato. In the *Timaeus*, Plato had postulated a creator who crafts the universe from an amorphous mass into a series of perfect geometric solids. Both the idea of the divine architect and a world of metaphysical geometries were maintained and developed by his many interpreters, particularly Proklos.

For Proklos, geometry was the visible image of a truth that could not otherwise be apprehended. Inspired by Euclid, he argued for an essential hierarchy that began with a point, which becomes a line, which can then bound a plane figure, and eventually project into three dimensions. According to this reasoning, the circle took precedence over all lines and shapes because it was simply a line that returned upon itself, without beginning or end. The same logic dictated that next came the semicircle (because it consists of only two lines), then the triangle (three), and finally the square (four). The circle, like the Christian God, embodied a unity and perfection that was a center for diffusion. Hagia Sophia seems to comply with this geometrical metaphysics, in the circle of the dome resting on semicircles of arches and triangles of pendentives over the square of the crossing. Geometry was not just for engineering, it was to make Hagia Sophia a metaphysical object that modeled the descent in stages from the one (God) to the many (the world of things).[131]

Poets and churchmen were privy to the conceptualization of the structure in terms of a symbolic geometry. It was literally their world. Chorikios actually says that he describes a church in the language of the mechanopoios: that the apse vault is the "fourth part of a hollow sphere" (*Laud. Marc.* 1.27), and the main dome a "hollow sphere equally divided" (1.19). One could dismiss such turns of phrase as self-serving credentialization, but it is still the language of an idea, at a geometrical remove from mundanity, and mediating with the grand design of the cosmos and the Platonic chain of being. The chain-of-command model that is automatically assumed for an architectural commission, from patron to designer to workmen, ignores the collaborative culture that involved the shared ideas of emperor, priest, mechanikos, mason, and even poet or hymnodist.

LANGUAGE AND THE LIMITS OF REPRESENTATION

The first dome of Hagia Sophia collapsed in 558 during an earthquake, and three inauguration performances celebrated its reconstruction and Hagia Sophia's rebirth in 563. First, a series of readings from scripture, second the dedicatory anthem (the kontakion), and lastly Paul the Silentiary's ekphrasis from the gallery. All enjoyed mutually supportive roles in unfolding the meaning of the building. The kontakion, in contemporary Greek, glossed preceding readings on Creation, the Incarnation, and the various sorts of tabernacle to demonstrate how the limitless divinity could indwell the physical building.[132] It concluded that, thanks to the mystery of the Eucharist, Hagia Sophia became a genuine residence for God as a spiritual apparition of which the material cosmos was an imperfect representation.

Ekphrases communicated the same themes in a more internalized way. Like Theodore Metochites's much later paean to the Chora, Paul the Silentiary declaimed in Homeric Greek, a language that had become archaic a full millennium before he wrote. Today's equivalent would be to hold forth within a newly unveiled building by, say Frank Gehry, using only Chaucerian English. (Elizabeth Barrett Browning actually translated Paul the Silentiary into Shakespearean English.)[133] Predictably, such willful anachronism, which other authors purposefully adopt, has often been dismissed as "turgid archaism" or marginalized as a court language of scholarly rarefaction.[134] But Homer was not revered just as a poet, but as an ancient sage. Like Vergil, he had long been scrutinized by exegetes seeking to exhume the proto-Christian allegory supposedly buried in his texts.[135] Furthermore, Paul the Silentiary's verse is heavily indebted to the pioneering poet Nonnos (c. 470 AD), who had resurrected Homeric metrics to compose not only the enormous *Dionysiaca* but also the equally long *Paraphrase of the Gospel of John*.[136] The linguistic archaism of such poetry consciously linked Byzantine culture with biblical prehistory in a retroactive synthesis, and Paul's neo-Homeric voice not only appealed to the heroic dimension of the epic, but invoked a timelessness that paralleled the incommensurability of the building at hand.[137]

In the collective effort to will Hagia Sophia into a spiritual vision, its success was most apparent when it became literally unsayable, a regular trick of such orations. As Aristeides had much earlier commented in his own *Laus Urbis:* "All men sing of and will sing of this city [Rome], but they detract from it more than if they were silent, in as much as in silence it can neither be magnified nor diminished from its existing state, but it remains to be known in an uncontaminated condition."[138] Chorikios says of the church of St. Stephen at Gaza (c. 540–542):

> The very church for which many crafts labored, stoneworkers, those knowing about slabs [of marble], and [the artists] through whom imitation [is accomplished] in colors, this solitary oration attempts to describe. And yet, I think it fit that the different skills came into rivalry with each other as in a public contest, concerning which one of them will execute its appropriate task more carefully; and apart from these considerations it is easy neither to view all of these things, since one's vision is directed from one object to another, nor to describe whatever one has seen. Thus, we fall short of the true merit [of the church] by twice as much, because the church surpasses vision while the vision surpasses description.[139]

The elaborate ekphrases of Hagia Sophia were not, as we have said, essays in grandiloquence of indecipherable privacy. They were inductive texts intended to assimilate the material construction to the divine order, reenacting the building verbally, constructing it anew with every declamation. If Hagia Sophia truly embodied the divine in a vision of the invisible and unvisionable, the measure of its success as a divine artifact was the ultimate imperfection of describing it. The very building itself was an imperfect approximation to the unimaginable divine abode of the cosmos, expressed with an unfailing grammar but only in architectural words. The divine was ungraspable, could only be glimpsed in reflection, and could not be constructed in any words. Language must fail, and this indescribability became the literary correlative to the object not made by human hands. The void of description could only be filled by the Word of God enunciated by the lectors in the ambo and the hymns that were offered up to the heavens by the choristers during the Mass.

Ekphrases of Hagia Sophia conclude that the reed is broken and the tongue undone right up to the bitter end of Byzantium. The last Greek ekphrasis by the last

Greek humanist, written in 1411 by Manuel Chrysolaras, is a disingenuously spontaneous tumble of cunningly premeditated thought. Summing up a tradition of speechlessness in the face of the ineffable, where reams of words must be expended to say nothing, each one of them must be quoted:

> But that famous church, to which they rightly gave the name of the Divine Wisdom—and truly it is not a product of human wisdom . . . I believe, in fact, that nothing similar has existed and nothing similar ever will exist in the world. For this reason I too, having simply brought it to mind, will abstain from any other consideration of it, because one can say no more about it that measures up to reality, indeed, one cannot say anything about anything [there]. Whoever has it in mind is incapable of thinking about anything else, and every expression, I must repeat, proves inadequate. And if one does begin, one must spend more time talking about it than all the remainder, without succeeding, however, in ever saying anything that gave it its proper due. . . . What discourse can ever raise itself to a height or grandeur equal to her height or grandeur? What discourse could ever reach her beauty? What discourse could reach perfection worthy of comparison with hers? What simplicity, I mean, would equal her simplicity? What words would succeed in effect and variety so as to equal her variety? Which expressions would equal the precision with which materials have been deployed in her every part, and rival their excellence? Or so well structured as to bear comparison with her harmonious structure? And even if we wished to choose only some part or detail, even an insignificant one, we would still be unable to speak of it in a worthy manner. We could not even express ourselves in an adequate manner about the doors, or the entrances, about the floor, the vestibule and the beauties that can be admired in front of the doors, about the columns, mosaics, the marble revetments of the walls, the glass, bronze, lead, iron, sapphires, gold, or even the glass of the mosaics, or about any other part that we could meticulously catalogue.[140]

8 Walking on Water

Cosmic Floors in Antiquity and the Middle Ages

When Hagia Sophia was completed and dedicated with great fanfare shortly after Christmas 537, Emperor Justinian could rightly say that he had outdone Solomon. Such was the splendor of the completed vessel that God Almighty might be tempted to descend among men and dwell within this place. Justinian's expenditure on the church was fabled: it was said that he had spent forty thousand pounds of silver on the sanctuary screen alone. No expense was spared on its construction, its gilt tesserae, liturgical furniture, lamps, silk hangings, chalices, and patens, or its acres of polychrome revetment.[1] The glaring exception seemed to be the point where the whole construction of faith met the earth's surface, the floor (see figs. 7.1, 8.1). Here, there were no vermiculated mosaics, no rainbow imbrications, no intricate tessellations, not a petrous pullulation in sight. Instead the floor presented an expanse of Proconnesian-marble flagstones, compartmented by thin green stripes, with any eye-catching and multicolored paving screened off behind the sanctuary barrier.

Yet the wonder of visitors was undiminished. Over more than a millennium, observer after observer would report that the combined undulations of its book-matched slabs suggested that the entire surface was a "frozen sea." The perduration of this perception reveals the enduring propriety of an extraterrestrial image that the faithful could read into the shifting matter below their feet. By "walking on water," they were reminded of the world's watery genesis, its apocalyptic destiny in a glacial purity, and that, all this while, God's throne sat "above the waters," gliding over a celestial sea.

FLOORS AS "SEAS," EAST AND WEST

The simplicity of the floor is all the more striking because Justinian could have chosen from a rich array of paving options. The floor eschews even the simplest geometric schemes, such as those that Rome's basilicas had borrowed from temples or civic basilicas, which were geometrically straightforward but enlivened by the richness of their counterchanged marbles.[2] In the Eastern empire, extremely rich floor mosaics had decorated the churches of Palestine, Jordan, and Syria, in the fourth century abounding in personified seasons and the creatures of earth and sea. Despite a gradual drift toward a puritanical aniconism between the mid-fourth and early fifth centuries, under Justinian a full-blown renaissance of the medium reclaimed the whole decorative repertoire of

FIG. 8.1. Nave paving, Hagia Sophia, 532–537 AD. Proconnesian marble. Istanbul.

antiquity, whether floors in churches or Justinian's own palace. Mosaic floors might have been impractical for the traffic that Hagia Sophia could expect, but the large marble slabs also offered a greater shimmer than any mosaic and, thanks to the veining of Proconnesian marble, the overall image of a frozen sea.

When the imperial marshal Paul the Silentiary recited his famous ekphrasis on the spot in 563, he compared the solea and ambo, the access-gangway and pulpit that pushed out into the nave, to a wave-lashed isthmus in a stormy sea, and ventured that those traversing the church found a safe harbor only in the liturgical destination of the sanctuary:

> And as an island rises amidst the waves of the sea, adorned with cornfields, and vineyards, and blossoming meadows, and wooded heights, while the travelers who sail by are gladdened by it and are soothed of the anxieties and exertions of the sea; so in the midst of the boundless temple rises upright the tower-like ambo of stone adorned with its meadows of marble, wrought with the beauty of the craftsman's art. Yet, it does not stand altogether cut off in the central space, like a sea-girt island, but it rather resembles some wave-lashed land, extended through the white-capped billows by an isthmus into the middle of the sea, and being joined fast at one point it cannot be a true island. Projecting into the watery deep, it is still joined to the mainland coast by the isthmus, as by a cable. . . . Here the priest who brings the good tidings passes along upon his return from the ambo, holding aloft the golden book; and while the crowd strives in honor of the immaculate God to touch the sacred book with their lips and hands, the countless waves of the surging people break around. Thus like an isthmus beaten by waves on either side, does this space stretch out, and it leads the priest who descends from the lofty crags of this vantage point to the shrine of the holy table.[3]

This nautical motif recurs in the folkloric account of the ninth century, the *Diegesis* or *Narratio,* which commends the pavement as "like the sea or the flowing waters of a river," and the ekphrasis of Michael the Deacon (c. 1140/50) more extravagantly returns to the theme of a sea dotted with islands, one of which is the ambo:

> The floor is like the sea, both in its width and in its form; for certain blue waves are raised up against the stone, just as though you had cast a pebble into water and had disturbed its calm. This sea has broken out into a gulf to eastward, and one wave having been, as it were, piled up against its predecessor, and another against the next (for thus also does it happen during floods, the ever-approaching wave never allowing itself to be broken by the contrary wind), the sacred Sphendone has been formed into steps, and one step is raised up above another, and the highest steps which curve in billows have been flooded over by an effusion of silver worth many talents.[4]

The *Narratio* also describes the paving as traversed by the rivers of paradise, meaning the bands of green Thessalian marble (verde antico) that compartment the nave.[5] Although this parallel tradition was disseminated as far as London and Russia, it proved comparatively shorter-lived.[6] Even Mehmet the Conqueror, on the day of Constantinople's fall (May 29, 1453), so admired this "sea in a storm" that he took a sword to a disobedient soldier trying to prize a slab from the floor. Cafer Çelebi's slightly later encomium of the same building (1493/94) also extolled its marble waves, as would several Ottoman poets after him, while the Florentine Bernardo Bonsignori (1498) is perhaps the last Westerner to repeat the observation, comparing the surface to watered silk, before the pavement submerged under Muslim prayer mats.[7]

Within Constantinople, the Apostoleion (c. 536–550) had an almost identical floor, and the rippling influence of these Justinianic floors can still be observed in the "pools" that nostalgically fill later Byzantine churches such as the Chora (Kariye Camii, c. 1316–21; see fig. 7.12), the Parekklesion of the Theotokos Pammakaristos (Fethıye Camii, c. 1310–14), and perhaps the eleventh-century Pantanasse Church, also in Constantinople but now destroyed.[8] There were probably others, but today the only surviving Byzantine church that shares with Hagia Sophia the distinction of an entirely Proconnesian floor is the Acheiropoietos in Thessaloniki, a mid-fifth century structure, but one whose paving could date from the mid-seventh (fig. 8.2).[9] Unlike the frozen sea at Hagia Sophia, the floor of the Acheiropoietos suggests a fossilized river flowing toward the sanctum, and we might think of it as a second Jordan.

However, the "seafloor" had already traveled west to fill the naves and crossings of Italian churches, in a variety of techniques. Highly descriptive sea scenes, with barks and darting fish, had featured in the floors of the early

FIG. 8.2. Paving, Acheiropoietos Church, mid-5th century or c. 620 AD. Proconnesian marble. Thessaloniki.

FIG. 8.3. Crypt floor, S. Savino, c. 1120–30. Mosaic. Piacenza.

fourth-century double basilica of nearby Aquileia and of S. Pudenziana, Rome (c. 384–399).[10] Abstracted versions also persisted, as in the crypt floor of S. Savino (c. 1120/30) at Piacenza, where zodiacal roundels bob about in a zigzag sea, populated by leaping sea life, including mermaids and sirens (fig. 8.3).[11] Mosaic waves also lapped the floors of eleventh- and twelfth-century Venetian churches such as S. Zaccaria and SS. Maria e Donato on Murano (1141), albeit in the disguise of interlinked crescent-shaped shields (or *peltae*).[12] This particular convention had already arrived in Grado, the seat of the Venetian patriarchate until as late as 1451, in the late sixth century: in the nave of S. Euphemia (579), ranks of peltae stream toward the altar (like the Acheiropoietos in Thessaloniki), wingtip-to-wingtip but facing in alternate directions (fig. 8.4). Overall these "painted marbles concealing the squalid earth" (as the dedicatory inscription says) merge into a ripple effect so evocative of waves that it has been compared with the furrowed surface of a tide-swept beach.[13] As it happens, in 1211 a German visitor to a Crusader palace in Beirut had voiced the same thought when he admired "a fine marble pavement that so well feigns water stirred by a light wind that, whoever steps over it, seems to be wading, since they leave no footprints above the sand depicted there."[14]

All these floors differed in materials and techniques, but the Proconnesian "seafloor" had also migrated to occupy the floors of two of the most ambitious churches built in Italy during the High Middle Ages, the Abbey Church of Montecassino and the palatine chapel of S. Marco, Venice. At Montecassino (1066–71) the "sea" was covered over in 1725 to 1729 and eradicated by Allied bombing in 1944, but the floor of S. Marco survives.[15] Laid sometime between 1110 and 1150, it pullulates with tessellated whorls and quincunxes that figure astral geometries, though even these shifting patterns could be considered seas, as a near-contemporary description of a similar pavement observes: "What is spread on the floor, and what clothes the whole space like a dress worked in colors might at first be called a sea, which, moving on all sides in the gentlest waves, is suddenly petrified."[16] Yet the jeweled waves of S. Marco only accentuate the hiatus of massive Proconnesian slabs under the crossing. These slabs, which were quarried specially for the occasion, have been collectively known as "il mare" (the sea) since at least the seventeenth century, but the tradition must be much older (fig. 8.5).[17] The pendentives of the dome above this "mare" contain personifications of the four rivers of paradise—Gyon, Euphrates, Phison, and Tigris—that empty their amphorae toward their feet, and so the floor.[18] Added to all this was, of course, the inescapable frisson that Venice was herself a city founded upon water, a fact that became painfully obvious with every seasonal *acqua alta*.

FIG. 8.4. Nave floor, S. Euphemia, 579 AD. Mosaic. Grado.

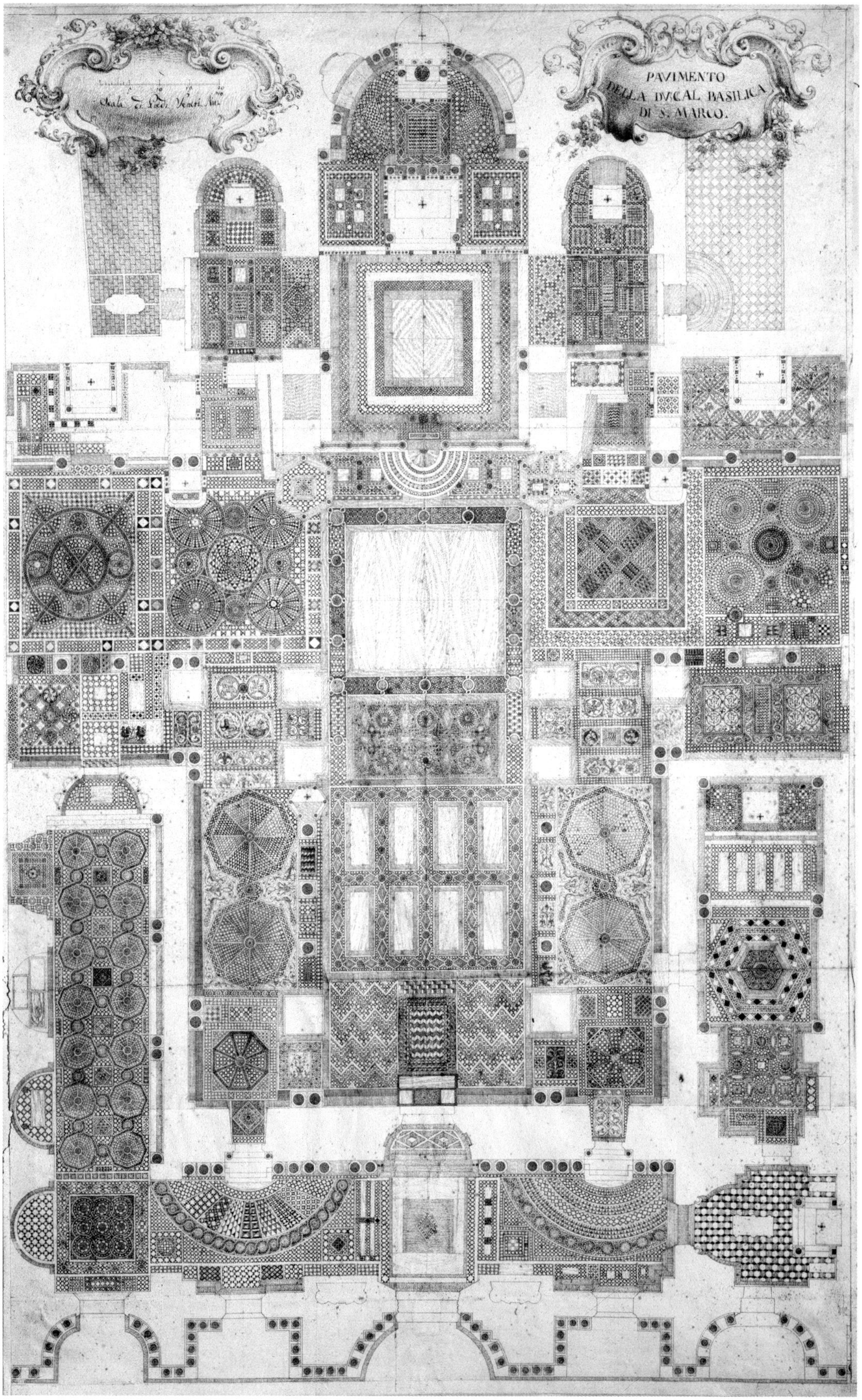

FIG. 8.5. Inlaid marble floor, with "Il mare" clearly visible at the crossing, S. Marco, 13th century. Various marbles. Venice. From Angelo Visentini, "Survey of the Pavement of the Basilica di San Marco," 1761. Archivio Storico della Procuratoria di San Marco, Venice.

The island quarries of Proconnesian marble were within easy seafaring reach of both Constantinople and Thessaloniki, on sea-lanes relatively immune to pirates, and had provisioned these cities and much of the Mediterranean with architectural marbles for centuries.[19] It might be argued, therefore, that the relatively unornamented floors of the Hagia Sophia and the Acheiropoietos were surfaced with this plentiful stone with little thought for any signification other than elegant utility. But there were far older and deeper associations to this marble, and marbles in general, that went straight to the heart of the matter. First was the still-lively perception that marbles might be liquid, and second the more particular kinship between marble and the sea.

MAR/MARMOR/MARMORA

The antique theories of geology that taught that marbles were of essentially watery origin still held, and in late antiquity they were still perceived as congelations of clammy vapors. When the poet Flavius Merobaudes, for example, eulogizes, c. 435/446 AD, a now-vanished baptistery (possibly that of S. Croce, Ravenna), he commends the marble font with the words "water wettens the marble pools . . . and burdens the crystals with rapidly flowing currents. The jewel, once liquid itself, now carries the liquid" (*Carm.* 2.5, 7–8: [*m*]*armoreos rigat unda lacus* . . . [*clar?*]*aque festinis onerat crystalla fluentis: / gemma vehit laticem, quae fuit ante latex*).

During the Medieval period, these geological perceptions received continued support from Arab science in the East, the same texts masquerading in the West as by Aristotle. Thus, in his *On the Congelation and Conglutination of Stones* (1021/23), the Arab physician Avicenna (980–1037) deduced from observation of alluvial formations (*conglutination*) and the growth of stalactites (*congelation*) that there must exist a lapidifying "mineral force" that freezes water.[20] Such a conclusion would have been considered especially authoritative, because a Latin epitome of Avicenna's treatise was often appended to Aristotle's *Meteorology 4* as an extra chapter ("*De mineralibus*"), and therefore attributed to "the Philosopher" himself.[21] Aristotle's word was itself sacrosanct because his Latin translators had infiltrated proto-Christian nuances into their translations, and the observations in "*De mineralibus*" continued to be disseminated by sainted writers such as Albertus Magnus as well (1220s).[22]

Significantly, Avicenna (pseudo-Aristotle) did not regard the transformation in question as a deposition of solids in water, but rather the actual metamorphosis of water itself into stone. The widely held belief that mountains could renew themselves by sweating out marble into their quarry scars also persisted, and the memory of Hierapolis, the city that had been supposedly built out of water (see fig. 3.4), endured in the West thanks to its brief description in Vitruvius, even though it had been abandoned after the tenth century.[23] In northern Europe, building stone was actually extracted from the deposits within the conduits of the aqueducts that had fallen into ruin. Charlemagne quarried this rather murky material to decorate his Palatine Chapel at Aachen, and between the eleventh and thirteenth centuries, more than a dozen churches were built using it in Cologne, whence it was also exported throughout the surrounding region (columns, altar slabs, and so on), and beyond, to England (Canterbury) and Sweden. A seventeenth-century Jesuit author even recounts how masons excavated the sinter from the cuniculus of the Aqua Virgo—that which passed over the Porta Maggiore (see fig. 3.5)—to make "alabaster" slabs, lamenting only that it could not take the high polish to which the masons were accustomed.[24]

These slabs were travertine, but such genuinely natural products fed wilder rumors. As late as 1491, an anonymous Ottoman writer found it necessary to refute the common belief that the porphyry columns in Hagia Sophia were formed from a frozen, water-based dye, and Filarete even cooked a piece of marble from a column at the Roman church of the Aracoeli to prove that it was not a water-based conglomerate.[25]

The presence of marine fossils only encouraged the perception that certain stones were petrified water, but literary tradition also invited the association enshrined in the word "*marmor*" itself. The Latin noun "marmor," which has resulted in all modern European derivatives ("marmo," "marbre," "marble," "marmor," and so on), descended from the Greek verb "*marmairein*," meaning "to glisten," and so on. "Marmairein" in turn was the iterative form of a verb whose Sanskrit root, "*mar*," implied motion. "Mar" had originally indicated the movement of the waves, and "*mar-mar*" the more agitated stirring (or murmuring) of the sea.[26] In poetry, this assonance proved pregnant with possibility. When Homer had spoken of "*hala marmara*" (*Il.* 14.273), he had simply meant the "shimmering sea," but when the Hellenophone Latin poet Ennius writes, about 184 BC, of a ship "skimming the calm sea's golden marble," a whole new realm of metaphor enters the poetic

FIG. 8.6. "Mouth of Truth," S. Maria in Cosmedin, 117/138 AD?. Phrygian marble. Rome.

FIG. 8.7. "Red Sea" Sarcophagus, mid-4th century. Red onyx. Museo Civico di Brescia.

consciousness.[27] There is no equivalent in English, but the nearest analogue is perhaps "the glassy sea." Catullus, Valerius Flaccus, Lucan, Lucretius, particularly Vergil, but many others often use "marmor" as a synonym for "mar" to imagine the sea's hard surface and hidden weight. Vergil fathomed marmor's depths by describing a calm in which "tides of marble smoothness meet the laboring oar,"[28] but the metaphor reaches fruition, and water again becomes stone, in the verses of Ovid. In wintry exile on the shores of the Black Sea, he writes that the chill is so fierce that "the ships, shut in by the cold, will stand fast in the marble surface and no oar will be able to cleave the stiffened waters."[29] Eventually this particular association became so much a mental habit that the emperor Julian, wintering in Paris (358–359 AD), will write home to describe a frozen Seine in Impressionist hues as huge sliding plates of Phrygian marble.[30]

With this in mind we can understand all the better why this marble was chosen to materialize the god Oceanus on the second-century drain cover, nowadays called the "Mouth of Truth" (fig. 8.6).[31] Geographically, the stream Oceanus encircled the world and marked its limits, but he was also a great cosmic power, the watery mass from which the Greek world was born. Thus, his disembodied visage and its staring eyes are the fulcrum of innumerable mosaic floors, and on some of these, and many sarcophagi, Ocean's hair and beard materialize out of piled waves. On the "Mouth of Truth," the deity's head is contained within the "birthmark" of the marble, so that when he reared his head in public it seemed a body of water itself, the face of the deep. Conversely, this mask must have made whatever court it once adorned a microcosm across which the waters streamed back into Ocean, just like those daily draining off the earth at large and from the sky above. Likewise, a mid-fourth-century sculptor who picked out an intensely red, onyx block with particularly gushing veining for a Christian sarcophagus in Brescia (fig. 8.7) understood the material paradox well enough to enlist this marble as the vehicle par excellence for images of the *Crossing of the Red Sea, Moses Striking the Rock, The Incredulity of St. Thomas,* and the *Flagellation.* When the Israelites had made their crossing, the sea had solidified into "a structure, created by a hanging wall of water, [which] held back the sea and kept it suspended in the air," Moses withdrew the water from rock, blood flowed from the wounds of Christ, and the column of his torture was spattered with his blood.[32]

CARYSTIAN AND PROCONNESIAN MARBLE

Ennius had written of waves of "golden marble," Homer of the "wine-dark sea," and rarely did ancient eyes regard the waves as "blue." More often than not, they were green, and one marble that regularly impersonated water was Carystian (*Cipollino*), from Euboea in Greece. Its marine veining, Statius reveled, was "the same color as the deep sea"; it was "the rock that vies with the gray-green sea," and that "joys to behold the waves." Even the quarry was "wavy."[33] As a result, Carystian was recurrently employed in bath complexes throughout the empire, private or public, and the desired effects are now nowhere more obvious than in the *frigidarium* of the Villa of the Quintili on the Via Appia (late second/early third century; fig. 8.8).[34] Here, Carystian ripples under the geometric floor, the columns are of the same marble, and if the now lost wall revetment were as well, one would have felt submerged before even entering the pool. Even in Nilotic mosaics, the river is sometimes indistinguishable from a shaft of Carystian marble.[35]

Analogously, the mosaic paving in the peristyle impluvium in the Maison du Char de Venus (after 317 AD) at Thuburbo Maius imitated large book-matched slabs of Carystian marble. This floor drains into a cistern below, which it therefore intimates, although one can only imagine the surface effects when the drain was plugged. Significantly, the frigidaria of nearby baths deploy the same paving.[36] Even where the marble was not so accurately imitated, its distinctive veining might be summarized geometrically, as in the enormous font of the Donatist Baptistery (late fourth/early fifth century) at Timgad in Algeria, where the chevrons of the mosaic steps evoke cascading veining (fig. 8.9).[37] In another example, at Hadrian's Villa at Tivoli, the columns and paving (and missing revetment?) of the ambulatory encircling the famous Teatro Marittimo (c. 118 AD) were again exclusively of Carystian marble, with the result that the emperor's hideaway would have originally floated in a tranquil "sea," not an ornamental pond.[38] In fact, since this is a ring canal, the sea in question was probably no less than Ocean himself.

By the sixth century, however, it was Proconnesian marble, the flooring of Hagia Sophia, that had largely supplanted Carystian in its power to epitomize the sea. Glistening white blocks of this marble had already been quarried from the Isle of Proconnesus for centuries, but the Byzantines sought out the faces streaked with dove-gray seams and enhanced its "ripple" veining by slicing the slabs across the bed and cutting columns on the bias (see fig. 7.11).

FIG. 8.8. Carystian marble floor panels and columns, Frigidarium, Villa of the Quintili, late 2nd/early 3rd century AD. Rome.

FIG. 8.9. Font, Baptistery, late 4th/early 5th century. Mosaic. Timgad, Algeria.

FIG. 8.10. Baptismal font, Baptistery, Hagia Sophia, 5th century. Proconnesian marble monolith. Istanbul.

Prokopios remarks on the "wavy lines of blue that mingle with the white," and Paul the Silentiary commends the "Bosporus Stone" that "gleams white but on whose white skin a blue vein winds a scattered path." But it is a twelfth-century poet who truly gushes, when describing a fountain whose "flowing water through its clarity took on the colors of the marble. Island marble decorated the base of the well; it was white but with a black hue in places and the black imitated the painter's art since the water seemed to be in constant motion and to billow up and almost burst out."[39]

Finally, by the time of the Latin conquest of Constantinople (1204), the equation of Proconnesian with water was consummated in the knowledge that the island from which these slabs hailed ("sea-girt Proconnesus," as Paul the Silentiary says) had given the encompassing sea, originally called Propontis, its popular name: "Marmara."[40]

In the Baptistery of Hagia Sophia, the font was hewn from a single massive block of Proconnesian marble that had been cut so that its blue-gray veins lay horizontally, like tidemarks in the stone (fig. 8.10). This vessel too originally lay under a gilded dome, and it was made a century before Justinian's Hagia Sophia, meaning that the effect of seeing it scintillating with real water ("the jewel, once liquid itself, carries the liquid") and immersing oneself in its transformative waters was both an intimate and widely shared experience before anybody came to think about how they should pave the nave of the new church.[41]

THE COSMIC FLOOR: THE MEADOW, THE OCEAN, AND THE GLASSY SEA

Marine images were often desirable for a church floor. Numerous late antique and medieval floor mosaics in both the eastern and western Mediterranean have as their template the ancient ideogram of the mythical Ocean encircling the world (see fig. 13.3).[42] In the eastern examples, the nave floor (representing the *oikoumene*, or inhabited world) is often bounded by a decorative border representing Ocean. In one instance, Ocean may even have been simulated in the real canals that surrounded the great cathedral at Edessa, Syria (c. 543–554 AD).[43] In each case, the devotee found the world at his feet, transforming the church (or imperial hideaway) into a model of the universe that put it on a cosmic footing. Moreover, if one pursued the cartographic suggestions of the floor to their logical conclusion, then the sanctuary occupied the position of paradise itself.[44]

To most contemporary eyes, the glittering Cosmati meadows that were strewn across S. Marco and Montecassino rendered the church an Eden of gems that flashed underfoot like the garden of God described by Ezekiel (28:13, 14). But a watery floor that imaged an entire sea, as did the expanse of Proconnesian marble in Hagia Sophia, promised to be the alpha and omega of all such premonitory materiality. For when its shimmering surface lay beneath an overarching dome of luminous gold, the whole construct became a simulacrum of God's "Separation of the Waters" in Genesis (1:2–8): "And the Spirit of God moved upon the face of the waters . . . and God made the firmament, and divided the waters which were under the firmament from the waters which were above the firmament. . . . And God called the firmament 'heaven.'"

These verses were fundamental for the Judaic view of the universe, and thereafter the Christian and Muslim versions.[45] Not only had God divided the waters, but he also sat "above the waters" since Creation. And on Judgment Day, Revelation tells us, his apocalyptic throne will finally be visible to all, resting upon "a sea of glass like to Crystal," or alabaster, a "sea of glass mingled with fire" (4:6; 15:2). In fact, this is exactly how we see him, above a shifting sea laced with flame flanked by horn-blowing apocalyptic angels, on the proscenium of S. Michele in Africisco (c. 545), Ravenna, mosaics that were installed within a decade of the inauguration of Hagia Sophia (fig. 8.11).[46]

In Hagia Sophia, observers could find both seas, the deep and the celestial. The two waters essentially compose the basso profondo underlying the dancing reflections of the materials. From remotest antiquity until the seventeenth century, few doubted that rock crystal was a form of ice that had been frozen by primordial cold. The

FIG. 8.11. Triumphal arch, S. Michele in Africisco, c. 545 AD. Ravenna. Mosaic. Bode Museum, Berlin.

FIG. 8.12. Enkolpion with Christ Pantokrator, 11th century; mount, 16th century. Rock crystal, gold, gemstones, and pearls, 2 3/8 × 2 3/8 in. (6.1 × 6 cm). Benaki Museum, Athens.

Byzantine poet Manuel Philes (c. 1275–c. 1345) penned verses to this effect that were inscribed on a rock-crystal relief of Christ Pantokrator, Christ the "all-powerful" (fig. 8.12):

> This stone is water, not really stone;
> He who freezes flowing water into ice
> Also freezes this into the nature of stone
> Lest the rock melt and flow away.[47]

It seemed as though light (the active principle of the Logos) was frozen into its very fabric. Thus, when marble, which was a more opaque cousin of crystal, was polished, it recovered this original light in a surface slick. The common resemblance of shimmer to wetness, the "wet look" that mosaics and marbles alike could achieve, therefore, pointed beyond the surface to a substratum of physical affinities. Mesarites will describe the shimmer of the glistening marbles in the Apostoleion by using the word "*hugrotēs*" ("wetness"); Niketas Magistros says that "the glitter of the marble" revetment in the Panagia Ekatontapiliani on Paros "exhibited such liquid refulgence as to surpass the brilliance of pearls"; and Michael the Deacon describes how the "countenance" of Hagia Sophia "flashes forth like liquid through gold which is everywhere. . . . The brightness of the gold almost makes the gold appear to drip down; for by its refulgence making waves to arise, as it were, in eyes that are moist, it causes their moisture to appear in the gold which is seen, and it seems to be flowing in a molten stream."[48]

The dome of heaven had "doors" and "windows" to let the waters from the supercelestial sea above fall to the earth (Genesis 7:11; Isaiah 24:18; Malachi 3:10). In Hagia Eirene (548) the inscriptions on the proscenium arch of the choir, and therefore below the dome, speak of the faith that pours from on high and includes the exclamation, "Who builds his upper chambers in the heavens and founds his vault upon the earth; who calls for the waters of the sea and pours them out upon the surface of the earth—the Lord is his name" (Amos 9:6).[49] The golden dome of Hagia Sophia too became a "shower of light," tumbling down in a luminous cascade, washing the walls and soaking the floor. The soldier-poet John Geometres (fl. 959–990) makes this sort of observation when he describes the columns of the Stoudios Church (454/463), Constantinople, melting back into their watery cradle in the earth and discharging over the floor in the process:

> The polished splendor of these stones
> Seems another sea without waves
> As though just now it has fallen calm.
> The light and luster of the columns,
> Their lovely sparkle, resembles
> A river glistening with dissolved
> Snow, which, almost another sea, flows
> Towards the glossy stones of the floor,
> Silently.

The floor instead was opus sectile, and the poet concludes:

> And the earth
> And its treasures? Look at the varied play
> Of its colors and the harmonious
> Designs. You see here a meadow
> Composed by art, clothed in flowers
> That time will not wither.[50]

The 562 kontakion of Hagia Sophia tells us, "In the beginning the firmament solidified in the midst of the waters . . . with a flowing substance . . . above it. . . . But here things are better and utterly wonderful: no flux, for the favor of God is the foundation on which rests the temple of God's Wisdom" (strophe 7); it also refers to the dome as the "*stereoma*," that is, "the firmament solidified in the midst of the waters." In the midst of this dome was an image of the Pantokrator, and it is as though the church had simply solidified from the flux at the command of this "all-powerful" one.

The glossy materials of the domed church allow it to be regarded as globally crystalline, impinged by the local color of earthly generation or the glow of heavenly ether. Divine light bleached the church's upper shell, paradisiacal landscapes (although rainbows or peacock wings, or other images of iridescence and multiplicity would do just as well) inhabited its walls, and cosmic waters pressed up against its floor. The symmetry of the whole construct is evoked by Claudian, who describes a crystal ball, or lens, as a miniature replica of the cosmos in his epigram "On a Crystal Enclosing Water" (*Anth. Pal.* 9.753; c. 390/395): "The snow-white crystal, fashioned by the hand of man, / Showed the variegated image of the perfect universe, / The heaven [i.e., spherical], clasping within it the deep-voiced sea."

Such liquid and crystalline light are also the themes that underlie the Islamic fountains that attempted to perform as microcosms of real water. An eleventh-century one in an Arab house in Cordoba is described thus: "From its head water fell in the form of a dome upon a floor of alabaster and marble; lights were set inside this 'dome' and were thus covered by it." A celebratory poem makes more obvious the microcosm when it surreptitiously asks:

> Tell me what is the torch upon the lamp
> That sprouts crystals onto a crystal base?
> A stream that will not kill fire in its midst,
> Its waters standing like a wall and missiles,
> A sky encrusted with an onyx skin
> Stretched over a ground of bedellium.[51]

The upper waters had been frozen at Creation into crystal, a fact that one knew because at night one could see the stars through the heavenly spheres, and this archetypal construct was born again in the letter of scripture. As the book of Job lays out, when God had *divided the waters,* his breath had crystallized the sky, and he had "shut in the sea with doors, when it brake forth, as if it had issued out of the womb" (38:8).[52] The simulacrum of the church therefore fixed the image of Creation in a material metaphor that literally enacted Job's words that "the waters are hid as with a stone, and the face of the deep is frozen" (38:30).

A church floor that was a frozen sea could, therefore, evoke at one and the same time the Creation and the Apocalypse, by recalling the ambience of God's throne room beyond human time and out of this world. In the beginning, God froze the waters, and when he renews the universe at the end of time, he will restore the earth's original luminosity, and its surface will no longer be dark and dull but will become a diaphanous mass as sleek as glass.[53] In this light, the brilliant, polished floor therefore becomes a mirror of the divine plan, and the importance of the words being spoken and the ritual taking place acquire ever greater resonance. The glacially white floors eventually recorded in various Constantinopolitan churches probably harbored similar ambitions.[54]

JUDAIC PRECURSORS AND ISLAMIC PROGENY

The patrons, architects, and theologians who thought "sea" when they surveyed a church floor would recall the "brazen sea" in the prototypical Temple of Solomon[55] or the waters that flowed from the temple in Ezekiel's vision, and the various pools and fountains of Hagia Sophia's long-lost atrium consciously prefigured the simulations of the interior.[56] Rabbinical commentaries on the third, Herodian temple make it clear that the visual association between marble and water was very well established in Judaic lore long before its Byzantine appropriation.[57] A Babylonian Talmud of the fourth century recounts that Herod "intended to overlay [the wall] with gold, but the Rabbis told him, 'Leave it alone for it is more beautiful as it is, since it has the appearance of the waves of the sea.'"[58] The identity of the marbles envisioned by the Amoraim (Talmudic scholars) is sealed by their use of the Greco-Latin word "marmar," and other texts refer to the stones of "*Perak Onsin,*" in which can be detected a corruption of "Proconnesus." Tziona Grossmark has shown that another, cryptic warning—"When ye arrive at the stones of pure marble, say not, Water, water!"—derived from the Hekhaloth, mystical accounts in which the visionary passed through seven heavenly palaces (Hekhaloth) to reach the throne of glory: one of the journeyers "stood at the entrance to the sixth palace and saw the splendor of the air of pure marble stones and he opened his mouth two times and said Water, water."[59] In other words, the material portended a heavenly vision, the supercelestial sea, and another medieval version of this story lays emphasis on this otherworldly dematerialization by describing the mirage of the "hundreds of thousands and millions of waves of water [that] stormed against him, and yet there was not a drop of water, only the ethereal glitter of the marble plates with which the (Sixth) Palace was tessellated."[60]

Nor were Islamic poets blind to the rainy allure of Proconnesian marble. A eulogy by the ninth-century poet Buhturi describes a Samarran palace, which was:

FIG. 8.13. "Foam of the sea," Great Mosque, 709–715 AD. Damascus.

> As if the glass walls of its interior
> Were waves beating upon the seashore;
> As if its striped marble, where its pattern
> Meets the opposite prospect [i.e., book-matching],
> Were streaky rain-clouds arrayed between clouds,
> dark and light,
> And striped, coming together and mingling.[61]

But these pan-Mediterranean water metaphors had no impact on mosque floors, both because they had to be covered with matting to accommodate the five daily prostrations toward Mecca, and because the mosque is essentially a prayer hall with no pretensions to housing the divine presence nor portraying the end of days. In the Great Mosque at Damascus it is the revetment surrounding the mihrab (fig. 8.13) that came to be known as "foam of the sea," not the floor.[62] In Islamic and Judaic lore, in midrashim and even the Qur'an, the image of the watery floor is more commonly found in descriptions of palace interiors. Because God's throne stood over a glassy floor, so did Solomon's, and when the queen of Sheba entered his palace, she too was fooled that its marble floor was a pool of water.[63] The crystal walls and watery floor of Solomon's palace would in turn exercise their rule over those sovereigns who wished to emulate the paragon of god-given kingship.[64] A glass pavement was even built, in Syria, in the late eighth or early ninth century, and the Qur'anic verses describing the glass floor frequently decorate the walls of medieval palaces with complex illusionistic or vitreous ornament, the Alhambra being an especial case in point.[65] "Glassy seas" and "Ocean floors" also existed in Western palaces, or at least poets imagined that they did. Baudri de Bourgueil (1045/46–1130), archbishop of Dol, describes the elaborately cosmographic floor in the audience hall of Adèle, comtesse de Blois: "It was completely covered by a glassy surface: the surface itself was called 'the glassy sea,' and its material was as clear as if not clearer than glass; lest the hostile feet of visitors crush it, it was supported by marble set below. The work was girdled by a fluid and green color, that you would think it the work of the sea in movement. This work took the name and the form of the ocean."[66]

Likewise the tenth/eleventh-century *Digenis Akritis*, a Byzantine ballad strongly infused with Islamic influences, imagines that the craftsman "paved the [palace] floor with onyx that had been so highly polished that onlookers thought it was water frozen into ice."[67]

THE SEA OF FAITH

The watery floor was not a medieval innovation, nor the monopoly of the Peoples of the Book. The Temple of Zeus at Olympia and the Parthenon had both had liquid floors, of oil and water, respectively (see figs. 2.8, 2.9). The nebulous alabaster floors of the Valley Temple of Khafre at Ghiza and Roman palaces (see fig. 5.8), and the virtualized fresco floors of the Minoans and Mycenaeans (see figs. 4.12. 4.13), all in their several ways also disputed the solidity of the earth underfoot. Their polished planes dissolved the floor and made the rock bottom drop out of the world of the spectator, who then trod a precarious line through extraterrestrial space. In the case of Hagia Sophia, still waters run deep. The conjunction of marble and sea explicitly collapses the antitypes of Creation, sea, and mountain into one horizon on which human finitude is set between lucid firmness and umbrous chaos.

Book-matching also meant that this stony sea emerged from an ordering that was innate. While the receptive spectator could read figurative images into these symmetries (exactly like Rorschach's inkblots), in this case they actually figured the image of a substance and element, water. This miracle of art erased the indices of facture, and helped sacralize the interior by assimilating the new creation to the product of artful nature, not the artwork of hands. No figural imagery appeared in Hagia Sophia's mosaics before the ninth century, so, in short, this was an aniconic floor for an aniconic church. The floor therefore avoided any of the pagan connotations of the personified "sea" which is, for example, the focus of the Church of the Apostles at Madaba in Jordan (578–579).[68] But although the imageless paving preordained no specific narratives, its received materiality and fundamental situation virtually predestined a proper range of reference.

For a start, the worshipper who headed across the waters to the sanctuary in Hagia Sophia, or Grado for that matter, retraced the steps of Peter in his march of faith across the stormy Sea of Galilee to meet Christ "the rock" (Matthew 14:29). Christ had quelled the chaotic waves of this sea just as his father had stayed those at Creation.[69] To slip beneath the surface would be to fall from God's favor, to await like Jonah reclamation from the depths, and several odes in the later Byzantine canon call upon salvation in paraphrases of Jonah's prayer in the belly of the whale.[70] Moreover, it was with this biblical event in mind that a ninth-century chronicler would have us believe that Galla Placidia, after shipwreck in 424, built a new church to St. John the Evangelist in Ravenna. Below mosaics depicting her perils on the open sea she had a pavement laid that was *undosum undique mare*—"everywhere a wavy sea"—as though it were "stirred by the winds, to produce the image of a tempestuous storm."[71] In S. Marco, pausing at the crossing is like waiting before the glassy sea of God's throne room. The same was once true for the abbey church of Montecassino, which hoped to surpass Hagia Sophia and whose floor was probably laid by Constantinopolitan craftsmen. As Alfanus of Salerno's poem on its splendors sings: "Here the green and porphyretic stones make the alabasters shine; and at the same time the Proconnesian paving matches these marbles to each other in such a way that this work may become a glassy sea." (Tellingly, he also claims that "Justinian's church [Hagia Sophia] would gladly trade places with you.")[72]

For others, Hagia Sophia's floor recalled the words of Job. The lector who chanted the kontakion (homiletic anthem) upon its second inauguration on Christmas Day 562, from the ambo in the middle of the "sea," told the masses, "In the beginning the firmament solidified in the midst of the waters . . . with a flowing substance, as it is believed to be, above it. . . . But here things are better and utterly wonderful: no flux, for the favor of God is the foundation on which rests the temple of God's Wisdom."[73] The lector sang in lucid, contemporary Greek, but when Paul the Silentiary lectured to the more select crowd of emperor and patriarch a few weeks later, from some gallery vantage overlooking the "sea," he declaimed Homeric hexameters. The linguistic shift reflected a revived republic of letters and the court culture that sponsored them, but it was also a timeless, epic voice rising to the heroic ambitions of the edifice, and one that simultaneously sought to harmonize Greek and biblical antiquity. Describing the tides of the populace that virtually assault the priest in their fervor to reach the Word (scripture, but at the place of the spoken word, the pulpit), Paul draws not only on Homer's figure of waves of Achaeans besieging Troy, but also the common sermonizing metaphor of the fertile island of the church as a "rock of faith" in a raging sea of sin. Likewise, when Paul speaks of the voyage of the faithful across the sea to the safe haven of the sanctuary, he calls upon a tradition as old as classical literature itself, referring not only to life's travails but even the struggles of literary composition.[74] It is all the more appropriate that Paul declaims in the voice of the *Odyssey*, for this was an epic that even classical commentators suspected lay beyond the domain of factuality (*exokeanismos*), since Odysseus's wanderings took place on that immense ocean that lay beyond earth's limits and touched the heavens. And when the lector of the kontakion describes the windows of Hagia Sophia as "spiritual luminaries fixed to the divine firmament," he adds that they enlighten "in the night of life those drifting about on the ocean of sin." Paul the Silentiary even adds that the church surpasses the Pharos of Alexandria (*Descr. S. Sophiae.* 921–933). We have every right to suspect that both authors juxtapose the still waters of the church interior with the peripheral territory of sin as an ocean in which even the best sailor may lose his course,[75] words that reverberated with Hagia Sophia's actual siting, a man-made mountain that dominated the straits of the Bosporus and the Sea of Marmara.

9 Relics, Tabernacles, Throne Rooms

Painting and Marbled Architecture in the Renaissance

Only in Byzantium had the marble quarries continued to function after the fourth century, but not for long. By the ninth century, emperors were cannibalizing old churches to build new ones and, by the early fourteenth, a jobbing painter at the Chora Church (c. 1316–21) would make a witty nod to the privations when he depicted a faux-marble panel dog-eared with an anomalous marble.[1] In Western Europe, in eighth-century Lucca, Tuscany, the same acknowledgment of standing conditions appears in the faux-revetment that once adorned the apse of SS. Giovanni e Reparata (fig. 9.1). Not only is this imagined assembly motley, but it is disharmoniously stapled together with iron cramps.

FIG. 9.1. Faux-revetment, SS. Giovanni e Reparata, 8th century AD. Lucca.

In Byzantium, there were sufficient stocks of ancient marbles for recycling that one could still pull off an image of cohesion. However, the only comparandum in Western Europe was S. Marco in Venice, and this was largely because it was built from spolia on an unprecedented scale, with boatload after boatload of marbles pilfered from all over the Adriatic, Aegean, and Constantinople itself (fig. 9.2).[2]

It was not until the mid-Quattrocento that marble revetment would gradually resume its ornamental role in chapels and churches on a scale that could even begin to rival antiquity, and emulation was largely the motivation for the revival. It is also in Italy, where the ruins still afforded abundant stocks for the taking, that the phenomenon originates. Before this revival, however, one occasionally finds some attempt made. One of the earliest instances is the chapel of the fourth-century bishop S. Giovenale in the aisle of his eponymous cathedral (originally constructed 1047–1145) at Narni, in Umbria (fig. 9.3). This chapel (also known as the Oratory of St. Cassius) faced into an extra aisle created in the thirteenth century to connect his tomb and oratory with the main church. At this time the chapel floor was cobbled together from a mishmash of Cosmati paving fragments, presumably from elsewhere in the cathedral, but, much more exceptionally, a tall marble façade was collaged together from sixth-century Cosmati transennae, other bits of liturgical furniture, stray slabs, and, over its doorway, a Paleo-Christian sarcophagus frontal.[3] It is a spolia patchwork, but there is for the first time in centuries a striving to achieve an architectural coherency, through the insistent articulation of pseudo-pilasters. It dimly looks back to antique walls like those in late antique palaces and Early Christian basilicas (see figs. 5.19, 6.2. 6.3, 6.5).

Rare as this coherency was, simulating marbles on architectural interiors was an unbroken tradition from antiquity, with varying degrees of know-how. Signs of the burgeoning revival of a coherent architecture in colored

FIG. 9.2. West façade, S. Marco, late 13th century. Venice.

marbles come with the frescoes by Giotto di Bondone (c. 1270–1337) in the dado of the Scrovegni Chapel in Padua (c. 1303–5; fig. 9.4).[4] Giotto not only replicates revetment but also book-matching, as well as simulating immediately identifiable lithotypes. Again, the painter's confidence in the technique was underpinned by the knowledge that fresco was a lithic meta-medium, made from earths and minerals, bonded with lime plaster, and petrified in unison.

At the other end of Europe, in Bohemia, we also find a couple of chapels with walls, again very unusually, completely covered in stones: the Chapel of St. Wenceslas (c. 1372–73) in the Cathedral of St. Vitus, Prague, and the Chapel of the Holy Cross (consecrated 1365) in the nearby fortress of Karlštejn (fig. 9.5). Their walls are not revetted (there is no skin of stone), but the colored stones have been embedded in a thick mortar. While both chapels undoubtedly remember St. John's Heavenly Jerusalem, they also immediately call to mind the arrays of gems on the reliquaries that traveled the length and breadth of Europe. The Chapel of the Holy Cross was, in fact, the heart of the castle treasury, where the bejeweled crown of the Holy Roman empire was stored.[5] However, this collage also readily recalls the practice of collecting stones as relics, which began with the origins of pilgrimage to the Holy Land in late antiquity, when travelers started identifying particular stones as contact relics of the Nativity, flight into Egypt, Passion, and Resurrection, as well as the travails of various martyrs and patriarchs. The examples are innumerable, and some itemized sacred history to an almost farcical extent. For example, one traveler mentions the "stone which Jesus used when he mounted the ass on Palm Sunday," and another stone in which "Saint Elizabeth hid her son Saint John the Baptist."[6]

Toward the eleventh century, some of these milestones in sacred history, such as a rock from the site of the Ascension, and that on which Christ had stood when delivering the Sermon on the Mount, began to make their way into Byzantium. Fragments from the Tomb of the Virgin, the Stone of Unction, and above all the Holy Sepulcher had also entered the imperial collections in Constantinople, where they became the subject of court poetry, and, after the sack of Constantinople in 1204, these made their way west to Paris.[7] The galleries of Hagia Sophia also contained various slabs and boulders that summarized the landscape of the absent holy land, and S. Marco in Venice followed suit.[8] Columns in the nave and narthex of S. Marco were said to come from Solomon's Temple or the Praetorium of Pontius Pilate; reliefs were said to have been carved by Christian sculptors in defiance of Diocletian, and so on. The baptistery was full of such things: the altar was a massive block of red granite, variously said to be the boulder on which Jesus had stood to preach in Tyre, the stone on which he

FIG. 9.3. Façade, Chapel of S. Giovenale, 13th century. Marble and Cosmatesque spolia. Co-cathedral di S. Giovenale, Narni.

FIG. 9.4. Giotto di Bondone, Dado, 1303–5. Fresco. Scrovegni Chapel, Padua.

FIG. 9.5. Chapel of the Holy Cross, Karlštejn Castle, consecrated 1365. Bohemia.

FIG. 9.6. Franciscus de Retza, *Defensorium Inviolatae perpetuaeque virginitatis castissime de genitricis Mariae*, 1487/90.

sat when he asked for water from the Samaritan, the block on which Abraham had tried to sacrifice Isaac, or the peak onto which Moses had climbed to receive the laws; there were the two slabs on which the baptist's decapitated head had landed (as one could tell from its dark red veining); there was even the stone from which Moses had brought forth water.[9] By the 1350s, when the baptistery was eventually decorated, the reputation and sanctity of S. Marco's miraculous stones and their Eastern origins was so consolidated that it became a leitmotiv of the new vault mosaics. As Debra Pincus has shown, the apostles are each shown baptized in fonts of identifiable marbles that together plot the routes of evangelism, the commercial reach of Venice, and thereby the legitimacy of the city's world mission.[10]

Putting contact relics to one side, marbles always arrived with some wondrous pedigree. By the ninth century, visitors to Hagia Sophia were convinced that its marbles had been ferried away from titanic palaces or fabled lands. The ninth-century *Narratio* began the trend and, six centuries later, Ottoman descriptions still relate that Justinian had removed stones from ancient temples in India, Arabia, Yemen, Maghrib, Persia, China, Turkistan, Byzantium, and Europe, or that a race of giants had found them at the ends of the earth.[11] In the West, as marble became an ever more precious commodity, spoliated stones assumed an increasingly mysterious character. Their original names had been lost as the quarries fell into oblivion, and with them the assurance that they were even natural. Marbles arrived in either tantalizing fragments or blocks of such dimensions that incredulous observers eventually concluded they had been quarried by giants and moved by angels. It is little wonder that the images perceptible in book-matching came to be regarded as miraculous apparitions too (see figs. 7.14, 7.15, 7.17). One that became quite famous was also in Venice: around 1222 to 1223, Albertus Magnus tells how masons discovered the head of a king when they split open a block "for a church," which all later writers took to be S. Marco (fig. 9.6).[12] Even the polish of marbles alone acquired mystical properties. In 1437, for example, the Portuguese voyager Pero Tafur recorded that by looking into two columns that had been brought to Pisa among the Crusader spoils of Jerusalem, "each one can foresee the evils which are to befall him and whatever he desires," in other words, scrying.[13] The Flemish manuscript *The Visions of the Knight Tondal* (1475),

FIG. 9.7. Simon Marmion, *Heavenly Wall*, in *Les Visions du Chevalier Tondal* (The Visions of the Knight Tondal), 1475. J. Paul Getty Museum, Los Angeles. 87.MN.141.

an originally eleventh-century story of a knight's itinerary from hell to heaven, concludes with an illumination of the heavenly walls of Revelation, "so bright and shiny that it would ward off any displeasure or trouble," and "inspire with consolation and joy the hearts of those who looked upon it" (fig. 9.7).[14]

FLESH AND BLOOD: THE COLUMN OF THE FLAGELLATION, THE HOLY SEPULCHER, AND OTHER RELICS

The most symbolically charged representations of marble appear in four interlocking categories of religious painting: representations of the Passion; the Virgin Annunciate; the Virgin as Queen of Heaven; and the Virgin and Child. In representations of the Passion, marbles are not simply narrative props but assume a powerful role as surrogate relics, whether that be the Column of the Flagellation, Christ's sepulcher, or the Stone of Unction. In all cases, it was not just the objective type of the relic (column, tomb, or slab) but its innate properties, its pattern and color, that clinched the identification. Once we have become more aware of this stone vocabulary and its modes of representation, we become alive to the implications of analogous stones in scenes of the Virgin. As we shall see, the columns and slabs that decorate Marian depictions also probe theological truths, rather than merely supplying authenticating detail.

Marble objects and veining assume particularly unambiguous roles in representations of the Passion. Zealous pilgrims, bent on finding the physical traces of Christ's earthly life, had identified numerous classical columns strewn across the Holy Land as *the* column on which he had been flagellated. The earliest testimony comes from Jerome, who reports (404 AD) that when Paula visited Jerusalem she was shown this column, still stained with the holy blood, and now reused in a church portico. By about 570 AD, the "Piacenza Pilgrim" writes that the real column, authenticated by the imprint of Christ's agonized fists, was in the Church of the Last Supper on Mt. Zion, where pilgrims came to kiss it and soak up its divine residue with bands of cloth. By the fourteenth century, yet another Column of the Flagellation was venerated in the Apostoleion, Constantinople, next to the column on which Peter had supposedly wept after his denial of Christ. By 1372 St. Bridget of Sweden noted that half of yet another column stood in a chapel near the Holy Sepulcher, Jerusalem, and the other half in S. Prassede, Rome.[15] It was only, in fact, when the latter relic was installed in S. Prassede, in 1223, that competition officially ended, though there are still many other contenders to this day. However, this tapering object (in reality an Egyptian offering stand) was bafflingly short to have accomplished its task, as well as unprepossessingly black and white (fig. 9.8). The presence of more evocative competitors, not to mention the demands of the artistic imagination, therefore did nothing to discourage medieval and Renaissance painters from depicting more gorily veined shafts when they came to scenes of Christ's torture (fig. 9.9).[16] In so doing they also remembered the bloodstains that pilgrims claimed to perceive in the rock at Golgotha, as well as yet another Column of the Flagellation nearby, "of blackish porphyry with certain natural, red markings that the people hold to be the bloodstains of Jesus Christ."[17]

Not only did these stains provide the necessary forensic evidence, but the Rock of Golgotha and Column of the Flagellation far outstripped any mere contact relic because they had soaked up the savior's blood and become consubstantial with it. Although they were not animate, they were to some degree incarnations. In the eyes of the evangelists, and also St. Paul, the most precious blood was synonymous with Jesus's Passion and death, the source of redemption (1 Peter 1:2.19; 1 John 1:7; Paul, Romans 3:25; Ephesians 1:7; Hebrews 9:10). Indeed, by the fifteenth century Dominican theologians argued that this blood remained united to the Word, and therefore enjoyed the potential to prompt miracles.

When artists came to represent such relics, a variety of mottled stones offered a ready resemblance to Christ's blood or tortured flesh. Porphyry, "so flaming . . . it was like blood gushing from a vein," immediately resembled clotted blood, and this had recommended it for portable altars (see fig. 6.23).[18] Thus, when Leo X (r. 1513–21) wished to commemorate the well in which Praxedis and Pudentiana had supposedly gathered the blood of Christian martyrs, again in S. Prassede, he capped it off with a porphyry rota.[19] Moreover, because porphyry was speckled with white, sometimes pink, granules, to

FIG. 9.8. Column of Flagellation, S. Prassede. Granito della Colonna (pegmatitic diorite). Rome.

FIG. 9.9. Cimabue, *The Flagellation of Christ*, c. 1280. Tempera on poplar wood, 9¾ × 7⅞ in. (24.7 × 20 cm). Frick Collection, New York.

simulate it with pigment seemed already to paint flesh and blood. Porphyry, indeed, bears more than a passing resemblance to Milanese salami, and rosso antico to bresaola. These are probably the stones the Florentine artist Filarete (c. 1400–c. 1469) remarked were "extremely varied and dappled with natural marks in such a way that many, with a mixture of white and red, appear quite like salted meat," and, more lightheartedly, Andrea del Sarto (1486–1530) would eventually capitalize on the resemblance by building a miniature temple out of blood sausages for a dinner party.[20]

A further motivation for depicting stones streaked with sanguine veins came from the bedrock of Christ's tomb in the Holy Sepulcher at Jerusalem, which pilgrims concurred was white veined with red.[21] Although Christ had not lain in a sarcophagus after death, pictorial and sculptural tradition supplied him with one in scenes of the lamentation over his body. The putative casket was frequently represented as a rich assembly of marbles and precious stones, fulfilling the prophecy of scripture, "And His sepulchre shall be glorious" (Isaiah 11:10: *et erit sepulcrum eius gloriosum*), and fusing the image of a reliquary casket with the various imperial sarcophagi that were known, and their later medieval imitations.[22] This must have been a subject of debate if Thomas Aquinas felt it necessary to put the arguments for and against Christ's interment in a sarcophagus, also citing Isaiah (*Summa.* III[a], q. 46–52).

This sarcophagus went through several iterations in painting and sculpture, but it is typically a white casket inlaid with three marble panels, often porphyry or

FIG. 9.10. Donatello, *Entombment*, 1445–46. Relief in bronze, limestone, colored marble, colored glass, and enamels, 52¾ × 74 in. (134 × 188 cm). S. Antonio, Padua.

serpentine or alabaster, but not only. Donatello's relief of the entombment in the Santo, Padua (1445–46), is an elaborate version (fig. 9.10).[23] Leon Battista Alberti even made a life-size replica of this putative casket when he came to simulate the Holy Sepulcher in the Cappella Rucellai (1457–67) at S. Pancrazio, Florence (fig. 9.11).[24] This is an altar, but liturgically altars are communion tables over tombs, and this one was also prepared to overlie the eventual grave of the patron, Giovanni Rucellai (1403–1481). Alberti repeated the formula on the walls of this cell, elaborated with contrasting roundels, the perimeter only broken by the figure of Christ's resurrected body ascending heavenward. The mental schema of Christ's marbled sarcophagus also informed tomb architecture in general, and it is no coincidence that triads of marble panels increasingly form the backdrops to Tuscan sarcophagi, beginning with Bernardo Rossellino's tomb for Leonardo Bruni (c. 1446–48) in S. Croce, Florence.[25]

However, an alternative and equally strong artistic tradition depicted Christ's sarcophagus as made from a marble that borrowed its bloody veining from the Rock of Golgotha and the cave tomb of the Holy Sepulcher. Pictorial play on the similarity between these stones and Christ's tortured flesh begins with at least Pietro Lorenzetti's panel *Entombed Christ as Man of Sorrows* (c. 1340–45; fig. 9.12a).[26] Lorenzetti cunningly draws (what would otherwise be) a marbleized frame into the field of the painting to confine Christ within a sort of anti-Mandorla outlined by the lip of the tomb. The painting is the right leaf of a diptych (the other showing the Madonna holding the swaddled infant close to her breast), and when one closed the diptych it became itself a casket, as the image disappeared from view below its verso, also marbled, in cream, red, and blue streaks that suggest Christ's flesh, veins, and wounds (fig. 9.12b). It is not impossible that the idea came to the artist from seeing portable altars (see fig. 6.23), and the material characterization was a departure from the Cosmatesque-type, porphyry-and-serpentine sarcophagus that Lorenzetti had preferred when he painted the *Deposition* (c. 1320–28) in the lower church of S. Francesco, Assisi.

Marbleized versos, spattered with blood or gushing with veining, soon became commonplace on panel paintings with analogous themes. Thus, Marco Zoppo's *Dead Christ with Saints John the Baptist and Jerome* (c. 1470) extends the sepulchral marbling onto the verso, which feigns a slab the color of dried blood, or the earth pigment

FIG. 9.11. Leon Battista Alberti, Cappella Rucellai, 1457–67. Interior, frescoes attributed to Giovanni da Piamonte imitating porphyry, serpentine and alabaster, marble altar. S. Pancrazio, Florence.

FIG. 9.12A–B. Pietro Lorenzetti, *Entombed Christ as Man of Sorrows.* c. 1340–45. Tempera on panel, 12⅝ × 20½ in. (32 × 52 cm); verso: faux-marble panel within raised and tooled border. Lindenau Museum, Altenburg.

has itself dried and faded to this same brown.[27] Another example is the verso of Jan van Eyck's *St. Francis Receiving the Stigmata* (c. 1428–29), which is speckled with red paint as though spattered with blood. In this case, the association could hardly have been lost on its original owner, the Bruges merchant Anselme Adorno (1424–1483), whose travel diary records every bloodstained rock he could find in the Holy Land and who completed the octagonal Jeruzalemkerk in Bruges in the 1450s to imitate the Holy Sepulcher.[28] The marbleized verso communicates a subtler embodiment in Antonio Leonelli da Crevalcore's *St. Francis as the Man of Sorrows* (c. 1490; figs. 9.13a–b). On the recto, the saint stands like the "Man of Sorrows" within Christ's tomb, the *orans* gesture manifesting all the more the divine gift of the stigmata. On the verso is painted a porphyry rota, within a serpentine ring, within a deep-brown porphyry (a rare sight outside of the quarry).[29] We might interpret this verso again as the lid of the tomb in which St. Francis stands on the recto, and just maybe Leonelli had seen Buggiano's Tomb of Giovanni di Bicci de' Medici and Piccarda Bueri (c. 1435) in the Old Sacristy, shielded by a huge table slab with a massive porphyry rota in the middle. But we would do better to recognize the verso for what it is, an abstracted materialization of the wounds of St. Francis, encircled by the Franciscan habit. The color of the stigmata had even been fixed by Franciscan decree. This signing was more specifically appropriate to this lithic painting thanks to the supreme authority of Bonaventura's 1263 *Legenda Maior,* which declared that not only had Francis been transformed into Christ (*Franciscus alter Christus*) by receiving the stigmata, but God had "painted" him with them. He was, in other words, as much an artwork as those miraculous images that appeared within marble: "St. Francis came down from the mountain . . . with him he bore a representation of Christ crucified which was not a work of an artist in wood or stone, but had been reproduced in the members of his body by the hand of the living God. . . . [Francis] did his best to conceal the sacred stigmata. However, it is for God to reveal his wonders for his own glory; he had impressed the stigmata on St. Francis in secret, but he publicly worked a number of miracles by them, so that their miraculous, though hidden, power might become clearly known."[30]

Such perceptions were nourished by other contact relics just as famous and just as incarnational as the Column of the Flagellation. The marble treads of the Scala Sancta at the Lateran Palace, reputedly the steps from Pilate's palace in Jerusalem, were dappled with the drops of blood that fell from Christ's body as he climbed to judgment after the

FIG. 9.13A–B. Antonio Leonelli da Crevalcore, *St. Francis as the Man of Sorrows,* c. 1490. Verso: faux porphyry and serpentine. Tempera on wood panel, 29⁵⁄₁₆ × 22¹⁄₁₆ in. (74.5 × 56 cm). Princeton Art Gallery.

FIG. 9.14. Balthasar Permoser, *Flagellated Christ*, 1728. Plassen limestone from the Untersberg massif near Salzburg, 31⅞ × 10 ⅝ × 9 13/16 in. (81 × 27 × 25 cm). Staatliche Kunstsammlungen, Dresden.

flagellation.[31] At S. Cristina, Bolsena, the altar was stained red in perpetuity by a host that spontaneously bled during Mass in 1263 when the priest, Peter of Prague, doubted the true presence in the Eucharist.[32] The Mass at Bolsena became a popular subject in Renaissance painting, immortalized by Raphael in the Vatican Stanze, but while the relic of the altar is more or less forgotten today, the story was still sufficiently memorable in the sixteenth century for it to be repeated about yet other bloody stones, this time in a Counter-Reform chapel in Rome, the Cappella Caetani in S. Pudenziana (1590–1603).[33] Of course, hard-stone chalices had long met with the same perceptions (see figs. 6.15, 7.20).

When artists came to represent such relics, marble veins were a facile metaphor for human ones, and to daub the dark streaks on the support was akin to painting in blood, sacred blood, such that painting became implicitly an act of piety as much as artistry. The painting was implicitly incarnational anyway, because marble was nature's divine painting. The same might apply to sculpture. Classical statues of Marsyas had capitalized on their veining to represent his flayed and tormented flesh (see fig. 3.7), and in the twelfth century Michael the Deacon had been beguiled by a marble slab in Hagia Sophia that "puts on the guise of living flesh, and whitish in color, displays all over itself what look like gaping veins of blood," such that "a statue of such material would be a plausible counterfeit of a man." Concurrently, the idea was architecturally at work in the chancel of Ávila Cathedral (see fig. 6.27). However, one must await the eighteenth century, and the *Flagellated Christ* carved by Balthasar Permoser (1651–1732), before a sculptor would revive this bruised and bloodied materiality and draw the material iconography of the Column of Flagellation into the bargain (fig. 9.14).[34] Permoser made three versions of the statue, but that which he carved at the end of his life (in 1728) hid a self-portrait on its back and, like a silent prayer, this pseudo-relic gave shape to his private desires for redemption.

BLOOD AND TEARS: THE STONE OF UNCTION AND MANTEGNA

Despite all the precious relics that the Venetians had managed to amass for the Treasury of S. Marco, and all the extraordinary pedigrees they attributed to the basilica's various marbles, two exceptional stones that would have crowned their collection eluded them: the Stone of Unction and the Stone of Nativity (or that of Ablution). The Stone of Unction was easily the more famous, the slab on which it was said that the dead Christ had been anointed after his deposition from the cross. A mottled, reddish stone with this identification had arrived in Constantinople in 1169–70, when Emperor Manuel I Komnenos carried it on his own back up the hill from the harbor to Our Lady of the Pharos, just as Christ had borne the cross to Calvary.[35] The Pharos was the palatine chapel, and it held the richest collection of relics from Christ's Passion anywhere, including the Crown of Thorns. There the stone remained until Manuel's death in 1180, when his widow, Eirene, had it moved to the emperor's burial chapel in the Pantokrator church, or rather the miniature church adjoining it that had been built as an imperial mausoleum. To commemorate both acts, a lengthy verse was now inscribed on or near the relic to elaborate on how it had been "rolled" (like the door of Christ's sepulcher) to Manuel's tomb to ensure his own Lazarus-like resurrection, and identifying Eirene with Mary Magdalene, shedding "tears like unguents" and "turned to stone with sorrow."

The Stone of Unction remained in the Pantokrator church in the utmost veneration until the sack of 1453, when it disappeared for good. Before it vanished, however, observer after observer marveled at the slab's mottling. In the mid-twelfth century, the akoluthia composed by George Skylitzes claims that the slab was originally a white stone that had been "dyed by the stream of blood flowing from God." Thereafter, figure was swapped for ground, and pilgrims and travelers were struck by wonder at the stone's miraculous watermark, the tears that the Virgin had shed at the Deposition and that were ever after trapped within its pores. The Greek historian John Kinnamos (c. 1180), the Frankish Knight Robert de Clari (1204), the Nestorian Rabban Sauma (1287), the Russian pilgrims Anthony of Novgorod (1348/49), Ignatius of Smolensk (1389), and Zosima the Deacon (1420–22), the Anonymous Armenian Pilgrim (1375/1434), the Spaniard Ruiz Gonzalez de Clavijo (1403), and the Burgundian Bertrandon de la Broquière (1422–23) all foreground the waxy or milky beads within the marble, which they earnestly believed to be the tears that the Virgin had showered on Christ's cadaver and spilled onto the slab.[36] Bertrandon de la Broquière, who was a court spy on a reconnaissance mission throughout the Middle East, provides the fullest description:

> In the beautiful church of the Pantokrator . . . one sees a stone, or slab, of various colors which Nicodemus had cut to place over his tomb, and which served him to lay down the body of Our Savior when he was taken down from the Cross. During this time the Virgin cried over the body; but her tears, instead of falling on Him, all fell on the stone, and one can still see them there. At first I thought that they were drops of wax, and stretched out my hand to touch them; then I bent down, so I could look horizontally and against the light, and they seemed to me like frozen drops of water. This is a thing that many people have been able to see, just like me.[37]

FIG. 9.15. Enkolpion of Stone of Unction. Breccia corallina. Vatopedi Monastery, Mt. Athos.

An *enkolpion* (a medallion worn around the neck by Orthodox bishops) at the Vatopedi monastery on Mt. Athos purports to enclose a fragment of this stone, and this fragment is a particularly subtle shade of the Asiatic marble known as *breccia corallina,* whose amber matrix contains clasts both opalescent and clouded, like pearls or salty tears (fig. 9.15). When Raban Sauma (1287) saw the stone "on which Mary wept," he added, presumably fooled by the polish, that "the place where her tears fell is still wet, and however much the wetness is wiped off, it returns again."[38] Innumerable pilgrims must have soon spread the Word to the West, where the Stone of Unction entered narratives of the Passion, eventually to be diffused by the fourteenth-century *Vita Christi* of Ludolphus of Saxony (c. 1295–1378), a commentary on the Gospels that became one of the most influential devotional writings of the late Middle Ages.[39] From the 1260s or 1270s the Stone of Unction also began to inform Franciscan imagery, in paintings that showed Christ's body laid horizontally on the Stone of Unction rather than on a sarcophagus.

The stone was especially familiar to Venetians, for the Pantokrator church was occupied by them during the Latin empire (1204–61), and the relic's teary identity would eventually be reaffirmed in Francesco Sansovino's translation of Ludolphus.[40] As the slab was famed far and wide, it seems inescapable that the tradition of seeing tears in its mottling generated Andrea Mantegna's depiction of spattered marble graining on the slab under Christ in his famous Brera *Lamentation over the Dead Christ* (1480s/90s?; fig. 9.16).[41] The same knowledge may have motivated the entire composition, for the drama of Mantegna's painting is best comprehended with the original relic in mind, and possibly its original staging in the Pantokrator as well.

This staging, unfortunately, remains somewhat vague because the imperial tombs were evicted from the church in 1453, the altars stripped, and the relics dispersed. However, it is clear that the slab was placed on a platform, somewhat as Carpaccio paints it in his own *Entombment* (c. 1505).[42] Moreover, as Cyril Mango notes, "It would be tempting to think that the images of

FIG. 9.16. Andrea Mantegna, *Lamentation over the Dead Christ*, 1480s/90s?. Tempera on canvas, 2¼ × 2¹³⁄₁₆ in. (5.8 × 7.1 cm). Brera, Milan.

FIG. 9.17. *Lamentation*, 1312. Fresco. Katholikon of the Vatopedi Monastery, Mt. Athos.

the Women at the Sepulcher and *Noli me tangere* which are mentioned in the Pantocrator *typikon* as being in an arch of the *herôon* [burial chapel] were located above the Stone of the Deposition." Thus are they shown in a fresco from Mt. Athos (fig. 9.17), where the aforementioned fragment of breccia corallina was kept as a relic (see fig. 9.15). It is also clear from at least one source that an image of the Virgin Mary was placed directly above the slab, reenacting the narrative of the relic, posing a model for the grieving empress, and perhaps even implying the Virgin's postmortem intercession on behalf of the deceased emperor.[43] In any case, the relic of the slab came alive and maternal agency was celebrated three times a year, when the Hodegetria (Constantinople's most famous icon of the Virgin) was processed to the Pantokrator for an all-night mass in the herôon that preceded the commemoration days of the emperor, his wife, and their son, all of whom were buried there.[44]

In Mantegna's painting, which may have been painted for his own devotion, it is therefore all the more striking that the sharply foreshortened Christ is also juxtaposed with the magnified and lachrymose heads of the distraught Virgin and Magdalene. Early scholars found the presence of these heads so discordant that they argued they were later additions, but for the purposes of the Stone of Unction, the importance of anguished expressions and gushing tear ducts overrode any need to represent the rest of their bodies.[45] Likewise, Mantegna's perspectival bravura privileges the slab. At one end stands an agate unguent jar (just like those spirited away from the Pantokrator to fill the Treasury of S. Marco), which references both the Magdalene and the Stone of Unction. All in all, the painting telescopes the narrative into an icon and "contains" the relic, while the protagonists reenact its history.

A SYMBOLIC PALINDROME: THE VIRGIN, THE STONES OF NATIVITY, ABLUTION, AND PROLEPTIC MARBLES

After the Shroud of Turin, the Stone of Unction was the supreme contact relic of Christ. He had lain upon it and he had risen from it. However, because the relic contained the tears of the Virgin it belonged as much to her as to Christ, petrifying for eternity the bond between human mother and divine son. It was not alone in this respect, for another tale was told of a wall in the grotto of the Nativity at Bethlehem: like the Rock of Calvary and that of the Holy Sepulcher, its earth was naturally red but spattered with white, which pilgrims explained were drops of the Virgin's milk that Christ had spat out as an infant.[46]

Travelers also record two other marble relics of Christ's Nativity, also red and white, that were still venerated in one of the galleries of Hagia Sophia long after the city had been conquered by the Ottomans. The first was said to be the stone on which he had been delivered, the Stone of Nativity, the second that on which he had been washed postpartum, the Stone of Ablution. By the seventeenth century, both stones had become praying stations for the faithful within the (now) mosque and were still used to cure sickly children, just as they apparently had before the Muslim conquest.[47]

While the Stone of Unction was long present in the Western imagination, the "stone on which Our Lord Jesus Christ was born" could not have been more so in the period after the Fall of Constantinople (1453), when the Venetians are said to have offered Mehmet thirty thousand ducats for it. As astronomical as the sum was, Mehmet replied that he would not part with it for even a hundred thousand.[48] It must be remembered that the Shroud of Turin, the only other relic of equal fame in its testimony to Christ's Passion, and one that had also been conserved supposedly in Constantinople, was purchased by the Savoy family in the same year, 1453.[49] No doubt Venice, and possibly the Savoy too, sought the legitimizing power of such extraordinary relics through the license to safeguard them.[50]

But knowledge of both relics seems to have recommended their quotation in Venetian representations of the Madonna and Child, especially in the oeuvres of Giovanni Bellini and Giovanni Battista Cima da Conegliano. Their quotation was instrumental in demonstrating Christ's future Passion in representations of his infancy. It was right and fitting that the Passion should cast its material shadow over such scenes, for the Virgin knew Christ's destiny from the cradle to the grave. Such foreknowledge is already pathetically perceptible in Pietro Lorenzetti's aforementioned *Entombed Christ* (c. 1340–45; see fig. 9.12a–b) that belonged to a diptych whose other leaf represents the Virgin and Christ Child.

Such mineral premonition is equally obvious in Castagno's frescoes in the refectory of S. Apollonia, Florence (c. 1445–50; fig. 9.18). In the scene of the Last Supper, one turbulent panel seems to prefigure the storm that will follow the Crucifixion, while the freely invented veining of that adjacent, and right over the head of Judas, suggests Christ's gaping wounds, and was perhaps influenced by Lorenzetti, whom he admired (see fig. 9.12b). Finally, when one looks at the wall as a whole,

FIG. 9.18. Andrea del Castagno, *Passion Cycle*, c. 1445–50. Refectory, S. Apollonia, Florence.

FIG. 9.19. Giovanni Bellini, *Lochis Madonna*, c. 1475. Tempera on panel, 19 × 13 in. (47 × 34 cm). Accademia Carrara, Bergamo.

it becomes obvious that the revetment of the enclosed space preannounces that of the paneled sarcophagi in the scenes of the Crucifixion, Deposition, and Resurrection in the register above. Indeed, it was at the Last Supper that Christ established the Eucharist and foretold his own sacrifice.

In Venetian painting of the Quattrocento and Cinquecento, these proleptic stones become thresholds to visions of the Madonna and Child. The Christ Child either stands on or lies prone upon them. It takes no great leap of the imagination to understand that the stones in question, which foreground so many of Giovanni Bellini's half-length Madonnas, and mediate between subject and viewer like a communion rail, were consciously premonitory props and metonyms for the Passion. They combine echoes of Christ's tomb or the Stone of Unction with the image of the near-identical stones of Nativity and Ablution. In Bellini's oeuvre perhaps the most satisfying example is the *Lochis Madonna* (c. 1475; fig. 9.19), but there are many others, and Cima da Conegliano is also full of such slabs and parapets, windowsills, and sarcophagus lips.[51] Had a portable altar ever been placed in front of one of these paintings, the implications of the materials as the site of sacrifice and redemption would have reached fever pitch.

Mantegna's *Presentation of Christ in the Temple* (fig. 9.20), variously dated between 1454 and 1466, was painted in roughly the same period as the abortive Venetian negotiations to purchase the Stone of the Nativity.[52] Strictly speaking, this is a presentation, not

FIG. 9.20. Andrea Mantegna, *Presentation of Christ in the Temple,* c. 1454/66. Tempera on canvas, 27⅛ × 34 in. (68.9 × 86.3 cm). Gemäldegalerie, Berlin.

a circumcision, and therefore centers on the theme of the Virgin's purification. However, as Sixten Ringbom points out, "Pseudo-Bonaventure made Simeon foretell the Passion having uttered *Nunc dimittis,*"[53] and the painting's foreboding stillness, the protagonists' expressions of apprehensiveness or restrained anguish, as well as the sepulchral gloom shrouding the background, all suggest the Passion. So does Christ's swaddling-cum-winding-sheet, and a red-stone frame encircles the ensemble like the rim of Lorenzetti's earlier diptych (see fig. 9.12a). Thanks to the function of the Constantinopolitan relics, where the Nativity ablution mirrored the Passion unction, the red stone in Mantegna's painting became materially the alpha and omega of the Christological life cycle.

This reading of Mantegna's painting seems undermined by the fact that when Giovanni Bellini copied

FIG. 9.21. Stone of Unction, 1820. Unidentified stone (probably veined Jerusalem limestone). Holy Sepulcher, Jerusalem.

it (in the 1470s?), he changed the frame from red to (green) verde antico.[54] But there was *another* Stone of Unction, installed in the Holy Sepulcher in Jerusalem at some point between 1327 and 1335, that displaced the Constantinopolitan one after the former was lost from sight. Pilgrims related that it was black sprinkled with red, gray, and white, or that it was "green porphyry," and "adorned all around with a checkerwork of white and red marbles."[55] This stone was badly damaged by a fire in 1820 and substituted or covered with another, that which pilgrims unwittingly venerate today as a true relic, and which is far more fleshy and sanguine-looking (fig. 9.21). The model was either the bedrock of Calvary, or the long tradition that had by now been established in painting, West and East.

THRONE ROOMS

But colored marbles were also the birthright of the Virgin, for after Christ's death and her own Assumption, she would become Queen of Heaven. Throughout European painting, the enthroned Virgin is almost always depicted surrounded by hallucinogenic marbles, often in a chamber that patently represents her celestial throne room. In Early Netherlandish painting, this location is evident from the saints who introduce the donor and the aerial perspective of the landscapes or city views beyond the crowned Virgin. The celestial penthouse is also frequently supported by columns and columnettes that resemble less true marbles than coral, crystal, or semitransparent stones halfway between alabaster and crystal.[56] The supernatural spectacle of such rooms was all the more apparent for the absence of chapels, churches, or palaces that were actually decorated with marbles.

More commonly, the heavenly audience hall was also contracted into the image of a throne, too cumbersome to be furniture, but ambitious enough to be architecture, often the pinnacle of towering steps that were an allusion to the biblical Throne of Solomon.[57] The examples are so countless that to pick any one is to shortchange the creativity of the genre, which was as enormously various as it was consistent in hewing to its theme. However, the version by Fra Antonio Falier da Negroponte (c. 1450) does at least have the merit of showing the Madonna on a micro-macro Throne of Wisdom that abbreviates a presbytery or entire church, swarming with putti real or carved, and inlaid with ethereal marbles and hard stones (fig. 9.22).[58] In the Pesaro Altarpiece by Giovanni Bellini

FIG. 9.22. Fra Antonio Falier da Negroponte, *Madonna and Child Enthroned,* c. 1450. Egg tempera and oil on panel, 9 ft. 10⅛ in. × 7 ft. 8 17⁄32 in. (300 × 235 cm). (Lunette attributed to Benedetto Diana c. 1510; marble aedicule added by Francesco Sansovino after 1534.) S. Francesco della Vigna, Venice.

FIG. 9.23. Giovanni Bellini, Pesaro Altarpiece, between 1472 and 1475?. Oil on panel, 8 ft. 7⅗₂ in. × 7 ft. 10½ in. (262 × 240 cm). Musei Civici, Pesaro.

(between 1472 and 1475?) we instead see a sort of compromise between throne room and throne, a double-seater apparently inspired by a palace window nook (fig. 9.23). Through a window (because the Virgin is also the *Fenestra Coeli?*) we glimpse a fortress, which signifies either the Heavenly Jerusalem or something altogether more arcane; opinion is very divided. Around this window various porphyries and hard stones interlace like a rosary (twelve per side, like the decades of prayer, plus the prayers before and after). These stones also rhyme the gems of her crown with the object that the frame inscribes. As the Venetian patriarch Lorenzo Giustinian recounted in his first sermon for the Feast of the Assumption, when Christ welcomed his mother as Queen of Heaven, she acceded to her honor with the words she had spoken when he became incarnate: "Ascend to the throne that I have prepared for you, take the crown set with gems" (Psalm 20:4).[59] On other occasions the only hint of the throne room is an intarsiated suppedaneum, a marbled floor, or treads up to the throne. In the vast majority of cases, as in Negroponte's painting (see fig. 9.22), the swirling psychedelic marbles depart from any identifiable lithotype, fantasy marbles that follow their own rules, and perhaps this is a sign of their heavenliness, that they are not subject to sublunar paradigms. A marble dais may span the entire composition under the feet of all protagonists to signal their heavenly location. Marble plinths fulfill this function even without the presence of the Holy Family, elevating saints above worldly things and making them into living statues.

THE COLUMNS OF THE ANNUNCIATION AND NATIVITY

The Virgin also had her own suite of columns to compete with the Column of the Flagellation, and they are her silent partners in a multitude of Annunciations. In fact, this army of columns has generated more mystification than revelation, not least because in legion examples the column apparently blocks the Archangel Gabriel's view of the Virgin. Michael Baxandall went so far as to remark, "If all Christian knowledge were lost, a person could well suppose that both figures, the Angel Gabriel and Mary, were directing some sort of attention to the column."[60] Baxandall, of course, meant this observation to be ironic, but close inspection of the Annunciation genre as a whole suggests that the interloping column is a frequent and willful intrusion.[61]

The motif is already present in Romanesque relief sculpture and illuminations, but it becomes a real protagonist in fifteenth-century painting. Among countless examples, the most instructive is Masolino's *Annunciation* (c. 1423–24) in the National Gallery of Art, Washington, DC (fig. 9.24).[62] Here, the arcade of the canopy overhanging Gabriel and the Virgin terminates in pendent bosses *except* for the one pinnacle that travels down to the ground as a column just so it can segregate the protagonists. In the long sequence of altarpieces that followed Masolino's, the rogue column may be interpreted as a totem or barrier or pivot, though none of these readings is mutually exclusive.

The stand-alone column might again be taken for a premonitory sign of Christ's future scourging, and the ultimate divine plan even at the moment of Incarnation. In some paintings this adumbration would explain the red-and-white veining chosen.[63] Moreover, the specter of the future Column of Flagellation haunted another, real one that pilgrim lore located at the scene of Christ's birth. To be more precise, there are actually three column-relics associated with the Virgin that medieval and early Renaissance travelers record. The Persian scholar Naser-e Khosraw (c. 1075/1100) relates that in a cave below a corner of Temple Mount, he saw a column next to a stone basin (supposedly the cradle of Jesus) and on it the marks

FIG. 9.24. Masolino, *Annunciation*, c. 1423–24. Tempera on panel, 58¼ × 45¼ in. (148 × 115 cm). National Gallery of Art, Washington, DC.

of the Virgin's fingers, where she had seized it during labor.[64] The legend was almost certainly known to the Crusaders, who captured Temple Mount in 1099 and converted it to Christian use. Exactly the same story was told about a second column, this time where we might more reasonably expect to find it, in the Grotto of the Nativity at Bethlehem.[65] This "birthing column" was certainly well known in the West because the Franciscans were gradually entrusted exclusive care of Christian sites in the Holy Land from 1309 onward, and the Pisan friar Giovanni de' Cauli inserts it into his account of the Nativity (1335/40), where he also records the stone there that Mary put within the crib. Thereafter, the column appears in fourteenth-century illuminations of his *Meditationes Vitae Christi* (fig. 9.25), and would become a standard feature of German Renaissance devotional prints.[66]

A third marble column more pertinent to the Annunciation was venerated at Nazareth itself, in the "Grotto of the Virgin" below the Church of the Annunciation, and this time pilgrims claimed that it was the column that the Virgin had clasped in fear at Gabriel's arrival.[67] This column (and yet another dedicated to Gabriel) had been installed by the Crusaders in 1102 to underpin the church, and were object examples for the pivotal pillar in Renaissance painting.[68] Furthermore, the manner in which the column bisects so many Annunciation scenes also happens to reflect the manner in which the Grotto of the Virgin at Nazareth is also divided into two chapels, dedicated to Gabriel and the Virgin, respectively.[69] This brings us to the column as a threshold or barrier that must be negotiated.

THE TABERNACLE OF THE VIRGIN

To appreciate this role, it must be recognized that the architecture surrounding the Virgin at the Annunciation is a vessel that must be infiltrated. The Virgin is almost always, it must be remembered, represented at the moment of conception, the moment when a shaft of divine light inseminates her womb. On a temporal level, the bisecting column is necessary, as Georges Didi-Huberman explains at length, because this is the moment when divinity and humanity, eternity and history, collide and then fuse.[70]

In contrast to the Old Testament, the New Testament had offered no archetypes of architecture that could contain God (the Mosaic Tabernacle, Solomon's Temple), but it did posit the body of Christ as the "New Temple" and the Virgin's womb as the vessel for his divinity. It is not surprising that vernacular tradition had first situated this in the Grotto of the Virgin at Nazareth, because caves and their springs had long been the place where things were born or came into being. The Gospels (Luke 1:26–38; Matthew 1:18–22) had not specified where the Annunciation had taken place, but Renaissance artists preferred to depict the event within an architecture that was permeable in sufficient degrees to express an outside and a within: a portico, a courtyard and garden (the *hortus conclusus*, a paradisiacal womb), or the bedchamber. The

FIG. 9.25. Tuscan artist, *Nativity*, c. 1423–24. From the *Meditationes Vitae Christi*. Bibliothèque nationale de France, Paris.

FIG. 9.26. Piero del Pollaiolo, *Annunciation*, c. 1470. Oil on poplar wood, 59¼ × 68 11/16 in. (150.5 × 174.5 cm). Gemäldegalerie, Berlin.

enveloping bedchamber was the most intimate room of the house that could include (in their times) a scrivener or prie-dieu, the places of thought and prayer. The bed was also where children were conceived and women gave birth. The bedchamber was therefore a ready metaphor for her womb and theological conceptions, and painters cunningly exploited the tumult of associations and affinities by depicting the room with varying degrees of enclosure and perforation.[71] Indeed, when the Virgin is shown receiving Gabriel in her bedroom, we know that the divine light has pierced the very penetralia of the sacred house, and so too the Virgin's womb.

Furthermore, theologians taught that the Virgin was not only the mother of Christ but his bride, *Mater Ecclesia*, and the earthly church their joint offspring. From at least the tenth century, the dedication liturgy implied that the physical church materialized this union when it broadcast verses that described the Apocalyptic Jerusalem as both "bride adorned for her husband" and "the Tabernacle of God" (Revelation 21:2–5).[72] More specific equations between her womb and built architecture had already been made for centuries. We will recall that Paul the Silentiary and Michael the Deacon had characterized Hagia Sophia as the "life-giving Queen," which "could be pregnant with many thousands of bodies," and that Venantius Fortunatus had said that St. Mary's in Bordeaux was "filled with light [because] the glittering hall is the image of Mary: she closed the light in her womb and this [church] encloses the day." The metaphor was reciprocal because, by virtue of being a terrestrial vessel for the covenant between man and God, the womb assumed a parity with both the Mosaic Tabernacle and Holy of Holies in Solomon's Temple, the place where the Ark of the Covenant was kept. Athanasius of Alexandria calls the Virgin a "templum" as early as the fourth century, and from the seventh she assimilates the House of Wisdom too.[73] The architectural analogies kept on multiplying through the Middle Ages. Albertus Magnus explains the Virgin as locus for the Incarnation with thirty-one architectural analogies that included *Maria thronus, Maria sedes, Maria cathedra, Maria habitatio, Maria cellula, Maria nidus,* and *Maria templum.* In the litanies, she came to be praised as *Templum Domini, Templum Sancti Spiritus* ("Temple of the Lord," "Temple of the Holy Spirit"), even *Domus Aurea* ("Golden House"), yet two identities were preeminent: *tabernaculum* and *ecclesia* ("tabernacle" and "church").[74]

The Virgin's bedchamber is identified as the Holy of Holies, for example, in Piero del Pollaiolo's *Annunciation* (c. 1470; fig. 9.26), thanks to the copious cherubim adorning the gilt wall hangings (1 Kings 6:29; 7:50). Cherubim were the only figurations that God had ordained for the Ark of the Covenant, but (along with seraphim) also

FIG. 9.27. Fra Angelico, *Annunciation*, 1425–26. Tempera on panel, 60⅝ × 76⅜ in. (154 × 194 cm). Museo de Prado, Madrid.

occupied the highest rank in the ninefold hierarchy of angels. They were nearest to God's essence. The rest of the house is marbled all over and assiduously attempts to recall ancient revetments, maybe those (now lost) of the portico antae of the Pantheon, Rome, which had been converted to new use as the Church of the Virgin (S. Maria ad Martyres; see fig. 10.15). Finally, the perspective of the fugitive corridor culminates in a brilliant-red column bisecting a bifora, which may recall the Column of the Flagellation.

The pictorial desire to assimilate the Virgin's house to the temple, church, or an all'antica house did not therefore serve archaeologizing accuracy. This would explain neither the loving, sometimes obsessive, fashion in which fictive marble interiors are depicted, nor the often searching painting of the marbles themselves. The latter is particularly true for Fra Angelico and Lippo Lippi, in whose works the Incarnation metaphor rewards every returning deliberation. A powerful and subtle example is Fra Angelico's *Annunciation* (1425–26), painted for the monastery of S. Domenico in Fiesole—that is, for Dominican friars, foremost doctrinal experts in Mariology (fig. 9.27). It is already remarkable that Fra Angelico made no effort to depict identifiable marbles in the way that Giotto had in the Scrovegni Chapel (see fig. 9.4), but the goal of representation is marble as a medium, not a lithotype. The floor is not really paved, there are no joins, in fact no surface. It is a single ground with rainbow eddies that arouse the age-old perception of marble as a liquid or vapor that is on the point of solidifying.[75] It is marble on a primordial plane. In this respect, it bears comparison with that earlier one at Westminster Abbey (see fig. 6.26), or rather its central roundel, which showed the "archetype" and "macrocosm." The swirling and nebulous veining of this roundel suggested *prima materia*, the melting pot of matter in the divine mind before it was separated into created substances, that is, before material things could be born. In its potential to represent paint, marble, and matter simultaneously, we might call it a primordial intermediality.

The vault of the portico that Fra Angelico represents is as celestial as that in the Mausoleum of Galla Placidia (see fig. 6.8), and the Virgin wraps herself in this heavenly mantle. But the depicted vault is an iconographical cloth—it is in "quotation marks"—representing human representation and outlining its semiotic limitations, while the blues and yellows of the floor mix heaven and light within their own nature. A roundel in the portico façade also presents us with Christ as a sculpture bust, and not coincidentally over the centralizing column. As for the Virgin's body, it could not be more spotless, yet her pink under-tunic is as much skin as the flesh that Christ assumed only for it to be bloodied at the scourging and Crucifixion. The pink and gold of Gabriel's robe more calmly embody the Incarnation of lux in flesh. Gabriel is bending toward the Virgin and has bypassed the corner column, so we know this is the very moment of conception, even without the ray of light from Godhead to maidenhead. Yet the angel's back foot still teeters on the brink, at the threshold between paradise lost (the expulsion from Eden) and regained (the Incarnation in the Virgin's womb). The labial slit in Gabriel's skirt figures the vagina not as a menstrual wound but as a gold-and-blue Mandorla, the almond-shaped, time-space portal in which the risen Christ or Virgin appears throughout medieval and early Renaissance art.

Medieval Byzantine poetry had articulated these ideas about stones more explicitly than in the West, by eulogizing white stones as venues for acheiropoietic images. Manuel Philes (c. 1290/1330) eulogizes reliefs of Nativity with the words, "I honor you, oh stone" because "you contain the uncontainable. You are truly unspotted because you represent the unspotted mother," or "Blessings upon you, sculptor's hand, on the stone from the mountain uncut by human hands, Him you delineate on unspotted stone, Who came forth from the unspotted Virgin."[76] By extension, veined, streaked, or nebulous stones offered a gamut of generative associations from the "living rock" to "marbles as meadows," not to mention Christ the rock in all his scriptural guises.[77]

ARCHITECTURE AND PAINTING: PIETRO LOMBARDO AND HIS SONS AT SANTA MARIA DEI MIRACOLI

Paintings of enthroned Virgins on marbled thrones became more reified once marble frames were made to encircle them, or the Madonna became mistress of her own chapel space. But the possibility that the Virgin could be a building, and that architecture could become like painting, came in the works of the Ticinese sculptor-architect Pietro Lombardo (1435–1515). It was Lombardo who also forged a style in which relief sculpture and architecture blended to create a super-medium, "pictorial architecture."

Sculpture was never foreign to architecture because the latter's materials were stone and clay, the primary materials of the sculptor, and because the theoretical paradigm for ordering architecture was the human body, also the primary subject of sculpture. Painting too had never been far from the architectural imagination because it could picture the world, and both arts were strongly assimilated by the invention and wide-scale adoption of one-point perspective at the beginning of the fifteenth century.

The possibilities of an architectural world pictured through one-point perspective were explored first by painters such as Masaccio, and then by sculptors such as Lorenzo Ghiberti and Donatello in a series of marble or bronze reliefs, often called "pictorial reliefs" because their backgrounds are either landscape or architecture (fig. 9.28). They are also known as "*rilievi schiacciati*" ("flattened reliefs"), and commentators agreed that such reliefs were the only place where painting and sculpture merged, because they required a command of illusion equal to painting to compress the representation of space into so shallow a field. Even Leonardo, who coined the term "*paragone*" ("contest") to pit the arts against each other and relentlessly insisted on the primacy of painting over sculpture, conceded that "low relief is a form of painting." He specifies, "This may be conceded as far as drawing is concerned, because it participates in perspective," eventually granting, "Low relief is a mixture of painting and sculpture." The rilievi schiacciati that Agostino di Duccio carved on the interior of the Tempio Malatestiano in Rimini also had pigmented backgrounds, and in one of them, *Ganymede as Aquarius* (fig. 9.29), the sculptor also exploited the veining of the stone to suggest a downpour, a relief that probably inspired a description, in the *Hypnerotomachia*, of another, representing the Winter Winds, for which "the consummate artist had carefully

FIG. 9.28. Donatello, *Miracle of the Mule,* 1447–50. Gilt bronze, 22 7/16 × 48 7/16 in. (57 × 123 cm). Basilica del Santo, Padua.

FIG. 9.29. Agostino di Duccio, *Ganymede as Aquarius,* 1449–57. White marble with vertical bluish veining (Proconnesian?). Cappella dello Zodiaco, Tempio Malatestiano, Rimini.

FIG. 9.30. Pietro Lombardo and Mauro Codussi, Façade, with reliefs of "The Healing of Anianus" and "The Conversion of Anianus" by Tullio and Antonio Lombardo, Scuola Grande di S. Marco, 1489–95. Venice.

chosen a marble that, beside its whiteness, was veined with black (in the appropriate places) such as to depict the dark, lightless and cloudy sky with its falling hail."[78]

Pietro Lombardo had served an apprenticeship in Padua in the 1450s, where he had absorbed firsthand the work of Donatello, as well as Mantegna, and from where he brought the "pictorial relief" to Venice. Lombardo and his sons began providing marble altarpieces for chapels, from relatively simple affairs such as the Cappella Gussoni in S. Lio (c. 1485), to more complex realizations such as the Bernabò Chapel in S. Giovanni Crisostomo (c. 1499–1502) and the Cappella Giustinian-Baoer in S. Francesco della Vigna (c. 1500).[79] Following the example of Donatello (see fig. 9.10), Pietro and his sons practiced "pictorial relief" and brought it still closer to painting by "colorizing" it with varicolored marble and hard-stone inlays (which became a family trademark), and amalgamized the tinted relief with architecture, on the façade of the Scuola Grande di S. Marco (1489–95), when they carved them at full scale with the building (fig. 9.30). On the lower story are inlays of false-perspective arcades that terminate in streaky marbles "painted by nature," to suggest cloud-riven skies. These illusionistic corridors frame sculptured scenes of holy miracles, which exploit the gamut of relief from almost freestanding, to bas-relief, to figures that are virtually engraved.

About the same time, Mantegna himself devised a series of pictures that fictionalized low-reliefs of statuesque figures in gilt bronze or marble and silhouetted against hard-stone backgrounds (fig. 9.31). Not only was

FIG. 9.31. Andrea Mantegna, *Samson and Delilah*, c. 1495–1500. Glue tempera on linen canvas, 19 × 15 in. (47 × 37 cm). National Gallery, London.

sculpture subsumed into painting in this (virtually) intermedial synthesis, but the faux-hard stones of the backgrounds were of themselves "paintings in stone." They evoke cameo, which was itself a remediation of painting in stone. But in a real cameo the grain would not run against and behind the white lamination in this way, and the filigree carving required here would have tested any sculptor. The brush beats the chisel.

A decade or so earlier, Pietro Lombardo's essays in thinking across the three media had found a volumetric outcome in the whole church of S. Maria dei Miracoli (1481–87), in Venice (figs. 9.32, 9.33).[80] This was not a parish church, but a pilgrimage one built solely to enshrine an image of the Virgin renowned for its recent miracle-working (hence the dedication). It has passed unnoticed that this was also the first church officially dedicated to the Immaculate Conception. Immaculism, the Franciscan-sponsored doctrine that the Virgin had been born free of original sin (immaculately conceived) and was therefore a "spotless" receptacle for Christ, had gathered strength from the twelfth century, met with repeated opposition, mainly from Dominicans, and it was only under Sixtus IV, a Franciscan and the reigning pope (1471–84), that it received papal sanction.

The outside of the church was extravagantly clad all over with the choicest marbles (pavonazzetto, bardiglio di Carrara, rosso di Verona, Istrian limestone, Proconnesian, cipollino, cipollino rosso, verde antico, porphyry, and serpentine), studded with jewels as it were, a precious vessel that was knowingly designed to be a foreign body within the mainly brick fabric of the city. Its only real point of comparison was with S. Marco (see fig. 9.2), which was itself so much a material apparition that contemporaries said it seemed "to have been made by angels rather than men."[81] The external revetment of the Miracoli was all book-matched slabs, outlined in bardiglio, and subdivided by bars of Veronese marble (in reality a limestone), as though mullions and transoms, and recalling Mantegna's own inventions (see fig. 10.2).

Inside the Miracoli, the material opulence continued. This was a prayer hall culminating in an elevated choir that exalted the thaumaturgic icon over the heads of the faithful and beckoned their prayers. In the nave were tall dados of marble, and above them revetment was simulated in paint. But in the choir the marbles not only ran up to the springing of the dome, but covered the spandrels and drum as well, a great innovation, and the entire proscenium arch in front was also revetted all the way to the barrel vault. The marble used was what the Venetians called "pavonazzetto" for its resemblance to the antique Phrygian marble that had acquired this modern name, but in reality it was quarried in the Apuan Alps in the far north of Tuscany. This marble had the virtue of being white streaked with purple, like its predecessor, but also flushed with gold. It was a marvelous combination for evoking the union of flesh, blood, and light. The mottling of the stone might suggest internal germination, and once the image was incubated, it could be revealed by splitting open the block, as in book-matching. Indeed, one exegete, the Dominican friar Franciscus De Retza (1343–1427), used precisely this metaphor when he revived the story told by Albertus Magnus of witnessing the head of a king in book-matched marbles to defend the doctrine of the Immaculate Conception (1487/90; see fig. 9.6).[82]

FIG. 9.32. Pietro Lombardo and others, Façade, S. Maria dei Miracoli, 1481–87. Venice.

FIG. 9.33. Pietro Lombardo and others, Nave and choir, S. Maria dei Miracoli, 1481–87. Venice.

FIG. 9.34. Giovanni Bellini, S. Giobbe Altarpiece, c. 1478–80. Oil on panel, 15 ft. 5 7⁄16 in. × 8 ft. 5 9⁄16 in. (471 × 258 cm). Formerly S. Giobbe, Venice, now Accademia, Venice (Photomontage of altar in frame by the author.).

The raised choir put the thaumaturgic painting of the Virgin—which was more an ensouled medium than a simple artwork—on a pedestal, so to speak, and enthroned her in a chamber suffused with the light that poured in through its large windows, and enveloped in a mirage of pullulating marbles. It was both tabernacle and throne room. The Virgin is presented, just as she is in the Office of the Immaculate Conception that Sixtus IV had approved as "like the rising dawn, beautiful as the moon, raised up like the sun" (Song of Songs 6:10); and once arrived and enveloped in the glow, the priest or communicant meets her image as the apocalyptic "woman clothed with the Sun" (Revelation 12:1).

In designing such an axial, perspectival, *di-sotto-in-su* conception of the choir, Pietro Lombardo employed the same visual and mental model that the painter Giovanni Bellini (c. 1430–1516) had invented to reconceptualize Venetian altarpieces, especially those dedicated to the Virgin, namely a perspectival projection of a heavenly throne room that simultaneously also figured her as a tabernacle. The signal example is Bellini's S. Giobbe Altarpiece (fig. 9.34), which current consensus dates to about 1478 to 1480, only two or three years before ground was broken for the Miracoli.[83] This set the Madonna against an archaizing backdrop of mosaic, intended to recall those at S. Marco.

FIG. 9.35. Workshop of Giovanni Bellini, or Vittore Carpaccio, *Annunciation*, c. 1487/89. Oil on canvas, both leafs, 88 3/16 × 41 5/16 in. (224 × 105 cm). S. Maria dei Miracoli, now Accademia, Venice.

To ascend the steep staircase in S. Maria dei Miracoli and enter the choir meant passing between statues of the announcing Gabriel and the Virgin annunciate on the newels of the balustrade. However diminutive these actors, passing between them meant to trip an invisible wire that sets alight the arena of the Incarnation, and the place of the eucharistic sacrifice. It is no coincidence that when it came to paint the shutters of the organ that once stood on the same elevated choir, the Bellini workshop, or young Carpaccio, made the equation crystal clear in the *Annunciation* (c. 1487/89; fig. 9.35).[84] Within the painting the shutters seem blown inward as if by a nuclear wind, and the marbling behind the sacred actors reproduces the faux-revetment of the actual nave of the Miracoli. Painting and architecture confront each other and resume a fruitful reciprocity between the historical event of Christ's Incarnation and the ongoing Transubstantiations in the Sacrament of multiple Masses.

10 Renaissance Chapels and Church Façades

Antiquarianism, Gems, and Sympathetic Magic

If colored marbles abound in Quattrocento painting, the same cannot be said for actual buildings. The examples are scattered, but the greatest concentration and the most impressive examples are in Venice. Venice had Byzantium in her blood, first as a client state, then trading partner, then conqueror, and finally as her successor. For a century, from the early Trecento, Venice expanded onto the terra firma, gaining Verona in 1337, Treviso and Bassano del Grappa in 1339, Vicenza in 1404, Padua in 1405, and Brescia, Udine, and Friuli in 1426. With Constantinople's fall in 1453, Venice filled the vacuum politically and culturally, and in 1462 she discarded the medieval title of *commune* in favor of the imperial title of "*dominium.*" Meanwhile, refugee Greek scholars flocked to Venice, Cardinal Basilios Bessarion (1403–1472) transferred his precious library of Byzantine manuscripts there, and the press of the humanist Aldus Manutius (1449/52–1515) began to publish the first texts in Greek. Among the early editions was Theophrastos's *On Stones,* bound in with Aristotle, and appearing in 1497.[1] (Pliny's *Natural History* had already been printed in Venice by Giovanni Spira back in 1469.)

The overwhelming influence of Byzantine icons is also manifest in Giovanni Bellini's divulgation and transformation of Constantinople's most revered example, the Hodegetria (see fig. 9.19), and from the 1470s Venetian altarpieces began to stage sacred history with backgrounds inspired by the Byzantinizing setting of S. Marco.[2] Between the 1490s and 1530s there also occurred a wave of neo-Byzantine architecture in which spoliated marbles were conspicuous components.[3] In fact, La Serenissima continued to scour the eastern Mediterranean for colored marbles well into the seventeenth century, and the ruins of the Istrian coast, Aegean Isles, Crete, Epirus, and the Peloponnese littoral all became her quarry.[4] Venice began to see herself as Old Rome, New Rome, and Jerusalem all at the same time, all cities that were rich in stones and their lore.[5]

The baton was passed in the paintings of Vittore Carpaccio, Giovanni Bellini, Giovanni Battista Cima da Conegliano, and others, all of whom imagined Eastern cities clad head-to-toe in precious stones, fusing Renaissance Venice with medieval Byzantium (fig. 10.1). Cima's acute knowledge of masonry construction shows that he had studied both S. Marco and rising edifices such as the Miracoli, including its marble pendentives, blending them all into one virtual continuum.[6] In essence,

Cima's painting fuses the compositional and illusionistic prototypes of Bellini with the archaeological acumen of Mantegna. Andrea Mantegna, for his part, found his wife (Nicosia Bellini, Giovanni Bellini's aunt) and pigments in Venice, and no doubt his marble literacy stemmed from there as well.[7] His depiction of colored stones deserves a long study unto itself, but one may defer to the rich contribution made by Roger Jones. Nonetheless, besides the Brera *Lamentation over the Dead Christ* and the *Samson and Delilah* (see figs. 9.16, 9.31), we can also single out his *Circumcision of Christ* (1464; fig. 10.2), staged in the Temple of Jerusalem, for its envisioning of a marble architecture that reeks of Venice, and whose only modern competitor was the doorway of Alberti's Tempio Malatestiano in Rimini made a decade earlier (see fig. 10.8).

The degree to which marble splendor became associated with Venice also emanates from the work of painters north of the Alps.[8] In Albrecht Dürer's *Madonna and Child,* painted after a Venetian sojourn in 1494–95 or 1505–7, the Madonna and Christ Child are as Nordic as the landscape visible through the window, but the Madonna is another Hodegetria of the type divulgated by Bellini, and the room

FIG. 10.1. Giovanni Battista Cima da Conegliano, *The Healing of Anianus,* 1498. Tempera on poplar panel, 67¾ × 53⅛ in. (172 × 135 cm). Gemäldegalerie, Staatliche Museen, Berlin.

FIG. 10.2. Andrea Mantegna, *The Circumcision of Christ*, 1464. Tempera on wood, 33⅞ × 16⁹⁄₁₆ in. (86 × 42 cm). Galleria degli Uffizi, Florence.

around her is shot through with marble veining as if it were carved from a single block (fig. 10.3).[9]

In religious architecture the longest shadow was cast by S. Marco, the most marbled building in the West, and more Byzantine than Byzantium (see fig. 9.2). When the Florentine architect and sculptor Filarete (Antonio di Pietro Averlino, c. 1400–c. 1469), who knew Constantinople only secondhand through his mentor, Francesco Filelfo (1398–1481), advocated the marbled interior of Hagia Sophia as a model for princely patronage, his model was really S. Marco:

> There are still many more marbles of this sort in Greece as we can understand from the things I have seen in Rome and also in Venice that came to us from Greece. . . . They are very beautiful because they are extremely varied and dappled with natural marks in such a way that many, with a mixture of white and red, appear quite like salted meat while many others are like clouds; others are varied in other ways such that different animals appear to lie within, made by nature and made in a strange way. One may see an example in Venice in the church of San Marco where there are many slabs in which there are within various things made by nature [and] amongst which . . . you will see an actual figure that you will say is painted and in the form of a hermit with a beard and a hairshirt, standing with his hands in such a way that he seems to be praying; exactly how this is made I do not know, but it was created by nature; and at the time those two slabs were sawn, as happens when an intarsia is made by human hand, then that thing appeared and when the two slabs were set together later they showed this figure. . . . And so I have understood that there are many similar [slabs] in Santa Sophia in Constantinople and with varied forms of men and other animals [in them]; such that, if you have any way to procure such beautiful things they would render great honor to the building because they would ornament it greatly. Just as the Venetians have done in their church, which is known to connoisseurs.[10]

Filarete's allegiance to Greekness also shines through the philo-Hellenic language of his *Trattato*, which contrasts with the strained Latinity of Alberti's own architectural writings.[11] The debt to Byzantine ekphrasis is more obvious in the Venetian Francesco Colonna's Greek-titled romance, the *Hypnerotomachia Poliphili* (Venice, 1499), which even cites the authority of the itinerant archaeologist Ciriaco d'Ancona (1391–1453/55) in its preface. The *Hypnerotomachia* is riddled with the most detailed descriptions of fantastic edifices clad in the richest marbles, and a few lines of its alliterative Venetian-cum-Byzantine rhapsody are more expressive of the Venetian fascination with exotic and veined marbles than any detailed further analysis. As the author himself says in the preface, "In order to understand [the book] one needs Greek and Latin no less than Tuscan and the vernacular." Indeed, the *Hypnerotomachia* is written in an idiosyncratic hybrid of

FIG. 10.3. Albrecht Dürer, *Madonna and Child,* 1495 or 1507. Oil on panel, 20 × 16 in. (50 × 40 cm). National Gallery of Art, Washington, DC.

Latin vocabulary imposed upon Italian syntax with a liberal sprinkling of Greek loan words, none of which makes for easy translation. Thus, describing a Temple of Zeus Aigiochoios, Colonna says:

> *Et secundo la requisitione dilla harmonia di lo aedificamento, cusì tutte le parte cum approbata elegantia constavano. Cum decente partiario dil coloramento degli marmori et vaga discriminatione, cum participamento allo obiecto gratioso, di Porphyrite, di Ophite, Numidice, Alabastritie, Pyropecile, Lacedaemonice, et candido di marmore varicosamente undulate, et Andracine digenerate cum bianchissime macule. Altre di multiplice coloramento confusamente commixto.*[12]
>
> Thus all the parts were composed with due elegance according to the harmony appropriate to the building; with the fitting distribution of the coloring and beautiful contrasts of the marbles agreeably partaking in the object: Porphyry, Ophite, Numidian marble, Alabaster, fiery granite, Lacedaemonian marble, and brilliant white marble varicosely veined, and cinnabar stained by the whitest mottle, and still others of multiple colors chaotically intermingled.

Moreover, the Fall of 1453 did not signify the end of eyewitness contact with Constantinople itself. Commerce continued under the Ottomans, artists such as Costanzo da Ferrara and Gentile Bellini briefly worked there, and humanists ventured eastward to tour the antiquities or seek out Greek manuscripts, among them Benedetto Dei, Niccolò and Bernardo Michelozzi, Bonsignore Bonsignori, and (the Constantinople-born but Venice-based) Janos Lascaris.[13] Hagia Sophia continued to fill the Western imagination through travelers' reports, Western chronicles, and circulating drawings. Ciriaco d'Ancona's drawings of Hagia Sophia (probably from the 1440s), periegetically organized like a Byzantine ekphrasis and extolling the "great temple notable for 104 porphyry, serpentine, and marble columns of various noble and remarkable stones," also made their way into the architectural consciousness through the copies of Giuliano da Sangallo (c. 1445–1516).[14] Those lucky enough to enter this church-turned-mosque were still exhilarated by its adornments, and some would come to the same conclusions as Prokopios, that "all its materials are the most valuable productions of nature, so that viewing it employs the thoughts of the spectator with delight and admiration."[15]

Venice was also within easy reach of some of the best marbled interiors to survive antiquity, at Ravenna. When Ambrogio Traversari, the Florentine friar, humanist, and probable ideator of Lorenzo Ghiberti's Porta del Paradiso (1425), visited Ravenna in 1433, he was euphoric about the ornament within and without, the huge columns and porphyry columnettes, and the shining, precision-cut marble slabs in which you could see the world reflected as if in a mirror. He had not seen anything better in Rome, he said.[16] When Antonio da Sangallo the Younger saw S. Vitale eighty years later (c. 1526), he too commented that "the fantasy is beautiful of stones sawn and opened like books that make beautiful forms accordingly."[17] Likewise, the Venetian humanists who investigated Ravenna and Aquileia were constantly struck by the *varium marmorum ornamenta* and *marmorea pavimenta et emblemata vermiculata.*[18]

All these ornaments left their mark on Venetian painting, but also on a batch of new chapels, in which painting (and mosaic), architecture, and sculpture enter into dialogue.[19] The earliest of these is the dogally-commissioned Cappella Foscari or Cappella dei Mascoli (1431) in S. Marco (fig. 10.4). This was the first chapel to be coherently revetted in the West since late antiquity. Its walls are sheathed in *cipollino rosso* (Iasian marble) and *greco fiorito*, all spolia, right up to the level of its vault mosaics.[20] On the altar wall, the slabs are laid symmetrically and knowingly so that the veining suggests a halo around the radiant altarpiece. This chapel was undoubtedly an inspiration to Pietro Lombardo at S. Maria dei Miracoli (see fig. 9.32), and it is in his work that the Venetian pursuit of an intermedial and pictorial architecture would culminate. However, other chapels continued to be extravagantly marbled, such as the Cappella Emiliana in S. Michele in Isola (1527–30), as well as a number of monumental altars such as the Garzadori Altar (c. 1500–1502) in S. Corona, Vicenza, and the Morosini Altar in S. Francesco della Vigna (1520s).[21] The jasper and porphyry inlays and columns of transparent alabaster on the Altar of the Sacrament that Lorenzo Bregno inserted into the apse of S. Marco in 1518 also demonstrate a delight in strident marble effects that was meant to evoke the opulence of Solomon's Holy of Holies.[22] Bregno's chapel, like the earlier Mascoli chapel, was equally inspired by the need to measure up to the building that housed it, S. Marco.

FIG. 10.4. Cappella Foscari (or dei Mascoli), S. Marco, 1431, Venice.

CHAPELS OUTSIDE VENICE

Outside Venice, there are few marbled chapels to pick from in the Quattrocento, and with one exception, they have none of the intermedial sophistication of the Lombardo, but tend to be more historicizing in their recollections of marble use. In Florence the venerable exempla of the Duomo Baptistery (1059–1128) and the Church of S. Miniato al Monte (1018/1207) offered the bichromatic model of revetment in green-and-white stone, privileging geometrical inlays (fig. 10.5). By contrast, Venetian use of marble was distinguished by a love of the slabs' innate pattern and palette; the procurators responsible for building S. Maria dei Miracoli did not care whether the marbles were Greek or Italian, just so long as they were veined.

In Florence, the Chapel of the Magi (c. 1444) in the Palazzo Medici and the highly influential Cardinal of Portugal Chapel designed by Antonio Manetti in S. Miniato al Monte, Florence (1460–68), have intricate and archaizing pavements mimicking Cosmati work. In the Chapel of the Magi, the use of porphyry is lavish and the several rotae consciously evoked the serial disks that carpeted imperial throne rooms from Constantine onward and that ran up the nave of Old St. Peter's like carpet runners. The Medici had already appropriated this symbolism for the floor tomb of Cosimo de' Medici the Elder in S. Lorenzo (1464).[23] Only the Chapel of Sigismondo Malatesta in S. Francesco, Rimini (1447–52), and the Cappella del Perdono in the Ducal Palace at Urbino (c. 1475/94) actually have revetted walls.[24] The Malatesta Chapel was probably designed by the sculptor and executive architect Matteo de' Pasti (1420–1467/68), and its dado is a motley assembly of pavonazzetto and Proconnesian spolia looted from nearby Ravenna.[25] Instead, the Cappella del Perdono (1472/80), almost certainly by Francesco di Giorgio Martini (1439–1501), applies a vision of an antique cella (like those in the Temple of Venus and Roma in Rome or the so-called "Temple of Diana" in Nimes) to the cult of the Virgin (fig. 10.6). Before entering the vestibule to this chapel and another, dedicated to the Muses, an inscription announces: "Whosoever seeks these sacred thresholds with a pure heart, seeks the shining kingdom of the Eternal City."[26]

The Cardinal of Portugal Chapel (fig. 10.7) represents a conscious attempt to reproduce a Paleo-Christian mausoleum with prototypes ranging from the Mausoleum of Galla Placidia in Ravenna (see fig. 6.8) to the ninth-century Chapel of S. Zeno at S. Prassede in Rome. Some material

FIG. 10.5. Apse, Baptistery, S. Giovanni Battista, revetted interior, 1059. Florence.

FIG. 10.6. Francesco di Giorgio Martini, attrib., Cappella del Perdono, Ducal Palace, 1472/80. Urbino.

cues are obvious: the cardinal's sarcophagus resembles the porphyry basin that still stood in front of the Pantheon; the *intarsie* on the altar paliotto mimic those in the Romanesque nave of S. Miniato itself. Moreover, some attempt was also made to imitate the mosaic and marble schemes of Paleo-Christian chapels and mausolea. The vault is covered with an iridescent skin of glazed tiles, and Alessio Baldovinetti tried to capture the effects of mosaic in the spandrel and lunette frescoes by applying gold leaf and varnishes that have since flaked off. The floor is an archaeologically correct copy of Cosmati paving, and its quincunx mirrors the tondi of the evangelists orbiting the Holy Spirit in the vault. Like the floor of Westminster Abbey (see fig. 6.26), the scheme probably aspires to some larger cosmic interpretation, here one whose end can be seen in its beginning. The wall revetment, on the other hand, is all'antica cladding that subtly extends the illusions of the frescoes into the chapel, as, for example, the wall panel of the Virgin's *hortus conclusus* that materializes into real revetment behind the episcopal throne.[27]

FIG. 10.7. Antonio Manetti, Antonio Rossellino, and others, Cardinal of Portugal Chapel, S. Miniato, 1460–68. Florence.

MARBLE CHURCH FAÇADES

Although colored marbles remained scarce in architecture for most of the Quattrocento, almost as many church façades were as revetted as chapel interiors. In most cases, the new façades reflected the antique architecture indigenous to the region, and the category "antique" might encompass Paleo-Christian and Byzantine buildings. This latitude of inclusion helped to accommodate the forms of classical architecture to a building typology foreign to it, the church. There was obviously some confusion too. On the one hand, Antonio Manetti could join visual style to historical period, when he explained (in his *Life of Brunelleschi*) that Tuscan Romanesque was Carolingian; on the other, from the 1300s on, Tuscan chroniclers could claim that the Florentine Baptistery (1059–1128) was actually an Augustan temple of Mars.[28]

The earliest of the new marble church façades were all designed by Leon Battista Alberti (1404–1472), who wrote the first systematic treatise on architecture (*De re aedificatoria,* 1443/52) since Vitruvius, and applied his vast knowledge of classical architecture and literature to the task. In practice, he also metabolized the idioms of the Duecento and Trecento architecture that he had studied during a youth spent between Genoa, Padua, and Bologna. He also knew Venice well, and S. Marco right down to the air vents in its pavement.[29]

The first façade that Alberti designed, around 1450, in the midst of composing his treatise, was on S. Francesco in Rimini, which became the first church façade of the Renaissance to reproduce an ancient structure convincingly, including its epigraphy (fig. 10.8).[30] The patron, Sigismondo Pandolfo Malatesta (1417–1468), was a cut-throat *condottiere,* but one who gathered a coterie of poets, humanists, and artists around him. The existing Church of S. Francesco was Gothic, but Alberti transformed it into a classical *templum* with a cocoon of arcades down its sides and a new façade, and from now on it would be known as the "Tempio Malatestiano." Malatesta's court poet and Graecist, Basinio Basini (1425–1457), extolled the result as a "marvellous Temple of Parian marble," and modern scholars have tended to be somewhat color-blind and associate the renovation

FIG. 10.8. Leon Battista Alberti, Façade, Tempio Malatestiano, c. 1450–c. 60. Rimini.

with Alberti's own comment that he would be "easily persuaded that purity and simplicity of color, as life, is most pleasing to the supreme Gods." However, Alberti actually goes on to praise lavish materials, and recommends, "It will be most convenient to clad the innards of the [church] with marble or glass, either as paneling or inlays."[31] Moreover, while the façade of the Tempio Malatestiano is predominantly built of white, Istrian stone, there are many colored refinements, combinations of stones that could be seen in Venice, the city that the masons hailed from. The blind arches are lined with gray limestone; the inscriptions on the side of the church are in the same material (to imitate bronze?); the stylobate is trimmed with red Veronese marble; the façade is studded with porphyry roundels; and the entire central bay is packed with porphyries and serpentines, despoiled from S. Apollinare in Classe, Ravenna.[32]

The ancient prototype for the façade was indeed the all-white Arch of Augustus (27 BC), a short walk away, from which Alberti quoted the engaged order with its broken architrave, cable-work shafts, archivolts, *clipei,* and stylobates. However, the façade was also to some extent indebted to the Arch of Constantine in Rome, which preserved a much more colored vision of monumental Roman architecture, and one that would inform Alberti's later design of the façade of S. Maria Novella, Florence (see fig. 10.11). The columns and pilasters of the arch were in yellow Numidian marble, the attic plinths were in green Carystian, the statues of Dacian captives were in purple-streaked Phrygian marble, and the fields around the tondi were inlaid with porphyry. Even the carcass of the arch was a fizzling mix of Proconnesian and Carrara, and the frieze (now missing) must have been inlaid in porphyry or serpentine. The Arch of Constantine preserved most of its inlays intact to Alberti's day, as is clear in Benedetto Bonfigli's contemporary fresco, the *Miracle of San Ludovico* (1454–61) in the chapel of the Palazzo dei Priori, Perugia.[33]

However, Alberti's probable point of inspiration for the porphyry roundels was the medallions on the side façades of S. Marco, for only there does one find oval insets in porphyry set vertically and horizontally. Others have also noted a similarity between the doorway to the Cella delle Reliquie within the church and the entrance to the baptistery at S. Marco.[34] In short, however Romanizing the façade of the Tempio, the Byzantine tradition transmitted to Venice and Ravenna still made itself felt. Indeed, the dedicatory inscriptions are also in Greek language and lettering. Sigismondo Malatesta even had stolen the body of Georgios Gemistos Plethon (1355–1452), the last philosopher of the last Byzantine court at Mystra, to be reinterred in one of the neo-Byzantine sarcophagi on the church's lateral arcade. Perhaps in the sudden vacuum left by the demise of Byzantium, Sigismondo saw himself as a new Constantine. Basini compares him to Augustus, and the Malatesta family had boasted that their forebears included Noah, the Trojan hero Tarchon, the Holy Roman Emperor Otto III, and the Roman general Scipio Africanus. The Riminese scholar Roberto Valturio (1405–1475), who wrote his *De re militari* (1446/55) for Sigismondo, and who is also interred in one of the church's external sarcophagi, compares him with every general he can think of from Alexander the Great onward, all of them mighty in arms, but builders of lavish temples in peacetime. He concludes that the Tempio Malatestiano is the most ancient building one can see in Rimini, in fact better than antiquity itself.[35]

The central bay is a complete innovation, and Basini and the itinerant scholar Porcelio Pandone (c. 1407–1485) admired its porphyry and "Parian."[36] The eye-catching tartan grid of statuary marble and serpentine and porphyry that fills the lunette above the portal, with alternating ovals and diamonds, is the first attempt to reconstruct the intricacies of antique revetment during the Renaissance, and reveals the welter of influences that Alberti metabolized. Porphyry and serpentine were like man and wife in the aristocracy of colored marbles, frequently used in combination on antique floors and revetment, and without this filler the portal would also have been hopelessly underscaled. However, the question remains as to where Alberti, a man so observant of classical remains and so erudite in his borrowings, could have found a precedent for this idea. About the time Alberti's façade was rising, Piero della Francesca also imagined an Early Christian basilica façade in his *Proofing of the True Cross* at Arezzo (1452/66), clad with implausibly large sheets of marble and studded with colored rotae, but no Early Christian façade of the sort actually existed. In truth, Alberti fused his knowledge of three sources: the late antique revetments in Rome and Ravenna (see figs. 6.1–6.3); the pronaos of the Pantheon, which, until at least the 1520s, still retained its paneling in porphyry and other granites (see fig. 10.15), and where there was also a large and blank lunette over the bronze portal in which Alberti may have imagined some revetment; and the altar bay of the Florence Baptistery (see fig. 10.5).

The façade of the Tempio Malatestiano had almost immediate influence on two façades added to preexisting buildings. The first is fairly absent from most accounts of the progression of Quattrocento Italian architecture, as it does not fit the old historiographical mold of progressive structuralism and supposedly ever-whiter (purer) architecture. This is the façade of the Oratory of S. Bernardino in Perugia, designed by the sculptor Agostino di Duccio in 1457 to 1461 (fig. 10.9).[37] The oratory façade inscribes a Romanesque portal (with canted frame, trumeau, and stacked aedicules) within a classical triumphal arch, sculptured all over, and covered with a highly polychromed blend of local pink-and-green limestones and surface paint from head to toe. Agostino di Duccio had only recently executed the reliefs within the Tempio Malatestiano (see fig. 9.29), and while it is clear that he took away with him the example of a triumphal arch façade and the particular Veneto-Byzantine love of external polychromy, the pictorial blend owes as much to the rilievo schiacciato, of which he was a master, that intermedium between painting and sculpture. In fact, one could easily imagine the same façade at a fraction of its scale, functioning as a sacrament tabernacle within a church. Finally, it is not only the classical moldings, triumphal arch, and pedimental sculpture that announce its studied antiquity but also the classical inscription, which records not the Christian dedication to S. Bernardino but only the year of inauguration and the city's Roman name, copied, even in its letter forms, from the city gate almost opposite and remodeled by Augustus in 40 BC: AVGVSTA PERVSIA MCCCCLXI.

FIG. 10.9. Agostino di Duccio, Façade, Oratory of S. Bernardino, 1457–61. Perugia.

The second immediate legacy of the Tempio Malatestiano was the façade that Mauro Codussi added (1468–77) to S. Michele in Isola, in the Venetian lagoon (fig. 10.10).[38] Codussi indubitably knew Alberti's façade firsthand, as he had worked for the Camaldolese friars at S. Apollinare in Classe, near Ravenna and on the road to Rimini.[39] The mostly Istrian stone of Codussi's façade initiated a Venetian trend for white architecture, and when viewed from the city it drifts across the lagoon like the Fata Morgana. Debarking at the island, however, the masonry becomes very insistent, thanks to the widely channeled joints and even to the point where the pilasters are not shafts but rather projections of the bossed wall, as though Codussi were positing stone-upon-stone as an antidote to Venetian veneer. By contrast, the attic is faced with Proconnesian sheets and trimmed with gray marble, as are the pediments. Presumably Codussi drew this trimming detail from Alberti's Tempio Malatestiano, for he also borrowed the trilobed superstructure (incomplete on the Rimini façade, but represented on Matteo de' Pasti's medal), the all'antica epigraphy in the friezes, the "*muro triumphato,*" and the porphyry rotae that orbit the oculus to compose a quincunx. Codussi's signature motif, the colored tympanum over the portal, was instead inspired by Venetian example, namely the exquisite porphyry tympanum on the door into the Madonna dell'Orto (before 1444), from which he also quoted the scalloped coving he would use to cap S. Michele's façade. The most telling detail of all, however, is the porphyry oval in the tympanum that had previously been used only in the portal lunette of the Tempio Malatestiano.

FIG. 10.10. Mauro Codussi, Façade, S. Michele in Isola, 1468–77. Venice.

FIG. 10.11. Leon Battista Alberti, Façade, S. Maria Novella, 1458–70. Florence.

The intricate intarsie of Alberti's next façade after the Tempio Malatestiano, on S. Maria Novella (1458–70), Florence, required inordinate amounts of marble and extraordinary craftsmanship (fig. 10.11). The result was the first marbled church façade in Florence since S. Miniato al Monte (1018/1207) and the acres of Trecento cladding on the flanks of the Duomo.[40] The patron, Giovanni Rucellai, had visited Rome for the Holy Year of 1450 and filled his journal with descriptions of the marble incrustations of Early Christian churches.[41] However, the fourteenth-century skin of S. Maria Novella's lower façade was the starting point for Alberti, its blind arcades copied from the Florentine Baptistery.[42] The baptistery remained an architectural icon of Florentine antiquity, and its marble facing was restored throughout the Quattrocento.[43] On the upper story Alberti instead quoted the geometrical intarsie of S. Miniato.

Alberti was unable to execute marble façades for his two other churches, S. Sebastiano and S. Andrea, both in Mantua. At S. Sebastiano, Ludovico Gonzaga's budget did not permit the façade to be built in imported stone, and the whole church took almost forty years to complete anyway (1460–99).[44] However, cleaning in the 1990s revealed that a square-and-octagon pattern had originally been painted on the plaster, of red and yellow panels, respectively, and reminiscent of the later Palazzo Schifanoia in Ferrara (figs. 10.12, 11.9).[45] While this must be the work of the local architect Luca Fancelli (c. 1430–c. 1502), who

FIG. 10.12. Leon Battista Alberti and Luca Fancelli, Façade, S. Sebastiano, 1460–99. Mantua. Reconstruction of façade polychromy.

supervised construction, and Alberti's façade underwent many compromises before it was semi-completed, other faux-revetments have been found on the medieval carcass behind the arcades that Alberti added to the flanks of the Tempio Malatestiano.[46]

Even more elaborate was the façade of S. Andrea, the most important church in Mantua after the Duomo, which was begun in the year of Alberti's death, 1472, and completed by Fancelli in 1493. This façade retained its very elaborate polychromy until a puritanical "restoration" of 1822–35 eradicated every trace.[47] Originally the capitals and bases of the giant order were in stucco mixed with brick dust (to imitate Veronese marble), as was the trim of the pilasters and the podium of steps. The pilasters of the dwarf order, the portals, and the giant order pedestals were all in gray-green stone (*noriglio* from the Trentino). Coffering, spandrels, and friezes were all painted, the latter as faux-intarsia, and liberally gilded. Finally, the niches and remaining wall surfaces were "quilted" with jewel-studded, faux-revetments. In many respects, the completed façade was the painted equivalent of Alberti's earlier S. Maria Novella, though employing a broader palette.

The most ambitious marbled façades in Venice were instead indebted to the example of S. Marco. Its influence on Lombardo's S. Maria dei Miracoli (1481–c. 87) is evident in the desire to sheathe the entire exterior with marbles; the book-matching and subdivision by contrasting marble bands drawn specifically from the Treasury wall of S. Marco; and the idea of a barrel-vaulted interior with revetted walls terminating (toward the *controfacciata*) in a façade with an oculus (see figs. 9.32, 9.33). The façade of the Scuola Grande di S. Marco (1489–95) could also hardly but draw on the imagery of S. Marco's façade, given the saintly patronage of the confraternity it housed (see fig. 9.30). Indeed the marbles even came from the same stable, the procurators of S. Marco.[48] Mauro Codussi, Pietro and Tullio Lombardo, and perhaps the Bellini all had a hand in the design of the façade, which blends palace and church, architecture and painting.[49] The influence of the exterior polychromy of S. Marco survived, unexpectedly, until the late sixteenth century, when Simone Sorella added a "Palladian" façade to Palladio's own S. Giorgio Maggiore, whose giant order of engaged columns were intended to have been monoliths in a very un-Palladian red limestone.[50]

Outside Florence and Venice, we must briefly mention two works by Giovanni Antonio Amadeo (c. 1447–1522): the façade of the Colleoni Mausoleum (1470–75) in Bergamo, and the Certosa of Pavia (begun in 1473, but executed mainly from 1492 onward).[51] The Colleoni Mausoleum is entirely clad in illusionistic rhomboids made from pink, white, and blue marbles (fig. 10.13).[52] The motif was common on antique paving (see fig. 4.29), which Amadeo may have known, though nearly identical revetment appears in paintings by Piero della Francesca.[53] Nonetheless, as Richard Schofield has suggested, the on-site prompt was probably Filarete's proposal for the façade of S. Vincenzo (now S. Alessandro), which faces

FIG. 10.13. Giovanni Antonio Amadeo, Colleoni Mausoleum, 1470–75. Bergamo.

the mausoleum and remained largely unexecuted, but for which he left a model and which he illustrated in his *Trattato,* and which was to some extent based again on S. Miniato al Monte.[54]

Alberti's colored classicism and the Venetian inheritance from Byzantium had little influence in Rome, the epicenter of ancient marbling. When the façades of the major churches of the *Renovatio Romae,* at S. Marco (begun 1466), S. Maria del Popolo (1472–77), and S. Agostino (1479–83), were faced in stone, it was not in marble but travertine, the material of antique engineering and the Colosseum. Only within the churches is there any evidence that antique revetments were being revived, and then in paint and under the impetus of émigré painters.

A series of chapels in S. Maria del Popolo decorated or inspired by Pintoricchio—the Cappella (di Domenico) Basso della Rovere (Cappella di S. Girolamo, 1477/79), Montemirabile (Cappella di S. Giovanni Battista, 1479/84), and Costa (after 1488)—had feigned columns and feigned revetment vaguely derived, like the shape of the chapels themselves, from those in the fifth-century Oratory of the Holy Cross at the Lateran (see fig. 6.3).[55] Pintoricchio painted more extensive faux-revetments in the later Cappella Eroli (1497) in the Cathedral of Spoleto as well.[56] The trend may have received further stimulus from the Cappella del Battista that Mantegna decorated in the Villa del Belvedere for Innocent VIII, in 1488/89–90. A description of the chapel published only a few years before it was destroyed (1780) reveals that most of the wall space was taken up by figurative scenes, but it also specifies corner pilasters fictively supporting the impost cornice, window reveals "paneled in faux-marble," and that there was a genuine quincunx floor of intarsiated marbles; finally, the cupola was "adorned . . . with many roundels, each connected to the other in the form of a grille."[57] This last feature was a leitmotif in the work of Mantegna and his followers, such that we can at least imagine the Vatican vault's appearance from that painted by Giovanni Pietro da Cemmo or Antonio Della Corna in the vestibule of the Palazzo Fodri in Cremona in c. 1490/1500.[58] In another foray, Mantegna decorated the diminutive Church of S. Maria della Vittoria (1495–96) in Vicenza with a decorative scheme that included faux-marble paneling on the nave piers and archivolts; after being secularized and converted to military use in 1877, only recently have fragments of this scheme been uncovered, leaving the possibility that other faux-revetments may also come to light, as one might expect from their depiction in his easel paintings.[59]

THE CHIGI CHAPEL: ARCHAEOLOGY AND REBIRTH *IN ASTRIS*

Some marble chapels continued to be built sporadically through the early Cinquecento. One of the least known is the great *Arca* (or sepulcher) for the remains of S. Bernardino da Siena in the basilica dedicated to the saint at L'Aquila, in the Abruzzi, designed by Silvestro dell'Aquila (c. 1450–1504) before 1500 and finished by his workshop in 1505.[60] This mausoleum is elaborately carved in white statuary marble, but its side walls are completely paneled with colored marble slabs. Also worthy of note are Giuliano da Sangallo's Gondi Chapel in S. Maria Novella, Florence (1503/8), and Lorenzo and Giambattista Bregno's Cappella del SS. Sacramento in the Duomo, Treviso (1501–7).[61] Giuliano's interest in antique *crustae* is clear from his drawings of the Basilica of Junius Bassus (see fig. 5.19) and the Oratory of S. Croce. The Gondi Chapel is, however, a tentative affair, with low-paneled walls whose effects depend on a simple contrast between dark and light stones. The far more monumental Cappella del SS. Sacramento instead copies the pictorial innovations of the Miracoli and Cappella Bernabò with head-to-toe "pavonazzetto" revetment (Apuan marble) marbling that enveloped the blessed sacrament in storm clouds of veining.

However, demand for the material and curiosity in its expressive potential was beginning to accelerate, at least on paper. In 1506 the Volterran humanist Raffaelle Maffei, whose brother would become the site manager at the Villa Madama, thought crustae were simply marbled plaster.[62] By 1521, however, the Milanese architect Cesare Cesariano (1475–1543), who also wrote an exceptionally long commentary on Vitruvius (1520), considered the creeping use of real revetment worthy of note, although one wonders exactly where he was looking.[63]

In Rome there is also nothing to compare with the colored revetments of northern Italian chapels before Raphael's Chigi Chapel in S. Maria del Popolo (fig. 10.14). Built in 1513–16 (but not actually completed until 1655), it remained a complete anomaly in the city until the Cappella Gregoriana in New St. Peter's was decorated in 1578–80 (see fig. 12.5).[64] The patron, Agostino Chigi, commissioned the chapel as a mausoleum for himself and his bride, Francesca Ordeaza, and it essentially couples Bramante's crossing at St. Peter's with the material paradigm of the Pantheon. Raphael must have had his own handiwork in mind when writing his well-known *Letter to Leo X* (c. 1514/15 or 1519/20), which stressed:

> There are modern buildings recently built which are now very well known, which, despite their novelty, have still neither attained total excellence nor the extraordinary expense that one sees and considers in the works of the ancients. To wit, although the architecture of our times has been much awakened and come very close to the manner of the ancients, as one may see in the many beautiful works of Bramante, nonetheless its ornaments are not of as precious a material as those of the ancients, who, it seems, put into effect at infinite expense whatever they imagined and whose will alone was enough to overcome any difficulty.[65]

Only in the Chigi Chapel did Raphael enjoy the financial wherewithal of Chigi, the richest man in Europe, to make good on his promise to revive the precious materials of antique architecture. It was natural that Raphael, who was a painter by training and a brilliantly inventive architect by inclination, should have cultivated an interest in colored marbles, and this penchant must have been encouraged by his 1515 papal appointment to procure marbles for St. Peter's. Although Raphael was authorized only to safeguard stones bearing inscriptions, scouring the ruins must have sensitized him further to the vast palette available as well as to antique schemes and techniques of revetment. In the Chigi Chapel, Raphael took the Pantheon as his model, though in a more rigorously archaeological manner than anyone before him, copying the marbled panel-work of the piers *in antis* around the main portal (fig. 10.15). It is therefore reasonable to assume that the most unusual feature of the chapel's incrustations, the raised-block revetment flanking the lateral pyramid-obelisks, derives from the same model.[66] This hypothesis receives a measure of support from alterations made by an unknown artist to Raphael's own drawing of the Pantheon interior, where opus isodomum revetment is visible in the exedrae above paneling with rotae, and so forth. No such revetment now survives, but Palladio did record the same detail, which must have still been in place when he visited Rome in 1541. Francesco Salviati's view of the chapel also shows that the unusual cladding was meant to fill the lunettes above the cornice level too.[67] The single-most expensive item in the chapel was the monolithic threshold of Aswan granite, costing six hundred scudi alone and the only colored stone step in the entire church. Even this may have mimicked the africano (Tean marble) monolith that provides a propitious welcome to the other rotunda.

Few theories have otherwise been advanced to explain Raphael's choice of marbles in the chapel. One of the most ingenious, that the classical names of the marble pun on the Sienese origins of the patron, does not tally with the utter inability of sixteenth-century observers to identify the geographical origins of almost any marble except porphyry.[68] Instead, as Ingrid Rowland observed, there is an overall chromatic transition from black-and-white floor, through the fiery Chigi tombs and their earthy backgrounds, the subdued greens and blues of the pendentive tondi of the *Four Seasons*, to the brilliant azure and golds of the illusionistic mosaic, which revived Paleo-Christian decorations and evoked the celestial domain of God and the zodiac.

The transitional elements linking the two zones are the twin cenotaphs on the side walls, whose proportions suggest pyramids but whose astragals identify them as obelisks. Since pyramids were both tombs and symbols of eternity, they were eminently suited to a funerary chapel, and pairing them had more than one Roman resonance. First, it was known that twin Aswan granite obelisks had flanked the entrance to the Mausoleum of Augustus. Secondly, medieval tradition held that St. Peter had been martyred at a spot equidistant between two antique pyramids, that of Cestius and the even bigger Meta Romuli near the Vatican (demolished in 1499). Thirdly, in the Trecento, two pyramids were still visible standing near the Vatican, for they are represented in contemporary altarpieces. Finally, the Piazza del Popolo itself had once contained two such *metae,* englobed in houses bracketing the mouth of the Via del Corso.[69] How these metae were faced we will never know, but the Meta Romuli at the Vatican may have been revetted in colored marbles, as we see from Pinturicchio's representation in the Borgia apartments (1492–94) and Filarete's on the bronze doors of St. Peter's (1434–45).

Yet Raphael did not follow these latter fanciful reconstructions and instead looked to the signature material of obelisks, Aswan granite, which Pliny and all Renaissance writers who followed him called "*Pyropoikilos*" (*HN.* 36.63: "fire-colored").[70] For sixteenth-century antiquarians, both pyramid and obelisk formally and etymologically evoked the *rogus,* or funeral pyre, of the emperor. Moreover, pyramids were said to take the shape of eternal flames, and an erroneous etymology that was often repeated derived the

FIG. 10.14. Raphael, Chigi Chapel, S. Maria del Popolo, 1513–16 (completed 1655). Rome.

word "pyramid" from the Greek "*pur*" ("fire"), the source of our own English "pyre." Likewise, the word "porphyry" was reckoned, again incorrectly, to enjoy the same derivation, and this stone was also perceived as intrinsically fiery.

Although Raphael substituted Chian marble (porta santa) for Aswan granite or porphyry, the material still assisted the conceit, because its red-and-gray veining could suggest both combustion and the funeral smoke wafting its way heavenward, up to the gold-and-blue mosaic constellations. There, in the "o'erhanging firmament," the "majestical roof fretted with golden fire," the evanescent soul would finally escape the shadowed world, traverse the ether, and enter the domain of the zodiac to mingle with the stars and be reborn *in astris*. This passage is implicit in the color and composition of the materials, which help fulfill Agostino Chigi's intention, recorded in the bull of the chapel's concession: "Desiring by a happy exchange, to turn earthly things into heavenly, and transitory into eternal" (*cupiens terrena in coelestia, et transitoria in aeterna felici commercio commutare*).[71]

THE HIGH ALTAR OF THE DUOMO, VICENZA (1534–36): NATURAL MAGIC AND MARBLES AS GEMS

For most of the sixteenth century, there were two initiatives at play in the revival of marble use, one expressly all'antica, the other tacitly neo-medieval. One sought to reclaim the syntax of ancient construction; the other clung to the ambient of the medieval reliquary and the thaumaturgy of gems, often at the cost of the visual unity of the container. Very rarely does one intuition take precedence over the other as clearly as it does in the Chigi Chapel.

Antiquarianism and reliquary imagery were inextricably coupled in the Cappella delle Reliquie in S. Apollinare

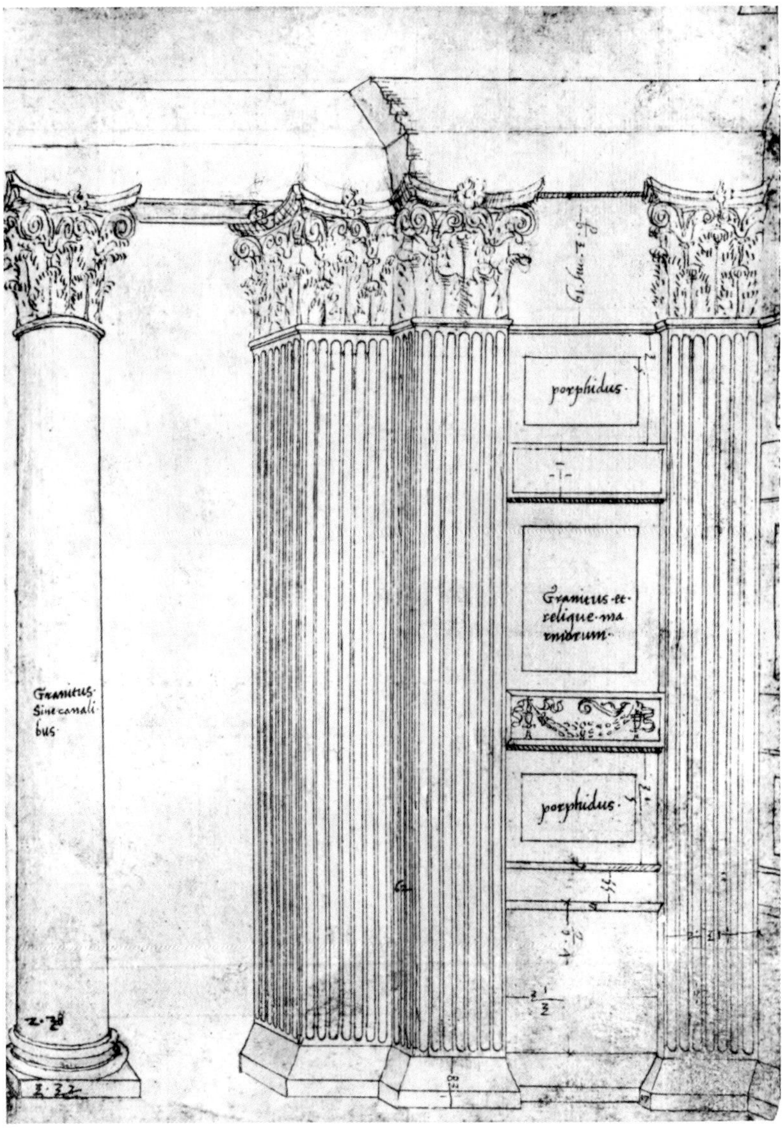

FIG. 10.15. Pantheon portico with revetments surviving in the 1520s, Codex Coner (Ashby 1904).

Nuovo, Ravenna, built in 1540 to host the bodies of St. John and St. Apollinaris himself (fig. 10.16).[72] Their hallowed remains were deposited in a marble casket held aloft by porphyry columns pilfered from the baldachin that Theoderic (493–526) had set up in the choir of the same church. From there also came the fret-sawn transennae, and the chapel walls included large panels of intarsiated chalices, again spolia from the earlier church (which in turn guaranteed the chapel's eventual mutilation in 1951, when most of these remains were restored to their original locations). The jeweled style of the cell continued on the chapel façade, whose frame was studded with gem-like inserts, and articulated with the best Ravennate spolia available: not only serpentine columnettes on alabaster columns but an extraordinary rarity, serpentine capitals.

An excellent example of the tension between the two strains, classicizing and medievalizing, is instead the High Altar of the Cathedral of Vicenza (1534–36), built by the local "Pedemuro Masters," Giovanni da Porlezza and Girolamo Pittoni da Lumignano, quite probably assisted by the youthful Palladio, to an overall design that has been attributed to Jacopo Sansovino (fig. 10.17).[73] Sansovino had, in fact, built a polychrome marble tabernacle for Cardinal Quiñones in the apse of S. Croce in Gerusalemme in Rome, which, like the Vicenza altar, was both a sacrament altar and a tomb.[74] Originally the Vicenza altar was freestanding, under the mouth of the choir, and its façade was a triumphal arch dedicated to the Corpus Christi. Its patron, Andrea Dell'Acqua, was to be buried at its foot. (He died in 1539, three years after the altar's completion.)[75]

FIG. 10.16. Cappella delle Reliquie, S. Apollinare Nuovo, 1540 (condition before 1951). Ravenna.

FIG. 10.17. Jacopo Sansovino?, Giovanni da Porlezza, and Girolamo Pittoni da Lumignano, High altar (Dell'Acqua Altar), Duomo, 1534–36. Vicenza.

This is the most ornate tabernacle altar of the Renaissance, as its façade is truly a constellation of glittering hard stones, granites, marble plaques, and polished globes. Moreover, while all the altar's recherché moldings and bases and capitals establish its credentials in the most up-to-date idiom of Renaissance classicism, the whole edifice is restored to the context of *ars sacra* by its overwhelming panoply of brilliant stones. This division between architectural membering and ornamental inlay in fact reflects separate competencies, for the altar was designed by an architect (possibly Sansovino), while the vibrant stones were installed by the Pedemuro Masters under the direction of the patron. Fine marbles cover all surfaces, even the pilasters.

When stones were formally segregated in this way, their exquisite hues and high polish gave them an affinity with gems. This affinity had been exploited in Paleo-Christian and Byzantine churches (see figs. 6.5, 6.6), and Suger had voiced the transformation that came about when stones were mounted on sacred objects: "This stone [porphyry] deserves to be enclosed in gems and gold. It was marble but in these [mounts] it has become more precious than marble."[76] By the sixteenth century, some commentators believed that, where size permitted, ancient architects had actually substituted gems for marbles, and Romance literature was already rife with fantastic descriptions of jeweled palaces and temples.[77] Moreover, as marble decorations became progressively more common in chapels during the sixteenth century, few contemporaries were slow to notice the analogy with gems, and similar sentiments were even voiced in the Ottoman world.[78] Contracts promoted the effect by commonly stipulating that "fine marbles have no flaws at all" and be keenly polished.

The gem analogy endured because such marbled tabernacles could conjure up both the Ark of the Covenant and the Heavenly Jerusalem. The stones of apocalypse, of course, were not marbles but hard stones. Yet, as Andrea Bacci would eventually point out in his 1597 treatise on the Papal Cope, early patristic commentators such as Epiphanius and Jerome had confused at least two gems with the stones onyx and lapis.[79] Moreover, ancient geology (as well as the ancient four-element theory, which regarded all matter as inter-transmutable compounds of the four elements under the influence of the four principles) implied that marbles and gems did not necessarily occupy separate physical categories, but stood at different degrees on a sliding scale of materiality. Marbles shared the same crystalline essence as gems and crystals, just with more earthy matter mixed in. And even if marbles could not enjoy the brilliant transparency of jewels or crystals, observers from Restoro d'Arezzo (1282) to Vincenzo Danti (1565) still esteemed the saturated color and hardness of porphyries and agates, thence their capacity for luster, which likened them to gems.[80]

Porphyry and serpentine were the only reliably antique names to have survived the oblivion of the rest—"alabaster" and "Parian" were really generic terms—and they were also the most prized stones because they were rare and precious, and because their hardness and saturated hues, and even the luster they acquired through high polish, assimilated them to gems. They were treated like jewels in practice as well. Granites were normally sawn by the gem-cutters or crystal-workers who stocked emery, not by the stonemasons who used sand. It is hardly a coincidence that it was a gem-cutter, Pier Maria Serbaldi da Pescia (1454/55–after 1522), who first recovered the art of carving porphyry. There is, in fact, no record of specialized *segatori di piere fine* in Venice before 1480, and this class of workers emerged with the greater use of marble facing.[81] The Venetians sprinkled precious porphyry and serpentine on church façades, such as S. Michele in Isola and the Miracoli, as though they were jewels, and this impersonation became clear-cut when they were also set in the form of the *crux gemmata* (see figs. 9.32, 9.33).[82]

But all writers could mention gems and marbles in the same breath anyway, because they belonged to a common metaphysical aristocracy. They could effect miracles, and they embodied light. In fact, on the Dell'Acqua Altar, the analogy of marbles with gems acquires particular propriety because it conjures up the tradition of sympathetic magic, which held that all precious and semiprecious stones were dormant vessels of occult powers instilled in them by astral rays, which could be catalyzed by proper arrangement and invocation.

These beliefs originated in those lapidaries of Hellenistic origin that were copied and recopied well into the seventeenth century. Between the third and fifth centuries AD, a group of lapidaries had been written in Greek, apparently in Alexandria, certainly under the influence of the gnostics, but which purported to be works by, among others, the legendary Hermes Trismegistus.[83] Each of these spun outlandish tales about the thaumaturgic properties of various stones, many of them unidentifiable or fictitious, and thereafter they formed the basis for a string of medieval lapidaries, among which the most influential remained those by Isidore of Seville (c. 560–636) and Marbode of Rennes (c. 1035–1123).[84]

Furthermore, between the Treatise of Theophrastus (315/314 BC) and that of Albertus Magnus (1254/61 AD), no systematic writing on mineralogy was produced in the West. Galen and Dioskourides, for example, had only described and classified the minerals used for medicinal purposes, and ever after most such treatises continued to be written by physicians. In other words, semiprecious stones were overridingly associated with curative or preventative agency, and hence thaumaturgy. Reinforcing the credibility of this tradition was Pliny and his Christian reception. Pliny's acceptance of "stone power" was all the more reliable for his dismissal of magic as fraud, and for Christians his word carried extra weight not only thanks to his divinization of nature, but also because his writings on stones came with the recommendation of St. Jerome.[85]

The tradition of writing lapidaries continued unabated into the sixteenth century, and the touchstone for Renaissance gem lore was the publication, in Venice in 1502, of Camillo Leonardi's *Speculum Lapidum*.[86] The

Speculum proved so popular that it was reprinted several times on its own or in compilations, and Lodovico Dolce even plagiarized it wholesale in an Italian translation of 1565.[87] Leonardi's book was itself, as he admitted, largely a compendium of earlier lapidaries, from Dioskourides (c. 40–90 AD) to Marbode, but it also organized the earlier catalogues of stones into a treatise that laid out the principles governing their formation and physical qualities, the "virtues" instilled in them by astral rays, and finally the means to draw out their potential by carving magical and zodiacal figures on them.[88]

Leonardi had distinguished between the composition of gems and marbles, but the distinction was always rather porous, as we can see from the earliest illustration of stars catalyzing precious stones (fig. 10.18). The illumination (c. 1023), in a manuscript of Hrabanus Maurus's *De universo* (c. 842) in Montecassino, shows undulating lines (astral forces) descending from a star-studded arc (the sidereal sphere) to animate two rocks, shaped like doughnuts and gripped by an individual who seems electrified by the experience.[89] There is no passage in Hrabanus's text to correspond with the image, and such "astrophysics" properly affects gems, not marbles; worse still, these stones are even labeled "common stones" (*de lapidibus vulgaribus*). Such confusion did not die with the Middle Ages, for we find the same phenomenon in a text written only five years after the Dell'Acqua Altar was finished, Pietro Contarini's 1541 poem "Argoa Voluptas" (Argoan Delights). This contains a lengthy ekphrasis on the façade of S. Marco, which assigns every marble column a Plinian label, which is incorrect in almost every case, and prioritizes their thaumaturgic powers as though they were gems or hard stones.[90]

FIG. 10.18. "On Common Stones" (*de lapidibus vulgaribus*), illumination in Hrabanus Maurus, *De universo*, c. 842. Lib. XVII, 3 (Cod. Cas. 132, fol. 418b).

The Dell'Acqua Altar in the Duomo at Vicenza (see fig. 10.17) was decorated with stones that the patron had assembled over a lifetime, and his will specifies that the masons "insert all the porphyries, serpentines, jaspers, alabasters, marbles and fine veined stones and of every sort as shall be provided, pieces large and small, round, and of every type." It is customary to interpret such bequests as posthumous aggrandizement, but Dell'Acqua's intellectual habits suggest otherwise. He was the author of the *Catena Evangelica*, a synchronic rewriting of the Gospels, and Lionello Puppi and Manfredo Tafuri have stressed his reputation within Venetian humanistic circles and his intimacy with Francesco Zorzi, Pietro Bembo, Andrea Gritti, Giangiorgio Trissino, and others.[91] Of these, Zorzi was the author of a famous *memoriale* prescribing modifications to Sansovino's design for S. Francesco della Vigna, designed to "tune" the building to the cosmos on the basis of a harmonic-proportional reasoning of obvious magical-hermetic inspiration.[92] It was also, tellingly, Zorzi who praised Dell'Acqua's *Catena Evangelica* as attaining a "Concord, inspired by the Holy and Harmonic Spirit," and Zorzi may even have recommended the altar's architect to Dell'Acqua. Finally, Zorzi had written another book that must have been well known to Dell'Acqua, *De Harmonia Mundi* (1525), which included a lengthy commentary on the stones of the Heavenly Jerusalem, tackling the questions: "Why is it said that this city was built out of gold?" "Why were the walls of Jasper?" and "What were the twelve stones in the foundation?" After a digression on the means by which stones possess supernatural powers, he eventually reveals the "real material of the heavenly city": light. Light was the "brilliant spirit," the "spiritual splendor or divine fire, vivifying and making all things glow," the *fundamental element* of the "city of light," where the angels and blessed will reside together.[93]

In this sense, the glittering display of the Dell'Acqua Altar is still in step with the anagogical materiality of Hypatius and Suger, where glitter and shimmer inspired dull minds to rise to truth. The altar's opulence, craftsmanship, and studied aniconicity also recalled the ark and the Heavenly Jerusalem, they honored God, and the variety and abundance of stones presented a mineral map of his creation.[94] Thus, the only sculptures on the altar are the two gilt-bronze angels that turn to face the door of the sacrament, just like the cherubim on the lid of the Ark. Likewise, it is more than a coincidence that the marble globes that are immured in this altar reappear in illuminations from the thirteenth through fifteenth centuries, encrusting the ramparts of the Heavenly Jerusalem (fig. 10.19). The stones on the Dell'Acqua Altar seem to radiate, or irradiate, from the host at the altar's heart, and their cosmic pretensions underlie their congregation within a great "fanlight" resembling a star-studded firmament. To some extent the model for these encrustations was another sacrament altar, that inserted by Lorenzo Bregno into the apse of S. Marco in 1518, from which our altar borrowed the design of some of its intarsiated whorls, a motif that originally held Christological connotations.[95]

The only area of the Dell'Acqua Altar that is conspicuously not "bejeweled" is the stepped plinth below the tabernacle door, of stridently veined and book-folded Egyptian alabaster. The stone may be a eucharistic accent, used to manifest light-filled cloud. Alabaster easily resembled sun-filled cloud, as texts from at least the 1300s mention, and pieces with particularly nebulous veining were called "*alabastro nuvoloso.*"[96]

Clouds had always escorted the theophany in scripture. Cloud was not only the carriage of divinity, but its veil, the sunscreen of his blinding brilliance, and hence the truest testament to God's presence (the *Shekinah*).[97] In the Old Testament, divine cloud had entered the ark in the desert and the Temple of Solomon at its consecration, and had appeared to Moses on Mt. Ararat. In the New Testament,

FIG. 10.19. Jean Colombe, *Heavenly Jerusalem, Apocalypse of the Dukes of Savoy*, c. 1490. El Escorial, Monastery Library of Saint Lawrence.

cloud indicated the presence of the entire Trinity, for "the Holy Ghost is sent in two ways, visibly and invisibly. . . . First, He appeared in the form of a bright cloud above Christ transfigured." Cloud thus readily signified the eventual Resurrection when, on the last day, "they will see the Son of man coming on the clouds of heaven with power and glory" (Matthew 24:30; Mark 13:26; Luke 21:27).

Furthermore, patristic writings could not make too much of the identification of Christ with cloud, and cloudiness with his flesh, and hence with the Incarnation. Hrabanus Maurus is the most succinct on the subject: the cloud tabernacle paralleled Christ-Ecclesia, and "allegorically speaking, in fact, cloud . . . signifies the incarnation of Christ."[98] This perceived similitude could and perhaps did provide an especial incentive to use alabaster on a eucharistic tabernacle. In the end, the altar also honors the donor, for his family arms appear three times, but considering that he was buried at the altar's foot, it is just as likely that this heraldic branding tied his posthumous destiny to the function of the funerary Mass and the daily Resurrection effected by the eucharistic rite.

Through the analogy of gem "astrophysics," the altar offered an image of transformation and a metaphor of divine agency at the most appropriate venue in the church, the high altar, the place of eucharistic sacrifice, and tabernacle for the consecrated host. The power of gems was innate but dormant. They were sleeping volcanoes, and only proper ritual could transform them and invoke their beneficent agency. The same could be said for the eucharistic host, and by the thirteenth century lapidaries had become quite explicit on the point, even beginning by comparing the power of gems to the act of Transubstantiation. The *Lapidario Estense,* in fact, opens with the following words:

> 1) According to proverb, in olden times one said that in stones, words and herbs reside the virtues. . . .
> 2) By virtue of the words that the priest says at the altar during the office of the Mass, the host, which appears to be lifeless bread, becomes the living body and the wine, which fills the chalice, becomes the living blood.
> 3) We also know that water and many other things become holy by the virtue of words, with the sign of the cross. . . .
> 12) Because of the virtues that reside in stones, he who holds them and carries them correctly receives many graces and escapes many dangers.[99]

11 Marble Mansions and Painted Palaces in Renaissance Italy

Dreams of palaces hewn from a single stone or strewn with jewels did not dissipate as the empire declined, but marble supplies rapidly dwindled. Aristocrats in the outlying provinces who hoped to live like the Dominus of the Porta Marina Domus at Ostia (see fig. 5.16) had instead to content themselves with ersatz substitutes. Many relied on painted plaster, such as the fifth-century patrician who decorated his residence at Constantia (Salamis, Cyprus). Others, such as the lord of the "House of the Consuls" (526/528 AD) at Apamea-on-the-Orontes (Syria), conjured up encrustations almost as elaborate as those at Ostia, but concocted from terra-cotta sherds, limestone chips, slate, and stucco.[1]

It was the same story in Western Europe, outside of Italy, where palaces continued to be built throughout the Middle Ages, but there was a scarcity of available marbles. An oblique nod to the tension between the ideal option and the sober reality comes from the *Libri Carolini,* a theological tract on images written about 790 to 792 (possibly by Theodulf, bishop of Orléans) at the Frankish court in Aachen: "If someone building a wooden house desires to adorn its walls with marble slabs or variegate them with multicolored pieces of glass, when he realizes that these same minerals cannot adhere to wood by nature, he discards them . . . and tries to finish it in wood anew" (3.30).

In Constantinople, however, stockpiles and spoliation meant that new mansions could still be enriched in the accustomed way, and ensured that the imperial palace remained a work in progress. The Emperor Theophilos (r. 829–842) added a richly marbled and mosaicked trikonchos and chrysotriklinion to the palace; and in the ninth or tenth century a sea frontage thick with spolia was added to the Boukoleon Palace.[2] Byzantine nobles apparently indulged their passion for marbles (roundly condemned already by St. Basil and Gregory of Nyssa in the fourth century) at the cost of true home comforts. When Liudprand, bishop of Cremona, visited Constantinople in 968 to 969, he moaned that it was all very well that his lodging was lofty, had ample windows, and was festooned with diverse marbles, but it kept out none of the elements, his bedding was a marble slab, and even his pillow was a stone.[3]

As time progressed, new imperial palaces were added along the city walls of Constantinople within intramural parks. One of them, the Blachernai Palace, was built in the eleventh century, extended in the twelfth, and by the early fourteenth it had effectively supplanted the Great

FIG. 11.1. Tekfur Sarayı, c. 1261/91. Istanbul. Photograph by Guillaume Berggren (1880s).

Palace.[4] The addition built by Andronikos II Palaiologos (r. 1282–1328) was again decked with marble floors, columns, micro-columns, and gold mosaic vaults.[5] More fortified palaces followed, such as the Tekfur Sarayı ("Palace of the Sovereign"), probably built in 1261/91 for Constantine Porphyrogenitus as a wing of the Blachernai (fig. 11.1), as well as the Mermerkule (c. 1390), a massive tower built entirely of solid marble blocks that stood at the point where the land walls turned to front the Sea of Marmara.[6] Still other extravagantly marbled palaces are known from literary sources almost until the twilight of Byzantium, such as the Palace (c. 1305/28) of the Grand Logothete Theodore Meteochites and the late fourteenth-century Palace of Theodore Palaiologos Kantakouzenos (uncle of Manuel II Palaiologos).[7]

In Byzantine literature, bedazzling palaces continued to prosper in newly composed romances, the earliest of which is the *Digenes Akrites* (originally twelfth century). In this hero's home, "every stone was polished with an art so wonderful that the surface became invisible. This house was built in very beautiful stones, placed so as to form, by the variety of their nuances, a kind of charming variegation."[8] Many details of this palace recur throughout the subsequent romances: a splendid bath, a building lit by a giant ruby, interiors decorated with shining marbles, precious stones and ores, and so on. Meanwhile, once Constantinopolitan fortresses and bastions had been repurposed as palaces, the motif of the enchanted castle entered Greek romance literature (throughout the Paleologan era, 1261–1453).

As the genre evolved, the castle also took on an increasingly allegorical character. Thus, in *Kallimachos and Chrysorroe* (1280/1340, attributed to Andronikos Komnenos, nephew of Michael VIII Palaiologos), after the hero slays the dragon, the Drakontokastron (Castle of the Dragon) metamorphoses into the Chrysokastron (Golden Castle), and "shines like a star, like the sun at midday." In *Livistros and Rodamne* (thirteenth century) there is the Argirokastron (Silver Castle), and in *Belthandros and Chrysantza* (fourteenth century), the equally opulent Erotokastron (Castle of Eros). The tradition culminates in the anonymous *Consolatory Fable about Bad and Good Fortune* and *To Chastity* by Theodore Meliteniotes (c. 1320–1393), who was no simple litterato but an archdeacon, comptroller-general of Hagia Sophia, and theologian in the patriarchal school of Constantinople.[9] His *To Chastity* is the climax of the allegorical trend, wherein all the material opulence serves not to intimate any earthly wealth but predicts the heavenly palace, for this castle finally transforms into a *heliokastron* (sun castle).

The shimmering abodes that had featured in late antique poetry were not forgotten in Western European literature either, but were sustained and nourished in the many iterations of the Hellenistic *Alexander Romance,* which was told and retold in an extraordinary range of languages. The entrancing description of the suites in the Palace of Candaces, with bright rooms that seemed as if the sun shone from within their very walls, can be found as far afield as fifth-century Armenia and thirteenth-century Ethiopia. The *Alexander Romance* was early translated into Latin and printed, and editions followed in modern European languages, with the last one in Greek published at Venice in 1529.[10] From the early twelfth century onward, the Byzantine tradition was also paralleled by Western romances, especially the hugely popular and much translated *Le Pèlerinage de Charlemagne* (before 1140). They too describe microcosmic throne rooms of shimmering materials, frequently stones, and culminate in the thirteenth-century descriptions of the crystalline Temple of the Grail. In the same tradition will follow Petrarch's *Africa* (1338/39), and even Francesco Colonna's *Hypnerotomachia Poliphili* (1499).[11]

The jeweled palaces of Western European literature may have inspired a representation on the Bayeux Tapestry (1070s) of a circular donjon, possibly the now-destroyed Haute Vielle Tower at Rouen (tenth/eleventh century), whose blocks are shown inset with colored stones (fig. 11.2).[12] In truth, hardly any Western European palace could remotely measure up to the literary visions. The exception that proves the rule is the enigmatic Castel del Monte built by Frederick II Hohenstaufen in Puglia, in the heel of Italy, in the 1240s (fig. 11.3). All its revetments and many of its columns were carted off to adorn the Bourbon palace at Caserta in 1757, but the remains are still arresting enough to make visitors feel they are tracing their steps through some illumination in a medieval manuscript.[13] Those columns (Byzantine spolia) that do remain, attached to the internal walls, were chosen for their strident veining, in some cases enhanced with red paint. The most eye-catching stone, however, is the locally

FIG. 11.2. Detail of Bayeux Tapestry, Haute Vielle Tower of Rouen?, c. 1070/77.

FIG. 11.3. Portal, Castel del Monte, 1240s. Lintel and jambs in blocks of breccia corallina. Puglia.

quarried breccia corallina (possibly from S. Lucia, near Castel Pagano), which resembles the ancient marble of the same name, that from which the Stone of Unction was supposedly hewn (see fig. 9.15). Monoliths of this lively stone were used for the castle portals and even the entrance façade, and the grand halls were once probably revetted all over with it.

VENICE AND THE EAST

The Venetians neither confined colored marbles to interiors nor limited them to churches, for they were the ornament of their most palatial homes as well. This was again a Byzantine legacy. Byzantine spolia had been used to adorn Venetian palaces in the late Middle Ages, but the building that most palpably propagated marble magnificence was the Palazzo Ducale (1340–65/67; 1424–c. 50), the only residence there to bear the name "palazzo," because it began as a *palatium* for the *dux* (doge), or Byzantine governor (fig. 11.4). Its rose-and-cream façade was the noblest expression of Venice as a princely republic, the alter ego to S. Marco's bejeweled façade, and wherever La Serenissima established administrative palaces on her terra firma, she equipped them with this stone uniform.[14]

Opinion has long been divided over the source of the Palazzo Ducale's diaper patterning, its façade as a whole, and the entire Venetian palace type, especially the characteristic feature of a piano nobile portico. Byzantine, Muslim, and even Persian forerunners have all been suggested.[15] However, Jürgen Schulz put an end to any idea that Romanesque Venetian palaces derived from exarchal Byzantium or Islamic sources, and argued that the type was a subset of an "upper-hall house" very common throughout Western Europe.[16] Nonetheless, the question stands as to whether the increasingly rich marble exteriors of Venetian palaces reflect Constantinopolitan prototypes. In truth, while the halls of palaces in the Byzantine capital were swathed in marbles, it is only an assumption that they were externally revetted.[17]

FIG. 11.4. Palazzo Ducale, c. 1424–50. Diaper pattern cladding in Istrian stone and Veronese marble (limestone). Venice.

Hadrian's Villa (117–128) at Tivoli may have been externally clad with colored marbles, and several external walls of the (probably) imperial villa (early fourth century AD) at Piazza Armerina, Sicily, conserve stretches of painted marbles.[18] Palaces "shining afar with checkered patterns of stone" must also have existed in mid-fifth-century Byzantium, for this is how Nonnos describes one (*Dion.* 18.63). Moreover, colored ceramic, multicolored stone, or mixed brick and stone are archaeologically attested on the façades of Byzantine buildings in the ninth and tenth centuries, and this practice spread through the Balkans, whence it may have become known in Venice.[19]

In emulating the past, misunderstanding and happy error also had their creative parts to play. Throughout the Middle Ages and into the Renaissance, pilgrims and itinerant antiquarians were liable to identify almost any hulking ruin as a palace, even the ones that had not been colonized for inhabitation. For example, Abbot Suger reminisced about the "Palatium Diocletiani" at Rome (in reality, the Baths of Diocletian), and in the *Hypnerotomachia Polifili,* Polifilo spends much time guessing the function of the various ancient buildings he encounters. One Roman ruin in particular, the so-called "Palazzo de Sesare" (actually the firewall of the Forum of Augustus), would inspire a whole new breed of rusticated palaces in Florence, beginning with that of Cosimo de' Medici (begun 1445).[20] Venetian aristocracy and artisans, for their part, may also have conjured up palaces from Byzantine piles, even contemporary descriptions of the late imperial palace, or the earlier Boukoleion Palace. Relations between Venetian and Byzantine palace design are, however, partially confirmed by the brick-and-stone polychromy of one tantalizing survivor of Constantinopolitan imperial palaces, the last one, the Tekfur Sarayı (c. 1261/91; see fig. 11.1).

Even if the models for the Palazzo Ducale were instead Islamic, we need not bypass Byzantium to reach them. Spheres of influence did overlap, trade exchanges continued across the Mediterranean, and there were long-standing and reciprocal transmissions of artists and their skills. The palace in the epic *Digenes Akrites* belongs to Basil the "Digenes Akrites" ("Two-Blooded Border Lord"), whose mother was Greek and whose father was Arab, and all his heroic feats straddle the borderlands. As early as circa 830, Emperor Theophilos had built the Palace of Bryas, Constantinople, in imitation of one in Baghdad. Moreover, several Islamic decorative practices, including glazed brickwork, would infiltrate Byzantine palace design again during the 1140s, and in cultural buffer zones such as Syria and Lebanon, Greeks and Arabs continued to work side by side.[21] In fact, in 1211 William of Oldenburg attributed the very success of the extraordinary marble and mosaic decorations (c. 1205/11) in the Beirut palace of the crusader John I of Ibelin (c. 1179–1236) to the friendly rivalry between Syrian, Saracen, and Greek artisans.[22] Conversely, Western, especially Venetian, influences are equally visible on the Tekfur Sarayı (machicolations, gargoyles, heraldic escutcheons, and so on), as they are also on the Paleologan Palace (c. 1250–1400) and other monuments of Byzantium's last capital, Mystra.[23]

The polychromy of palace façades was no innovation in the West, but the will to clad them so lavishly in colored marbles is noticeable in Venice as nowhere else in Italy, or in Europe for that matter. The penultimate Byzantine emperor, John VIII Palaeologus (r. 1425–48), was mesmerized by the city's beauty on his official visit in 1438; the diplomat Philippe de Commynes (1495) was struck by the Istrian stone façades of the palaces, with their "large slabs of porphyry and serpentine on their fronts"; and the pilgrim Pietro Casola (1494) recalled, "It was hard to believe the quantity of marbles of every kind and color, and so well worked that it is astounding," and even acclaimed Venice "the promised land."[24] Cardinal Bessarion himself had already said that Venice was "*quasi altera Byzantium.*"

The Fourth Crusade (1202–4) may have provided the impetus for marble display on Venetian palaces, for one chronicle claims that after 1204, Doge Enrico Dandolo had sent back marbles from Constantinople to decorate the façade of his son's palace, which tradition holds to be the Ca' Farsetti (1200–1208/9).[25] This may be a Venetian triumphalist myth, but well into the fifteenth century palaces such as the Ca' Morosini (now Sagredo) and the Ca' d'Oro continued to incorporate reused Byzantine bas-relief panels and roundels. Seventeenth-century authorities even believed, erroneously, that the Palazzo Ducale itself was built from the marbles that La Serenissima had collected from its supremacy over Greek lands (rather than newly quarried Istrian and Veronese stones). Just as important in spreading the taste for marbled façades must have been the hallucinatory example of S. Marco itself, more Byzantine than Byzantium.[26]

The taste for extraordinary polychromy continued through the fifteenth century and characterized palace fronts from the famous Ca' d'Oro (carved 1425–29, erected and finished 1429–33), to the Ca' Contarini de Corfù, to the equally famous Ca' Dario (c. 1487/94; fig. 11.5).[27] All these

FIG. 11.5. Ca' Dario, c. 1487/94. Venice.

families held long associations with Constantinople, both before and after her fall, and thereby access to her antiquities and marbles too. The Contarini, one of whose members (Pietro) wrote the earliest surviving ekphrasis of the façade of S. Marco, fought in a line regiment in the final desperate defense of Constantinople, while Giovanni Dario (1414–1494) was ambassador to Bayezit II in 1479–87.[28]

No other Venetian palace façade would ever match the polychrome exhibitionism of the Ca' Dario, but even after a foreign (i.e., Tuscan or Roman) architectural language had taken root, while Istrian stone prevailed in the new classicizing whiteness, tradition maintained that porphyry and serpentine rotae must continue to stud palace façades like huge jewels (fig. 11.6). Whether this taste has any real continuity with Byzantine practice is hard to say, though when Venetian, or Venetian-trained, artists such as Cima da Conegliano and Carpaccio depicted Eastern palaces, they were always sheathed in precious stones and projected back onto them the image of contemporary Venice (see fig. 10.1).

THE VENETO AND LOMBARDY

Under Venetian influence, the taste for marble façades, sometime just painted, extended across northern Italy. Francesco di Giorgio Martini recommends decorative reticulate bossing and imbricated tessellations, which he calls "feathered" (*spiumati*), recalling the iridescence of a bird's hackles.[29] Such tessellation was employed at least once, to clad the stair added to the Palazzo della Ragione (the basilica) at Vicenza in 1495/96, where the courses also alternate in color.[30] Amadeo was more ambitious on the façade of the Colleoni Mausoleum (1470–75), where he resuscitated the perspectival-lozenge motif in colored stones for the first time since antiquity (see figs. 4.27, 10.13).

Cesare Cesariano's 1521 commentary on Vitruvius also offers a fascinating insight into the state of architectural knowledge across Lombardy, the Veneto, and the Po valley. When discussing masonry types (Vitr. *De. Arch.* 2.8), Cesariano says of opus reticulatum (fig. 11.7):

> Amongst these walls in opus sectile are these which are reticulate: they were invented by the Persians and were

FIG. 11.6. Mauro Codussi and Lombardo workshop?, Ca' Contarini delle Figure, 1504–46. Venice.

used for the magnificence of the incrustations that could be made with various colored, well-polished, and lucid stones: as they did in the Palace of Mausoleus: Unable to find the varied and colored stones appropriate to mosaic-work: some like the Egyptians and the Arabs used terracotta sherds and painted and glazed their surfaces in the same way that they make the vases called "Damascene": which we still use and make just like the Egyptians of Damascus [i.e., Mamluks].[31]

Cesariano's grasp of antique construction was as pitiful as his geography, but his spectacular error offers a keyhole into wider misunderstandings and may reflect the authority of Venetian construction itself. It is telling that Cesariano associated this sort of decoration with the Near East, and his only real conduit for this association could have been Venice. Cesariano not only mistook opus sectile for opus reticulatum, but he also commented that glazed brickwork was a good substitute where stone revetment was unavailable. The painter-architect Giovanni Battista Caporali copied his errors, and these were further disseminated across Europe by those who plagiarized his illustrations.[32] Cesariano, it seems, was unable to stifle his predilection for marbles, and his illustration of *opus tetradorum* (a type of brick wall) again shows a marbled wall, while his discussion of bricks wanders to the subject of marble paneling on the contemporary Duomo in Milan and the antique Porta del Popolo.[33] In fact, the only constant in Cesariano's garbled mélange of antique masonry techniques is that he cannot conceive of a wall

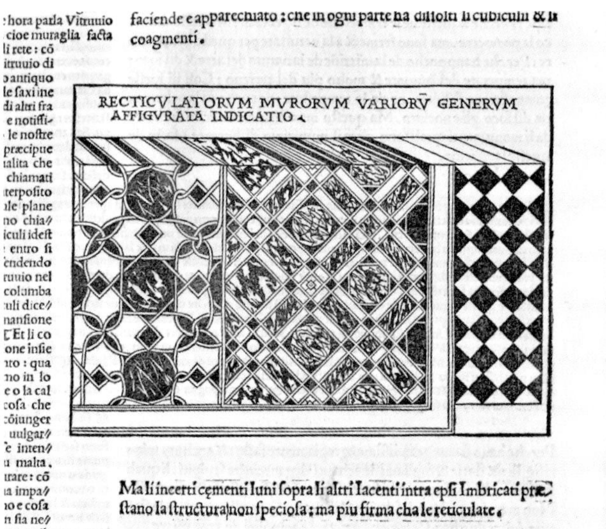

FIG. 11.7. Cesare Cesariano, "Opus Reticulatum," in Vitruvius, *De Architectura* (1521), lib. II, fol. 38v. Only the vertical checkerboard strip at right resembles true opus reticulatum (made with tufa cones).

as finished, whatever the fabric of its bearing walls, until it has received a veneer of stone, marble, or some simulation. Cesariano was himself adept at painting imitation stone and marble, and one of his few architectural works, the atrium of S. Maria presso S. Celso, Milan (after 1513), features colored marble revetment and bronze capitals.[34]

FAUX-MARBLE MANSIONS AND JEWELED PALACES

Today, a handful of Romanesque, Gothic, and early Renaissance palaces survive in Venice with their painted façade adornments intact, but many paintings (especially by Carpaccio) demonstrate that far more palace fronts were once adorned with painted patterning, including the diaper pattern of the Palazzo Ducale. Not content with simply painting the brick or stucco, the Venetians also doctored the stones to upgrade them to precious marbles. On the Ca' d'Oro, for example, Giovanni di Francia streaked the Istrian stone battlements to imitate marble (*ombrizadi a muodo de marmoro*), and pink Veronese marbles were oiled for an artificial polish the stone could not achieve alone. From the sixteenth century, fine marmorino intonacos were being devised by laying a render of Istrian stone chips over a pancake of brick dust, and then varnishing them with soap, wax, or boiled linseed oil before polishing, all the better to imitate the effects of marble. Sometimes Veronese marble dust or colored earths were mixed in as well.[35]

Nor was this desire for colored marble palace façades confined to Venice. We find elaborate designs for them in Urbinate intarsia work and painting,[36] and the poet Alberto Avogadro da Vercelli extols (1459–65) a material panoply on the façade of the Palazzo Medici in Florence, where such decorations plainly neither exist nor ever could have:

> If you wonder what binds together the stones,
> it is tenacious bitumen that joins white marble to the building,
> but the façade has no marble; what the soft bitumen fastens
> has stones of brilliant and triple color.
> A stone worthy of shining alabaster occupies the topmost part;
> porphyry stone the right-hand side;
> the left is of the stone which our ancestors called serpentine in the vulgar tongue.[37]

In northern Italy, a city such as Brescia could still count on the same quarries used in antiquity for a brilliant white stone that never seemed to weather. But the many other centers of the terra firma had no local quarry, and the expense of importing stone in bulk was reserved for communal commissions. In Ferrara, or Mantua, or Verona, where brick was the staple building material, marbles were instead often simulated in paint, and fragmentary examples survive on Quattrocento and Cinquecento palace façades there.[38] Thus, in 1471 Luca Fancelli informs Ludovico Gonzaga that for the outside of the Domus Nova of the Palazzo Ducale in Mantua, "I will have [the walls] plastered and I will make it appear that there are marbles, porphyry, alabaster, and serpentines so that it will represent a great thing."[39]

Marbles were also deemed an expression of nobility. Giovanni Rucellai praised the "nobility" (*gentilezza*) of variegated slabs and eulogized the intarsie of Early Christian monuments when he visited Rome for the Jubilee of 1450.[40] Filarete actually adopted class distinctions for stones, likening them to "noblemen, citizens and peasants." Gems were the upper class of course, but Parian, porphyry, serpentine, and alabaster did not trail far behind.[41] A century later (1567) Daniele Barbaro would still agree that "marbles approach the honor of gems, for their beauty and their grace, and especially those noble marbles, which through the variety of their colors, or their great beauty, refinement, splendor, or transparency give rise to wonder; marbles like Parian, Porphyry, Serpentine, Alabaster and other similar veined marbles, or granites."[42] Thanks to this age-old distinction, porphyries and marbles ennobled the base materials of construction, and so stucco façades must also receive stone window frames and doorframes (or stucco-over-peperino imitations) not just for wear but as the trimmings of nobility. Even a palace façade completely fashioned out of travertine, such as that of the Palazzo Riario (Cancelleria, 1489–1510), still had window frames of white marble set within it to show that Raffaele Riario belonged to the crème de la crème (fig. 11.8).[43]

FIG. 11.8. Piano-nobile window, Palazzo Riario (Cancelleria), 1489–1510. Rome.

With this perceived kinship between marbles and gems in mind, we can better understand the motivation for using them on palace façades. The jewelry that princes wore did not simply display wealth but, *decus et tutamen*, also protected the bearer thanks to the celestial virtues that the individual gems attracted. Virtually all literature on gems since antiquity dilated, again and again, upon the virtues that were latent in gems thanks to astral rays, and this sort of "astrophysics" was common knowledge to nobles and their courtiers. Hundreds of Renaissance lapidaries circulated, and aristocrats often preserved lavish examples, such as the *Lapidario Estense*, in their palace libraries or wrote out their own abridgments. Such pastimes must have made patrons particularly alive to stone connoisseurship, and they often had themselves portrayed with their jewelry of prophylactic or thaumaturgic minerals.

The metaphor of a jeweled façade may underlie the frontage of the Palazzo Schifanoia in Ferrara, enlarged by Borso d'Este (1413–1471) in 1465–69, and the façade decoration of fictive coffered intarsie imitating alabaster, porphyry, and other marbles (fig. 11.9).[44] This conclusion seems inescapable for the Schifanoia not just because Borso's predilection for gems was excessive enough to earn the hypocritical censure of Pius II, but also because of the murals that Borso commissioned from Francesco del Cossa within the building (1470). As Aby Warburg's famous study demonstrated, the *salone* walls were divided into three registers with the decans (the rulers of ten-degree slices of the zodiac) and zodiacal signs occupying an upper tier, and the beneficent effects of their conjunction on the well-governed Ferrara illustrated below.[45]

Façades like the Schifanoia, even to the same pattern, must have been more common than today's miserable tally. An almost identical fictive revetment appears on the walls

of a salone in the Castello Carrarese in Padua (decorated in the 1370s), where hexagonal panels alternate with quadrangular ones that loosely recall jewelry mountings for table diamonds.[46] Cima da Conegliano also illustrates just such a palace in the background of his *Theseus at the Court of Minos* (c. 1495–97).[47] In the 1490s Mantegna himself repeated the signature faux-revetments of his earlier *Camera Picta* (1465–74) on the exterior of his own townhouse in Mantua opposite the equally bejeweled S. Sebastiano (see fig. 10.12).[48] These lost murals set the trend for several other palaces in Mantua, such as the chain mail of porphyries and serpentines that once covered a palace in the Via dei Fratelli Bandiera (fig. 11.10). Another Mantuan palace copied the same faux-revetments on its exterior, as did the Palazzo Maggi al Fontanone in Brescia, and their linked disks are again more reminiscent of jewelry than structural cladding. Still another palace, built slightly later (1500/1530) and also in Mantua, has similar faux-intarsie and even a "panel" of interlaced rotae like that on the Ca' Dario, except with a

FIG. 11.9. Façade decoration, Palazzo Schifanoia, 1465–69. Ferrara.

FIG. 11.10. Via dei Fratelli Bandiera 17, c. 1480/1500. Mantua (reconstruction: 2009).

diamond at its center.[49] The jewel metaphor might also account for the Venetian use of rotae and panels in real porphyry, or alabaster, that not only stud façades but appear as clipei suspended off them by ribbons, almost like campaign medals, but set with marbles that seem huge rubies, emeralds, and agates (see fig. 11.6).[50]

Painters had long preceded architects in reclaiming the material opulence of ancient architecture, on canvas, in fresco, and also in manuscript illuminations. Throughout the Quattrocento, deluxe manuscript codices of ancient texts and chorales were lavishly illuminated with an architecture that was both classicizing and fantastically opulent—especially the "frontispiece," the medieval Latin word ("*frontespitium*") given to a title page because it was the façade to the contents (fig. 11.11).[51] Illuminators enjoyed far more liberty in imagining their houses of knowledge on vellum than could ever be achieved in construction and, conversely, real buildings assumed the modalities of the illuminated page, both in their carved ornaments and the frescoes that began to invade their external walls. It is futile to find buildings that were archetypes or give precedence to one over the other, codices or palace façades. One is imagined as a building, and the other as an illumination, and their common genealogy is actually in romance literature and the fairy-tale palaces described there.

A good example is the stable block that Ludovico Sforza il Moro had frescoed at Vigevano toward the end of the century (c. 1490/94), with a design that has been attributed to Bramante (fig. 11.12). The medievalizing tracery under the windows acknowledges the Trecento terracotta details of the Castello Visconteo (begun 1345) that dominates the compound. The columns and architrave of the grisaille wall are in the same all'antica tradition as the Mantuan buildings and the frieze ornaments. They, again, derive from illuminated manuscripts, such as those images with which Giovan Pietro da Birago decorated printed editions of Giovanni Simonetta's life of Francesco Sforza (*Sforziada*). However, on the stable block at

FIG. 11.11. Henry of Amsterdam, Frontispiece from a gradual, 1486. Biblioteca Communale Malatestiana, Corale D, fol. 1r. after 1513.

FIG. 11.12. Bramante?, Painted façade, stables, Castello di Vigevano, c. 1490/94. Vigevano.

FIG. 11.13. Gian Cristoforo Romano or Tullio Lombardo, Porta gemmea, Castello, 1522–24. Mantua.

Vigevano, most of the wall is again covered with a reticulate pattern of diamond bosses, like a jeweled quilt.[52]

The gem metaphor most likely subtends Isabella d'Este's apartment in the Castello at Mantua, at whose heart was a priceless collection of jewels, medals, and cameos. Here, porphyry- and serpentine-inlaid portals, carved by Tullio Lombardo with stones provided by the Marchesa, graced her *studiolo.* Stepping up and out of this room into the *grotta,* the courtier would pass through the *porta gemmea,* carved by Gian Cristoforo Romano or Tullio Lombardo in 1522–24 (fig. 11.13). The marble *specchiature* in its white marble jambs visibly pun on both gems and wood paneling; beyond lay the grotta, with paneled walls and a neo-Cosmatesque, porphyry-and-hard-stone floor (1522), and here the courtier or VIP might linger and pore over actual cameos and Isabella's impressive collection of gems.[53]

DIAMONDS ARE FOREVER

Faux-marble façades were more common in cities such as Ferrara than today's raw-brick sobriety would suggest. Duke Ercole I d'Este, lord of Ferrara, had the courtyard walls (c. 1479–81) and piazza façade (c. 1492) of the Palazzo del Corte decorated with "checkered walls painted in the guise of green, white and red marbles," the colors of his heraldry.[54] However, when his brother, Sigismondo d'Este, built the Palazzo dei Diamanti in 1492 for his illegitimate son (Ercole di Sigismondo d'Este), illusion became fact and the potential of gem symbolism subsumed the whole building (fig. 11.14). The encomium that Ercole's court humanist, Giovanni Sabadino degli Arienti (c. 1443–1510), penned five years later (1497) spells this out, comparing Ercole d'Este to Augustus in familiar terms: "Just as Caesar Augustus said of himself, that he had found a Rome of brick and would leave it in marble, so too will your Highness, by virtue of your magnificence, be recognized by posterity with the greatest glory: that you found a Ferrara of painted brick, and of marble, but you will have left your doughty riches carved in diamonds. For which reason one can already make this judgment: that this, your city, gleams more than oriental gems and will be among the most wondrous cities in the World."[55]

Many other diamond-point palaces were built in Italy and Spain, and where the stone was not white it would have been stuccoed over. The exception is the palace in Naples that Novello de Sancto Lucano designed for Roberto Sanseverino, prince of Salerno, in 1470, which was clad in the local, dark volcanic stone. However, the humanist Giovanni Pontano comments (1486/98) that the prince was criticized for not using marble, and that his son Antonello often mentioned his own intention to remedy the matter.[56] The diamond wall also traveled as far afield as Moscow, where the émigré architects Marco Ruffo and Pietro Antonio Solari built the Palace of Facets (Granovitaya Palata, 1487–92) for Ivan III within the Kremlin, as the czar's main reception and coronation hall. The diamond wall may even be found in Scotland, in the courtyard of Crichton Castle, where Francis Stewart, fifth Earl of Bothwell, had a diamantine-clad range built (c. 1585) after seeing the Palazzo dei Diamanti in Ferrara.

No classical precedents exist for such jagged skins, which were seemingly invented by Bartolomeo Bon at Venice in the 1450s, and the default position has been to regard them as gentrified fortification.[57] The battered base of the Palazzo dei Diamanti does suggest an element of fortification, but otherwise the palace was the flagship of a vast new and unwalled quarter (the *Addizione Erculea*) in a self-confident city. Any allusion to fortification should have been quite alien, save for the sort of enchanted and crystalline castle that had been the theme of so much romance literature, East and West. Precision diamond faceting itself, made possible by the advent of the horizontal cutting wheel, was not even a century old when these façades were built. And fifty years after the fact, Serlio still knew that the architects who used such pyramidal bossing "wished to imitate worked diamonds [*diamanti lavorati*]".[58]

When Paolo Cortesi's later tract *De Cardinalatu* (c. 1510) refers to palace façades—one of the few texts that does—it prescribes that they be imposing enough to subdue a lawless mob, specifying that they should "dazzle the eyes of the people" (given more to the senses than reason) "by its dignified splendor."[59] The Palazzo dei Diamanti may be set in the tradition of the heliokastra of the late Byzantine novel and the Grail Palace of Western romance, but the most heavenly of all was the apocalyptic Jerusalem, which, as everyone knew, was an edifice of gems (see fig. 10.18). In fact, such diamantine walls may trace their origins to jewel-encrusted, heavenly mansions in medieval mosaic, or those shimmering walls that medieval mosaicists expressed with a diagonalizing, diamond-point convention (see fig. 2.29). Indeed, diamantine *sgraffito* on late Quattrocento palaces is hardly distinguishable, and in at least one instance, the side façade (Via delle Fogge) of the Loggia del Consiglio (1476–93) in Verona, such diamond-point cladding was painted, but the stones are colored like rubies or amethysts or sapphires, suggesting that the wall is truly gemmed (fig. 11.15).[60]

Sabadino degli Arienti recognized that the Palazzo dei Diamanti had been encrusted with not just any marble, but one that resembled diamond. Moreover, the apexes of the 8,500 blocks are not all perpendicular to the wall, but some ranks point downward and others up, such that the building's overall surface is faceted in gentle arcs, and may shimmer all the more.[61] Pliny had praised diamond (Greek, *adamas*) above all other stones, as did Renaissance lapidaries, because of its "in-domitable virtue" (*a-damas arete*), and diamonds were a favorite of princely and imperial treasuries. For exactly the same reasons, diamond rings became ever more popular, especially in the bond of marriage.[62]

Confirmation of the diamond-point conceit comes from the building contracts of a palace that lost no time in emulating the Ferrarese model. The Palazzo Raimondi (façade, 1495–96) in Cremona was executed by Giovan Pietro da Rho and Bernardino da Lera to the design (so

FIG. 11.14. Biagio Rossetti?, Palazzo dei Diamanti, begun 1492. Ferrara.

FIG. 11.15. Façade, Via delle Fogge, Loggia del Consiglio, 1476–93. Verona.

FIG. 11.16. Eliseo Raimondi, Façade, Palazzo Raimondi, 1495–96. Cremona.

say the façade inscriptions) of the patron, Eliseo Raimondi (fig. 11.16).[63] The Veronese-red pilasters are offset with faceted blocks of Brescian marble—a limestone that never dims because it never discolors—that, as the mason's contract notes, must portray "a thousand marble diamonds well carved, worked and fashioned, in the very image of the existing diamond of the aforesaid Lord Eliseo."[64]

In Filarete's aristocracy of stones, diaphanous ones were the most noble because "the noble Lord must be splendid and clear without a single blemish even though he is handled and touched by many." However, Filarete reserved diamond for the temporal-spiritual pope because it was both the hardest and most luminous of stones. The pope, he says, "must strike other lords when it is necessary, and he ought to reflect the virtue in himself as does the diamond."[65] The diamond more generally stood for eternity, and when coupled with another emblem, such as feathers, unwavering virtue. The diamond had been a d'Este device since the fourteenth century, such that the supporters of Ercole d'Este called themselves "diamanteschi," and there was a Camera dei Diamanti in the *corte* at Ferrara whose now-lost murals may, as Thomas Tuohy has suggested, have resembled those that still survive in Vignola.[66]

The prickly skin of the Palazzo dei Diamanti reminded the public of the prince's pellucid virtue by allowing him to bare his teeth in public, so to speak. When diamantine masonry reappears on fortresses for the same reasons, it is also because diamond is so hard that no earthly weapon can dent it. This conceit probably underlies the design by Antonio da Sangallo the Younger (1484–1546) of the enceinte and bastions of the Fortezza da Basso in Florence (c. 1534–37; fig. 11.17). Here the diamond-point masonry alternates with ball-embossed blocks. Together, they allude to the Medicean arms and *impresa,* the ball and the diamond (ring), but they also suggest that any cannonball would simply have been punctured on their diamond points.[67]

INSIDE THE PALACE: *VESTIBULA* AND *ATRIA*

Alberti prohibited the use of even white marbles (Hymettan or Parian) in the home, scoffing that "such things belong in Temples," but others remembered that the Palace of Mausolus had been clad in Proconnesian marble, and cherished the otherworldly impact of palaces carpeted in porphyries, alabasters, and serpentines.[68] Some patrons, inspired either by the knowledge that Phengite had figured in the Domus Aurea (a description

FIG. 11.17. Antonio da Sangallo the Younger, Fortezza da Basso, c. 1534–37. Florence.

FIG. 11.18. Unknown painter, Bibliotheca Graeca, c. 1454 and later. Vatican Palace, Rome.

repeated verbatim in the *Hypnerotomachia Poliphili*) or the continuing popularity of the *Alexander Romance*, wished to be bathed in just as much golden light as their ancient forebears. In 1457, for example, Giovanni Rucellai extolled the "windows made of alabaster in place of glass" in the salone of Giordano Orsini's Roman palace, the Palazzo Montegiordano, and his son, the great antiquarian Bernardo, emulated the example in the Orti Oricellari in Florence.[69] Cosimo de' Medici reserved marbles for his palace chapel, but an anonymous poet extols his "figures and inlays of serpentine and of alabaster and porphyry and marble," and a 1459 description by Niccolò de' Carissimi da Parma also praises "a garden made with the finest of polished marbles with diverse plants," which, predictably, "seems a thing not natural but painted," such that whoever visited the palace saw "things that were more celestial than earthly. . . . In sum, it is believed that there is no other earthly paradise in the world than this."[70]

It was beyond most patrons' purses to use real marbles on the palace interior, so imitation again substituted. In Rome, the earliest intact remains date to the pontificate of Nicholas V (1447–55), who had a colored-marble peristyle simulated on the walls of the Bibliotheca Graeca in the Vatican Palace (c. 1454), which has been attributed to Andrea del Castagno (though this seems unlikely) under the direction of Alberti (fig. 11.18).[71] The dating is disputed, but Alberti did write in *De re aedificatoria* (9.4; 1443/52), "For the plaster revetment [*crustationes*] of walls there is no pictorial composition more pleasing and admirable than that which represents colonnades of stone." He next mentions the colonnades of the Palatine, and perhaps he had seen traces of Second Style painting somewhere. Whatever the case, the room initiated an influential trend.

In the adjoining Bibliotheca Latina, Melozzo da Forlì portrayed the next pontiff, Sixtus IV (r. 1471–84), surrounded by four of his nephews (including Girolamo Riario) and the humanist Platina in a completely marbled library (fig. 11.19). The illusionary back wall of this scene, and the actual wall surfaces around Melozzo's fresco, were painted an emerald green almost certainly because another court humanist, Francesco Patrizi, had scoured classical and medieval authorities (primarily Vitruvius and Isidore) to recommend that library "walls must be in polished stucco and ornamented a green color, for all things green benefit the eyes and restore keen vision." Isidore even specified the most suitable marble for libraries, Carystian (cipollino).[72]

FIG. 11.19. Melozzo da Forlì, *Sixtus IV Appoints Platina Prefect of the Vatican Library*, 1471/84. Fresco, 12 ft. 1 21/32 in. × 10 ft. 4 in. (370 × 315 cm). Bibliotheca Latina, Vatican Palace, Rome, now Pincoteca Vaticana.

After the Bibliotheca Graeca and Bibliotheca Latina, monumental faux-marble interiors prospered in Rome. Around 1476/81 an accomplished painter, possibly Melozzo again, painted the Sala dei Notabili (or Sala della Piattaia) in the palace of Raffaelle Riario's uncle, Girolamo (now Palazzo Altemps) (fig. 11.20).[73] Girolamo's palace was the gathering place for artists such as Melozzo and prominent humanists (including Luca Pacioli, who commended marble cladding for the ideal building) to convene and discuss mutual artistic interests.[74] In the Sala dei Notabili the fiction of an engaged colonnade against an opus isodomum wall of possibly alabaster blocks was presumably meant to imitate some part of the Roman house, such as the vestibulum or atrium (what Cortesi called the "Aula"), components whose true appearance and function were as yet imperfectly understood.[75]

Alberti had also recommended that seigneurial residences "should have stately porticoes, and handsome courts with everything else in imitation of a public edifice, that tends to dignity or ornament" (*De re Aed.* 9:1, 157v/158r), and such porticoes should enclose the principal unit of the house, the cortile, whose social and physical functions rendered it "a public Market-place to the whole house," for from it "derives all the advantages of communication and light" (5:17, 87). The bosom of the house was to act like a public forum.

In the best-preserved examples, suites of faux-marble rooms suggest that the chain of fictions expressed a progression through the antique house, from illusionistically open courts to closed rooms, in which the fictive paneling definitely seals the perimeter. These ideas were at work in Pintoricchio's frescoes in the Palace of Domenico della Rovere (1490s), the Belvedere of Innocent VIII (c. 1487–c. 90), the Sala del Mappamondo in the Palazzo Venezia (1489/91), and partly in the Borgia Apartments as well (1492–94).[76] In the Palace of Domenico della Rovere, the contrast is particularly evident in the great salone (fig. 11.21), whose (heavily restored) walls simulate files of freestanding columns *en ressault* bearing a tall attic, quite possibly derived from the Forum of Trajan or Forum of Nerva—both thought to be palaces—while the cardinal's audience room feigns an extravagant floor-to-ceiling revetment (fig. 11.22).[77] Pintoricchio's preoccupation with antique models is anyway revealed by his grotteschi and ceiling designs, both inspired by the rediscovery of the Domus Aurea in the 1470s. All of these simulations portrayed an idealized antique wrought from long effort in the libraries. Theirs was a vision of the antique epitomized in texts such as the *Hypnerotomachia Poliphili* (Venice, 1499) that reiterated and compounded their primary sources, right down to the spelling mistakes.[78]

FIG. 11.20. Melozzo da Forlì?, Sala dei Notabili, Palace of Girolamo Riario (Palazzo Altemps), c. 1476/81. Rome.

FIG. 11.21. Pintoricchio, Salone, Palace of Domenico della Rovere, 1490s. Rome.

FIG. 11.22. Pintoricchio, Audience room, Palace of Domenico della Rovere, 1490s. Rome.

RAPHAEL, PERUZZI, AND OTHERS

From these beginnings, and the sweeping reform of architecture that Donato Bramante (1444–1514) effected in Rome in the first decade of the sixteenth century, the material tradition was decisively revitalized in the hands of Raphael (1483–1520), who coined a more archaeologically rigorous version that his pupils, Giulio Romano (c. 1499–1546) and Giovanni da Udine (1487–1564), then took with them to northern Italy. Testament to this detailed knowledge of marbles and hard stones is the *Ceres* or *Allegory of Abundance* (c. 1516), possibly executed by Giulio Romano, that served as a sliding cover for Raphael's *Holy Family with Saint Elizabeth and the Infant Saint John* (fig. 11.23).[79] This pseudo-revetment protected the devotional image in transit by reclaiming the marbled versos of Trecento and Quattrocento religious paintings (see fig. 9.12b) and the marble panoply that filled various earlier Marian paintings (for example, see figs. 9.22, 9.23, 9.26, 9.27).

Raphael advocated real colored marble architecture in his *Letter to Leo X* (c. 1514/15 or 1519/20) and vetted the ruins for building materials. The importance of materials to his project to revive Rome is also reflected in the

FIG. 11.23. Workshop of Raphael (attr. Giulio Romano), *Ceres (or Abundance)*, c. 1516. Oil on walnut, 14 15⁄16 × 12 3⁄16 in. (38 × 31 cm). Musée du Louvre, Paris.

works of his collaborator, the antiquarian Andrea Fulvio (c. 1470–1527), which typified a lingering schizophrenia about marbles. Thus, his epic poem, the *Antiquaria Urbis* (1513), presents a potted history of Rome from the Republic to its downfall, and then the age of the popes, in which he echoes Livy (*AUC.* 7.29.5) to lament that the Republic had been proceeding nicely:

> Until the state was turned towards wanton luxury;
> Until private and public buildings were paved
> With marble floors which exceeded ancient custom,
> With their cut marbles and "painted" inlay everywhere,
> They were ennobled by the outward splendor of the stones.
> There stood tawny Onyx gleaming with glassy brilliance,
> And Specular Stone from Arabia, and Ophite famed for its markings,
> Synnadian and Thasian stone, and green Lacaenian marble,
> And purple granite, and that which rocky Carystos offers
> And those that the inexhaustable Cyclades send from their high cliffs
> And every sort of stone that shines forth in the world;
> The hands of craftsmen ornamented every "painted" floor and revetment
> And made each proud and splendid through their skill.[80]

By contrast, Fulvio's slightly later *Antiquitates Urbis* (1527), which he declared the philological companion to Raphael's archaeological research and painted reconstructions, became the first guidebook on Rome to contain an appendix on colored marbles.[81]

Raphael himself made good on his promise to revive the medium's architectural use only in the Chigi Chapel (see fig. 10.14), but his workshop resumed the practice of simulating revetment on the palace interior. One of their earliest essays in this medium may have been the decorations of the Sala Vecchia degli Svizzeri (or Camera dei Paramenti, c. 1514–15; mostly destroyed) in the Vatican, for according to Vasari's *Vite,* Giovanni da Udine decorated the walls with "certain panels imitating veined stones of various types and similar to the ancient revetment that the Romans used to make in their baths, temples, and other places, as one can see in the Pantheon and the portico of St. Peter's."[82] The location of these antique prototypes at thresholds must have been equally influential, because the Sala Vecchia was the first room in the sequence of Vatican apartments now famous as the Stanze, and this room's liminal nature was spelled out in inscriptions on the ceiling.[83]

Giovanni's decorations were destroyed about 1558 but can be reliably imagined from drawings of the granite and porphyry panels of the Pantheon pronaos before their despoiling (see fig. 10.15), as well as Giulio Romano's later decoration of the Sala dei Cavalli (c. 1526–28) in the Palazzo del Tè, Mantua (fig. 11.24). In the Sala dei Cavalli, it is again probably no coincidence that the room's simulated pilaster order was derived from the very public pronaos of the Pantheon, for the sala was the only large and unambiguously public space in the palazzo, and the venue for its balls and banquets.[84] The techniques of antiquity were just as much on their material minds. Giovanni da Udine had rediscovered how to make a rock-hard and brilliant white stucco by integrating marble dust, and neither he nor Raphael can have been unaware of Vitruvius's statement, "The ancients who first used polished surfaces, originally imitated the variety and juxtapositions of marble veneers" (*De Arch.* 7.5.1).

Other artists followed suit. Almost contemporaneously, the Sienese painter-architect Baldassare Peruzzi (1481–1536), who made meticulous sketches of antique revetment with great archaeological acumen, frescoed the Sala delle Prospettive at Agostino Chigi's suburban villa, the Farnesina (1515–17), with a panoply of fictive marbles (fig. 11.25), a scheme soon mirrored in the Palazzo della Valle-de Rustici (c. 1526–34).[85] Through Peruzzi's illusionism, the Sala delle Prospettive on the piano nobile became a richly marbled court with deep loggias of monolithic africano columns overlooking the brick quarters of medieval Rome. Like a latter-day Augustus, Agostino Chigi surveyed Rome's medieval shambles from this aerial dining room, which feigned Vitruvius's choicest option, the Oecus Cyzicenus.

The Villa Chigi (Farnesina) was a pleasure park and occasional residence of Chigi's mistress, the courtesan Imperia Cognati (and after her death in 1512, of his second mistress, Francesca Ordeaza). This is why so much romantic imagery suffuses the Farnesina's decorations, and why in Egidio Gallo's poem (1511) on the gardens, Venus descends from Olympus to visit the villa and departs commanding spring to remain eternally present.[86] Thus, the faux-marbles of the Sala delle Prospettive materialize the Palace of Venus as it had been described in so much classical and neo-Latin poetry, including that of Egidio Gallo. Eventually life imitated art. After dining here in 1519 with Leo X and twelve cardinals, Chigi descended the stairs to

FIG. 11.24. Giulio Romano, Sala dei Cavalli, Palazzo del Tè, c. 1526–28. Mantua.

FIG. 11.25. Baldassare Peruzzi, Sala delle Prospettive, Villa Chigi (Farnesina), 1515–17. Rome.

the Loggia of Cupid and Psyche and finally married his long-suffering mistress, Francesca Ordeaza.

Peruzzi's exquisitely faked revetment breathes life into a cultural program announced in the verses of the *Suburbanum Augustinii Chisii* (1512) by Blosius Palladius (Biagio Pallai). Blosius belonged to a sodality that included Raphael and that gathered around the poet and antiquarian Johann Goritz (Corycius), and would eventually edit the group's serial poetic outpourings as the *Coryciana* (1524). In his own poem on the Farnesina, Blosius draws freely on Martial's, Statius's, and Sidonius's praise of baths to extol the Farnesina, but acknowledges the true and contemporary state of affairs when he asks:

> Oh, should I not speak of you stones? And you marbles strewn throughout,
> You marble stairs, and you marble revetments?
> And should I not sing of you, hot baths, that roll forth diaphanous vapors and yourself consist of Parian marble?
> Were Barbarian peoples still now trading,
> Then hither Paros, hither Numidia, and hither all those famed in marble,
> Hither all would drag their quarried rocks;
> And the porticoes would be thick with columns, and marble would fill every room;
> Both citrus and cedar beams would breathe upon the ceilings;
> Nor would the Barbarian wish his mountains to diminish in trees and rocks for any [other] place.
> So also, though no ships [now] bring foreign marbles,
> We see these that you have gathered on all sides from far and wide,
> Either dug up from the earth or broken from the ruins of the ancients.[87]

One visitor in 1511 says that the villa was beautiful in every way, but it was the marbles that passed all comparison. Even so, the furnishings must have remained incomplete, as 117 slabs of veined marbles and alabasters awaited use in a storeroom.[88]

Raphael's boldest domestic commission, the Villa Madama (1519–25), for Cardinal Giulio de' Medici, cousin of the reigning pontiff, Leo X, remained unfinished thanks to the artist's premature death in 1520 and the Sack of Rome in 1527. What remains is a mere fragment of the projected whole, which, had it been completed, would have fulfilled Raphael's most studied attempts to re-create a Roman villa in all its authenticity. A long ekphrasis, a "building in verse" penned in 1519 by one of its cultural engineers, the poet Francesco Sperulo (1463/66–1528/31), promised that no part of the finished villa would remain unadorned by exotic marbles, including the image-bearing variety from Paros (Pliny. *HN.* 36.4.14):

> Lo! Let Libyan marble gleam on the first threshold, and bear the bronze doors and shape the great vestibule, to tease visitors with its spectacle and detain the eyes. Let the wonders begin from this spot. This marble was discovered buried on the Capitoline Hill, and had once adorned the Thunderer's high threshold. Let the palace everywhere reflect variegated Synnadian marble and mottled Ophite. May the great halls rest on these columns that the hardy farmer has just unearthed (for they also work with the turned plough) in the Gardens of Lucullus, and let them be polished, for they are cut from that same marble, whence, once split, emerged (oh wonder!) the image of Silenus.[89]

Sperulo tells us that "heavy marbles excavated from the Roman ruins" were being transported to the site, and that Parian columns had been plowed up on the Pincio, in the Gardens of Lucullus. More extraordinarily, he reveals that Raphael's attempt to resurrect the Roman villa was so concerted that the site of the "House of Scaurus" (*domibus Scauri*) was being excavated in search of the thirty-eight-foot columns that he had installed in his own atrium back in 58 BC. If this excavation ever took place, it was doomed to failure for, apparently unbeknownst to them, Augustus had transferred Scaurus's columns to the Theater of Marcellus in 17 BC (Asc. *Scaur.* 23). Scaurus was still the man to beat, although until now something of a forbidden apple. It is equally glaring that the columns for the atrium were recovered from the Gardens of Lucullus, another patrician notorious for his wanton *luxuria.* Sperulo's text is of a rather different tenor than the medieval palace descriptions with which we began. It is studiously archeological, drawing not only on Statius, but also Vitruvius, and clearly Raphael himself, in a collaboration between the artist, his patron, and literary companions.

This cultural project continued, on a smaller scale, at the Casino Turini (now Villa Lante; 1518/21–22; 1523/24–31), commissioned from Raphael's pupil Giulio Romano by Baldassare Turini da Pescia, the papal datary (a highly profitable office). Here, Giulio's workshop frescoed the salone to cornice-level with imitation revetment that was partly based on the Pantheon, and partly on other revetments, now lost, all very authentic in feel (fig. 11.26).[90]

FIG. 11.26. Workshop of Giulio Romano, Salone, Casino Turini (Villa Lante), 1523/24–31. Rome.

Externally the attic was also painted with faux-marbles.[91] All the marbles are clearly recognizable (Numidian, Tean, Phrygian, and Egyptian alabaster—premier stones, and all favorites of the classical poets), and it is quite likely that Martial's eulogy of the marbles in the Baths of Etruscus (*Ep.* 6.42.8–21) particularly inspired the decoration of the sala in a villa that enjoyed its own *stufetta* (now disappeared). The will to produce this imaginary splendor was no doubt fueled by the fact that the casino was built off antique foundations that were firmly believed to belong to the Villa of Julius Martialis described by his cousin, the poet Martial (*Ep.* 4.64). Indeed, Turini had one verse, "You can take the measure of all Rome from here" (*hinc totam licet aestimare Romam*), inscribed in the loggia.[92]

ROMAN PALACE FAÇADES

As for the exterior of the Roman Cinquecento palace, stucco was the preferred material, and that fine-grade mixture of lime and marble dust that was considered virtually a marble itself (as the Venetian patrician Alvise Cornaro will write some decades later, "Stucco transforms into stone because it is made from stone" [*stucco si converte in sasso perché è fatto di sasso*]).[93] The only palace in Rome to deploy colored marbles externally was the Palazzo Lante (begun c. 1515), whose design is attributed to Giuliano da Sangallo, but is probably by Baccio Bigio.[94] This palace vaunts a huge portal and stringcourse friezes in porta santa, panels below the ground floor windows in the same material, and its courtyard boasts cipollino columns where other palaces only use granites. But this palace remains an anomaly.

When Giulio Romano came to build his own house in Mantua (c. 1540–44) he eschewed a faux-marble façade in favor of a rusticated surface yet, as Vasari records, it was also "fantastic, all worked up in colored stuccoes."[95] This façade has yet to be cleaned, but it is quite possible that the rusticated blocks themselves were painted in subtle alternating tints, just as his follower Giovan Battista Bertani (1516–1576) painted them in the Cortile della Cavallerizza (1556) at the Palazzo Ducale in Mantua (fig. 11.27).[96] This latter effect surely seems beholden to the shuffling of white Istrian and pink Veronese stones on the Roman amphitheater in Verona, whose brand of rustication it also recalls, not to mention the fact that one range of the cortile is a transparent screen just like the amphitheater's ruined perimeter.[97] This hypothesis raises the tantalizing possibility that the arcaded frontage

FIG. 11.27. Giovan Battista Bertani, Cortile della Cavallerizza, Palazzo Ducale, 1556. Painted stucco over brick. Mantua.

of Giulio's Mantuan home derives from the same model, or recollects (amphi)theaters that had been adapted to inhabitation, most notably the Palazzo Savelli (Theater of Marcellus) in Rome. As for Giovanni da Udine, when he also left Rome to return temporarily to Udine in 1534, he made his home in an old townhouse, whose façade he now transformed into his own antique abode with panels of faux-marbles, which still survive.[98]

However, the Renaissance palace that most stridently resurrected the colored marble domus, and in real marbles, was not in Italy, but Spain. The Palace of Charles V in Granada (1527; fig. 11.28) was built by the Toledo architect Pedro Machuco (1485–1550) to adjoin and overbear the Nasrid jewel that is the Alhambra (1238–c. 1391). The entire design was so precocious in its reception of Roman trends, especially its rustication, that the name of Giulio Romano has been mentioned as its progenitor.[99] The external façades were built in the local golden sandstone, but a

FIG. 11.28. Pedro Machuco, Palace of Charles V, 1527. Granada, Spain.

more polychrome vision came to the fore in the great circular colonnade at its heart, which took its cue from the unexecuted one at the Villa Madama. Its double-story colonnades were made (except for the architraves) from a colorful conglomerate stone discovered in a quarry called El Turro in the Sierra Nevada. It was enlivened by its clasts (of marble, quartzite, and schist) cemented in a matrix of calcium carbonate, although this has proven so friable over time that these inclusions have loosened and begun to fall out. The local masons called the stone *"Almendrado"* ("almond-stone") or *"puddinja* ['pudding'] *del Turro,"* recalling a Spanish delicacy called *"turrón"* (*"torrone"* in Italian), a hard, white nougat with almonds in it.

BRICK AND MARBLE

Augustus might have substituted a Rome of brick with one of marble, but when Sixtus IV refounded the city's brick industry in the 1470s, he was acclaimed as a new Augustus for the opposite reason, that he had revived ruined Rome in brick. These freshly baked, yellow bricks stood from all the generations of previous medieval buildings with their recycled and tattered ancient bricks, and the new fair-faced brickwork could claim an antique pedigree not too far below marble. Various treatises commended the durability, smooth finish, and polychromy of antique, fair-faced brick monuments such as the exceptional "Temple of Deus Redicolus" near the Via Appia (possibly the Tomb of Annia Regilla). Like the blocks in stone masonry, the bricks of these precursors were most lauded when their joins were so tight that the wall appeared to be all of one piece.

In the everyday terminology of Italian artisans, the distinction between brick and stone was anyway blurred. It was common to call clay *"pietra viva"* ("living stone") and brick *"pietra cotta"* ("cooked stone"). One nobleman (Alessandro Farnese) would unwittingly echo the Mesopotamians centuries earlier, when he remarked, "Stone is of two sorts, one natural, the other artificial, which is brick" (*la pietra è di doi sorti naturale et artificiale qual'è il mattone*).[100] Renaissance artists more than once played on this material metamorphosis, especially in two icons of late Renaissance architecture, the façade of the Palazzo Farnese, Rome, and the courtyard of the Convento della Carità, Venice.

The Palazzo Farnese façade offers a sophisticated display of dichromatic brickwork (fig. 11.29). Its brick-red, rhomboidal patterns in a cream field seem descendants of the diaper façade of the Palazzo Ducale in Venice.[101] The bricklayers often hailed from the north, and the foreman at Palazzo Farnese, Bartolomeo Baronino, came from Casal Monferrato in Piemonte. On the second story, red-brick lozenges of various dimensions fill the walls between the windows, and in the panels between the stringcourses, some of these meet to form X-patterns. The coordination of the patterning with the architecture, and the fact that the bricks must have been sorted by color before laying, preclude any accident or caprice. The patterning was, however, abandoned in later stages of construction, as becomes clear from its disintegration into "butterfly kisses" on the far right of the palace frontage. The fact that it also breaks off abruptly several feet below Michelangelo's crowning cornice, at the point where he decided to raise the height of the façade when he took over the project in 1546, strongly suggests that the design was the initiative of his predecessor, Antonio da Sangallo the Younger, and must therefore have been carried out between 1541 and 1546. The brick patterns of the Palazzo Farnese influenced a clutch of slightly later palaces in

FIG. 11.29. Façade brickwork, Palazzo Farnese, 1541/46. Rome.

Rome, yet its diaper patterning is far more varied than any of its followers, and begins to resemble the more variable symmetries of book-matched marbling. In the tympana of the window aedicules the irregular patterns can only suggest veining, and they are inset with roundels and rose patterns, in the manner of intarsia (opus sectile).

Andrea Palladio (1508–1580) also sought to simulate the effects of marble in the engaged columns of the courtyard at the Convento della Carità (1561–62). He had fabricated a type of brick whose epidermis of mixed clays purposefully imitated the blotchy veining of Veronese marble, the material that trims the interior. These bricks were knowingly juxtaposed with capitals, bases, and keystones in Istrian marble to emulate the color contrasts of ancient tombs (figs. 11.30, 11.31). Palladio himself makes no comment on the artifice other than to say that the building was made from *pietra* and *pietra cotta,* but two centuries later Scipione Maffei (1675–1755) recognized and admired the conceit.[102] In England, about the same time that Maffei was writing, streaky clay tiles were being used to surface the floor of a country house grotto (1762–65) at Goldney Hall, Clifton, so that it looked both marbly and watery, and by the 1780s a French architect would revive, on an industrial scale, the same process of mixing clays explicitly to obtain "the happy accidents of marble veining" (*les hazards heureux des veines du Marbre*).[103]

FIG. 11.30. Andrea Palladio, Courtyard, Convento della Carità, 1561–62. Venice.

FIG. 11.31. Engaged brick column, courtyard, Convento della Carità, 1561–62. Venice.

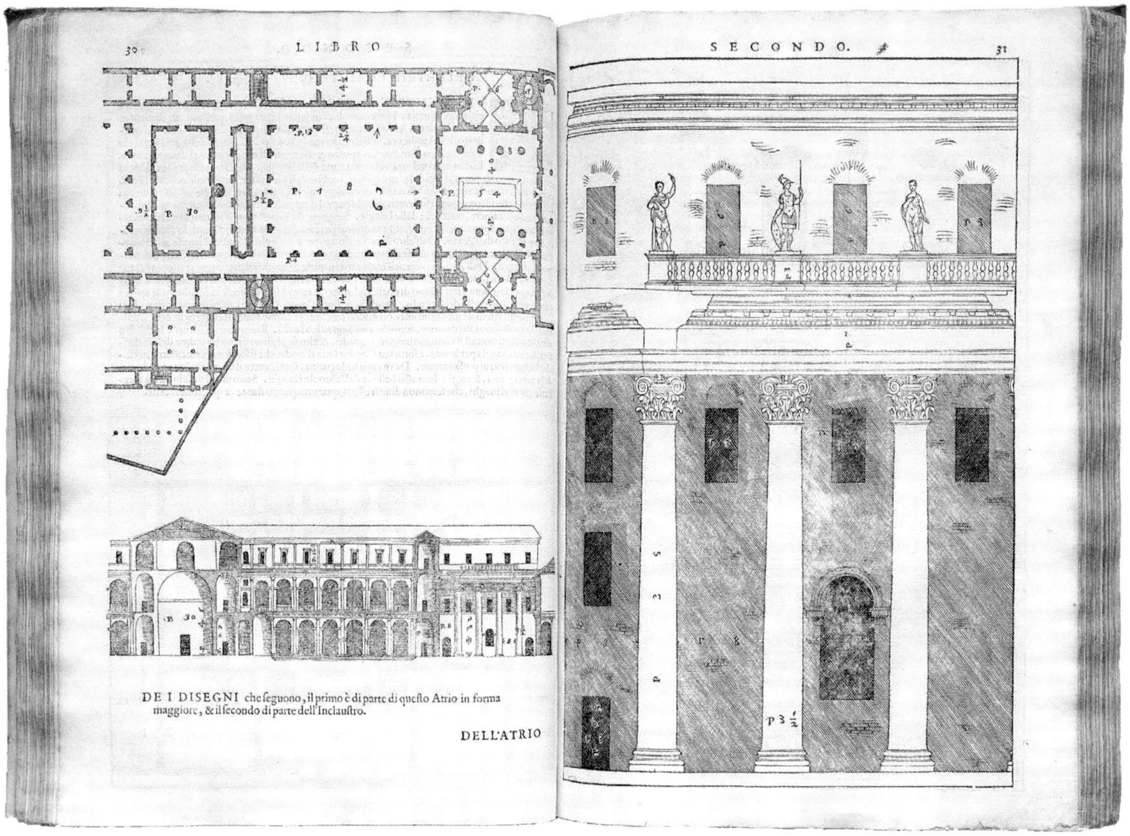

FIG. 11.32. Andrea Palladio, "Quattro Libri," Convento della Carità, 1570.

Palladio's Carità was one of his most considered attempts to revive the typology of the ancient house, on its most monumental scale because it must serve the communal demands of a populous, modern convent. The most imposing space, the monumental atrium (fig. 11.32), was unfortunately consumed by a conflagration in 1630. However, before it was, Sir Henry Wotton (1568–1639), the English ambassador to Venice, then writing his own free translation of Vitruvius, noted that the columns were made from "bricks, having first beene formed in a *Circular Mould,* and then cut before their burning into foure quarters or more, the sides outwards joyne so closely, and the points concenter so exactly, that the *Pillars* appeare one *entire Peece.*" Finally, the bricks were coated with a reddish glaze to augment their similitude to marble.

It is therefore legitimate to ask whether Palladio, in resurrecting the "House of the Ancients" in the guise of a convent, sought to rival, as Raphael abortively had thirty years earlier, the first feat of marble use in the ancient house, the columns of Aemilius Scaurus. It was possible to resurrect the palatial magnificence that Scaurus's house had epitomized only in image, for columns of this magnitude could not yet be quarried. Palladio's technical coup made a virtue of necessity and, in the process, as Wotton enthused, demonstrated "how in truth wee want rather *Art* then stuffe, to satisfie our greatest *Fancies.*"[104]

12 From Gems to Cloud Architecture

Reinventing Marbling in Early Modern Rome

In Rome, the desire for churches and chapels in colored marbles resurfaced in the mid-Cinquecento. In 1550 the Tuscan painter-architect, artists' biographer, and theorist Giorgio Vasari (1511–1574) took a tentative step when he designed wall tombs against colored marble fields in the Cappella del Monte in S. Pietro, in Montorio, Rome, although the walls ended up revetted only in white marble.[1] In the 1560s the revival really began to gather steam. The immensely wealthy cardinal Alessandro Farnese had the architect Giacomo Barozzi da Vignola (1507–1573) design richly marbled fireplaces and tabletops for his family palace with marbles culled from ancient Ostia and the Baths of Caracalla; foremost is the so-called "Farnese Table," which combined beguiling slabs of a misty alabaster with confronted peltae (Amazon shields) and other mixtilinear devices in *pietre dure* that, all together, took their cue from ancient revetment panels and opus sectile floors (fig. 12.1).[2] This table also heralded an industry in inlaid marble tabletops that would survive into the nineteenth century.

FIG. 12.1. Giacomo Barozzi da Vignola and Giovanni Mynardo, Farnese Table, 1565–73. Marble, hard stones, Egyptian alabaster, 37½ in. × 12 ft. 5¼ in. × 66¼ in. (95.3 × 379.1 × 168.3 cm). Metropolitan Museum of Art, New York.

FIG. 12.2. Giovanni Marchesi da Satrio, High altar tabernacle (now Tomb of Cardinal Francesco Pisani), S. Marco, before 1570. Rome.

The little-known Lombard mason and architect Giovanni Marchesi da Satrio (active 1536–78) provides an even more sophisticated glimpse of things to come in the Altar of S. Carlo Borromeo in the Duomo of Milan and the High Altar of S. Marco (before 1570), Rome; the latter was never installed and its *scenae frons* of stridently particolored stones, originally conceived to stage the Eucharist, was instead repurposed as a stage for mortal entries and exits on the Tomb of Cardinal Francesco Pisani in the same church (fig. 12.2).[3] This altar belonged to the same genealogy as the Dell'Acqua Altar in Vicenza (see fig. 10.17), as it too was descended from Francesco Sansovino's sacrament altar in the apse of S. Croce in Gerusalemme. Marchesi da Satrio's altar also resembles the sort of highly inlaid and enormously expensive marble cabinets that would begin to appear only a few years later. Polychrome projects accelerated in earnest, however, during the pontificate of Pius V (r. 1566–72), who commissioned tomb monuments in Rome, Naples, and Piedmont. Chief among these was the grand Tomb of Paul IV (1566) that Pirro Ligorio (c. 1512/13–1583) designed for the Carafa Chapel in S. Maria sopra Minerva (fig. 12.3), with its colored columns, dados, sarcophagus, and wall all encircling an equally polychrome figure of the pope (by

FIG. 12.3. Pirro Ligorio and others, Tomb of Paul IV, Carafa Chapel, S. Maria sopra Minerva, 1566. Rome.

FIG. 12.4. Matteo Nigetti and others, Cappella dei Principi, S. Lorenzo, begun 1604, lower story revetted in 1613–24. Florence.

Giacomo Cassignola) against a background intarsiated with multicolored peltae.[4] This tomb inspired, in turn, the equally impressive sepulcher that Pius V built himself (but never used) in the vast new Church of S. Croce (1567–71) in his hometown, Bosco Marengo, in Piedmont. Colored marbles appear throughout this church, beginning at the main portal, but the Sepulcher of Pius V was "arranged with such mastery that it would be the envy of any excellent painting," one observer remarked.[5]

Meanwhile, a colored marble revival was also underway in Florence. The Cavalcanti Chapel (1560–62) in S. Spirito, by an unknown designer, was clad in a variety of colored marbles, which the chapel inscriptions bragged had arrived from as far afield as "Egypt and England."[6] In the 1560s, Cosimo I de' Medici had asked Giorgio Vasari to design a third Medici Chapel at S. Lorenzo in Florence, which would be "large, and similar to that which Michelangelo already made there [the New Sacristy], but all in various veined marbles and mosaic."[7] Nothing came of it for the moment, but when the sculptor-architect Giovanni Antonio Dosio (1533–1611) returned from Rome in 1575, where he had studied Raphael's Chigi Chapel (see fig. 10.14) and designed intarsiated tabletops, he initiated a long sequence of polychrome chapels with his Cappella Gaddi in S. Maria Novella (1575–76) and Cappella Niccolini in S. Croce (begun 1579).[8] Dosio was a skilled pragmatist, successfully negotiating the acquisition of enormous quantities of marble from Rome despite skyrocketing costs, and his sensitivity to the new lithic palette is reflected in drawings with delicate washes that would become a staple in the architect's repertoire. In the Cappella Gaddi the best antique marbles available (porphyry, africano, porta santa) alleviated the Michelangelesque bones in the Florentine signature stone, gray pietra serena. The Florentine poet Antonfrancesco Grazzini (Il Lasca) promptly crowed that now Rome and Venice would have to yield precedence to Florence in such beauty and splendor.[9] Dosio's chapel initiated a long sequence of colored marble chapels and altars in Florence, the most opulent (and gaudy) of them the Cappella dei Principi at S. Lorenzo, which had been planned by Vasari in 1561/68, but which was not begun until the late 1580s, in competition with the papal Cappella Gregoriana (see below), and then to an unwieldy decorative program dictated by the grand dukes Francesco I (r. 1574–87) and Ferdinando I Medici (r. 1587–1609) (fig. 12.4).[10]

THE CAPPELLA GREGORIANA (1571–80)

The real watershed in colored marble construction, however, came in Rome with the construction of the Cappella Gregoriana (1571–80; fig. 12.5).[11] This was the first of the four corner chapels to be completed in the rising new St. Peter's, and only a chapel in name, for it was the size of a large church. It had been built in 1571–77, and every square inch was decorated with marble and mosaic in 1578–80. In a cavernous stucco-and-travertine basilica rising from

FIG. 12.5. Giacomo della Porta, Cappella Gregoriana, St. Peter's, built 1571–77, decorated 1578–80. Rome.

a brick floor, with none of the gilding nor extravagant marbling it would receive over the next two centuries, and no plans as yet to execute them, the Cappella Gregoriana emerged from the shadows as a mesmerizing oasis of color. The stakes were high because this pile was planned to be the mausoleum of the chapel's patron, Pope Gregory XIII (r. 1572–80); it must enshrine a twelfth-century wonder-working image of the "Madonna del Soccorso"; it was the basilica's sacrament chapel; and it was also to entomb the body of the Greek father St. Gregory Nazianzenus below the altar. The latter decision sent a clear signal of papal policy to proselytize or, if possible, reunite the Eastern Church with the Vatican.

The Cappella Gregoriana was decorated in a period of intense Christian antiquarianism. It incorporated columns from the old Constantinian basilica, and its vaults were the first in St. Peter's to receive mosaic, seeking material guidance from Paleo-Christian buildings such as the crumbling Oratory of the Holy Cross (see fig. 6.3). As if to acknowledge the fact that the chapel looked back to Christian antiquity, the Cappella Gregoriana also became the subject of ekphrasis, which found a new lease of life in the verse and prose compositions by the literati Lorenzo Frizolio (c. 1530–1597) and Ascanio Valentino (dates unknown), respectively.[12]

Frizolio was an erudite and accomplished poet, who knew the mini-ekphrasis (374 AD) that Gregory Nazianzenus himself had written on the church his own father had built (after 328/329) as well as translations of Prokopios into Latin (1537, 1538) and Italian (1547). Indeed, Frizolio produced the most expansive ekphrasis of marbles since Byzantium. In one passage, we can see just how alive the topos of the meadow still was, the metaphor of textiles to describe revetments, and also the tradition of the lapidary:

> Do you think that the sawn marbles,
> Which shimmer with divers colors,
> Can be understood with our discourse?
> These ones vie with the driven snow in their brilliance:
> These green ones surpass emeralds by far,
> And these yellow ones resemble Chrysoliths
> and the Pyropian flames in their polished glow.
> And do not these ones surpass black ebony, or jet?
> Yet others are distinguished by their innate blots or
> sprinkled droplets,
> These marvelous things are joined amongst themselves
> by art,
> And the sheen of the images by the new weavings of
> their markings.
> With their light the floor radiates, and the wall resplends,
> Just as the various fields laugh with expanses of every
> sort of color
> On Mount Timolus in Cilicia, or in the valleys of
> Sicilian Enna;
> Just as marigolds, violets, white privet,
> Narcissus flowers, or blue cornflowers shine, budding
> like gems in their garden beds;
> Just as woven hangings glitter, when the varicolored
> weft weaves in and out, stitched with Arabian
> patterns.
> As you look at these things, also marvel at the
> rosettes that burn in the lofty coffers, and do not
> ask for more.[13]

Ascanio Valentino was instead a prose stylist, and he too left no stone unturned in his attempt to demonstrate that the chapel had no peers in antiquity. All his detailed information plainly came from official sources, probably the architect Giacomo Della Porta (1532–1602) himself, and he attempted to identify every marble, describe every technique, and resurrect the specialist vocabulary of Pliny and Vitruvius. He took a more geological tack when it came to the small panels of amethyst that decorated the altar frame:

> Looking into the Amethyst, the image of some river enters the mind; which, discharged from the heights of the mountains, flows in a fine line, twisting and turning through the harsh lands; then bursts forth abundantly, just as clashing waves do; swells and billows; and dashing against the crags, catches up the surge in its assault, and washes over towns, whose idols [εἴδωλα] topple before the eyes. . . .

He continues:

> On the stylobates at every side, three slabs of the most beautiful alabaster possible have been set in place: in one of these [slabs] the eyes of the mind may perceive with the utmost pleasure the semblance of water, and in another [slab] that of earth; and as though water that has sprung in the mountain ridges is then snatched downhill, bends into a second course when opposed by the rocky obstacles, the gullies submerged under the billowing; just as they do when they irrigate plowed soil; how they flow together in one place, dispersed by the mass of rocks, and spread over the whole earth. In

another [slab] mountains are divided amongst themselves in equal intervals, and cascade into hills one below the other, some of them greater in height, while others appear more pinched because of the windings of their peaks.[14]

One will not find a similar passage in any contemporary text, not even Frizolio, and it signals a new material awareness of marble. The Cappella Gregoriana was contemporary with the same pope's institution of a mineralogical museum (Metallotheca) in the Vatican Palace under the direction of the papal physician, Michele Mercati (1541–1593).[15] In this gallery of nature, minerals were sorted in cupboards or cabinets, which sometimes look very much like altars themselves (fig. 12.6). Mercati's museum still followed the Aristotelian division of minerals into "things that can be mined" and "things that can be quarried" (that is, metals and stones). Modern geological classifications remained unformulated, and stones still seemed to owe their color and veining to a varying cocktail of air, earth, moisture, and colored ores. It was mottled and veined marbles, what Italians called "mixed marbles" (*marmi mischi*), that best seemed to confirm antique doctrine that all stones were "congelations" of subterraneous "exhalations," replenishing themselves in the core of an earth constantly channeled by generative streams. For example, Cesare Cesariano had commented (1521) that the quarries offered "various mixtures of stones mottled in varied disorder, complex and painted by nature like something compressed in a perfectly balanced and tempered congelation," while Vasari says that such stones are "called 'mixed' from the mixing of different stones that have frozen together and been made whole by time and the harshness of the waters."[16]

FIG. 12.6. Frontispiece, Armarium I (de lapidibus), *Metallotheca*.

Analogously, Valentino itemizes the marbles of the Cappella Gregoriana as here more earthy, there more aqueous, and alternately describes their veining as mountain peaks poking through cloud, or clotted masses traversed by snowy rivulets. It is a language that sometimes evokes modern satellite photography, but at heart one that regards the stone as being only in a temporary stasis before it will disassemble and return to its elemental origins.

PROVIDENTIAL IMAGES, HEAVEN ON EARTH, AND MARBLES AS GEMS

At the same time, it is striking that in among all the Ciceronian prose, Valentino decided to use a Greek word once, "εἴδωλα" (*eidōla*), a term invariably used to mean false idols throughout the Septuagint and New Testament. The subtext, which requires some prompting now, lies in the distinction that the church fathers drew between idols, which are carved by hand, and images naturally figured in the living rock, which are acheiropoieta sanctioned by the almighty.[17]

The reason that Valentino could draw such ethical conclusions from the images he perceived in the slabs of the Cappella Gregoriana was the still-vigorous religious mentality that any such marble images were heaven-sent. When the naturalist Ulisse Aldrovandi (1522–1605) wrote his *Musaeum Metallicum* in the 1580s, he devoted an entire chapter to marbles, which, for all its rigorous taxonomy, still concludes by grouping them according to their apparent figurations: "anthropomorphic," "dog-like," "shell-like," "fish-like," even "Mackerel-like" marbles.[18] Which marble went with which classical name had been long forgotten, and when marbles acquired new monikers during the Middle Ages, they regularly derived from their mimetic propensities. *Brocatello* resembled brocade, *lumachella* shells, *Grecco scritto antico* cursive Greek texts, and so on.[19] Even though Aldrovandi's mimetic marbles were obviously fossils, he persisted in regarding them as natural artifacts "painted" by *Natura Pictrix*, at the behest of the *Artifex Divinus* himself.

Seventeenth-century naturalists such as Fabio Colonna (1616) and Nicolas Steno (1669) would eventually contest the view that fossils were any kind of art form, but when the great Jesuit polymath Athanasius Kircher (1601/2–1680) published his *Mundus Subterraneus* in 1665, his argument hardly differed from Aldrovandi's (several of his plates were, in fact, copies) (fig. 12.7). A series of other observers, a great many north of the Alps, continued to publish notices on recently unearthed "figured stones." The Academia Naturae Curiosorum in Leipzig, for example, published a series of papers on the subject from the 1670s through to the middle of the eighteenth century.[20] And when Carl Lange published his *Historia Lapidum* in 1708—the product of pan-European correspondence—he too explained the generation of minerals in terms of terrestrial liquors and he still called fossils, as had Aldrovandi, "*lapides picti.*"[21]

While scholars of this ilk considered figured stones curiosities and games of nature, for others the "image made by chance" meant an image made by providence. These prodigies were, as Kircher explained, "made properly by Nature . . . because the Operation of Nature always depends on the concourse of all things by provident cause," and "if the phenomena of Nature are consequential, then they must be anticipated by cause, reflections of the Divine archetype modified by the laws of Nature."[22] Kircher echoed the opinion of a public predisposed to see freak images as pregnant with ominous foreboding.

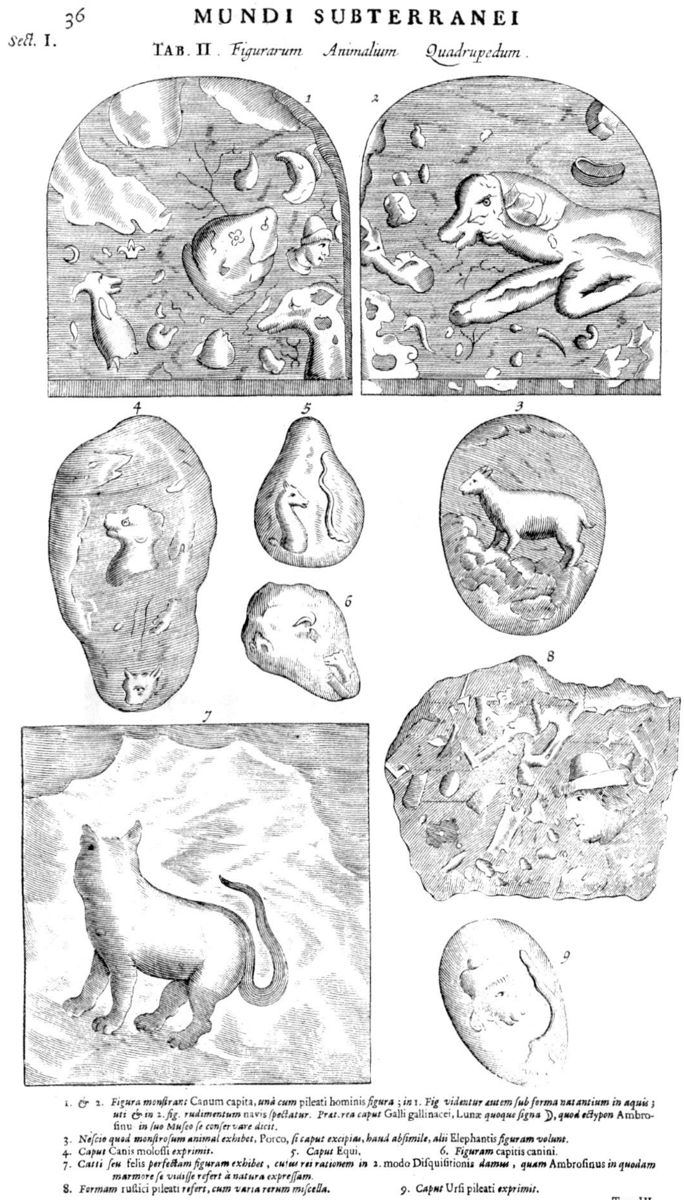

FIG. 12.7. Athanasius Kircher, Animal-stones, *Mundus Subterraneus*, 1665.

Thus, when the felling of an old apple tree near Harlem in 1628 revealed the image of a friar, the urbane Peter Paul Rubens found both the correct precedent in Pliny and recorded that the neighboring Dutch feared that this image portended a return to Catholic subjugation.[23] Guidebooks well into the eighteenth century continued to note, and occasionally depict, the various monks and Madonnas apparently encoded in church marbles, and one Frenchman could claim, for example, to discern a praying friar in the apse of St. Peter's.[24]

Furthermore, if fossils were not the fingerprints of original Creation, they must be the products of nature's ongoing creativity. Apparently, nature was continuously updating herself, and when one Venetian collector wrote a short treatise, as late as 1753, about the only agate he owned, he could identify it as "a Bavarian Duke" with some certainty (fig. 12.8).[25] Nature, he said, could only have generated the stone portrait very recently, because the ancients did not have moustaches and goatees, nor wear lace collars or the *Toison d'or* (the chivalric Order of the Golden Fleece). The same logic led him to remark, without a shadow of irony, that while the ancients had seen Pans and Apollos in tree roots and gems, today everybody unearthed Christs and Madonnas. Similarly, in 1665 the German historian Peter Lambeck claimed to discern the script B. XRISTO R. S. XXX ingrained in a *patera* of Oriental

FIG. 12.8. Michele Lazzari, Agate of "Bavarian Duke," 1753.

agate, from which he was able to decipher "BEATORI ORBIS CHRISTO, REGI SEMPITERNO, TRI-UNO CRUCIFIXO" ("To the Blesser of the World, Christ, the Everlasting King, the Tri-Unine Crucified One").[26] Lambeck was apparently unmoved by the fact that in a vessel of Egyptian or Indian origin and possibly Greek manufacture, Nature had chosen to express herself in Latin.

However, it was the innocence of the unlettered, discovering nature's secrets in the realm of untamed nature, that best guaranteed the authenticity of such prodigies. The Chilean Alonso de Ovalle's *History of Chile and Its Missions* (1646) reported several cases that were instrumental in Jesuit conversions, included an apparition of the Virgin on a cliff face and a miraculous tree in the form of the crucified Christ.[27] These images converted the natives by their guileless veracity and were proofs of divine favor for Jesuit activity in Chile. The same held for similar prodigies nearer home. On August 2, 1660, a certain Peter Bücher supposedly found a miraculous Virgin and Child engraved in stone in the iron mines of the canton of Uri in Switzerland. The local noble Konrad von Sonnenberg immediately wrote a tract pronouncing it the city's "palladium," and had the stone set up on the high altar of the Capuchin church of Gotteswald in the woodlands overlooking Lucerne.[28] The timing and imagery of the find were, however, suspicious. The Virgin stands on a crescent moon at the vortex of a brilliant, spiraling sun, the typical iconography of the *Immacolata,* and Von Sonnenberg's book was published in 1661, the year that Alexander VII issued a bull ratifying the doctrine of the Immaculate Conception. The implication, perhaps intended by Von Sonnenberg, was that the stone was a Swiss portent for the success of the Immaculists' cause.

Quite a few stones like Von Sonnenberg's had been brought in from the wilds, framed, and set up on altars. Aldrovandi, Kircher, and others mention several in churches all over Italy and Europe. Some also had liturgical functions, such as a rare, early Cinquecento *pace* (a utensil the devotee might kiss just before the communion) from S. Maria in Aracoeli, Rome, illustrating Augustus's apocryphal vision on the spot (fig. 12.9). Its inscription declares: "It is not the earth that has given this [stone], more precious than a gem, but God Almighty sent it from the Highest Heaven. As the prophets, seers and holy Sibyls sing, the image of the Virgin is revealed in this small marble."[29]

The tradition of identifying marbles with gems, which made the celestial metaphor possible, also showed no signs

FIG. 12.9. *Vision of Augustus*, c. 1515/20? Pace. S. Maria in Aracoeli, Rome.

of flagging. Aldrovandi, Andrea Bacci, Alonso Chácon, Mercati, Agostino del Riccio, and François la Rue (all of whom wrote treatises on marbles in the 1570s and 1580s) resumed the long textual tradition detailing the thaumaturgic powers of precious stones that spanned from Pliny to Camillo Leonardi and Lodovico Dolce.[30] Men such as Bacci quite explicitly quoted the Hermetic texts on the astral influence and sympathetic attractions of stones, to explain that gems and marbles bore a common soul. They were, as Bacci says, "the noblest and purged materials from the bowels of the earth, where the rays of sun and stars fuse to excite their potentialities and generate the most perfect things of celestial form, and consequently miraculous effect." As Giambattista della Porta writes, like plants, "the great power of precious stones, *if they have been cut at the proper hour and season,* comes from the heavens . . . thus they are more lively and vigorous and the countenance of the stars shines forth in them more distinctly."[31]

The affinity of marbles with jewels also justified their use in papal chapels as reflections of the Heavenly Jerusalem. In the Cappella Sistina (1585–90), the marbles were also spolia from buildings such as the Oratory of the Holy Cross (see fig. 6.3), and so in themselves Paleo-Christian relics in the Sistine representation of Rome as a Christian capital. Inaugural orations continued to tout extravagant chapels as apocalyptic visions and, as marbled chapels and churches proliferated, visitors would even claim that the whole city seemed a heaven on earth.[32]

THE NEW MARBLE CHAPELS OF ROME

The Cappella Gregoriana introduced a succession of papal funerary chapels that imitated or vied with its polychrome skin: the Cappella Sistina (1585–90) and Cappella Paolina (1605–15), both in S. Maria Maggiore, and the Cappella Clementina (marbling 1601–2) in St. Peter's. To the same period belongs Giacomo della Porta's extensive revetting of the transept of St. John in Lateran (1597–1601).

Over the next three hundred years, these papal commissions would set the standard for innumerable patrician chapels, and then whole churches, throughout the city. Already under Gregory XIII (r. 1572–85), demand for marble had increased so dramatically that a series of proclamations were issued prohibiting their excavation from the ruins.[33] By the century's close, agents were writing to out-of-town painters to apprise them: "According to the present custom of almost all the architects here in Rome, the altar must be ornamented with four columns of the finest marble, with other refinements in alabaster and other highly prized stones, so that as many honors as possible will be paid to the altar."[34] Marbles began to receive honorable mention in pastoral visits, and in time whole chapels were outfitted with them.[35]

Yet the process of re-marbling Rome was more gradual than the material's present ubiquity suggests. Religious communities might begin the century openly hostile to this luxury and finish the century basking in its light. In 1625 the Franciscan friars of S. Isidoro reserved marble use exclusively for the high altar of their new church and forced prospective patrons to content themselves with gilded plaster in the chapels. They were doubtless keen to forestall any charge of worldliness at the very moment in which their neighbors, the Capuchins at S. Maria della Concezione—brothers but rivals—were very publicly obstructing any papal attempt to introduce marbles into their own church.[36] However, by the 1660s, when the friars at S. Isidoro agreed for Bernini to clad their flagship chapel, the Chapel of the Immaculate Conception, they knew marbles were necessary for a chapel to be really *signorile.* Not long before, the new expectations were summed up by Bernini's testy rebuttal to the elders of Rieti (1654), when they suggested using plain stucco rather than marble to clad the new Chapel of S. Barbara he was to build them in the cathedral: "To make a chapel worthy of this city it would be better to wait for alms, or some other handout, than make something fit for Capuchins."[37]

Cost was a decisive factor. The prior of the equally meek Trinitarians at S. Carlo alle Quattro Fontane remarked that, had funds been available, he would have made his little church richer than Solomon's Temple, with a pavement made of emeralds and precious stones, since exterior splendor expressed an interior love of God, and nothing could be excessive when it came to God's house.[38] S. Carlino remained glaringly white, however, and took decades to complete (1634–75), in dramatic contrast to S. Maria della Vittoria, half a mile up the same street. When Cardinal Cornaro decided to shower the best-quality marbles on the new family chapel there (see figs. 12.21–23), it took only five years (1647–51) and cost 12,089 scudi, more than the entire complex of S. Carlino (11,678.53 scudi), though only one-fifth the cost of Bernini's later S. Andrea al Quirinale on the same street (56,030 scudi) (see figs. 12.29, 12.30). Similarly, the revetments of the Spada Chapel (1663–79), in S. Maria in Vallicella, cost approximately 12,000 scudi, about twelve times the cost of the actual structure, not to mention consuming fourteen years from quarrying to installation (1665–79). The earlier Nero di Nero Chapel (1600), in the same church, proved so expensive to clad that the bookkeeper eventually gave up recording the expenses.[39]

TABERNACLES AND THE CONFESSIO OF S. CECILIA

Marbles and semiprecious stones were also showered on the eucharistic tabernacle, which had become the focus of the Counter-Reform Church and the high altar. The most materially allusive forebear was the Ark of the Covenant in Solomon's Temple, but the form that became standard, the small centralized temple (*tempietto*), materialized celestial architecture and surrogated Christ as the "New Temple," the "greater and more perfect tabernacle, not made with hands" (Hebrews 9:11). A tabernacle such as that on the Lateran Altar of Sacrament (c. 1600) combines these allusions with a pedestal inlaid with hard stones, which plainly miniaturizes the foundations of the Heavenly Jerusalem (fig. 12.10).[40] Altars dedicated to the

FIG. 12.10. Pier Paolo Olivieri, Tabernacle, Altar of the Sacrament, St. John in Lateran, c. 1600. Rome.

FIG. 12.11. Giacomo Della Porta and Gaspare Guerra, Confessio, S. Cecilia, 1599–1603. Rome. Condition before the remodeling of 1899–1900.

Virgin were often accorded the same honors, because she was the perfect vessel for Christ and reigned as Queen of Heaven. Innumerable altars surrounded her image with a ring of precious stones as though with a heavenly crown. Cult statues were regularly festooned with necklaces and bracelets of semiprecious stones, and in certain parts of Italy young women donate their jewels to the Virgin to this day.

The other favored locale for colored revetments and paving was the *confessio,* the sunken pit in front of the high altar, normally accessible by stairs, which communicated with the tomb of a martyr (Early Christian) below the altar by means of a grated window (*fenestella confessionis*), and which was archaizingly reinstituted from the reign of Sixtus V (1585–90) on. The most imaginative confessio, designed by Giacomo Della Porta and Gaspare Guerra for S. Cecilia, Rome, was begun in 1599 (figs. 12.11, 12.12).[41] The body of the third-century martyr Cecilia had been "providentially" rediscovered, miraculously incorrupt, during the church's renovation in advance of the holy year of 1600. Paolo Camillo Sfondrato, the titular cardinal, spared no expense on restaging the event with a spectral statue and lavish display of marbles. In its original form (altered in 1899–1900), Della Porta elevated a balustrade over the wall separating the raised choir from the nave and moved the stairs to the rear, out of sight. There could be no clearer division between the nave, with its brick floor, and the ensemble of marbles and mosaic that loomed above it. In the spirit of the archaizing restorations that Cardinal Baronius had carried out in other medieval churches just a few years earlier, the screen wall in S. Cecilia mimicked medieval compounds such as that in the nave of S. Clemente and just possibly the revetments in late antique catacomb cubicula.[42]

The screen wall also frames reliefs of Cecilia and companion martyrs, gilt-bronze appliqués on panels of black marble (see fig. 12.12). This colorist segregation of figure and ground accentuated the idea of bas-relief as a halfway house between sculpture and painting. However, black limestones and marbles were generically called "*pietra di paragone*" ("touchstone") because they were used for assaying gold, especially coins, where the pure ore could be distinguished from an alloy by comparing the traces they left when rubbed side by side on the slab. The mount therefore became a touchstone of the absolute value of the sculpted artifact.

At the heart of the ensemble was Cecilia's black marble niche tomb, housing the recumbent statue that Stefano

FIG. 12.12. Giacomo Della Porta and Gaspare Guerra, Confessio, S. Cecilia, 1599–1603. Rome. Condition after the remodeling of 1899–1900.

Maderno supposedly carved limb-for-limb and fold-for-fold from the disinterred body. The unblemished marble replicated the incorruptibility of Cecilia's cadaver, and the sculptor chose what he thought was a block of "Parian" marble for the purpose (though it is actually Pentelic). Not coincidentally, when Aldrovandi illustrated Parian marble in his treatise, he selected an antique bust of the Republican heroine Lucretia because she was the "leading light of chastity" (*praecipuum castitatis lumen*).[43] Maderno's shining statue lies between the tomb and the altar, between the place of death and that of transcendence. She evokes the white-clad martyr who lies beneath the apocalyptic altar (in Revelation 6:9–11) and the effigy elides the transition between life and life eternal by recording both the historical body that defies corruption and the diaphanous soul that awaits reunion with it.

Just as Cecilia's incorruptibility is not circumscribed by human time, the gleaming marbles that ring her are the currency in a transaction between this world and the next. In the *Passio* (1600) written by Antonio Bosio, the antiquarian spelunker often called the "Columbus of the Catacombs," Cecilia avows that martyrdom "is not to lose youth, but only to exchange it; to discard mud for gold; pebbles for sparkling diadems; and a dark hovel for a great and spacious palace built of gold and precious stones . . . to abandon something that is perishable in exchange for something which knows no end . . . to rid oneself of vile stones which are trampled underfoot so as to accept a precious stone that sparkles in a royal diadem."[44]

The mesmerized observer was, in fact, still subject to an anagogical program and, until the mid-seventeenth century, the confessio was one of the few places where one found colored marble ensembles within the new churches. So the pilgrims who trod the clay floors of the nave knew that the stones in the vision that met their eyes were literally *not of this earth.* Indeed, Onofrio Panvinio, another great antiquary of Paleo-Christian remains, called the confessio the "Gates of Paradise."

PAINTING *IN* AND *ON* STONE: THE *PARAGONE* AND "ETERNAL PAINTING"

Bosio characterizes the marbles at S. Cecilia as both "gems" and "stones of many colors, which are so closely matched and joined, that they seem to have imitated the most elegant kind of painting" (*lapidibus multiplicium colorum, qui ita apte conferti, coagmentatique sunt, ut elegantissimae picturae speciem aemulati videantur*). Marbles, as we know, had always been considered natural paintings, and since the Quattrocento they were actively posited as gardens of potential cognition for artists in search of pictorial invention. Leonardo, Piero di Cosimo, and Sandro Botticelli all counseled painters to study colored stones for compositional inspiration. Botticelli apparently flung paint-loaded sponges at walls in search of the same creative serendipity, and Leonardo was so entranced by the random beauty of the patterns that he sought to reproduce them by binding goose quills into bundles that he called "*mistioni.*"[45]

For the artist, the idea of "painting in stone" was also as much a matter of practice as of theory. The journey from raw product to medium was so integral to the painter's art that when Vasari came to publish his *Vite* (1550), he began them with colored stones and only then progressed to painting techniques, and only after that to biography. It is symptomatic of our present detachment from the transformational properties of stones that his prolegomenon is still generally published as a separate book. In the most important surviving medieval treatise on the painter's trade, *Il libro dell'arte* (c. 1400), Cennino Cennini describes the skill, handed down from father to son, of prospecting the raw material from the "wrinkles" of the earth; back in the studio began the daily grind of mineral into pigment, the tabletop "alchemy" (Cennini's word) that made stone into painting, when the colored earths were crushed between porphyry or serpentine slabs and mullers.[46] In

FIG. 12.13. Anonymous Roman painter, *Conversion of St. Paul*, c. 1600?. Oil on calcite alabaster. Private collection.

fact, if artists had few problems accurately simulating these exquisite granites, it is because these were slabs on which they also made tracing paper, providing another opportunity to scrutinize their markings.

It may well be the diurnal habit of the grinding slab that taught painters to appreciate marbles as potential supports for painting.[47] By 1530 Sebastiano del Piombo had devised a technique for painting in oils on stone, wherein the colors "once dry, unified with the marble as though they had petrified."[48] Many painters, including Titian, Jacopo Bassano, and Vasari, followed suit and were proud to claim that they too had created a "painting that was virtually eternal," an opinion seconded by humanists such as Benedetto Varchi.[49] Toward the end of the century, the stones were used not just as hardy supports, but their found veining became an integral part of the painting. Lapis lazuli, amethyst, agate, alabaster, *pietra paesina,* or *albarese,* any manner of veined marbles—and paragone (the generic term for any black marble)—were all used. Lapis and amethyst were ideal for brilliant skies, reddish alabasters for scenes such as the *Crossing of the Red Sea* (artist unknown), alabaster and agate for scenes of revelation filled with heavenly cloud, and so on (fig. 12.13). Paragone was perfect for nocturnes, but the word itself invited comparison between the two art forms. Now "the skill of the artist played with the art of nature," fulfilling the suggestions of Nature the painter, and the human painter found himself in partnership with the stone, and almost a spectator to the collaboration. Collectors marveled at the results, "in themselves noble because they are eternal," and because "one cannot say whether they are treasures of Nature or miracles of Art, [for they are] precious in their materials, marvelous in their artifice."[50] The effects were so treasured that Antonio Carracci would replicate the veins of alabaster when he painted the *Virgin Appearing to St. Francis,* although its true support is really copper.[51] Paintings on marble became so familiar that novelty turned to expectation. Thus, although Athanasius Kircher applauded the High Altar of S. Sebastiano fuori le Mura as a work on marble "wherein clouds and the bright parts of the sky can be seen wonderfully feigned," the striated sky was actually tempera painted on plaster.[52]

Vasari says that Sebastiano not only made paintings on stone, but surrounded "them with ornaments of other veined stones that, once polished, made the most beautiful accompaniment." Furthermore, when paintings on marble were set into altars, as on the High Altar of S. Maria della

FIG. 12.14. Francesco Mazenta?, High altar, S. Maria della Passione, 1600/1619. Milan. The marble panels may have been painted by Giulio Cesare Procaccini and Gian Battista Crespi.

Passione in Milan (c. 1600), the old lore of gems fused with the theme of divine painting (fig. 12.14).[53]

By the 1580s, painting *on* stone turned to painting *in* stone, at first in the Florentine Opificio delle Pietre Dure (then known as the Galleria dei Lavori), and then in the workshops from Prague to Naples that turned out elaborate tabletops and easel pieces that were jigsaws of hard stones judiciously selected to match directly the colors and brushstrokes of painted cartoons.[54] In the late seventeenth century, Filippo Baldinucci would even pose a paragone between painting and these stone inlays (*commessi*), vaunting the greater skill required to make them:

> Whereas in painting the learned painter skilfully mixes and lays on his colours . . . and thoroughly blends them to create multiple tertiary colours, however he sees fit, and quite different from their natural pigmentation . . . [the *commettitore*] cannot dissolve his material, nor blend one colored stone with another to make a third of his own choosing, but must exploit the color of the stone just as nature has formed it. . . . He must seek out even the most minute markings that nature

> has fashioned by herself for what serves his own ends . . . and must above all be so experienced in the art of painting, that whenever he observes the markings of stones or gems he will recognize how each of them may serve to outline or shade the composition at hand, even the deepest darks and the half-tones.[55]

The Florentine commessi were a conscious revival of ancient opus sectile. However, Ferdinando I proudly claimed that a "new way of representing images and portraits of persons in marble inlays was our invention, not in the ordinary way of mosaics, but with another more ingenious artifice, the appearance and portraits of the persons with natural colors . . . the portrait composed in stone that seems as lifelike as if it had been painted with a brush."[56] Indeed, not only was the Cappella dei Principi (fig. 12.4) adorned with altarpieces in pietra dura reproduced from the cartoons of renowned painters, but it was intended for it to display pietra dura portraits of the Medici as well. The Medici workshops also produced deluxe gifts in the medium as touchstones of the new art (fig. 12.15).[57] When, in 1621, Michelangelo Buonarotti the Younger—great-nephew of the artist—came to deliver a eulogy of Cosimo II (1590–1621), he could even make the disputable claim that it was the very incrustations of the Cappella dei Principi that had led artists to "find a way of making painting eternal."[58]

Eventually, by the 1630s, the Sienese painters Michelangelo Vanni and Niccolò Tornioli would transform the metaphor of "painting in stone" into a new technology whereby images could be chemically instilled *within* marble slabs, as though they had been produced by providential nature.[59] This was an exceptional episode, but buon fresco was still a living art and, during the course of the seventeenth century, *scagliola* (a composite paste of plaster of Paris, glue, and various natural pigments) was also developed to imitate marble and marble inlay.

All this interest in marble as a form of painting was rekindled not only by the new genre of paintings on stone, and all the other innovations, but also by a revival of mosaic and opus sectile, both of which were also regarded as "paintings in stone." Vasari singled out glass and marble mosaic as "a painting capable of withstanding water, wind, and sun for an eternity," and mosaic had never been fully abandoned in Venice, unlike Rome, because it withstood the damp in a way that paint could not. But its use was intermittent and the technique lived on mainly thanks to the necessity of maintaining the vast mosaics at S. Marco. If Vasari was right, it took an artist of the stature of Titian, who provided cartoons for new mosaics in S. Marco, to reinstate the medium in the vanguard of artistic endeavor.[60]

The remains of antique opus sectile were also important because, alongside the ruined frescoes of the Domus Aurea and a few other remains, they provided one of the few sure ways to glimpse the colors of ancient painting and showed how to coordinate an entire wall all'antica. Thus, when Giovanni Battista Della Porta (1542–1597) revetted the Cappella Caetani in S. Pudenziana, to the designs of Carlo Maderno, in 1595 (fig. 12.16), he competed with the fourth-century Basilica of Junius Bassus (see fig. 5.19), which was still substantially intact and only a stone's throw away.[61] Antiquarians in Rome wondered at the skill of workers in opus sectile, while antiquarians in Venice marveled how the book-matched slabs of S. Marco seemed naturally painted and made the wall appear all of a piece.[62] All this renewed interest is even echoed in strictly religious commentaries. Thus, Juan de Piñeda (1558–1637)—no less than the censor of the Spanish Index of Prohibited Books—dared challenge St. Jerome's Vulgate translation by arguing that the walls of Solomon's Holy of Holies were *not* covered by painted fabrics, but that the stones were so variegated that they *seemed* painted: "What is called *lapis variegatus* in the original, from the word *Rakam,* which means *to variegate,* and even *to paint,* or *to interweave with varied handiwork and divers colours* . . . has little to do with fabrics, but rather stone. . . . I therefore trace [the word] more to the variation of the stones themselves, than painted coverings, which would conceal the beauty and variety of the precious stones."[63]

Moreover, as had been the case in antiquity and Byzantium, "painting" encompassed far more media than we would understand by the term today. By the second decade of the seventeenth century, the papal physician and art critic Giulio Mancini (1559–1630) would comment that mosaic, opus sectile, intarsia, marquetry, even tapestry were all "species of painting . . . because color can be expressed and imitated by different materials, which, with their colors, start to imitate and express the coloring that we desire, and from this it arises that there are as many types of paintings in color as the amount of colored materials that we employ for such imitations." Mancini adduces several modern examples but also, notably, the Basilica of Junius Bassus (see fig. 5.19) as an ancient example of "painting in stone."[64]

FIG. 12.15. Romolo di Francesco Ferrucci (del Tadda), from a cartoon by Jacopo Ligozzi, *Pope Clement VIII, 1600–1601.* Inlaid panel, marble, hard stones, mother of pearl, 38 3/16 × 26 3/4 in. (97 × 68 cm) with frame. J. Paul Getty Museum, Malibu.

FIG. 12.16. Wall revetment, Cappella Caetani, S. Pudenziana. Executed by the mason Giovanni Battista della Porta to the designs of Carlo Maderno. Rome.

THE PARAGONE AND "PAINTED" ARCHITECTURE: THE FAMILY CHAPELS OF VIRGILIO SPADA (1631–32, 1654–57)

During the sixteenth century, another factor in popularizing the colored marble medium was the accumulating theorization of the paragone, the long-standing debate by practitioners and theorists over the relative superiority of sculpture or painting.[65] Especially pertinent among the commonplaces of this debate was the allegation made by sculptors, for example Tullio Lombardo in 1526, that painting was inferior to sculpture because its materials were not as durable as marble. Painters countered that this owed nothing to art and everything to the material, but nonetheless this was a grave deficit for any art of commemoration. Thus, in 1537 the Florentine noble Giovan Battista Strozzi (called Filippo il Giovane, 1489–1538) even willed (unsuccessfully) that his family chapel in the apse of S. Maria Novella be completely marbled, whatever the cost to Filippo Lippi's famous frescoes there, precisely "because [painting] is not very durable by its nature," whereas marble revetment would be "perpetual."[66]

Until now, architecture had generally stood on the sidelines of the paragone debate, disqualified by its inability—with the exception of the man-made grotto—to imitate from nature or life, with observers wavering between regarding it as the mother of the arts or their poor sister. However, already in the 1540s, the latest protagonists of the paragone had broached the question by arguing that because architecture is "necessary" (unlike painting and sculpture), "sculptures and paintings are made to decorate buildings rather than the reverse" (Benedetto Varchi) and, conversely, that the fine arts were necessary to architecture because they "mix in a little delicacy to its essence, which is in truth mechanical" (Vincenzo Borghini), meaning a skill or trade—construction—more than a true "art." By the end of the century, the issue was short-circuited by the painter-turned-treatise-writer Gian Paolo Lomazzo, who discussed architecture in equal terms and identical categories to painting, and in whose *Idea of a Temple of Painting* (1590), as the title suggests, the criteria of painting became the building blocks of a conceptual temple.[67]

"Painting in stone" had become an open ambition in sculpture too, and Vasari approvingly observes that Ligorio's Tomb of Paul IV (see fig. 12.3) enclosed "a statue made in pieces . . . that represents that pope, with a mantle of veined brocatello, and the frieze and other things are also made of veined marbles of various colors, which renders it marvelous. And so we see . . . that through colors sculptors approach sculpture by imitating painting."[68] With this in mind, it becomes all the more understandable why painters should be concurrently celebrating compositions on marble as a species of "painting that was virtually eternal."

By the end of the century, painters and architects were also conjointly becoming ever more sensitive to the potential of colored marble revetment to amplify the altarpiece beyond its frame, and to decorate the chapel or chancel walls without the busy competition of grotteschi or figural stuccoes. Thus, in the new Oratorian Church of S. Maria in Vallicella in Rome, an *équipe* of artists picked the most appropriate palette of marbles to decorate the Marian cycle of chapels (1586–1621) dedicated to the *misteri gaudiosi, misteri dolorosi,* and *misteri gloriosi,* according to the theme of each chapel. Accordingly, as devotees progressed around the church, they passed from the festive colors around Barocci's *Presentation* (marbling 1589–92, altarpiece 1603) to the somber rust-reds and funereal blacks surrounding Caravaggio's *Deposition of Christ* (1602–4).[69] By the 1650s, compilers of ecclesiastical Latin not only found "Cappella" to be synonymous with a richly encrusted chapel, but defined any decorative motifs as "emblemata"—the Latin word that denoted those mosaic panels that had always imitated easel painting.[70] The idea that "painting in stone" could be carried out on an architectural scale was gathering strength, enough to tempt a new intermedial art form, "architecture as painting."

Unfortunately, the first building consciously to propose this concept, and on a grand scale, remained on the drawing board, the Spada Chapel in S. Andrea della Valle, Rome. The ideator, Virgilio Spada (1596–1662), was a career cleric and able administrator, but also a dilettante architect with a keen eye on both the practicalities and poetics of construction.[71] He relied on trained architects to make the drawings, perfect them, and execute the design, but the ideas always originated from him. In 1631 Spada was busy planning the decorations of the family chapel in S. Andrea della Valle and, from the outset, he wished to invent a decor radically different from the other rich and recently completed chapels (1603–16) in the church. As these already fulfilled clear categories of decoration—all marbling, all painting, and statuary-and-painting—Spada decided that his own chapel must be marbled, but that he would eliminate any painting and absorb it into the realm of sculpture by substituting bas-relief for canvas. The entire chapel was to be sheathed in black marble, except for narrative reliefs on the side walls, and "the altar and the ornament of the altar, which, being a glorious sepulchre, should be extremely beautiful and may consequently be adorned with every type of stone of various colors," reasoning that the altar deserved to be as glorious as Christ's sepulcher (Isaiah 11:10: "His Sepulchre shall be Glorious").

Spada extols black marbles for their lugubrious propriety, but since all black marbles were generically called "paragone" (the term Spada uses too), it would be extraordinary if the paragone debate did not lurk at the back of his mind. Some years later his protégé, the architect Francesco Borromini (1599–1667), made the pun inescapable when he described the coal-black portal to the Roman Oratory as *"ornata di pietra di paragone, lavoro senza paragone in Roma"* ("adorned with paragone-stone, a work without peer in Rome").[72] One of Spada's own drawings for the chapel betrays his preoccupations, because a bas-relief is shown hanging from sculpted cords just like an easel painting in a drawing room. The scheme was never executed, but the paragone became the more overt conceptual armature for Spada's next scheme.

Wanting something "out of the ordinary," he "came to the idea that the marble composition should form a perspectival view, with the hope that the variety of colors of the marbles could represent the subject no less than the colors of painters." He even called in the famous Bolognese *quadratura* painter Girolamo Curti ("Il Dentone," 1575–1632) to design the scheme, but he died soon after. With inlaid tabletops in mind, Spada also specified that the prospective painter paint a full-size cartoon on which masons could assemble a jigsaw of marbles to match the colors. Finally, he stressed that the illusionistic architecture must be inherently "painted," and that no "animals, trees, men" or any other figural elements intrude into the composition. Any such subjects must be confined to separate bas-reliefs in white marble. Again this scheme was never built, and when Spada returned to the project after a thirty-year gap (1662), he died before any start could be made. However, a drawing does survive

FIG. 12.17. Virgilio Spada?, Design for side wall inlaid with marbles, Spada Chapel, S. Andrea della Valle, 1662. Archivio di Stato. Rome.

FIG. 12.18. Cosimo Fanzago, Francesco Borromini, and others, under the direction of Virgilio Spada, Spada Chapel, S. Girolamo della Carità, 1654–57. Rome.

for this scheme (fig. 12.17), one that he proudly showed to Alexander VII, and one that shows how much Spada was now indebted to the side reliefs of Bernini's later Cornaro Chapel (see fig. 12.23).

Spada's only lasting contribution to the genre was instead the family chapel he had executed in S. Girolamo della Carità (1654–57) with help from Borromini and Cosimo Fanzago (1591–1678), but again hewing to his own symbolic agenda (fig. 12.18).[73] On its walls Neapolitan *marmisti* lavished the techniques of marquetry, even passing the stones over hot irons to tone them as was usual with wood marquetry. Spada specified that he wished "to represent the chapel as an upholstered room because in ancient times chapels were called *cubicula sanctorum* ["bedrooms of the saints"]."[74] The whole concept hinges upon the pun contained in this conscious archaism, and Spada's simulated damasks gave shape to the "sleep of death" within a sacred bedroom, truly a "chapel of repose."[75] The chapel was to be dressed with marble inlays based on the embroidered silk pilaster hangings of S. Maria in Vallicella, the mother church of Spada's affiliate order, the Oratorians. Spada's *concetto* found few followers in Rome, although the metaphor of revetment as brocade became extremely diffuse in Naples and Sicily, and its eighteenth-century acme is the Gesuiti in Venice (marblework, c. 1714–35). Even in Spada's triumph of meticulous intarsia, however, he would still point out that the veneers facing the proscenium arch were "uniformly made from marble with the veining placed horizontally as though the chapel had been excavated from a block of similar marble."

FIG. 12.19. Girolamo Rainaldi and others, Altar, Cappella Paolina, S. Maria Maggiore, 1611–13. Rome.

ARCHITECTURE AS PAINTING: GIANLORENZO BERNINI AND THE CORNARO CHAPEL (1647–52)

Spada's early designs for a pictorial marble chapel at S. Andrea della Valle had come to nothing, but their promise was fulfilled by Bernini in a string of projects from the 1630s and 1640s. His own inspiration was the opulent altar (1611–13) in the Cappella Paolina in S. Maria Maggiore, which had been designed by a committee including the architect Girolamo Rainaldi, the gentleman architect Giovanni Battista Crescenzi, the painter Antonio Tempesta, the goldsmith Pompeo Targone, and the pope himself (fig. 12.19).[76] This was to house the miraculous icon of the Virgin, known as the Salus Populi Romani, which tradition held St. Luke had painted himself.

To hypostasize this icon's supernatural character, it was made to float within an aedicule in a heavy gilt-bronze frame borne by gilt-bronze angels, against a field of lapis lazuli slabs, which the contract stipulated should be "set together in such a way that they feign the air." Contemporaries likened the whole ensemble to a window onto a star-spangled sky.[77] In an early design, the suggestion of parting clouds also appeared in the stridently veined sheets of alabaster sheathing the wall behind the aedicule columns, but in execution great pains were instead taken to match the white mottle in the tarsie of the lapis sky to suggest wisps of cloud and vapor.[78]

Bernini's enthusiasm for the altar is recorded in a dialogue of 1633, and its influence is tangible in the two important chapels he subsequently designed at St. Peter's, the Cappella del Santissimo Sacramento (1634) and Cappella della Cattedra Petri (1637), to host the blessed sacrament and St. Peter's throne, respectively. To highlight the otherworldly nature of these focal objects, Bernini devised alabaster backdrops of clouds parted by divine light.[79] He repeated the formula on the reliquary balconies

FIG. 12.20. Gianlorenzo Bernini, Reliquary balcony (Veil of Veronica), St. Peter's, 1640–41. Rome.

of the crossing piers (1640–41), which housed the basilica's most precious relics and were the loges for their Easter ostension (fig. 12.20).[80] In each case the artifice signified a paradigm shift, for the focus of each display was an object of otherworldly provenance (and a fragment of divine time). Bernini repeated the formula on the high altar at SS. Domenico e Sisto (1636–c. 40), whose inlays of incandescent clouds also encircle a miracle-working Madonna, but all these early experiments in pictorial marbles would culminate in the Cornaro Chapel of 1647–52 (fig. 12.21).[81]

The patron of the Cornaro Chapel, Cardinal Federico Cornaro (1579–1653), had come to Rome for the conclave of 1644, upon the death of Urban VIII, with his eyes set on the papacy. When this bid failed, he turned his gaze to posterity instead and commissioned a burial chapel from Bernini. Bernini's brief was threefold: he must commemorate the five Cornaro cardinals (and one doge) who had lived over the previous hundred and fifty years; he must make manifest Federico's desire for postmortem redemption; and he must celebrate St. Theresa of Avila, a newly canonized mystic famed for her visions and spiritual tracts.

As St. Theresa floats on a cloud, collapsed in ecstasy, an angel poises to thrust an arrow into her entrails; she is consumed by its fiery tip and feels herself transported from this world to spiritual union with God (fig. 12.22). She occupies an ovular tempietto that evokes the saint's description (in *The Interior Castle,* 1577) of the soul "as resembling a castle, formed of a single diamond or a very transparent crystal, and containing many rooms, just as in heaven there are many mansions . . . [and] in the center, in the very midst of them all, is the principal chamber in which God and the soul hold their most secret intercourse." This tempietto is also the junction of two axes. Along the vertical, the resurrection of the flesh is proleptically enacted: reanimated skeletons wriggle through illusionistic manholes in the chapel floor, from the burial vault below, into the light that bursts through the clouds parting the vault, the light that is cast by the supernal essence who will remain unseen until the last syllable of recorded time. This vector is intersected by a horizon of mediation between the temporal and the eternal. At the junction, Theresa momentarily dies to the world, while to either side the Cornaro cardinals convene for a family reunion impossible in life; they emerge from the sort of spectatorial boxes that VIPs occupied at canonizations (see fig. 12.20), but their fugitive perspectives tell us that they too haunt heavenly mansions (fig. 12.23).

At the apex of the chapel, angels brandish a banderole with the inscription "*nisi coelum creassem ob te solam crearem*" ("If I had not created heaven I would have to create one just for you"), words that Jesus himself had supposedly uttered to Theresa in her delirium. The spoken challenge for Bernini was to re-create the heaven of her prayers in the chapel, and the Carmelites of S. Maria della Vittoria would congratulate him on his success. In the foreword to a compendium of Theresa's writings, published in 1652 to celebrate the completion of the chapel and the beneficence of Cardinal Cornaro, they say that the second creator ("Cornaro *qua* Bernini") "has been able not only to render stones animate and colors eloquent, but also to transfer the sovereign kingdom from heaven to earth. . . . If God had not created a heaven for others, he professed, he would have created it for Theresa alone; and Theresa, besides that [heaven] which she bids you to relish through her intercessions once this mortal life has run its course, opens up to you another in which to rejoice while still living." As the Tridentine Catechism (1566) instructs us, it is the privilege of saints to witness the full mysteries that remain secret to the common devotee, and it was Bernini's task to make her vision our vision, a heaven for all to share.[82]

FIG. 12.21. Gianlorenzo Bernini, Cornaro Chapel, S. Maria della Vittoria, 1647–52. Rome.

FIG. 12.22. Gianlorenzo Bernini, *St. Theresa in Ecstasy*, 1647/50. Cappella Cornaro, S. Maria della Vittoria, Rome.

To do so, he removed the "fourth wall" on the drama in Theresa's interior castle, simultaneously dilating the space beyond the confines of the chapel wall by cantilevering the tempietto beyond the chapel and out over a side alley.[83] The suspension of disbelief in the vision was also assured by the suspension of the effigy, for *St. Theresa in Ecstasy* is the first statue ever made without a statuary base. Deflated, she expires on a cloud bank, and the whole mass of stone balances, like an angel on a pin, atop a travertine pier invisible from the church crossing. This antigravity mode makes the artist a conjuror, and the miracle of the moment is achieved by a miracle of art.

Bernini had inherited from his father, Pietro (1562–1629), not only his extraordinary facility in carving, but also his personal engagement with the paragone. Pietro used paintings as models for his sculpture and "translated" them into relief, overtly straining to rival the art in tours de force such as the massive *Assumption* he carved for S. Maria Maggiore in 1606–10 (fig. 12.24). The multiplicity of planes, degrees of polish, varieties of texture, chiaroscuro carving, anamorphic foreshortening, and impression of movement were all testament to the fusion of sculpture's capacity for materialization (to manifest things "as they are") with painting's capacity for illusionism (to represent things "as they seem").[84] Following his father's example, Gianlorenzo also strove against the proscriptions of the paragone, that sculpture could never compete with the effects of painting, by creating a genre of "sculpted painting" that would defy a material as unyielding as stone.[85] And the paragone weighed enough on Gianlorenzo's mind for him to write a comedy (1635) in which a painter's studio and a sculptor's appeared onstage side by side, and all the parts were played by his own studio-hands.[86]

This preoccupation with the paragone in the Cornaro Chapel is immediately betrayed by his calculating use of bas-relief and flattened relief (rilievo schiacciato), that halfway house between painting and sculpture, whose ambiguous allegiance he accentuated by making the reliefs material hybrids. Thus, on the altar *paliotto* a gilt-bronze relief of the *Last Supper* sparkles against a ground of lapis lazuli, in the genre of appliqués executed fifty years earlier at the Confessio of S. Cecilia (see fig. 12.12). Likewise, on the side walls the sharply foreshortened effigies of the Cornaro are carved in white silhouette against a colored stucco background (*stucco lustro* imitating gray *bardiglio* marble) (see fig. 12.23).[87] At the same time, Bernini's ability "to carve color" into colorless marble by virtue of his chisel alone was lauded by an (anonymous) poet in a sonnet probably written for the chapel's inauguration:

> *To the spectator.*
> *On the marvelously carved vestments of the same cardinals.*
> Look how, under their purple garments,
> The great souls and mighty prayers of the fathers ennoble the theater.
> Behold, their brilliance is visibly reddened by a secret dye,
> When the holy purples appear in Parian marble.
> But why do you wonder that an artist has carved colors,
> When the soul of all these heroes is carved here too?[88]

But these are merely supplements to the main event. Only the dialectics of the paragone explain why the central effigy of *St. Theresa in Ecstasy* transgresses conventional artistic boundaries. Like every sculptor who chose to walk in the footsteps of Michelangelo, Bernini knew that sculpture in stone was the art of subtraction, not addition, and all the effects of Theresa and the angel are extorted from a single block of stone. Yet the achievement is illusory because, although the figures are freestanding, they are not in the round. As in a painting, only what meets the eye is given form; Theresa has no right arm and she is as hollow as a tin mold. Yet the figure group is not a bas-relief either, as it has no backing slab; Theresa, mobile and pedestal-less, traverses her own dedicated volume, in her own light. Bernini has given the earlier idea of relief-as-painting an extra dimension by transforming "sculpted painting" into "built painting."

Theresa's cloud escalator itself belongs to painting, derived from scenes such as Giovanni Lanfranco's *Vision of St. Theresa Receiving the Rosary from the Virgin Mary* (fig. 12.25). The discrete lighting, which makes the sculpture a tableau, instead comes from works such as Caravaggio's *Conversion of St. Paul* (1601), where hiding the light-source offstage only augmented the mystery of its origin.[89] Seventeenth-century thought still differentiated between the invisible light of God's essential nature (lux, or "uncreated light"), and its product, the light we see, the light that makes the world intelligible at all (lumen, or "created light"). To this end, Bernini shunted daylight down a concealed flue over the tempietto, and it was (once) filtered through an ocherish-glass laylight to tinge the air around Theresa with an amber glow. This colored plasma seemed to materialize in the gilded spokes fanning out behind the statue, because when lux enters the realm of the senses it must become a *thing* in the swim of lumen; yet these spokes are themselves dissipated by

FIG. 12.23. Gianlorenzo Bernini, Reliefs with Cornaro Cardinals, side walls, Cornaro Chapel. Rome.

the immaterial glints of their burnished gold. Bernini becomes a second Vulcan, the divine smith who had forged the lightning bolts of Jupiter.

The dialectic between painting and sculpture also saturates the heavenly environment that envelops all the sculptures. At its completion, the chapel was a brilliant mirage of color in a church whose nave and vault were originally plain white and whose floor was clay (fig. 12.26). As an (anonymous) observer remarked, when Bernini later (1652–61) completed Raphael's Chigi Chapel (see fig. 10.14), "in the disposition of colors and markings [of the marbles] Painting becomes the companion of Architecture, as both collude with Nature" (*in colorum, et macularum dispositione Architecturam Pictura concomitata est, cum utraque colludente Natura*).[90] In the new coalition of the arts at the Cornaro Chapel, the absence of an altarpiece and any color at the heart of the chapel even necessitated a painterly approach to the remainder. The whole chapel is now a canvas, in which light and dark veining "bleed" from one stone into another like the brush trails of wet-on-wet painting. Moreover, the main trabeation, the backgrounds of the Cornaro reliefs, and the tympanum of the aedicule are not actually marbled but utterly deceptive imitations in stucco lustro; not only this, but the marble panels of the walls, which seem single slabs of exhilarated veining, are in reality pieced together from dozens of shards. So "painting in stone" was a practice, not just a concept. Even the columns have inserts, artificial xenoliths, to improve their veining. This was all necessary because the chapel was predominantly clad with spolia, and choice specimens were increasingly harder to find, but even in antiquity these mosaics of stone camouflage had been an art in their own right (see fig. 4.34). A year or two after the Cornaro Chapel was completed, Cosimo Fanzago would fashion

FIG. 12.24. Pietro Bernini, *Assumption of the Virgin Mary*, 1606–10. Carrara marble. S. Maria Maggiore, Rome.

FIG. 12.25. Giovanni Lanfranco, *Vision of St. Theresa Receiving the Rosary from the Virgin Mary*, between c. 1615 and c. 1618. S. Giovanni a Capo le Case, now Convento delle Carmelitane Scalze, Via della Nocetta.

FIG. 12.26. Guidobaldo Abattini (attr.), *Cappella Cornaro*, c. 1651. Oil on canvas, 5 ft. 6 in. × 3 ft. 11 in. (168 × 120 cm). Staatliches Museum Schwerin.

FIG. 12.27. Cosimo Fanzago, Paliotto, Cappella di S. Antonio, S. Isidoro, c. 1654. Alabastro di pecorella. Rome.

a paliotto for another chapel, from *alabastro di pecorella* ("lamb's fleece alabaster"), in a consummate performance of the technique (fig. 12.27). The cross blossoms as if from book-matching, but close inspection reveals that there are no orthogonal seams, and that the whole thing is again a jigsaw of hundreds of fragments. One friar acutely concluded, "It exhibits a contest between Natural Art and Artful Nature" (*naturalis artis et artificiosae naturae certamen exhibet*).[91]

The only contemporary comment on the Cornaro Chapel's revetment comes from another sonnet by our anonymous poet:

> *To the same Bernini.*
> *On the stones, and especially the mottled ones, that are marvelously arranged in the same Chapel of Saint Teresa.*
> Your stones, which will make their mark on the theater
> with their varied renown, cheer themselves with proud prayers.
> And the mottled pile is not bashful of the shining mass,
> If a shapely plague bursts forth from the precious shadows,
> When you command the blots to serve sacred splendor.[92]

The last couplet oxymoronically characterizes the marble veining as a "shapely plague" (*formosa lues*), which seems a discordant note in an encomium until one realizes that allusions to the blotching of contagion were art-critical shorthand for the disgregation of form by shadow in Caraveggesque painting.[93] Critics used the term pejoratively, but for artists such as Bernini the *macchia* ("*macula*" in the poem) invoked the gestalt of blots and stains that reached back to Leonardo and suggested unmediated invention, fantasy, and, of course, painterliness. Thus, Bernini passingly remarked to Chantelou (1665) that in large-scale works one should proceed by composing in masses (*macchie*), but when the art theorist (and Bernini's biographer) Filippo Baldinucci (1624–1697) published his theoretical lexicon of art in 1681, he specifically related the macchia to marbles. The macchia summed up, he said, "the quality of some drawings, and sometimes paintings too, made . . . in such a way that it almost seems that they were made not by the hand of an artist, but had appeared by themselves on the sheet or canvas, and they say 'this is a beautiful *macchia*.' In varicolored stones *macchia* signifies that color, which appears to overlay the background color; and one therefore calls such stones 'mottled' [*macchiate*]. . . . In resemblance to these stones, one calls *macchie* also those various colors that artfully mottle sheets that are called 'marbled' [*marezzati*]."[94]

Gian Paolo Lomazzo had opened his influential *Idea del Tempio della Pittura* (1590) by exalting painting to the heavens because God himself painted nature. Tasso seconded the tenet in his *Le sette giornate del mondo creato* (1592/94) and, more pertinently, Anselm De Boodt's highly influential 1609 treatise on gems explained the generation of streaky marbles with the metaphor that nature had mixed them from the elements just as painters mixed primaries on their palettes.[95] Bernini also paints as much with the elements as he does with stones, at one moment tailoring the marbles to form pictures (opus sectile), at another allowing the veins of the stones to picture themselves: in other words, moving between "natural art" and "artful nature."

To put this in visual terms, the Cornaro Chapel features every artifice of inlay, from the bone-upon-bone skeletons in the floor; to the black-and-yellow marbles that imitate hanging cloths; to the jaspers that are sliced and bedded to capture the sheen of velvet cushions; and, par excellence, to the sham side doors (see fig. 12.23), where a very costly alabaster slums it in the guise of olive root, thereby subjugating the preciousness of the material to the primacy of mimesis (or *materiam superabat opus;* Ov. *Met.* 2.5). But when the marbles discard their artful fictions, as they do on the altar wall that links the heaven-bound Theresa and descending heaven on high, they instead foreground an image of their own "geological" nature (see fig. 12.21). In 1650, modern geological classifications—sedimentary, igneous, metamorphic—remained unformulated, and the generation of stone was still vividly imagined in terms of the four elements (earth, water, air, fire). The apparent substance of the Cornaro Chapel is a flux of elemental compounds for which architecture is a scaffold. Thus, only the bones of the chapel are white: the capitals, bases, architraves, frames, and string moldings. Bernini shadows Vincenzo Scamozzi, who had stressed (1615) that these tectonic elements, and any figural reliefs too, must always be white because the natural veining of "mixed marbles" would otherwise disgregate the architecture: "However much [granites and veined marbles] create beauty by their vivacity and difference in color, in the end they will appear broken and cracked by their criss-cross of markings, here light and varicolored, there dark . . . and one should never make [reliefs and architectural elements] in 'mixed stones' with varied markings because it is clear that such diversity brings the greatest confusion to the work."[96]

FIG. 12.28. Sebastiano Serlio, *Rustic Gate XXIX*, Libro Estraordinario, 1584.

Yet Bernini's shuffling of monochrome tectonics with "animate" marbles seems to take a leaf from Sebastiano Serlio's earlier proposition of a "mixed" style in architecture (fig. 12.28). Explaining his design for a gateway, Serlio argued, "A single style will be made from a mixture, representing something that is partly the work of nature and partly the work of artifice: because the columns that are bound by rustic stones, and the architrave and frieze interrupted by voussoirs, reveal the work of nature, but the capitals and parts of the columns, and so also the cornice and pediment, represent the work of hands."[97] Serlio's architecture is incised into the rough stone, but rude nature fights back and the bestial maws in the spandrels threaten to consume the portal. In Bernini's confection the marbles interweave instead without dissent, as their luminous veining wanders and the stones seem to contain the seeds of their own dispersion. As nature in action (*natura naturans*), the Cornaro altar wall progresses from an ashy and earthy floor, through a muddy plinth, into a medley of autumnal hues and gassy veils, and up to the vault, where the clouds part and we look beyond the crystal ether at a pure and golden light, that is lux, light increate, God himself.

However, divine light is not only illumination, it is fecundating, it is heat. Its scanty survival in matter still retains that spark of divinity, which will ignite Theresa's heart with the fire of love (her *incendium cordis*). This blaze already figures at the chapel gate where a particularly "fiery" slab of *breccia di Sciro* is juxtaposed with a black-lacquered door on which gilded hearts visibly flame. It flares up in the flame-like folds of drapery that engulf Theresa's body, and recurs throughout the chapel in that counterchanging between black and orange that evokes ash and ember. Even the angel seems a heavenward-spiraling plume of vapor.[98] The spiritual fire that consumes Theresa is, as Kircher expounds in his mammoth tract on spiritual and physical light, the *Ars Magna Umbrae et Lucis* (1646), "the furnace of divine light . . . that purifies body and soul . . . through the hardy exercises of contemplation . . . the more to approach the divine light . . . so that, once purged and rendered incandescent by this light, [the devotee] in turn acquires the force to heat and illuminate."[99] To borrow from the terminology of modern thermodynamics, the divine light catalyzes a "phase change" (alchemical "sublimation") within the materials, from the solid earth to the dispersing vapors of the superplastic heavens, and from the absolute zero of the tomb to the white heat of lux. As lux had ever been an "increate" and originary essence, it was remote from the tribulations of sublunar matter (birth-growth-decay-regeneration) and immutable.

In other words, the "material imagination" of premodern geology supplies the visible register for the transition from death to resurrection and earth to ether. Thus combined, the stones reacquire a state that can only be imagined, where the material returns to its thermoplastic state and matter is transformed before the eyes. The middle ground in the transformation, from earth through air to ether (the gamut of substance), is the image of cloud that dominates the intermediate zone of the chapel; from the chiseled cumulus below Theresa's feet to the swirling patterns in the wall slabs of *alabastro nuvoloso* on either side of the aedicule. As we know, cloud was the sign of Judeo-Christian theophany, and it would come to fill Early Modern painting, whether altarpieces or domes or stage sets.[100] And it was also the stuff of dreams. Marble is a medium not just a material, and alabastro nuvoloso especially: if we look at a panel in the painting-on-stone genre, the veining materializes as cloud at the top and trodden earth at the bottom, such that the stone is literally the middle ground for the subject, the conversion of St. Paul on the road to Damascus, that is a conversion by light (see fig. 12.13).[101]

This concern with cloud also makes sense of Bernini's innovations in stucco beyond illusionism for its own sake. First at the Cappella Pio (1644) in S. Agostino, and later more confidently in the Cornaro Chapel (see fig. 12.21), Bernini had Guidobaldo Abbatini (1600–1656) fresco glories of angels on stucco clouds that roll apart the vault and spread over the architecture and bas-relief. The painter Giovanni Battista Passeri called this illusionism a "new deceptive artifice," which "by means of certain parts in relief makes appear effectively real that falsity which is feigned" (*un arteficio nuovo, che inganna, e con certi riporti di rilievo ha fatto parere vero effettivo quel falso, che è finto*).[102] What he fails to say is that, by giving truth to this fiction, Bernini gives volume to what is most insubstantial: cloud, light, and angels (which are creatures of light, and who only assume substance as condensed air).[103] Furthermore, Bernini's subversion deconstructs the picture plane by converting illusionist recession into physical intrusion. For he endows the deceptions of painting with a physical *rilievo* superior to those real reliefs it overlies, while the illusion of almost infinite depth remains an achievement of the brush and not the trowel. And if Bernini makes painting physically (though in essence immaterially) penetrate the space of architecture, it only complements painting's simultaneous inherency in the marbles themselves.

It remains briefly to consider what the implications of this dialectic between illusionism and material image held for the chapel as an ensemble. It has long been asserted that Bernini's special contribution was to achieve a "unity of the arts," or Gesamtkunstwerk, or *bel composto* ("beautiful composite"). But the arts had always worked in untroubled partnership or unison. Bernini's only recorded statement on the "bel composto" was that carving drapery had been "the greatest prize of his chisel, with which he had overcome the difficulty of rendering marble as malleable as wax, and that he had thereby known how to combine in a certain way painting and sculpture."[104] The eye does not simply apprehend form but also perceives substance (the "material imagination," or elemental imagination), and neither alone, and it is craft or art that mediates any transformation. Bernini indeed emphasizes that it was the transformation of the material (treating cold stone as if it were warm wax) and thence the implicit transformation of facture (from carving to modeling) that allowed sculpture to become painterly. It is not just the eye that judges, but the hand that knows. In other words, the matter of representation was also the matter at hand, for *materia* homologously indicated both a physical substance and an "immaterial" theme or argument. The same duality is at play when the artist's son, Domenico Bernini, tells us, "In the matter of architecture, the Cavaliere made it a rule for his pupils that it was first necessary to reflect on its matter, thereafter the invention, the arrangement of parts, and finally give them the perfection of grace and delicacy."[105] In other words, the matter of representation was also the matter at hand. The Carmelite brothers did, in fact, proclaim that Bernini had "rendered stones animate and colors eloquent."

The matter of Theresa's transverberation is her "striking through" by lux. This striking through makes her angelic encounter a second Annunciation—the precedent for Bernini's sculptural composition—whose own matter is the Incarnation of divine essence in mortal substance, a historical event that is enacted on the altar below in the daily Transubstantiation. We might conclude that the creative imagination was so embodied in matter that ideas could only surface in substance, rather than simply being reified as material objects—that the dialogue between matter and vision was in fact an analogue of that between the worldly and the divine.

Nonetheless, scholars have called the Cornaro Chapel an "enormous picture" or "three-dimensional picture," a picture that comes into sharp focus when we compare the well-known painting (c. 1649–52) of the chapel probably executed by Abbatini (the man who painted the vault) with a perspectivally adjusted photo (see figs. 12.26, 12.21).[106] This pictorial vision does not contradict Bernini's pronouncement that "architecture corresponds more to sculpture as relief, than to painting as a work of illusions and shadows," because architecture and painting had shared the same cognitive armature since the early Renaissance, perspectivity.[107] Between painting and architecture lay the deceptions of scenography, which created whole worlds onstage and, through the ephemera of the *Quarant'ore* (the forty hours' devotion of the Holy Sacrament), the heavenly otherworld as well.[108] In the Cornaro Chapel, painting is the reconciling vision for all the arts, because it is the means of picturing them together, but also because everyone agreed that it was the least haptic and therefore the most visionary of the arts (Leonardo had called it "a miracle").

However, the truth of the work of art and its capacity to sustain the vision could not remain hostage to the precarious veracity of illusionism alone. Although Baldinucci asserted that Bernini "was the first to attempt to unite architecture with sculpture and painting in such a manner that together they make a beautiful composite [bel composto]," in the mouth of Bernini's son Domenico, the same phrase instead referred to the artist's capacity "to unite together the fine arts of sculpture, painting, and architecture, in such a way that he formed *within himself* a marvelous composite [*maraviglioso composto*]."[109] The distinction is important because the ultimate cohesion of the arts in Bernini's hands lay not in their mere conjunction, but their ability to impersonate and perform as each other. That is, sculpture behaves (through the ambiguity of rilievo) like painting, colored architecture ranges between the two, and painting steps outside its frame to become statue. We can no longer decide whether sculpture is engulfed by painting or emerges from it. In fact, the distinction itself has become an idle question, surrendering to that autonomous "intermediality" that verifies the vision-world created, endows it with its own ontology, and thereby nullifies any issue of truth or deception.

CLOUD ARCHITECTURE: S. ANDREA AL QUIRINALE AND OTHER WORKS BY BERNINI

Bernini would indulge the potential of elemental images and metamorphic marbles within a decade, in the Church of S. Andrea al Quirinale (1658–71; fig. 12.29).[110] His understanding of the potential of marbling had been broadened at St. Peter's, where two newly discovered marbles, one foreign (*rosso di Francia* from Languedoc) and the other homegrown (cottanello from the Sabina), permitted him to clad the nave (1649–50) on a scale unknown since Byzantium.[111] Europe had never seen the like of them, one erudito claimed, and the pope had outdone Constantine. The new expanses of marbling even made the earlier Cappella Gregoriana look like "the work of ebanists," Virgilio Spada said.[112]

At S. Andrea al Quirinale, Bernini again exploited cottanello, and also Sicilian jasper (*diaspro di Sicilia*). Jasper was the apple of Bernini's eye ("of all modern stones, this is the most beautiful," he said), from which he formed the socle of the Cathedra Petri (1657–66), and which he now urged upon the Jesuits at S. Andrea. These newly quarried marbles could sheathe the church all over and provide monolithic columns of monumental dimensions, and Bernini went to the luxury of having them cut in thick blocks, not thin veneers, so that their faces could be dressed into a curve and the walls of the oval-plan church appear seamless.[113]

FIG. 12.29. Gianlorenzo Bernini, Church interior, S. Andrea al Quirinale, 1658–71. Rome.

With the money of Prince Camillo Pamphilj and unlimited supplies of cottanello and Sicilian jasper, Bernini reduced the graded coloristic progression from earth to ether, which had characterized the Cornaro Chapel, to a binary at S. Andrea: the interior is divided between a drum of rose-and-rust marbles and the gilded albumen of the vault. The two worlds intersect as the sunburst of ribs in the dome "earths" and hardens into the white pilasters overlaying the drum, rather like the spokes that spread behind the statue of St. Theresa. Underlying these pilasters are piers of cottanello, a roseate stone interspersed with wispy white veining that seems to condense light and earth (fig. 12.30). One must say "condensation," because to contemporary eyes few marbles projected the images of one element alone: all stones were images of elemental transmutation, and their elements always came to visibility suspended in precarious balance or promiscuous flux.

Thus, according to Kircher, the Aswan granite obelisk that Bernini was erecting in Piazza Navona for the Pamphilj pope in 1650 figured a solidified sunbeam (as "obelisk" means "finger of the sun"), and its granite embodied everything under the sun (the world as the commingling of the four elements). Likewise, a canon in far-off London observed that the great alabaster rota at Westminster Abbey (see fig. 6.26) "has within it the colors of the four elements of this world, that is of fire, air, water and earth."[114] Kircher goes on to explain how the obelisk's flared form gave shape to the trickle-down of lux into the world, and his diagram shows how a single point of light (God) first engenders the world as prima materia before that divides into the four elements, which disperse to the four corners

FIG. 12.30. Cottanello columns, S. Andrea al Quirinale. Rome.

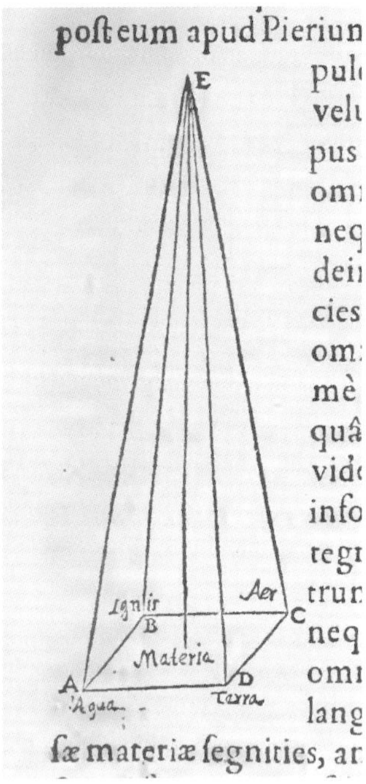

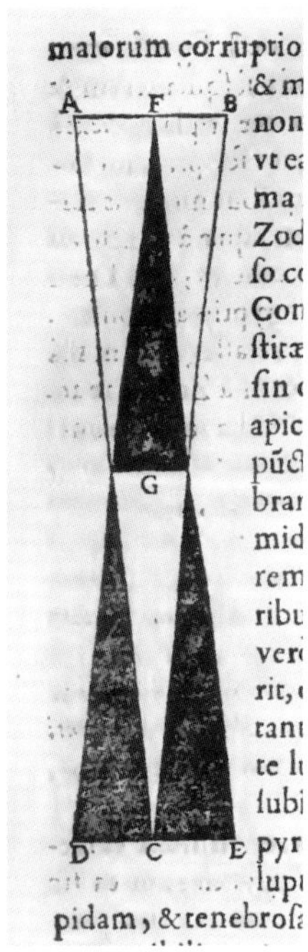

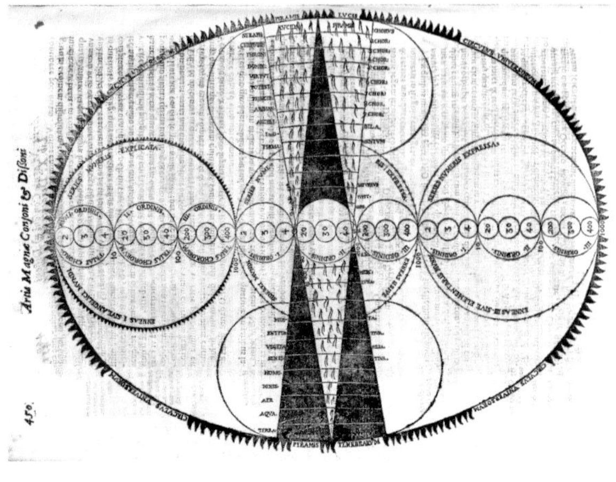

FIG. 12.31A–C. Athanasius Kircher a) pyramid of the diffusion of light into matter (*Obeliscus Pamphilius*, 1650); b) "Pyramid of Light and Pyramid of Darkness within the Circuit of the Universe," (Obeliscus Pamphilius, 1650); c) "Hierarchy of the Elements and Angelic Choirs" (*Musurgia Universalis*, 1650).

like a compound in a centrifuge (fig. 12.31a). Kircher finally marries form and matter in a diagram of intersecting light and dark pyramids (fig. 12.31b). The light pyramid represents the descent of divine light into dull matter, the dark one the degrees to which that same matter aspires to its brilliant origin. This works for matter and humans alike, for humans are body-soul composites of heaven and earth, and the light pyramid therefore stands in for the soul, and the dark one the body. The base of the pyramid of light resides in the divine realm beyond the zodiac and touches earthly things only at its point, and the pyramid of matter rests on the earth and only grazes the orbit of immortal things (fig. 12.31c). In short, the world is an admixture that lies between utter darkness and blinding light; all existence is an equilibrium; and fickle matter hangs in the balance of the four elements that decide the heat, weight, and inherent direction of each and every substance.

Bernini had no need to read Kircher's book, as the four elements and the Aristotelian world system still belonged to common lore. However, a figurative confirmation comes from Raphael's Chigi Chapel in S. Maria del Popolo, which Bernini was seeing to completion in the same years (1655–62) that he was building S. Andrea al Quirinale (1658 on). Here, the combustive conversion of matter is represented by the ruddy-and-gray marbles of pyre-like pyramids that smolder and fume up toward the blue-and-gold dome that contains the ether, zodiac, and mansions of the Lord (see fig. 10.14).

The fascination of cottanello and its wholesome power to suggest the admixture of the elements is also explained by Borromini, Bernini's archrival, in his account of another sepulcher, that of the prelate Clemente Merlini (1655–58; fig. 12.32):

> What stands out in this small work is its strength, tricking the eye that it is a single stone which has been carved whole in the quarry. The work's efficacy is also conspicuous because it is not susceptible to the injuries of time, as are inlays or panels of diverse and subtle little stones: it teaches, as Vitruvius says, that Nature prevails over ornament and, in the end, encloses within itself the qualities of the subject for whom it was made, which in this case is Monsignor Merlini Auditor of the Sacra Rota Romana, whose fortitude and constancy in administering justice with rectitude was combined with the purity and innocence of his morals. Therefore, a perfect architect would only have erred had he not chosen a red and white stone, wherein the red signified that prelate's fortitude, and the white his innocence.[115]

Domenico Bernini would have said that Merlini was a "marvelous composite," Shakespeare that his "elements were well mixed." Finally, two elements are privileged above others: as Cleopatra says, "I am fire and air; my other elements / I give to baser life" (*Antony and Cleopatra* V, ii, 288–89).

In Kircher's mind, lumen is no simple emanation of lux but an extension of it (*productio*), and therefore luminosity is not a quality of a material but evidence of the operation of lux within it.[116] So too the darkening palette of the marbles as they near the ground on the altar wall of the Cornaro Chapel charts the descent of lux into the world, and its progressive dimming within matter. This pictorial shading does not simply represent light cast on a surface (lumen). That the vivifying lux infuses the substance itself is further apparent at S. Andrea al Quirinale, where the blue mosaic below the Gloria downwardly diminishes in saturation (fig. 12.33) and the effect inverts a real sky, which is saturated at its zenith, and resembles more the diffusion of light one might perceive at the bottom of a swimming pool.

If the white flecks and veins of the marbles in S. Andrea al Quirinale can suggest the entrapment or infusion of light in dull matter, then white stucco should conversely materialize light's emancipation and emission (the desire of disembodied lux to return to its divine home). The souls of the Cornaro cardinals are white, as are the angels on the proscenium arch, as also is the swooning, out-of-body St. Theresa. The stucco whiteness of St. Andrew's soul, which either emerges from cloud or is

FIG. 12.32. Leonardo Sormanzi, to the design of Francesco Borromini, Tomb of Monsignor Clemente Merlini, S. Maria Maggiore, c. 1655–58. Rome.

distilled from the nebulous marbling, makes all the more sense when one remembers that Guillaume Courtois's high altarpiece (1668; see fig. 12.33) illustrates a passage in *The Golden Legend* and the liturgy for St. Andrew's feast day (November 30). As Andrew hung on the cross "drenched in brilliant light" (*luce perfusus*), he offered up a final prayer that he might cast off his earthly body as though it were an old and heavy coat, before "mounting, in this light . . . to the realms of light" (*cum luce . . . pergit ad lucis atria*).[117] The transformation for which he prayed is arguably the concetto for the whole church. Like sfumato, a smoky "cloud-stone" resolved the paradox of presenting an insubstantial heaven through worldly materials, and the stony cloud that envelops the interior both elevates and grounds us. Whether we look into the dome or the apse, the same holds. We are either in a cave looking out, or the clouds are parting to disclose the divine revelation.

S. Andrea is not a tableau like the Cornaro Chapel, for in this arena we are as much protagonists as spectators (we walk onstage, so to speak). Nor does it have the painterly effects of the Cornaro Chapel, or any of its knowing cues to the paragone. Nonetheless, the marbles do figure some foggy substance, which readily evokes vapor, if not cloud. Even the slabs of jasper he picked are uncharacteristically hazy, and in the frieze and tympanum, cumulus and cirrus patterns become more obvious thanks to painted (or stucco lustro) simulations. Clouds were as indexical of God's presence as smoke is of fire, because without them his unfiltered essence, lux, would have been fatal in its effulgence. Furthermore, protean clouds, like mottled marbles, were also a major venue for providential images, even whole castles in the air: one Jesuit describes the shifting fata morgana he witnessed off the coast of Calabria, on the Feast of the Assumption: aerial porticoes that transformed into aqueducts, and the next minute whole castles, which soon dissipated into single towers, then into a piazza of colonnades, which morphed into a windowed façade, which mutated into pinewoods and cypress groves, before melting away with a gust of wind.[118] The dissolving luminescence and veining of marble had always allowed the liturgical drama to push beyond tangible boundaries, but when its innate image was the utterly liminal cloud, the senses could participate in a material whose identity was ultimately as elusive as the essence it sought to impersonate. Beholders who occupied the center of such a permeable perimeter could assure themselves that they really nestled in a spiritual embrace.

FIG. 12.33. Bernini and Guillaume Courtois, Choir and altarpiece, *Martyrdom of St. Andrew*, S. Andrea al Quirinale, 1668. Rome.

It is also through the image of cloud that the whole evanescent construct reverts to the "atmospheric" dome painting of Giovanni Lanfranco (1582–1647), and before him Correggio (1489–1534), because the cloud had become one of the most important weapons in the Baroque pictorial arsenal. Although Walter Benjamin would rail against Baroque painters who "did not so much transfigure the world . . . as cast a cloudy sky over its surface," artists such as Lanfranco actually drew their nebulous filters over Baroque domes precisely in order to intimate God's proximity to create a heaven just for us.[119] The book of Job (22:14) puts it nicely: "Thick clouds are a covering to him, that He seeth not; and He walketh on the vault of the heavens."

"A painter who has not done a cupola can hardly be called a painter at all," Bernini remarked, and he followed the example of dome painters by setting the transcendent St. Andrew on the horizon between marble drum and stucco sky.[120] For his flying St. Andrew is nothing more than a translation into sculpture of a figure like the Virgin in Lanfranco's dome of S. Andrea della Valle (1625–28; fig. 12.34). It is as though Bernini had merely subtracted the nave and transepts, and one walks straight into the heavenly drum. A later modello for a vault by Bernini's preferred painter, Baciccio (1639–1709), summarizes the concetto of S. Andrea al Quirinale in two dimensions as a threshold between brilliant epiphany and umbrous world (fig. 12.35). St. Ignatius assumes the same pose as Andrew, the three-quarter profile that implies that only the saint receives the full vision of the Godhead, as the catechism stipulated, and makes him our intermediary. With

FIG. 12.34. Giovanni Lanfranco, *Assumption of the Virgin*, 1625–28. S. Andrea della Valle, Rome.

FIG. 12.35. Baciccio, *Apotheosis of St. Ignatius*, c. 1685. Bozzetto. Galleria Nazionale di Arte Moderna, Rome.

this distancing in mind, we may reinterpret the drum of S. Andrea al Quirinale (that is, the corps of the church below the dome), which is somber but flecked with light, as underbelly, as shadowed cloud, even—with Kircher's obelisk in mind—as a substance cast by the Gloria of the vault. And from the same perspective, we may also regard the transverse oval plan of the church—which has none of the anamorphic benefits that attend a longitudinal oval (namely, the capacity to appear circular from its entry point)—as a projection of the celestial circle into a sublunar oval. As all geometers knew, a continuity between the divine figure and its terrestrial reflection, between circle and oval, existed in projection because both circle and ellipse (which an oval approximates) are conic sections. They are slices of a continuum. This last fact is symbolically tantalizing because it coincides with an emanatistic view of Creation, of medieval origin, in which God is the apex from whom all created things are descending radiations and concretions of light in various degrees.

In two chapels exactly contemporary with S. Andrea al Quirinale, Bernini again figured clouds in marble, in the Fonseca Chapel at S. Lorenzo in Lucina (1662–65) in cottanello, and in the De Sylva Chapel at S. Isidoro (1660–63) more overtly in alabastro nuvoloso (fig. 12.36).[121] In the latter, marble clouds fill the marble paliotto, and the putti bearing Maratta's Immacolata emerge from them; fresco clouds extend along the side walls behind the De Sylva tombs, where hovering putti toy with skulls, flowers, and upturned and downturned torches. In toto, all these fictive clouds coalesce into a divine fog that rubs its back against the picture plane. The unitary implication is that there is no wall, rather the dimensionless hinterland of the *al di là* from which the allegories emerge leaning over a *diaspro* parapet. We occupy the center and are sheltered by a baldacchino. This is again a funerary chapel to host daily masses for the souls of the departed, and this eschatological meaning comes into sudden focus in the revelation on the paliotto, and its allusion to Matthew, that on the final day "they shall see the sign of the Son of man

FIG. 12.36. Bernini, Cappella De Sylva, S. Isidoro, 1660–63. Rome.

coming in the clouds."[122] Since this paliotto appears below Maratta's Immacolata crushing the serpent of original sin, the juxtaposition is a symbolic summa, what could be termed a "brief history of time."

Bernini had substituted the bijou world of the Cinquecento chapel and its gemmy marbles reflecting celestial virtues with a "cloud architecture" that brought heaven to earth. In the process he also abandoned the "eternal paintings" of intarsiated clouds with which he had begun (see fig. 12.20), in favor of a whole building made from what seemed petrified vapors. One is reminded of Galileo's earlier dismissal of Tasso's poetry as conceptually discontinuous as marquetry, and his praise of Ariosto's as enjoying a sweet and sfumato fluency; while Ariosto's poetry flowed, in Tasso's you could really see the joins.[123]

CONCLUSION

Bernini's various innovations in "sculpted painting" and painterly—but also elemental—architecture echo through the works of his successors. Within a decade of the Cornaro Chapel, Melchiorre Cafà (1636–1667), whom even Bernini acknowledged was as dexterous with the chisel as himself, would build the High Altar of S. Caterina da Siena a Largo Magnanapoli (c. 1662–64; fig. 12.37). Cafà's shimmering *St. Catherine in Glory* (re)immerses Bernini's Theresa within an intarsia background obviously borrowed from the "eternal paintings" in the reliquary balconies of St. Peter's (see fig. 12.20).[124] However, Cafà's backdrop of lapis and alabaster is made not from the choicest specimens (as in the Cappella Paolina) but impure blocks (the lapis is streaked with yellow veins), and they have not been honed to a level surface although they are polished. The result is an impasto that evokes cloud and vaporous sky, not only in palette but also texture. Bernini's pupil, Carlo Fontana, also took the idea of "painting in stone" when he executed the Ginetti Chapel in S. Andrea della Valle (1670–75). This was built within the same shell that Virgilio Spada had abortively sought first to clad completely in pietra di paragone and white bas-reliefs, and then with massive quadratura inlays (see fig. 12.17). Like Virgilio's lost chapel, in Fontana's version there are no paintings, the altarpiece is a massive relief, and the side walls feature white marble effigies of the Ginetti in eternal adoration silhouetted against great slabs of black stone. Fontana not only marbled the pendentives (like S. Maria dei Miracoli two centuries earlier) but went one step further by giving Rome its first marbled cupola, capped by a lantern with "a beautiful clump of clouds, but made without any work of the brush . . . [only] from pure alabaster, pieced together [*or* conceived] in such a way that in the same place it deceives and consoles the eye" (*un bel groppo di nuvole, ma senz'opra di Pennello . . . da puro alabastro, congegnato in modo, che nel punto stesso delude, e consola la vista*).[125]

The Cornaro Chapel also influenced those painter-architects who conceived chapels and their altarpieces as one optical whole. In the choir of S. Maria Maddalena de' Pazzi in Florence (1679–85), Ciro Ferri painted the altarpiece and designed the chapel around it. When it was unveiled to universal praise in 1685, the artist boasted that he was "not surprised at all . . . for the entire time he was painting he had the Chapel, the colors, luster, and harmony of its stones before his eyes," such that "outside its setting the painting would have lost as much as it had acquired by being placed there."[126] The whole ensemble was also visible through the proscenium of a nave with white walls, and this view must have been conceived much like a painting itself (fig. 12.38).

FIG. 12.37. Melchiorre Cafà, *St. Catherine in Glory,* c. 1662–64. S. Caterina da Siena a Largo Magnanapoli, Rome.

FIG. 12.38. Ciro Ferri, Choir, S. Maria Maddalena de'Pazzi, 1679–85. Florence.

Bernini's influence spread beyond the Alps, first to Bavaria. The brothers and painter-architects Cosmas Damian Asam (1686–1739) and Egid Quirin Asam (1692–1750) were trained in Rome, where they metabolized the latest innovations, especially the work of Bernini. In the oval vestibule of the Asamkirche (1733–46), their private church-sized chapel in Munich, an embossed and gilded sun sprinkles its rays like molten droplets onto a cloudy backdrop (fig. 12.39) and, in their mind's eye, the Asam brothers returned to the reliquary balconies of St. Peter's (see fig. 12.20). However, Bernini's most enduring influence was through S. Andrea al Quirinale. The kaleidoscopic cloud architecture of this church is the governing metaphor, at a generation's remove, behind much of the *Stuckmarmor* (or stucco lustro) cladding of Bavarian churches and others that occupy a swathe of Europe from Einsiedeln to Prague.[127] The clearest translation is the Jesuitenkirche in Vienna (1703–5) by the émigré Italian painter-architect Andrea Pozzo, but more painterly reinterpretations are found in the Bavarian pilgrimage churches, most notably the Abbey Church at Ottobeuren (1737–66) by Johann Michael Fischer (1743–72; fig. 12.40), the Vierzehnheiligen near Bad Staffelstein by Balthasar

FIG. 12.39. Egid Quirin Asam and Cosmas Damian Asam, Narthex ceiling, St. John Nepomuk, 1733–46. Munich.

FIG. 12.40. Johann Michael Fischer, Basilica of Sts. Alexander and Theodore, Benedictine Abbey, 1737–66. Ottobeuren.

Neumann, and the Wieskirche (1745–54) at Steingaden by Dominikus and Johann Baptist Zimmermann.

Now the columns are clad with marbles that cannot be found in nature. The pastel clouds and ever more Tiepolesque palettes of the altarpieces have dilated beyond their frames to drift across the walls, suffuse the columns, and fill the volume. This "marble" is Stuckmarmor, anhydrite that has been leavened with diluted bone-glue, colored with earth pigments, and the dough kneaded into a marbleized loaf, which is sliced into veneers that, once hardened, can be ground and polished. This technique descends from that which the artisans of Nuzi used to make a goblet for the wife of Thutmosis III four millennia earlier (see fig. 1.5). The Bavarian craftsmen veined and mottled their "marbles" with complete freedom, according to their fantasy, as masters of their medium, mindful of its place in a unified artistic world. The clouds they imagined were not carved or modeled, nor were they simply applied pigment, but they were painted in stone.

Epilogue

The history of architecture in colored marbles does not stop in eighteenth-century Bavaria, but this is a convenient juncture at which to discontinue the narrative of this particular book. The themes of chthonic-celestial matter (an elemental bridgehead between heaven and earth), the tropological density of signifying stones, the image within the stone, and "natural painting" all ceased to hold the pivotal importance they had in preceding ages.

This is not to say that marbles diminished in popularity. In fact, their use multiplied. The first half of the eighteenth century saw increasingly delicate and colorist palettes deployed in church and chapel interiors, not only in Rome, but also northern Italy, where many new marbles were quarried to vie with the old ones, and then farther afield. By the nineteenth century, it had become increasingly common to sheathe new and old churches in marble revetment, and eventually palaces and civic buildings as well, but often leading to a dampening uniformity rather than a scintillating vitality. Simultaneously, colored marble lost the powerful dialectical role it had once enjoyed as it gradually became an ornamental adjunct to an architecture that increasingly (from Charles Perrault on and, more dogmatically, from the time of Marc-Antoine Laugier) stressed structural fundamentalism. When the ardent Neoclassical theorist of architecture Francesco Milizia (1725–1798) discussed colored marbles in his *Principles of Civil Architecture* (1785), he did endorse the "diversity of marbles," and stressed, "The diverse coloring of marbles requires of the architect the intelligence to combine them in a work, such that a painterly harmony [*accordo pittoresco*] results . . . for which reason it suits the architect to learn from a good painter a harmonious palette [*accordo de' colori*]." However, he adds only as a dismissive afterthought that architects should exploit "that little which nature gives us."[1] Milizia's conclusions are aestheticizing recommendations for a tasteful decor, no more, no less. What has also dissipated along the way is the intermedial communion of the arts, of which the paragone "debate" (it seems to have been less a case of dueling arts than a ritualized academic argument) was only a self-conscious theoretical moment. Aesthetics (the criteria of a modern philosophy of "beauty") and "taste" begin to rule, and the secret life that made a slab of marble a "painting in stone" was ebbing away. Architecture was also becoming a distinct profession, one that still had, and has, a foundation in the fine arts; but those selfsame arts were devolving into academic categories, and architecture was slowly ceasing to be a field in which artists, whether they be painters or sculptors, might be called upon to work and add their medial perspectives.

The idea that stones held magical properties and consonances with the astral realm also fell foul of the Enlightenment (though popular belief in their therapeutic and astral character never died out completely, resurfacing strongly in the late twentieth century with "New Age" philosophies and natural remedies and alternative medicines). Likewise, heaven-sent images in stones began to be recognized only as "images made by chance"—or as actual fossils—from the early seventeenth century onward, and definitively so in the eighteenth century. The taming of the marble medium was in lockstep with the advent of geology and geochemistry as analytical sciences, and the narrowing focus on the rational faculty and empirical data contrasted with and contested the mythopoetic vision of the old macrocosm-microcosm.

The rampant imagining of the world in terms of an elemental tetrad of earth, fire, air, and water that had remained basic to the understanding of matter from the fourth century BC was displaced by quantitative chemistry in the eighteenth century and definitively superseded by the periodic table in the nineteenth. After Antoine Lavoisier published (1789) a list of thirty-three chemical elements, grouped according to gases, metals, nonmetals, and earths, a succession of nineteenth-century chemists proceeded to arrange the ever-increasing number of elements in sequence according to their atomic weight, culminating in Dmitri Mendeleev's decision (1869) to arrange them "periodically" as well.[2] While the whole world that had encompassed ancient and medieval humanity could be entirely comprehended in terms of a fluctuating medium of four elements, today the periodic table numbers 118 elements and counting, only 98 of which exist in nature, the balance (99–118) being synthesized in laboratories or reactors, and some stable for a mere instant (tennessine, element 117, was only synthesized in 2010, and its most stable isotope has a half-life of about eighty milliseconds).

The nineteenth century did, however, see a revival in colored stones and architectural polychromy, and the most significant episodes took place, strangely perhaps, in England. This renewed appreciation of marble poetics came with the Gothic revival, whose loosely shared precepts were codified in the three-volume *The Stones of Venice* (1851–53) by John Ruskin, a sequel to his bestselling *The Seven Lamps of Architecture* (1849), which was itself a summary of many ideas already in the air. Reactionary, moralizing, and abhorring the technologies of the Industrial Revolution, Ruskin yearned, like Augustus Pugin before him, for a "truer" architecture, rooted in intelligent craft, truth to materials, and a structural sensitivity that aspired to the sublimity of nature. All these qualities he found and endorsed in the English Gothic, especially the Early Decorated mode. However, as English architecture had never enjoyed the marble spectrum of its continental cousins, and the Reformation had vandalized church fittings and whitewashed their murals, Ruskin headed for Venice to revel in its tapestry of materials and ornamental libido.

Meanwhile, the holdings of the Museum of Practical Geology (rebuilt 1845–49, inaugurated 1851) and the new Oxford University Museum of Natural History (1855–60) reopened eyes to the mineral wealth of the British Isles. In Oxford, each column of the double-story arcades enclosing the exhibition hall was fashioned from a different polishable British stone (Scottish granites on the west, metamorphic rocks on the north, igneous rocks on the east, and English and Welsh marbles on the south), selected by the keeper of the museum, the geologist John Phillips. The ensemble was a lithic library of the whole United Kingdom. On the one hand, the Gothic was now acquiring new associations with modern science, technology, and progress; on the other, fossil-hunting curates and weekend geologists were fascinated by the formations that illustrated "intelligent design."

The succession of stylistic phases in a Gothic church might be compared to geologies formed over time, and the popularity of brickwork with different-colored courses possibly employed the metaphor of geological stratification to supplement the old and well-known analogy between Gothic and natural forms. As Ruskin explained in *The Stones of Venice:*

> It is perfectly natural that the different kinds of stone used in [a wall's] successive courses should be of different colours; and there are many associations and analogies which metaphysically justify the introduction of horizontal bands of colour, or of light and shade. They are, in the first place, a kind of expression of the growth or age of the wall, like the rings in the wood of a tree. . . . They are valuable in their suggestion of the natural courses of rocks and beds of the earth itself.[3]

Interest continued to grow. In 1855 George Edmund Street published *Brick and Marble in the Middle Ages: Notes of Tours in the North of Italy* (1855), urging English architects to examine the particolored fabrics of (northern Italian) medieval churches for structural expression. In the same period, William Butterfield was busy building the Church of All Saints, Margaret Street (1850–59), in London. Outside black-and-red patterned brickwork banded with stone, inside it was a potpourri of diaper tiling, geometrical patterned brick, painted tile, stained glass, brass, gilded timberwork, and marble columns and columnettes. Unnoticed, unless one knelt to receive the sacrament, was the marble communion rail, for which Butterfield had chosen a particularly fossiliferous stone, either in memory of Purbeck marble or to remind the faithful that such stones were fingerprints of the Creation. In "Natural Colour in Sacred Architecture," published in *The Ecclesiologist* in 1861, one member of the Oxford Architectural Society (unwittingly?) echoed Prokopios when he wrote that "natural stone must predominate [in a church] . . . in order that we may feel, and in truth we shall feel, that the edifice recalls to our minds the Great Architect Himself. The works of nature remind us of the GOD of nature."[4]

It was a more self-conscious Byzantine revival that inspired the provisioning and use of marble when John Francis Bentley came to design Westminster Cathedral (1895–1903), the monument to the full fruition of Catholic emancipation in Anglican England. This was to be built in a neo-Byzantine style meant to diverge from the Gothic and Anglican Westminster Abbey at the other end of Victoria Street, not to mention the neo-perpendicular of the adjacent Houses of Parliament (1840–76). It was a foregone conclusion that when it came to alleviating the interior of this striped brick behemoth with marble and mosaic to make it as authentically Byzantine as possible, Bentley would turn to the stonemason and marble supplier William Brindley (1832–1902). Already in the 1880s, alive to the increasing use of marble in English architecture as well as buildings such as Charles Garnier's Opéra (1861–75) in Paris, Brindley had singlehandedly prospected and rediscovered the ancient quarries of Carystian marble (cipollino), Chian (porta santa), Thessalian (verde antico), and even porphyry (which he was unable to export) as well as exploited the already discovered quarries in the Mani (rosso antico). Brindley reported all his findings in the *Journal of the Royal Institute of British Architects* as well as *The Builder*, was elected a fellow of the Geological Society (1888), and finally collaborated with the painter Sir Lawrence Alma-Tadema, whose meticulous reconstructions of ancient marble interiors are the most accurate of the nineteenth century, on their joint *Marbles: Their Ancient and Modern Application* (1907).[5]

Bentley reclaimed the spirit of Byzantium not simply in material outlay but in the symbolism of the floor he designed, which proved, however, too expensive to execute. Under the influence of the Arts and Crafts architect and theorist William Richard Lethaby (1857–1931), he set out to salvage the idea of the spiritual sea that had pervaded the floor of Hagia Sophia (see figs. 7.1, 8.1). This Victorian reception had germinated with George Edmund Street, who had remembered the old tradition that the floor of Hagia Sophia seemed a stormy sea, but had gone so far as to conclude that the peaks and valleys of S. Marco's much-settled floor were intentional; that "instead of being laid level and even [the floor] swells up and down as though its surface were the petrified waves of the sea, on which those who embark in the ship of the church may kneel in prayer with safety, the undulating surface serving only to remind them of the stormy sea of life." Street's words had enough impact on Lethaby for him to include a whole chapter, "On Pavements as Seas," in his recently published *Architecture, Mysticism and Myth* (1891).[6] And

FIG. 13.1. John Francis Bentley, Unexecuted design for marble floor, Westminster Cathedral, 1895.

Bentley now vowed to fill the whole nave ("*navis*" means "ship") of Westminster Abbey with Proconnesian pools darting with every variety of fish that was "promised to St. Peter's net" (fig. 13.1). Nothing came of that, but the young architect Robert Weir Schultz (1860–1951)—who was an aficionado of Lethaby, had participated in an extensive photographic survey of Greece and Turkey, and published several texts on Byzantine art—did execute a watery floor for the Chapel of St. Andrew at Westminster in 1910–15 (fig. 13.2). Most of this floor is taken up by book-matched paneling, but this is skirted by a meander, signifying Ocean as the stream that encircles the earth and marks its limits, along which marble fish and crabs zodiacally swim, to clinch the particular aquatic nuance of the chapel, its dedication to St. Andrew as "fisher of souls." The design may now seem an overly literal reclamation of tradition, but in the year of its inauguration, 1915, an unforeseen prototype was unearthed at Nikopolis (Actium) in western Greece (fig. 13.3). The transept floor of the Basilica of Doumetios (between c. 525 and c. 550) is also encircled by highly realistic, mosaic streams with fish, waterbirds, and even fishermen, and conveniently the inscription spells out, "Here you see immense and splendid Ocean that holds in its grasp the earth."[7]

FIG. 13.2. Robert Weir Schultz, Chapel of St. Andrew, Westminster Cathedral, 1910–15. London.

FIG. 13.3. Basilica of Doumetios (Hagios Demetrios), Nikopolis (Actium), between c. 525 and c. 550 AD. Mosaic floor in the north wing of the transept.

The same years of the early twentieth century would witness a less conscious return to the Byzantine aesthetic of architecture as aniconic painting in the work of the architects Adolph Loos (1870–1933) and Ludwig Mies van der Rohe (1886–1969). Both were sons of stonemasons, and both used highly veined marbles to color and pattern their architecture without recourse to subsidiary ornamentation. Loos produced a serious of interiors (and one exterior, the Looshaus, Vienna, 1909–12) that were swimming in marble cladding (and wood veneers) of the most striking veining and saturated color, often book-matched, sometimes (for example, the apartment lobby of the Looshaus) in conjunction with mirrors that replicated the symmetries to infinity. As is well known, this approach was indebted to Gottfried Semper (1803–1879), who advocated for cladding and extolled surface, taking a strong position in favor of ancient architectural polychromy as well as theorizing the textile origins of architectural form (Ruskin had also spoken of the "wall-veil"). Semper even cited, somewhat incongruously, Pliny's passage "We have begun to paint in stone," to support his view that Roman architects emphasized the lines of stone and structure by carving and painting.[8] A contemporary Swedish journalist remarked on the phantomatic automata that he perceived in the book-matched slabs of the Looshaus, and it seems little coincidence that the Rorschach test (1921) was being developed (from 1912) even as Loos was designing his buildings. However, it is also worth remembering a parallel course in modern psychology, Carl Jung's formulation of the "collective unconscious," sometimes called the "objective psyche" (and first proposed in his 1916 essay, "The Structure of the Unconscious"). Jung theorized that underlying Freud's psyche (Jung called it the "personal unconscious") exists a deeper unconscious, which he likened to a well or lake, made up of instincts and archetypes (fundamental images, symbols, or forms) inherited from our ancestors, which are repressed by the conscious mind. For all the enduring faith in perceptual psychology (and now attempts to chart neural responses to art), Jung's collective unconscious is at the least a serviceable trope for

FIG. 13.4. Ludwig Mies van der Rohe, Barcelona Pavilion, 1928–29 (here the reconstruction of 1983–86). Cipollino, onyx doré, Tinian marble, travertine, chrome, and tinted glass. Barcelona.

the continuity of material and imagistic affordances that attach to the substrate of marble.

Book-matching also recurs throughout Mies's oeuvre, beginning with the Barcelona Pavilion (1928–29), richly dressed in cipollino, onyx doré, Tinian marble, and travertine (fig. 13.4). Here the cladding has ceased to be mere surface and is released into the realm of objecthood as freestanding wall. Mies later claimed that he had come across the stridently veined onyx doré, from Algeria, only by chance (or providence?) as he was planning the project, and that he had acknowledged the blessing by making the slab the vertical module for the building (as if the building were again an unfolding repetition derived from the dimensional symmetries of book-matching). Mies's choice of book-matched marbles also looked back to his youth at Aachen, where, as a schoolboy at the Domschule, he had admired the cathedral's brand-new expanses of marble cladding that were put up (1902–13) in the famous Palatine Chapel of Charlemagne, to "restore" it, make it look more imperial, and (contemporary observers commented) to produce a "painterly effect."[9]

The return to the idea of an inherently painted architecture would come full circle in 1949, when the German émigré and newly appointed curator of architecture and design at the Museum of Modern Art (MoMA), New York, Peter Blake constructed a model of an "Ideal Museum" that assembled maquettes of Jackson Pollock paintings in lieu of the onyx and travertine slabs that had divided the Barcelona Pavilion (fig. 13.5). Pollock was in fact eager to work as a muralist, and the overall concept for this painting-building derived from a more recent incarnation of the Barcelona Pavilion, Mies's unbuilt *Museum for a Small City* (1942), which Mies had conceived in a series of collages wherein the paintings (most prominently, Picasso's *Guernica,* itself a wall-scale canvas) bridged between the near-invisible planes of floor and ceiling. Blake too envisioned a building in which Pollock's paintings would be "suspended between the earth and the sky, and set between mirrored walls so as to extend into infinity. Beyond these floating canvases would be the marshes and the inlets . . . the relentlessly horizontal landscape of that end of Long Island." Arthur Drexler,

FIG. 13.5. Jackson Pollock and Peter Blake examine Blake's model of an "Ideal Museum" at the Betty Parsons Gallery, New York, 1949.

who succeeded Blake in 1951, congratulated Blake for comprehending that Pollock's painting seemed to end only "because there was no canvas left for more." Yet again, almost like the continuously unfoldable patterns of book-matching (see fig. 7.10), one long wall in Blake's model terminated on both ends not with opaque partitions but mirrors. The overall result, Drexler concluded, was "a re-integration of painting and architecture wherein painting is the architecture."[10]

Ruskin also found a continuator or revivalist in the shape of Adrian Stokes (1902–1972), the British painter and art theorist who published *The Quattro Cento* (1932) and *Stones of Rimini* (1934), filled with observations he had been mulling over since a trip to northern Italy in 1925, as well as a trip to Greece to soak up its temples. Stokes fell in love with stone as an essence and disavowed all the apparatus of style, history, and even contemporary observation to focus on the medium itself, beginning from its elements:

> In all stonework typically Mediterranean there is somewhere expressed the identification or mutual consummation of limestone and water, there is expressed water made solid, permanent, glowing instead of glassy, set in space and brightened by the dripping rains. . . . Then we may understand how mere marble men and women could be works of art and could be deities, why the waters of springs were gathered deep and clear in marble shrines. . . . The external thing, the objective, concrete, thing in its form of the glowing limestone, became for Mediterranean man the symbol of realized expression. Time or succession appeared to be summed in stone form as homogeneous concrete matter.[11]

Many pages on the symbiosis of stone and water follow, in which there is much poetry and much wisdom, before he engages Quattrocento sculpture, particularly relief, more particularly low relief (rilievo schiacciato), and still more particularly Agostino di Duccio (see figs. 9.29, 10.9). Duccio's sculpture, he explained, is "carved to flower," peeled away by the chisel to release the life compacted in the stone and to surface the fantasies that the artist read into its depths: "The reliefs are for the most part low, yet their forms possess many values of sculpture in the round: while the quickened mass of a human shape between wind-strewn films of drapery, the delicious torture of hair and clothing by an unseen, evocative wind upon the outer and intermediary surfaces of a relief, give to its body the effect of vitality, of that stone blossom we prize so high. . . . Agostino gave back to marble its primeval eddies."[12] But Stokes was both behind the times and ahead of them, both too Victorian and too Postmodern. *The Stones of Rimini* was well reviewed by Ezra Pound, T. S. Eliot thought it contained "a lot of meat, but . . . a fair amount of suet as well," while Kenneth Clark balked at its engagement with psychoanalysis (Stokes had begun treatment with Melanie Klein in 1929). It sold less than two hundred copies in the first ten years, and would receive little attention again for another fifty.

I want to draw this brief sketch of the "afterlife" of colored marbles to a close with two contemporary works by two living and radically different artists, Stephen Cox and Marina Abramović (both born 1946), works that illustrate the interaction between a material imagination and the "collective unconscious" of the material itself. Cox began his career as a minimalist, in the mold of Donald Judd, working with largely humble and industrial materials. In the 1980s he was awakened to the world of colored marbles by reading Vasari's preface to the *Lives of the Artists* and the recent publication of *The Critical Writings of Adrian Stokes* (1978), which included *The Stones of Rimini*. Both books became itineraries for the quarries, and Cox worked his way from Puglia to Verona, aiming to carve in every stone that Vasari had mentioned. Emboldened, he sought out the grail of all ancient marbles, porphyry, in the remote mountaintop quarries of Jebel Dukhan; then those at Wadi Hammamat, for its seaweed-colored breccia flecked with multicolored clasts; then to Nubia for the bluish-gray, diorite-streaked quartzite/gneiss that had lain hidden under the desert sands since Khafre (c. 2520–2494 BC); and finally to the mottled gray granite of Mahabalipuram in southeast India, from whose outcrops had been hewn vast reliefs and whole temples ex uno lapide beginning in the seventh century AD. All in all, Cox comes to his work with a lithic palette that is virtually unparalleled and a deep knowledge of the quarries and cultures. Nonetheless, the sculptor's unexpected account of the making of one recent artwork reads like a medieval myth and resumes many of the themes discussed throughout this book. This is the altar (2005) for the Chapel of St. Anselm (late twelfth to early fourteenth century) in Canterbury Cathedral (fig. 13.6). As Cox says:

> The altar for Anselm required a "leap of faith." When I make "holy objects" I set out on a pathway with no preconceptions outside of the "construction," in this case that of a box. The altar was to be a smaller version of the earlier and much bigger *Interior Space* [Houghton Hall, Norfolk, 2005]. What happened, splitting the block in the workshop, was the discovery of an angled line of quartz that I then decided would suggest a perspectival pathway through the slot into the "beyond." On the back of the box the book-matched slabs [of Hammamat breccia] were paired differently so the quartz vein was now inverted [to make a V-shape] and suddenly suggested a verdant "valley" in this greenish stone that is scattered with multi-coloured stones like flowers.
>
> [The altar was a gift to Canterbury from the Regione Autonoma Valle d'Aosta, the birthplace of St. Anselm, and was therefore made from the indigenous marble *Verde Vecchia Chiesa d'Aosta*.] When commissioned to make the Canterbury altar I was looking for the Alpine valley landscapes where Anselm found God, and so the "calcite" veins in this new stone. From a 20-ton block with no obvious patterning in the veins, I processed it into manageable and hopefully "fruitful" sections, some of which were sawn into 10 cm slabs and pieces for the altar cut from them. When I asked my assistant, who was standing on the other side of the slabs laid out on the studio floor, if he thought the lines were representative of a landscape, he said "well from this side they look like the outstretched arms of Christ." A gift from God indeed! As for the slot it now became "The Cross."[13]

Marina Abramović is not a sculptor, but a performance artist, one who has always consciously avoided creating permanent artworks and their materials. However, in 2013 she began collaborating with Adam Lowe, and the artists and technicians of Factum Arte (Madrid), to materialize the themes of presence and absence that recur throughout her oeuvre. One of the first results to be unveiled was *Five Stages of Maya Dance* (2018; figs. 13.7a–c), of which the artist confessed, "At this point of my life, facing mortality, I have decided to capture my performance in a more permanent material than just film and photography. I chose alabaster based on its history and properties (luminosity, transparency . . .)."

Factum Arte sourced the alabaster from Fuentes de Ebro, near Zaragoza in Aragon, where this particularly

FIG. 13.6. Stephen Cox, St. Anselm. Verde Vecchia Chiesa d'Aosta. Altar, Chapel of St. Anselm, Canterbury Cathedral, 2005.

milky variety (a cousin of the lapis specularis of antiquity) has been quarried since at least the Middle Ages. Lowe had Abramović 3-D scanned while striking histrionic poses expressing a gamut of emotions ("anger, jealousy, shyness, shame, contemplation"), and then five blocks of the alabaster were drilled (using CNC milling) to form "negatives" of the scans, the profiles of face and arms creating valleys and gullies in the stone.[14] When these blocks have LED screens set behind them, wherever the block has been most thinned, it becomes most translucent. The five alabaster portraits were unveiled at Masterpiece London (2018), the annual marquee show of antique and contemporary dealers, where after passing through the lobby the visitor entered a penumbrous bottleneck, a small theatral space, by passing through a curtain of mist onto which was projected another image of Abramović's upturned head. Face-on, the alabaster images also projected vivid if spectral images of photographic clarity intended to "merge performance, light, and sculpture . . . something material yet immaterial at the same time," and because of the non-contact methodologies employed, something between the image-not-made-by-human-hands and frames from a film. As one moves away from the slabs and turns to see them from the side, her face decomposes into submarine valleys and trenches or "something that looks like a Chinese landscape" (as Lowe put it). The resulting artwork, like a medieval church window, is a materialization of light; and in a Renaissance or Baroque paragone, the "painting in stone" has been figured in the whiteness by subtraction and relief; but the fleeting image that the material conjures then decomposes, once the viewer ceases to confront it, to hint at celluloid and the more contemporary immateriality of digital media.

Cox and Abramović (with her collaborator, Lowe)—there are others—rescue colored marbles and translucent stones from the more familiar abasement they have suffered in multiple corporate lobbies and back-lit bars. Each artist aspires to coerce a miraculous presence from the cold stone, and each is as much alive to history as he or she remains militantly modern; but otherwise the artists' paths diverge. Abramović's self-portraits absorb the cosmic nature of the material into personal psychology and individual finitude. The stone rebounds like a poltergeist from the indifferent screen of a digital universe before dissolving back into a landscape of peripheral vision. Cox's altar is more introspective, a secretive object that quietly beckons in-reading and insight, and whose minimalism and Modernist "truth to materials" strike uncanny resonances with the medieval mason, and thereby revive age-old claims to an innate and shared natural order.

FIG. 13.7A–C. Marina Abramović, *Five Stages of Maya Dance*, 2018. Masterpiece, London.

Sat Prata Biberunt

Abbreviations

Alberti: Leon Battista Alberti
De re aed.: De re aedificatoria
Ambr.: Ambrose
De ob. Theod.: De obitio Theodosii Consolatio
De ob. Valent.: De obitu Valentiniani consolatio
Amm. Marc.: Ammianus Marcellinus
Res Gest.: Res Gestae
Anth. Lat.: Anthologia Latina
Anth. Pal.: Anthologia Palatina
Ap. Rhod.: Apollonius Rhodius
Argon.: Argonautica
Apollod.: Apollodorus mythographus
Bibl.: Bibliotheca
Apul.: Apuleius
Met.: Metamorphoses
Aquinas: Thomas Aquinas
Summa: Summa Theologica
Ar.: Aristophanes
Ach.: Acharnenses
Nub.: The Clouds
Arist.: Aristotle
Cael.: De Caelo
Col.: De Coloribus
Metaph.: Metaphysica
Mete.: Meteorologica
Poet.: Poetica
Aristid.: Aelius Aristeides
Orat.: Orationes
Asc.: Asconius
Scaur.: Commentary on Cicero, *In Scaurum*
Ath.: Athenaeus
Deip.: Deipnosophistae
August.: Augustine
De civ. D.: De civitate Dei
Aus.: Ausonius
Mos.: Mosella
Avien.: Avienus
Orb. Terr.: Periegesis seu Descriptio Orbis Terrarum
Boeth.: Boethius
Consol.: De Consolatione Philosophiae
Caes.: Julius Caesar
B Civ.: Bellum Civile
Cassiod.: Cassiodorus
Var.: Variae
Catull.: Catullus
Carm.: Carmina
Choric.: Chorikios
Laud. Marc: Laudatio Marciani
Cic.: Cicero
Att.: Epistulae ad Atticum
Div.: De Divinatione
Fam.: Epistulae ad familiares
Leg.: De Legibus
Mur.: Pro Murena
Nat. D.: De natura deorum
Scaur: Pro Scauro
Tusc.: Tusculanae disputationes
Verr.: In Verrem
CIL: Corpus Inscriptionum Latinarum, 1863–
Claud.: Claudianus
Carm.: Carmina
Ep. ad. Ser.: Epistula ad Serenam (Carm. min. 31)
Cons. Hon.: De consulatu Honorii
Epith.: Epithalamium de Nuptiis Honorii Augusti
Eutrop.: In Eutropum
Col.: Columella
De Agr.: De Agricultura
De Arb.: De Arboribus
Coripp.: Corippus
Laud. Just.: In laudem Justini Augusti minoris
Dio Cass.: Dio Cassius
Hist. Rom.: Historia Romana
Diod. Sic.: Diodorus Siculus
Bibl. Hist.: Bibliotheca historica
Dion. Hal.: Dionysius of Halicarnassus
Antiq. Rom.: Antiquitates Romanorum
Enn.: Ennius
Ann.: Annales
Ennod.: Ennodius
Carm.: Carmina
Eur.: Euripides
Ion: Ion
Euseb.: Eusebius
Hist. Eccl.: Historia Ecclesiastica
MP: De martyribus Palaestinae (Martyrs of Palestine)
Vit. Const.: Vita Constantini
Fest.: Festus
Apogr.: Apographs
Frontin.: Frontinus
Aq.: De aquae ductu
Gal.: Galen
Nat. Fac.: De naturalibus facultatibus
Gell.: Aulus Gellius
NA: Noctes Atticae
Hdt.: Herodotos
Hist.: Historiae
Hero: Hero of Alexandria
Stereom.: Stereometria
Hes.: Hesiod
Theog.: Theogony
Hippoc.: Hippokrates
Epid.: Epidemiae
Hom.: Homer
Il.: Iliad
Od.: Odyssey
Hor.: Horace
Ars P.: Ars Poetica
Carm.: Carmina or *Odes*
Hyg.: Hyginus
Fab.: Fabulae
Isid.. Isidorus
Etym.: Etymologiae
Isoc.: Isokrates
Joseph.: Josephus
AJ: Antiquitates Judaeorum
BJ: Bellum Judaicum
Julian.: Julianus Imperator
Ep.: Epistulae
Mis.: Misopogon
Juv.: Juvenal
Sat.: Satirae
Livy.: Livy
AUC: Ab urbe condita libri
Long.: Longus
Daph.: Daphnis and Chloe
Luc.: Lucan
Phars.: Pharsalia
Lucian: Lucian
Dom.: De Domo
Hipp.: Hippias seu Thermae
Sacr.: De Sacrificiis
Salt.: De Saltatione

Lucr.: Lucretius
Rer. nat: De rerum natura
Macrob.: Macrobius
Sat.: Saturnalia
Manil.: Manilius
Astron.: Astronomica
Mart.: Martial
Ep.: Epigrammata
Merobaud.: Flavius Merobaudes
Carm.: Carmina
Narratio: Narratio Sanctae Sophiae
Nonn.: Nonnos
Dion.: Dionysiaca
Olymp.: Olympiodoros
Com. in Meteora: Commentaria in Meteorologica Aristotelis
Ov.: Ovid
Ars Am.: Ars Amatoria
Fast.: Fasti
Her.: Heroides
Met.: Metamorphoses
Pont.: Epistulae ex Ponto
Tr.: Tristia ex Ponto
Paul.: Paulus Silentiarius
Descr. Amb.: Descriptio ambonis Sanctae Sophiae
Descr. S. Sophiae: Descriptio Sanctae Sophiae
Paus.: Pausanias
Desc.: Description of Greece
Pers.: Persius
Satir: Satirae
Petr.: Petronius
Sat.: Satyricon
PG: J.-P. Migne, ed., *Patrologia Graeca* (Paris: Garnier, 1857–1905)
Philo: Philo Judaeus
Leg.: Legatio ad Gaium
Leg. alleg.: Legum Allegoriae
Quis haer.: Quis rerum divinarum haeres sit
Philostr.: Philostratos
VA: Vita Apollonii
VS: Vitae Sophistarum
Phot.: Photios
Hom.: Homilies
Pind.: Pindar
Isth.: Isthmian Odes
Nem.: Nemean Odes
Ol.: Olympian Odes
Pyth.: Pythian Odes
PL: Jacques-Paul Migne, ed., *Patrologia Latina*, 217 vols. (Paris: Migne, 1844–55)
Pl.: Plato
Criti.: Critias
Leges: Laws
Ti.: Timaeus
Plaut.: Plautus
Asin.: Asinaria
Most.: Mostellaria
Pseud.: Pseudolus
Pliny (E): Pliny (The Elder)
HN: Naturalis Historia
Pliny.: Pliny (The Younger)
Ep.: Epistulae
Plut.: Plutarch
Amat.: Amatorius
De fac.: De facie in orbe lunae
Mar.: Marius
Mor.: Moralia
Per.: Pericles
Pomp.: Pompey
Popl.: Publicola
Quaest. conv.: Quaestiones convivales
Quaest. Rom.: Quaestiones Romanae
Sull.: Sulla
Vit.: Vitae Parallelae
Procop.: Prokopios
Aed.: De Aedificiis
BP: Bellum Persicum
Prop.: Propertius
El: Elegiae
Prudent.: Prudentius
c. Symm.: Contra Symmachum
Cath.: Cathemerina
Perist.: Peristephanon
Psych.: Psychomachia
Psell.: Michael Psellos
Orat. min.: Oratioria minora
Quint.: Quintilian
Inst.: Institutio oratoria
Rutil.: Rutilius Namatianus
De red.: De redito suo
Sedul.: Sedulius Scottus
Carm.: Carmina
Sen. (E): Seneca (The Elder)
Controv.: Controversiae
Sen.: Seneca (The Younger)
Ben.: De Beneficiis
Ep.: Epistulae
Ira: De ira
Phae.: Phaedra
QNat.: Quaestiones Naturales
Serv.: Servius
Ad Aen.: Commentary on Vergil, *Aeneid*
Ad Verg. G.: Commentary on Vergil, *Georgics*
SHA: Scriptores Historiae Augustae
Helioglab.: Heliogabalus
M. Ant.: Marcus Antoninus
Pert.: Pertinax
Sev.: Severus
Sid. Apoll.: Sidonius Apollinaris
Carm.: Carmina
Epist.: Epistuale
Stat.: Statius
Silv.: Silvae
Theb.: Thebais
Strab.: Strabo
Geog.: Geographia
Suet.: Suetonius
Aug.: Divus Augustus
Cal.: Gaius Caligula
Domit.: Domitianus
Iul.: Divus Iulius
Ner.: Nero
Tib.: Tiberius
Tit.: Divus Titus
Symm.: Symmacus
Ep.: Epistulae
Tac.: Tacitus
Ann.: Annales
Hist.: Historiae
Tert.: Tertullian
Apol.: Apologeticus
De resurrect.: De resurrectione carnis
Them.: Themistios
Orat.: Orationes
Theocr.: Theokritos
Id.: Idyllia
Theophr.: Theophrastos
De Lap.: De Lapidibus
Val. Max.: Valerius Maximus
Facta: Facta et dicta memorabilia
Varro: Varro
Antiq.: Antiquitates rerum divinarum
Rust.: De Re Rustica
Vell. Pat.: Velleius Paterculus
Hist.: Historiae
Ven. Fort.: Venantius Fortunatus
Carm.: Carmina
Mart.: Vita Sancti Martini
Verg.: Vergil
Aen.: Aeneid
G.: Georgics
Vitr.: Vitruvius
De Arch.: De Architectura
Xen.: Xenophon
An.: Anabasis
Por.: Poroi
Zeno.: Zenobius

Notes

INTRODUCTION

1. Gnoli 1988 (1st ed. 1971; 3rd ed. 2018); Borghini 1990.
2. Studies are legion, but the regular conference acts of ASMOSIA (Association for the Study of Marble and Other Stones in Antiquity), AISCOM (Associazione Italiana per lo Studio e la Conservazione del Mosaico), and the catalogue by De Nuccio and Ungaro (2002) deserve special mention.
3. The following were originally published in the 1930s and 1940s in French: Gaston Bachelard, *The Psychoanalysis of Fire* (Boston: Beacon, 1964); Gaston Bachelard, *Water and Dreams: An Essay on the Imagination of Matter* (Dallas: Pegasus Foundation, 1983); Gaston Bachelard, *Air and Dreams: An Essay on the Imagination of Movement* (Dallas: Dallas Institute of Humanities and Culture, 1988); Gaston Bachelard, *Earth and Reveries of Will: An Essay on the Imagination of Matter* (Dallas: Dallas Institute of Humanities and Culture, 2002).

CHAPTER 1. A MEDIUM FORETOLD

1. Uruk: Boehmer 1990. Exhaustive documentation in Eichmann 2007. Eridu: Safar et al. 1981, 66, 81, 240, figs. 118–19. For the construction process at Uruk, see Artefacts, "(Re-) Constructing the Stone-Cone Building in Uruk," https://vimeo.com/54015188.
2. Vanstiphout 2003, 58–59, ll. 38–54; Electronic Text Corpus of Sumerian Literature, "Enmerkar and the Lord of Aratta," http://etcsl.orinst.ox.ac.uk. The four Sumerian narrative poems, collectively known as the *Matter of Aratta*, originated about 2112/2004 BC but date in their present form to about 2017/1763 BC. See Electronic Text Corpus of Sumerian Literature, "The Sumerian King List: Translation," http://etcsl.orinst.ox.ac.uk.
3. Adda Seal: British Museum 89115. For the Mari mural, see Parrot 1958b, 76–78, fig. 57, pls. XVII, XIX.1, color pl. E. The deity is identified as Sin and the room as a chapel in Nunn 1988, 74, 83.
4. For the cone mosaics at Telloh, Tell al-'Ubaid, Ur, Tell Uqair, Tell Brak, and Buto (as well as Uruk and Eridu), see Jordan 1932; Van Buren 1946, 323–45; Moorey 1994, 309–10, 339, with further bibliography.
5. Tite and Bimson 1989.
6. Legrain 1937, no. 1498; Oppenheim et al. 1970, 10–15, 50–51; Nicholson 2012, 11–15.
7. Recipe (c. 1124–1103 BC) for *pappardilû*-stone and *dusû*-stone in Oppenheim 1966, 29–35, 42–43. See also, on BM 89159 (chert or chalcedony, dyed to resemble agate), Sollberger 1987; discussion and other examples in Merrillees 2005, 144–45, 147 (VI.1.4., VI.2.1.6).
8. Beretta 2009, 3–4.
9. Postgate 1997.
10. On obsidian, in a broad literature, see Cauvin et al. 1998.
11. Campbell Thompson 1938, 6; K. Foster 1991, 391–92.
12. Paynter and Tite 2001.
13. de Garis Davies 1935, pl. ix; Shortland 2000.
14. Metropolitan Museum of Art, New York, inv. no. 26.7.1175. See Dan Barag in Oppenheim et al. 1970, 139–40, 181–82, cat. nos. 14–15, figs. 13, 14, 92; a more definitive analysis in Lilyquist and Brill 1993, 9–12, figs. 10–14; Lilyquist 2003, 151, cat. no. 104. See also the glass fragments from the Tomb of Amenhotep II (r. 1427–1401 BC) in Lilyquist and Brill 1993, 34, fig. 19.
15. *Enmerkar and the Lord of Aratta*, ll. 397–434: Vanstiphout 2003, 78–81; Electronic Text Corpus of Sumerian Literature, "Enmerkar and the Lord of Aratta: Translation," http://etcsl.orinst.ox.ac.uk. Van Buren 1956.
16. Alabaster consists of gypsum (calcium sulfate, $CaSO_4$), whereas travertine is a rock composed of calcite (calcium carbonate, $CaCO_3$). Harrell 1990 has argued that "travertine" is the only geologically correct term, whereas Klemm and Klemm 1991 advocate "calcite-alabaster."
17. Van Haarlem 1997.
18. Foster 2008, 179. See also Landgráfová 2006a and b.
19. Friedman 1998; Caubet and Pierrat-Bonnefois 2005, 27; Nicholson 2012.
20. Lauer 1936, 34–38; Friedman 1996, 342–43; Friedman 1998, 66, 72, 180–81 (cat. nos. 17–20).
21. Pettazoni 1954, 40–41, with bibliography; Metevelis 2000, with further bibliography.
22. Horowitz 1998, 166–68. For the bowl-shape, see the hymns to Šamaš and Marduk: Ebeling 1953, 14–15; Lambert 1960, 135, ll. 154–55; Lawson Younger 2012.
23. Kramer 1956, 76.
24. Hymns to the temples of Utu in Larsa and Sippar: Sjöberg and Bergmann 1969, 27 (hymn 13, l. 169), 45 (hymn 37, l. 475), nn87, 137.
25. *Enûma Eliš. l.* 5.121; trans. in Foster 1996, 351–402.
26. Erich Ebeling, *Keilschrifttexte aus Assur religiösen Inhalts* (Leipzig: J. C. Hinrichs, 1915–19, 1920–23), no. 307, ll. 30–33. The formula is repeated in the late Babylonian compendium in the Louvre, AO 8196 (iv 20–22). See Horowitz 1998, 3–19, esp. 10–14; Livingstone 1986, 79–88; Huxley 1997.
27. The Mesopotamian lapidary, the *Abnu-šikinšu* (14–15), says, "The stone whose appearance is like *ḫulālu*-stone and *muššaru*-stone is named *luludānītu*-stone. / The stone whose appearance is red covered with white and black patches is named *luludānītu*-stone." *Muššaru*-stone has been identified as a red-and-white-banded agate (sardonyx), and texts suggest that it was a variety of ḫulālu-stone: Schuster-Brandis 2008, 429 (no. 65, luludanītu, lulladanītu), 433 (no. 76, muššaru); Schwemer and Abusch 2008, 471; André-Salvini 1999, 391n27. Other evidence suggests that ḫulālu-stone may refer specifically to onyx, a black-and-white-banded chalcedony: Schuster-Brandis

2008, 436. For "jasper" (ašpû), see the chalcedony pendant (on which this name is inscribed) given by the Elamite king Šilhak-Inšušinak (r. c. 1150–1120 BC) to his daughter, Bar-Uli: British Museum inv. no. 113886; Sollberger 1965. For elmēšu as amber, see Landsberger 1967, 190–98; Bodi 1991, 88–94. For saggilmut as amethyst, see Black 2001. See also Schuster-Brandis 2008, 401–2 (no. 16, ašpû), 409–10 (no. 29, elmēšu), 440–41 (no. 88, sangil(i)mud).

28. For Vardanes, see Philostr. *VA* 1.25.3. For Ḳosrow II: Nikephoros in Mango 1990, 56–57. Ferdowsī in Bertels 1960–71, 9.223.3577–79; Ṭaʿālebī in Zotenberg 1900, 699. That the Medici dome was to be clad in lapis is often repeated (e.g., Baldini et al. 1979), but I have yet to see a documentary source.
29. For the White Temple and its dating: Eichmann 2013; van Ess 2013. For a video reconstruction, see Artefacts, "Anu Ziggurat and the White Temple of Uruk," https://vimeo.com/148629834.
30. Keš temple hymn, 20–40, 13–15; see Electronic Text Corpus of Sumerian Literature, "The Kes Temple Hymn," http://etcsl.orinst.ox.ac.uk.
31. Temple Hymn 17, on the Badtibira Temple of Dumuzi at Emush.
32. Aruz and Wallenfels 2003, 200, cat. no. 128, with bibliography.
33. Prichard 1969, 573–76.
34. Winter 1994. For Gudea's Cylinder A, col. 28, 1–2, see Electronic Text Corpus of Sumerian Literature, c.2.1.7, 757–758, http://etcsl.orinst.ox.ac.uk.
35. Woolley and Moorey 1982, 165–66.
36. Grayson 1976, 18 (no. 55).
37. For the Ninhursag columns (BM 115328, BM 116760), see Hall and Woolley 1927, 17, 40, figs. 3, 100–103, pls. IV.1–6, XXXIV.3, XXXV.6–7, XXXVII; Giovino 2007, 182–85.
38. Tablet IX, 173–90, see George 2003, 1:672–75, and commentary at 2:867–68.
39. On glazed brick generally: Nunn 1988, 309–22; Crespo 2013, 212–13.
40. Köcher 1959, 237; Oppenheim et al. 1970, 17. Schuster-Brandis (2008, 57) translates "gišnugallu" as "limestone," but see Mitchell and Middleton (2002, 94–95) for "calcite-alabaster."
41. Marzahn 1995, 29–30. The compendium Catalogue of the Babylonian Section, tablets in the University Museum, Philadelphia, 6060, equates various minerals with deities, among them lapis-lazuli with DIL.BAT (Ištar: Livingstone 1986), 176–79.
42. George 2011, 160, 168, 169. The E-temen-anki was damaged beyond repair in the Persian period (c. 484 BC) and leveled by Alexander the Great (331 BC) for a reconstruction halted by his death: George 2005–6.
43. Edzard 1987; K. Foster 1991, 391–95; Maul 1997; van de Mieroop 2003, 262–64. For the late Babylonian map of the world (British Museum, tablets, 92687), see Horowitz 1988.
44. Cylinder A, 110–23, 132–46: ETCSL t.2.1.7, Electronic Text Corpus of Sumerian Literature, "The Building of Ningirsu's Temple," http://etcsl.orinst.ox.ac.uk. For other references to the gods using a "Tablet of the Stars of Heaven" (*dub.mul.an*) said to be made of lapis lazuli, see Horowitz 1998, 166–68.
45. James and van der Sluijs 2008. For Khorsabad, see Place 1867–70, 1:141–48, 2:79, 3:pls. 36–37. For Borsippa, see Allinger-Csollich 1998.
46. Foster 1996, 796, cited in van de Mieroop 2003, 267, 273.
47. Ezekiel 1:22, 26–27; 10:1. Kingsley (1992) deems the Babylonian origins of this transmission "subconscious," forgetting that Ezekiel knew the real E-temen-anki.
48. The operative noun is "סַפִּיר" (*cappiyr*), in the Septaugint translated as "σάπφειρος" (Latin, *Sappirus*), but which means "lapis lazuli." As Pliny says (*HN.* 37.38–39.119–20), "[Sapphire] also shines with gold specks. 'Sapphires' are also cerulean blue" (*aurum punctis conlucet. Caeruleae et sappiri*). Exodus 24:10.
49. Enoch 1 18:6–10: "I saw at the end of the earth the firmament of the heaven above. And I proceeded and saw a place that burns day and night, where there are seven mountains of magnificent stones, (7) three towards the east, and three towards the south. And as for those towards the east, one was of colored stone, and one of pearl, and one of jacinth, and those towards the south of red stone. (8) But the middle one reached to heaven like the throne of God, of alabaster, and the summit of the (9,10) throne was of lapis lazuli." Isaiah 54:11–12: "Behold, I will set your stones in antimony, and your foundations I will lay in lapis lazuli. / Moreover, I will make your battlements of rubies, And your gates of crystal, And your entire wall of precious stones." Tobit 13:16–17: "The gates of Jerusalem will be built with sapphire and emerald, and all your walls with precious stones. The towers of Jerusalem will be built with gold, and their battlements with purest gold. (17) The streets of Jerusalem will be paved with rubies and stones of Ophir." In the Aramaic text, "Description of the New Jerusalem" (2nd/1st century BC) from the Dead Sea Scrolls at Qumran, all buildings, windows, walls, and streets are covered with white stones (limestone?), alabasters, onyx, sapphires, rubies, and pure gold: García Martínez 1994, 129–35.
50. The most detailed study of the stones in Revelation 21:18–21 is Reader 1981.

CHAPTER 2. A MEDIUM FULFILLED

1. Wright 1978, 134, 228–36; Wright 1987 (who dogmatically argues for political symbolism); Santillo Frizell 1998; Nelson 2003; Nelson 2006; Fitzsimons 2006; Mason 2007; Hitchcock et al. 2016.
2. Shaw 2006, passim, esp. 222–23. For stone vessels made of conglomerate and breccia, and the painted imitation of conglomerate and breccia on ceramics, see Warren 1969, 127–28, 130, 172–173.
3. Blackwell 2014, esp. 475.
4. Higgins et al. 1968. Cooper (1988, 71) instead thinks the red-and-green stone is from the hills to the northeast of Nauplion, in the Argolid. See Warren 2012.
5. All discussions of the façade depend solely on the survey by Thiersch (1879), and the very detailed observations and diagrams in Perrot and Chipiez 1894. See also Wace et al. 1921–23, 342–49, with other bibliography; Robertson 1941. Several reconstructions have been proposed: Thomas Leverton Donaldson in Stuart and Revett 1830, 4:32, pl. V; Perrot and Chipiez 1894; Fletcher 1924, 70, F; Wace 1949, fig. 51; Mylonas 1966, fig. 114; Higgins et al. 1968, fig. 1. Younger (1987) has also argued that an older gypsum frieze (c. 1475 BC) of a galloping and standing bull was set into the lower half of this triangle (British Museum A56, A57).
6. Perrot and Chipiez 1894, 641–47; Wace et al. 1921–23, 359–61. For the date, see Mason 2013.
7. The "Tomb of Clytemnestra" was concealed by a Hellenistic theater cavea, but not the "Treasury of Atreus." Pausanias (9.36.4–5, 9.38.2) also described the Mycenean tomb at Orchomenos as the "Treasury of Minyas."
8. For Minoan examples, see the engaged half-columns (of unknown date) on the north entrance from the Great Court at Phaistos, and fluted columns in the Little Palace at Knossos (c. 1450–1400 BC): Evans 1913/14, 62, fig. 76. Some stucco tomb façades built on later Macedonian tumuli, for example, the *Tomb of the Judges* at Lefkadia (325/300 BC), also apparently represent palace façades.
9. The essential study of the sarcophagus remains Long 1974. For the dating: La Rosa 1999.
10. de Polignac 1995; Morgan 1990; Sourvinou-Inwood 1993; Wright 1994; Mazarakis Ainian 1997; Mazarakis Ainian 2016, 15–30, with further bibliography.
11. At Selinous (Temple E1, c. 580 BC, and Temple C), Syracuse (Temple of Apollo, c. 565 BC?), Paestum ("Basilica," c. 550–530/520), and Foce del Sele (Temple of Hera I, c. 500).
12. Barletta 2001, 1–20, 126–56, with bibliography; Barletta 2009; Wilson Jones 2014, 63–81, with further bibliography.
13. Vitr. *De Arch.* 4.2.5–6. For the date of *De Architectura:* Baldwin 1990. Vitruvius gives his sources at 7.praef.12, 14.
14. Respectively, Sapirstein 2016; Donderer 2005.
15. On debate over Egyptian influence, most conveniently, see Wilson Jones 2014, 94–110, with bibliography.
16. Boardman 1999; Skon-Jedele 1994, 1663–1953; Möller 2000.

17. Diod. Sic. 1.48; Richter 1970, 27–29; Kyrieleis 1996, esp. 110–15; Boardman 1978, 18–24; Kokkorou-Alewras 1995, 87–88, cat. 18; Gruben 1997, esp. 267–93. The Apollo is approximately the same size as the statue of Ramesses II (without crown) from Memphis (1279/1213 BC). Hdt. 2.130.2; 131.1, 3; 143.2, 4 (bis); 149.2, 153; 175.1; 176.1 (bis); 4.152.4.
18. For temples and landscape, see Scully 1962, 1–8.
19. Heidegger 1975, 40–43.
20. For the formal development of the krepidoma, see Martin 1965, 322–56; for a table of krepidomai (one to four steps), see 336–45. For the origins and development of the stepped altar, see Hoffmann 1952; Hoffmann 1953, with references.
21. Hom. *Il.* 14.185. For other glosses of leukos, see Ferrini 1998.
22. See Lapatin 2001, 22–42, for earliest examples; 42–60 for archaic Greek ones; 63–90 for Pheidias's Athena Parthenos and Zeus Olympios.
23. Lucian. *Sacr.* 11. Cf. Quint. *Inst.* 12.10.9; Livy. 45.28.5; Cic. *Verr.* 2.5.187, 2.4.72. See also Lapatin 2010.
24. See Giuliano 2004–5, there attributed to Euphranor, c. 350 BC.
25. Dietrich 1983; Smith 1988.
26. Clay 1981–82. See also Vernant 1991.
27. Pind. *Ol.* I; Verg. *G.* 3.8; Hyg. *Fab.* 83; Ov. *Met.* 6.404–11; Lucian. *Salt.* 54; Serv. *Ad Aen.* 6.603. On skin and ivory: Caubet 1989.
28. Pliny. *HN.* 28.34. This relic had disappeared by the time Pausianias visited the sanctuary, c. 174/175 AD (Paus. 5.13.4).
29. Häger-Weigel 1997; Despinis 2004; Lapatin 2001, 61.
30. "Morgantina Goddess": approx. 7 ft., 2⅜ in. (2.2 meters) high, "wackestone" (limestone) and Parian marble, pigment traces on limestone, from Morgantina, now Museo Archeologico, Aidone. Probably Persephone from the monumental sanctuary of Demeter at S. Francesco Bisconti: Marconi 2011, with bibliography. For the provenance and lithotypes, see *Summary of Proceedings from a Workshop Held at the Getty Villa on May 9, 2007*, esp. 23–28, http://www.getty.edu. For Selinous, see Marconi 1994, 138–40, 244nn10–17, with bibliography.
31. For pōros, see Wycherley 1974.
32. For marmaros and marmarein, see Ciani 1974, 143–44, 149–50, 159–60; also Neer 2010, 73–76.
33. For the Naxian quarries, see Kokkorou-Alevras 1992; Tambakopoulos and Maniatis 2017.
34. Pind. *Isth.* 6.73 (480 BC). "Bronze-dominating Naxian whetstone." On archaic Naxian sculpture, see Kokkorou-Alewras 1995.
35. For the Terrace of the Lions, see Santerre 1959, esp. 33–36. For the colossus, see Kokkorou-Alewras 1995, 87–88, cat. 18; Gruben 1997, esp. 267–93 (for the Colossus at Delos), 293–300 (Colossal Kouros at Apollonas).
36. Courbin (1980, 1987) argues that the "oikos" was the temple; Santerre (1984) and Gruben (1997) that it was a hestiatorion; Mazarakis Ainian (1997, 180–81), a temple-hestiatorion; Roux (1979), a temple-treasury; Lambrinoudakis (2005) notes that cult and banquet buildings are indistinguishable in the archaic phases of sanctuaries on Naxos itself; Coarelli (2016, 49–57) is adamant that it was a temple. Later still (560–550), the oikos received a columnar portico also in marble.
37. Paus. 5.10.3 (Byzes); Ohnesorg 1993, 65–66, 141, no. 1, pls. 6, 37.2.
38. Morris discovered a marble roof tile on Mt. Zas, Naxos, speculating that an undiscovered shrine to Zeus may have occupied the peak: Morris 2007, 100.
39. Lambrinoudakis 1991; Ohnesorg 2005, with further bibliography. For the Temple of Demeter at Sangri, see Ohnesorg 1993, 67–73; Lambrinoudakis et al. 2002. An ear survives from the statue at Yria: Lapatin 2001, nos. 34, 60, fig. 146.
40. D'Onofrio 2008, with previous bibliography; Sturgeon 2006.
41. Ohnesorg 1993, 9, 53–59, 65; De Siena et al. 2002. For Parian imports into Magna Graecia, see Lazzarini 2007.
42. Hoepfner (2001) has postulated translucent roofs for the temples of Artemis in the Delion on Paros (c. 490–480 BC), "Temple of the Athenians" on Delos (c. 421–417 BC), the Parthenon (447–438 BC), and Apollo Epikourios at Bassai (c. 429–427 BC). These are all refuted, though only on structural grounds, in Ohnesorg 2011.
43. Ike and Otaki 1998.
44. The Nashville Parthenon was designed by Russell E. Hart in collaboration with the renowned archaeologist William B. Dinsmoor, but apparently the idea of a translucent roof was resisted: Wilson 1937, 62–64.
45. Thomas 1991, 2:630.
46. Hennemeyer 2015, 26–27, with bibliography.
47. For the Parthenon: Stevens 1955, esp. 267–70; Graham 1963, 80–81; Boardman 1967, 509; Lapatin 2001, 79, 85–86, n239. For Olympia: Morgan 1952, 316–18; Barry 2007, 639–40.
48. Interview with Richard Wilson, *Guardian*, April 4, 2003. First installed in Matt's Gallery, London (1987); reinstalled in the Saatchi Gallery, St. John's Wood (1991); in the Saatchi Gallery, South Bank (2003): this version in the exhibition *Space Shifters*, Hayward Gallery, September 26, 2018–January 6, 2019.
49. For the transmitted light, see Ike and Otai 1998, 238. For the doors, see Pope and Schultz 2014. For the glaze: Jenkins and Middleton 1988, esp. 198. Note that Pliny (*HN.* 36.55.177) records that Panainos, Pheidias's brother, stuccoed the interior of the Temple of Athena at Elis with a plaster blended with milk and saffron.
50. Martin 1965, 422–32 (who claims the practice is attested from the 7th century BC).
51. For this stone and its quarries, see Peschlow-Bindokat 1990; Lazzarini 2016.
52. Philostratos (*VA.* 2.20, 24) describes a temple in India made from "conch-like stone" (*lithos konchuliatēs*). Marshall (1951, 1:222–29) identified this with the Greco-Indian Temple C at Jandial, near Taxila (eastern Punjab), and reasoned that he meant a stucco of powdered shell (in place of lime); Marshall is followed by Ḍār (1984, 43–53), but contested by Bernard (1996, 507–12).
53. For the Paros quarries, see Higgins and Higgins 1996, 180; Korres 2000, 61–82; Aliprantes 1996, 64–78, 150–56.
54. In the *Gilgameš* epic, tablet XII, 29–30, it is said of Ereškigal, queen of the Netherworld, "her gleaming shoulders are not draped in a garment, her breast, uncovered, [is] like a jar of stone": see George 2003, 1:731, and commentary at 2:901. See Philodemos (c. 110–c. 40/35 BC), epigram 9, l. 3, "τὰ λύγδινα κώνια μαστῶν" ("the Lygdian-marble cones of her breasts"): Sider 1997, 95–96. See also *Anth. Gr.* 5.48, l. 3 (Rufinus, possibly 3rd century): "δειρὴ λυγδινέη καὶ στήθεα μαρμαίροντα" ("her Lygdian-marble neck and gleaming breasts"). Lygdian marble is a synonym for the whitest marble, and therefore possibly Parian: See Diod. 2.52.9: "ἡ Παρία λύγδος" ("white marble of Paros"). See also Lucilius (c. 180–103/102 BC) frg. 887f.: "hic stare papillas / pectore marmoreo" ("here stand out the nipples from the marble breast").
55. Hor. *Carm.* 1.19.5; Petr. *Sat.* 126 ("*Parium marmor extinxerat*"). Poseidippos describes Aphrodite as "οἷά τε λύγδου / γλυπτήν" ("as though carved of white marble"): see Austin and Bastianini 2002, 162–63 (no. 126), ll. 3–4.
56. The treasuries of the Klazomenai and the Knidians (before 544 BC); of the Siphnians (c. 530–525 BC, built from the profits of their gold and silver mines); the Treasury of the Massalians (c. 530 BC) and "Doric Treasury" (c. 490–460 BC) in the precinct of Athena Pronaia; the treasuries of the Athenians (500 BC), Cyrenaeans (c. 373–338 BC), and Syracusans (414 BC): Tomlinson 2000; Partida 2000; Palagia and Herz 2002.
57. Neer 2004, with previous bibliography.
58. Herodotos (5.62) gives the date 514/513 BC, and Philokhoros 508/507 BC: Felix Jacoby, *Die Fragmente der griechischen Historiker*, 3 vols. (Berlin: Weidmann, 1923–59), no. 328, frag. 115.
59. *Hymn to Pythian Apollo*, 3.441–50, in *The Homeric Hymns and Homerica*, trans. Hugh G. Evelyn-White (London: W. Heinemann, 1920), 355, 357. Burkert (1979) suggested this hymn was composed by Kynaithos of Chios for the Delian-Pythian festival on Delos in 522 BC.

60. For the 4th-century reconstruction of the temple, see Amandry and Hansen 2010, 389, 409, 432. On the chariot of Helios, see Laroche and Jacquemin 1986.
61. Livy. 31.26.11; Strab. 9.1.23; Thompson 1980; Korres 1995; Abraldes 1996.
62. Wescoat 2018.
63. Bonamici 1989; Bonamici 1991.
64. Livy. 40.51.3. For opus marmoratum: Vitr. *De Arch.* 7.3.6, 7.6.1; Varro. *Rust.* 1.59.3; Pliny. *HN.* 36.55.176–77; Bettini 2001, 86, with bibliography. The Assisi inscription ("CIL XI, 8021") dates to 100/50 BC. The Brescia inscription has been dated to 90/50 BC (Fausto Zevi), the Augustan era (Albino Garzetti), and c. 27/28 BC (Alberto Valvo): Valvo 2008, 137–40, with bibliography.
65. For example, "Temple A" (of Juturna? or Feronia?) at Largo Argentina, Rome, progressively patched and resurfaced from the mid-1st century BC to c. 80 AD: Ziolkowski 1986. The Temple of Portunus in the Forum Boarium dates from 80–70 BC, but its frieze probably dates to the late 1st or 2nd century. In the mid-4th century the fire-damaged Temple of Apollo, Antioch, was restored with "columns of baked brick and plaster, encasing them in marble-stucco" (Julian. *Ep.* 29).
66. Livy. 42.1.3; Val. Max. 1.1.20; La Rocca 1996; Ruga 1996.
67. Gros 1973. For an overview of Augustan temple construction, see Gros 1976.
68. Mars in Circo (built by Hermodoros for Decius Iunius Brutus Callaicus, c. 133 BC), and Neptune (built by Cnaeus Domitius Ahenobarbus, c. 130/115 BC): Zevi 1976; Bernard 2010, with bibliography.
69. Strong and Ward-Perkins 1960; Heilmeyer and Rakob 1973, esp. 8, 28–29. Ziolkowski (1988) argues the temple was dedicated by Mummius; Coarelli (1988, 60–105; 164–204) that it was dedicated to Hercules Victor Olivarius by M. Octavius Herrenus (or Hersennus), 100/90 BC. For the marble, see Gorgoni et al. 2002.
70. Plut. *Pomp.* 42.4 (theater); Pliny. *HN.* 36.48 (Mamurra); Cic. *Att.* 4.16.7 (Saepta Iulia). See Fant 1999, 277–78. The Temple of Castor and Pollux (6 AD), Arch of Titus (81–82 AD), and Arch of Septimius Severus (203/212 AD) in the Forum Romanum, and the white marbles of the Pantheon are, for example, all Pentelic.
71. For example, Apollo Sosianus (34/29 BC?), Mars Ultor (42–42 BC), and Concordia (10 AD): Viscogliosi 1996, esp. 160ff.; Gasparri 1979, 39, 42, 57–58, 65–69, 77–78, 128, 130–31, 134, 140; Ganzert 2000, esp. 50–53, 97–110, figs. 30, 31, 33, 63–77.
72. Aristid. *Orat.* 26.6–7 (155 AD); trans. Behr 1981, 2:74. For the marbles involved: Bruno et al. 2002.
73. For the glass orb, see Bejarano Osorio 2002.
74. Serv. *Ad Verg. G.* 3.82. See André 1949, 25–31 (*albus*); 31–38 (*candidus*); 39–42 (*niveus, lacteus, marmoreus, argenteus, eburneus*); J. Clarke 2003, 50–66 (*candeo/candesco/candidus/candor*). For what follows, see the far more expansive discussion in Barry 2011a.
75. For example, Verg. *Aen.* 1.354 (the ghostly visage of Dido's dead husband, Sychaeus). See also, for example, Ov. *Heroides* 1.14 (Penelope pales at Hector's name) and "pallida mors" (Hor. *Carm.* 1.4.13). André 1949, 139–47 (*pallor/pallere/pallidus*); J. Clarke 2003, 120–23, 316 (*palleo/pallesco/pallidus/pallor*). For "rubeus," see Bradley 2004.
76. André 1949, 43–52 (*ater*), 52–59 (*niger*), 60–63 (*furvus, piceus, aquilus, coracinus*).
77. Museo Nazionale Romano, Rome, inv. no. 56230: Giuliano 1979, 274–77 (cat. no. 170), with bibliography.
78. Kreikenbom 1992, 213–15, cat. III 93, with bibliography. According to Martial and Statius, Domitian's side, breast, ear, and feet were sacred: Sauter 1934, 105–16.
79. Mart. *Ep.* 6.13.1–4: "quis te Phidiaco formatam, Iulia, caelo / vel quis Palladiae non putet artis opus? /candida non tacita respondet imagine lygdos / et placido fulget vivus in ore décor." The poem defies translation: "caelo" means "chisel" and "heaven" (because Julia is carved and divinised): "respondere" can mean both "responds" and "corresponds" (i.e., both "answers" and "matches"); "imago" can mean both "effigy" and "likeness." "Lygdus" means "Parian."
80. Antium Girl (Museo Nazionale Romano, Rome, inv. no. 50170): Lauter 1971; Borbein 1973, 134–37; Lucilla de Lachanal in Giuliano 1979, 274–77 (cat. no. 121).
81. *Anth. Pal.* 6.317: "Πραξιτέλης ἔπλασε Δαναὴν καὶ φάρεα Νυμφῶν / λύγδινα, καὶ πέτρης Πᾶν' ἐμὲ Πεντελικῆς."
82. Varro in Pliny. *HN.* 36.4.14. Lychnites was not quarried before the Hellenistic period: Leibdaros and Tsaimou 2000.
83. "Pectore candidus," meaning "pure-hearted" (Ov. *Pont.* 4.14.43); "candidus Hylas repellebat improbam Naida" (Petr. *Sat.* 83); "candida humanitate" (Petr. *Sat.* 129.11). Other examples in André 1949, 37–38. Parian as "*purus*" ("pure," "undefiled"), Hor. *Carm.* 1.19.6; as "*clarus*" ("bright," "shining"), Sen. *Phaedra.* 797.
84. Ov. *Met.* 15.845–50. Cf. Pliny. *HN.* 2.23.94; Suet. *Iul.* 88.
85. On Augustus as Apollo, see Suet. *Aug.* 94.4–5; 70.1. On the temple, see Zink 2008, with bibliography; on the House of Augustus, see Iacopi and Tedone 2006.
86. Prop. 2.31.1–3, 9–12; Ov. *Met.* 2.1–4. For an extended gloss of the latter passage: Brown 1987.
87. On the portico: Pensabene 1997, 151–92. The columns were giallo antico, pavonazzetto, and porta santa; the precinct paving was also Luna marble. Note that "radiare" ("to beam," "radiate") literally means "to furnish with spokes."
88. Ov. *Ars Am.* 3.113. See Gros 1976, 41–42, citing Prop. 2.31.1–2, 4.1.5; Ov. *Ars Am.* 3.9.43–44; Ov. *Fast.* 1.223–24. Augustus restored eighty-two temples in 28 BC alone: *Res Gestae Divi Augusti* 20.
89. Homolle 1890, 496–99; Bousquet 1938, 389–90; Martin 1965, 1:431. On ganosis, see Vitr. *De Arch.* 7.9.3–4; Pliny (*HN.* 33.40.122) says of Punic wax, "sic et marmora nitescunt."
90. Rutil. *De red.* 1.95–96. *Versus de Verona* (795/806 AD), 13, 15, 17: Godman 1985, 180.
91. Pliny. *HN.* 36.4.14. Michele di Pietro, in Pietrasanta, to Michelangelo, in Rome, May 13–16, 1518: "e vale marmi mostrono di volere mostrare la luna ne' pozo": Barocchi and Ristori 1965–83, 2:6.
92. The critical text is in Austin and Bastianini 2002, 148–51 (no. 118), ll. 3, 12: "Παρνησοῦ νιφόεντος νὰ πτύχ[α]ς ἤ παρ' Ὀλύμπωι . . . φήμη τις νιφόεντ' οἰκία τοῦ Παρίου. Note the repetition of νιφόεις" ("snowy," "snowclad"). Austin's own translation of l. 12 is instead, strangely, "an oracle to the snow-white house of the man from Paros."
93. Joseph. *BJ.* 5.222–24, in *Josephus: The Jewish War*, books IV–VII, trans. H. St. J. Thackeray (New York: W. Heinemann, 1928), 71. Cf. Joseph. *AJ.* 15.11.3.
94. Fischer and Stein 1994; Kaufman 2002; Demetrokalles 1994–98. Even Aliprantes (1996) persists in the error that the temple was Parian marble.
95. For the identification of the temple, see Warland 2003, 130–32. The mosaic was restored in 1929–31, but identical conventions are visible in the *Annunciation* in the same mosaics (Wilpert 1916, 3:pl. 56). The diagonal shading on gold backgrounds of the roof tiles suggests gilt bronze.
96. Ferrandus, *Sancti Fulgentii Ruspensis Vita*, ch. 13 (*PL* 65, col. 131A).
97. A selection only in an ever-burgeoning literature: Bankel and Liverani 2004; Brinkmann et al. 2007; Liverani and Santamaria 2014.
98. Summitt 2000, 281–97.

CHAPTER 3. ANCIENT GEOLOGY, LIVING ROCK, AND *EX UNO LAPIDE*

1. On the four elements in Egypt, Beretta (2009, 14–15) cites Diodorus Siculus (60/30 BC), Diogenes Laertius (3rd century AD), and Hecateus of Abdera (4th century AD). Oddly, he omits Sen. *QNat.* 3.14. For Greek presocratic texts on the four elements, see Cerri 1998.
2. Pl. *Ti.* 47e–51b; Arist. *Mete.* 1.339a 11–339b 2, 1.340b 19–28, 2.359b 29–360a 13, 3.378a 14–378b 6, 4.378b 10–25, 4.379b 10–26, 4.388a 10–15. Sokolowski 1970; Bolzan 1976; Longrigg 1985; Bostock 1995. See also the lively and incisive discussion in Bucklow 2009, 49–58.
3. Caley and Richards 1956; Eichholz 1965; Halleux 1974. For the date of *On Stones*, see Eichholz 1965, 8–11; Agricola 1556.
4. Arist. *Mete.* 1.341^{b}ff., 3.378^{a}15ff., discussed in Eichholz 1949. Olympiodoros (*Com. in*

Meteora. 6.4–6) mentions Aristotle's *περὶ μετάλλων*. See also Düring 1944; Horne 1966; Wilson 2013, 48–72, 271–77.

5. Eichholz 1965, 15–47. Aristotle's own commentators had difficulty distingushing between the efficient and material causes: Eichholz 1965, 42–44.
6. Eichholz 1965, 36–37.
7. Higgins and Higgins 1996, 13–14; Wickens 1986; Siebert 1990; Larson 2001, 226–67; Sporn 2010; Sporn 2013; van Opstall 2013.
8. Halleux 1970; Plumpe 1943.
9. Cic. *Verr.* 2.1.147–48; Alchermes 1994, 167n2.
10. Boson 1914, 43; Eliade 1978, 8, 33, 40–52; Pearce 2014.
11. Pliny. *HN.* 34.49.164–65. Pseudo-Aristotle (*de Mirabilibus Auscultationibus*, 93) relates that the mines on Elba stopped producing copper and started producing iron instead; Halleux 1970, 20. For life-giving air, see Paparazzo 2011, 92–93.
12. Papirius Fabianus cited in Pliny. *HN.* 36.24.125; Restoro d'Arezzo in Morino 1997, bk. 6, ch. 8; Capece 1546, 195; Milizia 1785, vol. 1, pt. 1, cap. XII ("Della combinazione de' vari marmi"), 335. Strab. 5.2.6; on Kleitarchos's dating, see Prandi 2012.
13. Inscriptiones Graecae, XII 55, 245. Bodnar 1973; Berranger 1983, 235–60; Edwards 1985, 800–814, no. 89; Bruno 2000, 91–94; Larson 2001, 179–81. For earlier quarries and their shrines, see Angliker 2019.
14. The dating of the invention of cameo is disputed: Plantzos 1996a; Plantzos 1996b. However, the earliest literary reference to a cameo may be in the *lithika* of Poseidippos (c. 310–c. 240 BC): Kosmetatou 2003.
15. On the Mouseion: Strab. 17.1.1–8; Ath. 5.196, 14.654; Fraser 1972, 305–55, 1:312–15, 2:467–69; di Pasquale 2005.
16. Cairo, Egyptian Museum, papyrus inv. 65445, ll. 142–54: "ἀσπάσιοι βα[σιλέως τοῦτο] δέχοισθε γέρας / ὃς καὶ λάινον [ἔργον ἐθ]ήκαιυ διψιλὲς οἴκωι / κτίσμα, πά[ρου λ]ευκὴν ἐκποδίσας σταγόνα, / εἰς ἡμίσφαιρο[ν τ]εύξας θέσιν· ἡ δὲ λυχνῖτις / ζωνὴ στυλοῦται πέζαν ἴωνι τύπωι· / ῥάβδου κοίλης ἔντος ἀποστίλβει δὲ συηνὶς / στικτὴ πρὸς πτερναῖς· κιόνος ἥδε θέσις. / ἡ δ᾽ ἀφ᾽ Ὑμηττοῦ πέτρος ἐρευγομένη πόμα κρήνης / ἐκδέχεται σπιλάδων ὑγρὰ διαινομένη. / εἰκόνα δ᾽ ὑμετέρην ἐτυπώσατο πίονι λύγδωι / πρηύνας, μέσσην δ᾽ ἥρμοσ[ε]ν Ἀρσινόην / σύγκληρον νύμφαις κατὰ πᾶν ἔτος. ἀλλ᾽ ἐπὶ πηγὴν / τήνδε μετ᾽ εὐνομίης βαίνετε Κρηνιάδες."

    I have omitted ll. 140–41: "θοινα . . . τε φλεγετ . . . / σιγηλου . . . ηρια καὶ Πτολεμ . . . ," as these lines are too fragmentary ("Feast . . . burning ? Silent . . . and of [?] Ptolemy"). Debate persists over the text and terminology, and the contributions have been many: for convenience, see Fernández Delgado and Pordomingo 2010, with previous bibliography. On the attribution to Poseidippos, see also Lehnus 2002. I follow the edition of Fernández Delgado and Pordomingo except for l. 144, where I retain: "Πά[ρου]." According to Lavagne (1988, 131–32), "κρήνη" indicates a fountain rather than a natural spring ("πηγή"), and "σπιλάς" an artificial rock.
17. Vanderpool 1953–54.
18. Papiri dell' Università degli Studi di Milano VIII.309, cols. I 1–IV 6: Bastianini et al. 2001, 28–45, 110–34. In the sprawling literature, I have found the following contributions most useful: Kramer 2002; Kosmetatou 2003; Kosmetatou 2004; Lelli 2004; Petrain 2005.
19. For the Ebers Papyrus, see Ghalioungui 1987. For the development of the lapidary, see Postgate 1997; Horowitz 1992; Schuster-Brandis 2008, passim. On the "περὶ λίθων" of Sotakos (Pliny. *HN.* 36.38.146) and "περὶ λίθων τιμίων" ("On Precious Stones") of "Zoroaster," see Halleux and Schamp 1985, xvi–xviii, xxii–xxiii, with references; Keyser and Irby-Massie 2008, 755, 851–52. Zachalias the Babylonian: Pliny. *HN.* 37.60.169. Soudines: Polyaenus *Strategmata* 4.20; Strab. 16.1–6; Pliny. *HN.* 9.56.115, 36.12.59, 37.9.25, 37.11.34, 37.24.90, 37.35.114, 37.50.133. The classical sources on lapidaries are largely reconstructed in Wellmann 1935.
20. Lelli 2004, esp. 136; Kramer 2002, 40–41; Petrain 2005, 340–43.
21. Theophr. *De Lap.* 6–7. Eichholz 1944; Caley and Richards 1956, 72–76; Eichholz 1965, 58–59, 92–93. Eichholz first gives "διάζωμα" (literally, "girdle") as "cornice," then "stone course" (his recourse to Ath. 5.206C, for the latter must be wrong, as tomb painting suggests alabaster in such cases).
22. Demetrios of Phaleron (c. 350–c. 280 BC) may have been chief librarian from 297 to 283 (Letter of Aristeas). Theophrastos's successor as scholiarch at the Lyceum, Straton of Lampsakos (c. 335–268 BC), served as tutor to Ptolemy II Philadelphus and himself wrote a "Περὶ τῶν μεταλλικῶν μηχανημάτων" (Diogenes Laertius, "On Machines for Mining," 5.58–59).
23. Although Page (1942, 1:451) remarked that "'having set free the white water-drop of Paros' must mean 'having quarried Parian marble'" he preferred to integrate the lacuna with the adverb "πά[ρος]" ("formerly") over the genitive "Πά[ρου]"; most editors follow him. "ἐκποδίζειν," a *hapax*, is the antithesis of "ἐμποδίζειν," "to bind [the feet]" (cognate with "impede"), and can be translated as "unleash" or "let run."
24. Rabbi Joshua bar Hanania. Rabbinical Midrash of Leviticus Rabba, 22, 27, cited in Grossmark 1998, 281.
25. *Descriptions*. 5.3, in *Philostratus, Imagines; Callistratus, Descriptions*, ed. and trans. Arthur Fairbanks (London: W. Heinemann, 1931), 393.
26. Lavagne 1988, 127–35.
27. Sear 1975, 88–92.
28. Szafranska 1989; Cenci 1992; Morel 1998; Rinaldi 1999; Heidet 2004.
29. Vanorio and Kanitpanyacharoen 2015.
30. Blackman and Grewe 2001.
31. Porta Maggiore: Coates-Stephens 2004, with bibliography. Claudium: Colini 1944, 137–62; Prandi 1953, 373–420.
32. Stat. *Silv.* 1.5.27–28; Rutil. *De red.* 1.97–99. For the converse idea of trumping nature by spanning her rivers, see Kleiner 1991. For Theodore II Lascaris: Mango 1972, 245.
33. Cassiod. *Var.* 4.51.3–4 (Theodoric to Symmachus). Mommsen (1894) dates this letter to 507/511 AD. The outer circuit had already collapsed in the early 5th century (*CIL* 6:1191, "THEATRUM POMPEI, [collapso] EXTERIORE AMBITU"). Shelley 1840, 2:120 (22 December 1818).
34. Schneider 2002, 88–91, 97.
35. Walters 2000.
36. Museo Nazionale Romano (Palazzo Altemps), Rome, inv. no. 8572: Palma and De Lachenal 1983, 1.5 (I Marmi Ludovisi nel Museo Nazionale Romano), 185–88 (cat. no. 79). Schneider 2002, 96.
37. Adams 1930, 10. On Pliny, see also the brief notes in Eichholz 1962. The classic reference book for technical literature on stones is Lenz 1861.
38. On Pliny, see Beagon 1992, 39–40, 92–102. On Lucretius, see Long 1997, 132–33, citing Lucr. 3.931–62; Rosenmeyer 2000.
39. *HN.* 36.1.1–2. For human subversion of nature, see Wallace-Hadrill 1990.
40. *Anth. Lat.* 162 (173): "Aerio victus dependet Marsya ramo / nativusque probat pectora tensa rubor / docta manus varios lapidem limavit in artus / arboris atque hominis fulget ab arte fides." Four life-size Marsyas in Phrygian marble survive in Florence, Rome, and Karlsruhe, and two statuettes in Worcester and Palermo. Two other statues (Taranto, Kos) are in white marble but bear traces of red paint: Weis 1992, app. A, 42, 106, 49, 31, 44, 20; Agnoli 2014. For the River Marsyas: Strab. 13.8.15, Xen. *An.* 2.7, Pseudo-Plutarch, *De fluviis*, 10.
41. Constantine Manasses (late 12th century) describes a relief of the "Cyclops tearing asunder Odysseus' companions and eating them . . . a stone suffused with red. . . . I marvelled at the skill and inventiveness of the craftsman, that, wishing to carve a slaughter and deceit, he cleverly devised the work in suitable fashion, and made its basic [material] of a colour matching the subjects of the carvings, in order that the stone not be engrained with spurious and alien tints, but should be bathed in blood from its core." Text in Sternbach 1902, cols. 83–5, ll. 1–6; trans. Maguire 1992, 139–40.
42. Stat. *Silv.* 1.5.22–23. See also Stat. *Silv.* 2.2.87–89. Catullus (*Carm.* 63.7) had earlier described Attis "staining the ground around with his shed blood."
43. *Anth. Lat.* 161 (172R): "De Daphne. / Frondibus et membris servavit dextera sollers / congruus ut sculptis posset inesse

color. / Dant mirum iunctae ars et pictura decorem, / ostendit varius cum duo signa lapis."

44. *Apollo and Daphne:* Princeton Art Museum, Gift of the Committee for the Excavation of Antioch to Princeton University y1965–219. *Triton:* Istanbul, Archaeological Museum, inv. 5247 T, unpublished.
45. Archaeological Museum, Aphrodisias: Rockwell 1991, 138–40.
46. Pliny. *HN.* 36.1.2–3, 37.9.23; Strab. 2.3.4.; Sen. *QNat.* 3.25.12; Diod. Sic. 2.52.1–2; Claudian in *Anth. Pal.* 9.753–54; Claud. *Carm.* 56–62. Bühler 1973, 22; Kalus 1991; Shalem 1994, 6.
47. Halleux 1974, 19–60. Isidore. *Etym.* 16.17.2 may be the first use of "metalla" to mean "metals" but, even in his time, "metalla" applied to glass tesserae too.
48. Alexander of Aphrodisias (appointed to his chair of philosophy between 198 and 209 AD) and Olympiodoros (495/505–after 565 AD): Eichholz 1965, 42–44.
49. Sheedy and Pike 2008.
50. Naxian "[τ]ô αϝὐτô λίθο ἐμὶ ἀνδριὰς καὶ τὸ σφέλας, Attic τοῦ αὐτοῦ λίθου εἰμὶ ἀνδριὰς καὶ τὸ σφέλας." Bruneau and Ducat 1983, 127–28; Kontoleon 1973; Courbin 1974; Boardman 1978, 51, caption to fig. 60; Chamoux 1990; Holtzmann 1996, 203n23; Wesenberg 2000; Di Cesare 2004; Queyrel 2014.
51. Bruneau 1988; Butz 2009. On mantic stones, see O'Bryhim 1996.
52. Angliker 2019.
53. Donohue 1988.
54. Paus. 8.37.3. For the sanctuary and statue group, see Dickins and Kourouniotis 1906/7; Jost 1985, 172–78.
55. Pliny. *HN.* 36.4.37. A summary history of the topos from Herodotos to Master Gregorius is given in Settis et al. 1999, 79–81. Pliny mentions two other sculptures "ex uno Lapide": *Winged Cupids Playing with a Lioness* (*HN.* 36.41) and *Chariot with Apollo and Diana* on the Arch of Octavian (*HN.* 36.36).
56. Claud., *Carm.* 7 [87]: "De quadriga marmorea. / Quis dedit innumeros uno de marmore vultus? / surgit in aurigam currus, paribusque lupatis / unanimi frenantur equi: quos forma diremit, / materies cognata tenet discrimine nullo. / Vir redit in currum; ducuntur ab axe iugales; / ex alio se quisque facit. quae tanta potestas! / una silex tot membra ligat ductusque per artem / mons patiens ferri varios mutatur in artus."
57. Stokes 1934, 43.
58. Vase no. 216791, Paris, Musée du Louvre, G372; see Cromey 1991, with bibliography. For Segni, see Cifarelli 2003, 68–72 (dating), 88–96 (construction). Magister Gregorius (*Narratio de Mirabilibus Urbis Romae,* 11): "The marvellous theater in the marble mountain at Heraclea . . . is carved in such a way, that all the chambers in the structure, all the seats in the cavea, and all the exit ramps and caverns were carved from a single solid stone." On the amphitheater at Cagliari, see Levi 1942.
59. "ἁρμογή": Polybius 6.18.1; Gal. *Nat. Fac.* 19.460; Joseph. *AJ.* 13.11.3; Pollitt 1974, 150–54. In painting and sculpture, the term also referred to the seamless meshing of one part with another. It is cognate with "ἁρμόζω" ("to fit together"). "ἁρμονία": Paus. 8.41.8, on the Temple of Apollo at Bassae; Paus. 8.8.8, on the blocks in fortifications; Diod. Sic. 2.8.2, on the construction of a bridge.
60. Amphion: Ap. Rhod. *Argon.* 1.745–51; Apollod. *Bibl.* 3.5.5; Paus. 9.5.6–8. The idea of harmony is more explicit in Prop.1.9.8–10, 4.2.3–4 ("saxa . . . agitata per artem / sponte sua in muri membra coisse ferunt"); Hor. *Carm.* 3.11.1–4; *Ars P.* 394–96; Honestus of Corinth (c. 30 AD), *Anth. Graec.* 9.250; also, by the same poet, *Anth. Graec.* 9.216. *Anth. Lat.* 210.6, Flavius Felix on the Baths at Alianae: "Ostendens pronis currere saxa iugis." Megara: Paus. 1.42.2; Ov. *Met.* 8.17.
61. Lugli 1957, 1:207–8; Martin 1965, 114–99; Orlandos 1966–68, 2:99–100; Coulton 1982, 46–47. For Middle Eastern origins of this technique, see Nylander 1970, 82–88, with bibliography.
62. For the cella wall of the Temple of Venus Genetrix in the Forum Iulium, dated to the Trajanic rebuilding, see Amici 1991, 88. For the Mausoleum of Plancus, see Fellmann 1957. For the cella wall of the Temple of Hercules Victor: Strong and Ward-Perkins 1960, 11–16, 32; Heilmeyer and Rakob 1973, pl. 17.1.
63. "Coniunctio lapidum": inscription (*CIL.* 8:14728) on a triumphal arch (379–383 AD): Lepelley 1979–81, 2:251 (where it is erroneously cited as no. 14729): "QUADRA[ATIS LAPIDIBUS OLIM] EXTRUCTUM . . . EORUNDEM LAPIDUM CONIUNCTIONIS"; Angilbert, *Carmen de Karolo Magno III,* v. 94 (790/814 AD), on Charlemagne's palace at Aachen ("saxa locat, solido coiungens marmora nexu"), see Schlosser 1896, 25, cat. 97.
64. "hic tantam faciem superesse sepulchri / perpetua novitate sui, sic stare nitentes / consensus lapidum, sic de radice levatos / in melius crevisse gradus, ut et angulus omnis / sic quasi mollitae ductus sit stamine cerae." Tomb of the Flavii, Colonia Flavia Cillium (Kasserine, Tunisia). For the full text, see Courtney 1995, 188–93. The monument is thoroughly analyzed in Groupe de recherches 1993.
65. Sen. *Ep.* 86.4, 90.10. See also *Ep.* 8.5–6; 110.14–18; 115.3–9. Seneca the Elder also reviles "that stone which is cut to cover walls with its thin veneer" (*Controv.* 2.1.12). Ovid, in his treatise on female cosmetics, also suggests this sort of beauty is only skin-deep: "Lofty halls are plated with gold, black earth lays hidden under set marble" (*Medic.* 7–8).
66. Magister Gregorius, *Narratio de Mirabilibus Urbis Romae,* 29: "Si lapis est unus, dic qua sit arte levatus, / Si lapides plures, dic ubi congeries."
67. DeLaine 2002, 219–20. See also Carettoni 1961.
68. *Orat.* 27: Keil 1958, 2:129–30; trans. Behr 1981, 101, 105.
69. McEwen cites the variants of Sextus Empiricus and Pomponius: McEwen 2003, 55–66.
70. Plut. *Peric.* 13.3: Bernadotte Perrin, trans., *Plutarch's Lives,* 11 vols. (London: W. Heinemann, 1914–26), 3:41. Hephaistos: Hom. *Il.* 8.539, 12.323, 17.444; *Od.* 7.94.
71. Taut 1936.
72. Thomas and Witschel 1992.
73. Cassiod. *Var.* 7.15.3: "quid dicamus columnarum iunceam proceritatem? moles illas sublimissimas fabricarum quasi quibusdam erectis hastilibus contineri et sub tanta aequalitate concavis canalibus excavatas, ut magis ipsas aestimes fuisse transfusas, ceris iudices factum, quod metallis durissimis videas expolitum, marmorum iuncturas venas dicas esse genitales, ubi dum falluntur oculi, laus probatur crevisse miraculis."
74. For the details and an alternative (rather pedestrian) explanation, see Shoe 1949, esp. 344–48, 349.
75. Boulter 1970; Harrison 1977; Palagia 2000. Skopas simulated the Erechtheion technique on the metopes of the later 4th-century Temple of Athena Alea at Tegea. Stern 1985; Lesk 2004, 427–28, 449, 510, 575, 583–84.
76. Cooper 1992, 8, 37, 98–119, 257–79 (roof), 296 (Ionic capitals, except for the pair at the end, which were limestone), 315 (Corinthian column and capital).
77. Cooper 1992, 245.
78. Coarelli 1985, 245–57, with bibliography; Middleton 1886a, pls. at 426/27 and 430/31.
79. Fant 1988, esp. 151–52.
80. Manil. *Astron.* 5.510–12: "Augusta nitent sacratis munera templis / aurea Phoebeis certantia lumina flammis / gemmarumque umbra radiantes lucibus ignes."
81. Gros 1976; Viscogliosi 1996, esp. 160ff.; Gasparri 1979, 39, 42, 57–58, 65–69, 77–78, 128, 30–31, 34, 40; Ganzert 1996; Ganzert 2000, esp. 50–53, 97–110, figs. 30, 31, 33, 63–77. Excavations of Agrippa's Pantheon revealed an opus sectile floor: Steinby 1993, 5:280–83. For Brescia, see Rossi 2014, 223–72, 283–84. See also the cella of the Capitolium at Pompei in a tinted engraving of 1824: Mielsch 2001, 13, fig. 5, after Mazois 1824.
82. Schneider 1986, esp. 115–25; Schneider 2001, 7–8; Cohon 1990; Lipps 2016. For a Phrygian marble throne accommodating a seated Augustus in white marble, see Cohon 1989.
83. Rakob 1979.
84. Petilä-Castrén 1987, 132; Hölscher 1988; Schneider 1986, 145–46; Fant 1993, 147; Fant 1999, 278–79.

85. Waelkens 1985; Schneider 1998.
86. Sid. Apoll. *Carm.* 5.34–41, "Panegyric on Maioranus" (c. 460 AD). A different number symbolism is at work in Sid. Apoll. *Carm.* 11.17–18: "Here is stone from five regions, giving five colours / Ethiopian, Phrygian, Parian, Punic and Spartan / purple, green, mottled, ivory and white."
87. On the quarries, see Fant 1989.
88. Tibullus 3.3.13; Hor. *Carm.* 3.1.41–46; Juv. 14.305–8; Ov. *Her.* 15.141–42; Aus. *Mos.* 48–54. For other literary references until the 10th century, see Robert 1962.
89. The literature is unwieldy, but see mainly Delbrück 1932; Deér 1959; de Blauuw 1991; Butters 1996.
90. Morton 2003, 63–67, 95, 111–13, 133–47, esp. 148; Baratte et al. 2004.
91. Robert 1940, no. 98 (ll. 11–12).
92. Yanni 1999, 52, 114.
93. See, for example, Peacock and Williams 1999; Isserlin 1998; Paton and Schneider 1999.
94. Schneider 1986, 137–38, 150–52; Schneider 2001, 6.
95. Pliny. *HN.* 36.1.1–2, 36.24.123–25. Cf. Beagon 1992, 41–42.
96. Ov. *Ars Am.* 3.125; Pliny. *HN.* 36.24.125.
97. In 333 AD, Barbarius Pompeianus declared that he had beautified Abella, Campania, "silicibus e montibus excisis non e dirutis monumentis advectis" (*CIL* X, 1199).
98. The 4th- and 5th-century legislation against spoliation is summarized in Janvier 1969. For an overview of earlier legislation from the Republic to Constantine, see Anguissola 2002.
99. When the *Praefectus Urbis* Lampadius robbed materials from standing buildings in Rome (Amm. Marc. 27.3.8–10).
100. *New History*, 5.24.5; Ridley 1982, 111–12.

CHAPTER 4. PAINTING IN STONE

1. Uffler 1971.
2. Fragments painted with curving bands in yellow, gray, red, and white, from the "Loom-Weight Deposit" at Knossos. See Fyfe 1902, 109, fig. 2; Evans and Evans 1921, 251, fig. 188a, b; Immerwahr 1990, 22, fig. 6f. For the dating (MM IIB, Protopalatial), see Hood 2005, 48–49. On Minoan use of lime, see Jones 2005, 199–209, table 13.1; on technique, 217–20, table 13.3. See also Noll et al. 1975, 93; Cameron et al. 1977, 147–57.
3. Blakolmer 1997.
4. Gale et al. 1988; Chlouveraki 2002; Chlouveraki 2006; Chlouveraki and Lugli 2009; Shaw 2015, 135–40.
5. Warren 1969, esp. 124–56, 182–91; Warren 1989. In general, see Bevan 2007.
6. For nodular and lenticular selenite and balatino, both calcium sulfate dihydrate, see Chlouveraki 2006, 1:132–35; 2:figs. 2.18–19, 3.10–11, 3.39–43, 3.64–65, 4.3–4, 7.84.
7. Betancourt et al. 1979, esp. 3–5, 14–16, pls. 2–5. The slip was probably made from iron-rich calcareous clay. Schiering first suggested that Vasiliki ware–inspired fresco: Schiering 1960, 17–20. See also Schiering 1965; Schiering 1987.
8. Haggis 2007, 731–37, figs. 5d, 13–16; Warren 1996.
9. For "Print ware" frescoes in the South-West House (MM IIA), see Koehl 2013, 170–71, figs. 1–2. For "splash painting," see Abramovitz 1980, 81–84; E. Davis 2007, 145–47.
10. For the composition of Minoan pigments: Jones 2005, 209–16, table 13.2. See also Cameron et al. 1977, 139–46, 157–60.
11. See a 1977 interview with Diego Rivera's assistant, Eric Halberstadt, discussing Rivera's fresco practice: SFMoMA, "Painting Diego Rivera's Frescoes," https://www.sfmoma.org. On lime production without kilns, see Russell and Dahlin 2007.
12. Estrin 2015.
13. For the pigments, see Rasmussen et al. 2004; Hejl and Tippelt 2005.
14. Shaw 2015, 19–20, 26, 113, 217; Hirsch 1975; Hirsch 1977a, 7–22; Shaw and Shaw 2006, 208–11.
15. Podium on the east side of Hall 64: Parrot 1958a, 105–6, 165; Parrot 1958b, 67–69, fig. 54, pl. 15. See also Winter 2000, 750–51; Niemeier and Niemeier 2000, 772–73; Koehl 2013, 172–73, esp. fig. 4 (unpublished fragment excavated 2004). Brief technical analysis of the frescoes is in Muller 1995. Winter and Koehl call the podium frescoes "monochromes," but it is only the excavation photographs that were in black and white. The dados of a passage leading into Court 31 of the "Royal Apartments" were also painted in imitation of a veined stone: Parrot 1958a, 165, pl. 39.2. Plinth near to the northwest corner of Court 31: Parrot 1958a, 166, fig. 187, pl. 39.1. Parrot 1958b, 67–69, 109–10, fig. 54, pl. 15.
16. Durand 1997–2000, 3.15–18 (nos. 855–56); Michel 1999.
17. Woolley 1955, 232, pl. XXXIX ("purplish-black . . . flecked with very small spots of red and green which give a marbled effect"); Niemeier 1991.
18. The "West House," Akrotiri, Santorini, rooms 4, 5: Televantou 1990; Doumas 1992, 30–31, 46, 49, figs. 14–17, 24, 49–56, 63–64; Televantou 1994, pls. 3–6, 9, 12–17, 19–20.
19. Bietak et al. 2012/13, 137–38, fig. 5; Bietak 2013; du Buisson 1935, frontis., 143, no. 2; von Rüden 2011, 32, 81–83, taf. 22, 23. See also Pfälzner 2013, 200–215; Niemeier and Niemeier 2000, 765–80. For the carbon-dating of Tel Kabri, see Cline et al. 2017.
20. Mycenae: Rodenwaldt 1919, esp. pl. 7; Lamb and Wace 1919–21, 193–95, pl. 55; Hirsch 1977b; Hirsch 1980; Rodenwaldt 1912, 2:23–29; Pylos: Lang 1966, 164–69. Tiryns: Hackl 1912. Gla (Room D): de Ridder 1894, 291. Egan 2016. See also a gridded floor imitating veined marble slabs at Trianda (Ialysos), on Rhodes, from the Middle Bronze Age: Marketou 2013.
21. Plaster fragments sponge-painted to imitate conglomerate stone in Building AA at Kommos: Shaw and Shaw 2006, 182 (no. 96), 224–25, pl. 2.38a, with copious other examples.
22. On timber reinforcing: Shaw 2015, 68–72, 85–88. On triple zones: Evely 1999, 68–69, 133–35. On orthostates: Shaw 1983; Shaw 2015, 122–24. See also the remarks in Dworakowska 1992.
23. "Inkrustationstil" was coined by Mau (1882, 11–23); "Structural Style" by Hinks (1933, 8); and "Masonry Style" by Dawson (1944, 58, 72).
24. For classical/Hellenistic precedents in Athens: Wirth 1931; Bruno 1969. Particularly important is the reconstruction (of the Hieron interior, c. 325 BC) in Lehmann 1964.
25. Guldager Bilde 1993. The earliest surviving example is a five-band wall of alternating white and "marble" blocks in the "House of Many Colors" at Olynthos (390/348 BC): Robinson and Mylonas 1946, 139, 182–206, 291, fig. 30, pl. 43. For other examples at Olynthos and Amphipolis: Andreou 1988, cat. nos. 23–24, 144–47. The literature on individual finds is now too vast to be cited here.
26. The best remains are in the houses of Dionysos, of Hermes, of the Trident, of the Masks, and of the Comedians: Bulard 1908; Chamonard 1922–24, 2:356–91; Alabe 1995; Westgate 2000b, 397–400, 402–4, 406–13. Only two houses have been fully published, the House of the Masks and the House of the Comedians: Chamonard 1933; Bezerra de Meneses 1970.
27. Excavated 1976–81, cursorily published without notes, excavation report, or drawings in Siganidou 1982. See also Ginouvès et al. 1993, 136–38, 172–73, 177, 180. See also the faux–Masonry Style paintings of the "tombe à ciste 22" (late 2nd century BC) at Amphipolis: Malama 2007, 118–19. Malama also cites tombs at Thessaloniki and Symbola.
28. Puglia: Tomb of the Pomegranate and Tomb of the Pilaster at Egnatia, the Hypogeum of the Palmieri at Lecce, and Tomb 7 of the Akropolis of Monte Sannace-Gioa del Colle: Steingräber 2000, 37–40, pls. 31.1–2, 32.1, 33.1–2. Tomb 8 at Monte Sannace also has faux-isodomic masonry. See also Mazzei 1995, 197–98; Mazzei 2002, 67–77. Two of the Pulian examples given by Tinè Bertocchi (1964, tav. 6) have dados with diagonal striping that suggest book-matched marbling. See also the Peristyle House (c. 300 BC) at Iaitas (Monte Iato), Sicily: Brem 2000, esp. 116–21.
29. Cell VII. Continuous dado zone in light gray with reddish veining; ocher fascia; two rows of faux-isodomic masonry with blocks painted black, yellow, and red. The blocks bend around the corners, unlike

real masonry; above this a Lesbian *kyma* in yellow on a gray background. The tympana are painted red: Francesco Roncalli in Buranelli 1987, 81, 107, figs. 26–29. A marble orthostate is also painted in the slightly later Tomb of the Garlands, Tarquinia: Steingräber 1988; Harari 2010.

30. Cavari and Donati 2015, 63–74, with bibliography. On the chemical composition: Cavari et al. 2015. Roman fresco technique would become ever more refined, requiring three or four layers of plaster for its effect: Vitr. *De Arch.* 7.3.
31. Iorio 2007.
32. In a broad literature, see Laidlaw 1985; Rosenberg 2010; Marcattili 2011, with copious bibliography.
33. Bruno 1969, 306–8, 317; Walter-Karydi 1998, 33–55; Coarelli 1983; Laidlaw 1985, 307ff.; Leach 2004, 61–63; Wallace-Hadrill 1994, 25–26.
34. Previously dated to 130/120 BC, now to 89/80 BC: Ohr 1991, 17–18, 35–36; Dobbins and Foss 2007, 159, 167–72. The *terminus ante* is a graffito scratched into the plaster in 78 BC: *CIL.* 4.1842. An 1879 scale model (Museo Archeologico, Naples) shows how veined and multicolored the much-faded plaster once was.
35. House of the Ceii: Michiel 1990, figs. 58–68. On external polychromy, see Santoro Bianchi 2001–2.
36. Leach 1993.
37. Mols 2005.
38. Vitr. *De Arch.* 7.5.1: "antiqui qui initia expolitionibus instituerunt imitati sunt primum crustarum marmorearum varietates et conlocationes, deinde coronarum et <silaceo>rum cuneorum inter se varias distributiones." The lacuna has been restored as either *siliculorum cuneorum* ("pod-shaped wedges") or *silaceorum cuneorum* (Fra Giocondo's interpolation, literally "ocherish wedges"). Rouveret and Romano argue the former, Eristov the latter: Rouveret 1985, 438; Romano in Gros et al. 1997, 2:1085n140; Eristov 1987, 121ff. At *De Arch.* 7.4.4, Vitruvius specifies, "In [winter triclinia] painted black and polished panels should be set above the dadoes, with ocherish or cinnabar blocks [*cuneis silaceis*] interposed between them."
39. Wesenberg 1975–76, 24ff; Elisa Romano in Gros et al. 1997, 2: 1084–85n139; Rouveret 2002, 109–11. For a summary of the controversy over whether Vitruvius means faux-ashlar (Masonry/First Style) when his text mentions only faux-revetment, see Leach 2004, 59–64.
40. For example, Room 12 in the acropolis palace at Jebel Khalid, the Hellenistic room in Beirut, and room at Tel Anafa: G. W. Clarke 2003, 18–19; Aubert and Eristov 2001, 213; Kidd 2001, 6, fig. 3. Pompei and Herculaneum: Mau 1882, 114ff.
41. Cairo, Egyptian Museum, Zenon papyri no. 3 59445, ll. 2–8, in Hunt et al. 1932–34, 1:no. 171; Nowicka 1984.
42. Calamiotou et al. 1983 (kind reference of Nikos Akamatis). Vitruvius (7.7–14) and Pliny (*HN.* 35.12–13) make it clear that the relative costs of pigments were well known to contemporary viewers; in fact, cinnabar was provided by the patron, not the painter. On the value of pigments, see Brecoulaki 2014.
43. Orlandos 1966–68, 1:151–53. Delos, House of Dionysos, House of the Sword: Bulard 1908, 113–14, 152, pls. 6A *a*, 8A *k*; Bruneau 1970, fig. 109. Excellent reproductions of Pergamon, Erythrai, and Knidos in Bingöl 1997, 89–92. Iasos: Tober 2011.
44. On the Villa Farnesina: Jones 2015, 118–28, with references.
45. Cic. *Div.* 2.48–49; Janson 1961.
46. Pliny. *HN.* 36.4.14. The "Relief of Adamas": Stuart 1816, ix; Reinach 1909–12, 2:360, no. 1.
47. For figural friezes at Delos, see Bruneau 1970, 151–93, pls. 21–26. See also Knidos, Hellenistic House 1, North Wall: Mellink 1969, 218, figs. 16, 19; Love 1971, 397–98, ill. 4. For Pompei, see Mau 1882, 210; Cassetta 2011. For other examples as far afield as Syria (Area 19, "House of the Painted Frieze," Jebel Khalid, c. 150 BC): Jackson 2009.
48. Pliny. *HN.* 33.39.117: "The old artists [Zeuxis, Polygnotos, etc.] used to paint with cinnabar the pictures that are even now called monochromes." Cf. *HN.* 35.5.15; 35.9.64; 35.29.11; 35.34.56, and Quint. *Inst.* 11.3.43. Bruno instead argues that the Delos monochromes on black derive from the friezes of the Erectheion (white marble figures over gray-blue Eleusinian stone), but even these could also represent dark cameo: Bruno 1985, 43–45.
49. The Pella image was first noticed in Laidlaw 1985, 33n28. The Pompeian image is noted in Eristov 1979, 697, 757 (fig. 4). The vine-leaf is from Cubiculum F, Villa of P. Fannius Synistor, Boscoreale (Metropolitan Museum, New York). Éva Dubois-Pelerin (in Barbet and Verbanck-Piérard 2013, 1:39) considers the image merely an "accident" during painting.
50. Mart. *Ep.* 8.55.6 (86/103 AD): "marmore picta Nomas"; Stat. *Silv.* 1.3.35–36 (c. 93/94 AD): "marmora picturata lucentia vena."
51. Ennod. *Carm.* 2.10.4–7 (c. 500/514 CE): "tamen arte regat natura figuras / viscera dum lapidum fingit imaginibus / candorem roseo perfundat docta robore / depingat sparsis congrua membra notis."
52. *TLL* 1900–, vol. 10, fasc. 14, 2081–84. For pictura as a medium or as variegation: Plaut. *Most.* 262; Cic. *Verr.* 2.4.1 ("nego ullam picturam neque in tabulis neque textilem fuisse"); Pliny, on tree foliage (*HN.* 16.95), on gems (2.207), on animals (7.7), snake markings (8.85), peacock colors (24.162); Stat. *Theb.* 6.58 ("picturatus . . . agger floribus"); Claud. *Ep. ad Ser.* 3 ("picturataeque volucres"). With regard to clothes the phrase might indicate either rich weave or figuration: Verg. *Aen.* 3.483 ("picturatas auri sebtemine vestes"); Lucr. 2.35 ("textilibus . . . picturis"). Note also that "pictura" is originally a future participle, not a noun. For pictura as procedure, see Quint. *Inst.* 2.18.2: "quam ποιητικήν appellamus, qualis est pictura." Properly speaking, a panel painting is a "*pinax*" or "*tabula picta*," though "pictura" is also used synecdochically: Plaut. *Asin.* 764; Cic. *Tusc.* 5.114; Sen. *Ep.* 5.34.14, 16.8.
53. Pliny. *HN.* 35.36.109, 110; 35.40.145.
54. Plin. *HN.* 36.9.55–56: "Augustean and, more recently, Tiberian marble, which were found in Egypt for the first time during the principates of Augustus and Tiberius respectively . . . in Augustean Serpentine [the markings] curl over like waves so as to form curls, while the Tiberian has scattered greyish-white spots which are not rolled into curls." "Marmor Augusteum" is possibly Metagabbro granite (*granito verde della sedia*), and "Marmor Tiberianum," quartz diorite (*granito bianco e nero*).
55. Vitruvius calls plaster surfaces "*politiones*" ("polishings"): Vitr. *De Arch.* 7.3.9; 7.5.1. Alabaster-grit has been detected in the frescoes from the House of Livia; calcite crystals have been detected in pigment from Cologne: Ling 1990, 200, 204 with bibliography.
56. Pliny. *HN.* 36.60.184. For the techniques: Vitr. *De Arch.* 1.7.
57. Moormann and Swinkels 1983. For an example of this tessellation on Delos: Westgate 2000a, 256–60. See also the marble emblema in the House of the Labyrinths: Guidobaldi and Vincenti 2005.
58. Laidlaw 1975. For the date of the decorations: Hoffmann and Faber 2009, 111.
59. Mau also saw a panel from which a cup seemed to emerge: Mau 1882, 51. The watercolor of the centaurs by Giuseppe Marsigli (Naples Archaeological Museum, inv. ADS 405; October 23, 1831) was first published in *Pompei* 1981, 137, 190. For the bird, see Parland 1913, pl. 1b.
60. Vitr. *De Arch.* 7.3.10. When the Temple of Ceres, near the Circus Maximus, was restored (in 27 BC), the wall reliefs, by Damophilus and Gorgasus, were cut out and enclosed in framed panels (Pliny. *HN.* 35.45.154): Bergmann 1995, 100, with bibliography.
61. Athens: Meritt 1970. Sikyon and Sparta: Ath. 6.253B, 13.577C; Paus. 3.15.8. Olympia: Paus. 5.21.17. Delphi: Roux 1952, 183–85.
62. The ornamental motifs are palmettes and lotus flowers: Lucrezia Ungaro in De Nuccio and Ungaro 2002, 114–22 (cat. nos. 62A–70); Ungaro 2004; Santamaria et al. 2004.
63. For the paintings by Apelles, Suet. *Aug.* 72.3: *Alexander with the Dioscuri and Victory* and *Alexander in a Chariot with a Figure of War.*

Parerga: Pliny. *HN.* 35.36.101. For pinacothecae and murals: Leach 2004, 132–55; Jones 2014, 298, 300–301; Jones 2015.

64. On opus interrasile, see Bonanni 1998, esp. 259 and n12; Quaranta and Bruto 2006. For the prototypes in painting, see Gasparri 1985, 61–87; Fusco 2010, 88. On opus sectile, see also Dohrn 1965; Gasparri 1985; Fusco 2010, 83–92; Mari and Fiore Cavaliere 2001, 433–34; Mari 2004, 292; Fusco 2006, 25–27.
65. *HN.* 35.1.2–3: "nunc vero in totum marmoribus pulsa . . . nec tantum ut parietes toti operiantur, verum et interraso marmore vermiculatisque ad effigies rerum et animalium crustis. (3) Non placent iam abaci nec spatia montes in cubiculo dilatantia: coepimus et lapide pingere."
66. Pliny. *HN.* 7.3, 9.52, 21.1. For example, the phrase is translated "to paint on stonework" in Carey 2003, 108. However, Michael Reeve kindly informs me that all the authoritative Mss. have the ablative, "lapide." Elsewhere (*HN.* 36.5.46) Pliny remarks, "Painting would not have been esteemed at all, let alone so highly, had marbles enjoyed any prestige at all" when they were first quarried.
67. *HN.* 35.1.2–3: "hoc Claudii principatu inventum, Neronis vero maculas, quae non essent in crustis, inserendo unitatem variare, ut ovatus esset Numidicus, ut purpura distingueretur Synnadicus, qualiter illos nasci optassent deliciae. Montium haec subsidia deficientium."
68. Barry 2017.
69. Apicius, *De re coquinaria,* 7.17 (especially *In ovis aplis*). While Apicius's cookery recipes are thought to have been compiled in the late 4th or early 5th century AD, Pliny mentions Apicius several times (*HN.* 8.209, 9.30, 10.133, 19.137, 19.143). Fusco (2006, 22, 23) translates "ovatus" as "giallo-uovo" ("yolk-colored"), but no such gloss is recorded in *TLL* 1900– (9.2, 1187), and Pliny uses the word only to mean "oval" (*HN.* 15.23.85, of berries, and *HN.* 16.8.21, of acorns).
70. Gregory of Nyssa, *In Ecclesiasten homilia* 3.4.321A (mid- or late 4th century), in *PG* 161 v. in 167, 44, cols. 653, 656. See also Stat. *Silv.* 2.2.86–89: the stone "which the vein of oriental Syene has speckled [i.e., Aswan granite], and in Synnas [Phrygia], where the white ground is divided by purple curves in the painted marble." Claudian (399 AD) calls Phrygia "precious for the marble painted with purple veins, which Synnada quarries" (*Eutrop.* 2.272–73).
71. Museo Capitolino, Rome, inv. no. 404: Fittschen and Zanker 1983, 3.1:118–19, cat. nos. 180, 118, figs. 209–10. A smaller-scale portrait bust is in the Museo Nazionale delle Terme, inv. 65190: Anderson and Nista 1989, 92–94.
72. Zigzag joins (1st–4th centuries AD) are analyzed only in terms of repair, rather than creative intervention, in Fogagnolo and Lugari 2008; Fusco 2006, esp. figs. 4–8.
73. Ostia, inv. 39898 a–b: Pensabene 1994, 53 (cat. no. 3), figs. 59–64. For other columns where individual tasselli simply conceal iron cramps or pins along shear fractures: Pensabene 1994, figs. 66–72, 268–71; Bonanni 1998, 270n101; Matthias Bruno in De Nuccio and Ungaro 2002, 528–30. However, in this case the technique is too meticulous to be economic, and far exceeded the requirements of repairing a damaged shaft. Adam instead considers the result a "marbre polychrome artificiel": Adam 1994, 126, seconded in Fusco 2006, 33–34.
74. Fusco 2006, 35–37, fig. 12.

CHAPTER 5. HOMES FIT FOR HEROES

1. Decorated tiles also embellished doors and furniture in the palace of Amenhotep III at Malkata, of Akhenaten at Tel el Amarna, of Ramesses II at Qantir, and the palaces of Ramesses III at Medinet Habu and Tel el Yahudieh: Hayes 1959, 245–47, 290, 332–38, 367. See also Landgráfová 2004, Landgráfová 2006b. Malkata: Lansing 1918, 12–13 (here they are said to be wall revetments, but they are obviously ceiling panels—cf. the roof slab of the side chamber of the "Treasury of Minyas" at Orchomenos).
2. Foster 1979, 12–15; Friedman 1998, 20n3, with references.
3. Hayes 1937; Caubet and Pierrat-Bonnefois 2005, 95–97, figs. 255–58; Delange 2015.
4. The phrase "wm.t n.t ḏꜥm" has been variously interpreted as referring to panels surrounding the throne, a pavilion or baldachin, or a doorway: Gnirs 2009, 15, bibliography cited at n9.
5. "Instructions of Amenemhat" 13: Quirke 2004, 127–29.
6. *Tale of Sinuhe,* 12.1–5. Papyrus Lansing: Erman and Lange 1925, 103–6; Lichtheim 1976, 173.
7. von Beckerath 1999, 6–10; Gundlach 2009; Gardiner 1964, 75; Osing 1972–92.
8. Beckman 2002; Pongratz-Leisten 2013.
9. Winter 1994; Crespo 2013, 201.
10. Tiglath-pileser III, "Royal Inscriptions of the Neo-Assyrian Period (RINAP) Project," 1, 47, r 23'b- r 33', http://oracc.museum.upenn.edu/rinap/corpus. See also Tadmore and Yamada 2011, 47.
11. Luckenbill 1924, 136, 197 (inscription of Tiglath-Pileser I); Reade 1979/80; Russell 1991, 115, 260, figs. 60ff; Crespo 2013, 116–19. Identification and provenance of the limestone and gypsum: Mitchell and Middleton 2002. Glazed brickwork of Sennacherib and Esarhaddon: Oppenheim et al. 1970, 16–17; Grayson 1976, 33 (no. 125), 63–70; Lackenbacher 1982, 110, 114–15. Colored stones: Marchetti 2009; Simkó 2015, 337–48.
12. Beam veneers: Nelson 2007. Hall of the Double Axes: Evans and Evans 1921, 3:323–24, fig. 225.
13. Goodison 2004, 342–49.
14. Plommer 1965.
15. Blegen and Rawson 1966, 65–66; Nelson 2007, 19.
16. Tiryns frieze: Schliemann 1886, xxiv, 214, 284–92, pl. IV; Moser von Filseck 1986.
17. Cook 2004.
18. Nylander 1970.
19. Louvre, Department des antiquites orientales, Sb 3382: de Mecquenem 1947, 35, fig. 15; Lilyquist and Brill 2013, 10, n45; Harper et al. 1992, 225; Caubet and Pierrat-Bonnefois 2005, fig. 292. For Achaemenid brickwork in general: Razmjou 2004.
20. Pliny. *HN.* 6.47.6: Vitr. *De Arch.* 2.8.10. On the Hekatomnid building program: Hornblower 1982, 294–332, esp. 302.
21. Jeppesen 2000, 95, 158–59 ("Appendix 3: Commander de la Tourrette's Report on the Demolition of the Maussolleion AD 1522"). For the Mausoleum fabric: Walker and Matthews 1997.
22. Ar. *Nub.* 814–15 (423 BC, revised 420–417 BC): "οὔτοι μὰ τὴν Ὁμίχλην ἔτ᾽ ἐνταυθοῖ μενεῖς, / ἀλλ᾽ ἔσθι᾽ ἐλθὼν τοὺς Μεγακλέους κίονας" ("My word, by the foggy air, you can't stay here! Go and eat the columns of Megakles!"). *Eryxias,* 394e: "ἢ τούτου μὲν καταφρονεῖν τοὺς ἀνθρώπους καὶ μὴ εἶναι ὠνητάς, τῆς δὲ κυπαρίττου τῆς ἐν τῇ οἰκίᾳ καὶ Πεντελικῶν λίθων πολλοὺς τοὺς δεομένους τε καὶ βουλομένους πρίασθαι" ("Or is it that men look down upon this [wisdom] and there are no buyers, while many, needing and desiring cypress wood and Pentelic marble for the house [of Poulytion], buy these?"). For the wealthy Poulytion: Isoc. 16.7; Pherekrates, frg. 64; Paus. 1.2.5.
23. For the development of the oikos: Walter-Karydi 1998; Tsakirgis 2016, with bibliography.
24. Kottaridi 2011, with bibliography.
25. Phoenix, frg. 6, ll. 9–12: Powell 1925, 235–36; Petrain 2005, 340–41. On the poet's dates, see Carluccio 1995. In Pliny, "*smaragdus*" (σμάραγδος) indicates not only emerald but a range of twelve green stones (*HN.* 37.16–17.66); he also implies (32.18.73) that pieces of green porphyry quarried on Mt. Taygetus were called "*smaragdi,*" and Theophrastos possibly equates them: Eichholz 1958. Phoenix of Colophon may use "σμαραγδίτης λίθος" precisely to distinguish it from the gem σμάραγδος. Pliny (36.55) calls this stone "marmor Lacedaemonium," and it is called "Λάκαινα λίθος" (Lucian. *Hipp.* 5, "οἶκος . . . Λακαίνῃ λίθῳ κεκοσμημένος"). Pausanias (3.21.4) states that there was no continuous mass of rock, only pebbles.
26. Pergamon: Zimmer 2010, 157–58.
27. Kallixeinos of Rhodes, *Περὶ Ἀλεξανδρείας 1,* in Ath. 5.206c. For the scanty archaeological remains of the palace: McKenzie 2007, 68–71. Faux-marble revetments so far discovered in Alexandrian homes (Sector M at Kom el-Dikka) date only from the

Roman period and later: Majcherek 2002, 42; Tkaczow 2014.

28. Adriani 1940, 44–45, pls. XII.2 and XIV.2 (Alexandria 25654). The tiles were found among architectural fragments from the 1st century BC.
29. On the "Zone Style" and alabaster imitation in the Necropolis of Mustafa Pasha, Alexandria: Adriani 1936, 113–32, figs. 50, 55–56, pl. XIV. Hypogaeum 2, room 1; Hypogaeum 5, rooms 1 and 4, Necropolis of Anfushi, Alexandria. See also the Necropolis of Kom El Shoquafa: Adriani 1963, pl. 69, fig. 232. Pagenstecher (1919, 179–80) considered the dark checkers to imitate faience, and was followed in this by Adriani 1936 and Venit 1988 and 2002.
30. Venit 1988: 89, nn67–68; Venit 2002, 77–85, 115; Venit 2015, 56–58.
31. It has been argued that this was wall revetment: Delbrück 1932, 36; Tkaczow 1995. The sole textual source, the Scriptores Historiae Augustae (Alex. Sev. 25.7), clearly indicates paving (*stravit*) and claims it was named after Alexander Severus (r. 222–235 AD), who "was the first [*sic*] to set 'Alexandrine marblework'—which is made of two kinds of stone, porphyry and Lacedaemonian marble—and he used this kind of marbling to ornament the streets of the Palace." However, the same text (SHA, *Helioglab.* 24.6) attributes the invention to the previous emperor, Elagabulus (r. 218–222 AD), who "used Lacedaemonian stone and porphyry to pave the streets in the Palace, which he called Antonine [streets]; these stones lasted down to within our own memory."
32. "λιθοστρώτου σμαραγδίτου λίθου καὶ πιννίνου καὶ παρίνου λίθου (בַּהַט־וָשֵׁשׁ וְדַר וְסֹחָרֶת רִצְפַּת)." The Hebrew terms are very different. "בהט" (*behaṭ*), probably from an unused root meaning "to glisten," generically indicates marble but was translated as "σμαραγδίτης λίθος," "green porphyry." "שש" (*shêsh*) refers to bleached fabrics, and therefore white stone, but was translated as "πίννινος λίθος," "a pearl-like stone," which I take to indicate "Egyptian alabaster" because of its translucency and veining. "סחרת" (*sôchereth*) seems to mean "border," and is often translated as "black," but "παρινος [alt. "πάριος] λίθος" means "Parian." The date and location of the translation of the Septuagint is given by *The Letter of Aristeas* (2nd or 1st century BC) and Philo of Alexandria (early to mid-1st century AD), followed by Josephus (37–c. 100 AD).
33. For samples of "alabaster" imitation, see Mulliez-Tramond 2010. For purple walls: Salvadori 1998, 203–25. In the wing of the House of the Griffins (1st century BC), discovered in the 1980s, wall paintings also seem to imitate huge slabs of porphyry: Iacopi 1991. For real and faux marble in Alexandrian houses and tombs: Nowicka 1969. Faux-marble toichobates also appear in Macedonian tombs: Miller 1993, 12, 34, 41.
34. For Ptolemaic, fragmentary alabaster pavements at Kôm ed-Dick and Kôm el Shokafa, as well as an alabaster/porphyry/granite floor from Wardiyan, see Daszewski 1982, 399ff., figs. 13–15. For the pavement from the Horti Lamiani: Cima and La Rocca 1986, 61–63.
35. Harrell et al. 2007.
36. For the "Tomb of Alexander": Bonacasa 1984–85; Adriani et al. 2000, esp. 101–4.
37. Lindsay 1913, 282; Gnoli 1988, 139. Gaggiotti (1989) points out that these are probably cocciopesto floors. Cf. Cic. *Leg.* 2.2.
38. When Varro speaks (c. 80–67 BC) of "lithostrota pavimenta et parietes incrustatos" in his *Taphê Menippou* (frg. 533), he probably means mosaic floors and stuccoed walls. See also Varro. *Rust.* 3.1.10.
39. On this marble, see Dworakowska 1990. Its identity as africano and provenance from Teos are now generally accepted, though Dworakowska sees a discrepancy between its price in the Edict of Diocletian and that of other Greek island marbles. But the edict dates from 301 AD, while the quarry was apparently worked out by c. 170 AD, so marmor luculleum may have commanded high prices precisely because it had become a rarity.
40. Metellus Scipio extortionately levied the columnarium in Syria in 49–48 BC: Caes. *B Civ.* 3.32; Cic. *Fam.* 3.8.
41. Seneca had been Nero's tutor, but his *Epistulae morales* were written after his retirement from court in 62, *Ep.* 94 and 115 perhaps in autumn 64, and his *De beneficiis* maybe at the same time: Griffin 1976, 399, 400.
42. Pliny. *HN.* 36.2.6. For the sort of architectural terra-cottas in question: Rystedt et al. 1993. Pliny (*HN.* 34.16.34) says, "The images of the gods dedicated in the temples were preferably wood or terracotta, right down until the conquest of Asia, whence came luxury." Cf. Cato's lament in Livy. 34.4.1–4.
43. Pliny. *HN.* 36.7. This is patently false—at the very least the Temple of Jupiter Stator built by Metellus (c. 143 BC) had marble columns.
44. Pliny disingenuously argues that in the good old days "columns were certainly used in temples, and not for the sake of splendor—such things were as yet unappreciated—but because they could not be made firmer otherwise" (*HN.* 36.5.45). Cf. discussion of columns in houses in Wallace-Hadrill 1988, 64–70. Seneca's Hippolytus says that the free spirit needs neither a thousand columns nor a golden roof to shelter him (Sen. *Phae.* 496–97).
45. Fant (1988, 153, nn33, 35) calculates 3,712,500 sesterces.
46. On the cost of transportation as the main factor in construction costs: Stanier 1953. The standard composite text of Diocletian's edict is in Giacchero 1974, 2:210–11. It is persuasively argued that the edict prices marble per square foot, not cubic foot: Corcoran and DeLaine 1994. According to the same estimates, materials and manufacture of fresco cost approximately 9 denarii per square foot; revetment in white marble approximately 45 *d.*/ft^2; in Numidian marble approx. 206 *d.*/ft^2. Flooring in mosaic would cost approx. 10 *d.*/ft^2, in white marble approx. 9 *d.*/ft^2, Numidian approx. 28 *d.*/ft^2. The opus sectile floor of the Curia Senatus alone would have cost 1,063,800 *d.*
47. Juvenal (14.86–95) describes how the imaginary Caetronius (or Cretonius) frittered away his fortune raising villas with marbles transported from Greece or farther afield that outdid the temples ("culmina villarum Graecis longeque petitis / marmoribus uincens Fortunae atque Herculis aedem"), only for his son to squander the remainder on new mansions of still-costlier marbles ("dum meliore novas attollit marmore villas").
48. Tac. *Ann.* 3.55. On civic status and the Roman house, see Wiseman 1987; Wallace-Hadrill 1988, 50 and passim.
49. Guidobaldi and Salvatori 1988.
50. Pliny. *Epist.* 9.39 (c. 108/109 AD). Elsewhere (*Epist.* 3.6.5; c. 97/102 AD), Pliny leaves the choice of marble for a statue base to his slave Annius Severus, so long as Pliny's name is engraved on it.
51. Thus, the poet Persius (34–62 AD) writes, "Now at last there are verses so fluent that no join jars critical fingernails as they glide across the smooth surface" (*Sat.* 1.63–65: "carmina molli / nunc demum numero fluere, ut per leve severos / effundat iunctura ungues?"). The commentary of Pseudo-Acro (c. 200 AD) on *Horace's Ars Poetica* (at l. 294), specifies "'to the nail': a phrase taken from marbleworkers, who test for joins in the marble with their fingernail" ("'ad unguem': tractum a marmorariis qui iuncturas marmorum ungue pertemptant"). Alberti, *De re aed.* 6.10, says of revetment, "Everything should be composed and fitted to the fingernail so that all parts of the work appear equally perfect" ("omnia ad unguem componentur atque delicabuntur ut cunctae partes operis aequaliter sese absolutas praebeant"); and in 7.12 says, "They brought the [architrave and jamb] together in a perfect joint" ("iunctiones ad unguem conterminarunt"). For the other meanings of "ad unguem," see D'Angour 1999. The expression was derived from the Greek "εἰς ὄνυχα"; see entries in Henry G. Liddell and Robert Scott, "A Greek-English Lexicon," at http://www.perseus.tufts.edu. The ingenuity that went into making the slabs run completely flush is unveiled in Ball 2002.
52. Symm. *Ep.* 1.12 (before 377 AD). On the Domus Symmachorum, see Carignani and Spinola 1995.
53. Philostr. *VS.* 556. An altar inscription calls Regilla "ΤΟ ΦΩΣ ΤΗΣ ΟΙΚΙΑΣ"

(Inscriptiones Graecae, III 1417; Ameling 1983, 2 [Inschriftenkatalog]: 160, no. 47).

54. Paus. 1.19.6 (Stadium, Athens); 7.20.6 (Odeum, Athens); 10.32.1 (Stadium, Delphi). In addition, he dedicated a nymphaeum at Olympia, an Odeum and court at Corinth, and possibly another fountain at Peirene: Abraldes 1996, 1: 65–67, 193–207; 2: 69–70, 335–39; 3: 69, 317–20. Some of these buildings did incorporate other marbles (114–15). Although the Athenian Odeum was built in Annia Regilla's memory, it does not share the introversion of Herodes's own home.

55. Apul. *Met.* 5.1. See also the Palace of the Sun in Ovid. *Metam.* 2–1.30.

56. Kantorowicz 1963. For the continuing medieval comparison of kings to the rising sun: Bührer-Thierry 2004 and esp. Baldwin Smith 1956, 45–50, 56–58, 130–40, 166–78.

57. The Domus Aurea was rigorously oriented to the cardinal points: Voisin 1987.

58. Suet. *Ner.* 31; Tac. *Ann.* 15.42.1. Cf. Cima and La Rocca 1986, 124–27. An attempt to reconstruct the revetment pattern is in Rocco 1988, esp. 128–30. For archaeological and literary remains of precious metals in architecture of the eastern Mediterranean: Vallois 1944–78, 1:290–310.

59. Lanciani 1901, 220, plan at 219. Cima and La Rocca 1986, 124–27. See also the fragments of agate veneers found in a domus on the site of the Villa Patrizi at Porta Pia: Roberta Belli Pasqua in Anderson and Nista 1989, 109, figs. 44–56. Earlier discoveries included a room on the Palatine sheathed in silver and another on the "Little Aventine" paved in agate and cornelian and the walls clad in gilt copper: Pietro Santi Bartoli in Fèa 1790, 1:233; Vacca 1594, no. 101. Lanciani (1901, 220) also recounts: "I have seen another apartment paved with large slabs of occhio di pavone, the walls of which were paneled with crusts of black slate covered with graceful arabesques in gold-leaf . . . a third hall with the floor made of segments of alabaster, framed in green enamel, around the walls of which were jets of water, four feet apart."

60. Carettoni 1949; Bastet 1971; Bastet 1972.

61. L'Orange 1942; Hemsoll 1990, esp. 28–33; Champlin 1998; Neverov 1986; Fernández Uriel 1992.

62. Suet. *Domit.* 14.4. The colonnades were of Numidian marble. Domitian is Juvenal's target when he disparages a "Dominus" who builds a banqueting hall, supported on Numidian columns to catch the winter sun (7.182–83). Plut. *Popl.* 15.5: "Your mania is to build; like the famous Midas, you desire that everything become gold and stone at your touch."

63. In the Aula Regia, the niches were filled with colossal statues of divinities in basalt. Only Domitian himself would have been enthroned: Clauss 1999, 119–32. For the Triclinium remains: Gibson et al. 1994; Gros 2001, 252–61. Triclinium apse: Ward-Perkins 1977, 108, 140. "Coenatio Iovis": SHA. *Pert.* 11.6.

64. Stat. *Silv.* 4.2.10–11, 18–31. On this passage: Vessey 1983; Coleman 1988, 88–93; Newlands 2002, 266–71.

65. Lucian. *Dom.* 3. Them. *Orat.* 18.223a; Dindorf 1961, 271. Pseudo-Codinus also extolls the rich marbles of the Palace of Lausus: Janin 1964, 123, 379.

66. *Anth. Pal.* 9.811, in *The Greek Anthology*, trans. William. R. Paton, 5 vols. (London: Heinemann, 1925), 3:437.

67. Coripp. *Laud. Just.* 1.97–102; Cameron 1976, 133 at ll. 97ff.

68. Coripp. *Laud. Just.* 1.30–200; "lux sacra palatia conplet" (1.363); "lux urbis et orbis" (1.250); "et lumen membris regalibus auxit" (2.90–91); "imitator Olympum / officiis Augusta domus. Sic Omnia clara / . . . luce corusca" (3.179–81); "atria . . . sole metallorum splendentia" (3.191–92); "adytis radiavit ab imis / inclita lux, et concistoria tota replevit. / Egreditur princeps" (3.211–13); "credunt aliud Romana Palatia caelum" (3.244); "iubar axe sereno / emicuit . . . clara coruscantem lux auxerat altera lucem . . . nova Roma nitebat, / certebantque ipsi iucunda palatio caelo" (4.96–102); "sedes, auro gemmisque superba, / lumen habens sine sole solum" (4.115–16); "sacri luminis instar / illius aspectus gemmas vincebat et aurum, / angelicis oculis exaequans sidera caeli . . . terruit incessu populous, oculisque refulsit: quails . . . sol aureus . . . mundumque inluminat omnem / lucis honore suae" (4.245–48).

69. Luxorius, *Carmina*, 90. Rosenblum 1961, 164–65, 250–51. Luxorius lived 480/490–c. 534 AD; Hilderic was Vandal king of North Africa 523–530 AD. See also Flavius Felix on the Alianas Baths rebuilt by the Vandal king Thrasamund (496–523) in *Anth. Lat.* 213.12: "Here the newborn day flashes through the brilliant marbles" (*hic geminata dies per candida marmora fulget*).

70. Yu Huan, *Weilue*, 30.39 (composed 239/265 AD); Fan Ye (based on Ban Yong, 125 AD), *Hou Hanshu*, 88.15 (chronicle of 6–189 AD, written in 398/445); Fang Xuanling, *Jin Shu*, 97.5–6 (chronicle of 265–420 AD, written in 648); Zhao Ying and Liu Xu, *Jiu Tangshu*, 198.6 (chronicle of 618–907 AD, compiled 941); Ouyang Xiu and Song Qui, *Xin Tangshu*, 221b.19, 25 (chronicle of 618–907 AD, compiled in 1044). Possibly descriptions of Aelana (Elat), Antioch-on-the-Orontes, Constantinople, or just possibly Rome: Hirth 1885, 40, 44, 51, 57, 71, 77, 92, 228–30 (on "*liu-li*" as "opaque glass," versus "*po-li*," "transparent glass"), 238–41; Sevillano-López 2015, 24–25, nn44, 48, 49. "*Liuli*" (琉璃) refers to both opaque glass and glass-like glaze, and "*shuijing*" (水精) can mean equally rock crystal and crystal glass: Brill 1991/92, 129–30.

71. Ibrahim et al. 1976, esp. 208–19 and drawings XVI–XIX, XXVII–XXX, XXXIV A–XXXIV B, LIII A–LIV. Over a hundred panels were recovered, which would have covered a total wall area of 1,615 ft.^{2} (150 m^2). Dated there to c. 370/375 AD, and to c. 360 AD in Brill 1999, 1:63.

72. Bacchelli et al. 1995; Volpe et al. 2004.

73. Examples from Rimini (before 257/258 AD), Corinth (c. 250/300 AD): Ortalli 2000, 516, 19–20, no. 183; Oliver 2001. For a wall panel with an aquatic scene (1st century AD): Harden 1987, 32–33 (cat. no. 10).

74. Nesbitt 1880. The central section of the Scenae Frons in the Theatre of Scaurus had glass cladding, probably mosaic: Pliny. *HN.* 36.24.114; Lavagne 1983. Statius describes walls "shining with figures in vitreous variety." SHA, *Firmus Saturninus Proculus et Bonosus*, 3.2: "It is said that [Firmus, general of Zenobia] fitted his house with glass quarries fixed [to the wall] with bitumen and other substances." Nesbitt (1880, 275–76) also refers to a now-destroyed room in the the Vigna Alfieri (see Lanciani's *Forma Urbis*, sheet 46, bottom left) where figurative glass opus sectile occupied the wall above a marble dado. Other fragments of glass opus sectile found in Egypt and Rome are catalogued in Ibrahim et al. 1976, 262–65.

75. Eschebach 1979, 87nn26–27. A (so far) unique example of slabs covering entire walls was the Oecus Cyzicenus in the House of the Dioscuri, all of which were salvaged after the eruption: Richardson 1955, 63–65, pl. 14.2 (marbles: alabaster, rosso antico, giallo antico, and africano). All revetments on public buildings were also salvaged. For antique tunneling into the ruins: Zevi 1991.

76. Room 18 (possibly the house of the very wealthy Marcus Nonius Balbus): Maiuri 1977, 68–71; Guidobaldi 1992, 74–75; Grandi and Guidobaldi 2005.

77. "Baths of Pasa Ilica": Conze 1912, 131–35, cat. no. 48 (who considers the revetment Hadrianic); Deubner 1939, 22–23, pl. 4; Hamamcioglu-Turan et al. 2013.

78. Bath house: Middleton 1892, 1:85, fig. 14; Molinari and Priester 1999, 558–59, fig. 24. Middleton's engraving (1892, 1:311) does not agree with the photograph of the apse as he "adds" details based on the imprint of revetment in mortar in a room in the House of the Vestals. A patch of surviving skirting and dado is illustrated in Middleton 1886b, pl. at 404/5.

79. Janiculum Villa: Filippi 2005. Apsidal hall, Domus B under Palazzo Valentino: Baldassari 2012, 50, 52–54, figs. 17–19, 23–24.

80. For example, Bruto and Vannicola 1990. See also Deubner 1939; Becatti 1961; Becatti 1969, 123–60; Guidobaldi 1992, 72–77; Dunbabin 1999, 261–68; Bitterer 2013.

81. For example, the so-called Tomba dei Custodi (5th century?) in the Catacomb of Pratextatus: Spera 1992, esp. 284–89, 295–99. Other significant survivals of opus sectile

revetment include the room behind the cubiculum of Pope Gaius in the Catacomb of Callixtus (late 4th century) and a cubiculum in the Catacombs of Sts. Peter and Marcellinus (4th century): Guidobaldi and Guiglia Guidobaldi 1983, 64–65, 112. Also Cubiculum 12: Reekmans 1988, 82, figs. 43, 46–48, 57; Spera 1995, 433–46. There are also several examples of faux-revetment, the best-preserved being the Hypogaeum in the Catacomb of the Via Dino Compagni (315/350 AD) and the Cappella Graeca in the Catacomb of Priscilla (mid-3rd century): Ferrua 1991. Color illustrations in Fiocchi Nicolai et al. 1999, figs. 62, 94, 97, 104, 109, 141. Figurative and vegetal crustae have also been recovered from Neapolitan cubicula: Novara 1998b.

82. Regio III, insula VII, no. 8. Becatti 1969; Frazer 1971; Brenk 1998, 530–31; Guidobaldi 2000. Comparable domus types can be found in Becatti 1948a; Becatti 1948b; Guidobaldi 1986; Pavolini 1986, 244, 252–69. See also Lind Hansen 1997. Boin (2013, 58–61, 68) instead agrees with Becatti that the building was a collegium.
83. The latest coinage discovered below the detritus dates to 393/394 AD (Becatti 1969, 42, nos. 14, 15, 67–71).
84. Excavations documented in Becatti 1969. The revetment was reconstructed at the exhibition *Aurea Roma,* Rome (2000), and from 2005 at the Museo dell'Alto Medioevo, EUR: Stella Arena and Carruba 2005. Floor: Becatti 1969, 21, 112–14.
85. The hall is 24.6 × 22 ft. (7.5 × 6.7 m); the exedra is 19.6 × 12.8 ft. (6 × 3.9 m). In the hall, the revetment reached a height of 25.6 ft. (7.8 m), in the exedra about 23 ft. (7 m). No revetment was found bridging the 8.2-ft. (2.5-m) gap between the floor and the surviving opus sectile (the shear line of collapse), as the artisans worked from the ceiling down and had only reached this point when the room collapsed. What we are missing is probably another border of opus sectile, an orthostate, and a skirting.
86. Becatti 1969. Identified as a philosopher in Zanker 1995, 315–19; Brenk 2001, 266–68.
87. Verità et al. 2008.
88. De Rossi 1871; Nesbitt 1880; Marucchi 1891; Hülsen 1927; Ashby and Lugli 1932; Cagiano de Azevedo 1967–68; Becatti 1969, 181–215; Paris 1990; Paris 1991; Basile 1991; Cima 2000. Drawings in Borsi 1985, 169–73; Whitehouse 2001, 138–65.
89. Unsawn Numidian blocks in the detritus bore inventory marks of Domitian (81–96 AD): Becatti 1969, 22–23. No doubt they came from the stockyards of Ostia or nearby Portus.
90. Schelkle 1954.
91. Roberts 1989a, 47. Note that the disjunctive interspersing of Latin texts with Greek words was often expressed by the metaphor of mosaic or opus sectile: McPhail 2003.
92. On aulae and triclinia in the late antique domus: Ellis 1991; Baldini Lippolis 2001, 47–72.
93. Ellis 1988; Ellis 1991, 118.
94. Van Dalen 1991; Tione 1999, 196–97. Reticulatum may also have occupied a venerable place in Ostia tradition, as even the old town walls (c. 82–79 BC), unusually, had been built from *opus quasi reticulatum* (irregular reticulate work): Calza 1953, 79–88, 101–2.
95. "Caseggiato delle Taberne finestrate" (regio IV, insula V, no. 18).
96. Lancha 1981, pl. LXXVIII (pre-220 AD). For Bacchic symbolism: Parrish 1995.
97. Lugli 1928, 65–76; Righi 1980; De Nuccio and Ungaro 2002, 174–75; Angelelli 2004, 89–91, pl. XV.2. Villamagna (near Anagni, Provincia di Frosinone): Umberto Tommasi in Giammaria 1999, 45, fig. 40; marble paving in cella vinaria at Villamagna, near Anagni: Fentress and Maiuro 2011. For the isodomum pavement in Chian tiles with Numidian coursing: Guidobaldi and Guiglia Guidobaldi 1983, 71–72, fig. 17E.
98. Pseudo-Aristotle, *Περὶ Κόσμου*, 398a (written between 350 and 200 BC); Apuleius, *De Mundo*, 26.
99. Cassiod. *Orat. Reliqu.* 2.12–13: Mommsen 1894, 483. L. Traube as *Orationum Reliquiae*, in *Monumenta Germaniae historica, Auctores antiquissimi*, 15 vols. (Berlin: Weidmann, 1877–1919), 12:483–84. I repropose the reading "cameras" (instead of "venas"), and integration "negatur" (instead of "noscatur") given in Chatelain 1903, 48.
100. Thalassius and Bassianus converted two temples in Phoenicia, possibly Tyre. Yet another noble, Theodulus, built himself a villa out of the remains of a demolished temple: Libanius ep. 1364 /N105, F724/ W636: Bradbury 2004, 42, 219–20.
101. On hunting dogs as apotropaic: Mitchell 2012b, with bibliography; as a comparandum for the harness, Mitchell (n3) adduces a 5th-century mosaic at Halikarnassos. Compositions of felines overpowering other animals also graced the aula of the domus of Junius Bassus (c. 335 AD) and the Domus sopra le Sette Sale (c. 350–400 AD): Becatti 1969, 188–91; Bianchi 2000; De Nuccio and Ungaro 2002, 168.
102. *Secundini Manichaei epistola ad Augustinum* (*PL* 42, col. 574; written 399 AD): "For I confess that the marbles of the Palace of the Anicii do not glow with such diligence nor so great industry as your writings shine with eloquence." Statius had compared brilliant marbles and mosaics of the baths of Claudius Etruscus to his friend's "shining talent and effort" (*Silv.* 1.5.63–64). Seneca had used the metaphor in the contrary sense to extoll the literary style of Fabianus Papirius, that he eschewed "fashionable polish" ("recentis politurae") and did not use "a variety of marbles" (*Ep.* 100.5–6: "desit sane varietas marmorum"). Seneca rejected oratory that was too polished (*Ep.* 115.2: "orationem politam"). For intellectual brilliance and persons who shine: Hor. *Carm.* 1.5.12; Mart. 10.89.3; Catull. 2.5, 61.189; *Anth. Pal.* 9.399. For the rich revetments (including glass paste) of the aula absidata and grand portico of the Palace of the Anicii: Jolivet et al. 1999; Jolivet et al. 2000.
103. Cassiod. *Var.* 7.5. Cassiod. *Var.* 3.10 ("marmora, quae de domo Pinciana"). For the revetment remains: Novara Piolanti 1998; Novara 1998c.
104. Krautheimer 1933–77, 4:72; Kelly 1987, 2:263–64, 278–80; Rickert 1998; Terry 1986, 152–53, figs. 19–20 ("bricks" in porphyry, serpentine, Phrygian, and Numidian).
105. Clothes in turn could be weighed down with jewels (to become crustae) or, as Claudian describes Stilicho's consular toga, embroidered with a "palace with columns of red marble" (*De consulatu Stilichonis,* 2.341: "rutilis hic pingitur aula columnis").
106. 7.53–55: Jeffreys 1998, 204–5.
107. Museo Romano Nazionale delle Terme. To my knowledge, undated, unprovenanced, and unpublished.
108. For evidence of possible support for the pagan rebels at Ostia, see Bloch 1945; Becatti 1969, 71.
109. Julius Valerius Alexander Polemius, *Res gestae Alexandri Macedonis* 3.22.802–27, 831–33, 849–60: Latin text in Rosellini and Calderan 1993, 157–60. The relevant section of the 2nd-century AD Greek version is in Kroll 1926, 119–20.

CHAPTER 6. MEDIEVAL SUBSTITUTIONS IN THE WESTERN CHURCH

1. Stern 1958, 208–12; Amadio 1986, 23–26 (cat. nos. 3–4), 38–39 (nos. 12–13), 40–41 (no. 14), 50–51 (no. 21), 89–90 (no. 65). So did S. Stefano Rotondo (c. 470); see Flavio Biondo in D'Onofrio 1989, 151: "marmoreis columnis et crustatis varij coloris marmore parietibus musivoque opera." Other survivors were S. Adriano (the Curia), S. Lucia in Selci, SS. Cosma e Damiano, S. Maria Maggiore, and SS. Sergio e Bacco.
2. Rickert 1998.
3. Giovenale 1929, 122–23, 128–39; Kelly 1987, 265–66 (cat. no. 43); Romano 1991, 34. The wall revetment of the central space (removed under Urban VIII) is recorded in Bibliotheca Apostolica Vaticana, Vat. Barb. Lat. 4333, fol. 69v, and Barb. Lat. 4424, fol. 31r.
4. Kelly 1987, 267–68, cat. no. 44; Johnson 1995; Romano 1996; Mackie 1997.
5. At S. Maria Maggiore "aedes fuit lastricata . . . in aula interius ubi supra portas desunt tabule lapidee": Onofrio Panvinio cited in Biasiotti 1915, 23. For Old St. Peter's, see Bosman 2004.
6. Krautheimer 1933–77, 2:256; Nordhagen 1962, 54–56; Guiglia Guidobaldi 2004; Guidobaldi 1997.

7. Mesnard 1935, 93–96, figs. 30–33, 36–38; Melograni 1990, 160–61.
8. Guidobaldi and Guiglia Guidobaldi 1983.
9. Lusuardi Siena 1990; Pejrani Baricco 1990, 298, fig. 4f.3a.; Pejrani Baricco 2000; Beghelli 2011, 168–69, n80.
10. Ebanista 1997.
11. Kostof 1965, 57–58; Novara 2000; Deichmann 1969–89, 3:94–95.
12. "Permittimus omnibus, ut qui volunt caedere habeant licentiam attributam. Fore enim arbitrantur, ut etiam complures saxorum nitentium venae in lumen perveniant": *Code of Theodosius* 10.19.2 (October 22, 363), in Pharr 1952, 283. Constantine had already allowed this in North Africa in 320, and Valens, Gratian, and Valentinian drafted a similar decree in 376 (Pharr 1952, 284).
13. Roberts 1989a, esp. 47. The topos of urban beauty, as manifested primarily in its marble monuments, assumed prominence in epideictic oratory from the 4th to 6th centuries, "at precisely the time that the ancient structure of the cities was gradually disintegrating": Saradi 1995, esp. 41–42. For 5th-century Rome in particular, see Bauer 2001.
14. Claud. *Epith.* 86–91. See also Stat. *Silv.* 1.2.145–53; Sid. Apoll. *Carm.* 2.418–21, 9.49–85; Claud. *VI Cons. Hon.* 51–52; Ennod. *Carm.* 1.4. On the genre: Pavlovskis 1965; Roberts 1989b.
15. Flood 2001, 25ff. For enduring tales of the "great pearl": Shalem 1997.
16. Nonn. *Dion.* 18.67–86, in *Nonnos, Dionysiaca*, trans. W. H. D. Rouse, 3 vols. (London: Heinemann, 1940), 2:69.
17. For the house-churches that became tituli in Rome: Guidobaldi 1989; Guidobaldi 1993. The Domus Sessoriana, the palace of Emperor Elagabulus, was converted into S. Croce in Gerusalemme already by the mid-4th century. For a case study: Brenk 1999.
18. Curia: Bartoli 1963, 58–60; Mancini 1966–67. Sangallo says, "anticamente li pilastri sono corintii antichi di fodere e fralli pilastri e foderato di marmi porfidi serpentini pezi grandi riquadrati di più sorti." In 1562 Ligorio stripped most of the revetment for reuse in the Palazzo Apostolico and the Palazzo dei Conservatori. But some sections survived until 1653, when Martino Longhi stipulated that the capomastro must "levare di opera e riporli dove vorrano li detti RR. Padri tutte le lastre di serpentini, porfidi et altri marmi e mischi, che sono dentro, e sopra la cima della tribuna vecchia": Barroero 1983, 199–201, with bibliography. A partial elevation of the revetment is in Codex Destailleur D, f. 7v; Kunstbibliothek, Berlin, inv. n. 4151: Tessari 1995, 112. Hall of the Praefectus Urbis: Lanciani 1882, 37–40. The revetments (illustrated by Pirro Ligorio; Bibliotheca Apostolica Vaticana, Vat. Lat. 3439, fol. 14), for which see Luschi 1998, esp. 447–49, with more bibliography.
19. Isid. *Etym.* 15.4.11. Paulinus of Nola, *Carmina*, 28.97, cited in Lavin 1962, 16–17, with more references.
20. Zeno. *Tract.* ix (*PL* 11, col. 417B); Maximus of Turin, *Homilia* LXI (*PL* 57, col. 371); cited in Kantorowicz 1963, 139–40.
21. Macrob. *Sat.* 1.18.11 (on the temple to Sebadius at Zilmissus, Thrace). The *Saturnalia* were composed c. 431.
22. There is no exclusive rule. For example, the Temple of Portunus by the Tiber had no windows, while the nearby "round temple" has two large windows flanking the portal. For the question of accessibility, most conveniently: Stambaugh 1978.
23. The principal collection of tituli is in de Rossi et al. 1861–1971. See also Leclerq 1907–53, esp. 850–905 for the early Medieval period. A representative collection of those tituli especially regarding divine light is analyzed in Borsook 2000. Such inscriptions were often gathered into syllogae and disseminated throughout Europe.
24. Fulgentius [468–533 AD], *Mithologiae*, I, proemium, cited in Kantorowicz 1963, 124.
25. See also the apse titulus at S. Maria in Domnica (c. 817–824): "This house was previously broken down into ruins. Now it glitters properly, decorated with variegated minerals. And behold its ornament shines forth like the sun in its orbit who chases the gloomy veils of hideous night."
26. "The day dies into the night . . . thus the exiled light mourns and yet is restored back to life again in the whole world, with its ornament, with delight with the sun, the same integral and whole, destroying death, her night" (Tert. *De resurrect.* 12); Ambrose's funeral oration for Valentinian (392): "I seem to see you leaving your body and, having dispelled the darkness of night, arise at dawn like the sun to approach God" (Ambr. *De ob. Valent.* 64).
27. For the consecration ceremony, when excerpts from Revelation are read: Muncey 1930. For texts from the 10th century on: Stookey 1969.
28. SS. Cosma e Damiano (526–30): "God's precious hall gleams with bright minerals, in which the precious light of Faith sparkles even more through the physician-martyrs, to the people the sure hope of salvation came, and this place grew because of that sacred honor": Lanciani 1882, 38. Ennod. *Carm.* 2.8.2–3, 8: Xystus "joining the light of life to the worth of the work, established this temple [a chapel attached to S. Lorenzo, Milan] . . . this man offered this temple which He Who will come consecrates."
29. Gage 1982.
30. Paulinus of Nola, *Carmina*, 28.183–84: "It is radiant from the source of piety in the centre and in a miraculous manner it simultaneously renews and is renewed": Goldschmidt 1940, 82–83. Translation mine.
31. "Shining with the skill of the artisans the night is expelled and burnishes the light into being. By artifice the day is joyfully imprisoned within a luminous prison, from which sad shadows have been evicted after a long sojourn": Borrel 1883, 50. On Avitus: Shanzer and Wood 2002, 3–27, with bibliography.
32. *Anth. Lat.* 111 (122R). See also Sid. Apoll. *Epist.* 2.2.4: "Within the heated chamber there is full day and this abundance of enclosed light forces all modest persons to feel themselves something more than naked." Eusebius characterizes the baptismal light when he speaks of the "pure souls that have been washed like gold by the Divine bath" (*Hist. Eccl.* 10.4.64).
33. For example, *Anth. Pal.* 1.3 (inscription in the apse of the Blachernae, Constantinople): "Justin the Elder made the splendid temple, of such brilliant beauty, dedicated to the Mother of God. A radiant light which conquered the dawn was given it by Justin the younger, who reigned after him." The apses were reconstructed by Justin II (565–578), although the church was not in reality built by Justin I (518–527).
34. *Ep.* 50 in Peiper 1883, 78. For the whole text (dedication of the church founded by Arigius, c. 500), with a slightly different translation of this passage: Shanzer and Wood 2002, 327.
35. Paulinus of Nola, *Carmina*, 27, ll. 383, 387–88; Goldschmidt 1940, 54–55. See also Ven. Fort. *Carm.* 2.10.11–16: "The resplendent hall is raised on marble columns / and because it remains pure, greater is the grace which shines within it. / Illuminated by glazed windows, it is the first to capture the rays and, thanks to the artist's hand, closes the day within its ark."
36. "Therefore our halls shine, Father, with Thy gifts of noble flames; their emulous light plays the part of day when it has gone, and night with torn mantle flees before it in defeat. But who would not discern that the swift light has its source on high and flows from God? . . . The lamps gleam out, that hang by swaying cords from every panel of the roof, and the flame, fed by oil on which it floats lazily, casts its light through the clear glass. You would believe that starry space stood over our heads. . . . Thou are the true light of our eyes, the true light of our minds; by Thee we see as in a glass within, a glass without": Prudent. *Cath.* 5.25–30, 141–46, 153–54, in *Prudentius*, ed. and trans. H. J. Thomson, 2 vols. (London: Heinemann, 1949), 1:41, 47.
37. Gatti Perer 1983, esp. 129–33, 185–99.
38. Amatus's *De Duodecim Lapidibus*, the consecration hymn of Montecassino, itemizes the apocalyptic stones: Lentini 1958.
39. See, though without reference to the material issues, Priester 1990. The 12th-century campanile of SS. Giovanni e Paolo is the most spectacular, with at least 36 *bacini* (ceramic bowls, often of Islamic provenance) and 83 *specchi* (marble panels).

40. For S. Pietro al Monte at Civate, see Colli 1981; Müller 2009, 246–59, with further bibliography. A much later representation of the same idea is Bartolomé Bermejo's *Christ Showing the Crucifixion to the Patriarchs in Heaven* (1462/65).
41. For example, Ven. Fort. *Mart.* 3.509–16 (c. 576 AD): "Oh sweet eyes, when you contemplate these happy figures, you have seen Zion and the array of its twelve gates, [you have seen Zion and its twelve splendid gates], that the Almighty loves over all the tabernacles of Jacob, those gates adorned with the varied splendour of gems, covered with chased gold leaf, artistically encrusted with emeralds, glittering with Chrysolite, snowy with shining pearls. Grace has spilled out all the treasures of love, and on each one shimmers a single topaz of a thousand colors. By these gates, that noble people of your kingdom, penetrate into the starry city, carried by their merits and they will dwell without end for ever and ever enrolled as citizens of heaven": Quesnel 1996, 71–72.
42. Terry 1986.
43. Columns with reticulate or gridded channels, inlaid with green-and-gold glass and amethyst or porphyry and serpentine, have been recovered at Hagios Polyeuktos, Hagios Ioannes, in Hebdomon (Istanbul Archeological Museum, inv. no. 3908), and Hagia Euphemia en to Hipprodomo (Istanbul Arch. Mus., inv. no. 5078): Harrison 1989, figs. 82, 83, 94; Naumann and Belting 1966, 54–56, 98–102; Mathews 1976, figs. 14–5; Mathews 1971, figs. 41, 44, 49, 50.
44. Ennod. *Carm.* 2.56.8–10: "vellera ceu Serum murice tincta feras, / qualiter inclusas comit lux hospita gemmis, / nix lapidis quotiens pulcrior arte rubet." Ennodius visited Constantinople on diplomatic missions (515, 517).
45. The column in the window above the triumphal arch in St. Catherine's, Sinai, imitates this sort of jeweled column in its mosaic pattern; Forsyth and Weitzmann 1973, pl. CXXIXa. Jeweled columns support the fantastic architecture depicted in the mosaic vault of Hagios Georgios, Thessaloniki, and divide the mosaics in the apse of S. Vitale, Ravenna. For the ceramic tiles: Gerstel and Lauffenburger 2001, cat. nos. A.44–48, A.63–64, C.1, C.3, C.9. See also examples from the Musée nationale de Céramique at Sèvres (inv. B 3, B 5) illustrated in Maguire 1997, pls. 21, 22. See, in the same volume, Jannic Durand, "Plaques de céramique byzantine des collections publiques françaises," 25–50. For the canon tables: Anderson 2001, 122–25.
46. Mark the Deacon, *Vita Porphyrii*, 1247.84 (compilation, c. 600, of texts originally written after 420). Avitus: Perrat and Audin 1957, 448–49, ll. 13–14 (dedication speech normally dated to 513/516); Ep. 50 (dedication of the church founded by Arigius, c. 500), in Peiper 1883, 78. Cf. *Hom.* 22, 24. Statius had already sung of "baths bejeweled with glistening marbles" (*Silv.* 1.5.12–13).
47. Rossi 1589, 58; Zaffagnini 1965, 29–31. The basilica was demolished in 1733 to make way for the new duomo.
48. Onians 1980, 7–11.
49. *Dom.* 9, in *Lucian: Works*, trans. A. M. Harmon, 8 vols. (London: Heinemann, 1913), 1:187. See also Theophilus (early 12th century): "You have embellished the ceilings or walls with varied work in different colors and have, in some measure, shown to the beholders the paradise of God, glowing with varied flowers, verdant with herbs and foliage . . . for the human eye cannot decide on which work to fix first its gaze; if it beholds the ceilings they glow like brocades; if one considers the walls they are a kind of paradise": Dodwell 1986, 63–64 (bk. 3, praef.). See also the description of the floor of Montecassino as meadow: *Chronica monasterii Casinensis.* 3.27, in Hoffmann 1980, 396.
50. Maguire 1987a: 25ff; Halton 1976; Cameron 1984; Loukaki 2013.
51. Roberts 1989a, 51–52. For buds as gems: *TLL* 1900–, 6.2:1753.78–54.32. Poetic anthologies were also considered strings of literary gems: Schur 2004.
52. Suger, *De admin*, 33.199, trans. Panofsky and Panofsky-Soergel 1979, 63–64.
53. Alexander 1952, 180; Mango 1972, 117. The symbolism of glitter, radiant materials, and its connection both to Heavenly Jerusalem and idealized images of the saints is discussed in Dahl 1978. See also the broad study by Pentcheva 2010.
54. "De fonte baptisterii S. Stephani, et aqua quae per columnas venit. / En sine nube pluit sub tectis imbre sereno, / Et coeli facies pura ministrat aquas. / Proflua marmoribus decurrunt flumina sacris, / Atque iterum rorem parturit ecce lapis. / Arida nam liquidos effundit pergula fontes, / Et rursus natis unda superna venit. / Sancta per aethereos emanat lympha recessus, / Eustorgi vatis ducta ministerio" (*Carm.* 2.149; *PL* 63, col. 0361A–B; PL trans. by Michael Roberts, e-mail to author). Ennodius says "iterum" ("again") because water from a rock repeats the biblical miracle of Moses; "rursus natis" ("the reborn") refers to those reborn through baptism. By "pergula" we may imagine a columnar structure such as that in the Baptistery of St. John in Lateran.
55. For example, S. Maria at Gazzo Veronese; see Lusuardi Siena et al. 1989, 176–77, figs. 87–88. The water drained into a tank, probably at the entrance, used for lustration: S. Lusuardi Siena and M. Sannazaro in Russo 2001, esp. 653 and n14.
56. Schwarzenberg 2004.
57. Poseidippos, Papiri dell' Università degli Studi di Milano, VIII.309. col. II 29–32 in Bastianini et al. 2001, 37, 122–23. Gregory Nazianzenus, *Orat*, 21.31.
58. Examples are common from the Hellenistic era on: Pseudo-Kallisthenes (*Alexander Romance*, 2.41), Agatharchides of Knidos (*On the Erythraean Sea*, 5.84a), Philostratos (*Apoll.* 3.46) and Heliodoros (*Ethiopica*, 8.11): Stoneman 1992, 104–6.
59. Pliny mentions a lion statue with eyes of *smaragdus: HN.* 37.17.66. Lucian of Samosata (c. 125–after 164 AD) describes a gem (*Lychnis*) on the forehead of a cult statue that lit a whole temple: *De Dea Syria*, 32. Benjamin of Tudela (1161/62) has the jewels in the Blachernae Palace illuminate it by night: Wright 1848, 75. According to Anthony of Novgorod (c. 1200), a gem on the forehead of Leo the Wise next to the main portal of Hagia Sophia illuminated the church by night: Khitrowo 1889, 90–91. Sapphires and rubies were said to illuminate also the Great Mosque at Damascus and the Dome of the Rock; see the many Islamic examples in Shalem 1997, 45–48. For other examples, see Adams 1930, 474–79; Ball 1938.
60. Hdt. 2.44; Theophr. *De Lap.* 4.25; Eichholz 1958.
61. Reinhold Lubenau (1587), Simeon of Lwov (1608/9), John Covel (1674), Grelot, and Veryard: Mango 1962, 121, 128, 123. See also Jacques Paul Babin (1672): Laborde 1848, 1:198; Veryard 1701, 341; Hammer-Purgstall 1822, 1:374; Lethaby and Swainson 1894, 262, fig. 62. On Hosios Loukas, see Guillet de Saint-Georges (1676) in Wheler 1682, 4:321. In this case, the panels are not panes but transenna panels below the (missing) glazing: Schultz and Barnsley 1901, 25, pls. 13, 14, 15; Chandler 1817, 2:63.
62. Stephen of Novgorod (1348/49) in Majeska 1984, 32.
63. The Icelander Saewulf, who never visited Athens, mentions the "oleum in lampade semper ardens, sed nunquam deficiens," and the *Liber Guidonis*, a 12th-century manuscript possibly from Ravenna, now in the library of Bourgogne, cites the "divinum lumen atque inextinguibile": Norre 1966, 45, 169–70n6, and see the Parthenon sources collected in the appendix. See also van der Vin 1980, 1:200.
64. Nicolas de Martoni in Le Grand 1895, 652.
65. Galland (1672–73) in Schefer 1881, 38; de Dreux in Pernot 1925, 155; Evliya Çelebi (c. 1667) in Norre 1966, 230; Magni 1688, 65; Guillet de Saint-Georges 1674, 195; Spon and Wheler 1724, 2:90–91.
66. See L'Orange 1974–75; Reutersward 1984; Reutersward 1991.
67. *De Templo*, 9.6; *PL* 50, cols. 1114B–C; Connolly and O'Reilly 1995, 25. The passage is still cited by Carlo Borromeo in his *Instructiones fabricae et supellectilis ecclesiasticae* (1577), ch. 8: Voelker 1980, 109.
68. Strophes 13–14: McVey 1993.
69. The saints were added to the tympana between 867 and the 880s: Mango 1962, 48–65, diagrams III–IV, pls. 57–89;

Salzenberg 1854, pl. IX. In the lowest register the ensconced saints stand in front of fictive transennae. Triplets of windows at Edessa (see previous note) were regarded with Trinitarian significance, as were those in the apse of Hagia Sophia: *Diegesis* 12 in Dagron 1984, 203.

70. Lucr. 4.75–83. Literary sources on velaria in Graefe 1979. Ambrose (395 AD) describes the deceased Theodosius as assumed into the "awnings" of Christ (*De ob. Theod.* 1). Gladiatorial games were outlawed in 438 by Valentinian III, but beast hunts continued until much later. Velaria were also used in the circuses that were unaffected by the ban.
71. Barry 2002; Charalampadis 2003.
72. L'Orange 1974–75, 197ff.; Asmussen 1986, 74–77, 79.
73. *HN.* 36.45.160–62; see also 3.30, 9.56.113, 36.45.162, 37.77.203; Petr. *Sat.* 68. For windowpanes in houses: Juv. 4.21; Sen. *Ep.* 90.25; Mart. 8.14; Pliny. *Ep.* 2.17, 21; Pliny. *HN.* 19.64. Horti Lamiani: Philo *Leg.* 364. Also known as "διοπτρίτης λίθος" (Strabo 12.2.10) and listed in the *Papyrus Graecus Holmiensis (c. 300 AD)* 3.39 in Lagercrantz 1913, 163. In general, see Guarnieri 2015, with bibliography. Latter-day equivalents are Muscovite (or "Muscovy-glass") and Isinglass-stone.
74. Du Cange 1954, 7:549, cols. 2–3; Courcelle 1967, 357–63.
75. *De Opificio Dei* 8 (*PL* 7, cols. 0038A). Lactantius is at pains to refute Lucretius's rehearsal of the Epicurean theory of decortications: the eyes are not holes into which the image-bearing rays pour, but the mind is the faculty that exploits the eyes for guided vision. Hence the stress on the lens of the eye as a diaphragm, like "specular stone."
76. Agricola recorded "Mary's Ice" (*Marieneis*) "in a certain old church of Marienburg": Agricola 1530, 456. In 1546 he also mentions "a church in Cosuicus, Saxony, and another in Merseburg, Thuringia": Bandy and Bandy 1955, 93. Leo III (795–816) restored the windows of St. Peter's, S. Paolo fuori le Mura (after the earthquake of April 30, 801), and the Lateran "ex metallo cyprino" (or "cypriano," or "cypsino"; *PL* 102, cols. 1005B, 1005D, 1017A; 128, cols. 1219, 1221, 1235).
77. For remains of these windows at S. Sabina, S. Prassede, and S. Giorgio in Velabro, see Muñoz 1924, 31–32; Muñoz 1926, 26–27; Franz 1958, 67; Pasquini 2002, 80–81. Leo of Ostia also describes "fenestras . . . gipseas" in the aisles at Montecassino (1066–71): Hoffmann 1980, 397 (3.28); Messineo 1998; Gianandrea et al. 2017.
78. Symphosius 68 in Ohl 1928, 98–99. "Specular. Perspicior penitus nec luminis arceo visus, / transmittens oculos intra mea membra meantes; / nec me transit hiems, sed sol tamen emicat in me." Leary (2011) instead argues that the riddle refers to glass.
79. Ps.-August. *Serm.* 245.4: (*PL*, suppl. II, col. 922). Eugenius I in Vollmer 1905, 261.
80. Gregory of Tours, *De gloria beatorum Martyrum*, cap. 58; PL 71, col. 759. Some of the panes discovered at S. Vitale, Ravenna, and assumed to be 6th century, are also yellow, though not dichroic: Bovini 1965. A dichroic quarry was found at Jarrow: Cramp 1970, 327n2, pl. LIVc, d; Cramp 2001, 74–75. For the history of Early Christian glass, see the excellent Dell'Acqua 2003, with copious bibliography.
81. Alberti, *De re aed.*, 7.12, f. 129. Vasari-Milanesi 1906, 1:204. Vasari is echoed in Baldinucci 1681, 90.
82. For example, Stephanos (c. 610/641 AD): "Your Phengites rekindling astounds all my vision, your shining radiance gladdens all my heart"; Sherwood Taylor 1937, 122–23.
83. Gnoli 1988, 220. Strabo (12.2.10) says that large pieces were exported; the *Souda* highlights its translucency without further ado. The grammarian John Tzetzes equates it with selenite (*Comm. on Lycophron*, 98), but his testimony is very late (c. 1150) and unlikely. *HN.* 36.46.163: "in Arabia quoque esse lapidem vitri modo tralucidum, quo utantur pro specularibus," citing a lost work by Iuba. For the Umayyad mosque at Sana'a, see Serjeant and Lewcock 1983, 335–37, figs. 18.35–18.38.
84. On the modern confusion over whether the mishkat is a niche or a window: Worrell 1948; Flood 1999, 331–34.
85. Fumi 1891, 187–93, 213, 214.
86. Albertini 1510, 12. See also Agostino Del Riccio (1597) in Gnoli and Sironi 1996, 185. Even modern studies (erroneously) persist in calling these "phengite": De Angelis d'Ossat 1929; Gurrieri et al. 1988, 134.
87. De Tournefort (2:455; 1717) mentions an alabaster window in the Armenian church of the castle at Ankara. Gurk: Reutersward 1984, fig. 11; Mora et al. 1977, pl. 13.
88. Ricci 1914, 62–63.
89. For Jarrow, see Cramp 1968, 16–19; Cramp 1970; Cramp 2000. Red-streaked, Saxon fragments were also uncovered at Uley: Heyworth 1993, 190. For S. Vincenzo al Volturno, see Dell'Acqua 1997, 33–41; Mitchell et al. 2001, 1:173–203; 2:211–18, 377–79. For Müstair, see Goll 2001, 93, fig. 5. For Liège, see Evison 1988, 215–19. For more detailed discussion of this form of glass production, though the examples are glassware not quarries: Evison 1990, 217–28. Green window glass streaked with dark-green veining, and one fragment of red streaked with black and green, were also recovered at Farfa: Newby 1991, 35.
90. For example, Heyworth 1993, 192. Price noted similar streaked glass in the Romano-British temple at Uley, and at Silchester and Wroxeter: Price 1993, 189.
91. "Vitrum sapphireum," blue panes made by smelting Roman perfume bottles or mosaic tesserae, were a desired commodity until at least Suger's time. Many of the window quarries found in early medieval layers at the Abbey of Farfa, 25 miles (40 km) northeast of Rome in the Sabina, were blue: Newby 1987, 262.
92. Myrrhine ware: Pliny. *HN.* 36.67.198; Del Bufalo 2016. "Chalzadonii" were pioneered by Angelo Barovier (1405–1460), a Murano glassblower whose skill earned him mention in Filarete's imaginary description of the church at Sforzinda: Zecchin 1987–90, 2:226, illustrations at 3:83, 379, 381.
93. Rubens Vase: Ross 1943. Sacro Catino: Pero Tafur in Letts 1926, 28, 229. It was said to come from Jerusalem, but was actually among the spoils of Caesarea in 1101: Calcagno 2001.
94. Gage 1982.
95. Robert De Clari, cap. 82, in Lauer 1924, 81–82; Weiss 1995, 319.
96. Barry 2010, with bibliography.
97. Hor. *Carm.* 2.18.1–5, 17–19. For other examples of this topos, see Pearcy 1977.
98. *Anth. Lat.* 443: "De divitiis et inhonesto animo / quod tua mille domus solidas habet alta columnas / quod tua marmoreo ianua poste nitet / aurea quod summo splendent laquearia tecto / imum crusta tegit quod pretiosa locum / atria quod circa dives tegit omnia cultus: / Hoc animos tollit nempe, beate, tuos? / aedibus in totis gemmae licet omnia claudant / turpe est, nil domino turpius esse suo."
99. Agathias [c. 532–579/582 AD], 5.3.11: trans. Frendo 1975, 138.
100. Niketas Choniates in Magoulias 1984, 263. This metaphor, "to put on a coat of stone," may be found in the *Iliad*.
101. As Theodora says to Justinian, quoting Isokrates, "Empire is a fair winding-sheet" (Procop. *BP.* 1.24.33–38; Isoc., *Archidamos*, 45). For the sarcophagi: Vasiliev 1948; Deér 1959, 134; Downey 1959b; Grierson 1962; Asutay-Effenberger and Effenberger 2006.
102. So says the eyewitness Harun-Ibn-Haya, c. 880: Izeddin 1941–46, 56–57. See also Vasiliev 1948, 159; van der Vin 1980, 491; Karlin-Hayter 1991; Dagron 2007.
103. *Life of St. John the Almoner, Archbishop of Alexandria*, chap. 17. Greek text in Festugière and Rydén 1974, 365; trans. Dawes and Baynes 1948, 228–29.
104. Mango (1984, 30) argues that Leontios's text was largely plagiarized from an anthology of c. 560–570. For the coronation liturgy: Brightman 1901; Yannopoulos 1991; Majeska 1997, 2–4, with further literature. However, no study of the Byzantine coronation mentions the episode of the stonecutters.
105. "The marble workers and tomb builders approached him, bringing marble and stone of various types to show him. They went up to him and said, 'What sort of tomb will your majesty order for himself?,' symbolically reminding him. 'You are a mortal man, and subject to corruption'": Majeska 1984, 110n185, 433–35, who claims that it was

copied from a Slavic translation of Leontios. The practice is not, however, mentioned in the Paleologan *Treatise on the Offices* (c. 1360/90): Verpeaux 1966, 252–73.

106. Passavanti (c. 1302–1357), prior of S. Maria Novella: Jacopo Passavanti, *Specchio della Vera Penitenza* (Florence: Bartolommeo Sermartelli, 1585): cap. IV; cited in Gnoli 1988, 66–67. Passavanti claims Isidore as his source, but I have found no such passage in his writings.

107. Adam of Usk in Thompson 1904, 90, 262. The ritual is not mentioned in earlier ordines for papal coronations, nor does Burchard mention it for the coronation of 1484, although there is a lacuna at this point in the manuscript: Celani and Thuasne 1907–14, 75.

108. Augustine castigates "all those who adore sculpted things" or "who adore stones" (*Enarrationes in Psalmos XCVI*, 7, 11). Cf. the early 9th-century apologia of Claudius of Turin, an anti-icon/anti-relic/anti-pilgrimage treatise: "Let them adore stones too, because when He was taken down from the cross he was placed in a sepulchre of stone, and the Apostle says of it: Christ Himself was the Stone; (but Christ is tropically, not properly, called stone, lamb and lion in terms of significance not substance)": *PL* 105, cols. 459–64, esp. 462.

109. Granite, rather than marble, columns were more plentiful: Van Stekelenburg 1992/93. As Manuel Chrysolaras writes in 1411: "Even Rome has suffered the fate . . . and even our own city [Constantinople] of becoming a mine and stone quarry for itself and, as it is said of this whole universe, of extracting nourishment from itself by feeding upon its own resources": trans. Cortassa 2000.

110. Dahl 1981.

111. Dahl 1981, 158.

112. Fernie 1980. For Cosmati examples, see Claussen 1987, figs. 58, 63, 70, 140, 150, etc.

113. Hamilton 1949.

114. Fernie (1994) traces the dissemination of this patterning to Lindisfarne (1120s), Selby (c. 1097–1123), Dumferline (1128–50), Norwich (1096–1119), Waltham Abbey (c. 1130s–50s), Kirkby Lonsdale (undated), and the crypt of York Minster (1150s). To this list may be added St. Lawrence at Pittington (12th century). Spiral columns also appear in the crypt of Canterbury Cathedral (begun 1096). The crypts of St. Peter in Utrecht (consecrated 1048) and the Abbey of Rolduc at Klosterrath (1107–8) all sport spiral and zigzag channeled columns, while that of St. Lebuinus at Deventer (11th century) has incised spiral and wave motifs: Kubach and Verbeek 1976, 3, pls. 561.2, 176.1, 80.1; Hardering 1998, 19–20, 76ff., esp. 101–11. The crypt of Lund Cathedral (1123) also has columns with spiral and more complex motifs. For the method of executing the spirals at Durham, see Bony 1990, 25–26. I refer to the Georgian use of veining on columns to indicate marbling, engaged piers framing apse in the Church of St. Nicholas at Nikorzminda (1010–14).

115. For example, the depictions of spina, circus, and temple revetments on the floor of the palestra or sphaeresterium; Carandini et al. 1982, pls. LVI, LVII. See also examples in Michaelides 1985.

116. Goulet 2001–2. The motif is considered to be "*finto bardiglio*" in Maiuri 1958, 1:134. It has been termed "*Inkrustationsnachahmung*" in Eschebach 1979, 93.

117. Epstein 1980–81, with bibliography. The patterns reappear in medieval, rock-cut, Byzantine-influenced churches in southern Italy, e.g., SS. Andrea e Procopio near Monopoli and the Madonna delle Sette Lampade near Mottola; Dell'Aquila et al. 1998, 211–13, 226–27.

118. Nordenfalk 1937; Rosenbaum 1955.

119. For an early Qur'an frontispiece, see Bothmer 1987. On the Italian appropriation of illuminated Byzantine manuscripts, see Nelson 1995.

120. Kahn 1991, 31, figs. 21–24. For traveling master masons and the connections between England and Italy in the 11th and 12th centuries, see Meredith 1980, ch. 1, with bibliography.

121. Pozzi and Ciapponi 1964, 1:367. The text leaves no doubt, speaking of a "silicato overo pavimento tessellato et sectilio vermiculariamente fabrefacto di pretiosi lapilli . . . sedili de diasprea materia . . . sepulchro . . . di petra pretiosa di iacyntho di colore vermiglio transparente." In the *Hypnerotomachia* illustrations, however, the reticulate patterns most likely represent simply pierced lattices.

122. England: Fernie 1984. Rome: Malmstrom 1975.

123. Canon Primus and Canon Sextus, Lorsch Gospels.

124. For St.-Savin (but not its faux-marbling): Favreau et al. 1999. For St.-Savin and St.-Hilaire-le-Grand: Camus 1989. Camus mentions identical remains at Ronceray d'Angers, Vieux-Pouzauges in the Vendée, and unspecified sites in Berry, Burgundy, and the Loire Valley. The choir in the church at Anzy-le-Duc (Saône-et-Loire) also has faux book-matched marbling on its piers that may be original: Oursel 1989, pl. 61.

125. Similar examples of English Romanesque polychromy are summarized in Fernie 2000, 279–80. A thorough historiography of the phenomenon appears in Autenrieth 1997. A good photo of the late 13th-century faux-masonry in St.-Éliphe at Rampillon—originally the chapel of the Hospitaliers de St.-Jean-de-Jérusalem—is published in Kimpel and Suckale 1990, 19 (fig. 15), 215 (fig. 19). The deviation from true of the coursing is quite noticeable. Numerous examples are also illustrated in Michler 1977.

126. The key passages are Chronicles and Kings, esp. 1 Chronicles 29:2–3, "ego autem totis viribus meis praeparavi inpensas domus Dei mei . . . lapides onychinos et quasi stibinos et diversorum colorum omnem pretiosum lapidem et marmor parium abundantissime." For the Hebrew terminology, see Grossmark 1998; Fischer 1999.

127. Isid. *Etym.* 16.5.8: "Parius candoris eximii, Lychnites cognomento." Prudentius (*Perist.* 12) calls the columns of S. Paolo fuori le Mura "Parian," although they were Phrygian. The inscription in S. Sofia at Benevento (c. 758–768) claims "HIC DUX ARICHIS PARIO DE MARMORE TEMPLUM CONSTRUXIT": Belting 1962, 183. Similarly, the chronicle (1055–82) of Rudolphus, abbot of St.-Trond (near Liège, Belgium), describes the new tomb of Lady Bertha as "Pariis lapidibus polita": Mortet 1911, 1:159. Gisilbert (1066) describes the Church of St. Amand-les-Eaux as having "columnas . . . de Pario Lapide": Lehmann-Brockhaus 1955, 1:no. 1631. Gerald of Wales says that in 1174 the archbishop showed Geoffrey Plantagenet a newly built hall "Pariisque structuris et marmore distinctam" (*De Gadridi Archiepiscopi Eboracensis*, rolls ser., XXI [1873], IV, 367). For Richard of Ware (d. 1283), see Lehmann-Brockhaus 1955, 1:no. 2854.

128. *Perist.* 2.481–82. Prudentius also urged his fellow citizens to "wash the marbles that are bespattered and stained with putrid blood you nobles. The statues must rest clean . . . let no debased usage pollute the monuments of art and turn it to sin" (c. *Symm.* 1.501–5). Likewise, Adelm of Malmesbury (beginning of the 8th century) will argue that Rome's brilliant pagan temples and execrable deities are destroyed and her new temples only "shine with the blood of Christ" (*De virginitate*, 560–566).

129. *De Templo*, *PL* 91, cols. 745C–746B. Trans. Connolly and O'Reilly 1995, 17. Angelomus Luxovensis, *Enarrationes in Libros Regum*, 3.5, *PL* 115, col. 411C ff.; *S. Eucherius, Commentarii in Libros Regum 3*, *PL* 50; Hrabanus Maurus, *Commentaria in Libros IV Regum 3.5* (834 AD), *PL* 109, col. 140A ff.; Claudius of Turin (840 AD), *XXX quaestiones super libros regum*, lib. 3, *PL* 104, esp. col. 733A ff. For an explicit allusion to the imitation of Solomon's Temple, see the description of reconstruction works at St.-Bertin (1138–1163), near St.-Omer: Mortet 1911, 129.

130. Hrabanus Maurus: *PL* 111, col. 464A. Rodulphus Glabrus (Raoul Glaber), *Historiarum Sui Temporis Libri Quinque; PL* 142, 651CD.

131. Bede, *Homilies*, 2.25 in Martin and Hurst 1991, 111, 267.

132. Krinsky 1967; Schuler 1999.

133. Gauzlin, abbot of Fleury-sur-Loire (from 1005 to 1029); Odilon of Mercoeur, abbot of Cluny (994–1049): Mortet 1911, 38, 128. Cf. Suet. *Aug.* 28.3.

134. Plumpe 1943, 8–9. The basic references are Ephesians 2:20–22: "Built upon a foundation of the apostles and prophets, Jesus Christ himself being the chief corner stone: In whom all the building, being framed together, groweth up into a holy temple in the Lord. In whom you also are built together into an habitation of God in the Spirit"; I Peter 2:4–6: "So come to him, to the living stone which was rejected by men but chosen by God and of great worth to him. You also, as living stones, must be built up into a spiritual temple, and form a holy priesthood to offer spiritual sacrifices acceptable to God through Jesus Christ. For you will find in scripture: I am laying in Zion a chosen corner-stone of great worth. Whoever has faith in it will not be put to shame." In the dedication prayer at the end of the *Exhortatio ad Virgines*, Ambrose (340–397) beseeches the lord to keep his eye upon the church, its altar, and "these spiritual stones, in everyone of which a sensible temple is consecrated unto thee," and this had also been the theme of Eusebius's panegyric of the church at Tyre (c. 317 AD): *Hist. Eccl.* 10.4.60–65. For the classical tradition from which these ideas derive: Ehrhard 1945.
135. The theologian Pierre de Roissy (early 13th century) says, "Squared stones denote the square of virtues possessed by saints, that is, moderation, justice, courage and prudence. Polished stone signify the saints who were honed by patience in the face of adversity . . . and the stones are joined by mortar, because the saints are linked by charity. For which reason St. Paul says to the Ephesians 'solicitous to keep the unity of the Spirit in the bonds of peace' [Ephesians 4:2–3]": *Manuale de Mysteriis ecclesiae* in Mortet and Deschamps 1929, 2:185. Cf. "Keeping the fabric of the pure temple, his body, in good repair, he renewed and perfected it, and polished all the stones of the spotless sanctuary, his breast, and made them all square, and with the plummet of exact living built them into a house of perfection"; Walteri Danielis, *Vita Ailredi abbatis Rievall* (c. 1170), in Powicke 1950, 47–48.
136. Phot. *Hom.* 10.4–5, trans. Mango 1958, 184–90. Cf. Federico Barbarossa marveled at the marble gate of Milan "so well made by the artisans' hands that . . . just like the works of the Roman empire, there appeared almost no joint in any part of it": *Rahewini gesta Friderici I*, lib. III, fol. 37, in Lehmann-Brockhaus 1955, 1:466, no. 2240. Additional sources in Greenhalgh 1989, 124–26. See also Evliya Celebi's praise of the Parthenon (c. 1667), that to descibe it "the tongue falls short and the pen is broken. The white polished marbles on the walls are of elephant size. What is noteworthy is that one, though he may not be knowledgeable and skilled in architecture, still cannot see the joints between these stones. You think that it is a great wall of one piece, forty cubits high, and it is so shiny and radiant that any minute thing flying in the air can be seen on it and especially the color of the face of the persons who gather there, the genuflections and the bowing to the ground during public prayers are reflected there. We are justified in saying that there are four big walls like mirrors, of one piece. And it is praiseworthy that there is no mortar, no gypsum, no concrete in these walls," Ahmet 1897–1938, 8:255; trans. Norre 1966, 235–36.
137. *De gestis regum Anglorum*, quoted in Mortet 1911, 338 (*PL* 179, col. 1364A).
138. For examples (without this observation), see Millet 1916, 228–44; Megaw 1931–32; Krautheimer and Ćurčić 1986, 379–82.
139. On the Cistercian Abbey of Silvanès, in the Rouergue: Hugo Francigena, *Tractatus de conversione Poncii de Larazio et exordio Salvaniensis monasterii*, in Mortet and Deschamps 1929, 48.
140. We might compare this with Paulinus of Nola's anti-epithalamium, in which Christ is subsituted for the bride, and pagan ornaments deemed superfluous for a Christian woman: Pavlovskis 1965, 165–66.
141. Pastor Hermas, *Similitude IX*, 4, 5–8. Greek text and French trans. Joly 1968, 288; Latin text in Tornau and Cecconi 2014.
142. Pastor Hermas, *Vision III*; Joly 1968, xx.
143. Pseudo-Heraclius, *De coloribus et de artibus Romanorum*, lib. III, 25, in Garzya Romano 1996, 6:22, 51, nn103–5: "How to prepare a column for painting. . . . If, however, you wish to feign marble, you can do so on a background of color, either brown or black or some other color, when the [sealing preparation and undercoat of the] column has dried. Then paint it in the sun." For the date and origin of this text, see Garzya Romano 1996, xxi–xxv.
144. *Reginaldi monachi Dunelmensis libellus de admirandis b. Cuthberti vitutibus*, fols. 154–57; Lehmann-Brockhaus 1955, 1:372, no. 1399.
145. Budde 1997; Favreau 2003; Méhu 2013; Hahn 2014.
146. Arnold 1890–96, 1: xxxii. For earlier marble imports, see Greenhalgh 1989, 129; for Ireland, Lynn 1984.
147. *Historiae Dunelmensis ecclesiae continuatio altera, usque ad episcopatum Hugonis de Puteaco*, fol. 168 in Lehmann-Brockhaus 1955, 1:378–79, no. 1408; *Liber Gaufridi sacristae de Coldingham de statu ecclesiae Dunhelmensis*, fol. 11 in Lehmann-Brockhaus 1955, 1:380, no. 1412.
148. Halsey (1980, 61) proposes that the materials were originally ordered for an abortive scheme to beautify the shrine of St. Cuthbert at the other end of the church.
149. Vellacott 1908; Drury 1948; Purcell 1967, 88–93; Leach 1975; Tummers 1980, 12–13; Blair 1991, 41–56, with more bibliography.
150. Badham 2009.
151. Rappoport 1995, 145, 181, and Russian and Ukrainian bibliography cited there. Rappoport cites materials found in the excavations at the Desiatynna (Tithe) Church in Kiev (c. 988), which the Laurent'ev and Ipat'ev Chronicles record was built by Greek masons. Marble was used for the architectural details and liturgical furnishings of the Sviata Sofiia Cathedral, Kiev (c. 1037).
152. Especially pyrophyllite shale: Rappaport 1995, 54, and bibliography cited there. Rappoport prevaricates on the nature of the "red marble" mentioned as paving the floors of the church at Suzdal (1233) and Rostov (1280).
153. Red limestone from the Gerecse Hills. The first major monument to exploit the material was the Porta Speciosa, the western portal of St. Adalbert's Cathedral in Esztergom, c. 1190: Marosi 1971; Marosi 1984, 32–34, 61, 63; Lövei 1992; Lövei 2001.
154. In 1307, a Purbeck "toumbe de marbre ben polye" was made for King Edward I: Wright 1866, 2:382. Blair (1991, 50–52) illustrates an episcopal effigy at Rochester Cathedral and mentions other examples.
155. Griffith-Jones and Park 2010; Draper 2001.
156. The fresco decorates a wall behind the 18th-century Cappella della Assunta and was not rediscovered until 1887: Wilpert 1916, 2:631, pls. 126–29, 131, 166, 208, 215, 243, 270.
157. *Vita Sancti Hugonis Episcopi Lincolnensis*, 879–888, in Garton 1986, 54: "Inspectus lapis iste potest suspendere mentes / ambiguas utrum jaspis marmorve sit; at si / jaspis, hebes jaspis; si marmor, nobile marmor. / Inde columnellae, quae sic cinxere columnas, / Ut videantur ibi quamdam celebrare choream. / Exterior facies, nascente politior ungue, / clara repercussis opponit visibus astra. / Nam tot ibi pinxit varias fortuna figuras, / ut si picturam similem simulare laboret / ars conata diu, naturam vix imitetur." I here follow British Museum MS Bib. Reg. 13. A. iv. Bodleian MS Laud. 515, which gives "*natura*" instead of "*fortuna*," and "*littus videatur arare*" instead of "*naturam vix imitetur*."
158. Henri d'Avranches, *Vita Sancti Hugonis Episcopi Lincolnensis*, 868–871, in Garton 1986, 54. Translation mine.
159. SI LECTOR POSITA PRUDENTER CUNCTA REVOLVAT / HIC FINEM PRIMI MOBILIS INVENIET; SPERICUS ARCHETYPUM GLOBUS HIC MONSTRAT MACROCOSMUM. See R. Foster 1991, with bibliography; Grant and Mortimer 2002.
160. On Piedra Sangrante, see Garcia-Talegon et al. 2016.

CHAPTER 7. HAGIA SOPHIA AND BYZANTIUM

1. General introductions to Byzantine marble use are in Paribeni 1989; Sodini 1994; Ousterhout 1999, 235–39.
2. Terry 1986.

3. The original aisle revetment is convincingly reconstructed in Deichmann 1969–89, 2.2, 121–34. The apse revetment—which many scholars still believe to be original—was executed in 1902–8 to the design of Corrado Ricci (on the models of the Euphrasius Basilica, S. Sabina, and the Chora), and modified in 1919 and 1923. The piers of the nave were re-revetted in stages by Lanciani in 1845–47, 1852–53, 1860–62, and 1867–69: Lombardini 1991–92; Bertelli 1992; Novara 1998a.
4. Built c. 500 but damaged by serious fire in 620. For Soteriou and Soteriou (1952) the damage was catastrophic; for Spieser (1984, 167–68), minimal. Cormack (1985a, 55–56) follows Soteriou, but considers the tribelon and nave revetment essentially original. For pre-fire photographs and meticulous survey drawings (1907–9), see Cormack 1969, esp. 19–21; Cormack 1985a, cat. nos. 7, 15, 16, 18, 21–28. The opus sectile is described in Diehl et al. 1918, 91–93; Asemakopoulou-Atzaka 1980, 88–101; Spieser 1984, 174–83.
5. Bouras 1982, esp. 63ff; Ousterhout 1992.
6. For the dating: Mylonas 1990. For the marbling: Schultz and Barnsley 1901, 27–29.
7. Lampakes 1889, 106; Millet 1899, 67–69.
8. Talbot Rice 1930, 52, figs. 2–3; Menna 2005, 138–44.
9. Emperor Leo V (813–820) had supposedly sent "workmen and excellent masters in architecture" to construct the Church of S. Zaccaria in Venice: Frothingham 1894, 36.
10. Tronzo 1997, with bibliography; Krönig 1965; Borsook 1990, 51–79, with bibliography.
11. Little of today's revetments on the Dome of the Rock survive from its construction (c. 691): Allen 1999, 206, 208–9. Some original revetment is visible in old photographs: Richmond 1924, figs. 4, 6, 7, 59, 60. None of the Damascene revetment is original: the mosque was reputedly despoiled by the Abbasids c. 750, comprehensively refurbished in 1068/69, re-veneered in 1300–1330, and gutted by fires in 1479 and 1893: Flood 2001, 67ff.
12. Wüstenfeld 1857, 147. See the meticulous description of Muḥammad ibn Aḥmad Ibn Jubayr (1145–1217), a scholar from Granada who visted Mecca on pilgrimmage in 1183: Broadhurst 1952, 78–91.
13. Marble from c. 1200 was imported from Constantinople to decorate churches in Kiev, Chernigov, and Timutorokan: Martin 1995, 63.
14. Literature on Hagia Sophia is vast but see mainly Mainstone 1988; Kleinbauer 1999. Still useful and lucid is Lethaby and Swainson 1894. For a survey of the older literature, see Theis 1999. For the building's afterlife: Necipoglu 1992; Nelson 2004; Fossati 1992. It has been impossible to engage with Schibille 2014 here, though in several respects her arguments apparently parallel several of my own. I have also been unable to consult Pentcheva 2017.
15. On Sinan's competition with Hagia Sophia: Necipoglu-Kafadar 1985. On S. Ivo: Connors 1996.
16. In particular Macrides and Magdalino 1988, 50; James and Webb 1991, esp. 6, 11; Webb 1999. The rule books of ekphrasis are available in Kennedy 2003.
17. Webb 1999, 11: definition based on those of Theon, "Hermogenes," Aphthonios, and Nikolaos (1st–5th centuries).
18. Webb 1999, 12.
19. Nicholaos Mesarites, 12.1, in Downey 1957, 867. Cf. Michael the Deacon, on Hagia Sophia: "The enduring newness in Christ of this building . . . can be observed with spiritual vision": ll. 15–16 in Mango and Parker 1960, 235.
20. Macrides and Magdalino 1988, 60 (on Paul the Silentiary).
21. Stat. *Silv.* 1.5.34–46. Martial (9.19) ridicules a poet, probably Statius, for literary prostitution when he versifies on the same complex: Hardie 1983, 133–34. Coleman (1988, 83–84) hypothesizes that Statius's ekphrasis of the banqueting hall of Domitian (*Silv.* 4.2) was delivered after or toward the end of the banquet. Lucian's declamation *On the Hall* is another good example. For speakers performing in front of temples: Pernot 1993. For the performative character of Byzantine ekphrasis: Macrides and Magdalino 1988; James 1996, 111–23; Magdalino 1997.
22. Although the lemma specifies that the descriptive section of the oration (and the later ekphrasis on the ambo) were given in the patriarchate, this included chambers in the upper stories of Hagia Sophia: Cormack and Hawkins 1977, 199–202. Kostenec and Dark (2011) instead think that Paul's poem was recited in the main audience chamber (*secretarium*) of the patriarchate. The authoritative edition of the text is De Stefani 2011.
23. Papalexandrou 2001.
24. Michael the Deacon's ekphrasis was bound with rhetorical set pieces probably collected together as an anthology of models: Krumbacher 1897, 470–76. Avitus's 6th-century dedication speeches probably also circulated as models: Wood 1986, 74–75.
25. *Hist. Eccl.* 10.4.45–62; Wilkinson 1982. On the Solomonic tradition, including Hagia Sophia, see Ousterhout 2010.
26. The Edessene Hymn was first published in Goussen 1925. A variant text and translation was produced by Dupont-Sommer 1947. Schneider (1941b) first suggested a reconstruction of the destroyed building on the basis of the hymn. A revised reconstruction and analysis is in Grabar 1947. A new edition of the Syriac text was provided and Grabar's errors rectified in McVey 1983. McVey's edition was strongly challenged by Palmer and Rodley (1988, esp. 52–57), with yet another reconstruction. Palmer's "idiosyncratic eclectic edition" was, in turn, tenaciously contested by McVey (1993) with a revised edition.
27. Palmer and Rodley 1988, 125–26, with review of the sources.
28. Frequently attributed to Romanos Melodios, the master of the genre. McVey (1993) has also argued for the kontakion's descent through late antique Syrian hymnody, via the Edessene Hymn (c. 543–554), back to Balai's *madrasha* (before c. 431).
29. Tursun Beg (c. 1487). Text: (in Latin characters) in Mertol Tulum 1977, 63–64. Italian trans. Pertusi 1976, 1:329–30.
30. Preger 1975, 74–108; French trans. Dagron 1984, 196–211, with excellent commentary; partial English trans., Mango 1972, 96–103. For redactions of the text, see Vitti 1986. For other tales, see Efthymiadis 2015.
31. "It is well put to represent the Roman Empire as one body composed, as it is right to say, of many limbs. We are therefore the head of this united body; for this holy head Wisdom is made the consort, to rule with me together in honour of the world entrusted to me, sitting in the same place": Coripp. *Laud. Just.* 2.195–200 (on Justin II, 565 AD), trans. Cameron 1976, 53.
32. Amm. Marc. 16.10.14. See also Pliny (*HN.* 36.24.101), that if we piled up Rome's buildings, we would see "such a grandeur towering above us as to make us think that some other world were being described, all concentrated in one place" (*universitate vero acervata et in quendam unum cumulum coiecta non alia magnitudo exurget quam si mundus alius quidam in uno loco narretur*).
33. It is ambiguous whether the "temple" in question is the church or Christ himself, so closely are they equated: McVey 1983, 96–97.
34. On the Hellenistic topos of Temple-House-Cosmos and its Early Christian appropriation: Festugière 1944–54, 2:233–38; Bornert 1966, 99ff.; Finney 1984.
35. Downey 1959a; Brzozowska 2012.
36. Meyendorff 1959; Meyendorff 1987; Fiene 1989; Deutsch 1990.
37. Pl. *Ti.* 44a. Ennod, *Vita Epifani*, 173.
38. Evans 1970, 8. The best summary is in Grumel 1926.
39. Michael the Deacon, 4.89–90, in Mango and Parker 1960, 237; Prudentius also calls the Church of Eulalia in Merida "motherly" (*Perist.* 3.192: *atria . . . alma*). Cf. Maximus Confessor, *Mystagogia 4; PG* 91, 672 A–C.
40. "Here you have founded a blessed temple for holy Mary, whence vanquished night has fled since the day never ends. Filled with light, the glittering hall is the image of Mary: she enclosed the light in her womb and this [church] encloses the day": Ven. Fort. *Carm.* 1.15.55–58, in Reydellet 1994, 1:35–36.
41. The equation of light with wisdom is elaborately pursued in the kontakion: Palmer and Rodley 1988, 146. Paul the Silentiary's ekphrasis also exploits the theme of light

because it was probably delivered on January 6, both Epiphany and the Feast of Lights, when Christ's baptism in the Jordan is commemorated: Macrides and Magdalino 1988, 63–64; Whitby 1985, 217, n18. December 31, 562, has also been suggested: Macrides and Magdalino 1988, 63–67.

42. McVey 1983, 99, with references. On Procop. *Aed.* 1.1.34, 46 and neo-Platonism, see Kaldellis 2013, 364–66.
43. Paris, Académie d'Architecture, III6 HP.DES.10/9.I. Drawn 1905/11. Translations of Paul the Silentiary, Prokopios, and "Codinus" (the *Narratio*) appeared in Lethaby and Swainson's *Sancta Sophia* of 1894 and were extremely influential though censured by some: Thomas 1982. Friedländer's definitive study of ekphrasis was published in 1912.
44. For the associations of gold, see the exhaustive Janes 1998.
45. Antoniades 1907–9, 3:101–6, fig. 617; Mainstone 1988, 209–10. Contested by Taylor (1996), who argues the dome sat on a fenestrated drum.
46. Michael the Deacon, 73–78, in Mango and Parker 1960, 237.
47. Schneider's excavations in the 1930s revealed traces of colored revetments on the atrium walls: Schneider 1941a, 23. The atrium columns were probably Proconnesian, as numerous pilgrims mention their whiteness.
48. Paul. *Descr. S. Sophiae.* 305–6. Strube 1973, 36; Kostenec and Dark 2011, 103–4. The external revetment likely disappeared in the 9th/10th centuries: Mainstone 1988, 104–5.
49. For technical analysis of the revetment: Lethaby and Swainson 1894, 234–63; Swift 1940, esp. 73–76; Antoniades 1907–9, 1:94–103, 2:19–21; Kleinert 1979, 7–44; Kleinert 1981; Asemakopoulou-Atzaka 1980, 97–102; Kelly 1987, 311–19 (cat. no. 75), 154–90, figs. 244–314.
50. Macrides and Magdalino 1988, 69; Paul. *Descr. S. Sophiae,* 379–80, 387–88, 391ff., 539ff., 620–46, 673–80; Agathias, 5.9.8, in Frendo 1975, 144. See Chorikios on St. Stephen at Gaza, "that each homeland of the most famous marbles presented [them] here as gifts"; on St. Sergius, Gaza, "let someone ennumerate for me the cities that pride themselves in the sumptuousness of the marble [they bring forth]: Proconnesus . . . Lacedaemonia . . . Caria . . . All of these places have contributed to the church gifts from [the resource] in which they glory. Thessaly too": Choric. *Laud. Marc.* I. 41–42, in Mango 1972, 63–64. Constantine instructs Macarius: "Concerning the columns and marbles [for the Holy Sepulcher] . . . these may be conveyed from every quarter, for it is fitting that the most wondrous place in the world should be adorned according to its worth": Euseb. *Vita Constantini,* 3.31, in Mango 1972, 12.
51. Underwood 1960, 212–13.
52. Paul. *Descr. S. Sophiae.* 549–50. The quarrying and shifting of large columns still commanded the respect of many Byzantine writers: Mango 1972, 7, 22n3, 31–32, 60, 61, 68–69.
53. L. 636 in Legrand and Reinach 1896, 55.
54. Stat. *Silv.* 2.2.90–91; Sid. Apoll. *Carm.* 5.38, 22.139. Likewise, Naser-e Khosraw (1047) admires the paving and balustrades of the Haram, Jerusalem, "of a flecked green marble that looks like a meadow with flowers in bloom": Thackston 1986, 24, 34. The Byzantines called serpentine "rue" (*pēganon*): Theophanes Continuatus, *PG* 109, col. 160D, cited in Gnoli 1988, 141. Russian pilgrims also refer to the Thessalian marble column-relics of Peter and Christ in the Apostoleion as "clover-colored": Majeska 1984, 300–301 (esp. nn70–79).
55. Delorme 1960, 332–36. For associations between colonnades and plane groves, see Vitr. *De Arch.* 5.11.4; Mart. 3.19.1–2; Prop. 2.32.11–13. The equation of column and trunk is also noticeable in Varro. *Rust.* 3.5.11–12. See also Bergmann 2002, esp. 99–100.
56. Ennod. *Carm.* 2.10.3 (c. 503–513). The "Palestra" at Hadrian's Villa was paved with large slabs (7.7 × 3.1 ft.; 234 × 95 cm) of Carystian, in contrast to the intricate opus sectile of the surrounding portico: Betori and Mari 2005, esp. 778.
57. Merobaud. *Carm.* 3.5–7 (c. 435/446 AD) in Clover 1971, 11, 28–29, 61.
58. Cornelison 2005, 95–113, with bibliography.
59. Long. *Daph.* 4.1–3. Cf. Col. *De Arb., De Agr.* 13.1.1. See Settis 2002, 42–47; Beck 2012.
60. Maffei 1986. Cf. Pliny. *HN.* 35.94; Petr. *Sat.* 83; Apul. *Met.* 2.4.
61. Regio V, Insula VII, nos. 1–2. Late 3rd-/early 4th-century remodeling of a mid-2nd-century peristyle building. Pensabene (1994, 371–74) sees the revetment as a 1930s restoration, but Laird (2000) notes that "identical paneling is sandwiched between the corner pilaster and a later wall in the northwestern corner of the peristyle."
62. Michelangelo later uses Thessalian ("cipollaccio") to imitate moss, lining a niche at the Belvedere "per ornamento di un fiume antico, acciò in questo campo fatto a guisa di scogli apparisca": Vasari-Milanesi 1906, 1:114; Ackerman 1954, 37, 58, 59, cat. 47. The destroyed niche (c. 1532/34), drawn by Heemskerck, contained a statue of Nile or Tiber: Hülsen and Egger 1916, 1:pl. 63.
63. Philagathos of Cerami, homily 27 (June 29, 1143/1146), on the Cappella Palatina in Palermo, in Rossi Taibbi 1969, 175. Philagathos modeled his description on circulating Byzantine ekphrases: Antonopoulou 2015.
64. Prudent. *Cath.* 5.113–14 (on heaven); Prudent. *Perist.* 3.191–95, 198–200 (on the church of the martyr Eulalia in Merida). In Greek lyric, the meadow of lilies could represent chastity, the meadow of hyacinths eroticism: Slings 1978, 38.
65. *Anth. Pal.* 1.10.60–65. See Mango and Sevcenko 1961; Connor 1999; Whitby 2006. For the scanty remains of the revetment, see Harrison 1986, 168–81.
66. Michael the Deacon, 110–12, in Mango and Parker 1960, 238. Nikolaos Mesarites, 37.5, in Downey 1957, 890. Themistios in Dindorf 1961, 219.
67. *Laudatio S. Theodori; PG* 46.737, in Mango 1972, 36–37.
68. "Quartered marble paneling is never found . . . in any Muslim monument after the Umayyad period": Creswell and Gautier-van Berchem 1969, 1.1:174, figs. 189–90.
69. Scholten 1929, 1:fig. 93. In 1047, at Ramallah, a Persian poet observed blocks being cut using Meccan sand and toothless saws "drawn along the length of the shaft, not across the grain, as with wood": Thackston 1986, 20.
70. Ousterhout 1987, 39–42. For a more detailed analysis of the pavement: Guiglia Guidobaldi 1999b, 321–58.
71. Herculaneum, House of the Relief of Telephus, Oecus 21. Late 2nd- or early 3rd-century mausoleum in Boult-sur-Suippe: Allag et al. 1988, esp. 97. Torah shrine, synagogue, Dura Europos (244 AD): Kraeling 1956, 54–56. Apse, Red Monastery, Sohag, Egypt: Bolman 2006, pls. 1, 5, 11, 16, 17, 19. Cubiculum (room 12a), House 2 (440/450 AD), Ephesus: Strocka 1977, figs. 275–79. Basilica, Tróia, Portugal: Nunes Pedroso 2001, pl. LXI. Temple façades in the Piazza Armerina mosaics (4th century): Carandini et al. 1982, pls. LVI, LVII. House of the Cascade, Utica: Cintas 1954, 149, fig. 173 (now destroyed). Tunisian mosaic floors: Michaelides 1985, 155–63, pls. 1–7; Trilling 1998, 119 and pl. 4; Balmelle and Ben Abed 2011, 96–99; Darmon 2011, 414–18.
72. Illustrated in De Nuccio and Ungaro 2002, 51, fig. 22. Book-matched Carystian revetment survives in room 31 ("Marmorsaal") of Hanghaus 2.
73. Oliver 2001, 360–61; Budetta 1987, 198–99, fig. 90 (stool); Mols 1999, 167–69, no. 13, figs. 88–93 (couch) and 182–83, no. 23, fig. 125 (stool).
74. *Orat.* 18.39 (*PG* 35, 1037). For the Mousikos, see Theophanes Continuatus in Mango 1972, 164. Built by Theophilus (829–842) in the Great Palace. Predictably, the passage continues, "When you see this building, you think it is a meadow abounding in various flowers."
75. Nikolaos Mesarites, 37.4, in Downey 1957, 890.
76. Alberti, *De re aed.* bk. 9.7, f. 170r, in Orlandi 1966, 839. Alberti, *De re aed.* VI.10.105v: "Maculae enim maculis et colores coloribus et talia talibus coaptanda sunt, ut alter alteri mutuo praestat gratiam."
77. Treu 1895, 28 (ll. 1029–38); Featherstone 2011, esp. 223.

78. Paul. *Descr. S. Sophiae.* 608–10, 348–50; Procop. *Aed.* 1.1.60; Choric. *Laud. Marc.* II. 40.

79. Cassiod. *Var.* 1.6.2: "magnitudini tuae studiosissime delegamus . . . de urbe nobis marmorarios peritissimos destinetis, qui eximie divisa coniungant, ut venis colludentibus illigata naturalem faciem laudabiliter mentiantur. De arte veniat quod vincat naturam: discolorea crusta marmorum gratissima picturarum varietate texantur, quia illud est semper in pretium, quod ad decorem fuerit exquisitum": Mommsen 1894, 16–17, 84–85 (for the date). See Kennell 1994, esp. 164–65.

80. Onians 1980; Trilling 1987; Trilling 1997; Trilling 1998, esp. 121–23; Gamboni 2002; Mitchell 2012a; Flood 2016.

81. Ibn Jubayr in Broadhurst 1952, 89.

82. *Anth. Pal.* 9.695: "ὁρᾷς τὸ κάλλος ὅσσον ἐστὶ τῆς λίθου / ἐν ταῖς ἀτάκτοις τῶν φλεβῶν εὐταξίαις." The lemma specifies "εἰς λίθον ἀκοίτονον," "on the stone *Akoitonos,*" apparently a phonetic spelling of "Aquitanus": Pontani 1978–81, 3:714n695. The stone is therefore Marmor Aquitanicum (bianco e nero antico, from *Aquitanum,* southwest Gaul), a black-and-white marble that Paul the Silentiary applauds (*Descr. S. Sophiae.* 340), as does Constantine Rhodius when he mentions "shining slabs of Aquitaine" (660).

83. For the images at Hagia Sophia, see Antoniades 1907–9, 1:94–97 (*genike*), 159–61 (*narthekos*), 2:14–21 (*kentrikou chorou*).

84. Kessler 1991–92.

85. Trilling 1998, 117–18, 126–27.

86. Illustrated in Bernia et al. 1648, 758, description at 755. See also Ricci 1905, 33; Agostino del Riccio in Gnoli and Sironi 1996, 95; Rossi 1589; Fynes Moryson (1590s) in Moryson 1907, 208; Hentzner 1617, 253; Churchill and Churchill 1732, 4:692; Barbaro 1739, 425. For other images in other, now lost, slabs: Torsani 1562, 64v; Panciroli and Gualtieri 1612, 48; Fabri 1664, 59; Pasolini 1678, 168–69; Malvasia 1678, 247.

87. Agricola 1556, 631. Illustrated in Bernia et al. 1648, 758 (description at 757–58). Crucifixion (Ruy González de Clavijo, 1403) in López Estrada 1999, 131–32.

88. Cantiga 342 in Kulp-Hill 2000, 416. This illumination at Escorial Ms T.I.1, fol. 44r.

89. Magdalino 1993, 119.

90. Oikos 9. Some scholars consider this passage (written c. 543–554) to refer to the actual acheiropoieton of Edessa: summary in McVey 1983, 100–101. But Drijvers (1998, 19–20) demonstrates that the first textual reference to the Edessene Image dates only from 787.

91. Dobschütz 1899, 37–39.

92. Guiglia Guidobaldi 1999a; Fobelli 2005.

93. Maguire 1994; Ćurčić 1992, 17–21, figs. 11–15; Dauterman Maguire et al. 1989, 19, fig. 18. Cf. the late 12th-century eulogy on "the newly built house of Leo Sikountênos at Thessaloniki," where "wondrous is the foundation of this house . . . in that its lower part receives support from above, for it is upheld by the symbols of virtue that are represented both inside and out" (Mango 1972, 225).

94. Cormack 1981, 133–35. The Apostoleion (dedicated 550) also seems to have received its first figural decorations only during the reign of Justin II (565–578). Several monophysite churches were aniconic, and some Nubian ones even favored faux-marble revetment over figural fresco: Qasr Ibrim (c. 590), Aksha (7th/9th century), Faras (630): Mundell 1977. But this rule was by no means exclusive, and Mundell provides several examples of figural decoration.

95. Liber Pontificalis in Duchesne and Vogel 1957, 172, 176.

96. Mango notes an alternative text edition that corrects "τὸ μὴ οὐχι" to "τὸ μὴ ὄν" ("the void"), presumably referring to the interstices between the tesserae: Mango 1958, 187n11.

97. For marble imitation in mosaic: Michaelides 1985; Darmon 2011.

98. Christ and the Hodegetria in aedicules flanking the templon; a large panel of *The Dormition of the Virgin* over the naos entrance: Underwood 1967–75, 1:164–71; 2:4–5, pls. 186–87.

99. For a summary of this tradition, see Close 1969.

100. *Carm.* 2.91 (*PL* 63, col. 352A): "Marmoribus opere sarsorio / visceribus lapidum permixta lege coactis / naturam faciunt artificum studia. / In solidum fractis riguerunt marmora membris / partibus in crustam colligitur genius. / Unam de variis speciem componere frustis / qui potuit, saxum duxit in obsequium." "Sarsorius" ("patch-work") indicates anything from mosaic to fabric, but Du Cange (1954, 6:313, col. 2) also defines "Opus sarsorium, ex quo colligitur ita appellari varias discolorum marmorum crustas invicem commissas, ut unum corpus et unam quasi picturam efficiant." Thus, St. Gregory of Tours writes (c. 577 AD) of the Church of St. Symphorianus built by Euphronius at Autun in the mid-5th century, "parietes ad altarium opere Sarsurio ex multo marmorum genere exornatos habet" (*Historia Francorum,* lib. 2, cap. 16; *PL* 71, col. 0215A).

101. On the history of the word "kosmos," see Finkelberg 1998, with bibliography.

102. Phot. *Hom.* 10.4, in Mango 1958, 185.

103. Choric. *Laud. Marc.* I. 25, in Förster and Richsteig 1929, 9.

104. *PL* 88, cols. 181, 380. McCrindle 1897, 146–48.

105. Literature on the eucharistic liturgy is unwieldy. For an introduction, see Schulz 1986, with extensive bibliography; on Hagia Sophia, Mathews 1971, esp. 138–49; Mainstone 1988, 226–34. An excellent summary of the Divine Liturgy is in Kallistos 1988.

106. Theodore of Mopsuestia (d. 428) in Tonneau and Devreese 1949, 485, 497; cited in Kallistos 1988, 9. Germanos of Constantinople (d. c. 733), *Hist. Myst.* 1, in Mango 1972, 141–42.

107. Russian Primary Chronicle in Cross and Sherbowitz-Wetzor 1953, 111.

108. Maximus Confessor (c. 580–662) and Nikolaos Kabasilas (c. 1322–c. 1391), cited in Mathews 1990, 210–13.

109. Paul the Silentiary refers to the ambo as central but "nodding" to the east (*Descr. Amb.* 50–51, 229, 232), and the substructures of Hagios Polyeuktos show that its ambo was very nearly on center: Harrison 1989, fig. 48.

110. On the liturgical functions of the ambo: Taft 1975, 312–13; Jakobs 1987.

111. Choric. *Laud. Marc.* II. 36. When the Pantheon was converted to Christian use during the Byzantine occupation (609 AD), the apse acquired a templon of porphyry columns that survived until 1711: Pasquali 1996, 32, 36n33, 41, 48n32, figs. 16, 17. Prokopios says of the Column of Theodora, "The column is purple, and it clearly declares before one sees the statue that it bears an Empress" (Procop. *Aed.* 1.11.9).

112. An analogous argument is made for the liturgical disposition in later Byzantine churches in Mathews 1990, 191–214. According to the *Mystagogia* (628–630) of Maximius Confessor, chs. 1–7, the binary of hierateion/naos (i.e., sanctuary/nave) expresses the dialectic between invisible, angelic heaven and the corporeal world of men; soul and body; and higher and lower faculties.

113. Hagia Sophia was begun 326 (Constantine), consecrated 360 (Constantius), rebuilt 404–415 (Theodosius II). As S. Paolo fuori le Mura, in Rome, was completed in the latter's reign, it offers the best clue to Hagia Sophia's original appearance. Apostoleion: Procop. *Aed.* 1.4.9–24. For the pre-9th-century building: Krautheimer 1964; Dark and Özgümüş 2002. The sources are collected in Janin 1953, 46–55. The Apostoleion was the resting place of the emperors and—crucially—Mango argues that the mausoleum of Constantine *preceded* the construction of the first Apostoleion: Mango 1990. Constantine Rhodius says that "either Anthemios or Isidoros" designed the Apostoleion (l. 640), although no other source mentions them: Janin 1953, 46–55.

114. For a bibliographical summary: Mainstone 1998, 282–84. Ćurčić (1992), for example, argues that Hagia Sophia evolved from a dome-on-pendentives typology. Others can find no precise precedent and argue that Hagia Sophia is an anomaly, e.g., Federico Bellini, "Cristologia e architettura in Santa Sofia di Constantinopoli," *Quaderni di storia dell'architettura* 2 (1999): 21–40; Kleinbauer 1999; Concina 2002, 59–63.

115. Procop. *Aed.* 1.4.9–24.

116. *Narratio,* 17, trans. Mango 1972, 99. The altar is also described by Kedrenos, Niketas

Choniates, Robert de Clari, and in the Souda: Dagron 1984, 242–43n139.

117. Ciggaar 1973, 339, 345–46.

118. Dagron (1984, 284) adduces the 6th- to 9th-century alchemical texts in Berthelot and Ruelle 1887–88.

119. Buckton 1984, 90–95, 129–40, 156–70, 189–90, 194–95, 252–57; cat. no. 5 (ewer); nos. 10, 11, 15, 16, 17, 23 (chalices); nos. 18, 25 (pattens); no. 35 (cruet). Several vessels probably came from the Pantokrator church, the seat of the Venetian *podestà* during the Latin domination (1204–61), but Hagia Sophia, the seat of the Venetian patriarch, is also possible. Unfortunately, the only inventory of the treasury of Hagia Sophia dates from 1396: Miklosich and Müller 1860–90, 2:566–70. However, Prokopios (*Aed.* 1.1.64) does say that Justinian endowed the church with gold and silver vessels and works in precious stones.

120. *Descr. Amb.* 50ff. Mango 1972, 92.

121. Porphyry urn, Hildesheim Cathedral, donated by Otto III (1020): Delbrück 1932, 196, pl. 101. Porphyry vase in S. Maria in Porto (Ravenna): Ambrogio Traversari to Niccolò Niccoli, 12 October 1433, in Baxandall 1971, 152–54. An identical vessel is mentioned by Rabban Sauma at the Pantokrator monastery in 1287: Brock 1969, 246, 250; Anonymous Russian (1389–91) in Majeska 1984, 152; Brock 1967, 87; Letts 1926, 142.

122. "The wine on top of hand is like a lamp / overflowing with rays and flames and radiance. / The beholder considers that, because of its unity with its / beaker, it exists there without any vessel": Abu'l-Fadl al-Kaskari (10th century), trans. Shalem 1994, 6, with other examples.

123. "A precious chalice out of one solid sardonyx . . . in which one [stone] the sard's red hue, by varying its property, so keenly vies with the blackness [*sic: "nigredo"* a lapsus for *"albedo"*?] of the onyx that one property seems to be bent on trespassing upon the other" (Suger, *De admin*, 34.207–8); Panofsky and Panofsky-Soergel 1979, 78–79.

124. John of Damascus describes the Eucharist as a divine "coal"; cf. Isaiah 6:6. Other sources describe the eucharistic transformation as a descent of fire: Kallistos 1988, 12–13. Aristotle explained that the color *oinops* (Homeric "wine-dark") occurred when sunlight intermingled with the dark juice of the berry, as when stirring wine in the krater (*Col.* 2.792b.7–8). Homer had called wine *"aithops"* (fiery), perhaps because when the juice ferments it visibly "boils": Schwarzenberg 2000, 21.

125. Inscription on a lost chalice, *Anth. Pal.* 9.819: "I, the mystic cup, by the flow of the Holy Spirit pour into the heart a stream of repentance," in *The Greek Anthology*, trans. William. R. Paton, 5 vols. (London: Heinemann, 1925), 3:441.

126. While The Piacenza Pilgrim (late 6th century?) mentions "the onyx cup He blessed at the Last Supper," Adamnan, less than a century later, describes it as a "silver cup holding the measure of a Gallic sextarius and with two opposite handles" (see Geyer 1898, 154, 173, 234, 305). The Sacro Catino in Genoa is of green glass, though long said to be emerald, but the chalice at Valencia in Spain is a cup of agate.

127. Buckton 1984, 169–70 (patten), 159–65 (chalice), both 10th/11th century. These phrases appear in the consecration formulas of both St. John Chrysostom and St. Basil: Hammond and Brightmann 1896, 1: 15ff.

128. Pappos 3.1022–24, in *Selections Illustrating the History of Greek Mathematics*, ed. and trans. I. Thomas, 2 vols. (London: Heinemann, 1939), 2:614–17. On mechanikoi, see Downey 1946; Meek 1952; McKenzie 2007, 327–28; Schibille 2009. See also the feats attributed to Ur-architects in Aus. *Mos.* 298–317. On mechanics and nature, see also Schiefsky 2007.

129. Agathias, 5.7.2–5.8.4, 5.6.3, in Frendo 1975, 142–43, 41. Among several studies, see Huxley 1959; and Knorr 1983.

130. Paul 1 Corinthians 3:10; Concina 2002, 63.

131. O'Meara 2004. On Anthemios and Isidoros and Neoplatonism, see also Kaldellis 2013.

132. Palmer (and Rodley 1988, 139–40) discerns Genesis 1; Exodus 25ff, 35ff; 1 Kings 8; 2 Chronicles 5–7; Psalms 23 and 65:5, 7; Proverbs 8–9 (Proverbs 8:24–29); Jeremiah 33:11; John 1; 1 Thessalonians 5:17; Hebrews 8; and Job 38. Malalas recounts that Psalm 23 ("open up the divine palace to our verses") was certainly sung: Macrides and Magdalino 1988, 55.

133. Barrett Browning 1863, 70–75.

134. For an overview: Bartusis 1995; Hunger 1969–70. Prokopios affected ignorance of contemporary terminology to maintain the authenticity of his classical voice in his *History of the Wars*, but not in his *On Buildings:* Cameron 1965. So does Paul the Silentiary: Chuvin 1997, 35.

135. Lamberton 1986, 78–82, 144–232.

136. Most conveniently, see the bibliography cited in Agosti 2003.

137. At Skiprou (Boetia, Greece), all inscriptions on the Church of the Virgin (9th century) are in contemporary Greek except those that commemorate the patron's eternal virtues, which are in Homeric verse: Papalexandrou 2001, 278–79. See also Papalexandrou 2007; and James 2007, in a growing bibliography. Paul's description also follows an internalized, ahistorical time loop, for his narrative begins with the dawn procession that had inaugurated the church and climaxes in night where she becomes a beacon of eternal illumination for mankind: Macrides and Magdalino 1988, 59–61, 73.

138. Aristid. *Orat.* 26.4, in Behr 1981, 73.

139. Choric. *Laud. Marc.* II. 25–26, in Förster and Richsteig 1929, 34–35; Litsas 1980, 140. Older versions in Abel 1931; Hamilton 1930–31; Downey 1950.

140. Italian trans. Cortassa 2000, 92–95.

CHAPTER 8. WALKING ON WATER

1. For more examples, expansive notes, and exhaustive bibliography, see Barry 2007. This did not, however, mention the baptisteries at Timgad and Hagia Sophia, medieval aqueduct quarrying, or Hagia Eirene, all discussed here.

2. Guidobaldi and Guiglia Guidobaldi 1983, esp. 19–58.

3. Fayant and Chuvin 1997, 146–49.

4. Mango and Parker 1960, 237, 239 (ll. 73–78, 156–74), 243.

5. See the commentary on the *Narratio* in Majeska 1978; and the text redactions in Vitti 1986, 462.10–16, 67.5–8 (a); 486.2–4, 487.7–11 (k); 503.16–20 (f); 561.1–5 (n); 578.23–28 (x); 598.14–21 (f); 616.23–25 (y). The other conspicuous exception to Proconnesian in the nave is the huge rota (or *omphalion*) with orbiting rotae that is a much later insertion, which has been dated to both the reign of Basil I (867–886) and the 12th century: Pedone 2012.

6. Stubbs 1876, 1:93–94; Anstruther 1851; Marichal 1960, 257, 259.

7. Levend 1958, 76–78; Borsook 1973, 173n95.

8. Apostoleion: Constantine Rhodius (931/944), in Legrand and Reinach 1896, 56, ll. 670–71; Nikolaos Mesarites (1198/1203), in Downey 1957, 890, 914. Chora Church: Underwood 1967–75, 1:17–20, pl. 9; Ousterhout 1987, 39–45, 66, 137–39, figs. 50–54; Guiglia Guidobaldi 1999b, 322–24, nn18–19. Pammakaristos: Belting et al. 1978, 20–21, figs. 6, 7, 9. Panatanassa: Brock 1967, 87, 90, 95. Majeska 1984, 377.

9. Diehl et al. 1918, 35–58, pls. 3–12. Synopsis of the dating in Kourkoutidou-Nikolaïdou 1989. See also Kleinbauer 1984, esp. 2:243.

10. See also the Oratorio della Pesca at Aquileia. Bisconti 1996, with vast bibliography. The pavement is attributed to the church foundation (and possibly the campaign of Siricius, 384–399) in Guidobaldi 2002, 1067–69, figs. 17, 18.

11. Kingsley Porter 1912a; Kingsley Porter 1912b; Cecchi Gattolin 1978. For its dating: Valla 1992. For its inscriptions: Trovabene 2001. The sea motif reoccurs in the contemporary apse mosaic of S. Michele in Pavia, flanking a central panel of a labyrinth: Peroni 1977, 718. Medallions also floated against wavy backgrounds in the contemporary Church of S. Tomaso at Reggio: Trovabene 1996. I suspect that the wavy tessellations on the 13th-century floor under the crossing of the duomo at Osimo also evoke waves: Barsanti 1996.

12. Barral i Altet 1985, 78.

13. Tavani 1999, 330. Other examples are the early 5th-century baptistery at Salona, the atrium at the "post-Attila" basilica at Aquileia, and the nave of the basilica di Via Madonna at Trieste, and possibly the Basilica Euphrasiana at Poreč: Mirabella Roberti 1975, 199–200.

14. Description of the Palace of John I of Ibelin at Beirut written by William of Oldenburg: Laurent 1864, 167. A similar conceit already appears in Auson. *Mos.* 10.53–54.
15. Bloch 1986, 44–52, fig. 27.
16. Nicephoros Kallistos Xanthopoulos (c. 1256–c. 1335) on a floor in the Palace of Andronikos Palailogos the Elder (1282–1328), Constantinople: Richter 1897, 368, no. 980.
17. Stringa 1610, 19. Loosely comparable paving is found at Hosios Loukas (1011/48), the Church of Sagmata (c. 1105), and the Church of the Dormition at Nicea (after 1065): Farioli Campanati 1997, 12. A Proconnesian field at the crossing is juxtaposed with more intricate opus sectile patterns also in the south church at Koutsovendis (c. 1090): Mango et al. 1990, 68. Large Proconnesian slabs also adorned the crossing of the church built by Alexios Apokaukos at Selymbria (Siliviri) c. 1325: Eyice 1978a, with bibliography.
18. Demus 1984, 1 (text), 194–95; 2 (plates), figs. 234, 327–29.
19. On Proconnesian quarrying and export: Barsanti 1989; Asgari 1992; Asgari 1995; Harper 1997. Cf. Sodini 1989.
20. Holmyard and Mandeville 1927, 46.
21. De Mély and Ruelle 1896–1902, 3:xxxiv.
22. *De mineralibus*, 1.1.3; Wyckoff 1967, 14–17.
23. Vincent of Beauvais's *Speculum naturale* 5.49, written c. 1250, very closely follows Vitruvius's description (*De Arch.* 8.3.10): Schuler 1999, 178.
24. Grewe 1991; Donati 1695, 257.
25. Addendum to the 9th-century (?) Greek *Chronicle of the History of Constantinople from Its Beginning until the End* (anonymous 1491 chronicle): Giese 1922, 93; Yerasimos 1990, 27. Filarete in Spencer 1965, 1:31–32; bk. III, fol. 17v–18r.
26. Schwarzenberg 2000, 22.
27. Gell. *NA*. 2.26.21–23 (Enn. *Ann.* 10.384–85).
28. *Aen.* 7.27. Cf. *Aen.* 6.729. Servius (*Ad Aen.* 7.718; 10.208) gives the gloss "marmore *mari*" (i.e., "marble" means "sea";). Cf. Lucr. 2.766–67.
29. Ov. *Tr.* 3.10.47–48. Cf. Valerius Flaccus, *Argonautica*, 6.568; Luc. 9.349; Catull. 63.88. The usage persists in medieval poetry; see *Novum Glossarium Mediae Latinitatis: Ab Anno DCCC usque ad Annum MCC* (Hafnia: Munksgaard, 1957–), *ad vocem*.
30. Julian. *Mis.* 341B. Late antique poets continue to match "mar" and "marmor": Avien. *Orb. Terr.* 56, 137, 169, 187, 206, 230, 245, 429, 492, 552, 635, 709, 714, 751, 775, 828, 1310.
31. Barry 2011b.
32. Avitus, *De transitu Maris Rubri*, 5.592–593. Cf. Prudent. *Cath.* 5.67–68. The early 5th-century *Heptateuch* of Cyprianus Gallus actually uses "marmor rubrum" (Exodus 434). The "Red Sea" sarcophagus (Museo Civico di Brescia, MR 5852) survived antiquity as the frontal of the high altar in the Church of S. Afra (now S. Angela Merici), Brescia: Marchini 2014.
33. Stat. *Silv.* 1.2.149–50 ("concolor alto vena mari"), 4.2.28 ("glaucae certantia Doridi saxa"), 2.2.92 ("gaudens fluctus spectare"), 1.5.19 ("undosa Carystos"). The perception that the sea was green explains Isidore's comment: "The Greek word 'marble' is called after its greenness" (Isid. *Etym.* 16.5.1: "marmor sermo graecus est a viriditate vocatus").
34. For the villa: Paris 2000, esp. 76–79. For the quarries: Lambraki 1980; Vanhove et al. 1995. Cf. Mart. *Ep.* 9.80–81: "extruit thermas / de marmore omni, quod Carystos invenit."
35. Museo Nazionale Romano (Palazzo Massimo), 171, c. 200–300 AD: Versluys 2002, 76–78, cat. no. 19, with bibliography.
36. Khader 1987, 61–91, esp. 70, pl. XXVII. Many other examples of such faux-Carystian floors survive from Tunisia, several transported to the Bardo Museum. Winter Baths (199 AD): Khader et al. 1985, 72–73, pls. XXVII, XXX, XXXII. Also the frigidarium of the public baths at Maktar.
37. Germain 1969, 123, with bibliography.
38. For the marbles: Ueblacker and Caprino 1985, 37–38. For the paving: Guidobaldi et al. 1994, 112–13, no. 39, pl. LXXVII.
39. Procop. *Aed.* 1.10.19; Paul. *Descr. Amb.* 150–153; see also *Descr. S. Sophiae.* 664–67. Eumathios Makrembolites (1140s), *Hysmine and Hysminias*, 1.5.7.
40. Pero Tafur in Letts 1926, 114. "Marmora" is also spelled "Marmara" and "Malmara." For the earliest surviving references: Gautier Dalché 1995; Villehardouin in Faral 1939, 2:292 (c. 1212 AD).
41. Dirimtekin 1965; Eyice 1978b.
42. Kitzinger 1951, 102–3; Maguire 1987b; Maguire 1987a, 21.
43. Palmer and Rodley 1988, 127, 134, with earlier bibliography.
44. For example, the Basilica at Tegea (Arcadia, late 5th century), the church at Khaldé (Lebanon, c. 450/500), SS. Cosmas and Damian at Gerasa (Jarash, Jordan, 533), and the basilica at Heraclea Lynkestis (Bitola, former Yugoslav republic of Macedonia, late 5th/early 6th century): Maguire 1987a, 24–26, 33–40.
45. *Qur'an*, 11.7/9, derived from Psalms 29:10. See O'Shaugnessy 1973, 212. For the ancient Near Eastern roots of the tradition, see Neiman 1969.
46. Grossmann 1973; Deichmann 1969–89, 1:220–25, fig. 211; 2.2:35–46, esp. 40–43; Effenberger 1989, esp. 60–64, n114.
47. Talbot 1999, 88; *Carm.* 86 in Miller 1855–57, 1:38.
48. Mesarites 37.4 in Downey 1957, 890, 914. Niketas Magistros, *Vita S. Theoctistae Lesbiae*, ch. 3, *Acta Sanctorum*, Nov. IV, 226; Mango 1972, 104. Michael the Deacon in Mango and Parker 1960, 235, 37.
49. George 1913, 48–50. The other inscriptions are from quotes from Psalm 65:4 and 6.
50. *PG* 106, col. 943; Cramer 1841, 4:306. The Stoudios columns are actually Thessalian marble (verde antico), but Paul the Silentiary calls even this marble "fresh green as the sea." On the floor, see Barsanti 2012.
51. Bargebuhr 1956, 211–12n60. These were also materials of Paradise: "A river flowed from Eden to water the garden . . . bedellium [crystal] and onyx stone are found there" (Genesis 2:10–12). Some alabasters were also compared to water, such as that from Hierapolis on the ambo of Hagia Sophia (which Paul the Silentiary extols): Mango 1972, 92. For the latter alabaster: Bruno 2002.
52. God "alone spreadeth out the heavens, and treadeth upon the waves of the sea," "by the breath of God frost is given: and the breadth of the waters is straightened," but also the "sky which is strong, and as a molten looking glass" (Job 9:8, 37:10, 37:18). Job is generally considered a text of Hellenistic date.
53. Cosmas Indicopleustes explains that "the heavens being on fire shall be dissolved, and the elements shall melt with fervent heat but [we look for] new heavens and the new earth" (Hebrews 6:20; 1 Peter 3:12) means "that with a great noise, as in the twinkling of an eye, all the elements being on fire as in a furnace and being thus purified, undergo the change for the better"; Cosmas Indicopleustes 7.290–91 in McCrindle 1897, 287–89. The crystalline transformation appears also in Thomas Aquinas, *Summa Theologica*, 91:3–4, based upon Isaiah 30:26.
54. The floor in S. Akakios gave the impression "that the whole church is coated with snow" (Procop. *Aed.* 1.4.25); that of the Kauleas monastery was "paved with white slabs [forming] a continuous translucent [surface], uninterrupted by any other colour": Leo the Wise (886–912), *Sermon 28*, trans. Mango 1972, 202. The pavement of the Pharos Chapel "was of a white marble so smooth and clear that it seemed to be of crystal": McNeal 1997, 103. A floor "en plaques rectangulaires de marbre blanc avec un cadre noir" was discovered in the excavation of "Basilica A" at Bayazit, Istanbul, in 1946 and immediately destroyed along with the entire complex: Firatli 1951, 167, fig. 1; Mamboury 1951.
55. 1 Kings, 7:23; Ezekiel 47:1–12. The appeal to Solomon's brazen "sea" is noted in Scheja 1962, 51. The *Narratio* (19) claims that Justinian originally wanted to sheathe the floor in silver. It also claims that Justinian made a fountain in imitation of Solomon's in the atrium and calls the floor below the altar, which was sheathed in silver, a "sea": Dagron 1984, 205, 243n142. Constantine Porphyrogenitos calls the altar a "little sea."
56. More difficult to locate and interpret is the "sea" he made "on the right side of the Gunaikitis . . . in which water collected to the depth of one span, and a gangway for the priests to walk over the pool"; Mango 1972, 101. See the various redactions of this

passage in Vitti 1986. Such immersion pools, and their Judaic and classical precedents, are analyzed in Pallas 1952, esp. 39–40, 146–56; Broilo 2009.

57. Grossmark 1998, esp. 277–78; Grossmark 2012.
58. Epstein and Simon 1948, Sukkah 51b, Baba Bathra 4a.
59. Epstein and Simon 1948, Hagigah 14b; Neusner 1999, Hagigah 2, 2–4; Elior 1982, 31. In another version, a rabbi warns his listeners not to confuse the alabaster pavement before God's throne with water: Strack and Billerbeck 1922–28, 3:798–99 on Ap. 4.6.
60. Scholem 1941, 52–53. Rosen-Ayalon (1989, 55) cites this passage to explain that the Proconnesian cladding within the Dome of the Rock specifically alluded to the Holy of Holies, as a suitable ambience for the Throne of God, when he will descend there at the end of time.
61. Scott-Meisami 2001, 73; Milwright 2005.
62. Muhammad ibn Sasra (1389/97) cited in Flood 2001, 67n46.
63. Grossmark 1998, 278; Yassif 1982, 50–56; the *Ma'ase Malkah Sheba* by Saadya ben Yosef (1702), cited in Silberman 1974, 70–71. The story is echoed in the Qur'an, *Sura* 27:44–46 ("she thought it a pool and uncovered her legs. [Solomon] said, 'It is a palace paved with glass'"). In a passage from the medieval Arabic travelogue *The City of Brass*, which could be a garbled account of a (possibly) 7th-century visit to an abandoned Roman city, the inner hall of the palace "was made of gleaming marble inlaid with precious stones, so that the spectator got the impression the floor was streaming water, and whoever walked upon it slipped. But the emir told the sheik to strew something on the floor, so they could cross it": Gerhardt 1963, 205.4, 48–49.
64. Gonzalez 2002, esp. 26, 30–48. For the afterlife of Solomon's "crystal palace," see Haag Bletter 1981.
65. Shiny green glass paving tiles were recovered in one of the public rooms of the Islamic Palace B at Raqqua-Rafiqua, in Syria: Weiss 1985, 517 (cat. 262). Plan in Saliby 1954. For a thorough study of Solomon's crystal palace and its influence on the Islamic palace: Flood 1993, 184–242.
66. Baudri de Bourgueil, *Carm.*, 134.728–35, in Tilliette 1998, 2:23; Barral i Altet 1987, 42.
67. *Digenis Akritis*. 7.56–58 in Jeffreys 1998, 204–5.
68. ΘΑΛΑССА ("Sea"): Lux 1968, pls. 14–35; Piccirillo and Alliata 1989, 96–107; Piccirillo et al. 1993, 106–7, figs. 78–95. In the middle of the floor of the Church of Bishop Sergius (late 6th century) at Umm al-Rasas, Jordan, there is instead a marine animal labeled "ΑΒΥССΟС" ("Abyss"): Hamarneh 1994; Gautier di Confiengo 2000, 450–51.
69. In fulfillment of Job 9:8: "Who treadeth upon the waves of the sea." See also the inscription that Venantius Fortunatus composed for the Oratory of Trasaric in Gaul, *Carm.*, 2.13.3–4: "This is the hall of Peter who locks the heavens with a key and under whose steps the sea stood firm as a stone" (*Haec est aula Petri caelos qui clave catenat / substitit et pelagus quo gradiente lapis*) in Reydellet 1994, 69. At l. 10, Venantius writes that the church "melts the shadows of the world and grasps the stars" (*tenebras mundi liquit et astra tenet*).
70. Ode 6, tone 3, oikos 1, orthros on Sunday morning: "Enveloped by the bottomless depths of my sins, I sense my life ebbing away. O Master, lift up Your hand and stretch it over to me; save me as You once saved Peter, O You who walk upon the waves." Ode 6, tone 3, oikoi 1 and 3, orthros on Tuesday morning: "I am drowning in the depth of sin. The sea of life is passing over me. But as Jonah came forth from the whale, so bring me up from the abyss of the passions and save me, O Lord . . . I am tossed on a storm of passions; but as You once ordered the waves to be calm of old and saved Your holy Disciples, O Christ Jesus, so extend Your hand to me and save me." Cf. ode 6, tone 6, orthros on Sunday morning. All in *Octoechos* 1999, 1:23–24, 230–31; 3:192–93. These prayers date between the 9th and 12th centuries.
71. *Tractatus Aedificationis et constructionis Ecclesiae sancti Johannis Evangelistae de Ravennae* in Colombo and Colombo 1942, 567–72. On S. Giovanni Evangelista: Deichmann 1969–89, 2.1:93–124, with comparative texts at 107–24. Cf. Psalm 65:5, 7: "O God of our salvation: who art the confidence of all the ends of the earth, and of them that are afar off upon the sea . . . which stilleth the noise of the seas, the noise of their waves."
72. *De situ, constructione ac renovatione coenobii Casinensis (Carm. 32)*, ll. 150–54, 188–89 (*PL* 147.1237C), in Lentini and Avagliano 1974, 176.
73. Oikos 7. Greek text in Trypanis 1968, 141–47 (no. 12); amended according to Palmer and Rodley 1988, 137n22, 140–44; Palmer 1990. Contemporary Syrian exegesis expresses the same ideas quite forcefully: McVey 1983, 114–15.
74. Bonner 1941; Brock 1995.
75. Oikos 9. The expression "sea of troubles" exists also in classical Greek: Aeschylus, *Seven against Thebes*, 758, and *The Persians*, 433.

CHAPTER 9. RELICS, TABERNACLES, THRONE ROOMS

1. "The Emperor Basil [r. 867–886] took away many pieces of marble and mosaic cubes from many churches on account of the Nea Ekklesia": Leo Grammaticus in Mango 1972, 181n1. John I Tzimiskes (r. 969–976) demolished the Oikonomeionex Baths in the Great Palace to build the Church of the Savior at the Chalke Gate in 971: Leo the Deacon in Mango 1959, 148. Isaak Angelos (r. 1185–95) removed marble slabs from the Great Palace to restore the church of St. Michael at Anaplous: Niketas Choniates in Mango 1972, 236–37; Ousterhout 1999, 140–45. Most of the revetment in the Kalenderhane Camii (late 12th century) had to be cut from column shafts: Kuban et al. 1997, 116–19. Chora mural: Underwood 1967–75, 3:pl. 536-b (the faux-spolia was first noted in Ousterhout 1999, with many observations on Byzantine spolia).
2. In a vast literature, see Barry 2010, with bibliography.
3. Pani Ermini 1998, with bibliography.
4. For the marbles that Giotto simulates, see Lazzarini 2008.
5. Möseneder 1981; Benesovská and Hlobil 1999, 44–58; Sedinová 1999; Sedinová 2000.
6. Leonardo Frescobaldi (1384) in Angelini 1944, 147, 148.
7. Lerou 2004, 177–79.
8. Majeska 1973.
9. Demus 1960, 121, 187–88; Lange 1964, 109–10 (cat. no. 39); Grabar 1976, cat. no. 123. The earliest record of the legend is from 1487.
10. Pincus 1990; Pincus 1996.
11. Semsüddin's 1480 chronicle: French trans. Dagron 1984, 196–211. Yerasimos 1990, 6–7, 27–28, 113–23; Necipoğlu 1992, 199–200. Cf. Evliya Çelebi (c. 1670) in Hammer-Purgstall 1846–50, 1:55; the anonymous 1491 chronicle, fol. 74, in Yerasimos 1990, 27.
12. *De mineralibus*, 2.3.1; Wyckoff 1967, xiii–xiv, 128.
13. Tafur in Letts 1926, 229. See also the *Roman des Sept Sages* and *Roman de Cléomadès* of c. 1285: Spargo 1934, 62, 64, 134–35. On scrying in medieval and later times: Delatte 1932, 13–132, 154–84.
14. Kren and Wieck 1990, 59, pl. 20.
15. Jerome: *PL* 22, col. 884 (*Ep.* 108.9). Piacenza Pilgrim: Wilkinson 1977, 83. The Piacenza Pilgrim is followed by Bede (*PL* 95, col. 362). Apostoleion columns: Majeska 1984, 300–301 (esp. nn70–79); Jørgensen 1954, 2:232–91; Vincent and Abel 1912–26, 2.3 (Jérusalem Nouvelle), 578.
16. Davanzati 1725, 227–32, 409–35; Vannutelli 1896; Gnoli 1988, 150–52; Del Bufalo 2001. Only during the Counter-Reform did artists begin to include this stubby column in their representations: Wisch 1991.
17. Bartolini and Caporali 1999, unpaginated in facsimile, but 42–43 in Italian trans.
18. Dante, *Purgatorio*, 11.101–2: "porfido . . . sì fiammeggiante, / come sangue che fuor di vena spiccia."
19. Francini et al. 1574, 58. Nicholas Muffel says something similar of a rota in St. Peter's: Wiedmann 1999, 49.
20. Filarete, lib. III, fol. 17r in Spencer 1965, 1:31. Vasari-Milanesi 1906, 6:610–11 (Andrea del Sarto).
21. Arculf (7th century): Adamnán, *De locis sacris*, I, 3.1, f. 232 trans. Wilkinson 1977, 96;

Giacomo da Verona (1335) in Röhricht 1895, 184; Anselme Adorno (1470/71) in de Groer and Heers 1978, 264.

22. The most famous post-antique imperializing sarcophagi are the Norman tombs in Monreale: Deér 1959. But it was an international phenomenon: e.g., the late 13th-century tombs of the English kings at Westminster Abbey: Binski 1995, 93ff.
23. White 1969a; White 1969b.
24. Heydenreich 1961, 222; Dezzi Bardeschi 1966.
25. For this tomb, see Markham Schulz 1977, 32–51, 99–104 (cat. no. 5), pls. 49–90.
26. Panel from a diptych, the left panel with the Madonna and Child against a gold ground. Altenburg, Lindenau-Museum: Oertel 1961, 68–69, nos. 47/48 (color pl. 5); Cämmerer-George 1966, 70–71.
27. Wood, painted surface 10⅜ × 8³⁄₁₆ in. (26.5 × 21 cm). National Gallery of Art, London, inv. no. 590: Dunkerton 1991, fig. 101.
28. Philadelphia Museum of Art, cat. 314: Asperen de Boer 1997. On the patron, see "Adorno, Anselmo," in *Dizionario biografico degli italiani* (Rome: Società Grafica Romana, 1960), 286–87.
29. Mundy 1977.
30. Habig 1983, 732.
31. Soresini 1672; Bambi 1775; Mazzucconi 1840. The "stains," in fact marble veins, are visible through brass portholes let into the nutwood treads that have protected the flags since 1723.
32. Barbier de Montault 1885, 3, 5, 7–8, 11, 13, 14, 15, 16, 18. The miraculous marbles are not mentioned before the 16th century: see documents in Lazzarini 1952; Leandro Alberti in degli Avanzi 2003, 2:70; and Adami 1734, 94.
33. Parlato 2009, 148–49, n33, with references. Another altar, conserved in the Lateran cloister, has the same distinction: Barbier de Montault 1885, 3.
34. Asche 1978, 115–17, cat. W. 01–03, pls. 336–42; Kanz 1994, esp. 238–39.
35. Graeve 1958, 227ff., nn23ff.; Mango 1970, 372–75; Drpić 2019, with more recent bibliography.
36. John Kinnamos in Brand 1976, 207; Robert de Clari in McNeal 1997, 112–13; Anthony of Novgorod in Erhard 1932; Clavijo in López Estrada 1999, 138–39, and Le Strange 1928, 83; Ignatius of Smolensk and Zosima the Deacon in Majeska 1984, 94, 186. Buondelmonti mentions the slab c. 1420: Gerola 1931, 276. The last eyewitness to the stone was a Greek pilgrim at some point after 1434: Mango 1970, n38.
37. de la Broquière in Schefer 1892, 160–61.
38. Raban Sauma in Brock 1967, 46–49.
39. Ludolphus does not explicitly mention the slab but says, "En ce lieu, dit-on, ses larmes sont encore apparentes": Lecoy de la Marche 1870, 168. For the many editions in several languages: Bodenstedt 1944, 16–23.
40. Janin 1944–45, 174–75; Sansovino 1585, pt. 2, 185. Sansovino thought the stone had been transferred to the Holy Sepulcher in Jerusalem.
41. Milan, Pinacoteca di Brera, 199. Variously dated to between 1464 and c. 1500, though Lightbown (1986, 421–23 [cat. no. 23], pl. X) prefers the 1480s/90s; Prater 1985; Martineau 1992, 155.
42. Choniates only says that Manuel was buried "in the *herôon*. . . . Next to the tomb there opens up a wide entrance, i.e. where the wall of the church forms an arch. . . . Beside it, placed on a pedestal a red stone, long as a human body, is exposed to veneration": Magoulias 1984, 125. See also Megaw 1965 (mistakenly referring to the relic as porphyry); Cormack 1985b, 194–214; Ousterhout 1999, 121–22; Ousterhout 2001.
43. Anonymous Armenian pilgrim (1375/1434) in Brock 1967, 87: "There is the stone on which they wrapped up Christ, above which appears the Mother of God." Mango 1970, 374.
44. Cormack 1985b, 209; Gaultier 1974. The typikon dates to 1136. Trans. Richard Jordan in Thomas et al. 2000, 2:725–81, esp. 754.
45. Fiocco, Rathe, Tietze-Conrat, Camesasca, and Arslan cited in Lightbown 1986, 423.
46. Barbier de Montault 1888, 486–87; de Mély 1890, 105; Didi-Huberman 1995, 110.
47. Reinhold Lubenau (1587), Pacifique de Provins (1622), and Evliya Çelebi (c. 1670): Sahm 1912, 143; Mango 1962, 120, 123; Hammer-Purgstall 1846–50, 1:64–65.
48. Babinger 1956, 8, 19, 21; Raby 1980, 99, n100.
49. Literature on the shroud is vast, but see Solero 1956, 138–41; Savio 1957, 80–130; Pugno 1961, 23–165; Wilson 1978, 219–23.
50. The Savoy had married into the last imperial family of Byzantium, the Palaiologoi, both the main line in Constantinople and the cadet branch at Monferrato: Oresko and Parrott 1997, 14, 20–21, 22. The Savoys' purchase of the shroud could, therefore, be regarded as supporting a dynastic claim to the throne of Constantinople as well.
51. Accademia Carrara, Bergamo, no. 724. On Bellini's half-length Madonnas, see Goffen 1975; Goffen 1989, 66ff. Examples by Cima in Humfrey 1983, cat. nos. 4, 5, 9, 16, 18, 25, 32, 39–40, 48, 52, 62–63, 68, 91, 93–94, 96–97, 102, 108, 111–12, 114, 116, 122, 126–27, 129, 137, 141, 151–52, 159, 168–71, 173, 175–78, 180, 185, 211, 214, 218, 228, 240.
52. Berlin-Dahlem, Staaliche Museum, 29: Ringbom 1965, 72–93. For bibliography and dating, see Lightbown 1986, 404–5, cat. 7, pl. IV.
53. Ringbom 1965, 74n7. For the iconography of the presentation, the fundamental study remains Shorr 1946.
54. Galleria Querini-Stampalia, Venice: Ringbom 1965, 77; Lightbown 1986, 405.
55. Felix Fabbri, fol. 117a, in Stewart 1893–97, VIII, 271; Petrucci and Petrucci 1972, 10; Elzear Horn (1724/44) in Hoade and Bagatti 1962, 123; Rachman-Schrire 2017. Bartolomé Bermejo's *Pietà* (c. 1465–66) may even have been based on an eyewitness account of the same slab: illus. in Young 1975, cat. no. A5, fig. 9. Cf. his cameo view of the same subject on a slightly later altarpiece (c. 1474–77): Young 1975, cat. no. A16, fig. 32.
56. For example, Jan Van Eyck, *Virgin and Child with Sts. Barbara and Elisabeth of Hungary and a Carthusian Monk* (c. 1441), Frick Collection, New York.
57. Piper 1873. Cf. Wormald 1961; Forsyth 1972.
58. On the history of this painting, see Eisler 2016.
59. On the painting, but little about the frame, see Wilson 1976; Chiappini Di Sorio 1986; Valazzi 1988; Goffen 1989, 123–37.
60. Baxandall 1972, 36.
61. For example, the Master of the Strauss Madonna (late 13th/early 14th century, Accademia, Florence). A superb investigation of the whole genre may be found in Arasse 1999.
62. Boskovits and Brown 2003, 466–71, with bibliography; Spencer 1955, 273–80. See also the Annunciation relief on the façade of St. Foye, Conques: Folda 1986, pls. 42, 43. Hours of the Virgin, Pierpont Morgan Library, New York, fol. 26v: Robb 1936, fig. 11.
63. For example, Masolino's *Annunciation* fresco in the Branda Castiglione Chapel, S. Clemente, Rome (c. 1428–1432)—red-and-white marble; Francesco di Giorgio Martini's *Annunciation* (Siena, Pinacoteca)—red-and-white marble; a stridently veined, red column is discreetly inserted into the left tondo of Filippino Lippi's *Annunciation* (1483, Pinacoteca, S. Gimignano). The many tropological interpretations of the column by Hrabanus Maurus are reviewed in Didi-Huberman 1995, 153–54. For a Romanesque example, see an *Annunciation* relief from S. Piero Scheraggio, Florence (c. 1180–1200), now in the Met Cloisters, New York.
64. Naser-e Khosraw in Thackston 1986, 26.
65. Frà Niccolò da Poggibonsi (1346–1350) in Bacchi della Lega and Biagatti 1945.
66. Ragusa and Green 1961, 32; Didi-Huberman 1995, 159–60; Bandmann 1970.
67. Anselme Adorno (1470/71) in de Groer and Heers 1978, 312.
68. Elzear Horn in Hoade and Bagatti 1962, 259–63.
69. Viaud 1910, 81–108; Bagatti 1984, 54–70; Folda 1986, 15–30.
70. Didi-Huberman 1995, 132–60.
71. See Didi-Huberman 1995, 132–60, for fascinating observations on the threshold, opened and closed doors, and other liminal motifs.
72. The main passage linking Mary with Jerusalem is in Isaiah 47ff. For a thorough but succinct review: Viller et al. 1937–95, 8: cols. 944–58; 10: cols. 416–21.
73. Livius and Vaughan 1893, 253–77; Cornelius 1930, 37–48. On the Virgin as tabernacle: Hirn 1912, 151–68. Albertus Magnus, *De laudibus Beatae Mariae Virginis,* in Borgnet and Borgnet 1898, esp. VIII, 2–4; IX, 1–8; X, *passim;* Didi-Huberman 1995, 70.

74. Rohault de Fleury 1878, 1:329ff.
75. Didi-Huberman (1995, 155) regards such material characterizations (i.e., stone that seems like cloud or fire) as evoking the "materials of 'dissimilitude' that Dionysius [the Areopagite] had electively proposed to represent the celestial mysteries" and as fluctuating between the varying tropological meanings of the figure.
76. Talbot 1999, 88–89.
77. Meier 1977.
78. In the ever-expanding literature on the paragone, see Pepe 1969; Collareta 1988; Fallay d'Este 1992; La Barbera Bellia 1997. *Hypnerotomachia* (Pozzi and Ciapponi 1964, 188): "el praestante artifice, electo solertemente el marmoro havea, che oltra la candidecia sua era venato (al requisito loco) de nigro, ad exprimere el tenebroso aere illumino, & nebuloso cum cadente grandine."
79. McAndrew 1980, 39–42, 47–50; Lieberman 1982, pls. 51, 52; C. Davis 2007; Markham Schulz et al. 2003. For a more detailed analysis of Pietro Lombardo's output, with sources, see Barry (forthcoming).
80. For the Miracoli in general: Lieberman 1986; Piana and Wolters 2003; Barry (forthcoming).
81. Felix Fabbri (1488), quoted in Fortini Brown 1995, 151.
82. de Retza 1487/90. On the book and its author: Purkart 1972.
83. Finocchi Ghersi 2004, 19–77, with bibliography. The recent redating (Wallace Maze 2013) of Giovanni Bellini's birth to 1424/28 (as opposed to 1435 or later) makes an early dating all the more likely.
84. Howard 1989, 685n13, 88, 89. Attributed to Giovanni Bellini and workshop (Folgolari 1908) or Carpaccio (Rearick 2003).

CHAPTER 10. RENAISSANCE CHAPELS AND CHURCH FAÇADES

1. Editio princeps: *Aristoteles et Theophrastus* (Venice: In Domo Aldi Manutii Romani et Graecorum Studiosi, 1497), tom. II, fols. 254–60.
2. Goffen 1975, 487–90.
3. For a nuanced view, see Morresi 1998.
4. Sperti 1996, 123. In 1304 the college asked the Capitano da Mar, Gabriele Dandolo, to look for fine marbles (*albis vergatis viridis porfiis*) among ruins in the Aegean Isles: Cecchetti 1886, 1:13, no. 99.
5. Pincus 1992; Fortini Brown 1995; Concina 1998, 114ff., with bibliography. Venetian writers were already comparing Venice to Athens, Sparta, Jerusalem, Rome, and Byzantium in the 1420s: Lorenzo de' Monaci, *De gestis, moribus et nobilitate civitatis Venetiarum* (1421–28), quoted in Gaeta 1980, 3.1:18. The theme of Venetian superiority over the models of ancient city would be resumed in the 15th century by Paolo Morosini and Francesco Filelfo, and systematized in the work of Bernardo Giustinian, whose *De Origine urbis venetiarum* was written between 1477 and 1489, but published posthumously in 1492.
6. See the marble vault pendentives in the *Virgin and Child with Sts. Sebastian, John the Baptist, Mary Magdalen and Roche, and Kneeling Confraternity Members* (c. 1486–88); alabaster spandrels in the *Virgin and Child with Sts. James and Jerome* (1489); interior with dome based on S. Marco in the *Virgin and Child with Sts. John the Baptist, Nicholas, Catherine, Apollonia, Francis and Peter* (1492–93); mosaic pendentives but marbled drum in the *St. John the Baptist with Sts. Peter, Mark, Jerome and Paul* (c. 1493–95); mosaiced pendentives and dome in the *Virgin and Child with Sts. Peter, Romuald, Benedict and Paul* (c. 1495–97); wall revetment based on the Miracoli in the *Virgin and Child with Sts. Michael and Andrew* (1490s); *Virgin and Child with Sts. John the Baptist, Cosmas, Damian, Apollonia, Catherine, and John the Evangelist* (1506–8) set in the marble and mosaic atrium of S. Marco: Humfrey 1983, 120–21 (cat. no. 81), 63–64 (cat. no. 60), 94–96 (cat. no. 37), 60–61 (cat. no. 57), 80–82 (cat. no. 12), 38–39 (cat. no. 19), and pls. 17, 31, 37, 52, 61. In his townscapes, marble is deployed on a very ambitious scale with large fields of book-matching and intarsia: Humfrey 1983, 99, 11–112 (cat nos. 44, 67).
7. Jones 1987. On Venetian influences in Mantegna's painting, see Fletcher 1993.
8. For example, Pius II in Totaro 1984, 1:566; Spesso 1997, 27; Giovanni di Tommaso Ridolfi in Jones 1988, 279–80; Alessandro di Filippo Rinuccini (1431–1494) in Calamai 1993, 127; Antonio da Crema (1486) in Nori 1996, 35 ("infiniti adornamenti di marmor fino"); Arnoldo von Harff (1497) in Reumont 1876, 395, 403, 404.
9. Hand and Mansfield 1993, 51–60. Dürer's 1494–95 Venetian trip is surmised on the basis of a dated drawing of the Alps. An alternative dating is c. 1506. Bellini's *Madonna Greca* (1476/77; Brera, Milan), *Contarini Madonna* (c. 1480; Accademia, Venice), and *Madonna degli Alberetti* (1487; Accademia, Venice) are the closest prototypes for Dürer's painting.
10. Spencer 1965, 1:31–32, lib. III, fols. 17r–18r. Translation mine.
11. Onians 1971; Grafton 2000, 320–21.
12. Pozzi and Ciapponi 1964, 1:48.
13. Raby 1980, 55–79; Raby 1987, with bibliography. Sigismondo Malatesta also abortively dispatched Matteo de'Pasti, medallist, miniaturist, and executive architect of the Tempio Malatestiano, to the court of Mehmed II. van der Vin 1980, 1:133–50, with bibliography. On Western admiration in the 15th century: Foschi 2002; Chieli 1995.
14. Smith 1987. For travelers' descriptions of Hagia Sophia after 1453, see Mango 1962, 117–40.
15. Gyllius 1632, 100 (written in 1544–47).
16. Dini-Traversari and McHus 1912, 49. See also Traversari's letter to Niccolò Niccoli of 12 October 1433: Cannetus 1759, 419–22; and his brief description of S. Vitale: Dini-Traversari and McHus 1912, 101. See also Castelli 1988.
17. "La fa[n]tasia è bella de priete segate e aperte come un libro e fanno varie forme conforme." Annotation on a plan of S. Vitale (c. 1526): Foschi 1995.
18. Spreti 1489; Candido 1521, bk. I.
19. A partial account of marbling in early Cinquecento Venice, which views it only as conspicuous consumption, is in Strupp 1993. For a systematic account of marble use 1380–1480, see Connell 1988. For a more sensuous account: Hills 1999, 23–46, 57–89.
20. Muraro 1961; Forlati 1975, 117–19; Wolters 1976, 277–78, with bibliography; briefly discussed in Smith 1995, 235–36, and Wolters 2000, 52–53.
21. Barbieri et al. 1956, 146–47; Mantese 1952, 3.2:961.
22. Markham Schulz 1991, 177–80; Stringa 1610, 23v.
23. MacKillop 1992, with bibliography.
24. De Zoppi 2004, with bibliography.
25. Hope 1992, 60–61, 82n113, 101n201. One hundred wagon-loads of porphyry, serpentine, and marble were taken from Ravenna, before 1449 and in 1454/58: Ricci 1925, 210–12, 38n21, 568ff (nos. IV–VI); Hope 1992, 66–67; Smith 1995, figs. 12, 15.
26. The seminal study of the Cappella del Perdono is Rotondi 1948. See also Wedepohl 2004, 160–75, 199–242, 251–65, 294–97, with further bibliography. Inscription: "*HAEC QUICUMQUE PETIT MONDO PIA LIMINA CORDE HIC PETIT AETERNI FULGIDA REGNA POLY.*"
27. Both Roman marble carvers and the Gamberelli workshop collaborated on the Cardinal of Portugal Chapel: Hartt et al. 1964; Koch 1996, with further bibliography. For the interaction of the various arts and illusions, see Shearman 1992, 76–78. The chapel became the model for Antonio Rossellino's Piccolomini Chapel in the Neapolitan Church of Monteoliveto: Pane 1975. The vault was also copied in the Martini Chapel in S. Giobbe, Venice: Ceriana 1996, with bibliography.
28. Burns 1971. For Giovanni Villani's *Chronica* (early 1300s), see Davis 1988, 43.
29. Alberti, *De re aed.* 1.8 (fol. 13v) in Rykwert et al. 1988, 22.
30. Bulgarelli 2008, 49–121; Bulgarelli 2016.
31. *Hesperis*, 13.345–46, in Drudi 1794, 288. Alberti, *De re aed.* 7.10 (fol. 126). Palladio paraphrases Alberti's comment on whiteness (*Quattro Libri*, IV, 2).
32. Limestone from Verucchio or Scorticata: Rodolico 1953, 219.
33. Arch of Constantine: Pensabene and Panella 1999, 40, 48, 63. Bonfigli: Garibaldi 1996.
34. Ricci 1925, 411; Smith 1995, 237–38, pls. 17–18.
35. The epigraphic model was the Temple of Castor and Pollux, Naples: Campana 1973–74, 90. On Pletho, see Woodhouse 1986, 374–75. Valturio 1534, 381–82.

36. Preudhomme 1539, 9–10, 77, 101. This revetment is not illustrated in the view of the façade (c. 1457/68) in Basinio's *Hesperis* (Oxford, Bodleian Library, MS Canon class. lat. 81, fol. 137r), but the building is shown in construction and the superstructure is also missing: Pächt 1951, figs. 5–6.
37. On the project and patronage: Cuccini 1990; Mercurelli Salari 1996. For a description of polychromy: Hesse 1992, 61–64.
38. Olivato and Puppi 1977, 18–45, 177–83. For the Albertian influences, Morresi 1998, 224–25.
39. Olivato and Puppi 1977, 23, 257.
40. Gurrieri et al. 1988; Dezzi Bardeschi 1970, esp. 18–19. The Duomo façade was begun by Arnolfo di Cambio in 1294 but left incomplete at his death in 1302; it was resumed in 1357 by Francesco Talenti but again left incomplete: Weinberger 1940/41, 72–74.
41. Perosa 1960, 68–69, 74.
42. Wittkower (1940–41, 7) considered the revetment between the Gothic tomb niches and attic as by Alberti. Kiesow (1962), followed by Heydenreich and Lotz (1974, 337n27), instead argue that everything below the attic level and within the terminal pier-and-column "bookends" is 14th century. On high medieval revetted decorations in Florence: Behne 1912; Rupp 1912; Paatz 1940.
43. Pietramellara 1973, 40–41; Corsucci et al. 1983, 55–61.
44. Begun 1460, vestibule built 1478–79 by Luca Fancelli, completed 1499 by Pellegrino Ardizoni. Calzona and Volpi Ghirardini 1994; Böckmann 2005.
45. Soggia and Zuccoli 1994.
46. Alternating panels of green and red were discovered during restoration (1999–2000): Iannucci 2006, 306–7, pl. 5.
47. Soggia and Zuccoli 1994. Inexplicably, these authors do not cite the groundbreaking article by Wolters 1986. See also Carpeggiani 2002; Bulgarelli 2006, esp. 128–43.
48. In 1489 Domenico di Piero, guardian grande of the Scuola Grande di S. Marco, bought marble panels from the procurators of S. Marco for the façade of his house: Sohm 1982, 121n7.
49. Paoletti 1929; Sohm 1982, 139–40; Schmidt Arcangeli 1996; Morresi 1998, 233–34; Schofield 2006.
50. Veronese "marble" from Valpollicella: Guerra 2001, 105, 110n77.
51. Borlini 1963; Shell 1993; Burnett and Schofield 1997; Burnett and Schofield 1998; Schofield 1999.
52. For the lozenge-pattern: Schofield 2002, esp. 75–76. A complete study of the chapel is in Kohl 2004, with bibliography.
53. On the wall behind the Virgin in Piero della Francesca's *Annunciation* scene in Arezzo, not to mention the *Flagellation* in Urbino.
54. Schofield 1992, 35–36; Spencer 1965, 2:bk. XVI, fol. 123r; Belotti 1940, 2:84–85; Crippa 1971, 21–30. The upper part of Filarete's façade is covered in a reticulate pattern that Schofield considers faux-opus reticulatum, but likely imitates the identical marble inlay on the façade of S. Miniato al Monte.
55. All remodeled in the late Cinquecento such that the original decorations are lacunary: La Malfa 2009, 44–51, 92–95, figs. 12, 84–87, with bibliography.
56. Bertini Calosso 1953.
57. Chattard 1767, 3:140, 141, 143–44.
58. Azzolini 1994, 70.
59. Bazzotti 2006; Gheroldi 2006.
60. Hardly studied, but see the information in Lopardi and Mucciante 1987, 47–53, with bibliography.
61. Gondi Chapel: Marchini 1939. The chapel at Treviso was built 1501–3; walls revetted in Brioni stone 1504; pediments in Pisan marble 1506; apse in yellow marble 1506–7; lunettes in yellow marble 1507: Biscaro 1899a; Coletti 1935, 291–93, 325, 333.
62. Maffei 1506, lib. XXVII, ccxci r–ccxcii v (*De Metallicis*), ccxcii v–ccxciii r (*De Pigmentis*), ccxciii r/v (*De Lapidibus*), ccxciii v– (*De Gemmis, De Crustationibus*).
63. Cesariano in Fiore et al. 1981, lib. II, fol. 37v.
64. The literature is large, but see mainly Shearman 1961; Rowland 1986, 700–711; Weil-Garris Brandt 1986; Fabjan 1999.
65. Shearman 2003, 1:500–545. For a bibliography: Di Teodoro 1994. Shearman dates the letter to 1514/15. The conventional date of 1519/20 is argued in Christof Thoenes, "La 'lettera' a Leone X," in Frommel and Winner 1986, 373–81.
66. Similar paneling can be found throughout Ostia, e.g., the House of Cupid and Psyche: Bruto and Vannicola 1990, 370, figs. 41, 42. According to Frommel (1994, 233), Varignana's use of such revetment on the façade of S. Petronio in Bologna was influenced by the Chigi Chapel.
67. Uffizi A 164: Shearman 1977. Palladio 1570, lib. 4: 82. Salviati: Uffizi 1193 E. The lateral pyramids were apparently inserted into the design at a late stage, as pentimenti suggest that the cornices were originally meant to traverse the fields between the corner piers: Shearman 1981.
68. Magnusson 1987. The threshold in rosso di Syéne (Aswan granite) would recall his Sienese origins; marmor Numidicum evoked Tuscany (because the marble was exported from Numidia on the River Tusca); marmor Chium would pun on his name ("Chigi" Latinized is "Chius").
69. Huskinson 1969; Levi and Levi 1974; Finch 1991; Parisi 2014.
70. For example, granito pyropecile: Pozzi and Ciapponi 1964, 1:48. Viel de Saint-Maux (1787, 2nd letter, 11) will still write that "une pierre pyramidale étoit l'emblême de la fiame divine."
71. Cugnoni 1880, 441.
72. The chapel is unstudied, but see Penni Iacco 2004, 100, 138. For the spolia: Vernia 2005.
73. Magrini 1848, 64–68, 99–104; Lampertico 1880; Zorzi 1958; Morresi 1996; Morresi 2000, 152–59 (cat. no. 24).
74. Giovannoni 1917.
75. Foscari and Tafuri 1983, pl. 25. In 2000 the altar was dismantled, cleaned, and reassembled against the back wall of the apse, a position to which it had been transferred following the Council of Trent. Until the 1560s, and between 1945 and 2000, it stood forward of the Duomo's choir on a podium of three or four steps.
76. Suger, XXXIVA; Panofsky and Panofsky-Soergel 1979, 79.
77. Agricola 1546, 319.
78. Borghini 1584, 602, 589. Ca'fer Efendi (1614/15) in Crane 1987, 71.
79. Bacci 1587, 1. A Latin version of Bacci's treatise was published in Frankfurt in 1603 (*De gemmis et lapidibus pretiosis*).
80. Restoro d'Arezzo in Morino 1997, 246–47.
81. Connell 1988, 144–45. See also Butters 1996, 144, 171.
82. For the motif in Early Christian and Byzantine art: Lipinski 1960.
83. De Mély and Ruelle 1896–1902, 2 (*Les lapidaires grecs*). For further comment, bibliography, and their medieval influence: Bianco 1992.
84. Isid. *Etym.* 16. For a wide-ranging but extremely compressed history of lapidaries from antiquity to the Middle Ages: Closs 1958. A lengthy list of medieval lapidaries is in Wyckoff 1967, 264–71 (app. B). A thorough review of sources and scholarly bibliography for the Medieval period is in Kitson 1978. On magic gems, see also Michel 2004.
85. Beagon 1992, 92–102, 106–8. Jerome, *In Esaiam XII–XVIII* (on Isaiah 54:11–14).
86. Leonardi 1502 (repub. Venice 1516, Augusta 1533, and Hamburg 1717). In 1610 it was republished as the preface to Arlensis de Scudalupis 1610. The latter was reprinted in 1717. Leonardi was a physician and astrologer from Pesaro. The *Speculum* is dedicated to Alexander VI's nephew, Cesare Borgia.
87. Dolce 1565; repub. Venice 1617.
88. De Bellis 1985.
89. Lib. XVII, 3 (Cod. Cas. 132, fol. 418b) in Amelli 1896, tav. CXVIII; Orofino 1994–2000, 2.2:81.
90. Contarini 1541, 25v–36r.
91. Mantese 1964; Puppi 1973, 2:237–38; Foscari and Tafuri 1983, 111n26.
92. Tafuri 1969, 18ff.
93. Zorzi 1525, 3:101r–2r. Canticle 3, tone 8, module 1 divided into five *Concentiae:* 1) "Why it is said that [the Heavenly Jerusalem] was built from gold"; 2) "Why the walls were of Jasper"; 3) "What the twelve stones of the foundation mean"; 4) "Digression on whence stones have their power"; 5) "What is the true material of the heavenly city."
94. Nagel calls the altar "the most 'abstract' altarpiece produced during the Italian Renaissance," regarding the absence of any painting as a conscious decision not to

detract from the real presence of the sacrament, the priority of representation: Nagel 2005, 406. See also the much fuller discussion in Nagel 2011, which incorporates material from the author.

95. Reutersward 1982.
96. The earliest text known to me is the contract (1364) for the tomb of Dante's son, in Treviso: Biscaro 1899b, 426, 430–31, no. X. The next relevant historical reference is in Filarete's *Trattato*, where he describes (almost certainly) alabasters as "chome dire nugholi": Spencer 1965, 1:31 (bk. III, fol. 17r).
97. Manning 1963; Luzarraga 1973; Otto 1994.
98. August. *Commentarii in Psalmos* in Ps. 98, 7 (*PL* 37, col. 1266) and *Commentarii in Psalmos* in Ps. 77, 13–16. Ambrose, *Commentarii in Psalmos* in Ps. 98, 5; *PL* 15, col. 318/19; Gregory the Great *Hom. VIII* in Ez. Lib I (*PL* 76, col. 868 B–C); Hrabanus, *In Num.* (*PL* 91, col. 364 A).
99. Tomasoni 1990, 20–22.

CHAPTER 11. MARBLE MANSIONS AND PAINTED PALACES IN RENAISSANCE ITALY

1. Constantia: Argoud et al. 1980, 28, pls. xxvi–xxvii. Apamea-on-the-Orontes: Balty 2004.
2. Mango 1965, 317–18; Mango 1997, with further bibliography.
3. Basil of Caesarea, *Sermon to the Rich* (c. 368 AD), 2; *PG* 31, col. 285; Gregory of Nyssa, *In Ecclesiasten homilia*, 3.4.320A–324A; *PG* 44, cols. 653–57. Liudprand, *Relatio de legatione Constantinopolitana ad Nicephorum Phocam*, 57, 13; *PL* 136, cols. 932, 916.
4. Macrides 2015, with bibliography.
5. *Nicephori Callisti Xanthopouli Historia Ecclesiastica libri XIII*, dedicatio, cap. 17, in *PG* 145, cols. 585B–86A; Failler 1997, 234–35.
6. Mango 1965, 330–36; Asutay-Effenberger (2007, 134–42) identifies the Mermerkule with the Πολίχνιον built by John V Palaiologos (1341–1391) around 1390. For a different identification, see Peschlow 2001, with further bibliography.
7. Theodoros Metochites, poem 19.155–226, in Featherstone 2000, 118–23; Guilland 1922. Johannes Chortasmenos marvels at the beautiful marble columns, columnettes, and domed roof in the palace of Theodore Kantakouzenos: Hunger 1969, 190–91.
8. *Digenis Akritas* (12th century), Trebizond manuscript version, 8.2732–35: Legrand 1902, 226.
9. *Kallimachos and Chrysorroe*, 868–69; *Belthandros and Chrysantza*, 246–483, esp. 339–442, 456–59; *Livistros and Rodamne*, 285–459. See Betts 1995; Cupane 1995; Miller 1858, esp. 1115–97 for an ekphrasis of precious stones; Cupane 1978; Cupane 2010; Cupane 2015.
10. Wolohojian 1969, 137; Wallis Budge 1933, 119–20; Holton 1974, 167. For the extremely complicated transmission of this text, see Stoneman 2003.
11. Schlauch 1932; Frankl 1960, 159–204; Petrarch, *Africa*, Palace of Truth (87–97).
12. Rowley 2016, 58–62.
13. About twenty columns, four in porphyry, two in verde antico, and the others white: Dell'Aquila 1970; Calò Mariani 1995. For the lithotypes, see Zezza 2005.
14. For example, Palazzo della Ragione, Vicenza, by the *ingenere* Domenico da Venezia and the *taiapria* Matteo di Giovanni, both Venetians (1451–60). The pre-Palladio diaper cladding of the basilica remains in situ at the terrace level: Barbieri 1968, pl. 4. At Udine, not only did the Palazzo Comunale (c. 1448–55) copy the cladding, but the piazza was also paved in 1482/84 "*iuxta quam in Venetiis salizatur*" ("just as they pave in Venice"), i.e., with alternating brick and stone courses: Battilotti 1991, esp. 22.
15. For Iranian models: Grube 1966; Howard 1991.
16. For "Veneto-Byzantine" palaces, see Howard 1980, 37–43. For the development of the Venetian palace: Goy 1989, chaps. 5, 8, 9; Arslan 1970; and esp. Schulz 2005, 12–21.
17. On Constantinopolitan palaces: de Beylié 1902; de Beylié 1903; Baldini Lippolis 1994.
18. An inscription from Tivoli (*CIL*. 14.3911) mentions "the painted façade of Hadrian's villa gleams" (*frontibus et pictis Aelia villa nitet*). "Pictus" could well refer to (naturally) painted stone.
19. Malalas does record that the Chalke gate of the Great Palace (538 AD) was "decorated with various kinds of marble and with mosaic work"; *Chronographia* 18.479, in Jeffreys et al. 1986, 285. On external revetment, see Reusche 1971; Pasadaios 1973; Canak-Medic 1981; Yenisehirlioglu 1982. Reusche deals with the Balkans, Romania, and Constantinople up to 1204; Pasadaios treats only brick polychromy and caprices in Constantinople.
20. Sinding-Larsen 1975; Tönnesmann 1984. See also Belli 1996.
21. Mango 1972, 160; Hunt 1984.
22. William of Oldenburg in Laurent 1864, 167.
23. Krautheimer and Ćurčić 1986, 473–75; Mango 1965, 334.
24. Laurent 1971, 212–14; Mandrot 1903, 2:208; Porro-Lambertenghi 1855, 9.
25. Chronicle of Stefano Magno cited in Tassini 1993, 7; Arslan 1970, 17. See also Schulz 2005, 165–86. The palace belonged to Renier Dandolo. The present marbling is 19th century. Other early palaces that retain original marble cladding are the Ca' Barzizza (upper façade early 13th century), Ca' da Mosto (before 1242), and Ca' Priuli-Bon (mid-13th century): Schulz 2005, figs. 29, 38, 40, 42.
26. Felix Fabbri (1441/42–1502): Hassler 1843–49, 3:403. Scamozzi 1615, 1:58. For the impact of S. Marco, see Schuller 2000.
27. For the Ca' d'Oro, see Connell 1988, 149–51; Goy 1992, 139–80, 227–34. For the Ca' Dario, see Connell 1988, 133–34; McAndrew 1980, 215.
28. Babinger 1961; Babinger 1962; Fortini Brown 1988, 64–66, 241–42; Calò 1992. Giovanni Dario was also a guardiano of the Scuola di S. Giovanni Battista and may perhaps have had a hand in the choice of revetment for its façade (begun c. 1479?): Ceriana 2003, 62n85.
29. Codice Torinese Saluzziano 148, fol. 20v, in Maltese and Maltese Degrassi 1967, 1:89, 259, tav. 36; Codex Ashburnham 361, fol. 20r, in Marani 1979, 1:20r. The caption also talks of "hornamento d'inchancellati a huso di ronbo fatti and hornamenti di piume e quadrati chantoni."
30. Built by Bernardino da Milano and Zanon Marchesini da Chiampo, advised by Pietro Lombardo; Barbieri 1968, 32–33, pls. 23, 24. The contract is in Zorzi 1926, 196–97n1.
31. Cesariano in Fiore et al. 1981, lib. II, fol. 38v; Wolters 2000, 47.
32. Caporali 1536, 56r. See, for example, Juan Gracian Alcalá de Henares's 1582 translation of Vitruvius, facsimile copy published in Pizarro Gómez and Mogollón Cano-Cortes 1999, 27.
33. Cesariano in Fiore et al. 1981, lib. II, fol. 34v.
34. Cesariano in Fiore et al. 1981, lib. II, fol. 38v. For Cesariano, see Gatti Perer and Rovetta 1996; Rovetta 2002, with bibliography.
35. Connell 1988, 150–51; Armani and Piana 1984, 50–51, fig. 11.
36. The doors (1470s/80s) of the Sala del Trono, in the Palazzo Ducale at Urbino, show two palazzi, one with a serpentine frieze and the other with a porphyry frieze: Trionfi Honorati 1983. These are presumably influenced by the depiction of the Praetorium of Pilate in Piero della Francesca's famous *Flagellation* (1460/70?) now in the Galleria Nazionale delle Marche, Urbino, and formerly in the sacristy of the Duomo.
37. Alberto Avogadro da Vercelli, *De religione et magnificentia illustris Cosmi Medices Florentini*, in Lami 1742, ll. 141–47. For similar poetry by Avogadro: Gombrich 1962.
38. Schweikhart 1973, cat. no. 27, pls. 50–53 (Loggia del Consiglio); cat. no. 73, pls. 143–46 (Corso Cavour 1); cat. no. 82, pls. 157, 159, 161 (Casa Mazzanti, Piazza dell'Erbe); Schweikhart 1977. The Loggia del Consiglio (1476–93) decorations are possibly by the Dalmatian painter Zuanno de Ragusa, active in Verona in 1482–92. The Casa Mazzanti frescoes were executed PATRIAE ORNAMENTO by the Mantuan painter Alberto Cavalli c. 1530. Some examples survive in Rome, e.g., the Palace of Prospero Santacroce (1480/1500): Garella 1992. For an exhaustive survey of a single city, Bergamo, see Rossi 2009.
39. "io lo farò nzessare et farò parerano de marmora e porfidi et alabastro e serpentini in modo rapresenterà una magna cosa": letter from Luca Fancelli to Ludovico Gonzaga, 7 July 1471, in Carpeggiani and Lorenzoni 1998, 159. Faded traces are still visible.
40. Perosa 1960, 68–69, 74.
41. Lib. III, fol. 17r, in Spencer 1965, 1:3.
42. Barbaro 1567, 84.

43. For the academic discussions behind the surface articulation of the Cancelleria façade, see Daly Davis 1989.
44. The painting is extensively preserved on a gable wall thanks to the addition that was built up against it: Varese 1978. The design is comparable with the studiolo ceiling at Gubbio, for which see Wilmering and Raggio 1999, 100–101, figs. 5.45, 5.46. The most complete study of the Schifanoia to date is Varese 1989.
45. Pius II in Gabel and Gragg 1959, 114; Warburg 1912.
46. Spiazzi 2005. The frescoes were discovered in 1992–2003.
47. Private collection, Zurich: Humfrey 1983, 167–68 (cat. no. 67), fig. 69a.
48. Sacchi 2000, 71–99.
49. Davies 1993, 82–83; Lampertico 1988. Palazzo Guerrieri at Via Cavour, 96: Zuccoli 1986, 33, figs. 27–30. Zuccoli also illustrates (fig. 31) the remains of more Mantegnesque rotae on the early 16th-century Torre della Podestà in the Piazza Broletto. See also Bazzotti et al. 2009. The façade frescoes of the Palazzo Maggi al Fontanone are now in the Pinacoteca Tosio Martinengo: Agosti 1993, 45, fig. 1.
50. For example, Mauro Codussi's Palazzo Corner-Spinelli (1490s), Palazzo Vendramin-Calergi (1490s); McAndrew 1980, 336–47, 348–57. Olivato and Puppi 1977, 148–61, 203–6, 221–26. Spoliated, antique clipei were visible at several points on the Adriatic coast: Rebecchi 1980. The Porta Aurea at Ravenna, which bore clipei, was not demolished until 1582.
51. Alexander 1994.
52. In general, see Germano Mulazzani, "La decorazione affrescata," in Bombi et al. 1988, 107–12, and illustrations at 92–97, reconstruction by Mario Castellani at 109, and construction history at 161–68.
53. Brown 1989–90; Signorini 1991. It can be inferred that Lombardo's lost pavement contained porphyry and hard-stone roundels from his letter to Isabella (12 January 1522), explaining that some roundels are bedded higher than their marble frames because they took far longer to wear down. Many years earlier (August 1507) Isabella had requested from Giorgio da Negroponto "un pavimento de profidi e serpentini et altre pietre amachiate de quelli pavimenti antiqui che se trovano qui a Roma": Luzio 1886, 6, no. 3. Isabella may have been inspired by the Camerino d'alabastro at Ferrara, but the Castello at Mantua already had a "chamerino intarsiato" by 1462, and this was to become Isabella's Grotta: Cantarelli and Brown 2002, 85, figs. 6–8.
54. Sabadino degli Arienti (1497), in Gundersheimer 1972, 51 (fol. 36r). Tuohy 1996, 53–95, 87–89, n187. Destroyed by fire in 1532.
55. Gundersheimer 1972, 75 (fols. 66r/v). For the Palazzo dei Diamanti, see Di Francesco 1991, with bibliography.
56. The façade was incorporated into the later church of the Gesù (1584): Gelao 1988, 17–19, with bibliography. Pontano in Tateo 1965, 95–96. For an excellent survey, see Ghisetti Giavarina 2007–8.
57. See the façade "de marmoro a diamante" of the Ca' del Duca, begun 1450s for the Corner family, probably by Bartolomeo Bon: Beltrami 1900. Diamantine bossing took over from the Trecento diaper cladding on the Rio frontage of the Palazzo Ducale when Antonio Rizzo built this wing in the 1480s. Gelao (1988, 12) claims—on what basis, I cannot fathom—that the bugnato of the Palazzetto Borgia at Bisceglie in Puglia predates 1450. Most other Puglian palaces in this style date from as late as the 1560s.
58. Serlio 1537.
59. D'Amico and Weil-Garris 1980, 88–89.
60. For Roman Quattrocento examples, see the palace on the Vicolo de' Matricciani and Vicolo del Governo Vecchio illustrated in Letarouilly 1840–53, 1:pl. 110. An extant example is in Via della Fossa 15–17.
61. Zevi 1960, 165–74, 260–75.
62. Pliny. *HN.* 37.15.55–61. For diamonds in treasuries, see McCrory 1997, 165–66. For diamond symbolism in rings, see Scarisbrick 1982.
63. Gussali 1912; Visioli 2001, esp. 59–69; Martinis 2014. Raimondi designed another palace (Contrada Bertesi, 8), also executed by Bernardo da Lera, and composed a treatise, *De Re Aedificatoria:* Arisi 1702–41, 1:399.
64. Contract, January 3, 1495, with Battista and Antonio de Molinari di Rezzato, in Visioli 1993, 84, 85–86. See also Coryate 1611, 322–24.
65. Lib. III, fol. 17v, in Spencer 1965, 1:32.
66. Tuohy 1996, 211–13, figs. 27 and 58. For a full study, see Torboli 2010.
67. Vasari-Milanesi 1906, 1:129. On the fortress, see Gurrieri and Mazzoni 1990.
68. *De re aed.* 9.1, fol. 158r.
69. Pozzi and Ciapponi 1964, 1:213; Rucellai in Perosa 1960, 76; Agostino del Riccio ("al giardino de' sigg. Rucellai in via della Scala sono degli alabastri che servono alla onorata Casa per finestre, usanza laudabile e bella"), quoted in Zobi 1853, 181.
70. Hatfield 1970, 233, 234, 246, 247. The Irish pilgrim Simon Fitzsimons on Cairene palace interiors (1322–23): "They seem nothing other than the house of God and the gate of heaven, since they are excellently decorated . . . with marble and other precious slabs are marvelously well and sumptuously paved": Hoade 1970, 26.
71. Yuen 1970. The *Bibliotheca Graeca* is instead dated to 1475/77 in Ruysschaert 1986, 104–5. However, see the correctives in Boyle 1991.
72. Patrizi 1471, lib. VIII, fol. xv; Aurigemma 1997; Bettini 2010. See Vitr. *De Arch.* 5.9.5 and Isid. *Etym.* 6.11.2–3, 16.5.15. Isidore's opinion is still echoed in Maffei 1732, tom. III, cap. IV, 190–91. Classical authors had also mentioned eye-soothing green gems in signet rings: Meadows 1945; Plantzos 1997; Trinquier 2002.
73. Attributed to Melozzo da Forlì (c. 1476–77), in Scoppola 1987, 197–203, esp. 200–202. As by Pedro Berraguete (c. 1480–81), in Benzi and Vincenti Montanaro 1997, 11–12.
74. Luca Pacioli, *De Divina Proportione* (Venice, 1509), in Bruschi et al. 1978, 92–93.
75. Pellecchia 1992.
76. The (destroyed) decorations in the Belvedere are described in Chattard 1767, 3:133–37. The Sala del Mappamondo was decorated 1489/91 for Cardinal Lorenzo Cybo, but heavily restored in 1922: Casanova 1992, 154–57, 251, 254, with bibliography. The lower walls of the Sala delle Arti Liberali in the Borgia Apartment were frescoed in imitation of marble revetted alcoves, some with quincunxes, but these are only visible in old photographs; Ehrle and Stevenson 1897, tav. LI.
77. An attempt to redate these frescoes to the early 1480s is unconvincing: La Malfa 2009, 81–87. Pintoricchio painted the Sala delle Lunette or Sala dei Profeti e degli Apostoli (audience room) and the Sala del Gran Maestro (Salone): Aurigemma and Cavallaro 1999, 46, 48, 175–82, 94–97. A large section of the end wall (the "spoglia Christi") of the Forum of Trajan, with engaged columns in giallo antico, was standing until 1526: Viscogliosi 2000, 90–91, 147–50, with bibliography.
78. For example, when Colonna cites the fabled column display in the House of the Gordians, he repeats Alberti's transcription (*De re aed.* 9.1, fol. 158r) from a corrupt manuscript of the *Historiae Augustae Scriptores* (Gord. 32, 2.), writing "*tisteas*" for "*carysteas*" and "*simiades*" for "*synnades*"; Pozzi and Ciapponi 1964, 2:105. See also the extremely knowledgeable identifications of marbles mentioned by Statius in Angelo Poliziano's commentary on the *Silvae:* Martinelli 1978, 235–38, 448–49, 655.
79. For this painting, see Fritz 1998, with bibliography; the sliding cover is attributed to Giulio and Giovanni in Jones and Penny 1983, 192.
80. Fulvio 1513, unnumerated, but at the printer's quire marks "Civ-Cii."
81. Fulvio 1527, lib. V, fol. C ("De marmorum generibus quibus urbs decorata est"). Fulvio claims (preface) that he accompanied Raphael in exploring the ruins, indicating to him the sites that Raphael would draw and paint ("priscaque loca tum per regiones exploras observari, quas Raphael Urbinas [quae honoris causa nomino] paucis an diebus quae ex ita decederet [me indicante] penicillo finxerat").
82. Vasari-Milanesi 1906, 6:554–55.
83. Shearman 1972, 8–9. This room also led into a fictive peristyle, the Sala dei Palafrenieri (or del Pappagallo), decorated by Raphael's shop in 1517/18; Shearman 1972, 369–424; Weddigen 2001.
84. For the Pantheon portico, see Nesselrath

2003, with bibliography. More restrained all'antica revetments are also simulated in the Camera di Ovidio (1527) and the Camera delle Imprese (c. 1527/30). Giulio used real marble (porta santa) for the doors in the Camera delle Aquile (1527/28); Belluzzi 1998, 1:109; 2:86–91, 112–17, 126, 162–69, 194, 330–35.

85. Wurm 1984, 87, 154, 160; Uguccioni 1989; Dacos 1998.
86. Egidio Gallo in Quinlan-McGrath 1989, esp. 42–56, 147–50; Luchterhandt 1996, esp. 228–30.
87. Quinlan-McGrath 1990, 122–23, ll. 108–21 (translation mine). The *Suburbanum Augustinii Chisii* is dated January 31, 1512. For Blosius Palladius (Biagio Pallai): Lesen 1926; Gnoli 1930, 3–19, 137–48; De Caprio 1981. The anthology of poetry he edited for Goritz is republished in Ijsewijn 1997. The case for Blosio's role in the "paternità culturale della villa" is strongly made in Bentivoglio 1987. See also Pagliai 1989. The marble floor in the Sala delle Prospettive dates only from 1929: Grazioli Medici 1992, 496.
88. Barbieri 2014, 71, with references.
89. Sperulo 1519; text, translation, and notes in Gwynne 2015, 2:98–149 (translation here slightly amended). On the villa, see most conveniently Frommel et al. 1984, 311–56. For Raphael's intensive study of antique texts (not just Vitruvius) and archaeological investigation: Burns 1984, esp. 391–94.
90. Lilius 1981, 1:121–30; 2:figs. 29, 33–35; Frommel in *Giulio Romano* 1989, 112–17, 292–93. The reliefs in the Sala were inserted in the early 19th century, but a drawing in Berlin (Kunstbibliothek Hdz 4151, cod. Destailleur D, fol. 119v) shows that they took the place of faux-rotae like those in the Pantheon: O'Gorman 1971.
91. A drawing by Raffaello da Montelupo (Lille, Musée Wicar, Fonds Wicar 767) specifies that the external attic was "dipinto come di marmi bianchi e misti" and the Vanocci sketchbook in Siena adds "quest'ultimo piano è dipinto come di marmi bianchi e misti" (Siena, Biblioteca Comunale, S. IV. 1).
92. Coffin 1979, 257–59, 261; Keller 1986; Kajava 2005.
93. Alvise Cornaro writing in 1550, in Archivio Segreto Vaticano, *Savi ed Esecutori alle Acque,* busta 986, filza 4, fol. 23, cited in Tafuri 1985, 242. See generally Forcellino 1990.
94. On the palace, see Marcucci and Torresi 1982; Marcucci and Torresi 1983, with previous bibliography. Attributed to Baccio Bigio (Caroline Elam, pers. comm.).
95. Vasari-Milanesi 1906, 5:549: "Fantastica, tutta lavorata di stucchi coloriti." For Giulio's townhouse: *Giulio Romano* 1989, 481–85.
96. The rusticated architecture painted on the walls of the bishop's villa at Quingentole, and attributed to Giulio, has its quoins and pilasters painted ocher and the infill of a dark, stony color (Amedeo Belluzzi in *Giulio Romano* 1989, 524–25).
97. Cortesi mentions palace façades built "in the manner of amphitheaters"; D'Amico and Weil-Garris 1980, 87.
98. Dacos and Furlan 1987, 160–61.
99. Adorni 2012, resuming an attribution made by Howard Burns and Manfredo Tafuri.
100. Maffei 1506, fol. 316r: "Romam denique ex lutea latericia sicut olim Augustus ex latericia marmoream reliquit." Farnese 1566/77, fol. 8r.
101. For a technical analysis, with extensive bibliography, see Montelli 1998, who adduces precedents in the courtyard of the Palazzo Ducale at Urbino (c. 1472) and Palazzo Ducale at Gubbio (c. 1476), and other examples include Villa Giulia (c. 1551–53/54 phase), Palazzo Mattei Caetani (after 1548), Palazzo Farnese at Caprarola (1562–68 phase), and Palazzo Grifoni in Florence (1563–73 phase). For the reception of classical brickwork, see 78–79 and notes. The pertinent documents are collected in Frommel 1973, 2:107–10.
102. For a technical analysis: Piana 2000. Veronese marble columns, pilasters, and architrave articulate the tablinum: Palladio 1570, 1:22; Maffei 1732, tom. III, cap. IV, 195: "mattoni di quella compositura, emulante il marmo."
103. Goldney Grotto: Savage 1989, 20–21. For the factory (called Argile-Marbre) of Léonard Racle (1736–1791) at Pont-de-Vaux, see *Mercure de France,* February 22, 1787, 180–82 (Slatkine anastatic repr., Geneva, 1974), 201–2. Racle is not remembered for any other endeavor: Thieme and Becker 1907–50, 27:545.
104. Wotton 1624, 44.

CHAPTER 12. FROM GEMS TO CLOUD ARCHITECTURE

1. Morrogh 1985b; Conforti 1993, 7–38, 129–42.
2. Raggio 1960; Riebesell 2003; Tuttle 2002, cat. nos. 81–82, 200–202.
3. Parlato 2014, 333–40; Parlato 2016, 155–58.
4. Naples: Caldwell 2011. Carafa Chapel: Cerasoli 1894; Pressouyre 1984, 181; Palmerio and Villetti 1989, 159–61.
5. Ieni 2002; Extermann 2016. The overall architect was Nanni di Baccio Bigio, but the tomb has been convincingly attributed to Giacomo Vignola: Frommel 2012, 38–40.
6. Pilliod 1989, 73–124, 190–219 (docs. 5–15); Pilliod 1991. Attributed to the sculptor Montorsoli in Morrogh 1985b, 315; Morrogh 1985a, 88, cat. no. 35. Alessandro Allori supervised payments in the chapel but does not claim it in his *Ricordi,* although it did influence his later design of the Cappella Antella (1600) in SS. Annunziata. See also Sicca 2002.
7. Vasari-Milanesi 1906, tomo VII, 712.
8. On Dosio, see Barletti 2011, with bibliography; Spinelli 1992; delli Santi 2000. Dosio drew a view into the Chigi Chapel (Uffizi 166 A) and a dimensioned elevation of one of its pyramids (Uffizi 3204 Av).
9. Acidini Luchinat 1991, 295–320.
10. For later Florentine chapels, see Martellacci 1995; Gambuti 1995; Bosman 2005. For the Cappella dei Principi: Baldini et al. 1979; Przyborowski 1982; Cresti 1988; Guarracino 2009; Morrogh 2017; Vaccaro and Vetri 2018.
11. For what follows, see Barry 2019a. On the chapel in general, see Zollikofer 2017, with ample bibliography.
12. Valentino 1583; Frizolio 1582.
13. Frizolio 1582, 417–34.
14. Valentino 1583, 31–32 (ll. 533–42, 547–62).
15. Accordi 1980; Cooper 1995.
16. Cesariano, *De Architectura,* lib. II, fol. 37v: "ancora sono trovate de varie mixtione de pietre e di varia confusione maculate, complexe e pincte da la natura como cosa infarcita in una coequale e temperate congelatione." Vasari-Milanesi 1906: "pietra chiamata mischio, dalla mescolanza di diverse pietre congelate insieme e fatto tutt'una dal tempo e dalla crudezza dell'acque." Vasari is repeated in Baldinucci 1681, 98.
17. The same motivation may underlie the slightly later Cappella di Renata di Francia (1590–91) in the Castello at Ferrara, which had no altarpiece, just strident marble paneling. See Marcolini 2001; Domenicali 2002; the entries by Carla Di Francesco, Rita Fabbri, Gian Carlo Grillini, and Giuliana Marcolini in Bentini and Borella 2002.
18. "*Marmor Ophites Anthropomorphites,*" "*Marmor Cynites,*" "*Marmor Nautilites,*" "*Marmor Scombriforme.*" Aldrovandi's museum was installed in the Palazzo Pubblico in Bologna in 1607, moving to the Istituto delle Scienze in 1742. For a 17th-century inventory of the minerals: Torricelli et al. 1993, 99–109, marbles at 114–17. For Bartolomeo Ambrosini's direction of the studio (1632–57): 61–65. Publication of Aldrovandi's complete works had been interrupted in 1621. For an introduction to Aldrovandi: Vai and Cavazza 2006.
19. See examples in Bugini et al. 2002.
20. A sample: Dolaeus 1677, Ledel 1696 (illustrated), De Muralto 1697–98 (illustrated), Scheuchzer 1697–98.
21. Lange 1708.
22. Kircher 1665, tom. II, lib. VIII, 48.
23. Letter from Peter Paul Rubens to Pierre Dupuy, 23 March 1628: Magurn 1955, 247. The Dutch engraving of the friar seen in the tree rings is published in Alpers 1983, 80.
24. Anonymous French guidebook manuscript 1676–77, published in Connors and Rice 1990, 26. Kircher mentions several other images, not all in St. Peter's, none of which can be presently identified: Kircher 1665, tom. II, lib. VIII, 32–33.
25. Lazzari 1753 (rare, this copy in Biblioteca Comunale degli Intronati, Siena, XC. I. 48).
26. Lambeck 1679, sec. XXVII, 25–26.
27. De Ovalle 1646, lib. VII, cap. XX; lib. I, cap. XXIII. The passage is paraphrased in Kircher 1665, tom. II, lib. VIII, 48.
28. Von Sonnenberg 1661. For the Sonnenberg family (though Konrad is not named) see the *Historische-Biographisches Lexikon der Schweitz,* band 6 (Neuenburg: Attinger, 1931), 447–48.

29. González-Palacios 1984, 1:151–52 ("Una Pace per i Medici"); 2:figs. 302, 303.
30. Bernia et al. 1648. Aldrovandi's library contained, as one might expect, copies of Marbode, Agricola, and Encelius: Ventura Folli 1993; Bacci 1587; Agostino Del Riccio in Gnoli and Sironi 1996; de Valles et al. 1588. For Chácon's treatise, see Barry 2019a.
31. Bacci 1587, 3–5. Della Porta 1560, 134; lib. IV, cap. 25.
32. Ostrow 1990, 253–56. See in particular "Ragionamento fatto dal M. R. P. Castello sopra le due Capelle in S. Maria maggiore nell'Ottava della Traslatione dell'Imagine della Verg. alli 3 di Febraro 1613," Archivio di S. Maria Maggiore, Fondo Cappella Borghese, Misc. III, fasc. 11, no. 52, in Ostrow 1996, 270–79. "Sogno occorso all'autore di ritornare alle patria all Ill. mo Sig.re Cardinal Borghese . . . , in Rime diverse di Massimiliano Caffarelli . . . con alcuni sonetti, et altre poesie fatte da diversi Sig.ri e altri virtuosi ingegni . . . all'Ill.mo e R.mo Cardinal Borghese . . . (1613)," Archivio Segreto Vaticano, FB I 449, 12r: cited in Hill 2001.
33. Gregory issued bandi in 1572, 1576, and 1578: Lanciani 1902–12, 4:51, 54–56; Tuena 1989.
34. Letter from Girolamo Bernieri da Correggio to Lavinia Fontana, 12 May 1599, cited in Galli 1940, 82–85.
35. Clement VIII specifically called for the use of marble during his pastoral visit of his diocese, Rome, in 1600: Beggiao 1978, 71. For the chapels: Ostrow 1990; Tuena 1990. For an anthology of contracts: Di Castro et al. 1994. An inconclusive appreciation appears in Kummer 1996.
36. Cleary 1925, 162; Harold 1931 (1st ed., 1662), 70; Pollak 1928, 1:165–72; Haskell 1963, 42–43.
37. Letter from Bernini to the committee for the construction of the Cappella di Santa Barbara in the Duomo at Rieti, 24 October 1654, cited in Aloisi 1998, 29: "sarebbe presto aspettare qualche limosina o altro assegnamento per poter fare una Cappella degnia di quella Città che fare una cosa da cappuccino."
38. Fra Juan di San Bonaventura, *Relazione*, 31, in Montijano García 1999, 60–61.
39. The Cornaro marbles were supplied by Gabriele Renzi, who was paid 5431 scudi, 24 baiocchi in 1647–53. The Teresa itself cost 2,000 scudi: Carloni 1999, 46nn22, 24; see also Barcham 1993. Spada Chapel: Butler 1994, 69. For a breakdown of the costs at S. Andrea al Quirinale, see Connors 1982, 21. Nero di Nero Chapel, document of July 1601: Tuena 1990, 97n49.
40. By Pompeo Targone: Freiberg 1995, pt. II, ch. 5; cat. no. III.H.4.
41. The marble setting was executed by Pompeo Targone in 1599–1603 under the direction of Giacomo della Porta and Gaspare Guerra. On S. Cecilia, see Kämpf 2015, with previous bibliography. On the confessio, see Ostrow 2009.
42. Turco 1997, esp. 74–86, with bibliography.
43. Bernia et al. 1648, 774.
44. Bosio 1600, 21–22.
45. McMahon 1956, 1:50–51, 59; Janson 1961; Berra 1999, with further bibliography; Kemp 1981, 343, fig. 100. Cennini in Broecke 2015.
46. Cennini in Broecke 2015, chaps. 36, 2, 115. Byzantine painters also ground pigment on porphyry: Partington 1934, 141. See the porphyry and serpentine slabs from Mantegna's studio mentioned in the inventory of his son Ludovico (1510) or those in the inventory of Durante Alberti (1613): Signorini and Radcliffe 1997, 57–58; Soergel Panofsky 1996, 369. The earliest authoritative Italian dictionary, the *Vocabolario degli Accademici della Crusca* (Venice: Giovanni Alberti, 1612), also defines "porfido" as a slab on which to grind powders.
47. Caillois 1970; Arduini 1995; Bellesi 1998; Bona Castellotti 2000; Acidini Luchinat and Chiarini 2000; Seifertová 2007. On the technique, see Collomb 2003; Collomb 2006.
48. Letter from Vittorio Soranzo to Pietro Bembo, 8 June 1530: Hirst 1981, 124.
49. Varchi 1549, 97.
50. Pona 1620, 17, 31; cited in Bona Castellotti 2000, 19.
51. Circa 1605, Musei Capitolini, Rome.
52. Kircher 1665, 2:bk. 8, 30; Barry 2017, 34–35, fig. 10.
53. The High Altar of S. Maria della Passione is unresearched. Some information is supplied by M. A. Zilocchi, in Bora 1981, 174–80. The marble panels may have been painted by Giulio Cesare Procaccini and Gian Battista Crespi, the altar possibly designed by Francesco Mazenta between 1600 and 1619.
54. For an overview: Zobi 1853; Giusti 1988; Giusti 2005, esp. 21–47; Rossi 2002.
55. Baldinucci 1846, 3:218–19.
56. Zobi 1853, 187 (letter from Ferdinando I de' Medici to Giovanni Maria Bardi, Count of Vernio, 10 October 1601).
57. Fusco and Hess 1994, with bibliography.
58. Buonarotti 1622, 33–34.
59. Barry 2017.
60. Vasari-Milanesi 1906, 1:144, 7: 466–67. On scagliola, see Fassinelli 1997, with bibliography.
61. Parlato 2009, with previous bibliography.
62. Braccius 1637, 6; Stringa 1610, 20v.
63. de Piñeda 1609, lib. V, cap. V, sec. XV, 87.
64. Mancini (1617/21) in Marucchi 1956–57, I, 18–19.
65. Ferrari and Papaldo 1999, xx–xxxii; Barocchi 2001.
66. The two wills of Giovan Battista Strozzi are in Niccolini 1847, 315–22 (August 1535), 323–31 (December 31, 1537).
67. Payne 2007; Barry 2009, 18–19.
68. Vasari-Milanesi 1906, 7:551.
69. Kummer 1987, 113–85; Barry 1996.
70. For example, Scorso 1655, 150–51, 158. Du Cange compares the word "exemplum" (in lieu of "emblema") with "the images in marbles inlaid in pavements": Du Cange 1688, *ad vocem*.
71. For Spada and his abortive chapel, see Barry 2019b, with bibliography and sources.
72. Borromini in Connors 1998, 75. Spada possibly coined the expression himself, as he largely ghost-wrote the *Opus*. For paragone the stone: Vasari-Milanesi 1906, 1:117. Del Riccio distinguished six types of paragone.
73. Heimbürger-Ravalli 1977, 82–113; Licoccia 1987; Lenzo 2008, with further bibliography.
74. Virgilio Spada, letter to Orazio Spada, n.d.; Archivio di Stato di Roma, Archivio Spada, vol. 490, ff. 168–74; Heimbürger-Ravalli 1977: 112–13n16. See also Martinelli (1660–63) in D'Onofrio 1969, 71; Giovanni Battista Passeri, *Vita di Francesco Borromini*, f. 173r, in Hess 1934, 363–64.
75. On this theme: Barry 1999.
76. Ostrow 1996, 118–83, esp. 42–67, with bibliography.
77. "compartiti in modo che parria aria finta" (contract of June 1610): Dorati 1967, 234; Ostrow 1996, 158–60. Pietro de Angelis, 1621: "instar fenestrae . . . ingens vero interceptum lapide lazuli incrustatur caeruleo colore, et auri micis velut stellis internittentibus coruscante," cited in Ostrow 1996, 329nn190, 191.
78. Jacob 1975, cat. nos. 274, 275.
79. D'Onofrio 1966, 133. St. Peter's, payments May–August 1634 for "l'incrostatura d'alabastro che fanno per la Cappella de Smo Crocefisso": Pollak 1928, 2:201–2 (nos. 600–607). Payments, July–August 1637, for the Cappella della Cattedra Petri, "commesso d'alabastro . . . con nuvole scure et chiare contronate, et spelndore nel mezzo di Giallo . . . fascia di marmo bianco nel mezzo dove va lo Spirito Sto con suoi raggi, and pare che ci fossero nuvole": Pollak 1928, 2:186–87 (nos. 533–35). For the building history of the Chapel of Crucifix and Cathedra Petri: Rice 1997, 182–86, 189–92, with bibliography.
80. Payments, February 1640–November 1641, for "fondo de mischi . . . con nuvole chiare et scure contornate per le figure con spelndore nel mezzo di Giallo antico": Pollak 1928, 2:487, 93–94, 97, 501 (nos. 1937–40, 78–81, 88d, 2006–8).
81. SS. Domenico e Sisto: Bernardini et al. 1991, 69–74, with earlier bibliography. Bernini designed but did not execute a similar scheme in the apse of S. Maria in Via Lata: Lavin 1980, 1:169–81, 2:87. See also Ostrow 1996, 180n254, 181–83, 333n255. The literature on the Cornaro Chapel is broad, but see Lavin 1980; Preimesberger 1986; Napoleone 1998; Carloni 1999; Barcham 2001, 329–91. For the drawings: Ilana Dreyer in Lavin 1981, 87–94. Previous studies of its marbling are Barry 1995; Napoleone 1996; Barry 2009; Barry 2016.
82. St. Teresa of Avila, *Castillo interior*, 1.2, 1.4, in Zimmerman 1921, 38, 40. Fra Tomás de Jesús, *Compendio dell'orazione mentale*, transcribed in Lavin 1980, 1:159, with discussion, 1:82–84, 139–40. *Cathechismus, ex Decreto Concilii Tridentini, Praefatio*, 1.

83. The necessary licence was conceded in 1647: Bentivoglio 1994, 23 (no. 120).
84. Ostrow 2004; Guido and Mancella 2014; Ostrow 2014.
85. Preimesbergher 1985. See also Bolland 2000; Ostrow 2004; Ostrow 2007; Barry 2009.
86. Montanari 2004, 305–8.
87. Napoleone 1998, 178. Colored stucco lustro was unusual in Rome but common in Venice, so the artisan responsible was perhaps Lazzaro Morelli, whose mother came from a long line of Venetian craftsmen: Carloni 1999, 45.
88. Montanari 2003. Translations mine. Bibliotheca Apostolica Vaticana, Ms. Chigi D.III.41, fol. 51v: "Spectatori. De vestibus eorundem cardinalium affabre exculptis. / Cerne ut nobilitant Tyria sub veste theatrum / ingentes animae votaque magna patrum. / En rubet arcano spectatus murice candor, / cum lapide in Pario purpura sacra patent. / Sed quid miraris fabrum sculpsisse colores / quando tot heroum sculpitur hic anima?"
89. Barry 2002; del Bravo 1983.
90. *Chisiani Sacelli descriptio addenda operi: Roma vetus, et recens auctore Alexandro Donato Senensi* (Bibliotheca Apostolica Vaticana, Mss. Chigi, N. 111. 78, fols. 421r–430v, in Strunck 2003, 142, 174).
91. Harold 1931, 70. This may have influenced the book-matched paliotto of the Cappella Farsetti, S. Maria Maddalena (formerly in the chancel until 1694).
92. Bibliotheca Apostolica Vaticana, Ms. Chigi D.III.41, fol. 51v: "Eidem Bernino. De lapidibus et praecipue de maculosis in eodem Sacello Divae Teresiae mirifice dispositis. / Nominibus variis insignitura theatrum / exhylarant sese votis tua saxa superbis, / nec maculosa pudet nitidæ molimina molis, / si formosa lues pretiosis emicet umbris / quando iubes sacro maculas servire nitori."
93. Sohm 2000.
94. Chantelou (10 October 1665) in Stanić 2001, 247. For the development of the term "macchia," see Lehmann 2015. Baldinucci 1681, s.v. "macchia." Cf. s.v. "ritrarre alla macchia."
95. Lomazzo in Ciardi 1973, 1:245. Tasso's verses (1592/94) were published posthumously in 1607: Petrocchi 1951, 26–27, 39, 266. De Boodt 1609, 24.
96. Scamozzi 1615, pt. 2, bk. 6, ch. 30, 141.
97. Serlio 1584, bk. 6, 133v.
98. For the drapery, see Petersson 1970, 101–2; Lavin 1980, 1:111; Hyde Minor 1989.
99. Abridged from Kircher 1646, 933–34.
100. Shearman 1987; Buccheri 2014.
101. Bona Castellotti 2000, 138–39 (cat. no. 78).
102. Passeri in Hess 1934, 237.
103. On the matter of angels, see Thomas Aquinas, *Summa Theologica*, I^a q. 51 a. 2 ad 3, a definition repeated in Early Modern treatises on angelology.
104. Bernini 1713, 149: "il pregio maggiore del suo Scalpello, con cui vinto haveva la difficultà di render' il Marmo pieghevole come la cera, & haver con ciò saputo accoppiare in un certo modo insieme la Pittura, e la Scultura."
105. Bernini 1713, 39: "il Cavaliere dava per documento a'suoi Discepoli in materia di Architettura, che prima bisognava riflettere alla materia, indi all'invenzione, poi all'ordinazione delle parti, e finalmente a dar loro perfezzione di grazia, e tenerezza."
106. Wittkower 1981, 33; Hibbard 1965, 134.
107. Nicodemus Tessin, "Osservationi dal discorso del Sig.or Cav.ro Bernini," in Kommer 1974, 158.
108. On the *Quarant'ore*, see Imorde 1997, with bibliography.
109. Baldinucci 1682, 67; Bernini 1713, 33. For a shrewd discussion of this issue, see Delbeke 2006.
110. For S. Andrea, see mainly Connors 1982; Frommel 1983; Sharon Cather in Lavin 1981, 194–207.
111. Tratz 1991–92.
112. Lualdi 1653, dedication. Letter from Virgilio Spada, undated but 1649, Archivio di Stato di Roma, Fondo Spada Veralli, tom. 18, cc. 30–38. Cottanello had been used in antiquity but was rediscovered by the mason and entrepreneur Santi Ghetti some time before 1648: Apolloni Ghetti 1977. Rosso di Francia had also been quarried in antiquity: Perrier 1996.
113. For Sicilian jasper, see Gagliardo Briuccia and Montana 1998; Napoleone 2001. For Bernini on jasper, see the letter of Leonardo Mascardi (29 October 1666) in Mercantini 1981, 42, doc. 14. For the dressing of the blocks at S. Andrea, see Terhalle 1996.
114. Kircher 1650, lib. I, cap. V, 49–51 ("De Materia Obeliscorum"), lib. II., cap. X, esp. 169 ("Distinctio II, De Pyramidis & Obelisci mysteriis"); Kircher 1646, frontis.; Kircher, *Musurgia Universalis*, 2:450; Kircher 1652–54, 2:2, 112; Fleet 1909, 114.
115. D'Onofrio 1969, 107–8. On the tomb, see Roca De Amicis 1987.
116. Kircher 1646, 33–34.
117. For the *Golden Legend*, see Sharon Cather in Lavin 1981, 200; Careri 1995, 91–96. For the text of the Mass, see Guéranger 1868–1903, 1:280.
118. Letter from Ignazio Angelucci to Athanasius Kircher, 22 August 1633, in Kircher 1646, 801.
119. Benjamin 2003, 79. On the development of Early Modern dome painting, see Morel 1984.
120. Chantelou in Stanić 2001, 242.
121. Negro 2002; Barry 2004b, with bibliography.
122. Cf. Matthew 24:30. Cf. Mark 13:26 and Luke 21:27; Otto 1994, 227ff.
123. Galileo Galilei, *Considerazioni al Tasso*, canto I, 1, in Albèri et al. 1842, 9:68.
124. Dating: Ventura 1991, 81. Cafà died in 1667 and the altar wall was completed in 1672: Bevilacqua 1993, 97n7. Titi's 1674 edition credits Cafà with the whole design: Contardi and Romano 1987, 1:146. Cafà's brother, Lorenzo, was indeed Malta's leading architect. Yet, while the central aedicule, derived from the Cornaro altar, may be a design by Melchiorre, the whole altar wall is close in style to Carlo Rainaldi (e.g., Gesù e Maria, Altar of S. Antonio da Padova, SS. Apostoli, etc.), with whom Cafà had collaborated elsewhere (S. Agnese in Agone, S. Maria in Campitelli).
125. Queba e Tuna 1687, 381. For the chapel, see Cavazzini 1999.
126. Lankheit 1962, 291, docs. 399, 412. For the chapel's construction, see Pacini 2003, with bibliography.
127. On Stuckmarmor, see predominantly Schiessl 1979.

EPILOGUE

1. Milizia 1785, tomo I, pt. I, cap. XII, "Della combinazione de' vari marmi," 334–35.
2. For the demise of the four-element theory, see Siegfried 2002.
3. John Ruskin, *The Stones of Venice* (London: Smith, Elder, 1851–53), vol. 1, chap. 26 ("The Wall Veil and Shaft"). For an illuminating discussion of the geological metaphor in Victorian architecture and its resonances with contemporary developments in the sciences and theology, see Hall 2000.
4. *Ecclesiologist* 19 (1861): 67–69, cited in Hall 2000, 83.
5. Alma-Tadema and Brindley 1907; Hardy 1994; Rogers 2008; Warren 2012.
6. Lethaby 1892, 168–83. Lethaby and Swainson 1894, 79–80.
7. Street 1855, 159–60; de L'Hôpital 1919, 1:126–29. For the full story, with references, see Barry 2007, 642–46.
8. Trilling 1994; Lenzo 2007; Semper 2004 (1860–63), 410–11; Chatterjee 2017.
9. Schulze 1985, 5; Cohen 2007, 13, 70; Belting 1984. The Aachen connection has been independently recognized by Flood 2016, 201–5.
10. Betty Parsons Gallery, New York, 1949. Drexler 1950; Blake 1993, 111–12; Lum 1999.
11. Stokes 1934, 87.
12. Stokes 1934, 105, 157.
13. Stephen Cox, e-mail to author, June 9, 2018.
14. Marina Abramović, Masterpiece London (June 28–July 4, 2018). Alabaster panels fabricated by Factum Arte, Madrid, installation and display designed by Skene Catling de La Peña, London/Madrid. Interviews and manufacturing details at Factum Arte, "Five Stages of Maya Dance," http://www.factum-arte.com; Ann Binlot, "Marina Abramovic on Why She's 'Under Pressure to Open Roads for Every Great Female Artist,'" Forbes, https://www.forbes.com; Gareth Harris, "Podcast Episode 38: Marina Abramovic and Michael Jackson," Art Newspaper, https://www.theartnewspaper.com.

Bibliography

PRIMARY SOURCES

ABEL 1931: Felix-Marie Abel, "Gaza au VIe siècle, d'après le rhéteur Choricius," *Revue biblique* 40 (1931): 5–31.

ADAMI 1734: Andrea Adami, *Storia di Volseno antica metropoli della Toscana* (Rome: Antonio de' Rossi, 1734).

AGOSTI 2003: Giovanni Agosti, ed., *Nonno di Panopoli. Parafrasi del Vangelo di San Giovanni. Canto Quinto* (Florence: Università degli Studi di Firenze, 2003).

AGRICOLA 1530: Georgius Agricola, *Bermannus, siue de re metallica* (Basel: Frobeniana Officina, 1530).

AGRICOLA 1546: Georgius Agricola, *De la generatione de le cose, che sotto la terra sono, e de la cause de' loro effetti e nature* (Venice: Michele Tramezzino, 1546).

AGRICOLA 1556: Georgius Agricola, *De re metallica libri XII* (Basel: H. Frobenius & N. Episcopius, 1556).

AHMET 1897–1938: Cevdet Ahmet, ed., *Evliyâ Çelebi Seyâhatnâmesi*, 15 vols. (Dersaadet'te [Istanbul]: Ikdam Matbaasi, 1897–1938).

ALBÈRI ET AL. 1842: Eugenio Albèri, Celestino Bianchi, and Vincenzio Viviani, eds., *Le opere di Galileo Galilei*, 15 vols. (Florence: Società Editrice Fiorentina, 1842).

ALBERTINI 1510: Francesco Albertini, *Memoriale di molte statue et picture nella città di Florentia* (Florence: 1510).

ALOISI 1998: Sabrina Aloisi, *La Cappella di Santa Barbara nel Duomo di Rieti. Una realtà dimenticata–Gianlorenzo Bernini* (Rieti: M. Rinaldi, 1998).

AMELING 1983: Walter Ameling, *Herodes Atticus*, 2 vols. (Hildesheim: Olms, 1983).

AMELLI 1896: Ambrogio Maria Amelli, ed., *Miniature sacre e profane dell'anno 1023 illustranti l'enciclopedia medioevale di Rabano Mauro riprodotte in 133 tavole cromolitografiche da un codice di Montecassino* (Montecassino: Tipo-Litografia di Montecassino, 1896).

ANGELINI 1944: Cesare Angelini, ed., *Leonardo Frescobaldi and Simone Sigoli, Viaggi in Terrasanta* (Florence: Le Monnier, 1944).

ANSTRUTHER 1851: Robert Anstruther, ed., *Radulfi Nigri Chronica: The Chronicles of Ralph Niger* (London: Smith, 1851).

ARISI 1702–41: Francesco Arisi, *Cremona literata* (Parma: Pazzoni & Monti, 1702–41).

ARLENSIS DE SCUDALUPIS 1610: Petrus Arlensis de Scudalupis, *Sympathia septem Metallorum ac septem selectorum Lapidum ad Planetas* (Paris: Charles Silvestre, David Gilles, Jean Petitpas, 1610).

ARNOLD 1890–96: Thomas Arnold, ed., *Memorials of St. Edmund's Abbey*, 3 vols. (London: Eyre & Spottiswoode, 1890–96).

ASHBY 1904: Thomas Ashby, "Sixteenth-Century Drawings of Roman Buildings Attributed to Andreas Coner," *Papers of the British School at Rome* 2 (1904): 1–88.

AUSTIN AND BASTIANINI 2002: Colin Austin and Guido Bastianini, eds., *Posidippi Pellaei quae supersunt omnia* (Milan: LED, 2002).

BACCHI DELLA LEGA AND BIAGATTI 1945: A. Bacchi della Lega and P. B. Biagatti O. F. M., eds., *Frà Niccolò da Poggibonsi, Libro d'oltremare (1346–1350)* (Jerusalem: Tipografia dei PP Francescani, 1945).

BACCI 1587: Andrea Bacci, *Le XII pietre pretiose, le quali per ordine di Dio nella Santa Legge adornavano i vestimenti del sommo Sacerdote, aggiuntevi il diamante, le margarite e l'oro poste da S. Giovanni nell'Apocalisse in figura della celeste Gierusalemme: Con un sommario dell' altre pietre pretiose* (Rome: Giorgio Martinelli, 1587).

BALDINUCCI 1681: Filippo Baldinucci, *Vocabolario toscano dell'arte del disegno* (Florence: Santi Franchi, 1681).

BALDINUCCI 1682: Filippo Baldinucci, *Vita del cavaliere Gio. Lorenzo Bernino, scultore, architetto, e pittore* (Florence: Vincenzio Vangelisti, 1682).

BALDINUCCI 1846: Filippo Baldinucci, *Notizie de' professori del disegno da Cimabue in qua*, 6 vols. (1681–1728; repr., Florence: V. Batelli, 1846).

BAMBI 1775: Gaspare Bambi, *Memorie sacre della cappella di Sancta Sanctorum e della scala del palazzo di Pilato detta volgarmente Scala Santa* (Rome: Generoso Salomoni, 1775).

BANDY AND BANDY 1955: Mark Chance Bandy and Jean A. Bandy, eds., *Georgius Agricola. De natura fossilium (Textbook of Mineralogy)* (New York: Geological Society of America, 1955).

BARBARO 1567: Daniele Barbaro, *I dieci libri dell'architettura* (Venice: Francesco de'Franceschi & Giouanni Chrieger, 1567).

BARBARO 1739: Tommaso Barbaro, *Il pellegrino geografo cronistorico da Napoli sino a Venezia* (Venice: Jacopo Tomasini, 1739).

BAROCCHI AND RISTORI 1965–83: Paola Barocchi and Renzo Ristori, eds., *Il carteggio di Michelangelo*, 5 vols. (Florence: Spes, 1965–83).

BARRETT BROWNING 1863: Elizabeth Barrett Browning, *The Greek Christian Poets and the English Poets* (London: Chapman & Hall, 1863).

BARTOLINI AND CAPORALI 1999: Gabriella Bartolini and Giulio Caporali, eds. and trans., *Bernhard von Breydenbach. Peregrinationes: Un viaggiatore del Quattrocento a Gerusalemme e in Egitto. Ristampa anastatica dell'incunabolo* (Rome: Vecchiarelli, 1999).

BASTIANINI ET AL. 2001: Guido Bastianini, Claudio Galazzi, and Colin Austin, eds., *Posidippo di Pella. Epigrammi (P.Mil.Vogl. VIII 309)* (Milan: LED, 2001).

BEHR 1981: Charles A. Behr, ed. and trans., *P. Aelius Aristides: The Complete Works*, vol. 2, *Orations XVII–LIII* (Leiden: Brill, 1981).

BERNIA ET AL. 1648: Marco Antonio Bernia, Bartolommeo Ambrosini, and Ulisse Aldrovandi, *Ulyssis Aldrovandi patricij Bononiensis musaeum metallicum in libros IIII distributum* (Bologna: Giovanni Battista Ferroni, 1648).

BERNINI 1713: Domenico Bernini, *Vita del cavalier Gio. Lorenzo Bernino* (Rome: Rocco Bernabò, 1713).

BERTELS 1960–71: A. E. Bertels, ed., *Shax-nāme: Kritičeskij Tekst*, 9 vols. (Moscow: Izdatel'stvo Nauka, 1960–71).

BERTHELOT AND RUELLE 1887–88: Pierre E. M. Berthelot and Charles E. Ruelle, eds., *Collection des anciens alchimistes grecs*, 4 vols. (Paris: G. Steinheil, 1887–88).

BETTS 1995: Gavin Betts, ed., *Three Medieval Greek Romances: Velthandros and Chrysantza, Kallimachos and Chrysorrhoi, Livistros and Rodamni* (London: Garland, 1995).

BORGHINI 1584: Raffaello Borghini, *Il Riposo* (Florence: Marescotti, 1584).

BORGNET AND BORGNET 1898: A. and E. Borgnet, eds., *Albertus Magnus. Opera omnia* (Paris: Vivès: 1898).

BORSOOK 1973: Eve Borsook, "The Travel of Bernardo Michelozzi and Bonsignore Bonsignori in the Levant (1497–1498)," *Journal of the Warburg and Courtauld Institutes* 36 (1973): 145–97.

BOSIO 1600: Antonio Bosio, *Historia Passionis B. Caeciliae Virginis, Valeriani, Tiburtii, et Maximi martyrum nec non Urbani, et Lucii Pontificum, et Mart. Vitae atque Pascalis Papae I Litera de eorumdem sanctorum corporum inventione, et in Urbe translatione . . .* (Rome: Stefano Paulini, 1600).

BRACCIUS 1637: Ignatius Braccius, *Phoenicis effigies in numismatis et in gemma quae in museo Gualdino asservatur et parietibus indicata et eiusdem avis vindiciae* (Rome: Vitale Mascardi, 1637).

BRADBURY 2004: Scott Bradbury, *Selected Letters of Libanius: From the Age of Constantius and Julian* (Liverpool: Liverpool University Press, 2004).

BRAND 1976: Charles M. Brand, ed., *John Kinnamos: Deeds of John and Manuel Comnenus* (New York: Columbia University Press, 1976).

BROADHURST 1952: R. J. C. Broadhurst, ed. and trans., *The Travels of Ibn Jubayr, Being the Chronicle of a Mediaeval Spanish Moor Concerning His Journey to the Egypt of Saladin, the Holy Cities of Arabia, Baghdad the City of the Caliphs, the Latin Kingdom of Jerusalem, and the Norman Kingdom of Sicily* (London: Cape, 1952).

BROCK 1967: Sebastian P. Brock, "A Medieval Armenian Pilgrim's Description of Constantinople," *Revue des études arméniennes* 4 (1967): 81–102.

BROCK 1969: Sebastian P. Brock, "Rabban Sauma a Constantinople (1287)," in *Mémorial. Mgr. Gabriel Khouri-Sarkis (1898–1968)* (Louvain: Imprimerie orientaliste, 1969), 245–53.

BROECKE 2015: Lara Broecke, ed. and trans., *Cennino Cennini's Il libro dell'arte: A New English Translation and Commentary with Italian Transcription* (London: Archetypes, 2015).

BRUSCHI ET AL. 1978: Arnaldo Bruschi, Corrado Maltese, Manfredo Tafuri, and Michele Bonelli, eds., *Scritti rinascimentali di architettura: Patente a Luciano Laurana, Luca Pacioli, Francesco Colonna, Leonardo da Vinci, Donato Bramante, Francesco di Giorgio, Cesare Cesariano, lettera a Leone X* (Milan: Il Polifilo, 1978).

BUONAROTTI 1622: Michelangelo Buonarotti the Younger, *Delle lodi del Granduca di Toscana Cosimo II di Michelagnol Buonarotti. Recitata dà lui nell'Accademia Fiorentina il dì 20. di Dicembre 1621* (Florence: Pietro Cecconcelli, 1622).

CALAMAI 1993: Andrea Calamai, ed., *Sanctissimo peregrinaggio del Sancto Sepolcro, 1474/in appendice Itinerario di Pierantonio Buondelmonti, 1468* (Pisa: Pacini, 1993).

CALEY AND RICHARDS 1956: Earle R. Caley and John C. Richards, eds. and trans., *Theophrastus on Stones* (Columbus: Ohio State University Press, 1956).

CAMERON 1976: Averil Cameron, ed. and trans., *Flavius Cresconius Corippus. In laudem Iustini Augusti minoris Libri IV* (London: Anthlone, 1976).

CAMPANA 1973–74: Augusto Campana, "Ciriaco d'Ancona e Lorenza Valla sull'iscrizione greca del tempio dei Dioscuri," *Archaeologia classica* 25/26 (1973–74): 84–102.

CANDIDO 1521: Giovanni Candido, *Commentariorum Aquileiensium libri octo* (Venice: Lorenzo Lorio, 1521).

CANNETUS 1759: Petrus Cannetus, ed., *Ambrosii Traversarii latinae epistolae* (Florence: Ex Typographio Caesareo, 1759).

CAPECE 1546: Scipione Capece, *De principiis rerum* (Venice: Paolo Manuzio, 1546).

CAPORALI 1536: Giovanni Battista Caporali, *Architettura con il suo comento et figure Vetruvio in volgar lingua* (Perugia: Iano Bigazzini, 1536).

CARPEGGIANI AND LORENZONI 1998: Paolo Carpeggiani and Anna Maria Lorenzoni, eds., *Carteggio di Luca Fancelli con Ludovico, Federico e Francesco Gonzaga, marchesi di Mantova* (Mantua: Arcari, 1998).

CELANI AND THUASNE 1907–14: Enrico Celani and Louis Thuasne, eds., *Johannis Burchardi diarium, sive rerum urbanarum commentarii, 1483–1506* (Città di Castello: S. Lapi, 1907–14).

CHANDLER 1817: Richard Chandler, *Travels in Asia Minor and Greece*, 3rd ed., 2 vols. (London: J. Booker & R. Priestley, 1817).

CHATELAIN 1903: Émile Chatelain, "Notes sur quelques palimpsestes de Turin," *Revue de philologie, de littérature et d'histoire anciennes* 27 (1903): 37–48.

CHATTARD 1767: Giovanni Pietro Chattard, *Nuova descrizione del Vaticano o sia della sacrosanta basilica di S. Pietro*, 3 vols. (Rome: Eredi Barbiellini, 1767).

CHURCHILL AND CHURCHILL 1732: Awnsham Churchill and John Churchill, eds., *A Collection of Voyages and Travels*, 6 vols. (London: John Walthoe, 1732).

CIARDI 1973: Roberto Paolo Ciardi, *Gian Paolo Lomazzo. Scritti sulle arti*, 2 vols. (Florence: Marchi & Bertolli, 1973).

CIGGAAR 1973: Krijnie N. Ciggaar, "Une description anonyme de Constantinople du XIIe siècle," *Revue des études byzantines* 31 (1973): 335–54.

CLEARY 1925: Gregory Cleary, *Father Luke Wadding and St. Isidore's College, Rome: Biographical and Historical Notes and Documents* (Rome: Tipografia del Senato, 1925).

CLOVER 1971: Frank M. Clover, ed. and trans., *Flavius Merobaudes* (Philadelphia: American Philosophical Society, 1971).

COLEMAN 1988: Katherine M. Coleman, *Silvae IV* (Oxford: Oxford University Press, 1988).

COLLI 1981: Agostino Colli, "L'affresco della Gerusalemme celeste di S. Pietro al Monte di Civate: Proposta di lettura iconografica," *Arte Lombarda* 58–59 (1981): 7–20.

COLOMBO AND COLOMBO 1942: Alessandro Colombo and Giuseppe Colombo, eds., *Anonymi Mediolanensis Libellus de situ civitatis Mediolani: De adventu Barnabe Apostoli et de vitis priorum pontificum Mediolanensium* (Bologna: Zanichelli, 1942).

CONNOLLY AND O'REILLY 1995: Seán Connolly and Jennifer O'Reilly, eds. and trans., *Bede: On the Temple* (Liverpool: Liverpool University Press, 1995).

CONNORS 1998: Joseph Connors, ed., *Francesco Borromini. Opus architectonicum* (Milan: Il Polifilo, 1998).

CONNORS AND RICE 1990: Joseph Connors and Louise Rice, *Specchio di Roma Barocca. Una guida inedita del XVII secolo* (Rome: L'Elefante, 1990).

CONTARDI AND ROMANO 1987: Bruno Contardi and Serena Romano, eds., *Filippo Titi. Studio di pittura, scoltura, et architettura, nelle chiese di Roma (1674–1763). Edizione comparata*, 2 vols. (Florence: Centro Di, 1987).

CONTARINI 1541: Pietro Contarini, *Argoa Voluptas* (Venice: Bernardino de Viani, 1541).

CORTASSA 2000: Guido Cortassa, ed., *Roma parte del cielo. Confronto tra l'Antica e la Nuova Roma* (Turin: Utet, 2000).

CORYATE 1611: Thomas Coryate, *Coryats crudities* (London: William Stansby, 1611).

COURTNEY 1995: Edward Courtney, ed., *Musa Lapidaria: A Selection of Latin Verse Inscriptions* (Atlanta: Scholars, 1995), 188–93.

CRAMER 1841: John A. Cramer, ed., *Anecdota græca e codd. Manuscriptis Bibliothecæ Regiæ Parisiensis*, 4 vols. (Oxford: Typographeum Academicum, 1841).

CRANE 1987: Howard Crane, ed. and trans., *Risale-I Mi'Mariye: An Early Seventeenth-Century Ottoman Treatise on Architecture* (Leiden: Brill, 1987).

CROSS AND SHERBOWITZ-WETZOR 1953: Samuel H. Cross and Olgerd P. Sherbowitz-Wetzor, eds. and trans., *The Russian Primary Chronicle: Laurentian Text* (Cambridge, MA: Mediaeval Academy of America, 1953).

CUGNONI 1880: G. Cugnoni, "Note al Commentario di Alessandro VII sulla vita di Agostino Chigi," *Archivio della Società Romana di Storia Patria* 3 (1880): 422–48.

CUPANE 1995: Carolina Cupane, ed. and trans., *Romanzi cavallereschi bizantini. Callimaco e Crisorroe. Beltandro e Crisanza. Storia di Achille. Florio e Platziaflore. Storia di Apollonio di Tiro. Favola consolatoria sulla cattiva e la buona sorte* (Turin: Utet, 1995).

D'AMICO AND WEIL-GARRIS 1980: John P. D'Amico and Kathleen Weil-Garris, "The Renaissance Cardinal's Ideal Palace: A Chapter from Cortesi's *De Cardinalatu*," in *Studies in Italian Art and Architecture, 15th through 18th Centuries*, ed. Henry A. Millon (Rome: L'Elefante, 1980), 45–123.

D'ONOFRIO 1969: Cesare D'Onofrio, *Roma nel seicento* (Florence: Vallecchi, 1969).

DAVANZATI 1725: Benigno Davanzati, *Notizie al pellegrino della basilica di Santa Prassede* (Rome: Antonio de' Rossi, 1725).

DAWES AND BAYNES 1948: Elizabeth Dawes and Norman H. Baynes, *Three Byzantine Saints: Contemporary Biographies* (Oxford: Basil Blackwell, 1948).

DE BOODT 1609: Anselm de Boodt, *Gemmarum et lapidum historia* (Hanover: Typis Wechelianis, 1609).

DE GROER AND HEERS 1978: Georgette de Groer and Jacques Heers, eds., *Itinéraire d'Anselme Adorno en Terre sainte: 1470–1471* (Paris: Éditions du centre national de la recherche scientifique, 1978).

DE L'HÔPITAL 1919: Winefride de L'Hôpital, *Westminster Cathedral and Its Architect*, 2 vols. (London: Hutchinson, 1919).

DE MÉLY AND RUELLE 1896–1902: Fernand De Mély and Charles E. Ruelle, *Les lapidaires de l'antiquité et du moyen age*, 2 vols. (Paris: Leroux, 1896–1902).

DE MURALTO 1697–98: Giovanni De Muralto, "De quibusdam lapidibus figuratis Helvetiae, ANC," *Miscellanea Curiosa Medico-Physica Academiae Naturae Curiosorum sive Ephemeridum Medico-Physicarum Germanicarum* Dec. III., no. An. V. et VI. (1697–98): 43–45.

DE OVALLE 1646: Alonso De Ovalle, *Historica Relacion del Reyno de Chile, y delas missiones, y ministerios que exercita en el la Compagnia de Iesus* (Rome: Francesco Cavalli, 1646).

DE PIÑEDA 1609: Juan de Piñeda, *De rebus Salomonis Regis libri octo* (Lyon: Horatius Cardon, 1609).

DE RETZA 1487/90: Franciscus de Retza, *Defensorium inviolate perpetueque virginitatis castissime de genitricis Mariae . . .* (Basel: Lienhart Ysenhut, 1487/90).

DE ROSSI ET AL. 1861–1971: Giovanni Battista de Rossi, Antonio Ferrua, and Angelo Silvagni, eds., *Inscriptiones Christianae urbis Romae septimo saeculo antiquiores*, Nova series (Rome: Pontificium Institutum Archaeologiae Christianae, 1861–88), and new series, 5 vols. (Rome: Libraria Doct. Befani, 1922–71).

DE STEFANI 2011: Claudio De Stefani, ed., *Paulus Silentiarius, Descriptio Sanctae Sophiae. Descriptio Ambonis* (New York: De Gruyter, 2011).

DE TOURNEFORT 1717: Joseph Pitton de Tournefort, *Relation d'un voyage du Levant*, 2 vols. (Lyon: Anisson & Posuel, 1717).

DE VALLES ET AL. 1588: Francisco de Valles, François La Rue, and Lievin Lemmens, *De iis, quae scripta sunt physicae in libris sive de sacra philosophia . . . De gemmis aliquot iis praesertim quarum Div. Joannes Apost. In sua Apocal. Meminit* (Lyon: François Le Fèvre, 1588).

DEGLI AVANZI 2003: Adriano Prosperi, ed., *Descrittione di tutta Italia di F. Leandro Alberti Bolognese aggiuntavi la Descrittione di tutte l'isole*, anastatic repr. of 1568 ed., publ. in Venice, ed. Lodovico degli Avanzi, 3 vols. (Bergamo: Leading, 2003).

DELLA PORTA 1560: Giambattista Della Porta, *Magiae Naturalis* (Antwerp: Christophor Plantinus, 1560).

DINDORF 1961: Wilhelm Dindorf, ed., *Themistii Orationes* (Hildesheim: Olms, 1961).

DODWELL 1986: C. R. Dodwell, ed. and trans., *Theophilus: The Various Arts. De Diversis Artibus* (Oxford: Clarendon, 1986).

DOLAEUS 1677: Johann Dolaeus, "Observatio LXVI. De lapidibus varias tum arborum, tum fruticum, tum animalium figuras exprimentibus," *Miscellanea Curiosa Medico-Physica Academiae Naturae Curiosorum sive Ephemeridum Medico-Physicarum Germanicarum* 6–7 (1675–76) [1677]: 92.

DOLCE 1565: Lodovico Dolce, *Libri tre ne i quali si tratta delle diverse sorti delle gemme, che produce la natura* (Venice: Gio. Battista Sessa e Marchio Sessa, 1565).

DONATI 1695: Alessandro Donati, S. J., *Roma vetus ac recens* (Amsterdam: Janssonio-Waesbergios & Joannem Wolters, 1695).

DOWNEY 1950: Glanville Downey, "Description of the Church of St. Stephen at Gaza by Choricius, Sections 37–46," in *The Dome: A Study in the History of Ideas*, ed. E. Baldwin Smith (Princeton: Princeton University Press, 1950), 155–57.

DOWNEY 1957: Glanville Downey, "Nikolaos Mesarites: Description of the Church of the Holy Apostles at Constantinople," *Transactions of the American Philosophical Society* 47, no. 6 (1957): 855–924.

DRUDI 1794: Lorenzo Drudi, ed., *Basini Parmensis Poetae Opera Praestantiora: Hesperis. Astronomicon. Meleagris. Argonautico* (Rimini: Typographia Albertiniana, 1794).

DU CANGE 1688: Charles Du Fresne Du Cange, *Glossarium ad scriptores mediæ et infimæ graecitatis* (Lyons: J. Posuel & C. Rigaud, 1688).

DU CANGE 1954: Charles Du Fresne Du Cange, *Glossarium mediae et infimae Latinitatis*, 7 vols. (Graz: Akademische Druck- u. Verlagsanstalt, 1954).

DUCHESNE AND VOGEL 1957: Louis Duchesne and Cyrille Vogel, eds., *Le Liber pontificalis* (Paris: E. Thorin, 1957).

DUPONT-SOMMER 1947: André Dupont-Sommer, "Une hymne syriaque sur la cathédrale d'Édesse," *Cahiers archéologiques* 2 (1947): 29–39.

DURAND 1997–2000: Jean-Marie Durand, ed. and trans., *Les documents épistolaires du palais de Mari*, 3 vols. (Paris: du Cerf, 1997–2000).

EICHHOLZ 1965: David E. Eichholz, ed. and trans., *Theophrastus: De Lapidibus* (Oxford: Clarendon, 1965).

ELIOR 1982: Rahel Elior, ed., *Hekhalot Zutarti* (Jerusalem: Universitah ha-'Ivrit, 1982).

EPSTEIN AND SIMON 1948: Isidore Epstein and Maurice Simon, eds., *The Babylonian Talmud* (London: Soncino, 1948).

ERHARD 1932: Marcelle Erhard, "Le livre du pèlerin d'Antoine de Novgorod," *Romania* 58 (1932): 44–65.

ERMAN AND LANGE 1925: Adolf Erman and Hans O. Lange, eds., *Papyrus Lansing, eine ägyptische Schulhandschrift der 20. Dynastie* (Copenhagen: A. F. Høst, 1925).

FABRI 1664: Girolamo Fabri, *Le sagre memorie di Ravenna antica* (Venice: Francesco Valuense, 1664).

FARAL 1939: Edmond Faral, ed., *Villehardouin. La Conquête de Constantinople*, 2 vols. (Paris: Belles lettres, 1939).

FARNESE 1566/67: Duke Alessandro Farnese, *Commentari di varie regole, e disegni di architettura civile, e militare, con altre istruzioni, e precetti di arte militare* (1566/77), Biblioteca Corsiniana, Rome, 663 (32 B 14).

FAYANT AND CHUVIN 1997: Marie-Christine Fayant and Pierre Chuvin, eds. and trans., *Paul le Silentiaire. Description de Sainte-Sophie de Constantinople* (Die: Die, 1997).

FÈA 1790: Carlo Fèa, *Miscellanea filologica, critica, e antiquaria*, 2 vols. (Rome: Pagliarini, 1790).

FEATHERSTONE 2000: Jeffrey M. Featherstone, ed., *Theodore Metochites's Poems "to Himself"* (Vienna: Verlag der Österreichischen Akademie der Wissenschaften, 2000).

FEATHERSTONE 2011: Jeffrey M. Featherstone, "Metochites's Poems and the Chora," in *The Kariye Camii Reconsidered*, ed. Holger A. Klein, Robert G. Ousterhout, and Brigitte Pitarakis (Istanbul: İstanbul Araştırmaları Enstitüsü, 2011), 213–37.

FESTUGIÈRE AND RYDÉN 1974: André J. Festugière and Lennart Rydén, eds., *Léontios de Néapolis. Vie de Syméon le fou et vie de Jean de Chypre* (Paris: Paul Geuthner, 1974).

FIORE ET AL. 1981: Francesco Paolo Fiore, Arnaldo Bruschi, and Adriano Carugo, eds., *De Architettura, Vitruvio traslato, commentato e affigurato da Caesare Caesariano* (Milan: Il Polifilo, 1981).

FÖRSTER AND RICHSTEIG 1929: Richard Förster and Eberhard Richsteig, eds., *Choricii Gazaei Opera* (Leipzig: B. G. Teubner, 1929).

FOSTER 1996: Benjamin R. Foster, *Before the Muses: An Anthology of Akkadian Literature,* vol. 1, *Archaic, Classical, Mature* (Bethesda, MD: CDL, 1996).

FRANCINI ET AL. 1574: Girolamo Francini, Andrea Palladio, Vincenzo Accolti, and Henry W. Poole, *Le cose merauigliose dell'alma citta di Roma . . .* (Rome: Vincenzo Accolto, 1574).

FRENDO 1975: Joseph D. Frendo, ed. and trans., *Agathias: The Histories* (Berlin: de Gruyter, 1975).

FRIZOLIO 1582: Lorenzo Frizolio, *Sacellum Gregorianum* (Rome: Domenico Basa, 1582).

FULVIO 1513: Andrea Fulvio, *Antiquaria Urbis* (Rome: Jacopo Mazocchi, 1513).

FULVIO 1527: Andrea Fulvio, *Antiquitates Urbis* (Rome: Marcello Silber, 1527).

GABEL AND GRAGG 1959: Leona C. Gabel and Florence A. Gragg, eds. and trans., *Memoirs of a Renaissance Pope: The Commentaries of Pius II* (New York: Putnam, 1959).

GARCÍA MARTÍNEZ 1994: Florentin García Martínez, *The Dead Sea Scrolls Translated: The Qumran Texts in English,* trans. Wilfred G. E. Watson (Leiden: Brill, 1994).

GARTON 1986: Charles Garton, ed. and trans., *The Metrical Life of Saint Hugh of Lincoln* (Lincoln: Honywood, 1986).

GARZYA ROMANO 1996: Chiara Garzya Romano, *I colori e le arti dei Romani e la compilazione pseudo-eracliana* (Bologna: Il Mulino, 1996).

GEORGE 2003: Andrew R. George, *The Babylonian Gilgamesh Epic: Introduction, Critical Edition, and Cuneiform Texts,* 2 vols. (Oxford: Oxford University Press, 2003).

GEROLA 1931: Giuseppe Gerola, "Le vedute di Costantinopoli di Cristoforo Buondelmonti," *Studi bizantini e neoellenici* 3 (1931): 249–79.

GEYER 1898: Paul Geyer, *Itinera hierosolymitana saeculi IIII–VIII* (Vienna: F. Tempsky, 1898).

GHALIOUNGUI 1987: Paul Ghalioungui, *The Ebers Papyrus: A New English Translation, Commentaries, and Glossaries* (Cairo: Academy of Scientific Research and Technology, 1987).

GIACCHERO 1974: Marta Giacchero, ed., *Edictum Diocletiani et collegarum de pretiis rerum venalium: In integrum fere restitutum e latinis graecisque fragmentis,* 2 vols. (Genoa: Istituto di Storia Antica e Scienze Ausiliarie, 1974).

GIESE 1922: Friedrich Giese, ed. and trans., *Die altosmanischen anonymen Chroniken [Tawarih-i-al-i-Otman]* (Breslau: Giese, 1922).

GNOLI AND SIRONI 1996: Raniero Gnoli and Attilia Sironi, eds., *Agostino Del Riccio. Istoria delle pietre* (Turin: Allemandi, 1996).

GODMAN 1985: Peter Godman, *Latin Poetry of the Carolingian Renaissance* (Norman: University of Oklahoma Press, 1985).

GOLDSCHMIDT 1940: Rudolf C. Goldschmidt, ed. and trans., *Paulinus' Churches at Nola* (Amsterdam: Noord-Hollandsche Uitgevers Maatschappij, 1940).

GOUSSEN 1925: Heinrich Goussen, "Über eine 'Sugitha' auf die Kathedrale von Edessa," *Muséon* 38 (1925): 117–36.

GRAYSON 1976: Albert K. Grayson, *Assyrian Royal Inscriptions II: From Tiglath-Pileser to Ashur-Nasir-Pal II* (Wiesbaden: Harrassowitz, 1976).

GROS ET AL. 1997: Pierre Gros, Antonio Corso, and Elisa Antonio, eds. and trans. and comm., *Vitruvio. De Architectura,* 2 vols. (Turin: Einaudi, 1997).

GUILLET DE SAINT-GEORGES 1674: Georges Guillet de Saint-Georges, *Athenes ancienne et nouvelle et l'estat present de l'empire des Turcs* (Paris: E. Michallet, 1674).

GWYNNE 2015: Paul Gwynne, *Patterns of Patronage in Renaissance Rome: Francesco Sperulo: Poet, Prelate, Soldier, Spy,* 2 vols. (Bern: Peter Lang, 2015).

GYLLIUS 1632: Petrus Gyllius, *De Constantinopoleos topographia libri IV* (Leiden: A. Elzevir, 1632).

HABIG 1983: Marion A. Habig, *St. Francis of Assisi: Writings and Early Biographies: English Omnibus of the Sources for the Life of St. Francis* (Chicago: Franciscan Herald, 1983).

HALLEUX AND SCHAMP 1985: Robert Halleux and Jacques Schamp, eds., *Les Lapidaires Grecs. Lapidaire Orphique, Kérygmes Lapidaires d'Orphée, Socrate et Denys, Lapidaire Nautique, Damigéron-Évax* (Paris: Belles lettres, 1985).

HAMILTON 1930–31: D. W. Hamilton, "Two Churches at Gaza," *Palestine Exploration Fund, Quarterly Statement* (1930–31): 178–91.

HAMMER-PURGSTALL 1846–50: Joseph von Hammer-Purgstall, ed. and trans., *Narrative of Travels in Europe, Asia and Africa in the Seventeenth Century by Evliya Efendi,* 2 vols. (London: W. H. Allen, 1846–50).

HAROLD 1931: Francis Harold, *Vita Fratris Lucae Waddingi,* 3rd ed. (Quaracchi: Ad Claras Aquas, 1931).

HASSLER 1843–49: Dietrich Hassler, ed., *Fratris Felicis Fabri evagatorium in Terrae Sanctae, Arabiae et Egypti peregrinationem,* 3 vols. (Stuttgart: Societas Literaria Stuttgardiensis, 1843–49).

HATFIELD 1970: Rab Hatfield, "Some Unknown Descriptions of the Medici Palace in 1459," *Art Bulletin* 52, no. 3 (1970): 232–49.

HENTZNER 1617: Paul Hentzner, *Itinerarium Germaniae, Galliae; Angliae; Italiae* (Breslaw: Johann Eyering & Johann Perfert, 1617).

HESS 1934: Jacob Hess, *Die Künstlerbiographien von Giovanni Battista Passeri* (Leipzig: Keller, 1934).

HOADE 1970: Eugene Hoade, *Western Pilgrims: The Itineraries of Fr. Simon Fitzsimons* (Jerusalem: Franciscan, 1970).

HOADE AND BAGATTI 1962: Eugene Hoade and Bellarmino Bagatti, eds., *Fr. Elzear Horn O. F. M., Ichnographiae monumentorum Terrae Sanctae (1724–1744)* (Jerusalem: Franciscan, 1962).

HOFFMANN 1980: Hartmut Hoffmann, ed., *Chronica monasterii Casinensis* (Hanover: Hahn, 1980).

HOLMYARD AND MANDEVILLE 1927: Eric J. Holmyard and Desmond C. Mandeville, eds., *Avicennae de Congelatione et Congluinatione Lapidum* (Paris: Paul Geuthner, 1927).

HOLTON 1974: David Holton, ed. and trans., *The Tale of Alexander* (Thessaloniki: Byzantines kai Neoellenikes Bibliotheke, 1974).

HÜLSEN AND EGGER 1916: Christian Hülsen and Hermann Egger, eds., *Die römischen Skizzenbücher von Marten van Heemskerck im Königlichen Kupferstichkabinett zu Berlin,* 2 vols. (Berlin: J. Bard, 1916).

HUNGER 1969: Herbert Hunger, ed., *Johannes Chortasmenos (ca. 1370–ca. 1436/37). Briefe, Gedichte und kleine Schriften. Einleitung, Regesten, Prosopographie, Text* (Vienna: Böhlau in Komm, 1969).

HUNT ET AL. 1932–34: Arthur S. Hunt, Denys L. Page, and Desmond M. Leahy, eds. and trans., *Select Papyri,* 2 vols. (London: W. Heinemann, 1932–34).

IJSEWIJN 1997: Jozef Ijsewijn, ed., *Coryciana* (Rome: Herder, 1997).

IZEDDIN 1941–46: M. Izeddin, "Un prisonnier arabe à Byzance au IXe siècle," *Revue des études islamiques* (1941–46): 41–62.

JEFFREYS 1998: Elizabeth M. Jeffreys, ed. and trans., *Digenis Akritis: The Grottaferrata and Escorial Versions* (Cambridge: Cambridge University Press, 1998).

JEFFREYS ET AL. 1986: Elizabeth Jeffreys, Michael Jeffreys, Roger Scott, and Brian Croke, eds. and trans., *The Chronicle of John Malalas* (Melbourne: Australian Association for Byzantine Studies, 1986).

JOLY 1968: Robert Joly, ed., *Hermas, Le Pasteur,* 2nd ed. (Paris: du Cerf, 1968).

JONES 1988: P. J. Jones, "Travel Notes of an Apprentice Florentine Statesman, Giovanni di Tommaso Ridolfi," in *Florence and Italy: Renaissance Studies in Honour of Nicolai Rubinstein,* ed. Peter Denley and Caroline Elam (London: Westfield College, 1988), 199–233.

KEIL 1958: Bruno Keil, ed., *Aelii Aristidis Smyrnaei quae supersunt omnia,* 2 vols. (Berlin: Weidmann, 1958).

KENNEDY 2003: George A. Kennedy, ed. and trans., *Progymnasmata: Greek Textbooks of Prose Composition and Rhetoric* (Leiden: Brill, 2003).

KHITROWO 1889: Sofiia P. Khitrowo, ed. and trans., *Itineraires russes en Orient* (Geneva: J.-G. Fick, 1889).

KIRCHER 1646: Athanasius Kircher, *Ars Magna Lucis et Umbrae* (Rome: H. Scheus, 1646).

KIRCHER 1650: Athanasius Kircher, *Obeliscus Pamphilius* (Rome: Ludovico Grignani, 1650).

KIRCHER 1652–54: Athanasius Kircher, *Oedipus Aegyptiacus*, 3 vols. (Rome: V. Mascardi, 1652–54).

KIRCHER 1665: Athanasius Kircher, *Mundus subterraneus* (Amsterdam: Janssonius & Weyerstraet, 1665).

KREN AND WIECK 1990: Thomas Kren and Roger S. Wieck, *The Visions of Tondal from the Library of Margaret of York* (Malibu: J. Paul Getty Museum, 1990).

KROLL 1926: Wilhelm Kroll, ed., *Historia Alexandri Magni (Pseudo-Callisthenes)*, vol. 1 (Berlin: Wiedmannsche Buchhandlung, 1926).

KRUMBACHER 1897: Karl Krumbacher, *Geschichte der byzantinischen Litteratur von Justinian bis zum Ende des oströmischen Reiches, 527–1453*, 2nd ed. (Munich: Beck, 1897).

KULP-HILL 2000: Kathleen Kulp-Hill, ed. and trans., *Songs of Holy Mary of Alfonso X, the Wise: A Translation of the Cantigas de Santa Maria* (Tempe: Arizona Center for Medieval and Renaissance Studies, 2000).

LABORDE 1848: Leon de Laborde, *Athènes aux XVe, XVIe et XVIIe siècles*, 2 vols. (Paris: J. Renouard, 1848).

LACKENBACHER 1982: Sylvie Lackenbacher, *Le roi bâtisseur: Les récits de construction assyriens des origines à Teglatphalasar III* (Paris: Éditions recherche sur les civilisations, ADPF, 1982).

LAGERCRANTZ 1913: Otto Lagercrantz, ed., *Papyrus graecus holmiensis (P. holm.); Recepte für Silber, Steine und Purpur* (Uppsala: Akademiska Bokhandeln, 1913).

LAMBECK 1679: Peter Lambeck, *Commentariorum de Augustissima Bibliotheca Caesarea Vindobonensi liber primus* (Vienna: Matthaei Cosmerovij, 1679).

LAMI 1742: Giovanni Lami, ed., *Deliciae eruditorum* (Florence: Pietro Caietano Viviani, 1742).

LANGE 1708: Carl Nicolaus Lange, *Historia lapidum figuratorum Helvetiae: Ejusque viciniae, in qua non solum enarrantur omnia eorum genera, species et vires aeneisque tabulis repraesentantur* (Venice: Jacopo Tomasini, 1708).

LAUER 1924: Philippe Lauer, ed., *Robert De Clari: La Conquête De Constantinople* (Paris: É. Champion, 1924).

LAURENT 1864: Johann C. M. Laurent, ed., *Peregrinatores medii aevi quatuor* (Leipzig: J. C. Hinrichs, 1864).

LAURENT 1971: Vitalien Laurent, ed., *Les Mémoires du Grand Ecclésiarque de l'église de Constantinople Sylvestre Syropoulos sur le concile de Florence (1438–1439)* (Paris: Centre national de la recherche scientifique, 1971).

LAZZARI 1753: Michele Lazzari, "Discorso Del Signor Michele Lazzari sopra le macchie naturali di una pietra, rappresentante un ritratto," in *Raccolta d' opuscoli scientifici e filologici, tomo quarantesimottavo a Sua Eccellenza Il Sig. Conte Ferdinando Scoti*, ed. Angelo Calogera (Venice: Simone Occhi, 1753), 267–87.

LAZZARINI 1952: Andrea Lazzarini, *Il miracolo di Bolsena: Testimonianze e documenti dei secoli XIII e XIV* (Rome: Edizioni di Storia e Letteratura, 1952).

LE GRAND 1895: Léon Le Grand, "Relation du Pèlerinage à Jérusalem de Nicolas de Martoni, notaire italien (1394–1395)," *Revue de l'orient latin* 3 (1895): 566–669.

LE STRANGE 1928: Guy Le Strange, ed. and trans., *Clavijo: Embassy to Tamerlane, 1403–1406* (London: Routledge, 1928).

LECOY DE LA MARCHE 1870: Albert Lecoy de la Marche, ed., *Vie de Jésus-Christ composée au Xve siècle d'après Ludolphe le Chartreux* (Paris: Hurtel, 1870).

LEDEL 1696: Samuel Ledel, "Observatio LXXXVIII. De lusu naturae in lapide," *Miscellanea Curiosa Medico-Physica Academiae Naturae Curiosorum sive Ephemeridum Medico-Physicarum Germanicarum* Dec. III, no. An. IV (1696): 188.

LEGRAND 1902: Émile Legrand, ed., *Les exploits de basile Digénis Acritas: Épopée byzantine. Publiée d'après le manuscrit de Grotta-Ferrata*, 2nd rev. ed. (Paris: J. Maisonneuve, 1902).

LEGRAND AND REINACH 1896: Émile Legrand and Théodore Reinach, eds., *Description des oeuvres d'art et de l'Eglise des Saints Apôtres de Constantinople: Poème en vers iambiques par Constantin le Rhodien, publié d'après le manuscrit du Mont-Athos* (Paris: Leroux, 1896).

LEHMANN-BROCKHAUS 1955: Otto Lehmann-Brockhaus, *Lateinische Schriftquellen zur Kunst in England, Wales und Schottland, vom Jahre 901 bis zum Jahre 1307*, 5 vols. (Munich: Prestel, 1955).

LENTINI 1958: Anselmo Lentini O. S. B., "Il ritmo 'Cives caelestis patriae' e il 'De duodecim lapidibus' di Amato," *Benedictina* 12, nos. 1–2 (1958): 15–26.

LENTINI AND AVAGLIANO 1974: Anselmo Lentini and Faustino Avagliano, eds., *I carmi di Alfano I arcivescovo di Salerno* (Montecassino: Badia di Montecassino, 1974).

LEONARDI 1502: Camillo Leonardi, *Speculum Lapidum* (Venice: Giovanni Battista Sessa, 1502).

LETTS 1926: Malcolm Letts, ed. and trans., *Pero Tafur: Travels and Adventures, 1435–1439* (London: Routledge, 1926).

LEVEND 1958: Agâh Sirri Levend, ed., *Türk Edebiyatinda Sehr-Engizler ve Sehr-Engizlerde Istanbul* (Istanbul: Istanbul Fethi Dernegi, 1958).

LICHTHEIM 1976: Miriam Lichtheim, *Ancient Egyptian Literature II: The New Kingdom* (Los Angeles: University of California Press, 1976).

LINDSAY 1913: Wallace M. Lindsay, ed., *Sexti Pompei Festi de verborum significatu* (Leipzig: Teubner, 1913).

LITSAS 1980: Fotios K. Litsas, "Choricius of Gaza: An Approach to His Work: Introduction, Translation, Commentary" (PhD diss., University of Chicago, 1980).

LÓPEZ ESTRADA 1999: Francisco López Estrada, ed., *Embajada a Tamorlán* (Madrid: Clásicos Castalia, 1999).

LUALDI 1653: Michelangelo Lualdi, *L'India Orientale soggettata al Vangelo* (Rome: Ignazio De Lazzari, 1653).

LUZIO 1886: Alessandro Luzio, ed., *Lettere inedite di Fra Sabba da Castiglione* (Milan: Bortolotti, 1886).

MAFFEI 1506: Raffaelle Maffei, *Commentariorum Urbanorum* (Rome: Ioannes Besicken, 1506).

MAFFEI 1732: Scipione Maffei, *Verona Illustrata* (Verona: Jacopo Vallarsi e Pierantonio Berno, 1732).

MAGNI 1688: Cornelio Magni, *Relazione della città d'Athene, colle prouincie dell' Attica, Focia, Beozia, e Negroponte, ne' tempi, che furono queste passeggiate da Cornelio Magni Parmegiano l'anno 1674* (Parma: G. Rosati, 1688).

MAGOULIAS 1984: Harry J. Magoulias, *O City of Byzantium: Annals of Niketas Choniates* (Detroit: Wayne State University Press, 1984).

MAGURN 1955: Ruth Saunders Magurn, ed. and trans., *The Letters of Peter Paul Rubens* (Cambridge, MA: Harvard University Press, 1955).

MAJESKA 1984: George P. Majeska, *Russian Travelers to Constantinople in the Fourteenth and Fifteenth Centuries* (Washington, DC: Dumbarton Oaks, 1984).

MALTESE AND MALTESE DEGRASSI 1967: Corrado Maltese and Livia Maltese Degrassi, *Francesco di Giorgio Martini. Trattati di architettura ingegneria e arte militare*, 2 vols. (Milan: Il Polifilo, 1967).

MALVASIA 1678: Carlo Cesare Malvasia, *Felsina pittrice: Vite de pittori bolognesi* (Bologna: Domenico Barbieri, 1678).

MANDROT 1903: Bernard de Mandrot, ed., *Mémoires de Philippe de Commynes*, 2 vols. (Paris: Picard, 1903).

MANGO 1958: Cyril Mango, ed. and trans., *The Homilies of Photius, Patriarch of Constantinople* (Cambridge, MA: Harvard University Press 1958).

MANGO 1972: Cyril A. Mango, ed. and trans., *The Art of the Byzantine Empire, 312–1453: Sources and Documents* (Englewood Cliffs, NJ: Prentice-Hall, 1972).

MANGO 1990: Cyril A. Mango, ed. and trans., *Nicephorus: Short History* (Washington, DC: Dumbarton Oaks, 1990).

MANGO AND PARKER 1960: Cyril Mango and John Parker, "A Twelfth-Century Description of St. Sophia," *Dumbarton Oaks Papers* 14 (1960): 233–45.

MARANI 1979: Pietro C. Marani, ed., *Trattato di architettura: Il Codice Ashburnham 361 della Biblioteca medicea laurenziana di Firenze* (Florence: Giunti Barbera, 1979).

MARTIN AND HURST 1991: Lawrence T. Martin and David Hurst, *Bede the Venerable: Homilies on the Gospels: Book Two, Lent to the Dedication of the Church* (Kalamazoo, MI: Cistercian, 1991).

MARTINELLI 1978: Lucia Cesarini Martinelli, ed., *Angelo Poliziano. Commento inedito alle Selve di Stazio* (Florence: Sansoni, 1978).

MARUCCHI 1956–57: Adriana Marucchi, ed., *Giulio Mancini. Considerazioni sulla pittura,* 2 vols. (Rome: Accademia Nazionale dei Lincei, 1956–57).

MAZZUCCONI 1840: Leonardo Mazzucconi, *Memorie storiche della Scala Santa e dell'insigne santuario di Sancta Sanctorum* (Rome: Ferretti, 1840).

MCCRINDLE 1897: J. W. McCrindle, ed. and trans., *The Christian Topography of Cosmas, an Egyptian Monk* (London: Hakluyt Society, 1897).

MCMAHON 1956: A. Philip McMahon, ed., *Leonardo da Vinci: Treatise on Painting (Codex urbinas latinus 1270),* 2 vols. (Princeton: Princeton University Press, 1956).

MCNEAL 1997: Edgar H. McNeal, ed. and trans., *Robert de Clari: The Conquest of Constantinople* (Toronto: University of Toronto Press, 1997).

MERTOL TULUM 1977: A. Mertol Tulum, *Tursun Bey. Târih-i Ebü'l-feth* (Istanbul: Baha Matbaas, 1977).

MIKLOSICH AND MÜLLER 1860–90: Franz Miklosich and Joseph Müller, eds., *Acta et diplomata graeca medii aevi sacra et profana,* 6 vols. (Vienna: C. Gerold, 1860–90).

MILIZIA 1785: Francesco Milizia, *Principi di architettura civile,* 3 vols. (Bassano: Remondini, 1785).

MILLER 1855–57: E. Miller, ed., *Manuelis Philae Carmina, ex codicibus Escurialensis, Florentinis, Parisinis et Vaticanis,* 2 vols. (Paris: Typographeum imperiale, 1855–57).

MILLER 1858: M. Miller, "Poëme allégorique de Meliténiote: Publié d'après un manuscrit grec de la Bibliothèque Impériale," *Notices et Extraits des Mss. de la Bibliothèque Imperiale et autres bibliothèques* 19, no. 2 (1858): 1–138.

MOMMSEN 1894: Theodor Mommsen, ed., *Cassiodori Senatoris variae* (Berlin: Wiedmann, 1894).

MONTANARI 2003: Tommaso Montanari, "A Contemporary Reading of Bernini's 'Maraviglioso Composto,'" in *Poetry on Art: Renaissance to Romanticism,* ed. Thomas Frangenberg (Donington: Shaun Tyas, 2003), 177–98.

MONTIJANO GARCÍA 1999: Juan María Montijano García, *San Carlo alle Quattro Fontane di Francesco Borromini: Nella Relatione della Fabrica di fra Juan de San Buenaventura* (Milan: Il Polifilo, 1999): 60–61.

MORINO 1997: Alberto Morino, ed., *Restoro d'Arezzo. La composizione del mondo* (Parma: Guanda, 1997).

MORTET 1911: Victor Mortet, ed., *Recueil de textes relatifs à l'histoire de l'architecture et à la condition des architéctes en France au moyen âge, XIIe–XIIIe siècles,* 2 vols. (Paris: Picard, 1911).

MORTET AND DESCHAMPS 1929: Victor Mortet and Paul Deschamps, eds., *Recueil de textes relatifs à l'histoire de l'architecture et à la condition des architéctes en France au moyen âge, XIIe–XIIIe siècles,* 2 vols. (Paris: Picard, 1929).

MORYSON 1907: Fynes Moryson, *An Itinerary, Containing His Ten Yeeres Travell through the Twelve Dominions of Germany, Bohmerland, Sweitzerland, Netherland, Denmarke, Poland, Italy, Turky, France, England, Scotland and Ireland* (Glasgow: James MacLehose, 1907).

MÜLLER 2009: Monika E. Müller, *Omnia in mensura et numero et pondere disposita. Die Wandmalereien und Stuckarbeiten von San Pietro al Monte di Civate* (Regensburg: Schnell & Steiner, 2009).

NEUSNER 1999: Jacob Neusner, ed. and trans., *The Tosefta* (Atlanta: Scholars, 1999).

NICCOLINI 1847: Giovanni Battista Niccolini, *Filippo Strozzi. Tragedia di G.-B. Niccolini corredata d'una vita di Filippo e di documenti inediti* (Florence: Felice Le Monnier, 1847).

NORI 1996: Gabriele Nori, ed., *Itinerario al Santo Sepolcro, 1486/Antonio da Crema* (Ospedaletto: Pacini, 1996).

OCTOECHOS 1999: *The Great Octoechos,* 4 vols. (Boston: Sophia, 1999).

OHL 1928: Raymond T. Ohl, "The Enigmas of Symphosius" (PhD diss., University of Pennsylvania, 1928).

ORLANDI 1966: Giovanni Orlandi, ed., *Leon Battista Alberti. De re aedificatoria,* 2 vols. (Milan: Il Polifilo, 1966).

OROFINO 1994–2000: Giulia Orofino, *I codici decorati dell'Archivio di Montecassino* (Rome: Istituto Poligrafico e Zecca dello Stato Libreria dello Stato, 1994–2000), 2.2 (I codici preteobaldiani e teobaldiani).

PAGE 1942: Denys L. Page, *Greek Literary Papyri,* 2 vols. (London: William Heinemann, 1942).

PALLADIO 1570: Andrea Palladio, *I quattro libri dell'Architettura* (Venice: Domenico de' Franceschi, 1570).

PALMER AND RODLEY 1988: Andrew Palmer and Lynn Rodley, "The Inauguration Anthem of Hagia Sophia in Edessa: A New Edition and Translation with Historical and Architectural Notes and a Comparison with a Contemporary Constantinopolitan Kontakion," *Byzantine and Greek Studies* 12 (1988): 117–67.

PANCIROLI AND GUALTIERI 1612: Guido Panciroli and Flavio Gualtieri, *Raccolta breve d'alcune cose più segnalate c'hebbero gli antichi, e d'alcune altre trovate da moderni* (Venetia: Bernardo Giunti & Giovanni Battista Ciotti, 1612).

PANOFSKY AND PANOFSKY-SOERGEL 1979: Erwin Panofsky and Gerda Panofsky-Soergel, eds. and trans., *Abbot Suger on the Abbey Church of St.-Denis and Its Art Treasures* (Princeton: Princeton University Press, 1979).

PASOLINI 1678: Serafino Pasolini, *Lustri ravennati* (Bologna: Giacomo Monti, 1678).

PATRIZI 1471: Francesco Patrizi, *De institutione rei publicae* (1464, presented 1471), Bibliotheca Apostolica Vaticana, Vat. Lat. 3084 (1471).

PEIPER 1883: Rudolf Peiper, ed., *Alcimi Ecdicii Aviti Viennensis Episcopi Opera quae supersunt* (Berlin: Weidmann, 1883).

PERNOT 1925: Hubert Pernot, ed., *Voyage en Turquie et en Grèce du R. P. Robert de Dreux, amônier de l'ambassadeur de France* (Paris: Belles lettres, 1925).

PEROSA 1960: Alessandro Perosa, *Giovanni Rucellai ed il suo Zibaldone* (London: Warburg Institute, 1960).

PERRAT AND AUDIN 1957: Charles Perrat and Aimable Audin, "Alcimi Ecdicii Aviti viennensis episcopi homilia dicta in dedicatione superioris basilicae," in *Studi in onore di Aristide Calderini e Roberto Paribeni* (Milan: Ceschina, 1957), 433–51.

PERTUSI 1976: Agostino Pertusi, *La Caduta di Costantinopoli,* 2 vols. (Milan: Mondadori, 1976).

PETROCCHI 1951: Giorgio Petrocchi, ed., *Torquato Tasso. Il mondo creato* (Florence: Le Monnier, 1951).

PETRUCCI AND PETRUCCI 1972: Armando Petrucci and Franca Petrucci, eds., *Viazo da Venesia al Sancto Iherusalem* (Rome: L'Elefante, 1972).

PHARR 1952: Clyde Pharr, *The Theodosian Code and Novels and the Sirmonidan Constitutions* (Princeton: Princeton University Press, 1952).

PIZARRO GÓMEZ AND MOGOLLÓN CANO-CORTES 1999: Francisco Pizarro Gómez and Pilar Mogollón Cano-Cortes, eds., *Los diez libros de arquitectura de Marco Vitruvio Polión según la traducción castellana de Lázaro de Velasco* (Cáceres: Ciclón: 1999).

POLLAK 1928: Oskar Pollak, *Die Kunsttätigkeit unter Urban VIII,* 2 vols. (Vienna: Benno Filser, 1928).

PONA 1620: Francesco Pona, *Sileno ovvero delle Bellezze del Luogo dell'Ill. mo Sig. Co. Gio. Giacomo Giusti* (Verona: B. Merlo, 1620).

PONTANI 1978–81: Filippo Maria Pontani, ed., *Antologia Palatina* (Turin: Einaudi, 1978–81).

PORRO-LAMBERTENGHI 1855: Giulio Porro-Lambertenghi, ed., *Viaggio di Pietro Casola a Gerusalemme* (Milan: Ripamonti Carpano, 1855).

POWELL 1925: John U. Powell, ed., *Collectanea Alexandrina* (Oxford: Clarendon, 1925).

POWICKE 1950: Frederick M. Powicke, *The Life of Ailred of Rievaulx* (London: Nelson, 1950).

POZZI AND CIAPPONI 1964: Giovanni Pozzi and Lucia A. Ciapponi, eds., *Hypnerotomachia Poliphili*, 2 vols. (Padua: Antenore, 1964).

PREGER 1975: Theodor Preger, ed., *Scriptores originum Constantinopolitanarum* (1901–7; repr., New York: Arno, 1975).

PREUDHOMME 1539: Christopher Preudhomme, ed., *Trium Poetarum Elegantissimorum, Porcelii, Basinii, et Trebani Opuscula* (Paris: Simon de Colines, 1539).

PRICHARD 1969: James B. Prichard, ed., *Ancient Near Eastern Texts Relating to the Old Testament*, 3rd ed. (Princeton: Princeton University Press, 1969), 573–76.

QUEBA E TUNA 1687: Lavinio Queba e Tuna, *Il fior fenice, cioè Marzio redivivo in Gio. Francesco cardinale Ginetti* (Venice: Antonio Pinelli, 1687).

QUESNEL 1996: Solange Quesnel, ed. and trans., *Venance Fortunat. Vie de Saint Martin*, vol. 4, *Oeuvres* (Paris: Belles lettres, 1996).

QUINLAN-MCGRATH 1989: Mary Quinlan-McGrath, "Aegidius Gallus, de viridario Augustini Chigii vera libellus: Introduction, Latin Text and English Translation," *Humanistica Lovaniensia* 38 (1989): 1–8, 10–99.

QUINLAN-MCGRATH 1990: Mary Quinlan-McGrath, "Blosius Palladius, Suburbanum Augustini Chisii: Introduction, Latin Text, and English Translation," *Humanistica Lovaniensia* 39 (1990): 93–155.

QUIRKE 2004: Stephen Quirke, *Egyptian Literature 1800* BC*: Questions and Readings* (London: Golden House, 2004).

RAGUSA AND GREEN 1961: Isa Ragusa and Rosalie Green, eds., *Meditations on the Life of Christ: An Illustrated Manuscript of the Fourteenth Century (Paris, BN, Ms. Ital. 115)* (Princeton: Princeton University Press, 1961).

REUMONT 1876: Alfredo Reumont, "Viaggio in Italia nel MCDXCVII del Cav. Arnoldo di Harff di Colonia sul Reno," *Archivio Veneto* 11 (1876): 393–407.

REYDELLET 1994: Marc Reydellet, ed., *Venance Fortunat. Poèmes*, vol. 1, *Livres I–IV*, 2 vols. (Paris: Belles lettres, 1994).

RICHTER 1897: Jean Paul Richter, *Quellen der byzantinischen Kunstgeschichte* (Vienna: C. Graeser, 1897).

RIDLEY 1982: Ronald T. Ridley, trans., *Zosimus: New History* (Canberra: Australian Association for Byzantine Studies, 1982).

RÖHRICHT 1895: Reinhold Röhricht, "Le pèlerinage du Moine Augustin Jacques de Vérone (1335)," *Revue de l'orient latin* 3, no. 2 (1895): 155–302.

ROSELLINI AND CALDERAN 1993: Michaela Rosellini and Roberto Calderan, eds., *Iuli Valeri Res Gestae Alexandri Macedonis Translatae ex Aesopo Graeco* (Leipzig: Teubner, 1993).

ROSENBLUM 1961: Morris Rosenblum, *Luxorius: A Latin Poet among the Vandals* (New York: Columbia University Press, 1961).

ROSSI 1589: Girolamo Rossi, *Historiarum Rauennatum libri decem* (Venice: Francesco de Franceschi, 1589).

ROSSI TAIBBI 1969: Giuseppe Rossi Taibbi, ed., *Omelie per i vangeli domenicali e le feste di tutto l'anno* (Palermo: Istituto Siciliano di Studi Bizantini e Neoellenici, 1969).

RYKWERT ET AL. 1988: Joseph Rykwert, Neil Leach, and Robert Tavernor, eds., *Leon Battista Alberti: On the Art of Building in Ten Books* (Cambridge, MA: MIT Press, 1988).

SAHM 1912: Wilhelm Sahm, ed., *Beschreibung der Reisen des Reinhold Lubenau* (Königsberg: Thomas & Oppermann, 1912).

SANSOVINO 1585: Francesco Sansovino, ed., *Vita di Giesu Christo nostro redentore, scritta da Landolfo di Sassonia del'ordine Certosino, et fatta volgare da M. Francesco Sansovino* (Venice: Altobello Salicato, 1585).

SCAMOZZI 1615: Vincenzo Scamozzi, *L'idea della architettura universale* (Venice: Expensis Auctoris, 1615).

SCHEFER 1881: Charles Schefer, ed., *Journal d'Antoine Galland pendant son séjour à Constantinople (1672–1673)* (Paris: E. Leroux, 1881).

SCHEFER 1892: Charles Schefer, ed., *Le voyage d'outremer de Bertrandon de la Broquière* (Paris: Leroux, 1892).

SCHEUCHZER 1697–98: Johann Jacob Scheuchzer, "De dendritis aliisque lapidibus, qui in superficie sua plantarum, foliorum, florum, figuras exprimunt," *Miscellanea Curiosa Medico-Physica Academiae Naturae Curiosorum sive Ephemeridum Medico-Physicarum Germanicarum* Dec. III, no. An. V. et VI. (1697–98): 57–80.

SCHLOSSER 1896: Julius Schlosser, ed., *Schriftquellen zur Geschichte der Karolingischen Kunst* (Vienna: Graeser, 1896).

SCORSO 1655: Francesco Scorso, S. J., *Latium Christianum elocutionis vocabula, et phrases, ad res ecclesiae christianae accomodata* (Palermo: Petride de Isola, 1655) (2nd ed. of *Latium Christianum sive loquendi formae ad explicanda christiana sacra latinis scriptoribus exquisitae* [1648]).

SEMPER 2004: Gottfried Semper, *Style in the Technical and Tectonic Arts, or, Practical Aesthetics*, trans. Harry Mallgrave and Michael Robinson (Los Angeles: Getty, 2004).

SERLIO 1537: Sebastiano Serlio, *Regole generali di architetura sopra le cinque maniere de gli edifici* (Venice: F. Marcolini da Forli, 1537).

SERLIO 1584: Sebastiano Serlio, *Tutte l'opere d'architettura* (Venice: Francesco de' Franceschi, 1584).

SHANZER AND WOOD 2002: Danuta Shanzer and Ian Wood, eds., *Avitus of Vienne: Letters and Prose* (Manchester: Liverpool University Press, 2002).

SHEARMAN 2003: John Shearman, ed., *Raphael in Early Modern Sources: 1483–1602*, 2 vols. (New Haven: Yale University Press, 2003).

SHELLEY 1840: Mary Shelley, ed., *Essays, Letters from Abroad, Translations and Fragments by Percy Bysshe Shelley*, 2 vols. (Philadelphia: Lea & Blanchard, 1840).

SHERWOOD TAYLOR 1937: F. Sherwood Taylor, "The Alchemical Works of Stephanos of Alexandria," *Ambix* 1, no. 2 (1937): 116–39.

SIDER 1997: David Sider, ed., *The Epigrams of Philodemos: Introduction, Text, and Commentary* (New York: Oxford University Press, 1997).

SJÖBERG AND BERGMANN 1969: Åke W. Sjöberg and Eugen Bergmann, *The Collection of Sumerian Temple Hymns* (Locust Valley, NY: J. J. Augustin, 1969).

SORESINI 1672: Giovanni Maria Soresini, *De Scala sancta Sanctorum in Laterano culta opusculum* (Rome: Varesius, 1672).

SPENCER 1965: John R. Spencer, *Filarete's Treatise on Architecture: Being the Treatise by Antonio di Piero Averlino, Known as Filarete*, 2 vols. (New Haven: Yale University Press, 1965).

SPERULO 1519: Francesco Sperulo, *Ad R.mum D. Iulium Medicem S. R. E. Presbyterum Cardinalem et Vicecancellarium humillimus servus Franciscus Sperulus Villam Iulium Medicam versibus fabricata mittit*, Bibliotheca Apostolica Vaticana, Vat. Lat. 5812 (March 19, 1519), fols. 2r–16v.

SPESSO 1997: Marco Spesso, ed., *Enea Silvio Piccolomini: Scritti di architettura* (Turin: Testo & Immagine, 1997).

SPON AND WHELER 1724: Jacob Spon and George Wheler, *Voyage d'Italie, de Dalmatie, de Grece, et du Levant: Fait aux années 1675 et 1676*, 2 vols. (The Hague: Rutgert Alberts, 1724).

SPRETI 1489: Desiderio Spreti, *De amplitudine, de vastatione et instauratione urbis Ravennae* (Venice, 1489).

STANIĆ 2001: Milovan Stanić, ed., *Chantelou. Journal de voyage du cavalier Bernin en France* (Paris: Macula L'insulaire, 2001).

STEWART 1893–97: Aubrey Stewart, *The Book of Wanderings of Brother Felix Fabri (circa 1480–1483)*, 2 vols. (London: Committee of the Palestine Exploration Fund, 1893–97).

STRACK AND BILLERBECK 1922–28: Hermann Strack and Paul Billerbeck, *Kommentar zum Neuen Testament aus Talmud und Midrasch*, 3 vols (Munich: Beck, 1922–28).

STREET 1855: George Edmund Street, *Brick and Marble in the Middle Ages: Notes of Tours in the North of Italy* (London: John Murray, 1855).

STRINGA 1610: Giovanni Stringa, *La Chiesa di San Marco; Capella del Serenissimo Principe di Venezia* (Venice: Francesco Rampazetto, 1610).

STUART 1816: James Stuart, *The Antiquities of Athens and Other Monuments of Greece Measured and Delineated*, vol. 4 (London: T. Bensley for J. Taylor, 1816).

STUART AND REVETT 1830: James Stuart and Nicholas Revett, *The Antiquities of Athens and Other Places in Greece, Sicily, etc.: Supplementary to the Antiquities of Athens* (London: Priestly & Weale, 1830).

STUBBS 1876: William Stubbs, ed., *Radulfi de Diceto decani Lundoniensis opera historica: The Historical Works of Master Ralph de Diceto, Dean of London*, 2 vols. (London: Longman, 1876).

TADMORE AND YAMADA 2011: Hayim Tadmor and Shigeo Yamada, *The Royal Inscriptions of Tiglath-Pileser III (744–727 B.C.) and Shalmaneser V (726–722 B.C.), Kings of Assyria* (Winona Lake, IN: Eisenbrauns, 2011).

TATEO 1965: Francesco Tateo, ed., *Giovanni Pontano. I trattati delle virtu sociali: De liberalitate, de beneficentia, de magnificentia, de splendore, de conviventia* (Rome: L'Ateneo, 1965).

THACKSTON 1986: Wheeler M. Thackston, *Naser-e Khosraw's Book of Travels (Safarnama)* (Albany, NY: Bibliotheca Persica, 1986).

THOMAS ET AL. 2000: John P. Thomas, Angela Constantinides Hero, and Giles Constable, eds., *Byzantine Monastic Foundation Documents: A Complete Translation of the Surviving Founders' Typika and Testaments*, 5 vols. (Washington, DC: Dumbarton Oaks, 2000).

THOMPSON 1904: Edward Maunde Thompson, ed., *Chronicon Adae de Usk, A.D. 1377–1421* (London: Oxford University Press, 1904).

TILLIETTE 1998: Jean-Yves Tilliette, ed., *Baudri de Bourgueil. Poèmes*, 2 vols. (Paris: Belles lettres, 1998).

TLL 1900–: *Thesaurus Linguae Latinae* (Leipzig: Teubner, 1900–).

TOMASONI 1990: Piera Tomasoni, ed., *Lapidario Estense* (Milan: Bompiani, 1990).

TONNEAU AND DEVREESE 1949: Raymond Tonneau and Robert Devreese, eds., *Les homelies catéchétiques: Reproduction phototypique du ms. Mingana Syr. 561 (Selly Oak Colleges' Library, Birmingham)* (Vatican City: Biblioteca Apostolica Vaticana, 1949).

TORNAU AND CECCONI 2014: Christian Tornau and Paolo Cecconi, eds., *The Shepherd of Hermas in Latin: Critical Edition of the Oldest Translation Vulgata* (Berlin: De Gruyter, 2014).

TORSANI 1562: Angelo Maria Torsani, *Orationes, qua de Umbriae, Romandiolaeq. celeberrimarum regionum Italiae urbiumque suarum praecipuarum laudibus agunt* (Venice: Giovanni Griffio, 1562).

TOTARO 1984: Luigi Totaro, ed., *I commentarii/Enea Silvio Piccolomini, Papa Pio II*, 2 vols. (Milan: Adelphi, 1984).

TREU 1895: Maximilien Treu, ed., *Dichtungen des Gross-Logotheten Theodoros Metochites* (Potsdam: Brandt, 1895).

TRYPANIS 1968: Constantine A. Trypanis, *Fourteen Early Byzantine Cantica* (Vienna: Böhlau, 1968).

VACCA 1594: Flaminio Vacca, *Memorie di varie antichità trovate in diversi luoghi della città di Roma* (Rome: Gaetano Zenobj, 1594).

VALENTINO 1583: Ascanio Valentino, *Sacelli Gregoriani descriptio* (Florence: Bartolomeo Sermartelli, 1583).

VALTURIO 1534: Roberto Valturio, *De Re Militari* (Paris: Christian Vechel, 1534).

VAN DER VIN 1980: J. P. A. van der Vin, *Travellers to Greece and Constantinople: Ancient Monuments and Old Traditions in Medieval Travellers' Tales*, 2 vols. (Leiden: Nederlands Historisch-Archaeologisch Instuut te Istanbul, 1980).

VANSTIPHOUT 2003: Herman J. L. Vanstiphout, ed. and trans., *Epics of Sumerian Kings: The Matter of Aratta* (Atlanta: Society of Biblical Literature, 2003).

VARCHI 1549: Benedetto Varchi, *Due lezzioni . . . Nella seconda si disputa quale sia più nobile arte la Scultura, o la Pittura . . .* (Florence: Lorenzo Torrentino, 1549).

VASARI-MILANESI 1906: Gaetano Milanesi, ed., *Giorgio Vasari. Le vite de'più eccellenti pittori, scultori ed architettori*, 6 vols. (Florence: G. C. Sansoni, 1906).

VERPEAUX 1966: Jean Verpeaux, ed., *Pseudo-Kodinos. Traité des Offices* (Paris: Éditions du Centre National de la Recerche Scientifique, 1966).

VERYARD 1701: Ellis Veryard, *An Account of Divers Choice Remarks, as Well as Geographical, as Historical, Political, Mathematical, Physical, and Moral; Taken in a Journey through the Low-Countries, France, Italy, and Part of Spain with the Isles of Sicily and Malta . . . as also, a Voyage to the Levant* (London: S. Smith and B. Walford, 1701).

VIEL DE SAINT-MAUX 1787: Charles-François Viel de Saint-Maux, *Lettres sur l'architecture des anciens, et celle des modernes, dans lesquelles se trouve développé le génie symbolique qui présida aux monumens de l'Antiquité* (Paris, 1787).

VITTI 1986: Evangelia Vitti, ed., *Die Erzählung über den Bau der Hagia Sophia in Konstantinopel: Kritische Edition mehrerer Versionen* (Amsterdam: Hakker, 1986).

VOELKER 1980: Evelyn C. Voelker, *Charles Borromeo's "Instructiones fabricae et supellectilis ecclesiasticae," 1577* (Ann Arbor: University Microfilms, 1980).

VOLLMER 1905: Frederick Vollmer, ed., *Fl. Merobaudis reliquiae: Blossii Aemilii Dracontii carmina, Eugenii Toletani Episcopi carmina et epistulae, cum appendicula carminum spuriorum* (Berlin: Weidmann, 1905).

VON SONNENBERG 1661: Konrad von Sonnenberg, *Vera et syncera Relatio, recenter inventi LAPIDIS Miraculosi, ipsa Die 25. Augusti Anno Dominicae Incarnationis 1660 in ferrifodinis montium Cantonis Vraniensis: Potentis Helvetiae: in quo Effigies IESV ET MARIAE Circumdata fulgentissimo Sole omnium Admiratione conspicitur, Neque invenire est Artificem, qui simile quid, in eiusmodi specie LAPIDIS, effingere attentet. Asservatur Summa cum Devotione in loco Miraculoso, vulgò SYLVA DEI Dicto ditionis Lucernensis* (Lucerne, 1661).

WALLIS BUDGE 1933: Ernest Wallis Budge, *The Alexander Book in Ethiopia: The Ethiopic Versions of Pseudo-Callisthenes, the Chronicle of Al-Makin, the Narrative of Joseph Ben Gorion, and a Christian Romance of Alexander: Translated into English from MSS. in the British Museum and Bibliothèque Nationale* (London: Oxford University Press, 1933).

WHELER 1682: George Wheler, *A Journey into Greece . . .*, 6 vols. (London: William Cademan, 1682).

WIEDMANN 1999: Gerhard Wiedmann, *Nikolaus Muffel. Descrizione della città di Roma nel 1452: Della indulgenze e dei luoghi sacri di Roma (Der ablas und die heiligen stet zu Rom)* (Bologna: Pàtron, 1999).

WILKINSON 1977: John Wilkinson, *Jerusalem Pilgrims before the Crusades* (Warminster: Aris & Phillips, 1977).

WOLOHOJIAN 1969: Albert Mugrdich Wolohojian, ed. and trans., *The Romance of Alexander the Great by Pseudo-Callisthenes, Translated from the Armenian Version* (New York: Columbia University Press, 1969).

WOODHOUSE 1986: Cristopher M. Woodhouse, *George Gemistos Plethon: The Last of the Hellenes* (Oxford: Clarendon, 1986).

WOTTON 1624: Henry Wotton, *The Elements of Architecture* (London: John Bill, 1624).

WRIGHT 1848: Thomas Wright, ed., *Early Travels in Palestine: Comprising the Narratives of Arculf, Willibald, Bernard, Sæwulf, Sigurd, Benjamin of Tudela, Sir John Maundeville, De la Brocquière, and Maundrell* (London: Henry G. Bohn, 1848).

WRIGHT 1866: Thomas Wright, *The Chronicle of Pierre de Langtoft, in French Verse, from the Earliest Period to the Death of King Edward I*, 2 vols. (London: Longmans, Green, Reader & Dyer, 1866).

WÜSTENFELD 1857: Ferdinand Wüstenfeld, ed. and trans., *Die Chroniken der Stadt Mekka. Band I. Die Geschichte und Beschreibung der Stadt Mekka von Al-Azraqī* (Leipzig: Brockhaus, 1857).

WYCKOFF 1967: Dorothy Wyckoff, ed. and trans., *Albertus Magnus, Book of Minerals* (Oxford: Clarendon, 1967).

YASSIF 1982: Eli Yassif, *Sipure Ben Sira bi-Yeme ha-benayim (The Tales of Ben Sira in the Middle Ages)* (Jerusalem: Hotsaat sefarim al shem Y. L. Magnes, ha-Universitah ha-Ivrit, 1982).

ZIMMERMAN 1921: Benedict Zimmerman, ed. and trans., *The Interior Castle, or The Mansions by St. Teresa of Avila*, 3rd ed. (London: Thomas Baker, 1921).

ZOBI 1853: Antonio Zobi, *Notizie storiche sull'origine e progressi dei lavori di commesso in pietre dure* (Florence: Stamperia Granducale, 1853).

ZORZI 1525: Francesco Zorzi, *De Harmonia Mundi Totius Cantica Tria* (Venice, 1525).

ZOTENBERG 1900: H. Zotenberg, ed. and trans., "Ġorar aḳbār molūk al-fors [of Ṯaʿālebī]," in *Histoire des rois des Perses* (Paris: Imprimerie Nationale, 1900).

SECONDARY SOURCES

ABRALDES 1996: Ana M. Abraldes, "Pentelethen: The Export of Pentelic Marble and Its Use in Architectural and Epigraphical Monuments" (PhD diss., University of California, Berkeley, 1996).

ABRAMOVITZ 1980: Katherine Abramovitz, "Frescoes from Ayia Irini, Keos. Parts II–IV," *Hesperia* 49, no. 1 (1980): 57–85.

ACCORDI 1980: Bruno Accordi, "Michele Mercati (1541–1593) e la Metallotheca," *Geologica Romana* 19 (1980): 1–50.

ACIDINI LUCHINAT 1991: Cristina Acidini Luchinat, "Tre madrigali e alcune osservazioni sulla Cappella Gaddi in Santa Maria Novella," *Studi di Storia dell'Arte* 2 (1991): 295–320.

ACIDINI LUCHINAT AND CHIARINI 2000: Cristina Acidini Luchinat and Marco Chiarini, eds., *Bizzarrie di pietre dipinte: Dalle collezioni dei Medici* (Cinisello Balsamo: Silvana, 2000).

ACKERMAN 1954: James S. Ackerman, *The Cortile del Belvedere* (Vatican City: Biblioteca Apostolica Vaticana, 1954).

ADAM 1994: Jean Pierre Adam, *Roman Building: Materials and Techniques* (Bloomington: Indiana University Press, 1994).

ADAMS 1930: Frank D. Adams, *The Birth and Development of the Geological Sciences* (London: Williams & Wilkins, 1930).

ADORNI 2012: Bruno Adorni, *Giulio Romano architetto. Gli anni mantovani* (Cinisello Balsamo: Silvana, 2012).

ADRIANI 1936: Achille Adriani, *Annuaire du Musée Greco-Romain (1933–34–1934–35): La Nécropole de Moustafa Pacha* (Alexandria: Whitehead Morris, 1936).

ADRIANI 1940: Achille Adriani, "Fouilles et découvertes. A). Alexandrie. IV. Nécropoles. A) Fouilles dans la nécropole orientale. II.–Nécropole de Ezbet El Makhlouf," *Annuaire du Musée gréco-romain* 3 (1940): 83–122.

ADRIANI 1963: Achille Adriani, *Repertorio d'arte dell'Egitto greco-romano*, ser. c, 3 vols. (Palermo: Fondazione "Ignazio Mormino" del Banco di Sicilia, 1963).

ADRIANI ET AL. 2000: Achille Adriani, Nicola Bonacasa, and Patrizia Minà, *La tomba di Alessandro: Realtà, ipotesi e fantasie* (Rome: Bretschneider, 2000).

AGNOLI 2014: Nadia Agnoli, *Marsia. La superbia punita* (Rome: De Luca, 2014).

AGOSTI 1993: Giovanni Agosti, "Su Mantegna 1 (All'ingresso della mostra 1992, Londra)," *Prospettiva* 71 (1993): 42–52.

ALABE 1995: Françoise Alabe, "Intérieurs de maisons hellénistiques: Les murs peints a Délos," *Revue archéologique*, no. 1 (1995): 191–97.

ALCHERMES 1994: Joseph Alchermes, "Spolia in Roman Cities of the Late Empire: Legislative Rationales and Architectural Reuse," *Dumbarton Oaks Papers* 48 (1994): 167–78.

ALEXANDER 1952: Paul J. Alexander, "Hypatius of Ephesus: A Note on Image Worship in the Sixth Century," *Harvard Theological Review* 45, no. 3 (1952): 176–84.

ALEXANDER 1994: Jonathan J. G. Alexander, ed., *The Painted Page: Italian Renaissance Book Illumination, 1450–1550* (Munich: Prestel, 1994).

ALIPRANTES 1996: Nikos C. Aliprantes, *Το Πάριο μάρμαρο στην ιστορία και στην τέχνη: κατά τους συγγραφείς, τις επιγραφές και τα μνημεία* (Athens: Demos Parou, 1996).

ALLAG ET AL. 1988: Claudine Allag, Brigitte Bardoux, and Daniel Chossenot, "La mort d'Adonis: Une peinture murale gallo-romaine à Boult-sur-Suippe (Marne)," *Bulletin de la Société archéologique champenoise* 81, no. 2 (1988): 93–107.

ALLEN 1999: H. R. Allen, "Observations on the Original Appearance of the Dome of the Rock," in *Bayt al-Maqdis: Jerusalem and Early Islam*, ed. Jeremy Johns (Oxford: Oxford University Press, 1999), 197–213.

ALLINGER-CSOLLICH 1998: Wilfred Allinger-Csollich, "Birs Nimrud II: 'Tieftempel'-'Hochtempel': (Vergleichende Studien Borsippa—Babylon)," *Baghdader Mitteilungen* 29 (1998): 95–330.

ALMA-TADEMA AND BRINDLEY 1907: Lawrence Alma-Tadema and William Brindley, "Marbles: Their Ancient and Modern Application," *Journal of the Royal Institute of British Architects* 14 (1907): 169–96.

ALPERS 1983: Svetlana Alpers, *The Art of Describing: Dutch Art in the Seventeenth Century* (London: John Murray, 1983).

AMADIO 1986: Adele A. Amadio, *I mosaici di S. Costanza: Disegni, incisioni e documenti dal XV al XIX secolo* (Rome: De Luca, 1986).

AMANDRY AND HANSEN 2010: Pierre Amandry and Erik Hansen, *Le Temple d'Apollon du IVe siècle* (Paris: De Boccard, 2010).

AMICI 1991: Carla M. Amici, *Il Foro di Cesare* (Florence: Olschki, 1991).

ANDERSON 2001: Jeffrey C. Anderson, "Tiles, Books, and the 'Church Like a Bride Adorned with Pearls and Gold,'" in Gerstel and Lauffenburger 2001, 119–41.

ANDERSON AND NISTA 1989: Maxwell L. Anderson and Leila Nista, eds., *Radiance in Stone: Sculptures in Colored Marble from the Museo Nazionale Romano* (Rome: De Luca, 1989).

ANDRÉ 1949: Jacques André, *Étude sur les termes de couleurs dans la lingue latine* (Paris: C. Klincksieck, 1949).

ANDRÉ-SALVINI 1999: Béatrice André-Salvini, "L'idéologie des pierres en Mésopotamie," in *Cornaline et pierres précieuses: La Méditerranée, de l'Antiquité à l'Islam: Actes du colloque organisé au Musée du Louvre les 24–25 novembre 1995*, ed. Annie Caubet (Paris: Documentation française, 1999), 391, 373–400.

ANDREOU 1988: Andreas Andreou, "Griechische Wanddekorationen" (PhD diss., Johannes Gutenberg-Universität, 1988).

ANGELELLI 2004: Claudia Angelelli, "Sectilia pavimenta minora e/o inediti della Villa di Domiziano," in *Studi e ricerche sul patrimonio archeologico del Parco Nazionale del Circeo*, ed. Valentina Livi and Roberto Righi (Sabaudia: Corpo Forestale, 2004), 83–95.

ANGLIKER 2019: Erica M. Angliker, "Worship and Stones on the Cycladic Islands: A Case Study of the Aniconic Cult of Apollo and Zeus," in *Worshipping and Stones*, ed. Bernard Gilhooly, Niamh Kelly, Sol Mallía-Guest, and A. David Newell (Leiden: Sidestone, 2019).

ANGUISSOLA 2002: Anna Anguissola, "Note alla legislazione su spoglio e reimpiego di materiali da costruzione ed arredi architettonici, I sec. a.C.–VI sec. d.C.," in *Senso delle rovine e riuso dell'antico*, ed. Walter Cupperi (Pisa: Scuola Normale Superiore di Pisa, 2002), 13–30.

ANTONIADES 1907–9: Eugenios M. Antoniades, *Ἔκφρασις τῆς Ἁγίας Σοφίας*, 3 vols. (Athens: Teubner, 1907–9).

ANTONOPOULOU 2015: Theodora Antonopoulou, "Philagathos Kerameus and Emperor Leo VI: On a Model of the Ecphrasis of the Cappella Palatina in Palermo," *Νέα Ῥώμη* 12 (2015): 115–27.

APOLLONI GHETTI 1977: Fabrizo M. Apolloni Ghetti, "Santi Ghetti Scalpellino e Imprenditore," *L'urbe* 40, nos. 3–4 (1977): 22–39.

ARASSE 1999: Daniel Arasse, *L'annonciation italienne: Une histoire de perspective* (Paris: Hazan, 1999).

ARDUINI 1995: Francesco Arduini, "Due inediti paragoni," *Labyrinthos* 14, nos. 27–28 (1995): 203–17.

ARGOUD ET AL. 1980: Gilbert Argoud, Olivier Callot, and Bruno Helly, *Une résidence byzantine "L'huilerie"* (Paris: E. de Boccard, 1980).

ARMANI AND PIANA 1984: Emanuele Armani and Mario Piana, "Primo inventario degli intonaci e delle decorazioni esterne dell'architettura veneziana," *Ricerche di Storia dell'Arte* 9, no. 24 (1984): 44–54.

ARSLAN 1970: Edoardo Arslan, *Venezia gotica. L'architettura civile gotica veneziana* (Milan: Electa, 1970).

ARUZ AND WALLENFELS 2003: Joan Aruz and Ronald Wallenfels, eds., *Art of the First Cities: The Third Millennium B.C. from the Mediterranean to the Indus* (New York/New Haven: Metropolitan Museum of Art and Yale University Press, 2003).

ARUZ ET AL. 2013: Joan Aruz, Sarah B. Graff, and Yelena Rakic, eds., *Cultures in Contact: From Mesopotamia to the Mediterranean in the Second Millennium B.C.* (New York/New Haven: Metropolitan Museum of Art and Yale University Press, 2013), 170–79.

ASCHE 1978: Sigfried Asche, *Balthasar Permoser: Leben und Werk* (Berlin: Deutscher Verlag für Kunstwissenschaft, 1978).

ASEMAKOPOULOU-ATZAKA 1980: Panagiota Asemakopoulou-Atzaka, *Ἡ τεχνική opus sectile στὴν ἐντοιχία διακόσμηση* (Thessaloniki: Kentron Vyzantinon Ereunon, 1980).

ASGARI 1992: Nusin Asgari, "Observations on Two Types of Quarry-Items from Proconnesus: Column-Shafts and Column-Bases," in *Ancient Stones: Quarrying, Trade and Provenance*, ed. Marc Waelkens,

Norman Herz, and Luc Moens (Leuven: Leuven University Press, 1992), 73–80.

ASGARI 1995: Nusin Asgari, "The Proconnesian Production of Architectural Elements in Late Antiquity, Based on Evidence from the Marble Quarries," in *Constantinople and Its Hinterland*, ed. Cyril Mango, Gilbert Dagron, and Geoffrey Greatrex (Aldershot: Variorum, 1995), 263–88.

ASHBY AND LUGLI 1932: Thomas Ashby and Giuseppe Lugli, "La basilica di Giunio Basso sull'Esquilino," *Rivista di Archeologia Cristiana* 9 (1932): 221–25.

ASMUSSEN 1986: Marianne Wirenfeldt Asmussen, "The Chapel of S. Zeno in S. Prassede in Rome: New Aspects on the Iconography," *Analecta Romana Instituti Danici* 15 (1986): 67–86.

ASPEREN DE BOER 1997: R. J. van Asperen de Boer, *Jan van Eyck: Two Paintings of Saint Francis Receiving the Stigmata* (Philadelphia: Philadelphia Museum of Art, 1997).

ASUTAY-EFFENBERGER 2007: Neslihan Asutay-Effenberger, *Die Landmauer von Konstantinopel-Istanbul: Historisch-topographische und baugeschichtliche Untersuchungen* (Berlin: de Gruyter, 2007).

ASUTAY-EFFENBERGER AND EFFENBERGER 2006: Neslihan Asutay-Effenberger and Arne Effenberger, *Die Porphyrsarkophage der oströmischen Kaiser. Versuch einer Bestandserfassung, Zeitbestimmung und Zuordnung* (Wiesbaden: Reichert, 2006).

AUBERT AND ERISTOV 2001: Catherine Aubert and Hélène Eristov, *L'habitat hellénistique de Beyrouth et son décor: Site du petit sérail* (Paris: Errance, 2001).

AURIGEMMA 1997: Maria G. Aurigemma, "Note sulla diffusione del vocabolario architettonico: Francesco Patrizi," in *Le due Rome del Quattrocento: Melozzo, Antoniazzo e la cultura artistica del '400 romano*, ed. Sergio Rossi and Stefano Valeri (Rome: Lithos, 1997), 364–79.

AURIGEMMA AND CAVALLARO 1999: Maria Giulia Aurigemma and Anna Cavallaro, *Il Palazzo di Domenico della Rovere in Borgo* (Rome: Istituto Poligrafico dello Stato, 1999).

AUTENRIETH 1997: Hans Peter Autenrieth, "Structures ornementales et ornements à motifs structuraux: Les appareils peints jusqu'à l'époque romane," in *Le rôle de l'ornement dans la peinture murale du Moyen Age*, ed. John Ottaway (Poitiers: Universite de Poitiers Centre national de la recherche scientifique, 1997), 57–72.

AZZOLINI 1994: Lidia Azzolini, *Palazzi del Quattrocento a Cremona* (Cremona: Turris, 1994).

BABINGER 1956: Franz C. H. Babinger, *Reliquienschacher am Osmanenhof im XV. Jahrhundert: Zugleich ein Beitrag zur Geschichte der osmanischen Goldprägung unter Mehmed II, dem Eroberer* (Munich: Bayerische Akademie, 1956).

BABINGER 1961: Franz Babinger, *Johannes Darius (1414–1494) Sachwalter Venedigs im Morgenland und sein griechischer Umkreis* (Munich: Verlag der Bayerischen Akademie der Wissenschaften, 1961).

BABINGER 1962: Franz Babinger, "Notes on Cyriac of Ancona and Some of His Friends," *Journal of the Warburg and Courtauld Institutes* 25 (1962): 321–23.

BACCHELLI ET AL. 1995: Barbara Bacchelli, Mariarosaria Barbera, Rita Pasqualucci, and Lucia Saguì, "Nuove scoperte sulla provenienza dei pannelli in opus sectile vitreo della collezione Gorga," in *Atti del II colloquio dell'AISCOM*, ed. Irene Bragantini and Federico Guidobaldi (Bordighera: Istituto Internazionale di Studi Liguri, 1995), 447–66.

BADHAM 2009: Sally Badham, "The Use of Sedimentary 'Marbles' for Church Monuments in Pre-Reformation England," *Church Archaeology* 11 (2009): 1–18.

BAGATTI 1984: Bellarmino Bagatti, *Gli scavi di Nazaret* (Gerusalemme: Tip. dei PP. Francescani, 1984).

BALDASSARI 2012: Paola Baldassari, "Archaeological Excavations at Palazzo Valentini: A Residential Area in the Shade of the [*sic*] Trajan's Forum," in *11th International Colloquium on Ancient Mosaics*, ed. Mustafa Şahin, Uludağ (Istanbul: Phoibos, 2012), 43–67.

BALDINI ET AL. 1979: Umberto Baldini, Anna Maria Giusti, and Annapaula Pampaloni Martelli, eds., *La Cappella dei Principi e le pietre dure a Firenze* (Milan: Electa, 1979).

BALDINI LIPPOLIS 1994: Isabella Baldini Lippolis, "Case e palazzi a Costantinopoli tra IV e VI secolo," in *Ravenna, Costantinopoli, Vicino Oriente*, ed. Raffaella Farioli Campanati (Ravenna: Girasole, 1994), 279–311.

BALDINI LIPPOLIS 2001: Isabella Baldini Lippolis, *La domus tardoantica: Forme e rappresentazioni dello spazio domestico nelle città del Mediterraneo* (Imola: University Press Bologna, 2001).

BALDWIN 1990: Barry Baldwin, "The Date, Identity, and Career of Vitruvius," *Latomus* 49, no. 2 (1990): 425–34.

BALDWIN SMITH 1956: E. Baldwin Smith, *Architectural Symbolism of Imperial Rome and the Middle Ages* (Princeton: Princeton University Press, 1956).

BALL 1938: Sydney H. Ball, "Luminous Gems, Mythical and Real," *Scientific Monthly* 47, no. 6 (1938): 496–505.

BALL 2002: Larry F. Ball, "How Did the Romans Install Revetments?" *American Journal of Archaeology* 106, no. 4 (2002): 551–74.

BALMELLE AND BEN ABED 2011: Abed Catherine Balmelle and Aïcha Ben Abed, "Pavements en opus sectile de Jebel Oust a effet de miroir," in *Marmoribus Vestita*, ed. Olof Brandt and Philippe Pergola (Vatican City: Pontificio Istituto di Archeologia Cristiana, 2011), 81–101.

BALTY 2004: Janine Balty, "L'opus sectile pariétal de la Maison aux Consoles d'Apamée," *Musiva e Sectilia* 1 (2004): 159–84.

BANDMANN 1970: Günter Bandmann, "Höhle und Säule auf Darstellungen Mariens mit dem Kinde," in *Festschrift für Gert von der Osten* (Cologne: DuMont Schauberg, 1970), 130–48.

BANKEL AND LIVERANI 2004: Hansgeorg Bankel and Paolo Liverani, eds., *I colori del bianco: Policromia nella scultura antica* (Rome: De Luca, 2004).

BARATTE ET AL. 2004: François Baratte, Neyla Ouertani, and Hassen Abdallah, "Un curieux ensemble sculpté de l'Afrique romaine: Les piliers de Meninx," in *Studi di archeologia in onore di Gustavo Traversi*, ed. Manuela Fano Santi (Rome: Bretschneider, 2004), 47–62.

BARBET AND VERBANCK-PIÉRARD 2013: Alix Barbet and Annie Verbanck-Piérard, eds., *La Villa Romaine de Boscoreale et ses fresques*, 2 vols. (Arles: Errance, 2013).

BARBIER DE MONTAULT 1885: Xavier Barbier de Montault, *Le miracle de Bolsène et le saint corporal d'Orvieto* (Lyon: X. Jevain, 1885).

BARBIER DE MONTAULT 1888: Xavier Barbier de Montault, "Inventaires de Saint-Pierre de Rome, II–IV," *Revue de l'art chrétien* 6 (1888): 486–87.

BARBIERI 1968: Franco Barbieri, *La Basilica palladiana* (Vicenza: CISA, 1968).

BARBIERI 2014: Costanza Barbieri, *Le "Magnificenze" di Agostino Chigi Collezioni e passioni antiquarie nella Villa Farnesina* (Rome: Accademia Nazionale dei Lincei, 2014).

BARBIERI ET AL. 1956: Franco Barbieri, Renato Cevese, and Liciso Magagnato, *Guida di Vicenza*, 2nd ed. (Vicenza: Eretenia, 1956).

BARCHAM 1993: William Barcham, "Some New Documents on Federico Cornaro's Two Chapels in Rome," *Burlington Magazine* 135, no. 1089 (1993): 821–22.

BARCHAM 2001: William L. Barcham, *Grand in Design: The Life and Career of Federico Cornaro, Prince of the Church, Patriarch of Venice and Patron of the Arts* (Venice: Istituto Veneto, 2001).

BARGEBUHR 1956: Frederick P. Bargebuhr, "The Alhambra Palace in the Eleventh Century," *Journal of the Warburg and Courtauld Institutes* 19 (1956): 192–258.

BARLETTA 2001: Barbara A. Barletta, *The Origins of the Greek Architectural Orders* (New York: Cambridge University Press, 2001).

BARLETTA 2009: Barbara A. Barletta, "The Greek Entablature and Wooden Antecedents," in *Koine: Mediterranean Studies in Honor of R. Ross Holloway*, ed. Derek B. Counts and Anthony S. Tuck (Oxford: Oxbow, 2009), 153–65.

BARLETTI 2011: Emanuele Barletti, ed. *Giovan Antonio Dosio da San Gimignano: Architetto e scultor fiorentino tra Roma, Firenze e Napoli* (Florence: Edifir, 2011).

BAROCCHI 2001: Paola Barocchi, "Die Wettstreit zwischen Malerei und Skulptur: Benedetto Varchi und Vincenzo Borghini," in *Ars et scriptura*, ed. Hannah Baader, Ulrich Müller Hofstede, Kristine Patz, and Nicola Suthor (Berlin: Gebr. Mann, 2001), 93–106.

BARRAL I ALTET 1985: Xavier Barral i Altet, *Les mosaïques de pavement médiévales de Venise, Murano, Torcello* (Paris: Picard, 1985).

BARRAL I ALTET 1987: Xavier Barral i Altet, "Poésie et iconographie: Un pavement du XIIe siècle décrit par Baudri de Bourgueil," *Dumbarton Oaks Papers* 41 (1987): 55–70.

BARROERO 1983: Liliana Barroero, ed., *Via dei Fori Imperiali. La zona archeologica di Roma: Urbanistica, beni artistici e politica culturale* (Venice: Marsilio, 1983).

BARRY (FORTHCOMING): Fabio Barry, "Santa Maria dei Miracoli (1481–89), the Immaculate Conception, and the Architectural Imaginary of Pietro Lombardo" (forthcoming).

BARRY 1995: Fabio Barry, "'I marmi loquaci': Painting in Stone," *Daidalos* 56 (1995): 106–21.

BARRY 1999: Fabio Barry, "'Pray to thy Father which is in secret': The Tradition of Coretti, Romitorii and Lanfranco's Hermit Cycle at the Palazzo Farnese," in *Barocke Inszenierung*, ed. Joseph Imorde, Fritz Neumeyer, and Tristan Weddigen (Emsdetten: Imorde, 1999), 191–221.

BARRY 2002: Fabio Barry, "Lux and Lumen: Real and Represented Light in Baroque Architecture," *Kritische Berichte* 4 (2002): 22–37.

BARRY 2004A: Fabio Barry, review of De Nuccio and Ungaro 2002, *Journal of the Society of Architectural Historians* 63, no. 3 (2004): 378–80.

BARRY 2004B: Fabio Barry, "New Documents on the Decoration of Bernini's Fonseca Chapel," *Burlington Magazine* 146, no. 1215 (2004): 396–99.

BARRY 2007: Fabio Barry, "Walking on Water: Cosmic Floors in Antiquity and the Middle Ages," *Art Bulletin* 89, no. 4 (2007): 627–57.

BARRY 2009: Fabio Barry, "Sculpture in Painting/Painting in Sculpture: c. 1485–c. 1660," in *Sculpture in Painting*, ed. Penelope Curtis (Leeds: Henry Moore Foundation, 2009), 13–19.

BARRY 2010: Fabio Barry, "Disiecta Membra: Ranieri Zeno, the Imitation of Constantinople, the Spolia Style, and Justice at San Marco," in *San Marco, Byzantium, and the Myths of Venice*, ed. Henry Maguire and Robert Nelson (Washington, DC: Dumbarton Oaks, 2010), 7–62.

BARRY 2011A: Fabio Barry, "A Whiter Shade of Pale: Relative and Absolute White in Roman Sculpture and Architecture," in *Revival and Invention: Sculpture through Its Material Histories*, ed. Sébastien Clerbois and Martina Droth (Oxford: Peter Lang, 2011), 31–62.

BARRY 2011B: Fabio Barry, "The 'Mouth of Truth' and the Forum Boarium: Oceanus, Hercules, and Hadrian," *Art Bulletin* 93, no. 1 (2011): 7–37.

BARRY 2016: Fabio Barry, "Im/material Bernini," in *Material Bernini*, ed. Evonne Levy and Carolina Mangone (Abingdon: Routledge, 2016), 39–67.

BARRY 2017: Fabio Barry, "'Painting in Stone': Early Modern Experiments in a Metamedium," *Art Bulletin* 99, no. 3 (2017): 30–61.

BARRY 2019A: Fabio Barry, "Between All'antica and Acheiropoieton: The Cappella Gregoriana in the Ekphraseis of Lorenzo Frizolio and Ascanio Valentino (1582, 1583)," in *Chapels in Roman Churches of the Cinquecento and Seicento: Form, Function, Meaning*, ed. Chiara Franceschini, Steven F. Ostrow, and Patrizia Tosini (Milan: Officina Libraria, 2019).

BARRY 2019B: Fabio Barry, "'Perché vorrebbe uscire dall'ordinario': Architecture and the *Paragone* in Virgilio Spada's Chapel Designs for Sant'Andrea della Valle (1630–34, 1662)," in *I colori del marmo*, ed. Cristiano Giometti and Cinzia M. Sicca (Pisa: University of Pisa, 2019).

BARRY AND GWYNNE (FORTHCOMING): Fabio Barry and Paul Gwynne, *Ars Artem Invocat: Architecture and Poetry in Early Modern Rome* (forthcoming).

BARSANTI 1989: Claudia Barsanti, "L'esportazione di marmi dal Proconneso nelle regioni pontiche durante il IV–VI secolo," *Rivista dell'Istituto Nazionale d'Archeologia e Storia dell'Arte* 12 (1989): 91–220.

BARSANTI 1996: Claudia Barsanti, "Il pavimento medievale del duomo di Osimo," *Atti del III colloquio dell'AISCOM*, ed. Federico Guidobaldi and Alessandra Guiglia Guidobaldi (Bordighera: Istituto Internazionale di Studi Liguri, 1996), 445–55.

BARSANTI 2012: Claudia Barsanti, "The Marble Floor of St. John Studius in Constantinople: A Neglected Masterpiece," in *11th International Colloquium on Ancient Mosaics*, ed. Mustafa Şahin (Istanbul: Phoibos, 2012), 87–98.

BARTOLI 1963: Alfonso Bartoli, *Curia Senatus: Lo scavo e il restauro* (Rome: Instituto di Studi Romani, 1963).

BARTUSIS 1995: Mark C. Bartusis, "The Functions of Archaizing in Byzantium," *Byzantinoslavica* 56, no. 2 (1995): 271–78.

BASILE 1991: Antonella Basile, "Nota sul restauro del pannello in opus sectile dalla Basilica di Giunio Basso con biga e fazioni del circo," *Bollettino di Archeologia* 7 (1991): 95–97.

BASTET 1971: Frédéric L. Bastet, "Domus Transitoria I," *Bulletin antieke beschaving* 46 (1971): 144–72.

BASTET 1972: Frédéric L. Bastet, "Domus Transitoria II," *Bulletin antieke beschaving* 47 (1972): 61–87.

BATTILOTTI 1991: Donata Battilotti, "Piazza Contarena a Udine. Uno spazio veneziano per la Serenissima," in *La Piazza, la chiesa, il parco: Saggi di storia dell'architettura (XV–XIX secolo)*, ed. Manfredo Tafuri (Milan: Electa, 1991), 9–55.

BAUER 2001: Franz Alto Bauer, "Beatitudo Temporum. Die Gegenwart der Vergangenheit im Stadtbild des spätantiken Rom," in *Epochenwandel?: Kunst und Kultur zwischen Antike und Mittelalter*, ed. Franz Alto Bauer and Norbert Zimmermann (Mainz am Rhein: Philipp von Zabern, 2001), 75–94.

BAXANDALL 1971: Michael Baxandall, *Giotto and the Orators: Humanist Observers of Painting in Italy and the Discovery of Pictorial Composition 1350–1450* (Oxford: Clarendon, 1971).

BAXANDALL 1972: Michael Baxandall, *Painting and Experience in Fifteenth Century Italy: A Primer in the Social History of Pictorial Style* (Oxford: Clarendon, 1972), 36.

BAZZOTTI 2006: Ugo Bazzotti, "La chiesa di Santa Maria della Vittoria e la pala di Andrea Mantegna," in Signorini 2006, 206–7.

BAZZOTTI ET AL. 2009: Guido Bazzotti, Stefano L'Occaso, and Francesca Vischi, eds., *Facciate dipinte nella Mantova di Andrea Mantegna* (Milan: Skira, 2009).

BEAGON 1992: Mary Beagon, *Roman Nature: The Thought of Pliny the Elder* (Oxford: Clarendon, 1992).

BECATTI 1948A: Giovanni Becatti, "Case ostiensi del tardo impero—I," *Bollettino d'arte* 33, no. 2 (1948): 102–28.

BECATTI 1948B: Giovanni Becatti, "Case ostiensi del tardo impero—II," *Bollettino d'arte* 33, no. 3 (1948): 197–224.

BECATTI 1961: Giovanni Becatti, "Incrostazione," in *Enciclopedia dell'arte antica classica e orientale* (Rome: Istituto Poligrafico dello Stato, 1961), 130–33.

BECATTI 1969: Giovanni Becatti, *Edificio con opus sectile fuori Porta Marina* (Rome: Istituto Poligrafico dello Stato, 1969).

BECK 2012: Thomas E. Beck, "Gardens as a 'Third Nature': The Ancient Roots of a Renaissance Idea," *Studies in the History of Gardens and Designed Landscapes* 22, no. 4 (2012): 327–34.

BECKMAN 2002: Gary Beckman, "'My Sun-God': Reflections of Mesopotamian Conceptions of Kingship among the Hittites," in *Ideologies as Intercultural Phenomena*, ed. Antonio Panaino and Giovanni Pettinato (Milan: Università di Bologna, 2002), 15–21.

BEGGIAO 1978: Diego Beggiao, *La visita pastorale di Clemente VIII (1592–1600): Aspetti di riforma post-Tridentina a Roma* (Rome: Pontificia Università Lateranense, 1978).

BEGHELLI 2011: Michelle Beghelli, "La basilica di San Giulio, la chiesa di San Lorenzo e la cristianizzazione della zona del Cusio: Fonti letterarie e archeologiche," in *Esglésies rurals a Catalunya entre l'Antiguitat i l'Edat Mitjana (segles V–X): Taula rodona, Esparreguera-Montserrat, 25–27 d'Octubre de 2007*, ed. Oriol Achón (Bologna: BraDypUS, 2011), 161–76.

BEHNE 1912: Adolf Behne, *Der Inkrustationsstil in Toscana* (Berlin: Ebering, 1912).

BEJARANO OSORIO 2002: Ana María Bejarano Osorio, "Una ampolla de vidrio decorada con la planta topográfica de la ciudad de Puteoli," *Mérida, excavaciones arqueológicas* 8 (2002): 513–32.

BELLESI 1998: Sandro Bellesi, *Stefano della Bella: Otto dipinti su pietra paesina* (Florence: Falciani, 1998).

BELLI 1996: Gianluca Belli, "Forma e naturalità nel bugnato fiorentino del Quattrocento," *Quaderni del Palazzo Te* 4 (1996): 9–36.

BELLUZZI 1998: Amedeo Belluzzi, *Palazzo Te a Mantova* (Modena: Panini, 1998).

BELOTTI 1940: Bortolo Belotti, *Storia di Bergamo e dei bergamaschi*, 2 vols. (Milan: Ceschina, 1940).

BELTING 1962: Hans Belting, "Studien zum Beneventanischen Hof im 8. Jahrhundert," *Dumbarton Oaks Papers* 16 (1962): 141–93.

BELTING 1984: Hans Belting, "Das Aachener Münster im 19. Jahrhundert: Zur ersten Krise des Denkmal-Konzepts," *Wallraf-Richartz-Jahrbuch* 45 (1984): 257–90.

BELTING ET AL. 1978: Hans Belting, Cyril Mango, and Doula Mouriki, *The Mosaics and Frescoes of St. Mary Pammakaristos (Fethiye Camii) at Istanbul* (Washington, DC: Dumbarton Oaks, 1978).

BELTRAMI 1900: Luca Beltrami, *La "Ca del Duca" sul Canal Grande ed altre reminiscenze sforzesche in Venezia* (Milan: U. Allegretti, 1900).

BENESOVSKÁ AND HLOBIL 1999: Klára Benesovská and Ivo Hlobil, *Peter Parler and St. Vitus's Cathedral 1356–1399* (Prague: Prague Castle Administration, 1999).

BENJAMIN 2003: Walter Benjamin, *The Origin of German Tragic Drama* (London: Verso, 2003).

BENTINI AND BORELLA 2002: Jadranka Bentini and Marco Borella, eds., *Il Castello Estense* (Viterbo: BetaGamma, 2002).

BENTIVOGLIO 1987: Enzo Bentivoglio, "La presenza di Baldassare Peruzzi nei lavori della casa di Blosio Palladio," in *Baldassare Peruzzi: Pittura, scena e architettura nel Cinquecento*, ed. Maria Luisa Madonna and Marcello Fagiolo (Rome: Enciclopedia Italiana, 1987), 193–204.

BENTIVOGLIO 1994: Enzo Bentivoglio, "Due libri di patenti dei 'Maestri di strade' di Roma degli anni 1641–45 e 1646–54. I Mss no 131 e no 142 dell'Archivio Doria Pamphili (II)," *Quaderni PAU* IV.8, 2, no. 2 (1994): 11–62.

BENZI AND VINCENTI MONTANARO 1997: Fabio Benzi and Caroline Vincenti Montanaro, *Palazzi di Roma* (Venice: Arsenale, 1997).

BERETTA 2009: Marco Beretta, *The Alchemy of Glass: Counterfeit, Imitation, and Transmutation in Ancient Glassmaking* (Sagamore Beach, MA: Science History, 2009).

BERGMANN 1995: Bettina Bergmann, "Greek Masterpieces and Roman Recreative Fictions," *Harvard Studies in Classical Philology*, Greece in Rome: Influence, Integration, Resistance 97 (1995): 79–120.

BERGMANN 2002: Bettina Bergmann, "Art and Nature in the Villa at Oplontis," *Journal of Roman Archaeology*, ed. C. Stein and J. H. Humphrey, supp. ser. no. 47 (2002): 87–120.

BERNARD 1996: Paul Bernard, "III. L'Aornos bactrien et l'Aornos indien. Philostrate et Taxila: Géographie, mythe et réalité," *Topoi* 6, no. 2 (1996): 475–530.

BERNARD 2010: Seth G. Bernard, "Pentelic Marble in Architecture at Rome and the Republican Marble Trade," *Journal of Roman Archaeology* 2, no. 1 (2010): 35–54.

BERNARDINI ET AL. 1991: Virginia Bernardini, Andreina Draghi, and Guia Verdesi, *SS. Domenico e Sisto* (Rome: Palombi, 1991).

BERRA 1999: Giacomo Berra, "Immagini casuali, figure nascoste e natura antropomorfica nell'immaginario artistico rinascimentale," *Mitteilungen des Kunsthistorischen Instituts in Florenz* 43, nos. 2/3 (1999): 358–419.

BERRANGER 1983: Danièle Berranger, "Le relief inscrit en l'honneur des Nymphes dans les carrieres de Paros," *Revue des études anciennes* 85, nos. 3–4 (1983): 235–60.

BERTELLI 1992: Fabio Bertelli, "Metodo storico alla prova nella ricostruzione del paramento lapideo dell'abside di San Vitale," *OPD Restauro* 4 (1992): 216–29.

BERTINI CALOSSO 1953: Achille Bertini Calosso, "Gli affreschi del Pintoricchio nel Duomo di Spoleto," *Rivista dell'Istituto Nazionale d'Archeologia e Storia dell'Arte* 2 (1953): 303–28.

BETANCOURT ET AL. 1979: Philip P. Betancourt, Thomas K. Gaisser, Eugene Koss, Robert F. Lyon, Frederick R. Matson, Steven Montgomery, George H. Myer, and Charles P. Swann, *Vasilike Ware: An Early Bronze Age Pottery Style in Crete: Results of the Philadelphia Vasilike Ware Project* (Göteborg: Paul Åström, 1979).

BETORI AND MARI 2005: Alessandro Betori and Zaccaria Mari, "Sectilia Pavimenta di Villa Adriana: Un primo aggiornamento a dieci anni dal Corpus," in *Atti del X colloquio dell'AISCOM*, ed. Claudia Angelelli (Tivoli: Scripta Manent, 2005), 777–92.

BETTINI 2001: Sergio Bettini, "'Opus tectorium,' 'opus albarium' 'gypsum': Note sullo stucco romano tratte dalle fonti antiche," in *L'arte dello stucco in Friuli nei secoli XVII–XVIII*, ed. Giuseppe Bergamini and Paolo Goi (Udine: Udine Musei, 2001), 75–86.

BETTINI 2010: Sergio Bettini, "In biblioteca: Sisto IV, il Platina e l'architettura dipinta nell'affresco di Melozzo da Forlì," *Saggi e memorie di storia dell'arte* 34 (2010): 55–66.

BEVAN 2007: Andrew Bevan, *Stone Vessels and Values in the Bronze Age Mediterranean* (Cambridge: Cambridge University Press, 2007).

BEVILACQUA 1993: Mario Bevilacqua, *Santa Caterina da Siena a Magnanapoli: Arte e storia di una comunità religiosa romana nell'età della Controriforma* (Rome: Gangemi, 1993).

BEZERRA DE MENESES 1970: Ulpiano T. Bezerra de Meneses, "Les peintures," in Bruneau 1970, 151–93.

BIANCHI 2000: Fulvia Bianchi, "Domus delle Sette Sale. L'opus sectile parietale dell'aula basilicale, studi preliminari," in *Atti del VI colloquio dell'AISCOM* (Ravenna: Girasole, 2000), 351–60.

BIANCO 1992: Ludmilla Bianco, *Le pietre mirabili: Magia e scienza nei lapidari greci* (Palermo: Sellerio, 1992).

BIASIOTTI 1915: Giovanni Biasiotti, "La Basilica di S. Maria Maggiore di Roma prima delle innovazioni del secolo XVI," *Mélanges d'archéologie et d'histoire de l'École française de Rome* 35 (1915): 15–40.

BIETAK 2013: Manfred Bietak, "The Impact of Minoan Art on Egypt and the Levant: A Glimpse of Palatial Art from the Naval Base of Peru-Nefer at Avaris," in Aruz et al. 2013, 188–99.

BIETAK ET AL. 2012/13: Manfred Bietak, Constance von Rüden, Johannes Becker, Johannes Jungfleisch, Lyvia Morgan, and Erico Peintner, "Preliminary Report of the Tell El-Dabca Wall Painting Project—Season 2011/2012," *Egypt and the Levant* 22/23 (2012/13): 127–43.

BINGÖL 1997: Orhan Bingöl, *Malerei und Mosaik der Antike in der Türkei* (Mainz: Zabern, 1997).

BINSKI 1995: Paul Binski, *Westminster Abbey and the Plantagenets: Kingship and the Representation of Power, 1200–1400* (New Haven: Yale University Press, 1995).

BISCARO 1899A: Gerolamo Biscaro, "Note storico-artistiche sulla Cattedrale di Treviso II. La Cappella del Santissimo," *Nuovo Archivio Veneto* 18 (1899): 179–97.

BISCARO 1899B: Gerolamo Biscaro, "La tomba di Pietro di Dante a Treviso," *L'arte* 2 (1899): 417–31.

BISCONTI 1996: Fabrizio Bisconti, "Considerazioni iconologiche sulla decorazione musiva dei cosidetti 'oratori' di Aquileia," in *Atti del III colloquio dell'AISCOM*, ed. Federico Guidobaldi and Alessandra Guiglia Guidobaldi (Bordighera: Istituto Internazionale di Studi Liguri, 1996), 273–86.

BITTERER 2013: Tobias Bitterer, "Marmorverkleidung stadtrömischer Architektur. Öffentliche Bauten aus dem 1. Jahrhundert v. Chr. bis 7. Jahrhundert n. Chr." (PhD diss., Ludwig-Maximilians-Universität, 2013).

BLACK 2001: Jeremy Black, "Amethysts," *Iraq* 63 (2001): 183–86.

BLACKMAN AND GREWE 2001: Deane R. Blackman and Klaus Grewe, "Excursus on Calx," in *Essays on Frontinus' De Aquis Urbis Romae*, ed. Deane R. Blackman and A. Trevor Hodge (Ann Arbor: University of Michigan Press, 2001), 109–15.

BLACKWELL 2014: Nicholas G. Blackwell, "Making the Lion Gate Relief at Mycenae: Tool Marks and Foreign Influence," *American Journal of Archaeology* 118 (2014): 451–88.

BLAIR 1991: John Blair, "Purbeck Marble," *in English Medieval Industries: Craftsmen, Techniques, Products,* ed. John Blair and Nigel Ramsay (London: Hambledon, 1991), 41–56.

BLAKE 1993: Peter Blake, *No Place Like Utopia: Modern Architecture and the Company We Kept* (New York: Knopf, 1993).

BLAKOLMER 1997: Fritz Blakolmer, "Minoan Wall-Painting: The Transformation of a Craft into an Art Form," in *TEXNH: Craftsmen, Craftswomen and Craftsmanship in the Aegean Bronze Age: Proceedings of the 6th International Aegean Conference: 6e Rencontre Égéenne Internationale, Philadelphia, Temple University (18–21 April 1996),* ed. Robert Laffineur and Philip P. Betancourt (Aegeum: Annales d'archéologie égéenne de l'Université de Liège, 1997), 95–105.

BLEGEN AND RAWSON 1966: Carl W. Blegen and Marion Rawson, *The Palace of Nestor at Pylos in Western Messenia,* vol. 1, *The Buildings and Their Contents* (Princeton: Princeton University Press, 1966).

BLOCH 1945: Herbert Bloch, "A New Document of the Last Pagan Revival in the West," *Harvard Theological Review* 38 (1945): 199–201.

BLOCH 1986: Herbert Bloch, *Monte Cassino in the Middle Ages,* 3 vols. (Cambridge, MA: Harvard University Press, 1986).

BOARDMAN 1967: John Boardman, "Water in the Parthenon?," *Gymnasium* 74 (1967): 508–9.

BOARDMAN 1978: John Boardman, *Greek Sculpture, the Archaic Period: A Handbook* (New York: Oxford University Press, 1978).

BOARDMAN 1999: John Boardman, *The Greeks Overseas: Their Early Colonies and Trade,* 4th ed. (London: Thames & Hudson, 1999).

BÖCKMANN 2005: Barbara Böckmann, "Il San Sebastiano di Leon Battista Alberti a Mantova. Progetto originale e modifiche successive," *Arte Lombarda,* 145, no. 3 (2005): 61–73.

BODENSTEDT 1944: Mary Immaculate Bodenstedt, *The Vita Christi of Ludolphus the Carthusian* (Washington, DC: Catholic University Press, 1944).

BODI 1991: Daniel Bodi, *The Book of Ezekiel and the Poem of Erra* (Freiburg/Göttingen: Freiburg Universitätsverlag and Vandenhoeck & Ruprecht, 1991).

BODNAR 1973: Edward W. Bodnar, "A Quarry Relief on the Island of Paros," *Archaeology* 26, no. 4 (1973): 270–77.

BOEHMER 1990: Rainer M. Boehmer, "Zur Funktion des Steinstifttempels in Uruk nach Befunden der Kampage 39," *Baghdader Mitteilungen* 21 (1990): 49–66.

BOIN 2013: Douglas Boin, *Ostia in Late Antiquity* (Cambridge: Cambridge University Press, 2013).

BOLLAND 2000: Andrea Bolland, "Desiderio and Diletto: Vision, Touch, and the Poetics of Bernini's Apollo and Daphne," *Art Bulletin* 82, no. 2 (2000): 318–22.

BOLMAN 2006: Elisabeth S. Bolman, "Late Antique Aesthetics, Chromophobia, and the Red Monastery, Sohag, Egypt," *Eastern Christian Art* 3 (2006): 1–24.

BOLZAN 1976: J. E. Bolzan, "Chemical Combination According to Aristotle," *Ambix* 23, no. 3 (1976): 134–44.

BOMBI ET AL. 1988: Giorgio Bombi, Marisa Laveroni, and Piero Lucca eds., *La biscia e l'aquila. Il castello di Vigevano: Una lettura storico-artistica* (Piacenza: Cassa di risparmio di Piacenza e Vigevano, 1988).

BONA CASTELLOTTI 2000: Marco Bona Castellotti, ed., *Pietra dipinta: Tesori nascosti del '500 e del '600 da una collezione privata milanese* (Milan: Motta, 2000).

BONACASA 1984–85: Nicola Bonacasa, "Un inedito di Achille Adriani sulla tomba di Alessandro," in *Giornate di studio in onore di Achille Adriani,* ed. Sandro Stucchi and Margherita Bonanno Aravantinos (Rome: Bretschneider, 1984–85), 5–19.

BONAMICI 1989: Marisa Bonamici, "Il marmo lunense in epoca preromana," in *Il marmo nella civiltà romana: La produzione e il commercio,* ed. Enrico Dolci (Carrara: Internazionale marmi e macchine, 1989), 83–114.

BONAMICI 1991: Marisa Bonamici, "Nuovi monumenti di marmo nell' Etruria settentrionale," *Archeologia Classica* 43 (1991): 795–817.

BONANNI 1998: Alessandro Bonanni, "Interraso Marmore (Plin. NH 35, 2): Esempi della tecnica decorativa a intarsio in età romana," in *Marmi Antichi II. Cave e tecnica di lavorazione. Provenienze e distribuzione,* ed. Patrizio Pensabene (Rome: Bretschneider, 1998), 259–92.

BONNER 1941: Campbell Bonner, "Desired Haven," *Harvard Theological Review* 34, no. 1 (1941): 49–67.

BONY 1990: Jean Bony, "The Stonework Planning of the First Durham Master," in *Medieval Architecture and Its Intellectual Context: Studies in Honour of Peter Kidson,* ed. Eric C. Fernie and Paul Crossley (London: Hambledon, 1990), 19–34.

BORA 1981: Giulio Bora, ed., *Santa Maria della Passione e il Conservatorio Giuseppe Verdi di Milano* (Milan: Silvana, 1981).

BORBEIN 1973: Adolf H. Borbein, "Die griechische Statue des 4. Jahrhunderts v. Chr. Formanalytische Untersuchungen zur Kunst der Nachklassik. Habilitationsschrift Marburg 1972/73," *Jahrbuch des Deutschen Archäologischen Instituts* 88 (1973): 43–212.

BORGHINI 1990: Gabriele Borghini, ed., *Marmi Antichi* (Rome: De Luca, 1990).

BORLINI 1963: Giampiero Borlini, "The Façade of the Certosa in Pavia," *Art Bulletin* 45, no. 4 (1963): 323–36.

BORNERT 1966: René Bornert, *Les commentaires Byzantins de la divine liturgie: Du VIIe au XVe siècle* (Paris: Institut français d'études byzantines, 1966).

BORREL 1883: E. -L. Borrel, "Étude sur l'Homélie prêchée par Saint Avit, au commencement du VIe siècle, dans la basilique de Saint-Pierre de Moutiers en Tarentaise (Savoie), à l'occasion de sa consécration," *Bulletin du Comité des travaux historiques et scientifiques* (1883): 46–55.

BORSI 1985: Stefano Borsi, *Giuliano da Sangallo: I disegni di architettura e dell'antico* (Rome: Officina, 1985): 169–73.

BORSOOK 1990: Eve Borsook, *Messages in Mosaic: The Royal Programmes of Norman Sicily (1130–1187)* (Oxford: Clarendon, 1990).

BORSOOK 2000: Eve Borsook, "Rhetoric or Reality: Mosaics as Expressions of a Metaphysical Idea," *Mitteilungen des Kunsthistorischen Institutes in Florenz* 44, no. 1 (2000): 2–18.

BOSKOVITS AND BROWN 2003: Miklós Boskovits and David Alan Brown, *Italian Paintings of the Fifteenth Century* (New York/Washington, DC: Oxford University Press and National Gallery of Art, 2003).

BOSMAN 2004: Lex Bosman, *The Power of Tradition: Spolia in the Architecture of St. Peter's in the Vatican* (Hilversum: Uitgeverij Verloren, 2004).

BOSMAN 2005: Lex Bosman, "Spolia and Coloured Marble in Sepulchral Monuments in Florence, Rome, and Bosco Marengo. Designs by Dosio and Vasari," *Mitteilungen des Kunsthistorischen Instituts in Florenz* 49 (2005): 353–76.

BOSON 1914: Giustino Boson, *Les métaux et les pierres dans les inscriptions assyro-babyloniennes* (Munich: Akademische Buchdruckerei von F. Straub, 1914).

BOSTOCK 1995: David Bostock, "Aristotle on the Transmutation of the Elements in *De Generatione et Corruptione* 1.1–4," *Oxford Studies in Ancient Philosophy* 13 (1995): 217–29.

BOTHMER 1987: Hans-Caspar Graf von Bothmer, "Architekturbilder im Koran; eine Prachthandschrift der Ummayyadenzeit aus dem Yemen," *Pantheon* 45 (1987): 4–20.

BOULTER 1970: Patricia N. Boulter, "The Erechtheion Frieze," *Antike Plastik* 10 (1970): 7–28.

BOURAS 1982: Charalambos Bouras, *Nea Moni on Chios: History and Architecture,* trans. David A. Hardy (Athens: Commercial Bank of Greece, 1982).

BOUSQUET 1938: Jean Bousquet, "Le temple d'Aphrodite et d'Arès a Sta Lenikà," *Bulletin de correspondence hellénique* 62, no. 2 (1938): 386–408.

BOVINI 1965: Giuseppe Bovini, "Gli antichi vetri da finestra della chiesa di S. Vitale," *Felix Ravenna* 91 (1965): 98–108.

BOYLE 1991: Leonard E. Boyle, "Sixtus IV and the Vatican Library," in

Rome: Tradition, Innovation and Renewal (Victoria: University of Victoria, 1991), 65–73.

BRADLEY 2004: Mark Bradley, "The Colour 'Blush' in Ancient Rome," in *Colour in the Ancient Mediterranean World*, ed. Liza Cleland and Karen Stears (Oxford: Archaeopress, 2004), 117–21.

BRECOULAKI 2014: Harikleia Brecoulaki, "'Precious Colours' in Ancient Greek Polychromy and Painting: Material Aspects and Symbolic Values," *Revue archéologique* 57, no. 1 (2014): 1–36.

BREM 2000: Hansjörg Brem, *Das Peristylhaus 1 von Iaitas: Wand- und Bodendekorationen* (Lausanne: Payot Lausanne, 2000).

BRENK 1998: Beat Brenk, "Ostia Tardoantica," *Rivista di archeologia cristiana* 74 (1998): 523–33.

BRENK 1999: Beat Brenk, "La cristianizzazione della Domus dei Valerii sul Celio," in *The Transformations of Urbs Roma in Late Antiquity*, ed. V. W. Harris (Portsmouth, RI: Journal of Roman Archaeology, 1999), 69–84.

BRENK 2001: Beat Brenk, "La christinatisation d'Ostie," in *Ostia: Port et porte de la Rome antique*, ed. Jean-Paul Descoeudres (Geneva: Georgs, 2001), 262–71.

BRIGHTMAN 1901: Frank E. Brightman, "Byzantine Imperial Coronations," *Journal of Theological Studies* 2 (1901): 359–92.

BRILL 1991/92: Robert H. Brill, "Some Thoughts on the Origin of the Chinese Word 'Boli,'" *Silk Road Art and Archaeology* 2 (1991/92): 129–36.

BRILL 1999: Robert H. Brill, *Chemical Analyses of Ancient Glasses*, 2 vols. (Rochester, NY: Corning Museum of Glass, 1999).

BRINKMANN ET AL. 2007: Vinzenz Brinkmann, Renée Dreyfus, and Ulrike Koch-Brinkmann, eds., *Gods in Colour: Polychromy in the Ancient World* (Munich: Stiftung Archäologie, Glyptothek München, 2007).

BROCK 1995: Sebastian Brock, "The Scribe Reaches Harbour," in *Bosphorus*, ed. Stephanos Efthymiadis, Claudia Rapp, and Dimitris Tsougarakis (Amsterdam: Hakkert, 1995), 195–202.

BROILO 2009: Federica A. Broilo, "'Cleanses the Sins with the Water of the Pure-Flowing Font': Fountains for Ablutions in the Byzantine Constantinopolitan Context," *Revue des études sud-est européennes* 47, no. 1–4 (2009): 5–24.

BROWN 1987: Robert Brown, "The Palace of the Sun in Ovid's Metamorphoses," in *Homo Viator: Classical Essays for John Bramble*, ed. Michael Whitby, Philip Hardie, and Mary Whitby (Bristol: Bristol Classical and Bolchazy-Carducci, 1987), 211–20.

BROWN 1989–90: Clifford M. Brown, "Tullio Lombardo and Mantua: An Inlaid Marble Pavement for Isabella d'Este's Grotta and a Marble Portal for the Studiolo," *Arte Veneta* 43 (1989–90): 121–30.

BRUNEAU 1970: Philippe Bruneau, *L'Îlot de la Maison des Comédiens*, Exploration archéologique de Délos, no. 27 (Paris: E. de Boccard, 1970).

BRUNEAU 1988: Philippe Bruneau, "Deliaca 55. Une idée de plus sur l'inscription archaïque de la base du colosse," *Bulletin de correspondance hellénique* 112 (1988): 577–82.

BRUNEAU AND DUCAT 1983: Philippe Bruneau and Jean Ducat, *Guide de Délos*, 3rd ed. (Paris: École française d'Athènes and E. de Boccard, 1983), 127–28.

BRUNO 1969: Vincent J. Bruno, "Antecedents to the Pompeian First Style," *American Journal of Archaeology* 73 (1969): 305–17.

BRUNO 1985: Vincent J. Bruno, *Hellenistic Painting Techniques: The Evidence of the Delos Fragments* (Leiden: Brill, 1985).

BRUNO 2000: Matthias Bruno, "The Results of a Field Survey on Paros," in *Paria Lithos: Parian Quarries, Marble and Workshops of Sculpture*, ed. Demetrius U. Schilardi, Stella Katsarou, Dora Katsonopoulou, and Carla McKinney Brenner (Athens: Demetrius U. Schilardi, 2000), 91–94.

BRUNO 2002: Matthias Bruno, "Alabaster Quarries near Hierapolis (Turkey)," in *Interdisciplinary Studies on Ancient Stone*, ed. Lorenzo Lazzarini (Padua: Bottega d'Erasmo, 2002), 19–24.

BRUNO ET AL. 2002: Matthias Bruno, Stefano Cancelliere, Carlo Gorgoni, Lorenzo Lazzarini, Paolo Pallante, and Patrizio Pensabene, "Provenance and Distribution of White Marbles in Temples and Public Buildings of Imperial Rome," in Hermann et al. 2002, 289–300.

BRUTO AND VANNICOLA 1990: Maria L. Bruto and Cinzia Vannicola, "Ricostruzione e tipologia delle crustae parietali in età imperiale," *Archeologia Classica* 42 (1990): 287–376.

BRZOZOWSKA 2012: Zofia Brzozowska, "The Church of Divine Wisdom or of Christ–the Incarnate Logos? Dedication of Hagia Sophia in Constantinople in the Light of Byzantine Sources," *Studia Ceranea* 2 (2012): 85–96.

BUCCHERI 2014: Alessandra Buccheri, *The Spectacle of Clouds, 1439–1650: Italian Art and Theatre* (Farnham: Ashgate, 2014).

BUCKLOW 2009: Spike Bucklow, *The Alchemy of Paint: Art, Science and Secrets from the Middle Ages* (London: Marion Boyars, 2009).

BUCKTON 1984: David Buckton, ed., *The Treasury of San Marco, Venice* (Milan: Olivetti, 1984).

BUDDE 1997: Michael Budde, *Altare Portatile: Kompendium der Tragaltäre des Mittelalters, 600–1600* (Werne an der Lippe: Budde, 1997).

BUDETTA 1987: Tommasina Budetta, "Attivita dell'ufficio scavi: 1985–1986," *Rivista di Studi Pompeiani* 1 (1987): 194, 196–99.

BUGINI ET AL. 2002: R. Bugini, L. Folli, and C. Ferrario, "A Glossary of Some Coloured Marbles–Comparison between Ancient and Modern Terms," in *Interdisciplinary Studies on Ancient Stone*, ed. Lorenzo Lazzarini (Padua: Bottega d'Erasmo, 2002), 177–82.

BÜHLER 1973: Hans-Peter Bühler, *Antike Gefässe aus Edelsteinen* (Mainz: Zabern, 1973).

BÜHRER-THIERRY 2004: Geneviève Bührer-Thierry, "Lumière et pouvoir dans le haut Moyen Âge occidental: Célébration du pouvoir et métaphores lumineuses," *Mélanges de l'École française de Rome. Moyen Âge* 116, no. 2 (2004): 521–56.

BULARD 1908: Marcel Bulard, *Peintures murales et mosaïques se Délos* (Paris: Fondation Eugène Piot, 1908).

BULGARELLI 2006: Massimo Bulgarelli, "Mantova: 1470 e dintorni. Alberti, Fancelli, Mantegna," in *Leon Battista Alberti e l'architettura*, ed. Massimo Bulgarelli, Arturo Calzona, Matteo Ceriana, and Francesco Paolo Fiore (Cinisello Balsamo: Silvana, 2006), 120–47.

BULGARELLI 2008: Massimo Bulgarelli, *Leon Battista Alberti, 1404–1472. Architettura e storia* (Milan: Electa, 2008).

BULGARELLI 2016: Massimo Bulgarelli, "Bianco e colori. Sigismondo Malatesta, Alberti, e l'architettura del Tempio Malatestiano," *Opus Incertum* 2 (2016): 48–57.

BURANELLI 1987: Francesco Buranelli, ed., *La tomba François di Vulci* (Rome: Quasar, 1987).

BURKERT 1979: Walter Burkert, "Kynaithos, Polycrates and the Homeric Hymn to Apollo," in *Arktouros: Hellenic Studies Presented to B. M. W. Knox*, ed. Glen W. Bowersock, Walter Burkert, and Michael C. J. Putnam (Berlin: De Gruyter, 1979), 53–62.

BURNETT AND SCHOFIELD 1997: Andrew Burnett and Richard V. Schofield, "The Medallions of the Basamento of the Certosa di Pavia: Sources and Influence," *Arte Lombarda* 120 (1997): 5–27.

BURNETT AND SCHOFIELD 1998: Andrew Burnett and Richard V. Schofield, "An Introduction to the Portrait Medallions on the Certosa di Pavia," in *The Image of the Individual: Portraits in the Renaissance*, ed. Nicholas Mann and Luke Syson (London: British Museum, 1998), 55–66.

BURNS 1971: Howard Burns, "Quattrocento Architecture and the Antique: Some Problems," in *Classical Influences on European Culture A.D. 500–1500*, ed. Robert R. Bolgar (London: Cambridge University Press, 1971), 269–87.

BURNS 1984: Howard Burns, "Raffaello e 'quell'antiqua architectura,'" in Frommel et al. 1984, 381–404.

BUTLER 1994: David Butler, "Orazio Spada and His Architects: Amateurs and Professionals in Late-Seventeenth-Century Rome," *Journal of the Society of Architectural Historians* 53, no. 1 (1994): 61–79.

BUTTERS 1996: Suzanne B. Butters, *The Triumph of Vulcan: Sculptor's Tools, Porphyry, and the Prince in Ducal Florence* (Florence: Olschki, 1996).

BUTZ 2009: Patricia A. Butz, "The Naxian Colossus at Delos: 'Same Stone,'" in *Asmosia VII: Proceedings of the 7th International Conference of Association for the Study of Marble and Other Stones in Antiquity*, ed. Yannis Maniatis (Athens: École française d'Athènes, 2009), 77–87.

CAGIANO DE AZEVEDO 1967–68: Michelangelo Cagiano de Azevedo, "La datazione delle tarsie della Basilica di Giunio Basso," *Rendiconti della Pontificia Accademia Romana di Archeologia* 40 (1967–68): 151–70.

CAILLOIS 1963: Roger Caillois, "D'étranges délires de marbre," *Connaissance des arts* 139 (1963): 36–43.

CALAMIOTOU ET AL. 1983: Maria Calamiotou, Maria Siganidou, and Sophokles E. Filippakis, "X-Ray Analysis of Pigments from Pella, Greece," *Studies in Conservation* 28 (1983): 117–21.

CALCAGNO 2001: Daniele Calcagno, "Il Sacro Catino specchio dell'identità genovese," *Xenia Antiqua* 10 (2001): 43–112.

CALDWELL 2011: Dorigen Caldwell, "A Neglected Papal Commission in Naples Cathedral: The Tomb of Cardinal Alfonso Carafa," *Burlington Magazine* 153, no. 1304 (2011): 712–17.

CALÒ 1992: Giuseppe Calò, *22 Dispacci: Da Costantinopoli al Doge Giovanni Mocenigo* (Venice: Corbo e Fiore, 1992).

CALÒ MARIANI 1995: Maria Stella Calò Mariani, "Castel del Monte: La veste ornamentale," in *Federico II. Immagine e potere*, ed. Maria Stella Calò Mariani and Raffaella Cassano (Venice: Marsilio, 1995), 305–12.

CALZA 1953: Guido Calza, ed., *Topografia Generale* (Rome: Libreria dello Stato, 1953).

CALZONA AND VOLPI GHIRARDINI 1994: Arturo Calzona and Livio Volpi Ghirardini, *Il San Sebastiano di Leon Battista Alberti* (Florence: Olschki, 1994).

CAMERON 1965: Averil Cameron, "Procopius and the Church of S. Sophia," *Harvard Theological Review* 58, no. 1 (1965): 161–63.

CAMERON 1984: Alan Cameron, "The Pervigilium Veneris," in *La Poesia tardoantica: Tra retorica, teologia e politica* (Messina: Centro di studi umanistici, 1984), 229–31.

CAMERON ET AL. 1977: M. A. S. Cameron, R. E. Jones, and S. E. Philippakis, "Scientific Analyses of Minoan Fresco Samples from Knossos," *Annual of the British School at Athens* 72 (1977): 121–84.

CÄMMERER-GEORGE 1966: Monika Cämmerer-George, *Die Rahmung der toskanischen Altarbilder im Trecento* (Strasbourg: P. H. Heitz, 1966).

CAMPBELL THOMPSON 1938: Reginald Campbell Thompson, "A Survey of the Chemistry of Assyria in the Seventh Century B.C.," *Ambix* 2, no. 1 (1938): 3–16.

CAMUS 1989: Marie-Thérèse Camus, "Capitelli e colonne dipinte. Antecedenti dell'XI secolo nella Francia dell'Ouest," in *Wiligelmo e Lanfranco nell'Europa romanica* (Modena: Panini, 1989), 141–49.

CANAK-MEDIC 1981: Milka Canak-Medic, *Actes du Congrès international d'études byzantines, 1976* (Athens: K. Michalas, 1981), 107–20.

CANTARELLI AND BROWN 2002: Vincenzo Cantarelli and Clifford M. Brown, "La torretta di San Nicolò del castello di San Giorgio a Mantova (circa 1395–1530). Fra ipotesi e certezze," *Quaderni del Palazzo Te* 10 (2002): 79–91.

CARANDINI ET AL. 1982: Andrea Carandini, Andreina Ricci, and Mariette de Vos, *The Villa of Piazza Armerina: The Image of a Roman Aristocrat at the Time of Constantine* (Palermo: S. F. Flaccovio, 1982).

CARERI 1995: Giovanni Careri, *Bernini: Flights of Love, the Art of Devotion*, trans. Linda Lappin (Chicago: University of Chicago Press, 1995).

CARETTONI 1949: Gianfilippo Carettoni, "Costruzioni sotto l'angolo sud-ovest della 'domus flavia.' Triclinio e ninfeo occidentale," *Notizie degli scavi di antichità* (1949): 48–79.

CARETTONI 1961: Gianfilippo Carettoni, "Il fregio figurato della Basilica Emilia," *Rivista dell'Istituto Nazionale d'Archeologia e Storia dell'Arte* 9 (1961): 16–21.

CAREY 2003: Sorcha Carey, *Pliny's Catalogue of Culture: Art and Empire in the Natural History* (Oxford: Oxford University Press, 2003).

CARIGNANI AND SPINOLA 1995: Andrea Carignani and Giandomenico Spinola, "Mosaici e pavimenti marmorei dai recenti scavi sul Celio," in *Atti del II colloquio dell'AISCOM* (Bordighera: Istituto Internazionale di Studi Liguri, 1995), 403–14.

CARLONI 1999: Livia Carloni, "La Cappella Cornaro in Santa Maria della Vittoria: Nuove evidenze e acquisizioni sulla 'men cattiva opera' del Bernini," in *Gian Lorenzo Bernini. Regista del Barocco. I restauri*, ed. Claudio M. Strinati and Maria Grazia Bernardini (Milan: Skira, 1999), 37–46.

CARLUCCIO 1995: Annamaria Carluccio, "Per Fenice di Colofone," *Rudiae* 7 (1995): 125–28.

CARPEGGIANI 2002: Paolo Carpeggiani, "Oltre l'Alberti: Storia e trasformazioni del Sant'Andrea in Mantova," in *Architettura: Processualità e trasformazione*, ed. Maurizio Caperna and Gianfranco Spagnesi (Rome: Bonsignori, 2002), 272–73.

CASANOVA 1992: Maria Letizia Casanova, *Palazzo Venezia* (Rome: Editalia, 1992).

CASSETTA 2011: Roberto Cassetta, "Un raro fregio figurato di I Stile dalla Casa del Naviglio (VI, 10, 11) di Pompei," in *Pittura ellenistica in Italia e in Sicilia*, ed. Gioacchino Francesco La Torre and Mario Torelli (Rome: Bretschneider, 2011), 473–79.

CASTELLI 1988: Patrizia Castelli, "'Marmi policromi e bianchi screziati': In margine alle valutazioni estetiche del Traversari," in *Ambrogio Traversari nel VI centenario della nascita*, ed. Gian Carlo Garfagnini (Florence: Olschki, 1988), 211–24.

CAUBET 1989: Annie Caubet, "Pygmalion et la statue d'ivoire," in *Architecture et poésie dans le monde grec, hommage à Georges Roux*, ed. Roland Étienne, Marie-Thérèse Le Dinahet, and Marguerite Yon (Lyon: Maison de l'orient, 1989), 247–54.

CAUBET AND PIERRAT-BONNEFOIS 2005: Annie Caubet and Geneviève Pierrat-Bonnefois, eds., *Faïences de l'antiquité: De L'Égypte à l'Iran* (Paris/Milan: Musée du Louvre and 5 Continents, 2005).

CAUVIN ET AL. 1998: Marie-Claire Cauvin, Alain Gourgaud, Bernard Gratuze, Nicolas Arnaud, Gérard Poupeau, Jean-Louis Poidevin, and Christine Chataigner, eds., *L'obsidienne au Proche et Moyen Orient: Du volcan à l'outil* (Oxford: Archaeopress, 1998).

CAVARI AND DONATI 2015: Fernanda Cavari and Fulvia Donati, "Rappresentazioni e composizione delle imitazioni marmoree nella pittura di I Stile dall'etruria romana," in *Antike Malerei zwischen Lokalstil und Zeitstil*, ed. Norbert Zimmermann (Vienna: Verlag der Österreichischen Akademie der Wissenschaften, 2015), 63–74.

CAVARI ET AL. 2015: Fernanda Cavari, Francesca Droghini, Marco Giamello, Cynthia Mascione, and Andrea Scala, "The Imitation of Coloured Marbles in a First Style Wall Painting from the Etruscan-Roman Town of Populonia (Li—Italy)," in *Asmosia X*, ed. Patrizio Pensabene and Eleonora Gasparini (Rome: Bretschneider, 2015), 55–62.

CAVAZZINI 1999: Patrizia Cavazzini, "The Ginetti Chapel at S. Andrea Della Valle," *Burlington Magazine* 141, no. 1156 (1999): 401–13.

CECCHETTI 1886: Bartolomeo Cecchetti, *Documenti per la storia dell'augusta ducale Basilica di San Marco in Venezia*, 2 vols. (Venice: Ongania, 1886).

CECCHI GATTOLIN 1978: Enrichetta Cecchi Gattolin, "I tessellati romanici della basilica di San Savino," in *La basilica di San Savino e le origini del Romanico a Piacenza*, ed. Roberto Salvini (Modena: Artioli, 1978), 115–51.

CENCI 1992: Elena Cenci, "L'antro delle ninfe. Significato ed evoluzione dall'antichità al cinquecento," in *Il giardino storico all'italiana*, ed. Francesco Nuvolari (Milan: Electa, 1992), 147–53.

CERASOLI 1894: Francesco Cerasoli, "Il monumento di Paolo IV nella chiesa della Minerva," *Studi e documenti di storia e diritto* 15 (1894): 131–34.

CERIANA 1996: Matteo Ceriana, "La cappella Corner nella chiesa dei Santi Apostoli a Venezia," in *All'ombra delle volte: Architettura del Quattrocento a Firenze e Venezia*, ed. Massimo Bulgarelli and Matteo Ceriana (Milan: Electa, 1996), 118–22.

CERIANA 2003: Matteo Ceriana, "L'architettura e la scultura decorativa," in Piana and Wolters 2003, 51–122.

CERRI 1998: Giovanni Cerri, "L'ideologia dei quattro elementi da Omero ai presocratici," *Aion* 20 (1998): 5–58.

CHAMONARD 1922–24: Joseph Chamonard, *Le Quartier du Théâtre: Étude sur l'habitation délienne à l'époque hellénistique*, 2 vols. (Paris: E. de Boccard, 1922–24).

CHAMONARD 1933: Joseph Chamonard, "Fouilles de Délos (juillet–septembre 1930)," *Bulletin de correspondance hellénique* 57 (1933): 98–169.

CHAMOUX 1990: François Chamoux, "L'épigramme du colosse des naxiens à Delos," *Bulletin de correspondance hellénique* 114 (1990): 185–86.

CHAMPLIN 1998: Edward Champlin, "God and Man in the Golden House," in *Horti Romani*, ed. Maddalena Cima and Eugenio La Rocca (Rome: Bretschneider, 1998), 333–44.

CHARALAMPADIS 2003: Constantine P. Charalampadis, "The Representation of the Uncreated Light (Lux Increata) in the Byzantine Iconography of the Transfiguration of Christ," *Arte Medievale* 2 (2003): 129–36.

CHATTERJEE 2017: Anuradha Chatterjee, *John Ruskin and the Fabric of Architecture* (Abingdon: Routledge, 2017).

CHIAPPINI DI SORIO 1986: Ileana Chiappini Di Sorio, *Giovanni Bellini: Incoronazione della Vergine "Pala di Pesaro"* (Venice: Arsenale, 1986).

CHIARINI 1970: Marco Chiarini, "Pittura su pietra," *Antichità Viva* 9, no. 2 (1970): 29–37.

CHIELI 1995: Francesca Chieli, "Riflessi della Grecità antica e bizantina nell'architettura italiana del quattrocento," *Quasar* 13–14 (1995): 119–21.

CHLOUVERAKI 2002: Stephania N. Chlouveraki, "Exploitation of Gypsum in Minoan Crete," in *Interdisciplinary Studies on Ancient Stone: Proceedings of the Sixth International Conference of the "Association for the Study of Marble and Other Stones in Antiquity," Venice, June 15–18, 2000*, ed. Lorenzo Lazzarini (Padua: Bottega d'Erasmo, 2002), 25–34.

CHLOUVERAKI 2006: Stephania N. Chlouveraki, "Gypsum in Minoan Architecture: Exploitation, Utilisation and Weathering of a Prestige Stone" (PhD diss., Institute of Archaeology, London, 2006).

CHLOUVERAKI AND LUGLI 2009: Stephania N. Chlouveraki and Stefano Lugli, "Gypsum: A Jewel in Minoan Palatial Architecture: Identification and Characterisation of Its Varieties," in *Asmosia VII. Proceedings of the 7th International Conference of Asmosia, Thassos, Greece, 25–20 September 2003*, ed. Yannis Maniatis (Athens: École française d'Athènes, 2009), 657–68.

CHUVIN 1997: Pierre Chuvin, "Paul le Silentiaire et la Description de Sainte-Sophie. Eclairages historiques," in Fayant and Chuvin 1997, 19–32.

CIANI 1974: Maria Grazia Ciani, *Φαος e termini affini nella poesia greca. Introduzione a una fenomenologia della luce* (Florence: Olschki, 1974).

CIFARELLI 2003: Francesco Maria Cifarelli, *Il tempio di Giunone Moneta sull'acropoli di Segni: Storia, topografia e decorazione architettonica* (Rome: Bretschneider, 2003).

CIMA 2000: Maddalena Cima, "Giunio Basso: La 'basilica' scomparsa. Il restauro delle tarsie dei Musei Capitolini," *Bollettino dei musei comunali di Roma* 14 (2000): 69–90.

CIMA AND LA ROCCA 1986: Maddalena Cima and Eugenio La Rocca, eds., *Le tranquille dimore degli dei: La residenza imperiale degli Horti Lamiani* (Venice: Marsilio, 1986).

CINTAS 1954: Pierre Cintas, "Nouvelles recherches à Utique," *Karthago* 5 (1954): 89–154.

G. W. CLARKE 2003: G. W. Clarke, *Excavating and Interpreting the Governor's Palace, Acropolis, Jebel Kahlid* (Canberra: Australian Academy of the Humanities, 2003).

J. CLARKE 2003: Jacqueline Clarke, *Imagery of Colour and Shining in Catullus, Propertius and Horace* (New York: Peter Lang, 2003).

CLAUSS 1999: Manfred Clauss, *Kaiser und Gott. Herrscherkult im römischen Reich* (Stuttgart: B. G. Teubner, 1999).

CLAUSSEN 1987: Peter Cornelius Claussen, *Magistri doctissimi Romani: Die römischen Marmorkünstler des Mittelalters* (Stuttgart: Franz Steiner, 1987).

CLAY 1981–82: Jenny Strauss Clay, "Immortal and Ageless Forever," *Classical Journal* 77, no. 2 (1981–82): 112–18.

CLINE ET AL. 2017: Eric H. Cline, Assaf Yasur-Landau, and Andrew Koh, "The Absolute Chronology of the Middle Bronze Age Palace at Tel Kabri: Implications for Aegean-Style Wall Paintings in the Eastern Mediterranean," *Journal of Ancient Egyptian Interconnections* 13 (March 2017): 43–47.

CLOSE 1969: Anthony J. Close, "Commonplace Theories of Art and Nature in Classical Antiquity and in the Renaissance," *Journal of the History of Ideas* 30, no. 4 (1969): 467–86.

CLOSS 1958: A. Closs, "Die Steinbücher in kulturhistorischer Überschau," *Graz Landesmuseum Joanneum Mineralogisches Mitteilungsblatt* 8 (1958): 1–34.

COARELLI 1983: Filippo Coarelli, "Architettura sacra e architettura privata nella tarda repubblica," in *Architecture et société de l'archaisme grec à la fin de la République Romaine* (Paris/Rome: CNRS and École française de Rome, 1983), 191–217.

COARELLI 1985: Filippo Coarelli, *Il Foro Romano. Periodo repubblicano e augusteo* (Rome: Quasar, 1985).

COARELLI 1988: Filippo Coarelli, *Il Foro Boario: Dalle origini alla fine della Repubblica* (Rome: Quasar, 1988).

COARELLI 2016: Filippo Coarelli, *I mercanti nel tempio. Delo: Culto, politica, commercio* (Athens: Scuola Archeologica Italiana di Atene, 2016).

COATES-STEPHENS 2004: Robert Coates-Stephens, *Porta Maggiore: Monument and Landscape* (Rome: Bretschneider, 2004).

COFFIN 1979: David R. Coffin, *The Villa in the Life of Renaissance Rome* (Princeton: Princeton University Press, 1979).

COHEN 2007: Jean-Louis Cohen, *Ludwig Mies van der Rohe*, 2nd ed. (Basel: Birkhäuser, 2007).

COHON 1989: Robert Cohon, "An Unpublished Fragment of Furniture from Augustan Rome," *Boreas* 12 (1989): 105–9.

COHON 1990: Robert Cohon, "Pavonazzetto Sculptures of Eastern Barbarians," *Journal of Roman Archaeology* 3 (1990): 264–70.

COLETTI 1935: Luigi Coletti, ed., *Catalogo delle cose d'arte e di antichità d'Italia: Treviso* (Rome: Libreria dello Stato, 1935).

COLINI 1944: Antonio M. Colini, *Storia e topografia del Celio nell'antichità* (Vatican City: Tipografia Poliglotta Vaticana, 1944).

COLLARETA 1988: Marco Collareta, "Le 'arti sorelle.' Teoria e pratica del 'paragone,'" in *La pittura in Italia: Il Cinquecento*, ed. Giuliano Briganti (Milan: Electa, 1988), 569–80.

COLLOMB 2003: Anne-Laure Collomb, "Les traités artistiques et la peinture sur pierre (XVIe–XVIIe siècles)," *Histoire de l'art* 52 (2003): 111–20.

COLLOMB 2006: Anne-Laure Collomb, *La Peinture sur pierre en Italie, 1530–1630* (PhD diss., Université Lumière Lyon 2, 2006).

CONCINA 1998: Ennio Concina, *A History of Venetian Architecture* (Cambridge: Cambridge University Press, 1998).

CONCINA 2002: Ennio Concina, *Le arti di Bisanzio. Secoli VI–XV* (Milan: Mondadori, 2002).

CONFORTI 1993: Claudia Conforti, *Giorgio Vasari architetto* (Milan: Electa, 1993).

CONNELL 1988: Susan Connell, *The Employment of Sculptors and Stonemasons in Venice in the Fifteenth Century* (New York: Garland, 1988).

CONNOR 1999: Carolyn L. Connor, "The Epigram in the Church of Hagios Polyeuktos in Constantinople and Its Byzantine Response," *Byzantion* 69, no. 2 (1999): 479–527.

CONNORS 1982: Joseph Connors, "Bernini's S. Andrea al Quirinale: Payments and Planning," *Journal of the Society of Architectural Historians* 41, no. 1 (1982): 15–37.

CONNORS 1996: Joseph Connors, "Borromini, Hagia Sophia and S. Vitale," in *Architectural Studies in Memory of Richard Krautheimer*, ed. Cecil L. Striker (Mainz am Rhein: Zabern, 1996), 43–48.

CONZE 1912: Alexander Conze, ed., *Altertümer von Pergamon*, vol. 1.1, *Stadt und Landschaft* (Berlin: Georg Reimer, 1912).

COOK 2004: Erwin Cook, "Near Eastern Sources for the Palace of Alkinoos," *American Journal of Archaeology* 108, no. 1 (2004): 43–77.

COOPER 1988: Frederick A. Cooper, "The Quarries of Mount Taygetos in the Peloponnesos, Greece," in Herz and Waelkens 1988, 65–76.

COOPER 1992: Frederick A. Cooper, *The Temple of Apollo Bassitas,* vol. 1, *The Architecture* (Princeton: American School of Classical Studies at Athens, 1992).

COOPER 1995: Alix Cooper, "The Museum and the Book: the Metallotheca and the History of an Encyclopaedic Natural History in Early Modern Italy," *Journal of the History of Collections* 7, no. 1 (1995): 1–24.

CORCORAN AND DELAINE 1994: Simon Corcoran and Janet DeLaine, "The Unit Measurement of Marble in Diocletian's Prices Edict," *Journal of Roman Archaeology* 7 (1994): 263–88.

CORDEZ 2013: Philippe Cordez, "Les marbres de Giotto astrologie et naturalisme à la Chapelle Scrovegni," *Mitteilungen des Kunsthistorischen Institutes in Florenz* 55 (2013): 8–25.

CORMACK 1969: Robin Cormack, "The Mosaic Decoration of S. Demetrios, Thessaloniki: A Re-Examination in the Light of the Drawings of W. S. George," *Annual of the British School at Athens* 64 (1969): 17–52.

CORMACK 1981: Robin Cormack, "Interpreting the Mosaics of S. Sophia at Istanbul," *Art History* 4 (1981): 131–49.

CORMACK 1985A: Robin Cormack, *The Church of Saint Demetrios: The Watercolours and Drawings of W. S. George* (Thessaloniki: Demos Thessalonikes, 1985).

CORMACK 1985B: Robin Cormack, *Writing in Gold: Byzantine Society and Its Icons* (London: George Philip, 1985).

CORMACK AND HAWKINS 1977: Robin Cormack and Ernest J. W. Hawkins, "The Mosaics of St. Sophia at Instanbul: The Rooms above the Southwest Vestibule and Ramp," *Dumbarton Oaks Papers* 31 (1977): 175–251.

CORNELISON 2005: Sally J. Cornelison, "When an Image Is a Relic: The St. Zenobius Panel from Florence Cathedral," in *Images, Relics, and Devotional Practices in Medieval and Renaissance Italy,* ed. Sally J. Cornelison and Scott B. Montgomery (Tempe, AZ: Medieval and Renaissance Texts and Studies, 2005), 95–113.

CORNELIUS 1930: Roberta D. Cornelius, *The Figurative Castle: A Study in the Medieval Allegory of the Edifice with Especial Reference to Religious Writings* (Bryn Mawr, PA: Bryn Mawr College, 1930).

CORSUCCI ET AL. 1983: Sergio Corsucci, Mauro Dalzocchio, and Leonardo Pedini, "Nuove Acquisizioni sulla fabbrica del Battistero di S. Giovanni in Firenze," in *Storia e restauro dell'architettura: Aggiornamenti e prospettive,* ed. Gianfranco Spagnesi (Rome: Centro di Studi per la Storia dell'Architettura, 1983), 55–61.

COULTON 1982: J. J. Coulton, *Greek Architects at Work: Problems of Structure and Design,* 2nd ed. (London: Granada, 1982).

COURBIN 1974: Paul Courbin, "L'inscription du colosse naxien de Délos," in *Mélanges helléniques offerts à Georges Daux* (Paris: E. de Boccard, 1974), 57–66.

COURBIN 1980: Paul Courbin, *L'oikos des Naxiens* (Paris: Fontemoing, 1980).

COURBIN 1987: Paul Courbin, "Le temple archaïque de Délos," *Bulletin de correspondance hellénique* 111, no. 1 (1987): 63–78.

COURBY 1927: Fernand Courby, *La terrasse du temple* (Paris: E. de Boccard, 1927).

COURCELLE 1967: Pierre Courcelle, *La consolation de philosophie dans la tradition littéraire. Antécédents et posterité de boèce* (Paris: Études Augustiniennes, 1967).

CRAMP 1968: Rosemary Cramp, "Glass Finds from the Anglo-Saxon Monastery of Monkwearmouth and Jarrow," in *Studies in Glass History and Design,* ed. R. J. Charleston, Wendy Evans, and A. E. Werner (London: Unwin, 1968), 16–19.

CRAMP 1970: Rosemary Cramp, "Decorated Window-Glass and Millefiori from Monkwearmouth," *Antiquaries Journal* 50, no. 2 (1970): 327–35.

CRAMP 2000: Rosemary Cramp, "Anglo-Saxon Window Glass," in *Glass in Britain and Ireland,* AD *350–1100,* ed. Jennifer Price (London: British Museum, 2000), 109–12.

CRAMP 2001: Rosemary Cramp, "Window Glass from the British Isles 7th–10th Century," in *Il colore nel medioevo. Arte, simbolo, tecnica. La vetrata in occidente dal IV all'XI secolo,* ed. Francesca Dell'Acqua and Romano Silva (Lucca: Istituto Storico Lucchese, 2001), 67–86.

CRESPO 2013: Johanna Tudeau Crespo, "Assyrian Building Practices and Ideologies According to the Assyrian Royal Inscriptions and State Archives" (PhD diss., University of Cambridge, 2013).

CRESTI 1988: Carlo Cresti, "La Cappella dei Principi: Un Pantheon foderato di pietre dure," in *Splendori di Pietre Dure,* ed. A. M. Giusti (Florence: Giunti, 1988), 62–73.

CRESWELL AND GAUTIER-VAN BERCHEM 1969: Keppel A. C. Creswell and Marguerite Gautier-van Berchem, *Early Muslim Architecture* (Oxford: Clarendon, 1969).

CRIPPA 1971: Giuliano Crippa, "Inediti sul Duomo di Bergamo dal Filarete al Fontana," *Bergamo Arte* 2, no. 5 (1971): 21–30.

CROMEY 1991: Robert D. Cromey, "History and Image: The Penelope Painter's Akropolis (Louvre G372 and 480/79 BC)," *Journal of Hellenic Studies* 111 (1991): 165–74.

CUCCINI 1990: Gustavo Cuccini, *Agostino di Duccio: Itinerari di esilio* (Perugia: Guerra, 1990).

CUPANE 1978: Carolina Cupane, "Il motivo del castello nella narrativa tardo-bizantina. Evoluzione di un'allegoria," *Jahrbuch der österreichischen Byzanzinistik* 27 (1978): 229–67.

CUPANE 2010: Carolina Cupane, "Traumpaläste Von Byzanz. Eine unbeachtete Ansicht von 'Constantinople Imaginaire,'" *Νέα Ῥώμη* 6 (2010): 407–39.

CUPANE 2015: Carolina Cupane, "Die Wirklichkeit der Fiktion. Palastbeschreibungen in der byzantinischen Literatur," in *Raumstrukturen und Raumausstattung auf Burgen in Mittelalter und Früher Neuzeit,* ed. Christina Schmid, Gabriele Schichta, Thomas Kühtreiber, and Kornelia Holzner-Tobisch (Heidelberg: Universitätsverlag Winter, 2015), 93–118.

ĆURČIĆ 1992: Slobodan Ćurčić, "Design and Structural Innovation in Byzantine Architecture before Hagia Sophia," in *Hagia Sophia from the Age of Justinian to the Present,* ed. Ahmet S. Çakmak and Robert Mark (Cambridge: Cambridge University Press, 1992), 16–38.

CURTIUS ET AL. 1890: Ernst Curtius, Karl F. W. Dittenberger, Wilhelm Dörpfeld, Adolf Furtwängler, Friedrich Graeber, Paul Graef, Josef F. M. Partsch, Karl Purgold, Georg D. K. Treu, Rudolf Weil, Richard Borrmann, and Friedrich Adler, *Olympia: Die Ergebnisse der von dem Deutschen Reich veranstalteten Ausgrabung,* 5 vols. (Berlin: Asher, 1890).

D'ANGOUR 1999: Armand J. D'Angour, "Ad Unguem," *American Journal of Philology* 120, no. 3 (1999): 411–27.

D'ONOFRIO 1966: Cesare D'Onofrio, "Note berniniane I. Un dialogo-recita di Gian Lorenzo Bernini e Lelio Guidiccioni," *Palatino* 10, no. 2 (1966): 127–34.

D'ONOFRIO 1989: Cesare D'Onofrio, *Visitiamo Roma nel Quattrocento. La città degli Umanisti* (Rome: Romana Società Editore, 1989).

D'ONOFRIO 2008: Anna Maria D'Onofrio, "L'apporto cicladico nella più antica plastica monumentale in Attica," in *La sculpture des Cyclades à l'époque archaïque: Histoire des ateliers, rayonnement des styles; actes du colloque international organise par l'ephorie des antiquités préhistoriques et classiques des Cyclades et l'École française d'Athènes (7–9 Septembre 1998),* ed. Yannos Kourayos and Francis Prost (Athens: École française d'Athènes, 2008), 201–62.

DACOS 1998: Nicole Dacos, "Perin del Vaga et trois peintres de Bruxelles au palais Della Valle," *Prospettiva* 91–92 (1998): 159–70.

DACOS AND FURLAN 1987: Nicole Dacos and Caterina Furlan, *Giovanni da Udine, 1487–1561* (Udine: Casamassima, 1987).

DAGRON 1984: Gilbert Dagron, *Constantinople imaginaire: Études sur le recueil des "Patria"* (Paris: Presses universitaires de France, 1984).

DAGRON 2007: Gilbert Dagron, "From the Mappa to the Akakia: Symbolic Drift," *From Rome to Constantinople–Studies in Honour of Averil Cameron,* ed. Hagit Amirav and Bas Ter Haar Romeny (Louvain: Peeters, 2007), 203–20.

DAHL 1978: Ellert Dahl, "Heavenly Images: The Statue of Ste. Foye of Conques and the Signification of the Medieval Cult Image in the West," *Acta ad Archaeologiam et Artium Historiam Pertinentia* 8 (1978): 175–92.

DAHL 1981: Ellert Dahl, "Dilexi decorem domus Dei: Building to the Glory of God in the Middle Ages," *Acta ad archaeologiam et artium historiam pertinentia* 1 (1981): 157–90.

DALY DAVIS 1989: Margaret Daly Davis, "'Opus isodomum' at the Palazzo della Cancelleria: Vitruvian Studies and Archaeological and Antiquarian Interests at the Court of Raffaele Riario," in Danesi Squarzina 1989, 442–57.

DANESI SQUARZINA 1989: Silvia Danesi Squarzina, ed., *Roma, centro ideale della cultura dell'Antico nei secoli XV e XVI: Da Martino V al Sacco di Roma 1417–1527* (Milan: Electa, 1989).

ḌĀR 1984: Saifurraḥmān Ḍār, *Taxila and the Western World* (Lahore: al-Waqar, 1984).

DARK AND ÖZGÜMÜŞ 2002: Ken Dark and Ferudun Özgümüş, "New Evidence for the Byzantine Church of the Holy Apostles from Fatih Camii, Istanbul," *Oxford Journal of Archaeology* 21, no. 4 (2002): 393–413.

DARMON 2011: Jean-Pierre Darmon, "Faux marbres en mosaïque, autant d'hommages rendus a l'opus sectile," in *Marmoribus Vestita*, ed. Olof Brandt and Philippe Pergola (Vatican City: Pontificio Istituto di Archeologia Cristiana, 2011), 408–34.

DASZEWSKI 1982: Wiktor A. Daszewski, "Die Fußboden-Dekoration in Häusern und Palästen des griechisch-römischen Ägypten," in *Palast und Hütte. Beiträge zum Bauen und Wohnen im Altertum von Archäologen*, ed. Dietrich Papenfuss and Volker M. Strocka (Mainz am Rhein: Zabern, 1982), 395–411.

DAUTERMAN MAGUIRE ET AL. 1989: Eunice Dauterman Maguire, Maggie J. Duncan-Flowers, and Henry Maguire, *Art and Holy Powers in the Early Christian House* (Urbana: University of Illinois Press, 1989).

DAVIES 1993: Paul Davies, "Quattrocento Palaces in Mantua and Ferrara," in *Mantegna and 15th-Century Court Culture*, ed. Francis Ames-Lewis and Anka Bednarek (London: Birkbeck College, 1993), 72–83.

DAVIS 1988: Charles T. Davis, "Topographical and Historical Propaganda in Early Florentine Chronicles and in Villani," *Medioevo e Rinascimento* 2 (1988): 35–52.

C. DAVIS 2007: Charles Davis, "La Cappella Bernabò in San Giovanni Crisostomo: Storia e imagine," in *Tullio Lombardo. Scultore e architetto nella Venezia del Rinascimento*, ed. Matteo Ceriana (Verona: Cierre Grafica, 2007): 216–78.

E. DAVIS 2007: Ellen Davis, "Brush Work," in *Κρινοι Και Λιμενοι: Festschrift in Honor of Joseph and Maria Shaw*, ed. Philip P. Betancourt, Michael C. Nelson, and Hector Williams (Philadelphia: INSTAP, 2007), 145–50.

DAWSON 1944: Christopher M. Dawson, *Romano-Campanian Mythological Landscape Painting* (New Haven/London: Yale University Press and Oxford University Press, 1944).

DE ANGELIS D'OSSAT 1929: Guglielmo De Angelis d'Ossat, "I marmi di Roma antica; Marmor Phengite," in *Atti del Io Congresso Nazionale di Studi Romani* (Rome: Istituto di Studi Romani, 1929), 1–7.

DE BELLIS 1985: Carla De Bellis, "Astri, gemme e arti medico-magiche nello 'Speculum Lapidum' di Camillo Leonardi," in *Il Mago, il cosmo, il teatro degli astri: Saggi sulla letteratura esoterica del Rinascimento*, ed. Gianfranco Formichetti (Rome: Bulzoni, 1985), 67–114.

DE BEYLIÉ 1902: Léon M. E. de Beylié, *L'habitation byzantine: Recherches sur l'architecture civile des Byzantins et son influence en Europe* (Grenoble: Falque & F. Perrin, 1902).

DE BEYLIÉ 1903: Léon M. E. de Beylié, *L'habitation byzantine: Les anciennes maisons de Constantinople* (Grenoble: E. Leroux, 1903).

DE BLAUUW 1991: Sible de Blauuw, "Papst und Purpur. Porphyr in frühen Kirchenausstungen in Rom," in *Tesserae: Festschrift für Josef Engemann*, ed. Ernst Dassmann (Münster: Aschendorff, 1991), 36–50.

DE CAPRIO 1981: Vincenzo De Caprio, "L'area umanistica romana (1513–1527)," *Studi Romani* 29 (1981): 321–35.

DE GARIS DAVIES 1935: Norman de Garis Davies, *Paintings from the Tomb of Rekh-Mi-Rē'atThebes* (New York: Plantin, 1935).

DE MECQUENEM 1947: Roland de Mecquenem, "Contribution à l'étude du palais achéménide de Suse," in *Archéologie Susienne*, ed. Roland de Mecquenem, Louis Le Breton, and M. Rutten (Paris: Presses universaitaires de France, 1947), 1–117.

DE MÉLY 1890: F. de Mély, "Les reliques du lait de la Vierge et la galactique," *Revue archéologique* 15 (Jan.–June 1890): 103–16.

DE NUCCIO AND UNGARO 2002: Matilda De Nuccio and Lucrezia Ungaro, eds., *I marmi colorati della Roma imperiale* (Venice: Marsilio, 2002).

DE POLIGNAC 1995: François de Polignac, *Cults, Territory, and the Origins of the Greek City-State*, trans. Janet Lloyd (Chicago: University of Chicago Press, 1995).

DE RIDDER 1894: André de Ridder, "Fouilles de Gha (Pl. X–XI)," *Bulletin de correspondance hellénique* 18 (1894): 271–310.

DE ROSSI 1871: Giovanni Battista De Rossi, "La basilica profana di Giunio Basso sull'Esquilino dedicata poi a S. Andrea ed appellata Catabarbara Patricia," *Bollettino di Archeologia Cristiana* (1871): 5–29, 41–64.

DE SIENA ET AL. 2002: A. De Siena, Stefano Cancelliere, and Lorenzo Lazzarini, "The Architectural Marbles of the Athenaion of Metapontum and Their Identification," in *Interdisciplinary Studies on Ancient Stone*, ed. Lorenzo Lazzarini (Padua: Bottega d'Erasmo, 2002), 303–8.

DE ZOPPI 2004: Giacomo De Zoppi, "La cappella del Perdono e il tempietto delle Muse nel Palazzo Ducale di Urbino. Analisi e proposta d'attribuzione a Francesco di Giorgio Martini," *Annali di Architettura* 16 (2004): 9–24.

DEÉR 1959: József Deér, *The Dynastic Porphyry Tombs of the Norman Period in Sicily* (Cambridge, MA: Harvard University Press, 1959).

DEICHMANN 1969–89: Friedrich W. Deichmann, *Ravenna: Hauptstadt des spätantiken Abendlandes*, 5 vols. (Wiesbaden: Franz Steiner, 1969–89).

DEL BRAVO 1983: Carlo del Bravo, "Sul significato della luce nel Caravaggio e in Gianlorenzo Bernini," *Artibus et Historiae* 4, no. 7 (1983): 69–77.

DEL BUFALO 2001: Dario Del Bufalo, "Il granito della Colonna," in *Le parole e i marmi: Studi in onore di Raniero Gnoli nel suo 70° compleanno*, 2 vols, ed. Raffaele Torella (Rome: Istituto Italiano per l'Africa e l'Oriente, 2001), 2:929–37.

DEL BUFALO 2016: Dario Del Bufalo, *Murrina Vasa: A Luxury of Imperial Rome* (Rome: Bretschneider, 2016).

DELAINE 2002: Janet DeLaine, "The Temple of Hadrian at Cyzicus and Roman Attitudes to Exceptional Construction," *Papers of the British School at Rome* 70 (2002): 219–20.

DELANGE 2015: Élisabeth Delange, "Le décor de palais de Séthi Ier. Porte(s) de Qantir," in *Trois monuments égyptiens du Nouvel Empire. La Chambre des Ancêtres, Les Annales de Thoutmosis III et le décor de(s) palais de Séthi Ier* (Paris: Khéops, 2015), 187–281.

DELATTE 1932: Armand Delatte, *La captoptromancie grecque et ses dérivés* (Paris: E. Droz, 1932).

DELBEKE 2006: Maarten Delbeke, "Gianlorenzo Bernini's Bel Composto: The Unification of Life and Work in Biography and Historiography," in *Bernini's Biographies: Critical Essays*, ed. Maarten Delbeke, Evonne Levy, and Steven F. Ostrow (University Park: Pennsylvania State University Press, 2006), 251–74.

DELBRÜCK 1932: Richard Delbrück, *Antike Porphyrwerke* (Berlin: Wilhelm de Gruyter, 1932).

DELL'ACQUA 1997: Francesca Dell'Acqua, "Ninth-Century Window Glass from the Monastery of San Vincenzo al Volturno (Molise, Italy)," *Journal of Glass Studies* 39 (1997): 33–41.

DELL'ACQUA 2003: Francesca Dell'Acqua, *"Illuminando colorat": La vetrata tra l'età tardo imperiale e l'alto Medioevo: Le fonti, l'archeologia* (Spoleto: Centro italiano di studi sull'alto Medioevo, 2003).

DELL'ACQUA 2008: Francesca Dell'Acqua, "Between Nature and Artifice: 'Transparent Streams of New Liquid,'" *RES* 53/54 (2008): 93–103.

DELL'AQUILA 1970: Carlo Dell'Aquila, "Le colonne di Castel del Monte," *Nel Mese* 4, no. 2 (1970): 51–53.

DELL'AQUILA ET AL. 1998: Franco Dell'Aquila, Aldo Messina, Nicola Amato, and Sergio Leonardi, *Le chiese rupestri di Puglia e Basilicata* (Bari: M. Adda, 1998).

DELLI SANTI 2000: Maurizio Delli Santi, "Il reimpiego dei marmi antichi a Firenze nel XVI secolo: La Cappella Niccolini in Santa Croce," in *Marmi antichi e pietre dure*, ed. Dario Del Bufalo (Lavello: Mario Congedo, 2000), 91–102.

DELORME 1960: Jean Delorme, *Gymnasion: Étude sur les monuments consacrés à l'éducation en Grèce (des origines à l'Empire Romain)* (Paris: Boccard, 1960).

DEMETROKALLES 1994–98: Georgios Demetrokalles, "Τα περί δήθεν χρήσεως Παριανού μαρμάρου κατά την ανοικοδόμησιν του Ναού του Σολομώντος," *Epeteris etaireias byzantinon spoudon* 49 (1994–98): 115–19.

DEMUS 1960: Otto Demus, *The Church of San Marco in Venice: History, Architecture, Sculpture* (Washington, DC: Dumbarton Oaks, 1960).

DEMUS 1984: Otto Demus, *The Mosaics of San Marco in Venice*, vol. 1, *The Eleventh and Twelfth Centuries*, 2 vols. (Chicago: University of Chicago Press, 1984).

DESPINIS 2004: Giorgios I. Despinis, "Zu Akrolithstatuen griechischer und römischer Zeit," *Nachrichten der Akademie der Wissenschaften zu Göttingen* 8 (2004): 245–301.

DEUBNER 1939: Otfried Deubner, "Expolitio. Inkrustation und Wandmalerei," *Mitteilungen des deutschen archäologischen Instituts, Römische Abteilung* 54 (1939): 14–41.

DEUTSCH 1990: Celia Deutsch, "Wisdom in Matthew: Transformation of a Symbol," *Novum Testamentum* 32, no. 1 (1990): 13–47.

DEZZI BARDESCHI 1966: Marco Dezzi Bardeschi, "Il complesso monumentale di S. Pancrazio a Firenze ed il suo restauro (nuovi documenti)," *Quaderni dell'Istituto di Storia dell'Architettura* 13, nos. 73–78 (1966): 19–25.

DEZZI BARDESCHI 1970: Marco Dezzi Bardeschi, *La facciata di Santa Maria Novella a Firenze* (Pisa: Nistri-Lischi, 1970).

DI CASTRO ET AL. 1994: Alberto Di Castro, Paola Peccolo, and Valentina Gazzaniga, *Marmorari e argentieri a Roma e nel Lazio tra Cinquecento e Seicento: I committenti, i documenti, le opere* (Rome: Quasar, 1994).

DI CESARE 2004: Riccardo Di Cesare, "Sull'Apollo dei Nassii a Delo e le iscrizioni della Base," *Eidola: International Journal of Classical Art History* 1 (2004): 23–60.

DI FRANCESCO 1991: Carla Di Francesco, ed., *Palazzo dei Diamanti. Contributi per il restauro* (Ferrara: Spazio libri, 1991).

DI PASQUALE 2005: Giovanni di Pasquale, "The Museum of Alexandria: Myth and Model," in *From Private to Public: Natural Collections and Museums*, ed. Marco Beretta (Sagamore Beach, MA: Science History, 2005), 1–12.

DI TEODORO 1994: Francesco Di Teodoro, *Raffaello, Baldassar Castiglione e la Lettera a Leone X* (Bologna: Nuova Alfa, 1994).

DICKINS AND KOUROUNIOTIS 1906/7: Guy Dickins and Konstantinos Kourouniotis, "Damophon of Messene (II)," *Annual of the British School at Athens* 13 (1906/7): 357–404.

DIDI-HUBERMAN 1995: Georges Didi-Huberman, *Fra Angelico: Dissemblance and Figuration* (Chicago: University of Chicago Press, 1995), 110.

DIEHL ET AL. 1918: Charles Diehl, Marcel Le Tourneau, and Henri Saladin, *Les monuments chrétiens de Salonique* (Paris: E. Leroux, 1918).

DIETRICH 1983: Bernard C. Dietrich, "Divine Epiphanies in Homer," *Numen* 30, no. 1 (1983): 53–79.

DINI-TRAVERSARI AND MCHUS 1912: Alessandro Dini-Traversari and Lorenzo McHus, *Ambrogio Traversari e i suoi tempi: Albero genealogico Traversari ricostruito. Hodoeporicon* (Florence: B. Seeber, 1912).

DIRIMTEKIN 1965: Feridun Dirimtekin, "Ayasofya Baptisteri/The Baptistery of Saint Sophia," *Türk Arkeoloji Dergisi* 12, no. 2 (1965): 54–87.

DOBBINS AND FOSS 2007: John J. Dobbins and Pedar W. Foss, eds., *The World of Pompeii* (New York: Routledge, 2007).

DOBSCHÜTZ 1899: Ernst von Dobschütz, *Christusbilder: Untersuchungen zur christlichen Legende* (Leipzig: Hinrichs, 1899).

DOHRN 1965: Tobias Dohrn, "Crustae," *Mitteilungen des Deutschen Archäologischen Instituts*, Römische Abteilung 72 (1965): 127–41.

DOMENICALI 2002: Enrica Domenicali, "La cappella Ducale del Castello Estense di Ferrara," *Analecta Pomposiana* 27 (2002): 333–66.

DONDERER 2005: Michael Donderer, "Das Heraion in Olympia und sein Säulenkranz," *Bulletin Antieke Beschaving* 80 (2005): 7–20.

DONOHUE 1988: Alice A. Donohue, *Xoana and the Origins of Greek Sculpture* (Atlanta: Scholars, 1988).

DORATI 1967: Maria Cristina Dorati, "Gli scultori della Cappella Paolina in S. Maria Maggiore," *Commentari* 18, nos. 2–3 (1967): 231–60.

DOUMAS 1992: Christos G. Doumas, *The Wall-Paintings of Thera* (Athens: Thera Foundation, 1992).

DOWNEY 1946: Glanville Downey, "Byzantine Architects: Their Training and Methods," *Byzantion* 18 (1946): 99–118.

DOWNEY 1959A: Glanville Downey, "The Name of the Church of St. Sophia in Constantinople," *Harvard Theological Review* 52, no. 1 (1959): 37–41.

DOWNEY 1959B: Glanville Downey, "The Tombs of the Byzantine Emperors at the Church of the Holy Apostles in Constantinople," *Journal of Hellenic Studies* 79 (1959): 33–51.

DRAPER 2001: Peter Draper, "Canterbury Cathedral: Classical Columns in the Trinity Chapel?," *Architectural History*, Essays in Architectural History Presented to John Newman, 44 (2001): 172–78.

DREXLER 1950: Arthur Drexler, "Unframed Space: A Museum for Jackson Pollock's Paintings," *Interiors and Industrial Design* 109, no. 6 (1950): 90–91.

DRIJVERS 1998: Jan J. W. Drijvers, "The Image of Edessa in Syriac Tradition," in *The Holy Face and the Paradox of Representation*, ed. Herbert L. Kessler and Gerhardt Wolf (Bologna: Nuova Alfa, 1998), 13–31.

DRPIĆ 2019: Ivan Drpić, "Manuel I Komnenos and the Stone of Unction," *Byzantine and Modern Greek Studies* 43, no. 1 (2019): 60–82.

DRURY 1948: G. D. Drury, "The Use of Purbeck Marble in Medieval Times," *Proceedings of the Natural History and Archaeological Society* 70 (1948): 74–97.

DU BUISSON 1935: Robert Du Mesnil du Buisson, *Le site archéologique de Mishrifé-Qaṭna* (Paris: E. de Boccard, 1935).

DUNBABIN 1999: Katherine M. D. Dunbabin, *Mosaics of the Greek and Roman World* (Cambridge: Cambridge University Press, 1999).

DUNKERTON 1991: Jill Dunkerton, *Giotto to Dürer: Early European Painting in the National Gallery* (New Haven/London: Yale University Press and National Gallery of Art, 1991).

DÜRING 1944: Ingemar Düring, *Aristotle's Chemical Treatise: Metereeologica, Book IV* (Göteborg: Elanders boktryckeri aktiebolag, 1944).

DWORAKOWSKA 1990: Angelina Dworakowska, "Once Again on Marble Luculleum," in *Marble: Art Historical and Scientific Perspectives on Ancient Sculpture* (Malibu: J. Paul Getty Museum, 1990), 253–62.

DWORAKOWSKA 1992: Angelina Dworakowska, "The Beginnings of Stone Quarrying on Crete, Some Reflections," in *Studia Aegaea et Balcanica in Honorem Lodovicae Press*, ed. Waldemar Chmielewski (Warsaw: Wydawnictwa Uniwersytetu Warszawskiego, 1992), 53–58.

EBANISTA 1997: Carlo Ebanista, "La 'basilica nova' di Cimitile. Resti del pavimento e della decorazione absidale in 'opus sectile,'" in *Atti del IV Colloquio dell'AISCOM*, ed. Rosa M. Bonacasa Carra and Federico Guidobaldi (Ravenna: Girasole, 1997), 645–67.

EBELING 1953: Erich Ebeling, *Die Akkadische Gebetsserie "Handerhebung"* (Berlin: Akademie, 1953).

EDWARDS 1985: C. M. Edwards, "Greek Votive Reliefs to Pan and the Nymphs" (PhD diss., New York University, 1985).

EDZARD 1987: Dietz-Otto Edzard, "Deep-Rooted Skyscrapers and Bricks: Ancient Mesopotamian Architecture and Its Imagery," in *Figurative Language in the Ancient Near East*, ed. Murray Mindler, Markham J. Geller, and John E. Wansbrough (London: School of Oriental and African Studies, 1987), 13–24.

EFFENBERGER 1989: Arne Effenberger, *Das Mosaik aus der Kirche San Michele in Africisco zu Ravenna*, 2nd ed. (Berlin: Evangelische, 1989).

EFTHYMIADIS 2015: Stephanos Efthymiadis, "Diegeseis on Hagia Sophia from Late Antiquity to Tenth-Century Byzantium," *Byzantinoslavica* 1–2 (2015): 7–22.

EGAN 2016: Emily C. Egan, "Textile and Stone Patterns in the Painted Floors of the Mycenaean Palaces," in *Woven Threads: Patterned*

Textiles in the Aegean Bronze Age, ed. Maria C. Shaw and Anne P. Chapin (Oxford: Oxbow, 2016), 131–47.

EHRHARD 1945: Arnold Ehrhard, "Vir Bonus Quadrato Lapidi Comparatur," *Harvard Theological Review* 38, no. 3 (1945): 177–93.

EHRLE AND STEVENSON 1897: Franz Ehrle and Henry Stevenson, *Gli affreschi del Pinturicchio nell'Appartamento Borgia del Palazzo apostolico Vaticano* (Rome: Danesi, 1897).

EICHHOLZ 1944: David E. Eichholz, "Theophrastus on ΠΟΡΟΣ (ΠΩΡΟΣ)," *Classical Review* 58, no. 1 (1944): 18.

EICHHOLZ 1949: David E. Eichholz, "Aristotle's Theory of the Formation of Metals and Minerals," *Classical Quarterly* 43, nos. 3–4 (1949): 141–46.

EICHHOLZ 1958: David E. Eichholz, "Theophrastus Περὶ λίθων, 25," *Classical Review* 8, nos. 3–4 (1958): 221–22.

EICHHOLZ 1962: David E. Eichholz, "References to a Theory of the Formation of Stones," in *Pliny: Natural History: Libri XXXVI–XXXVII* (London: William Heinemann, 1962), x–xv.

EICHMANN 2007: Ricardo Eichmann, *Uruk: Architektur I: Von den Anfängen bis zur frühdynastischen Zeit* (Rahden: Marie Leidorf, 2007).

EICHMANN 2013: Ricardo Eichmann, "Frühe Grossarchitektur der Stadt Uruk," in *Uruk: 5000 Jahre Megacity*, ed. Nicola Crüsemann, Margarete van Ess, Friedhelm Pedde, and Judith Ramadan (Petersberg: Michael Imhof, 2013), 117–27.

EISLER 2016: Colin T. Eisler, "Fra Antonio Da Negroponte Falier's 'Madonna' and the First Imperial Venetian Style," *Artibus et Historiae* 37, no. 73 (2016): 71–89.

ELIADE 1978: Mircea Eliade, *The Forge and the Crucible: The Origins and Structure of Alchemy*, trans. Stephen Corrin, 2nd ed. (Chicago: University of Chicago Press, 1978).

ELLIS 1988: Simon P. Ellis, "The End of the Roman House," *American Journal of Archaeology* 92, no. 4 (1988): 565–76.

ELLIS 1991: Simon P. Ellis, "Power, Architecture, and Decor: How the Late Roman Aristocrat Appeared to His Guests," in *Roman Art in the Private Sphere: New Perspectives on the Architecture and Decor of the Domus, Villa, and Insula*, ed. Elaine K. Gazda (Ann Arbor: University of Michigan Press, 1991), 117–34.

EPSTEIN 1980–81: Ann Wharton Epstein, "The Fresco Decoration of the Column Churches, Göreme Valley, Cappadocia; A Consideration of Their Chronology and Their Models," *Cahiers archéologiques* 29 (1980–81): 27–45.

ERISTOV 1979: Hélène Eristov, "Corpus des faux-marbres À Pompei," *Mélanges de l'École française de Rome. Antiquité* 91 (1979): 693–771.

ERISTOV 1987: Hélène Eristov, "Peinture romaine et textes antiques: Informations et ambiguïtés. A propos du 'Recueil Millet,'" *Revue archéologique* (1987): 109–23.

ESCHEBACH 1979: Hans Eschebach, *Die Stabianer Thermen in Pompeji* (Berlin: de Gruyter, 1979).

ESTRIN 2015: Seth Estrin, "Living Surfaces: The Materiality of Minoan Wall Paintings," in *Beyond Iconography: Materials, Methods, and Meaning in Ancient Surface Decoration*, ed. Sarah Lepinski and Susanna McFadden (Boston: Archaeological Institute of America, 2015), 109–25.

EVANS 1913/14: Arthur Evans, "The 'Tomb of the Double Axes' and Associated Group, and the Pillar Rooms and Ritual Vessels of the 'Little Palace' at Knossos," *Archaeologia or Miscellaneous Tracts Relating to Antiquity* 65 (1913/14): 1–94.

EVANS 1970: David B. Evans, *Leontius of Byzantium: An Origenist Christology* (Washington, DC: Dumbarton Oaks, 1970).

EVANS AND EVANS 1921: Arthur Evans and Joan Evans, *The Palace of Minos: A Comparative Account of the Successive Stages of the Early Cretan Civilization as Illustrated by the Discoveries at Knossos* (London: Macmillan, 1921).

EVELY 1999: Doniert Evely, *Fresco: A Passport into the Past: Minoan Crete through the Eyes of Mark Cameron* (Athens: British School at Athens, Goulandris Foundation, Museum of Cycladic Art, 1999).

EVISON 1988: Vera I. Evison, "Vieux-Marché, Place Saint-Lambert, Liège-the Glass," in *Les fouilles de la Place Saint-Lambert à Liège*, vol. 2, *Le vieux-marché* (Liège: Université de Liège, 1988), 215–19.

EVISON 1990: Vera I. Evison, *Red Marbled Glass, Roman to Carolingian* (Amsterdam: Association pour l'histoire du verre, 1990).

EXTERMANN 2016: Grégoire Extermann, "Marmi romani in Piemonte. Pio V e il cantiere di Santa Croce a Bosco Marengo," in *Splendor Marmoris. I colori tra Roma e l'Europa da Paolo III a Napoleone III, ed.* Grégoire Extermann and Ariane Varela Braga (Rome: De Luca, 2016), 69–86.

EYICE 1978A: Semai Eyice, "Encore une fois l'église d'Alexis Apocauque à Sélymbria," *Byzantion* 48 (1978): 406–16.

EYICE 1978B: Semai Eyice, "Le baptistère de Sainte Sophie d'Istanbul," in *Atti del IX congresso internazionale di archeologia cristiana* (Vativan City: Pontificio Istituto di Archeologia Cristiana, 1978), 257–73.

FABJAN 1999: Barbara Fabjan, "Fabio Chigi, la Cappella di Santa Maria del Popolo," in *Gian Lorenzo Bernini. Regista del Barocco. I restauri*, ed. Claudio M. Strinati and Maria Grazia Bernardini (Milan: Skira, 1999), 47–56.

FAILLER 1997: Albert Failler, "Pachymeriana Novissima," *Revue des études byzantines* 55, no. 1 (1997): 221–46.

FALLAY D'ESTE 1992: Lauriane Fallay d'Este, *Le paragone: Le parallèle des arts* (Paris: Klincksieck, 1992).

FANT 1988: J. Clayton Fant, "The Roman Emperors in the Marble Business: Capitalists, Middlemen or Philanthropists?," in Herz and Waelkens 1988, 147–58.

FANT 1989: J. Clayton Fant, *Cavum Antrum Phrygiae: The Organization and Operations of the Roman Imperial Marble Quarries in Phrygia* (Oxford: BAR, 1989).

FANT 1993: J. Clayton Fant, "Ideology, Gift and Trade: A Distribution Model for the Roman Imperial Marbles," in *The Inscribed Economy: Production and Distribution in the Roman Empire in the Light of Instrumentum Domesticum*, ed. William V. Harris (Ann Arbor: University of Michigan, 1993), 145–70.

FANT 1999: J. Clayton Fant, "Augustus and the City of Marble," in *Archéomatériaux: Marbres et autres roches*, ed. Max Schvoerer (Talence: CRPAA, 1999), 277–80.

FARIOLI CAMPANATI 1997: Raffaella Farioli Campanati, "Il pavimento di San Marco a Venezia e i suoi rapporti con l'Oriente," in *Storia dell'arte marciana: I mosaici*, ed. Renato Polacco (Venice: Marsilio, 1997), 11–19.

FASSINELLI 1997: Anna Maria Fassinelli, *Scagliola, l'arte della pietra di luna* (Rome: Editalia, 1997).

FAVREAU 2003: Robert Favreau, "Les autels portatifs et leurs inscriptions," *Cahiers de civilisation médiévale* 46 (2003): 327–52.

FAVREAU ET AL. 1999: Robert Favreau, François Jeanneau, Yves-Jean Riou, Marie-Thérèse Camus, Yves Christe, Marie-France de Christien, Isabelle Dangas, Bernard Brochard, and Dominique Simon-Hiernard, *Saint-Savin. L'Abbaye et ses peintures murales* (St.-Aignan-de-Grand-Lieu: Connaissance et promotion du patrimoine de Poitou-Charentes, 1999).

FELLMANN 1957: Rudolf Fellmann, *Das Grab des Lucius Munatius Plancus bei Gaëta* (Basel: Verlag des Institutes für Ur- and Frühgeschichte der Schweiz, 1957).

FENTRESS AND MAIURO 2011: Elizabeth Fentress and Marco Maiuro, "Villa Magna near Anagni: The Emperor, His Winery and the Wine of Signia," *Journal of Roman Archaeology* 24 (January 2011): 333–69.

FERNÁNDEZ DELGADO AND PORDOMINGO PARDO 2010: Pardo José Antonio Fernández Delgado and Francisca Pordomingo Pardo, "Pcair. 65445.140–154 a la luz del nuevo papiro de Posidipo," in *Dic Mihi, Mvsa, Virvm: Homenaje al Profesor Antonio López Eire*, ed. Francisco Cortés Gabaudán and Julián Víctor Méndez Dosuna (Salamnaca: Ediciones Universidad de Salamanca, 2010), 179–90.

FERNÁNDEZ URIEL 1992: Pilar Fernández Uriel, "Nero, Alter Apollo: La divinización del 'Princeps' en la ideología neroniana," in *Héroes, semidioses y daimones*, ed. Jaime Alvar, Carmen Blánquez, and Carlos G. Wagner (Madrid: Ediciones Clásicas, 1992), 159–73.

FERNIE 1980: Eric C. Fernie, "The Spiral Piers of Durham Cathedral," in *Medieval Art and Architecture at Durham Cathedral* (London: British Archaeological Association, 1980), 49–58.

FERNIE 1984: Eric C. Fernie, "The Use of Nave Supports in Romanesque and Early Gothic Churches," *Gesta* 23, no. 2 (1984): 107–17.

FERNIE 1994: Eric C. Fernie, "The Architectural Influence of Durham Cathedral," in *Anglo-Norman Durham: 1093–1193*, ed. Davis W. Rollason, Margaret Harvey, and Michael Prestwich (Woodbridge, Suffolk: Boydell, 1994), 269–79.

FERNIE 2000: Eric C. Fernie, *The Architecture of Norman England* (Oxford: Oxford University Press, 2000).

FERRARI AND PAPALDO 1999: Oreste Ferrari and Serenita Papaldo, *Le sculture del Seicento a Roma* (Rome: Bozzi, 1999).

FERRINI 1998: Maria F. Ferrini, "Τὸ μὲν A λευκὸν τὸ δὲ B ἄνθρωπος. Λευκός e μέλας nella filosofia Greca," *Quaderni Urbinati di Cultura Classica* 59, no. 2 (1998): 59–80.

FERRUA 1991: Antonio Ferrua, *The Unknown Catacomb: A Unique Discovery of Early Christian Art* (New Lanark: Geddes & Grosset, 1991).

FESTUGIÈRE 1944–54: André M. J. Festugière, *La révélation d'Hermès Trismégiste*, 4 vols. (Paris: Lecoffre, 1944–54).

FIENE 1989: Donald M. Fiene, "What Is the Appearance of Divine Sophia?," *Slavic Review* 48, no. 3 (1989): 449–76.

FILIPPI 2005: Fedora Filippi, ed., *Palazzo Altemps. I colori del fasto. La domus del gianicolo e i suoi marmi* (Rome: Electa, 2005).

FINCH 1991: Margaret Finch, "Petrine Landmarks in Two Predella Panels by Jacopo di Cione," *De Artibus et Historiae* 12, no. 23 (1991): 67–82.

FINKELBERG 1998: Aryeh Finkelberg, "On the History of the Greek ΚΟΣΜΟΣ," *Harvard Studies in Classical Philology* 98 (1998): 103–36.

FINNEY 1984: Paul C. Finney, "TOPOS HIEROS und christlicher Sakralbau in vorkonstantinischer Überlieferung," *Boreas* 7 (1984): 193–225.

FINOCCHI GHERSI 2004: Lorenzo Finocchi Ghersi, *Il Rinascimento veneziano di Giovanni Bellini* (Venice: Marsilio, 2004).

FIOCCHI NICOLAI ET AL. 1999: Vincenzo Fiocchi Nicolai, Fabrizio Bisconti, and Danilo Mazzoleni, *The Christian Catacombs of Rome: History, Decoration, Inscriptions*, trans. Cristino C. Stella and Lori-Ann Touchette (Regensburg: Schnell & Steiner, 1999).

FIRATLI 1951: Nezih Firatli, "Découverte de trois églises byzantines à Istanbul," *Cahiers archéologiques* 4 (1951): 167, fig. 1.

FISCHER 1999: Moshe L. Fischer, ". . . 'On Parian Pillars' (the Book of Esther 1:6): Bible, Midrash and Real Marble in the Ancient Near East," *Journal of Jewish Studies* 50, no. 2 (1999): 235–45.

FISCHER AND STEIN 1994: Moshe L. Fischer and Alla Stein, "Josephus on the Use of Marble in Building Projects of Herod the Great," *Journal of Jewish Studies* 45, no. 1 (1994): 79–85.

FITTSCHEN AND ZANKER 1983: Klaus Fittschen and Paul Zanker, *Katalog der römischen Porträts in den Capitolinischen Museen und den anderen kommunalen Sammlungen der Stadt Rom*, 6 vols. (Mainz am Rhein: Zabern, 1983).

FITZSIMONS 2006: Rodney D. Fitzsimons, "Monuments of Power and the Power of Monuments: The Evolution of Élite Architectural Styles at Bronze Age Mycenae" (PhD diss., University of Cincinnati, 2006).

FLETCHER 1924: Sir Banister Fletcher, *A History of Architecture on the Comparative Method for Students, Craftsmen, and Amateurs*, 7th rev. and enlarged ed. (London: B. T. Batsford, 1924).

FLETCHER 1993: Jennifer Fletcher, "Mantegna and Venice," in *Mantegna and 15th-Century Court Culture*, ed. Francis Ames-Lewis and Anka Bednarek (London: Birkbeck College, 1993), 17–25.

FLETE 1909: John Flete, *The History of Westminster Abbey*, ed. J. Armitage Robinson (Cambridge: Cambridge University Press, 1909).

FLOOD 1993: Finbarr Barry Flood, "Palaces of Crystal, Sanctuaries of Light: Windows, Jewels and Glass in Medieval Islamic Architecture" (PhD diss., Edinburgh, 1993).

FLOOD 1999: Finbarr Barry Flood, "Light in Stone: The Commemoration of the Prophet in Umayyad Architecture," in *Bayt al-Maqdis: Jerusalem and Early Islam*, ed. Jeremy Johns (Oxford: Oxford University Press, 1999), 311–59.

FLOOD 2001: Finbarr Barry Flood, *The Great Mosque of Damascus: Studies on the Makings of an Umayyad Visual Culture* (Leiden: Brill, 2001).

FLOOD 2016: Finbarr Barry Flood, "'God's Wonder': Marble as Medium and the Natural Image in Mosques and Modernism," *West 86th* 23, no. 2 (2016): 168–219.

FOBELLI 2005: Maria L. Fobelli, "La Santa Sofia di Costantinopoli nell'età di Giustiniano: Sistemi decorativi e strategia delle immagini," in *Medioevo: Immagini e ideologie*, ed. Arturo C. Quintavalle (Milan: Electa, 2005), 90–99.

FOGAGNOLO AND LUGARI 2008: Stefania Fogagnolo and Alessandro Lugari, "Un'antica metodologia di lavorazione nelle lastre in opus sectile: Il tipo di giunzione a zig-zag," in *Atti del XIII Colloquio dell'AISCOM*, ed. Claudia Angelelli and Federica Rinaldi (Tivoli: Scripta Manent, 2008), 513–24.

FOLDA 1986: Jaroslav Folda, *The Nazareth Capitals and the Crusader Shrine of the Annunciation* (University Park: Pennsylvania State University Press, 1986).

FOLGOLARI 1908: Gino Folgolari, "Le portelle dell'organo di S. Maria dei Miracoli di Venezia," *Bollettino d'Arte* 4 (1908): 1–32.

FORCELLINO 1990: Antonio Forcellino, "La diffusione dei rivestimenti a stucco nel corso del XVI Secolo," *Ricerche di Storia dell'Arte* 41–42 (1990): 23–52.

FORLATI 1975: Ferdinando Forlati, *La Basilica di San Marco attraverso i suoi restauri* (Trieste: Lint, 1975).

FORSYTH 1972: Ilene H. Forsyth, *The Throne of Wisdom: Wood Sculptures of the Madonna in Romanesque France* (Princeton: Princeton University Press, 1972).

FORSYTH AND WEITZMANN 1973: George H. Forsyth and Kurt Weitzmann, *The Monastery of Saint Catherine at Mount Sinai: The Church and Fortress of Justinian* (Ann Arbor: University of Michigan Press, 1973).

FORTINI BROWN 1988: Patricia Fortini Brown, *Venetian Narrative Painting in the Age of Carpaccio* (New Haven: Yale University Press, 1988).

FORTINI BROWN 1995: Patricia Fortini Brown, "Renovatio or Conciliatio? How Renaissances Happened in Venice," in *Language and Images of Renaissance Italy*, ed. Alison Brown (Oxford: Clarendon, 1995), 127–54.

FOSCARI AND TAFURI 1983: Antonio Foscari and Manfredo Tafuri, *L'armonia e i conflitti: La chiesa di San Francesco della Vigna nella Venezia del '500* (Turin: Einaudi, 1983).

FOSCHI 1995: Silvia Foschi, "S. Vitale in Ravenna: Censura e damnatio memoriae," *Ravenna, Studi e Ricerche* 2 (1995): 62–78.

FOSCHI 2002: Silvia Foschi, "Santa Sofia di Costantinopoli: Immagini dall'Occidente," *Annali di Architettura* 14 (2002): 7–33.

FOSSATI 1992: Gaspare Fossati, *Gaspare Fossati, 1809–1883* (Rancate: Fidia edizione d'arte, 1992).

FOSTER 1979: Karen Polinger Foster, *Aegean Faience of the Bronze Age* (New Haven: Yale University Press, 1979).

K. FOSTER 1991: Karen Polinger Foster, "Ceramic Imagery in Ancient Near Eastern Literature," in *Materials Issues in Art and Archaeology II: Symposium Held April 17–21, 1990, San Francisco, California, U.S.A.*, ed. Pamela B. Vandiver, James Druzik, and George S. Wheeler (Pittsburgh: Materials Research Society, 1991), 389–413.

R. FOSTER 1991: Richard Foster, *Patterns of Thought: The Hidden Meaning of the Great Pavement of Westminster Abbey* (London: Jonathan Cape, 1991), with bibliography.

FOSTER 2008: Karen Polinger Foster, "Minoan Faience Revisited," in *Vitreous Materials in the Late Bronze Age Aegean*, ed. Caroline M. Jackson and Emma C. Wager (Oxford: Oxbow, 2008), 173–86.

FRANKL 1960: Paul Frankl, *The Gothic: Literary Sources and Interpretations through Eight Centuries* (Princeton: Princeton University Press, 1960).

FRANZ 1958: Heinrich Gerhard Franz, "Transennae als Fensterverschluss. Ihre Entwicklung von der frühchristlichen bis zur islamischen Zeit," *Istanbuler Mitteilungen* 8 (1958): 65–81.

FRASER 1972: Peter M. Fraser, *Ptolemaic Alexandria*, 2 vols. (Oxford: Clarendon, 1972).

FRAZER 1971: Alfred Frazer, review of Becatti 1969, *American Journal of Archaeology* 75 (1971): 319–24.

FREIBERG 1995: Jack Freiberg, *The Lateran in 1600: Christian Concord in Counter-Reformation Rome* (Cambridge: Cambridge University Press, 1995).

FRIEDMAN 1996: Florence Dunn Friedman, "Notions of Cosmos in the Step Pyramid Complex," in *Studies in Honor of William Kelly Simpson*, ed. Peter Der Manuelian (Boston: David Brown, 1996), 337–51.

FRIEDMAN 1998: Florence Dunn Friedman, ed., *Gifts of the Nile: Ancient Egyptian Faience* (New York: Thames & Hudson, 1998).

FRITZ 1998: Michael P. Fritz, "'Una tavoletta doviziosa.' Gedanken zu einem Frühwerk Giulio Romanos in Paris," in *Diletto e maraviglia: Ausdruck und Wirkung in der Kunst von der Renaissance bis zum Barock*, ed. Christine Göttler (Emsdetten: Imorde, 1998), 238–51.

FROMMEL 1973: Christoph L. Frommel, *Der Römische Palastbau Der Hochrenaissance*, 2 vols. (Tübingen: Wasmuth, 1973).

FROMMEL 1983: Cristoph L. Frommel, "S. Andrea al Quirinale: Genesi e struttura," in *Gian Lorenzo Bernini architetto e l'architettura europea del Sei-Settecento*, ed. Marcello Fagiolo and Gianfranco Spagnesi, 2 vols. (Rome: Istituto Della Enciclopedia Italiana, 1983), 1:211–53.

FROMMEL 1994: Christoph L. Frommel, "Il progetto di Domenico Aimo da Varignana per la facciata di San Petronio," in *Una basilica per una città: Sei secoli in San Petronio*, ed. Mario Fanti and Deanna Lenzi (Bologna: Fabbriceria di San Petronio, 1994), 223–39.

FROMMEL 2012: Cristoph Luitpold Frommel, "Pio V, la chiesa di Bosco Marengo e l'architettura della controriforma," in *L'immagine del rigore. Committenza artistica di e per Pio V a Roma e in Lombardia*, ed. Luisa Giordano and Gianpaolo Angelini (Como: Ibis, 2012), 17–78.

FROMMEL AND WINNER 1986: Christoph L. Frommel and Matthias Winner, eds. *Raffaello a Roma* (Rome: L'Elefante, 1986).

FROMMEL ET AL. 1984: Christoph L. Frommel, Stefano Ray, and Manfredo Tafuri, eds. *Raffaello architetto* (Milan: Electa, 1984).

FROMMEL ET AL. 2002: Christoph L. Frommel, Luisa Giordano, and Richard V. Schofield, eds. *Bramante milanese e l'architettura del Rinascimento lombardo* (Venice: Marsilio, 2002).

FROTHINGHAM 1894: Arthur L. Frothingham, "Byzantine Architects in Italy from the Sixth to the Fifteenth Centuries," *American Journal of Archaeology* 9 (1894): 32–52.

FUMI 1891: Luigi Fumi, *Il duomo di Orvieto e i suoi restauri: Monografie storiche condotte sopra i documenti* (Rome: Società Laziale Tipografico-Editrice, 1891).

FUSCO 2006: Riccardo Fusco, "Neronis Maculae," *Marmora: An International Journal for Archaeology, History and Archeometry of Marbles and Stones* 2 (2006): 21–39.

FUSCO 2010: Riccardo Fusco, "Luminescenza, lucentezza e trasparenza. Note sullo stile dei rivestimenti parietali in marmo di età neroniana," in *Atti del X Congresso dell'AIPMA*, ed. Irene Brigantini (Naples: Università degli Studi di Napoli "L'Orientale," 2010), 83–92.

FUSCO AND HESS 1994: Peter Fusco and Catherine Hess, "A Rediscovered 'Commesso' Portrait," *Burlington Magazine* 136, no. 1091 (1994): 68–72.

FYFE 1902: Theodore Fyfe, "Painted Plaster Decoration at Knossos: With Special Reference to the Architectural Schemes," *Journal of the Royal Institute of British Architects* 10 (1902): 107–31.

GAETA 1980: F. Gaeta, "Storiografia, coscienza nazionale e politica culturale nella Venezia del Rinascimento," in *Storia della cultura veneta*, ed. Girolamo Arnaldi and Manlio Pastore Stocchi (Vicenza: Neri Pozza, 1980), 3.1 (*Dal primo Quattrocento al concilio di Trento*), 1–91.

GAGE 1982: John Gage, "Gothic Glass: Two Aspects of a Dionysian Aesthetic," *Art History* 5 (1982): 36–57.

GAGGIOTTI 1989: Marcello Gaggiotti, "Pavimenta Poenica marmore Numidico constrata," in *L'Africa romana*, ed. Attilio Mastino (Sassari: Edizioni Gallizzi, 1989), 215–21.

GAGLIARDO BRIUCCIA AND MONTANA 1998: Valentina Gagliardo Briuccia and Giuseppe Montana, *I marmi e i diaspri del Barocco siciliano: Rassegna dei materiali lapidei di pregio utilizzati per la decorazione ad intarsio* (Palermo: Flaccovio, 1998).

GALE ET AL. 1988: Noel H. Gale, H. C. Einfalt, Hans-Wolfgang Hubberten, and Richard E. Jones, "The Sources of Mycenaean Gypsum," *Journal of Archaeological Science* 15 (1988): 57–72.

GALLI 1940: Romeo Galli, *Lavinia Fontana: Pittrice, 1552–1614* (Imola: Paolo Galeati, 1940).

GAMBONI 2002: Dario Gamboni, *Potential Images: Ambiguity and Indeterminacy in Modern Art* (London: Reaktion, 2002).

GAMBUTI 1995: Alessandro Gambuti, "Pietre e marmi di pregio per adornare la chiesa quale 'immagine del cielo sulla terra,'" in *Architetture di altari e spazio ecclesiale: Episodi a Firenze, Prato e Ferrara nell'età della Controriforma*, ed. Carlo Cresti (Florence: Pontecorboli, 1995), 121–36.

GANZERT 1996: Joachim Ganzert, *Der Mars-Ultor-Tempel auf dem Augustusforum in Rom* (Mainz am Rhein: Zabern, 1996).

GANZERT 2000: Joachim Ganzert, *Im Allerheiligsten des Augustusforums: Fokus "oikoumenischer Akkulturation"* (Mainz am Rhein: Zabern, 2000).

GARCIA-TALEGON ET AL. 2016: Jacinta Garcia-Talegon, A. C. Iñigo, and Eloy Molina Ballesteros, "Silicified Granites (Bleeding Stone and Ochre Granite) as Global Heritage Stones Resources from Avila (Central of Spain)," *Geoscience Canada* 43, no. 1 (2016): 53–62.

GARDINER 1964: Sir Alan Gardiner, *Egyptian Grammar: Being an Introduction to the Study of Hieroglyphs*, 3rd rev. ed. (London: Oxford University Press, 1964).

GARELLA 1992: Luciano Garella, "Pitture sopra le facciate delle case di Roma. Ritrovamenti a Palazzo Santacroce ed al Palazzo Curti," *Alma Roma* 33 (1992): 124–37.

GARIBALDI 1996: Vittoria Garibaldi, ed., *Un pittore e la sua città: Benedetto Bonfigli e Perugia* (Milan: Electa, 1996).

GASPARRI 1979: Carlo Gasparri, *Aedes Concordiae Augustae* (Rome: Istituto di Studi Romani, 1979).

GASPARRI 1985: Carlo Gasparri, "Appunti sull' 'opus sectile' del Palatino," *Studi Urbinati di storia, filosofia e lettere* 58, no. B3 (1985): 61–67.

GATTI PERER 1983: Maria Luisa Gatti Perer, *"La dimora di Dio con gli uomini" (Ap 21,3): Immagini della Gerusalemme celeste dal III al XIV secolo* (Milan: Vita & Pensiero, 1983).

GATTI PERER AND ROVETTA 1996: Maria Luisa Gatti Perer and Alessandro Rovetta, eds., *Cesare Cesariano e il classicismo di primo Cinquecento* (Milan: Vita e Pensiero, 1996).

GAULTIER 1974: Paul Gaultier, "Le typicon du Christ sauveur Pantocrator," *Revue des études byzantines* 32 (1974): 1–145.

GAUTIER DALCHÉ 1995: Patrick Gautier Dalché, *Carte marine et portulan au XIIe siècle: Le "Liber de existencia riveriarum et forma maris nostri Mediterranei" (Pise, circa 1200)* (Rome: École française de Rome, 1995).

GAUTIER DI CONFIENGO 2000: Edoardo Gautier di Confiengo, "La catechesi figurata dei mosaici della chiesa del vescovo Sergio ad Umm al-Rasas di Giordania," *Studium Biblicum Franciscanum* 50 (2000): 439–52.

GELAO 1988: Clara Gelao, "Palazzi con bugnato a punta di diamante in terra di Bari," *Napoli nobilissima* 27, nos. 1–2 (1988): 12–28.

GEORGE 1913: W. S. George, *The Church of Saint Eirene at Constantinople* (Oxford: Oxford University Press, 1913).

GEORGE 2005–6: Andrew R. George, "The Tower of Babel: Archaeology, History and Cuneiform Texts," *Archiv für Orientforschung* 51 (2005–6): 75–95.

GEORGE 2011: Andrew R. George, "A Stele of Nebuchadnezzar II," in *Cuneiform Royal Inscriptions and Related Texts in the Schøyen Collection*, ed. Andrew R. George (Bethesda, MD: CDL, 2011), 153–69.

GERHARDT 1963: Mia Irene Gerhardt, *The Art of Story-Telling: A Literary Study of The Thousand and One Nights* (Leiden: Brill, 1963).

GERMAIN 1969: Suzanne Germain, *Les mosaïques de Timgad. Étude descriptive et analytique* (Paris: Éditions du Centre national de la recherche scientifique, 1969).

GERSTEL AND LAUFFENBURGER 2001: Sharon E. J. Gerstel and Julie A. Lauffenburger, eds., *A Lost Art Rediscovered: The Architectural Ceramics of Byzantium* (Baltimore: Walters Art Museum, 2001).

GHEROLDI 2006: Vincenzo Gheroldi, "La decorazione interna della chiesa di Santa Maria della Vittoria a Mantova: Indagini stratigrafiche, analisi chimiche, ricostruzioni techniche," in Signorini 2006, 220–29.

GHISETTI GIAVARINA 2007–8: Adriano Ghisetti Giavarina, "Il bugnato a punte di diamante nell'architettura del Rinascimento italiano," *Lexicon* 5–6 (2007–8): 9–26.

GIAMMARIA 1999: Gioacchino Giammaria, *Villamagna* (Anagni: Istituto di Storia e di Arte del Lazio Meridionale, 1999).

GIANANDREA ET AL. 2017: Manuela Gianandrea, Manuela Annibali, and Laura Bartoni, eds., *Il convento di Santa Sabina all'Aventino e il suo patrimonio storico-artistico e architettonico* (Rome: Campisano, 2017).

GIBSON ET AL. 1994: Sheila Gibson, Janet DeLaine, and Amanda Claridge, "The Triclinium of the Domus Flavia: A New Reconstruction," *Papers of the British School at Rome* 42 (1994): 67–100.

GINOUVÈS ET AL. 1993: René Ginouvès, Giannes M. Akamates, and Manoles Andronikos, eds., *La Macédoine de Philippe II à la conquête romaine* (Paris: CNRS, 1993).

GIOVANNONI 1917: Gustavo Giovannoni, "Un'opera sconosciuta di Jacopo Sansovino in Roma," *Bollettino d'Arte* 11, nos. 3–4 (1917): 64–81.

GIOVENALE 1929: Giovanni Battista Giovenale, *Il Battistero lateranense nelle recenti indagini della Pont. commissione di archeologia sacra* (Rome: Pontificio Istituto di Archeologia Cristiana, 1929).

GIOVINO 2007: Mariana Giovino, *The Assyrian Sacred Tree: A History of Interpretations* (Fribourg: Vandenhoek & Ruprecht, 2007).

GIULIANO 1979: Antonio Giuliano, ed., *Museo Nazionale Romano. Le sculture* (Rome: De Luca, 1979).

GIULIANO 2004–5: Antonio Giuliano, "Un Volto in Avorio," *Rendiconti della Pontificia Accademia Romana di Archeologia* 77 (2004–5): 289–308.

GIULIO ROMANO 1989: *Giulio Romano* (Milan: Electa, 1989).

GIUSTI 1988: A. M. Giusti, *Splendori di pietre dure: L'arte di corte nella Firenze dei Granduchi* (Florence: Giunti, 1988).

GIUSTI 2005: Anna Maria Giusti, *L'arte delle Pietre Dure da Firenze all'Europa* (Florence: Lettere, 2005).

GNIRS 2009: Andrea M. Gnirs, "In the King's House: Audiences and Receptions at Court," in *4. Symposium zur ägyptischen Königsideologie: Egyptian Royal Residences*, ed. Rolf Gundlach and John H. Taylor (Wiesbaden: Harrasowitz, 2009), 13–43.

GNOLI 1930: Domenico Gnoli, "Orti letterari nella Roma di Leone X," *Nuova Antologia: Rivista di lettere, scienze e arti* 269 (1930): 3–19, 137–48.

GNOLI 1988: Raniero Gnoli, *Marmora Romana*, 2nd ed. (Rome: L'Elefante, 1988).

GOFFEN 1975: Rona Goffen, "Icon and Vision: Giovanni Bellini's Half-Length Madonnas," *Art Bulletin* 57, no. 4 (1975): 499–505.

GOFFEN 1989: Rona Goffen, *Giovanni Bellini* (New Haven: Yale University Press, 1989).

GOLL 2001: Jürg Goll, "Frühmittelalterliche Fenster Gläser aus Müstair und Sion," in *Il colore nel medioevo. Arte, simbolo, tecnica. La vetrata in occidente dal IV all'XI secolo*, ed. Francesca Dell'Acqua and Romano Silva (Lucca: Istituto Storico Lucchese and Scuola Normale Superiore di Pisa, 2001), 87–98.

GOMBRICH 1962: Ernst H. Gombrich, "Alberto Avogadro's Descriptions of the Badia of Fiesole and of the Villa of Careggi," *Italia Medioevale e Umanistica* 5 (1962): 217–29.

GONZALEZ 2002: Valérie Gonzalez, *Le piège de Salomon. La pensée de l'art dans le Coran* (Paris: Michel, 2002).

GONZÁLEZ-PALACIOS 1984: Alvar González-Palacios, *Il tempio del gusto: Roma e il Regno delle Due Sicilie: Le arti decorative in Italia fra classicismi e barocco*, 2 vols. (Milan: Longanesi, 1984).

GOODISON 2004: Lucy Goodison, "From Tholos Tomb to Throne Room: Some Considerations of Dawn Light and Directionality in Minoan Buildings," in *Knossos: Palace, City, State*, ed. Gerald Cadogan, Eleni Hatzaki, and Antōnēs Vasilakēs (London: British School at Athens, 2004), 339–50.

GORGONI ET AL. 2002: Carlo Gorgoni, Maria Grazia Filetici, Lorenzo Lazzarini, Paolo Pallante, and Patrizio Pensabene, "Archaeometry of Two Important Marble Monuments of the Republican and Early Imperial Periods in Rome: The Tempio Rotondo and the Pyramid of Cestius," in Hermann et al. 2002, 308–15.

GOULET 2001–2: Crispin C. Goulet, "The 'Zebra-Stripe' Design: An Investigation of Roman Wall Painting in the Periphery," *Rivista di studi pompeiani* 12–13 (2001–2): 53–94.

GOY 1989: Richard J. Goy, *Venetian Vernacular Architecture: Traditional Housing in the Venetian Lagoon* (Cambridge: Cambridge University Press, 1989).

GOY 1992: Richard J. Goy, *The House of Gold: Building a Palace in Medieval Venice* (Cambridge: Cambridge University Press, 1992).

GRABAR 1947: André Grabar, "Le témoinage d'une hymne syriaque sur l'architecture de la cathédrale d'Edesse au VIe siècle et sur la symbolique de l'édifice chrétien," *Cahiers archéologiques* 2 (1947): 41–67.

GRABAR 1976: André Grabar, *Sculptures byzantines du moyen age* (Paris: Picard, 1976).

GRAEFE 1979: Rainer Graefe, *Vela erunt: Die Zeltdächer der römischer Theater und ähnlicher Anlagen* (Mainz am Rhein: Philipp von Zabern, 1979).

GRAEVE 1958: Mary A. Graeve, "The Stone of Unction in Caravaggio's Painting for the Chiesa Nuova," *Art Bulletin* 40, no. 3 (1958): 227ff.

GRAFTON 2000: Anthony Grafton, *Leon Battista Alberti: Master Builder of the Italian Renaissance* (Bath: Allen Lane, 2000).

GRAHAM 1963: J. W. Graham, "Acropolis and Parthenos: New Models in the Royal Ontario Museum," in *Parthenos and Parthenon*, ed. G. T. W. Hooker (Oxford: Clarendon, 1963), 77–83.

GRANDI AND GUIDOBALDI 2005: Monica Grandi and Federico Guidobaldi, "La decorazione in Opus Sectile della Casa del Rilievo di Telefo a Ercolano," in *Atti del X colloquio dell'AISCOM*, ed. Claudia Angelelli (Tivoli: Scripta Manent, 2005), 183–98.

GRANT AND MORTIMER 2002: Lindy Grant and Richard Mortimer, eds., *Westminster Abbey: The Cosmati Pavements* (Farnham: Ashgate, 2002).

GRAZIOLI MEDICI 1992: Priscilla Grazioli Medici, *Medici: Marmorari Romani* (Vatican: Tipografia Poliglotta Vaticana, 1992).

GREENHALGH 1989: Michael Greenhalgh, *The Survival of Roman Antiquities in the Middle Ages* (London: Duckworth, 1989).

GREWE 1991: Klaus Grewe, "Aquädukt-Marmor. Kalksinter der römischen Eifelwasserleitung als Baustoff des Mittelalters," *Bonner Jahrbücher des Rheinischen Landesmuseums in Bonn* (1991): 277–343.

GRIERSON 1962: Philip Grierson, "The Tombs and Obits of the Byzantine Emperors," *Dumbarton Oaks Papers* 16 (1962): 1–63.

GRIFFIN 1976: Miriam T. Griffin, *Seneca: A Philosopher in Politics* (Oxford: Clarendon, 1976).

GRIFFITH-JONES AND PARK 2010: Robin Griffith-Jones and David Park, eds., *The Temple Church in London: History, Architecture, Art* (Woodbridge: Boydell, 2010).

GROS 1973: Pierre Gros, "Hermodorus et Vitruve," *Mélanges de l'École française de Rome. Antiquité* 85, no. 1 (1973): 137–61.

GROS 1976: Pierre Gros, *Aurea Templa. Recherches sur l'architecture religieuse de Rome à l'époque d'Auguste* (Rome: École française de Rome, 1976).

GROS 2001: Pierre Gros, *L'architecture Romaine deu debut du III siècle av. J.-C. á la fin du Haut-Empire*, vol. 2, *Maisons, palais, villas et tombes* (Paris: Picard, 2001).

GROSSMANN 1973: Peter Grossmann, *S. Michele in Africisco zu Ravenna. Baugeschichtliche Untersuchungen* (Mainz am Rhein: Zabern, 1973).

GROSSMARK 1998: Tziona Grossmark, "'Shayish' (Marble) in Rabbinic Literature," in *Marble Studies: Roman Palestine and the Marble Trade*, ed. Moshe L Fischer (Constance: UVK Universitätsverlag Konstanz, 1998), 274–83.

GROSSMARK 2012: Tziona Grossmark, "Marble as Building Material in Rabbinic Literature," *Mediterranean Chronicle* 2 (2012): 61–78.
GROUPE DE RECHERCHES 1993: Groupe de recherches sur l'Afrique antique, ed., *Les Flavii de Cillium. Étude architecturale, épigraphique historique et littéraire du Mausolée de Kasserine (CIL VIII, 211–216)* (Rome: École française de Rome, 1993).
GRUBE 1966: Ernst Grube, "Elementi islamici nell'architettura veneta del medioevo," *Bollettino del Centro Internazionale di Studi di Architettura Andrea Palladio* 7, no. 2 (1966): 231–54.
GRUBEN 1997: Gottfried Gruben, "Naxos und Delos. Studien zur archaischen Architektur der Kykladen," *Jahrbuch des Deutschen Archäologischen Instituts* 112 (1997): 261–416.
GRUMEL 1926: Venance Grumel, "Léonce de Byzance," in *Dictionnaire de théologie catholique* (Paris: Letouzey & Ané, 1926), 400–426.
GUARNIERI 2015: Chiara Guarnieri, ed., *Il vetro di pietra. Il lapis specularis nel mondo romano dall'estrazione all'uso* (Faenza: Carta Bianca, 2015).
GUARRACINO 2009: Monica Guarracino, *Le pietre di Livorno: Transito e lavorazione delle pietre dure per la Cappella dei principi di Firenze nel XVII secolo* (Livorno: Sillabe, 2009).
GUÉRANGER 1868–1903: Prosper Guéranger, *The Liturgical Year,* trans. Laurence Shepherd, 15 vols. (Dublin: J. Duffy, 1868–1903).
GUERRA 2001: Andrea Guerra, "Quel che resta di Palladio. Eredità e dispersione nei progetti per la chiesa di San Giorgio Maggiore a Venezia," *Annali di Architettura* 13 (2001): 93–110.
GUIDO AND MANCELLA 2014: Sante Guido and Giuseppe Mancella, "Pietro Bernini e l'Assunzione della Beata Vergine nella Basilica di Santa Maria Maggiore. Note di tecnica artistica e restauro," in *Figure, liturgia e culto, arte. Ricerche sall'archivio della Basilica Papale di Santa Maria Maggiore,* ed. Michal Jagosz (Rome: Lisanti, 2014), 297–319.
GUIDOBALDI 1986: Federico Guidobaldi, "L'edilizia unifamiliare nella Roma tardoantica," in *Società romana e impero tardoantico,* ed. Andrea Giardina (Rome: Laterza, 1986), 165–237.
GUIDOBALDI 1989: Federico Guidobaldi, "Chiese Titolari di Roma nel Tessuto Urbano Preesistente," in *Quaeritur inventus colitur,* 2 vols. (Vatican City: Pontificio Istituto di Archeologia Cristiana, 1989), 1:383–96.
GUIDOBALDI 1992: Federico Guidobaldi, "L'opus sectile pariétal à Rome et en Italie," *Dossiers de l'archéologie* 173 (1992): 72–77.
GUIDOBALDI 1993: Federico Guidobaldi, "Roma. Il tessuto abitativo, le domus e i tituli," in *Storia di Roma,* ed. Lellia Cracco Ruggini, Andrea Giardina, Arnaldo Momigliano, Aldo Schiavone, and Andrea Carandini (Turin: Einaudi, 1993), 3:60–83.
GUIDOBALDI 1997: Federico Guidobaldi, "Gli scavi del 1993–95 nella basilica di S. Clemente a Roma e la scoperta del battistero paleocristiano. Nota preliminare," *Rivista di Archeologia Cristiana* 73 (1997): 459–91.
GUIDOBALDI 2000: Federico Guidobaldi, "La decorazione in opus sectile dell'aula," in *Roma aurea. Dalla città pagana alla città cristiana,* ed. Serena Ensoli and Eugenio La Rocca (Rome: Bretschneider, 2000), 251–62.
GUIDOBALDI 2002: Federico Guidobaldi, "Osservazioni sugli edifici romani in cui si insediò L'Ecclesia Urbis," in *Ecclesiae Urbis,* ed. Federico Guidobaldi and Alessandra Guiglia Guidobaldi (Vatican City: Pontificio Istituto di Archeologia Cristiana, 2002), 1033–71.
GUIDOBALDI AND GUIGLIA GUIDOBALDI 1983: Federico Guidobaldi and Alessandra Guiglia Guidobaldi, *Pavimenti marmorei di Roma dal IV al IX secolo* (Vatican City: Pontificio Istituto di Archeologia Cristiana, 1983).
GUIDOBALDI AND SALVATORI 1988: Federico Guidobaldi and A. Salvatori, "The Introduction of Marbles in Late Republican Rome: The Evidence from Mosaic Pavements with Marble Insertions," in Herz and Waelkens 1988, 171–75.
GUIDOBALDI AND VINCENTI 2005: Federico Guidobaldi and Valentina Vincenti, "Emblemata in sectile entro tappeti musivi come indizio di livello qualitativo ed elemento di datazione," in *Atti del X Colloquio dell'AISCOM,* ed. Claudia Angelelli (Tivoli: Edizioni Scripta Manent, 2005).
GUIDOBALDI ET AL. 1994: Federico Guidobaldi, Fulvia Olevano, Andrea Paribeni, and Daniela Trucchi, *Sectilia pavimenta di Villa Adriana* (Rome: Istituto Poligrafico dello Stato, 1994).
GUIGLIA GUIDOBALDI 1999A: Alessandra Guiglia Guidobaldi, "I mosaici aniconici di S. Sofia di Costantinopoli nell'età di Giustiniano," in *La mosaïque gréco-romaine VII,* ed. Mongi Ennaifer and Alain Rebourg (Tunis: Institut national du patrimoine, 1999), 691–702.
GUIGLIA GUIDOBALDI 1999B: Alessandra Guiglia Guidobaldi, "La decorazione pavimentale Bizantina in età Paleologa," in *L'arte di Bisanzio e L'Italia al tempo dei Paleologi, 1261–1453,* ed. Antonio Iacobini and Mauro Della Valle (Rome: Argos, 1999), 321–58.
GUIGLIA GUIDOBALDI 2004: Alessandra Guiglia Guidobaldi, "La decorazione marmorea dell'edificio di Santa Maria Antiqua fra tarda antichità e alto medioevo," in *Santa Maria Antiqua al Foro Romano cento anni dopo,* ed. John Osborne, J. Rasmus Brandt, and Giuseppe Morganti (Rome: Campisano, 2004), 49–66.
GUILLAND 1922: Rodolphe Guilland, "Le palais de Théodore Métochite," *Revue des études grecques* 35 (1922): 82–95.
GULDAGER BILDE 1993: Pia Guldager Bilde, "The International Style: Aspects of Pompeian First Style and Its Eastern Equivalents," in *Aspects of Hellenism in Italy: Towards a Cultural Unity?,* ed. Pia Guldager Bilde, Inge Nielsen, and Marjatta Nielsen (Copenhagen: Museum Tusculanum, 1993), 151–77.
GUNDERSHEIMER 1972: Werner L. Gundersheimer, *Art and Life at the Court of Ercole I d'Este: The De Triumphis Religionis of Giovanni Sabadino degli Arienti* (Geneva: Librairie Droz, 1972).
GUNDLACH 2009: Rolf Gundlach, "Horus in the Palace: The Centre of State and Culture in Pharaonic Egypt," in *4. Symposium zur ägyptischen Königsideologie: Egyptian Royal Residences,* ed. Rolf Gundlach and John H. Taylor (Wiesbaden: Harrassowitz, 2009), 45–68.
GURRIERI AND MAZZONI 1990: Francesco Gurrieri and Paolo Mazzoni, *La Fortezza da Basso, un monumento per la città* (Florence: Ponte alle Grazie, 1990).
GURRIERI ET AL. 1988: Francesco Gurrieri, Luciano Berti, and Claudio Leonardi, *La Basilica di San Miniato al Monte a Firenze* (Florence: Cassa di Risparmio di Firenze, 1988).
GUSSALI 1912: Emilio Gussali, "La casa di un umanista architetto. Il palazzo Raimondi a Cremona," *Rassegna d'arte* 12 (1912): 181–85.
HAAG BLETTER 1981: Rosemarie Haag Bletter, "The Interpretation of the Glass Dream—Expressionist Architecture and the History of the Crystal Metaphor," *Journal of the Society of Architectural Historians* 50, no. 1 (1981): 20–43.
HACKL 1912: Rudolf Hackl, "Die Fußboden," in Rodenwaldt 1912, 222–37.
HÄGER-WEIGEL 1997: Elisabeth Häger-Weigel, *Griechische Akrolith-Statuen des 5. und 4. Jhs. v. Chr.* (Berlin: Köster, 1997).
HAGGIS 2007: Donald C. Haggis, "Stylistic Diversity and Diacritical Feasting at Protopalatial Petras: A Preliminary Analysis of the Lakkos Deposit," *American Journal of Archaeology* 111 (2007): 715–75.
HAHN 2014: Cynthia Hahn, "Portable Altars (and the Rationale): Liturgical Objects and Personal Devotion," in *Image and Altar 800–1300: Papers from an International Conference in Copenhagen 24 October–27 October 2007,* ed. Poul Grinder-Hansen (Copenhagen: University Press of Southern Denmark, 2014), 45–64.
HALL 2000: Michael Hall, "What Do Victorian Churches Mean? Symbolism and Sacramentalism in Anglican Church Architecture, 1850–1870," *Journal of the Society of Architectural Historians,* 59, no. 1 (2000): 78–95.
HALL AND WOOLLEY 1927: Harry R. Hall and Leonard Woolley, *Al-'Ubaid: A Report on the Work Carried Out at Al-'Ubaid for the British Museum in 1919 and for the Joint Expedition in 1922–3* (London: Oxford University Press, 1927).
HALLEUX 1970: Robert Halleux, "Fécondité des mines et sexualité des pierres dans l'antiquité gréco-romaine," *Revue belge de philologie et d'histoire* 48, no. 1 (1970): 16–25.
HALLEUX 1974: Robert Halleux, *La problème des métaux dans la science antique* (Paris: Belles lettres, 1974).

HALSEY 1980: Richard Halsey, "The Galilee Chapel," in *Medieval Art and Architecture at Durham Cathedral* (London: British Archaeological Association, 1980), 59–73.

HALTON 1976: Thomas Halton, "The Coming of Spring: A Patristic Motif," *Classical Folia* 30, no. 2 (1976): 150–64.

HAMAMCIOGLU-TURAN ET AL. 2013: Mine Hamamcioglu-Turan, Yalcin Arisoy, Ayhan Nuhoglu, and Yusuf Percin Erturan, "Protection of Archaeological Remains in the Yortanli Dam Reservoir in Turkey," *International Journal of Architectural Heritage* 7, no. 6 (2013): 653–72.

HAMARNEH 1994: Basema Hamarneh, "I mosaici del complesso di S. Stefano: Proposta di lettura," in *Umm al-Rasas, Mayfa'ah I: Gli scavi del complesso di Santo Stefano*, ed. Michele Piccirillo and Eugenio Alliata (Jerusalem: Studium Biblicum Franciscum, 1994), 231–40.

HAMILTON 1949: Robert W. Hamilton, *The Structural History of the Aqsa Mosque: A Record of Archaeological Gleanings from the Repairs of 1938–1942* (London: Oxford University Press, 1949).

HAMMER-PURGSTALL 1822: Joseph von Hammer-Purgstall, *Constantinopolis und der Bosporos, örtlich und geschichtlich Beschrieben*, 2 vols. (Vienna: Anton Strauss, 1822).

HAMMOND AND BRIGHTMANN 1896: Charles E. Hammond and Frank E. Brightmann, eds., *Liturgies Eastern and Western* (Oxford: Clarendon, 1896).

HAND AND MANSFIELD 1993: John O. Hand and Sally E. Mansfield, *German Paintings of the Fifteenth through Seventeenth Centuries* (Washington, DC/New York: National Gallery of Art and Cambridge University Press, 1993).

HARARI 2010: Maurizio Harari, "La Tomba dei Festoni di Tarquinia e alcuni problemi di pittura greca tardoclassica," *Bollettino di Archeologia Online*, vol. 1 (2010): 56–68.

HARDEN 1987: Donald B. Harden, *Glass of the Caesars* (Milan: Olivetti, 1987).

HARDERING 1998: Klaus Hardering, *Die Abteikirche von Klosterrath: Baugeschichte und Bedeutung. De Abdijkerk te Rolduc: Bouwgeschiedenis en betekenis* (Utrecht: Clavis, 1998).

HARDIE 1983: Alex Hardie, *Statius and the Silvae: Poets, Patrons and Epideixis in the Graeco-Roman World* (Liverpool: Cairns, 1983).

HARDY 1994: Emma Hardy, "Farmer and Brindley, Craftsmen and Sculptors, 1850–1930," in *The Victorian Society Annual 1993*, ed. Peter Howell (London: Victorian Society, 1994), 4–17.

HARPER 1997: James G. Harper, "The Provisioning of Marble for the Sixth-Century Churches of Ravenna: A Reconstructive Analysis," in *Pratum Romanum: Richard Krautheimer zum 100. Geburtstag*, ed. Renate L. Colella, Meredith J. Gill, Lawrence A. Jenkins, and Petra Lamers (Wiesbaden: Reichert, 1997), 131–48.

HARPER ET AL. 1992: Prudence O. Harper, Joan Aruz, and Françoise Tallon, eds., *The Royal City of Susa: Ancient Near Eastern Treasures in the Louvre* (New York: Metropolitan Museum of Art and Abrams, 1992).

HARRELL 1990: James A. Harrell, "Misuse of the Term 'Alabaster' in Egyptology," *Göttinger Miszellen, Beiträge zur Ägyptologischen Diskussion* 119 (1990): 37–42.

HARRELL ET AL. 2007: James A. Harrell, Maarten Broekmans, and Dorothy Godfrey-Smith, "The Origin, Destruction, and Restoration of Colour in Egyptian Travertine," *Archaeometry* 49 (2007): 421–36.

HARRISON 1977: Evelyn B. Harrison, "Alkamenes' Sculptures for the Hephaisteion: Part I, the Cult Statues," *American Journal of Archaeology* 81, no. 2 (1977): 137–78.

HARRISON 1986: R. Martin Harrison, *Excavations at Saraçhane in Istanbul* (Princeton: Princeton University Press, 1986).

HARRISON 1989: R. Martin Harrison, *A Temple for Byzantium: The Discovery and Excavation of Anicia Juliana's Palace-Church in Istanbul* (Austin: University of Texas Press, 1989).

HARTT ET AL. 1964: Frederick Hartt, Gino Corti, and Clarence Kennedy, *The Chapel of the Cardinal of Portugal, 1434–1459, at San Miniato in Florence* (Philadelphia: University of Pennsylvania Press, 1964).

HASKELL 1963: Francis Haskell, *Patrons and Painters: A Study in the Relations between Italian Art and Society in the Age of the Baroque* (New York: Harper & Row 1963).

HAYES 1937: William C. Hayes, *Glazed Tiles from the Palace of Ramesses II at Kantir* (New York: Metropolitan Museum of Art, 1937).

HAYES 1959: William C. Hayes, *The Scepter of Egypt: A Background for the Study of the Egyptian Antiquities Part II: The Hyksos Period and the New Kingdom (1675–1080 B.C.)* (New York: Metropolitan Museum of Art, 1959).

HEIDEGGER 1975: Martin Heidegger, *Poetry, Language, Thought*, trans. Albert Hofstadter (New York: Harper & Row, 1975).

HEIDET 2004: Sandrine Heidet, "Pierres ponces, concrétions calcaires . . . Le décor à 'rocailles' des parois et des voûtes en Italie et en Gaule à l'époque romaine," in *Les roches décoratives dans l'architecture antique et du Haut Moyen Âge*, ed. Pascale Chardon-Picault, Jacqueline Lorenz, Pierre Rat, and Gilles Sauron (Paris: Comité des travaux historiques et scientifiques, 2004), 289–97.

HEILMEYER AND RAKOB 1973: Wolf-Dieter Heilmeyer and Friedrich Rakob, *Der Rundtempel Am Tiber in Rom* (Mainz am Rhein: Philipp von Zabern, 1973).

HEIMBÜRGER-RAVALLI 1977: Minna Heimbürger-Ravalli, *Architettura, scultura e arti minori nel Barocco italiano: Ricerche nell'Archivio Spada* (Florence: Olschki, 1977).

HEISENBERG 1908: August Heisenberg, *Grabeskirche und Apostelkirche, Zwei Basiliken Konstantins; Untersuchungen zur Kunst und Literatur des ausgehenden Altertums*, 2 vols. (Leipzig: Hinrichs, 1908).

HEJL AND TIPPELT 2005: Ewald Hejl and Gerold Tippelt, "Prehistorical Pigment Mining on Santorini's Neighbouring Island Anafi (Cyclades, Greece)," *Austrian Journal of Earth Sciences* 98 (2005): 22–33.

HEMSOLL 1990: David Hemsoll, "The Architecture of Nero's Golden House," in *Architecture and Architectural Sculpture in the Roman Empire*, ed. Martin Henig (Oxford: Oxford University Committee for Archaeology, 1990), 10–38.

HENNEMEYER 2015: Arnd Hennemeyer, "The Temple Architecture and Its Modifications during the 5th Century BCE," in *New Approaches to the Temple of Zeus at Olympia*, ed. András Patay-Horváth (Newcastle upon Tyne: Cambridge Scholars, 2015), 16–38.

HERMANN ET AL. 2002: John J. Hermann, Norman Herz, and Richard Newman, eds., *Asmosia 5: Interdisciplinary Studies on Ancient Stone. Proceedings of the Fifth International Conference of the Association for the Study of Marble and Other Stones in Antiquity, Museum of Fine Arts, Boston, 1998* (London: Archetype, 2002), 240–49.

HERZ AND WAELKENS 1988: Norman Herz and Marc Waelkens, eds., *Classical Marble: Geochemistry, Technology, Trade* (Dordrecht: Kluwer, 1988).

HESSE 1992: Sabine Hesse, *Die Fassade der Oratorium San Bernardino in Perugia. Ein Beitrag zum Werk des Agostino di Duccio* (Göppingen: Albert-Ludwigs-Universität Freiburg, 1992).

HEYDENREICH 1961: Ludwig H. Heydenreich, "Die Cappella Rucellai von San Pancrazio in Florenz," in *De artibus opuscula XL: Essays in Honor of Erwin Panofsky*, ed. Millard Meiss (New York: New York University Press, 1961), 1:219–29.

HEYDENREICH AND LOTZ 1974: Ludwig H. Heydenreich and Wolfgang Lotz, *Architecture in Italy, 1400 to 1600* (Harmondsworth: Penguin, 1974).

HEYWORTH 1993: Mike Heyworth, "Analysis of the Window Glass," in *The Uley Shrines*, ed. Ann Woodward and Peter Leach (London: English Heritage, 1993), 189–92.

HIBBARD 1965: Howard Hibbard, *Bernini* (Harmondsworth: Penguin, 1965).

HIGGINS AND HIGGINS 1996: Michael Denis Higgins and Reynold Higgins, *A Geological Companion to Greece and the Aegean* (Ithaca, NY: Cornell University Press, 1996).

HIGGINS ET AL. 1968: Reynold A. Higgins, S. E. Ellis, Richard Hope Simpson, and R. H. Higgins, "The Façade of the Treasury of Atreus at Mycenae," *Annual of the British School at Athens* 63 (1968): 331–36.

HILL 2001: Michael Hill, "The Mavellous Facades of S. Susanna (1603) and S. Maria della Vittoria (1626) in Rome," *Proceedings of the 18th Annual Conference of the SAHANZ* (Darwin: *SAHANZ*, 2001), 231–37.

HILLS 1999: Paul Hills, *Venetian Colour: Marble, Mosaic, Painting and Glass 1250–1550* (New Haven: Yale University Press, 1999).

HINKS 1933: Roger P. Hinks, *Catalogue of the Greek, Etruscan, and Roman Paintings and Mosaics in the British Museum* (London: British Museum, 1933).

HIRN 1912: Yrjö Hirn, *The Sacred Shrine: A Study of the Poetry and Art of the Catholic Church* (London: Macmillan, 1912).

HIRSCH 1975: Ethel S. Hirsch, "Infrared Photography and Archaeology, Painted Floors at Gournia," *Archaeology* 28 (1975): 260–66.

HIRSCH 1977A: Ethel S. Hirsch, *Painted Decoration on the Floors of Bronze Age Structures on Crete and the Greek Mainland* (Göteborg: P. Åström, 1977).

HIRSCH 1977B: Ethel S. Hirsch, "The Restorations of the Painted Courtyard Floor at Mycenae," *Antike Kunst* 20, no. 1 (1977): 54–56.

HIRSCH 1980: Ethel S. Hirsch, "Another Look at Minoan and Mycenean Interrelationships in Floor Decoration," *American Journal of Archaeology* 84, no. 4 (1980): 453–62.

HIRST 1981: Michael Hirst, *Sebastiano del Piombo* (Oxford: Oxford University Press, 1981).

HIRTH 1885: Friedrich Hirth, *China and the Roman Orient: Researches into Their Ancient and Mediaeval Relations as Represented in Old Chinese Records* (Shanghai: Kelly & Walsh, 1885).

HITCHCOCK ET AL. 2016: Louise A. Hitchcock, Anne P. Chapin, Emilia Banou, and James H. Reynolds, "The Conglomerate Quarry at the Mycenaean Site of Vapheio-Palaiopyrgi in Laconia," *Hesperia* 85 (2016): 65–90.

HOEPFNER 2001: Wolfram Hoepfner, "Der parische Lichtdom," *Antike Welt* 32 (2001): 491–506.

HOFFMANN 1952: Herbert Hoffmann, "Antecedents of the Great Altar at Pergamon," *Journal of the Society of Architectural Historians* 11, no. 3 (1952): 1–5.

HOFFMANN 1953: Herbert Hoffmann, "Foreign Influence and Native Invention in Archaic Greek Altars," *American Journal of Archaeology* 57, no. 3 (1953): 189–95.

HOFFMANN AND FABER 2009: Adolf Hoffmann and Andrea Faber, *Die Casa del Fauno in Pompeji (VI 12)*, vol. 1, *Bauhistorische Analyse. Die stratigraphischen Befunde und Funde der Ausgrabungen in den Jahren 1961 bis 1963* (Wiesbaden: Reichert, 2009).

HÖLSCHER 1988: Tonio Hölscher, "Historische Reliefs," in *Kaiser Augustus und die verlorene Republik*, ed. Mathias Hofter (Mainz am Rhein: Zabern, 1988), 384–86, cat. nos. 214–16.

HOLTZMANN 1996: Bernard Holtzmann, "Note Sur Le Colosse Des Naxiens," *Topoi* 6, no. 1 (1996): 199–204.

HOMOLLE 1890: Théophile Homolle, "Comptes et inventaires des temples Déliens en l'année 279," *Bulletin de correspondence hellénique* 14, no. 2 (1890): 389–511.

HOOD 2005: Sinclair Hood, "Dating the Knossos Frescoes," in Morgan 2005, 45–81.

HOPE 1992: Charles Hope, "The Early History of the Tempio Malatestiano," *Journal of the Warburg and Courtauld Institutes* 55 (1992): 51–154.

HORNBLOWER 1982: Simon Hornblower, *Mausolus* (Oxford: Clarendon, 1982).

HORNE 1966: R. A. Horne, "Aristotelian Chemistry," *Chymia* 11 (1966): 21–27.

HOROWITZ 1988: Wayne Horowitz, "The Babylonian Map of the World," *Iraq* 50 (1988): 147–65.

HOROWITZ 1992: Wayne Horowitz, "Two Abnu Šikinšu Fragments and Related Matters," *Zeitschrift für Assyriologie und Vorderasiatische Archäologie* 82, no. 1 (1992): 112–22.

HOROWITZ 1998: Wayne Horowitz, *Mesopotamian Cosmic Geography* (Winona Lake, IN: Eisenbrauns, 1998).

HOWARD 1980: Deborah Howard, *The Architectural History of Venice* (London: B. T. Batsford, 1980).

HOWARD 1989: Deborah Howard, "The Church of the Miracoli in Venice and Pittoni's St. Jerome Altar-Piece," *Burlington Magazine* 131, no. 1039 (1989): 684–92.

HOWARD 1991: Deborah Howard, "Venice and Islam in the Middle Ages: Some Observations on the Question of Architectural Influence," *Architectural History* 34 (1991): 59–72.

HÜLSEN 1927: Christian Hülsen, "Die Basilica des Iunius Bassus und die kirche S. Andrea cata Barbara auf dem Esquilin," in *Festschrift für Julius Schlosser zum 60. Geburtstage*, ed. Leo Planiscig and Arpad Weixlgärtner (Zurich: Amalthea, 1927), 53–67.

HUMFREY 1983: Peter Humfrey, *Cima da Conegliano* (Cambridge: Cambridge University Press, 1983).

HUNGER 1969–70: Herbert Hunger, "On the Imitation (MIMHSIS) of Antiquity in Byzantine Literature," *Dumbarton Oaks Papers* 23/24 (1969–70): 17–38.

HUNT 1984: Lucy-Anne Hunt, "Comnenian Aristocratic Palace Decoration: Descriptions and Islamic Connections," in *The Byzantine Aristocracy, IX to XIII Centuries*, ed. Michael Angold (Oxford: BAR, 1984), 138–57.

HUSKINSON 1969: J. M. Huskinson, "The Crucifixion of St. Peter: A Fifteenth-Century Topographical Problem," *Journal of the Warburg and Courtauld Institutes* 32 (1969): 135–61.

HUXLEY 1959: George L. Huxley, *Anthemius of Tralles: A Study of Later Greek Geometry* (Cambridge, MA: Greek, Roman and Byzantine Studies, 1959).

HUXLEY 1997: Margaret Huxley, "The Shape of the Cosmos According to Cuneiform Sources," *Journal of the Royal Asiatic Society* 7, no. 2 (1997): 189–98.

HYDE MINOR 1989: Vernon Hyde Minor, "Shapes of the Invisible: Bernini's Fiery Angels in Saint Peter's," *Artibus et Historiae* 10, no. 19 (1989): 149–55.

IACOPI 1991: Irene Iacopi, "Casa dei Grifi. Nuove pitture," *Bollettino di Archeologia* 7 (1991): 83–87.

IACOPI AND TEDONE 2006: Irene Iacopi and Giovanna Tedone, "Bibliotheca e Porticus ad Apollinis," *Mitteilungen des deutschen archäologischen Instituts. Römische Abteilung* 112 (2006): 351–78.

IANNUCCI 2006: Anna Maria Iannucci, "L'ultimo restauro del Tempio Malatestiano," in *Leon Battista Alberti e l'architettura*, ed. Massimo Bulgarelli, Arturo Calzona, Matteo Ceriana, and Francesco Paolo Fiore (Cinisello Balsamo: Silvana, 2006), 304–11.

IBRAHIM ET AL. 1976: Leila Ibrahim, Robert L. Scranton, and Robert Brill, *Kenchreai: The Panels of Opus Sectile in Glass* (Leiden: Brill, 1976).

IENI 2002: Giulio Ieni, "La chiesa conventuale di Santa Croce e alcuni arredi fissi," in *Santa Croce di Bosco Marengo*, ed. Fulvio Cervini and Carlenrica Spantigati (Alessandria: Cassa di Risparmio di Alessandria, 2002), 52–65.

IKE AND OTAKI 1998: Kozo Ike and Toru Otaki, "Natural Lighting Method of the Parthenon: An Experimental Study for Transparency of Pentelic Marble Tile," *Journal of Architecture, Planning, and Environmental Engineering* (*Architectural Institute of Japan*) 504 (1998): 235–43.

IMMERWAHR 1990: Sara A. Immerwahr, *Aegean Painting in the Bronze Age* (University Park: Pennsylvania State University Press, 1990).

IMORDE 1997: Joseph Imorde, *Präsenz und Repräsentanz: Oder die Kunst, den Leib Christi auszustellen* (Emsdetten: Imorde, 1997).

IORIO 2007: Vincenza Iorio, "I rapporti tra Delo e Pompei nel II secolo a.c. Aspetti religiosi e circolazione dei temi figurativi," in *Circulación de temas y sistemas decorativos en la pintura mural antigua*, ed. Carmen Guiral Pelegrín (Calatayud: Gobierno de Aragón, 2007), 59–66.

ISSERLIN 1998: Raphael M. J. Isserlin, "A Spirit of Improvement? Marble and the Culture of Roman Britain," in *Cultural Identity in the Roman Empire*, ed. Ray Laurence and Joanne Berry (London: Routledge, 1998), 125–55.

JACKSON 2009: Heather Jackson, "Erotes on the Euphrates: A Figured Frieze in a Private House at Hellenistic Jebel Khalid on the Euphrates," *American Journal of Archaeology* 113, no. 2 (2009): 231–53.

JACOB 1975: Sabine Jacob, *Italienische Zeichnungen der Kunstbibliothek Berlin: Architektur und Dekoration 16. bis 18. Jahrhundert* (Berlin: Mann, 1975).

JAKOBS 1987: Peter H. F. Jakobs, *Die frühchristlichen Ambone Griechenlands* (Bonn: R. Habelt, 1987).

JAMES 1996: Liz James, *Light and Colour in Byzantine Art* (Oxford: Clarendon, 1996).

JAMES 2007: Liz James, "'And Shall These Mute Stones Speak?' Text as Art," in *Art and Text in Byzantine Culture*, ed. Liz James (Cambridge: Cambridge University Press, 2007), 188–206.

JAMES AND VAN DER SLUIJS 2008: Peter James and Marinus Anthony van der Sluijs, "Ziggurats, Colors, and Planets: Rawlinson Revisited," *Journal of Cuneiform Studies* 60 (2008): 57–79.

JAMES AND WEBB 1991: Liz James and Ruth Webb, "To Understand Ultimate Things and Enter Secret Places: Ekphrasis and Art in Byzantium," *Art History* 14, no. 1 (1991): 1–17.

JANES 1998: Dominic Janes, *God and Gold in Late Antiquity* (Cambridge: Cambridge University Press, 1998).

JANIN 1944–45: Raymond Janin, "Les sanctuaires de Byzance sous la domination latine," *Revue des études byzantines* 2 (1944–45): 134–84.

JANIN 1953: Raymond Janin, *La géographie ecclésiastique de l'empire byzantin* (Paris: Institut français d'études byzantines, 1953).

JANIN 1964: Raymond Janin, *Constantinople byzantine: Développement urbain et répertoire topographique* (Paris: Institut francais d'études byzantines, 1964).

JANSON 1961: Horst W. Janson, "The 'Image Made by Chance' in Renaissance Thought," in *De artibus opuscula XL: Essays in Honor of Erwin Panofsky*, ed. Millard Meiss (New York: New York University Press, 1961), 254–66.

JANVIER 1969: Yves Janvier, *La législation du bas-empire romain sur les édifices publics* (Aix-en-Provence: Pensée Universitaire, 1969).

JENKINS AND MIDDLETON 1988: Ian D. Jenkins and A. P. Middleton, "Paint on the Parthenon Sculptures," *Annual of the British School at Athens* 83 (1988): 183–207.

JEPPESEN 2000: Kristian Jeppesen, *The Maussolleion at Halikarnassos: Reports of the Danish Archaeological Expedition to Bodrum*, vol. 4, *The Quadrangle: The Foundations of the Maussolleion and Its Sepulchral Compartments* (Aarhus: Aarhus University Press, 2000).

JOHNSON 1995: Mark J. Johnson, "The Fifth-Century Oratory of the Holy Cross at the Lateran in Rome," *Architectura* 25 (1995): 134–47.

JOLIVET ET AL. 1999: Vincent Jolivet, Henri Broise, and Marco Rossi, "Rome: Pincio (Jardins de Lucullus). Chronique du chantier 1998," *Mélanges de l'École française de Rome. Antiquité* 111 (1999): 481–86.

JOLIVET ET AL. 2000: Vincent Jolivet, Henri Broise, and Martine Dewailly, "Rome: Pincio (Jardins de Lucullus). Chronique du chantier 1999," *Mélanges de l'École française de Rome. Antiquité* 112 (2000): 432–53.

JONES 1987: Roger Jones, "Mantegna and Materials," *I Tatti Studies* 2 (1987): 71–90.

JONES 2005: R. E. Jones, "Technical Studies of Aegean Bronze Age Wall Painting: Methods, Results, and Future Prospects," in Morgan 2005, 199–224.

JONES 2014: Nathaniel Jones, "Ancient Painted Panels, Terminology and Appearance," *Mnemosyne* 67 (2014): 295–304.

JONES 2015: Nathaniel Jones, "Temple Inventory and Fictive Picture Gallery: Ancient Painting between Votive Offering and Artwork," in *Museum Archetypes and Collecting in the Ancient World*, ed. Maia Wellington Gathan and Donatella Pegazzano (Leiden: Brill, 2015), 118–28.

JONES AND PENNY 1983: Roger Jones and Nicholas Penny, *Raphael* (New Haven: Yale University Press, 1983).

JORDAN 1932: Julius Jordan, *Dritter Vorläufiger Bericht über die von der Notgemeinschaft der Deutschen Wissenschaft in Uruk unternommenen Ausgrabungen* (Berlin: Verlag der Akademie du Wissenschaften, 1932).

JØRGENSEN 1954: Johannes Jørgensen, *Saint Bridget of Sweden*, trans. Ingebord Lund, 2 vols. (London: Longmans Green, 1954).

JOST 1985: Madeleine Jost, *Sanctuaries et cultes d'Arcadie* (Paris: Librairie philosophique J. Vrin, 1985).

KAHN 1991: Deborah Kahn, *Canterbury Cathedral and Its Romanesque Sculpture* (London: Harvey Miller, 1991).

KAJAVA 2005: Mika Kajava, "La Villa Lante al Gianicolo e la Villa di Giulio Marziale," in *Villa Lante al Gianicolo. Storia della Fabbrica e cronaca degli abitatori*, ed. Tancredi Carunchio and Simo Örmä (Rome: Palombi & Partner, 2005), 11–18.

KALDELLIS 2013: Anthony Kaldellis, "The Making of Hagia Sophia and the Last Pagans of New Rome," *Journal of Late Antiquity* 6, no. 2 (2013): 347–66.

KALLISTOS 1988: Bishop Kallistos of Dioklea, "The Meaning of the Divine Liturgy for the Byzantine Worshipper," in *Church and People in Byzantium*, ed. Rosemary Morris (Chester: Bemrose, 1988), 7–28.

KALUS 1991: Ludvik Kalus, "Rock-Crystal Talismans against Drought," in *Jewelry and Goldsmithing in the Islamic World*, ed. Na'ama Brosh (Jerusalem: Israel Museum, 1991), 101–4.

KÄMPF 2015: Tobias Kämpf, *Archäologie offenbart: Cäciliens römisches Kultbild im Blick einer Epoche* (Leiden: Brill, 2015).

KANTOROWICZ 1963: Ernst H. Kantorowicz, "Oriens Augusti–Lever du Roi," *Dumbarton Oaks Papers* 17 (1963): 117–35.

KANZ 1994: Roland Kanz, "Balthasar Permosers Christus an der Geißelsäule," in *Pinxit/sculpsit/fecit. Kunsthistorische Studien: Festschrift für Bruno Bushart*, ed. Bärbel Hamacher and Christl Karnehm (Munich: Deutscher Kunstverlag, 1994), 226–41.

KARLIN-HAYTER 1991: Patricia Karlin-Hayter, "L'adieu à l'empereur," *Byzantion* 61, no. 1 (1991): 123–26.

KAUFMAN 2002: A. F. Kaufman, "Identifying Stone Remains: The Second Temple at Jerusalem as a Case Study," in *Interdisciplinary Studies on Ancient Stone*, ed. Lorenzo Lazzarini (Padua: Bottega d'Erasmo, 2002), 437–42.

KELLER 1986: Fritz-Eugen Keller, "Bemerkungen zur Villa Suburbana des Baldassare Turini (Villa Lante)," in Frommel and Winner 1986, 349–52.

KELLY 1987: Kristin A. Kelly, *Motifs in Opus Sectile and Its Painted Imitation from the Tetrarchy to Justinian*, 2 vols. (Ann Arbor: UMI, 1987).

KEMP 1981: Martin Kemp, *Leonardo da Vinci: The Marvellous Works of Nature and Man* (London: Dent, 1981).

KENNELL 1994: Stefanie A. H. Kennell, "Hercules' Invisible Basilica (Cassiodorus, Variae I, 6)," *Latomus* 53 (1994): 159–75.

KESSLER 1991–92: Herbert L. Kessler, "'Pictures Fertile with Truth:' How Christians Managed to Make Images of God without Violating the Second Commandment," *Journal of the Walters Art Gallery* 49–50 (1991–92): 53–65.

KEYSER AND IRBY-MASSIE 2008: Paul T. Keyser and Georgia L. Irby-Massie, eds., *The Encyclopedia of Ancient Natural Scientists: The Greek Tradition and Its Many Heirs* (Abingdon: Routledge, 2008).

KHADER 1987: Aïcha Ben Abed Ben Khader, *Thuburbo Majus. Les mosaiques dans le quartier ouest* (Tunis: Institut national d'archéologie et d'art, 1987).

KHADER ET AL. 1985: Aïcha Ben Abed Ben Khader, Mongi Ennaifer, Marie Spiro, Margaret A. Alexander, and David Soren, *Thuburbo Maius. Les mosaiques de la region des grands thermes* (Tunis: Institut National d'Archéologie et d'Art, 1985).

KIDD 2001: Benton Kidd, "Technique and Composition of the Tel Anafa Stucco," *MUSE* 33–35 (2001): 4–13.

KIESOW 1962: Gottfried Kiesow, "Die gotische Südfassade von S. Maria Novella in Florenz," *Zeitschrift für Kunstgeschichte* 25 (1962): 1–12.

KIMPEL AND SUCKALE 1990: Dieter Kimpel and Robert Suckale, *L'architecture gothique en France 1130–1270*, trans. Françoise Neu (Paris: Flammarion, 1990).

KINGSLEY 1992: Peter Kingsley, "Ezekiel by the Grand Canal: Between Jewish and Babylonian Tradition," *Journal of the Royal Asiatic Society* 2, no. 3 (1992): 339–46.

KINGSLEY PORTER 1912A: Arthur Kingsley Porter, "San Savino at Piacenza: I. History and Structure," *American Journal of Archaeology* 16, no. 3 (1912): 350–67.

KINGSLEY PORTER 1912B: Arthur Kingsley Porter, "San Savino at Piacenza II. Ornament. Conclusions," *American Journal of Archaeology* 16, no. 4 (1912): 495–517.

KITSON 1978: Peter Kitson, "Lapidary Traditions in Anglo-Saxon England: Part I, the Background; the Old English Lapidary," *Anglo-Saxon England* 7 (1978): 9–60.

KITZINGER 1951: Ernst Kitzinger, "Studies on Late Antique and Early Byzantine Floor Mosaics, I: Mosaics at Nikopolis," *Dumbarton Oaks Papers* 6 (1951): 83–122.

KLEINBAUER 1984: W. Eugene Kleinbauer, "Remarks on the Building History of the Acheiropoietos Church at Thessaloniki," in *Actes du Xe Congrès international d'archéologie chrétienne,* 2 vols. (Vatican City: Pontificio Istituto di Archeologia Cristiana, 1984), 2:241–57.

KLEINBAUER 1999: W. Eugene Kleinbauer, *Saint Sophia at Constantinople. Singulariter in Mundo* (Dublin, NH: Bauhan, 1999).

KLEINER 1991: Fred S. Kleiner, "The Trophy on the Bridge and the Roman Triumph over Nature," *L'antiquité classique* 60 (1991): 182–92.

KLEINERT 1979: Annette Kleinert, *Die Inkrustation der Hagia Sophia: Zur Entwicklung der Inkrustationsschemata im römischen Kaiserreich* (Münster: University of Münster, 1979).

KLEINERT 1981: Annette Kleinert, "Die Inkrustation der Hagia Sophia: Zur Entwicklung der Inkrustationsschemata im römischen Kaiserreich," *Das Münster* 34 (1981): 245–346.

KLEMM AND KLEMM 1991: Dietrich D. Klemm and Rosemarie Klemm, "Calcit-Alabaster oder Travertin?," *Göttinger Miszellen, Beiträge zur Ägyptologischen Diskussion* 122 (1991): 57–70.

KNORR 1983: Wilbur Knorr, "The Geometry of Burning-Mirrors in Antiquity," *Isis* 74, no. 1 (1983): 53–73.

KOCH 1996: Linda A. Koch, "The Early Christian Revival at S. Miniato al Monte: The Cardinal of Portugal Chapel," *Art Bulletin* 78, no. 3 (1996): 527–55.

KÖCHER 1959: Franz Köcher, "Ein spätbabylonischer Hymnus auf den Tempel Ezida in Borsippa," *Zeitschrift für Assyriologie und vorderasiatische Archäologie* 53 (1959): 236–40.

KOEHL 2013: Robert B. Koehl, "The Near Eastern Contribution to Aegean Wall Painting and Vice Versa," in Aruz et al. 2013, 170–79.

KOHL 2004: Jeanette Kohl, *Fama und Virtus: Bartolomeo Colleonis Grabkapelle* (Berlin: Akademie, 2004).

KOKKOROU-ALEVRAS 1992: Georgia Kokkorou-Alevras, "Τα ἀρχαία λατομεία μάρμαρου τής Ναξίου," *Archaiologike Ephemeris* 131 (1992): 101–27.

KOKKOROU-ALEVRAS 1995: Georgia Kokkorou-Alevras, "Die archaische naxische Bildhauerei," *Antike Plastik* 24 (1995): 37–138.

KOMMER 1974: Björn R. Kommer, *Nicodemus Tessin der Jüngere und das Stockholmer Schloss* (Heidelberg: C. Winter, 1974).

KONTOLEON 1973: Nikolaos M. Kontoleon, "Ἀνδριὰς καὶ τὸ σφέλας," in *Études Déliennes* (Athens: E. de Boccard, 1973), 239–51.

KORRES 1995: Manolis Korres, *From Pentelicon to the Parthenon* (Athens: Melissa, 1995).

KORRES 2000: Manolis Korres, "Υπόγεια Λατομεία της Πάρου," in Schilardi 2000, 61–82.

KOSMETATOU 2003: Elizabeth Kosmetatou, "Poseidippos, 'Epigr.' 8 AB and Early Ptolemaic Cameos," *Zeitschrift für Papyrologie und Epigraphik* 142 (2003): 35–42.

KOSMETATOU 2004: Elizabeth Kosmetatou, "On Large Gemstones," *Zeitschrift für Papyrologie und Epigraphik* 146 (2004): 81–84.

KOSTENEC AND DARK 2011: Jan Kostenec and Ken Dark, "Paul the Silentiary's Description of Hagia Sophia in the Light of New Archaeological Evidence," *Byzantinoslavica* 3 (2011): 89–94.

KOSTOF 1965: Spiro Kostof, *The Orthodox Baptistery of Ravenna* (New Haven: Yale University Press, 1965), 57–58.

KOTTARIDI 2011: Angeliki Kottaridi, "The Palace of Aegae," in *Brill's Companion to Ancient Macedon: Studies in the Archaeology and History of Macedon, 650 BC–300 AD*, ed. Robin J. Lane Fox (Leiden: Brill, 2011), 297–333.

KOURKOUTIDOU-NIKOLAÏDOU 1989: Eftychia Kourkoutidou-Nikolaïdou, *Acheiropoietos: The Great Church of the Mother of God* (Thessaloniki: Institute for Balkan Studies, 1989).

KRAELING 1956: Carl H. Kraeling, *The Synagogue: The Excavations at Dura-Europos* (New Haven: Yale University Press, 1956).

KRAMER 1956: Samuel N. Kramer, *History Begins at Sumer* (Garden City, NY: Doubleday, 1956).

KRAMER 2002: Bärbel Kramer, "Il rotolo di Milano e l'epigramma ellenistico," in *Un poeta ritrovato–Posidippo di Pella, giornata di studio, Milano, 23 nov. 2001* (Milan: LED, 2002), 33–45.

KRAUTHEIMER 1933–77: Richard Krautheimer, *Corpus Basilicarum Christianarum Romae* (Vatican City: Pontificio Istituto di Archeologia Cristiana, 1937–77).

KRAUTHEIMER 1964: Richard Krautheimer, "Zu Konstantins Apostelkirche in Konstantinopel," in *Mullus: Festschrift Theodor Klauser,* ed. Alfred Stuiber and Alfred Hermann (Munster: Aschendorff, 1964), 224–29; repr. in *Studies in Early Christian, Medieval, and Renaissance Art* (London: University of London Press, 1971), 27–34.

KRAUTHEIMER AND ĆURČIĆ 1986: Richard Krautheimer and Slobodan Ćurčić, *Early Christian and Byzantine Architecture* (Harmondsworth: Penguin, 1986).

KREIKENBOM 1992: Detlev Kreikenbom, *Griechische und römische Kolossalporträts bis zum späten Ersten Jahrhundert nach Christus* (Berlin: de Gruyter, 1992).

KRINSKY 1967: Carol Herselle Krinsky, "Seventy-Eight Vitruvius Manuscripts," *Journal of the Warburg and Courtauld Institutes* 30 (1967): 36–70.

KRÖNIG 1965: Wolfgang Krönig, *Il Duomo di Monreale e l'architettura normanna in Sicilia* (Palermo: S. F. Flaccovio, 1965).

KUBACH AND VERBEEK 1976: Hans E. Kubach and Albert Verbeek, *Romanische Baukunst an Rhein und Maas: Katalog der vorromanischen und romanischen Denkmäler,* 3 vols. (Berlin: Deutscher Verlag für Kunstwissenschaft, 1976).

KUBAN ET AL. 1997: Y. Doğan Kuban, Cecil L. Striker, and Albrecht Berger, *Kalenderhane in Istanbul: The Buildings, Their History, Architecture and Decoration: Final Reports on the Archaeological Exploration and Restoration at Kalenderhane Camii, 1966–1978* (Mainz: Philipp von Zabern, 1997).

KUMMER 1987: Stefan Kummer, *Anfänge und Ausbreitung der Stuckdekoration im Römischen Kirchenraum (1500–1600)* (Tübingen: Wasmuth, 1987).

KUMMER 1996: Stefan Kummer, "Antiker Buntmarmor als Dekorationselement römischer Kirchen im 16. Jahrhundert," in *Antike Spolien in der Architektur des Mittelalters und der Renaissance,* ed. Joachim Poeschke (Munich: Hirmer, 1996), 329–46.

KYRIELEIS 1996: Helmut Kyrieleis, *Der große Kuros von Samos* (Bonn: In Kommission bei R. Habelt, 1996).

L'ORANGE 1942: Hans P. L'Orange, "Domus Aurea, der Sonnenpalast," *Serta Eitremiana SO Supplement* 11 (1942): 68–100.

L'ORANGE 1974–75: Hans Peter L'Orange, "Lux Aeterna. L'adorazione della luce nell'arte tardoantica e alto-medievale," *Rendiconti della Pontificia Accademia Romana di Archeologia* 47 (1974–75): 191–202.

LA BARBERA BELLIA 1997: Simonetta La Barbera Bellia, *Il paragone delle arti nella teoria artistica del Cinquecento* (Bagheria: Cafaro, 1997).

LA MALFA 2009: Claudia La Malfa, *Pintoricchio a Roma. La seduzione dell'antico* (Cinisello Balsamo: Silvana, 2009).

LA ROCCA 1996: Eugenio La Rocca, "Le tegole del tempio di Hera Lacinia ed il tempio della Fortuna Equestre: Tra spoliazioni e restauri in età tardo-repubblicana," in Spadea 1996, 89–98.

LA ROSA 1999: Vincenzo La Rosa, "Nuovi dati sulla tomba del sarcofago dipinto di H. Triada," in *Ἐπὶ πόντον πλαζόμενοι. Simposio italiano di Studi Egei dedicato a Luigi Bernabò e Giovanni Pugliese Carratelli, Roma, 18–20 febbraio 1998,* ed. Vincenzo La Rosa, Dario Palermo, and Lucia Vagnetti (Rome: Scuola Archeologica Italiana di Atene, 1999), 177–88.

LAIDLAW 1975: Anne Laidlaw, "A Reconstruction of the First Style Decoration in the Alexander Exedra of the House of the Faun," in *Neue Forschungen in Pompeji und den Anderen vom Vesuvausbruch 79 N. Chr. Verschutteten Stadten*, ed. Bernard Andreae and Helmut Kyrieleis (Recklinghausen: Bongers, 1975): 39–52.

LAIDLAW 1985: Anne Laidlaw, *The First Style in Pompeii: Painting and Architecture* (Rome: Bretschneider, 1985).

LAIRD 2000: Margaret L. Laird, "Reconsidering the So-Called 'Sede degli Augustali' at Ostia," *Memoirs of the American Academy at Rome* 45 (2000): 41–72.

LAMB AND WACE 1919–21: Winifred Lamb and A. J. B. Wace, "Excavations at Mycenae," *Annual of the British School at Athens* 24 (1919/20–1920/21): 185–209.

LAMBERT 1960: Wilfred G. Lambert, *Babylonian Wisdom Literature* (Oxford: Clarendon, 1960).

LAMBERTON 1986: Robert Lamberton, *Homer the Theologian: Neoplatonist Allegorical Reading and the Growth of the Epic Tradition* (Los Angeles: University of California Press, 1986).

LAMBRAKI 1980: Anna Lambraki, "Le Cipolin de la Karystie. Contribution a l'étude des marbres de la Grèce exploités aux époques romaine et paléochrétienne," *Revue archéologique* (1980): 31–62.

LAMBRINOUDAKIS 1991: Vassilis Lambrinoudakis, "The Sanctuary of Iria on Naxos and the Birth of Monumental Greek Architecture," in *New Perspectives in Early Greek Art*, ed. Diana Buitron-Olive (Washington, DC/Hanover, NH: National Gallery of Art and University Press of New England, 1991), 172–88.

LAMBRINOUDAKIS 2005: Vassilis Lambrinoudakis, "A New Early Archaic Building on Naxos: Some Thoughts on the Oikos of the Naxians on Delos," in *Architecture and Archaeology in the Cyclades: Papers in Honour of J. J. Coulton*, ed. Marina Yeroulanou and Maria Stamatopoulou (Oxford: Archaeopress, 2005), 79–86.

LAMBRINOUDAKIS ET AL. 2002: Vassilis Lambrinoudakis, Gottfried Gruben, Aenne Ohnesorg, Manolis Korres, Themistokles Bilis, Maria Magnisali, and Eva Simandoni-Bournia, "Naxos—Das Heiligtum von Gyroula in Sangri. Eine neugefundene, drei Jahrtausende alte Kultstätte der Demeter," *Antike Welt* 33 (2002): 387–408.

LAMPAKES 1889: Georgios Lampakes, *Χριστιανικὴ ἀρχαιολογία Μονῆς Δαφωίου* (Athens: Alexandros Papageorgios, 1889).

LAMPERTICO 1880: Fedele Lampertico, "Aurelio dell'Acqua e l'istituzione della mensa aureliana," *Archivio Veneto* 20 (1880): 255–73.

LAMPERTICO 1988: Matteo Lampertico, "La facciata dipinta nell'antica contrada Innocentini," *Civiltà mantovana* 22 (1988): 11–40.

LANCHA 1981: Janine Lancha, *Recueil general des mosaiques de la Gaule III. Narbonnaise 2* (Vienne: CNRS, 1981).

LANCIANI 1882: Rodolfo Lanciani, "Degli antichi edificii componenti la chiesa di SS. Cosma e Damiano," *Bullettino della Commissione Archeologica Comunale di Roma* 10 (1882): 29–54.

LANCIANI 1901: Rodolfo Lanciani, *New Tales of Old Rome* (London: Macmillan, 1901).

LANCIANI 1902–12: Rodolfo Lanciani, *Storia degli scavi di Roma e Notizie intorno le collezioni romane di antichità*, 7 vols. (Rome: Loescher, 1902–12).

LANDGRÁFOVÁ 2004: Renata Landgráfová, "Faience Inlays from the Funerary Temple of Khentkaus II," *Zeitschrift für Assyriologie und vorderasiatische Archäologie* 131 (2004): 134–55.

LANDGRÁFOVÁ 2006A: Renata Landgráfová, *Abusir XIV: Faience Inlays from the Funerary Temple of King Raneferef: Raneferef's Substitute Decoration Programme* (Prague: Czech Institute of Egyptology, Faculty of Arts, Charles University in Prague, 2006).

LANDGRÁFOVÁ 2006B: Renata Landgráfová, "Faience Inlays and Tablets of Egyptian Blue," in *Abusir IX: The Pyramid Complex of Raneferef: The Archeology*, ed. Miroslav Verner (Prague: Czech Institute of Egyptology, 2006), 485–86.

LANDSBERGER 1967: Benno Landsberger, "Akkadisch-hebräische Wortgleichungen," in *Hebräische Wortforschung. Festschrift Zum 80. Geburtstag Von Walter Baumgartner*, ed. Benedikt Hartmann (Leiden: Brill, 1967), 176–204.

LANG 1966: Mabel L. Lang, *The Palace of Nestor at Pylos in Western Messenia*, vol. 2, *The Frescoes* (Princeton: Princeton University Press, 1966).

LANGE 1964: Reinhold Lange, *Die byzantinische Reliefikone* (Recklinghausen: Aurel Bongers, 1964).

LANKHEIT 1962: Klaus Lankheit, *Florentinische Barockplastik; Die Kunst am Hofe der letzten Medici, 1670–1743*, 2 vols. (Munich: Bruckmann, 1962).

LANSING 1918: Ambrose Lansing, "Excavations at the Palace of Amenhotep III at Thebes," *Metropolitan Museum of Art Bulletin* supp., *The Egyptian Expedition 1916–17* 13, no. 3, (1918): 8–14.

LAPATIN 2001: Kenneth D. S. Lapatin, *Chryselephantine Statuary in the Ancient Mediterranean World* (Oxford: Oxford University Press, 2001).

LAPATIN 2010: Kenneth D. S. Lapatin, "New Statues for Old Gods," in *The Gods of Ancient Greece: Identities and Transformations*, ed. Jan N. Bremmer and Andrew Erskine (Edinburgh: Edinburgh University Press, 2010), 126–51.

LAROCHE AND JACQUEMIN 1986: Didier Laroche and Anne Jacquemin, "Le char d'or consacré par le peuple rhodien," *Bulletin de correspondance hellénique* 110, no. 1 (1986): 285–307.

LARSON 2001: Jennifer Larson, *Greek Nymphs: Myth, Cult, Lore* (Oxford: Oxford University Press, 2001).

LAUER 1936: Jean-Philippe Lauer, *Fouilles à Saqqarah, la pyramide à degrés. L'architecture*, vol. 1, *Texte* (Cairo: L'Institut francais d'archeologie orientale, 1936).

LAUTER 1971: Hans Lauter, "Neues zum Mädchen von Antium," *Mitteilungen des Deutschen Archäologischen Instituts, Athenische Abteilung* 86 (1971): 147–61.

LAVAGNE 1983: Henri Lavagne, "'Luxuria inaudita': Marcus Aemilius Scaurus et la naissance de la mosaïque murale," in *Mosaïque: Recueil d'hommages à Henri Stern* (Paris: Éditions recherches sur les civilisations, 1983), 259–64.

LAVAGNE 1988: Henri Lavagne, *Operosa Antra. Recherches sur la grotte à Rome de Sylla à Hadrien* (Rome: École française de Rome, 1988).

LAVIN 1962: Irving Lavin, "The House of the Lord: Aspects of the Role of Palace Triclinia in the Architecture of Late Antiquity and the Early Middle Ages," *Art Bulletin* 44, no. 1 (1962): 1–27.

LAVIN 1980: Irving Lavin, *Bernini and the Unity of the Visual Arts*, 2 vols. (New York: Oxford University Press, 1980).

LAVIN 1981: Irving Lavin, ed., *Drawings by Gianlorenzo Bernini, from the Museum der Bildenden Künste, Leipzig* (Princeton: Princeton University Art Museum, 1981).

LAWSON YOUNGER 2012: K. Lawson Younger, Jr., "Another Look at an Aramaic Astral Bowl," *Journal of Near Eastern Studies* 71, no. 2 (2012): 209–30.

LAYARD 1849–53: Austen Henry Layard, ed., *The Monuments of Nineveh*, 2 vols. (London: John Murray, 1849–53).

LAZZARINI 2007: Lorenzo Lazzarini, "Indagini archeometriche sui marmi bianchi della statuaria e architettura della Magna Grecia," *Marmora: An International Journal for Archaeology, History and Archeometry of Marbles and Stones* 3 (2007): 21–52.

LAZZARINI 2008: Lorenzo Lazzarini, "I finti marmi di Giotto agli Scrovegni (Padova, Italia)," *Marmora: An International Journal for Archaeology, History and Archeometry of Marbles and Stones* 4 (2008): 131–40.

LAZZARINI 2016: Lorenzo Lazzarini, "La calcarenite gialla dei templi di Selinunte: Cave, caratterizzazione, problemi di conservazione," in *Selinunte: Restauri dell'antico*, ed. Valentina Nicolucci (Rome: De Luca, 2016), 145–52.

LEACH 1975: Rosemary Leach, *An Investigation into the Use of Purbeck Marble in Medieval England* (Hartlepool: E. W. Harrison, 1975).

LEACH 1993: Eleanor W. Leach, "The Entrance Room in the House of Iulius Polybius and the Nature of the Roman Vestibulum," in *Functional and Spatial Analysis of Wall Painting*, ed. Eric M. Moormann (Leiden: Babesch, 1993), 23–28.

LEACH 2004: Eleanor W. Leach, *The Social Life of Painting in Ancient Rome and on the Bay of Naples* (Cambridge: Cambridge University Press, 2004).

LEARY 2011: T. J. Leary, "'The Winter Does Not Pass through Me, but the Sun Sparkles within Me': A Literary Window on Roman Glass?" *Journal of Glass Studies* 53 (2011): 235–37.

LECLERQ 1907–53: H. Leclerq, "Inscriptions," in *Dictionnaire d'archéologie chrétienne et de liturgie*, ed. Fernand Cabrol, 15 vols. (Paris: Letouzey & Ané, 1907–53), 7, cols. 850–1089.

LEGRAIN 1937: Léon Legrain, ed., *Business Documents of the Third Dynasty of Ur* (London: University of Pennsylvania Press, 1937).

LEHMANN 1964: Phyllis Williams Lehmann, "The Wall Decoration of the Hieron in Samothrace," *Balkan Studies* 5 (1964): 277–86.

LEHMANN 2015: Claudia Lehmann, "Bernini's Macchia," in *A Transitory Star: The Late Bernini and His Reception*, ed. Claudia Lehmann and Karen J. Loyd (Berlin: De Gruyter, 2015), 95–116.

LEHNUS 2002: Luigi Lehnus, "Posidippean and Callimachean Queries," *Zeitschrift für Papyrologie und Epigraphik* 138 (2002): 12–13.

LEIBDAROS AND TSAIMOU 2000: Rousetos Leibdaros and Konstantina G. Tsaimou, "Νέα στοιχεία για την υπόγεια εκμετάλλευση του λυχνίτη στην Πάρο," in Schilardi 2000, 83–90.

LELLI 2004: Emanuele Lelli, "I gioielli di Posidippo," *Quaderni Urbinati di Cultura Classica* 76, no. 1 (2004): 127–38.

LENZ 1861: Harald O. Lenz, *Mineralogie der alten Griechen und Römer* (Gotha: E. F. Thienemann, 1861).

LENZO 2007: Fulvio Lenzo, "I marmi di Loos," in *Adolf Loos, 1870–1933: Architettura, utilità e decoro*, ed. Richard Bösel and Vitale Zanchettin (Milan: Electa, 2007), 115–30.

LENZO 2008: Fulvio Lenzo, "La Cappella Spada in San Girolamo della Carità. Una 'stanza adobbata' per le ambizioni di un cardinale," *Römische Historische Mitteilungen* 50 (2008): 383–428.

LEPELLEY 1979–81: Claude Lepelley, *Les cités de l'Afrique romaine au Bas-Empire*, 2 vols. (Paris: Études Augustiniennes, 1979–81).

LEROU 2004: Sandrine Lerou, "L'usage des reliques du Christ par les empereurs aux XIe et XIIe siècles: Le Saint Bois et les Saintes Pierres," in *Byzance et les reliques du Christ*, ed. Jannic Durand and Bernard Flusin (Paris: Association des amis du centre d'histoire et civilisation de Byzance, 2004), 159–82.

LESEN 1926: A. Lesen, "Blosio Palladio e il cenacolo letterario di Gian Goritz," *Terra sabina* 4 (1926): 37–44.

LESK 2004: Alexandra Lesk, "A Diachronic Examination of the Erechtheion and Its Reception" (PhD diss., University of Cincinnati, 2004).

LETAROUILLY 1840–53: Paul Marie Letarouilly, *Les édifices de Rome moderne*, 3 vols. (Liege: D. Avanzo, 1840–53).

LETHABY 1892: William R. Lethaby, *Architecture, Mysticism, and Myth* (London: Percival, 1892).

LETHABY AND SWAINSON 1894: William R. Lethaby and Harold Swainson, *The Church of Sancta Sophia, Constantinople: A Study of Byzantine Building* (London: Macmillan, 1894).

LEVI 1942: Doro Levi, "The Amphitheatre in Cagliari," *American Journal of Archaeology* 46, no. 1 (1942): 1–9.

LEVI AND LEVI 1974: Annalina Levi and Mario Levi, "The Medieval Map of Rome in the Ambrosian Library's Manuscript of Solinus (C 246 Inf.)," *Proceedings of the American Philosophical Society* 118, no. 6 (1974): 576–94.

LICOCCIA 1987: Isabella Licoccia, "Roma: S. Girolamo della Carità," *Ricerche di Storia dell'arte* 31 (1987): 43–45.

LIEBERMAN 1986: Ralph E. Lieberman, *The Church of Santa Maria dei Miracoli in Venice* (New York: Garland, 1986).

LIGHTBOWN 1986: Ronald W. Lightbown, *Mantegna: With a Complete Catalogue of the Paintings, Drawings and Prints* (Oxford: Phaidon, 1986).

LILIUS 1981: Henrik Lilius, *Villa Lante al Gianicolo. L'architettura e la decorazione pittorica*, 2 vols. (Rome: Bardi, 1981).

LILYQUIST 2003: Christine Lilyquist, *The Tomb of Three Foreign Wives of Tuthmosis III* (New York/New Haven: Metropolitan Museum of Art and Yale University Press, 2003).

LILYQUIST AND BRILL 1993: Christine Lilyquist and Robert H. Brill, *Studies in Early Egyptian Glass* (New York: Metropolitan Museum of Art, 1993).

LIND HANSEN 1997: Sanne Lind Hansen, "The Embellishment of Late-Antique Domus in Ostia and Rome," in *Patron and Pavements in Late Antiquity*, ed. Signe Isager and Birte Poulsen (Odense: Odense University Press, 1997), 111–24.

LING 1990: Roger Ling, *Roman Painting* (Cambridge, Cambridge University Press, 1990).

LIPINSKI 1960: Angelo Lipinski, "La 'Crux gemmata' e il culto della Santa Croce nei monumenti superstiti e nelle raffigurazioni monumentali," *Felix Ravenna* 30 (1960): 5–62.

LIPPS 2016: Johannes Lipps, "Statuen kniefalliger Orientalen aus Rom und ein Dreifuß im Olympieion von Athen," *Mitteilungen des deutschen archäologischen Instituts. Römische Abteilung* 122 (2016): 203–52.

LIVERANI AND SANTAMARIA 2014: Paolo Liverani and Ulderico Santamaria, eds., *Diversamente bianco: La policromia della scultura romana* (Rome: Quasar, 2014).

LIVINGSTONE 1986: Alasdair Livingstone, *Mystical and Mythological Explanatory Works of Babylonian and Assyrian Scholars* (Oxford: Oxford University Press, 1986).

LIVIUS AND VAUGHAN 1893: Thomas Livius and Herbert Vaughan, *The Blessed Virgin in the Fathers of the First Six Centuries* (London: Burns & Oates, 1893).

LOMBARDINI 1991–92: Nora Lombardini, "Corrado Ricci e i restauri alla basilica di San Vitale di Ravenna: Il caso della ricomposizione delle tarsie marmoree dell'abside," *Quasar* 6–7 (1991–92): 92–99.

LONG 1974: Charlotte R. Long, *The Ayia Triadha Sarcophagus: A Study Of Late Minoan and Mycenaean Funerary Practices and Beliefs* (Göteborg: P. Åström, 1974).

LONG 1997: Anthony A. Long, "Lucretius on Nature and the Epicurean Self," in *Lucretius and His Intellectual Background*, ed. K. A. Agra, M. H. Koenen, and P. H. Schrijvers (North-Holland: Koninklijke Nederlandse Adademie van Wetenschappen, 1997), 125–40.

LONGRIGG 1985: James Longrigg, "Elements and After: A Study in Presocratic Physics of the Second Half of the Fifth Century," *Apeiron: A Journal for Ancient Philosophy and Science* 19, no. 2 (1985): 95–113.

LOPARDI AND MUCCIANTE 1987: Antonella Lopardi and Graziella Mucciante, *La Basilica di San Bernardino* (L'Aquila: Soprintendenza BAAAS, 1987).

LOUKAKI 2013: Marina Loukaki, "Ekphrasis Earos. Le topos de la venue du Printemps chez des auteurs Byzantins," *Parekbolai* 3 (2013): 77–106.

LOVE 1971: Iris C. Love, "A Preliminary Report of the Excavations at Knidos," *American Journal of Archaeology* 76, no. 4 (1971): 397–98, ill. 4.

LÖVEI 1992: Pál Lövei, "A tömött vörös mészkö–'vörös márvány'–a középkori magyaroszági müvészetben," *Ars Hungarica* 20, no. 5 (1992): 3–28.

LÖVEI 2001: Pál Lövei, "'Virtus, es, marmor, scripta.' Red Marble and Bronze Letters," *Acta Historiae Artium* 42, nos. 1–4 (2001): 39–56.

LUCHTERHANDT 1996: Manfred Luchterhandt, "Im Reich der Venus: zu Peruzzis Sala delle Prospettive in der Farnesina," *Römisches Jahrbuch der Bibliotheca Hertziana* 31 (1996): 207–44.

LUCKENBILL 1924: Daniel D. Luckenbill, ed. and trans., *The Annals of Sennacherib* (Chicago: University of Chicago Press, 1924).

LUGLI 1928: Giuseppe Lugli, *Forma Italiae, Regio I. Latium et Campania*, vol. 1.2, *Ager Pomptinus. Circeii* (Rome: Danesi, 1928).

LUGLI 1957: Giuseppe Lugli, *La tecnica edilizia romana, con particolare riguardo a Roma e Lazio*, 2 vols. (Rome: Giovanni Bardi, 1957).

LUM 1999: Eric Lum, "Pollock's Promise: Toward an Abstract Expressionist Architecture," *Assemblage* 39 (1999): 62–93.

LUSCHI 1998: Licia Luschi, "Gli antichi edifici della basilica dei SS. Cosma e Damiano: Osservazioni su disegni ligoriani," in *XLIII Corso di cultura sull'arte ravennate e bizantina*, ed. Raffaella Farioli Campanati (Ravenna: Girasole, 1998), 429–52.

LUSUARDI SIENA 1990: Maria Silvia Lusuardi Siena, "Tarsie parietali del coro," in *Milano capitale dell'impero romano, 286–402 D.C.*, ed. Gemma Sena Chiesa (Milan: Silvana, 1990), 133–34.

LUSUARDI SIENA ET AL. 1989: Silvia Lusuardi Siena, Cinzia Fiorio Tedone, Marco Sannazaro, and Maria Motta Broggi, "Le traccie materiali del cristianesimo dal tardo antico al mille," in *Il Veneto nel medioevo. Dalla "Venetia" alla Marca Veronese*, ed. Andrea Castagnetti and Gian Maria Varanini (Verona: Banca popolare di Verona, 1989), 87–328.

LUX 1968: U. Lux, "Die Apostel-Kirche in Madaba," *Zeitschrift des Deutschen Palästina-Vereins* 84 (1968): 106–29.

LUZARRAGA 1973: Jesús Luzarraga, *Las tradiciones de la nube en la Biblia y en el judaismo primitivo* (Rome: Biblical Institute, 1973).

LYNN 1984: C. J. Lynn, "Some Fragments of Exotic Porphyry Found in Ireland," *Journal of Irish Archaeology* 2 (1984): 19–32.

MACKIE 1997: Gillian Mackie, "The Santa Croce Drawings: A Re-Examination," *RACAR* 24, no. 1 (1997): 1–13.

MACKILLOP 1992: Susan MacKillop, "Dante and Lumen Christi: A Proposal for the Meaning of the Tomb of Cosimo de' Medici," in *Cosimo "il Vecchio" de' Medici, 1389–1464*, ed. Francis Ames-Lewis (Oxford: Clarendon, 1992), 245–301.

MACRIDES 2015: Ruth Macrides, "The 'Other' Palace in Constantinople: The Blachernai," in *The Emperor's House: Palaces from Augustus to the Age of Absolutism*, ed. Michael Featherstone, Jean-Michel Spieser, Gülru Tanman, and Ulrike Wulf-Rheidt (Boston: De Gruyter, 2015), 159–68.

MACRIDES AND MAGDALINO 1988: Ruth Macrides and Paul Magdalino, "The Architecture of Ekphrasis: Construction and Context of Paul the Silentiary's Poem on Hagia Sophia," *Byzantine and Modern Greek Studies* 12 (1988): 47–82.

MAFFEI 1986: Sonia Maffei, "Le imagines di Luciano: Un 'patchwork' di capolavori antichi," *Studi classici e orientali* 36 (1986): 155–60.

MAGDALINO 1993: Paul Magdalino, *The Empire of Manuel I Komnenos, 1143–1180* (Cambridge: Cambridge University Press, 1993).

MAGDALINO 1997: Paul Magdalino, "In Search of the Byzantine Courtier: Leo Choirosphaktes and Constantine Manasses," in *Byzantine Court Culture from 829 to 1204*, ed. Henry Maguire (Washington, DC: Dumbarton Oaks, 1997), 141–65.

MAGNUSSON 1987: Cecilia Magnusson, "The Antique Sources of the Chigi Chapel," *Konsthistorisk Tidskrift* 56, no. 4 (1987): 135–39.

MAGRINI 1848: Antonio Magrini, *Notizie storico descrittive della Chiesa Cattedrale di Vicenza* (Vicenza: G. Tramontini, 1848).

MAGUIRE 1981: Henry Maguire, *Art and Eloquence in Byzantium* (Princeton: Princeton University Press, 1981).

MAGUIRE 1987A: Henry Maguire, *Earth and Ocean: The Terrestrial World in Early Byzantine Art* (University Park: Pennsylvania State Press, 1987).

MAGUIRE 1987B: Henry Maguire, "The Mantle of the Earth," *Illinois Classical Studies* 12 (1987): 221–28.

MAGUIRE 1992: Henry Maguire, "Byzantine Art History in the Second Half of the Twentieth Century," in *Byzantium: A World Civilization*, ed. Angeliki E. Laiou and Henry Maguire (Washington, DC: Dumbarton Oaks, 1992), 119–56.

MAGUIRE 1994: Henry Maguire, "Magic and Geometry in Early Christian Floor Mosaics and Textiles," *Jahrbuch der österreichischen Byzantinistik* 44 (1994): 265–74.

MAGUIRE 1997: Henry Maguire, *Materials Analysis of Byzantine Pottery* (Washington, DC: Dumbarton Oaks, 1997).

MAINSTONE 1988: Rowland Mainstone, *Hagia Sophia: Architecture, Structure and Liturgy of Justinian's Great Church* (New York: Thames & Hudson, 1988).

MAIURI 1958: Amedeo Maiuri, *Ercolano: I nuovi scavi (1927–58)*, 2 vols. (Rome: Istituto Poligrafico dello Stato, 1958).

MAIURI 1977: Amedeo Maiuri, *Herculaneum* (Congleton: Old Vicarage, 1977), 68–71.

MAJCHEREK 2002: Grzegorz Majcherek, "Alexandria. Kom El-Dikka. Excavations, 2000/2001," *Polish Archaeology in the Mediterranean* 13 (2002): 31–43.

MAJESKA 1973: George P. Majeska, "St. Sophia in the Fourteenth and Fifteenth Centuries: The Russian Travelers on the Relics," *Dumbarton Oaks Papers* 27 (1973): 71–87.

MAJESKA 1978: George P. Majeska, "Notes on the Archaeology of St. Sophia at Constantinople: The Green Marble Bands on the Floor," *Dumbarton Oaks Papers* 32 (1978): 299–308.

MAJESKA 1997: George P. Majeska, "The Emperor in His Church: Imperial Ritual in the Church of St. Sophia," in *Byzantine Court Culture from 829 to 1204*, ed. Henry Maguire (Washington, DC: Dumbarton Oaks, 1997), 1–11.

MALAMA 2007: Pénélope Malama, "Le décor pictural des tombes récemment mises au jour à Amphipolis," in *Peinture et couleur dans le monde grec antique*, ed. Sophie Descamps-Lequime (Milan/Paris: 5 Continents and Musée du Louvre, 2007), 109–19.

MALMSTROM 1975: Ronald E. Malmstrom, "The Colonnades of High Medieval Churches at Rome," *Gesta* 14, no. 2 (1975): 37–45.

MAMBOURY 1951: Ernest Mamboury, "Les fouilles byzantines à Istanbul," *Byzantion* 27 (1951): 433–37.

MANCINI 1966–67: Adele Mancini, "La chiesa medioevale di S. Adriano nel Foro Romano," *Atti della Pontificia Accademia Romana di Archeologia* 39 (1966–67): 191–245.

MANGO 1959: Cyril A. Mango, *The Brazen House: A Study of the Vestibule of the Imperial Palace of Constantinople* (Copenhagen: I kommission hos Munksgaard, 1959).

MANGO 1962: Cyril A. Mango, *Materials for the Study of the Mosaics of St. Sophia at Istanbul* (Washington, DC: Dumbarton Oaks, 1962).

MANGO 1965: Cyril A. Mango, "Constantinopolitana," *Jahrbuch des Deutschen Archäologischen Instituts* 80 (1965): 305–36.

MANGO 1970: Cyril A. Mango, "Notes on Byzantine Monuments," *Dumbarton Oaks Papers* 23/24 (1970): 369–75.

MANGO 1984: Cyril A. Mango, "A Byzantine Hagiographer at Work: Leontios of Neapolis," in *Byzanz und der Westen. Studien zur Kunst des Europäischen Mittelalters*, ed. Irmgard Hutter (Vienna: Österreichische Akademie der Wissenschaften, 1984), 25–42.

MANGO 1990: Cyril A. Mango, "Constantine's Mausoleum and the Translation of Relics," *Byzantinische Zeitschrift* 83 (1990): 51–62.

MANGO 1997: Cyril A. Mango, "The Palace of the Boukoleon," *Cahiers archéologiques* 45 (1997): 41–50.

MANGO AND SEVCENKO 1961: Cyril A. Mango and Ihor Sevcenko, "Remains of the Church of St. Polyeuktos at Constantinople," *Dumbarton Oaks Papers* 15 (1961): 243–47.

MANGO ET AL. 1990: Cyril Mango, Ernest J. W. Hawkins, and Susan Boyd, "The Monastery of St. Chrysostomos at Koutsovendis (Cyprus) and Its Wall Paintings. Part I: Description," *Dumbarton Oaks Papers* 44 (1990): 68.

MANNING 1963: Eugène Manning, "Le symbolisme de la nubes (nephélè) chez Origène et les pères latins," *Recherches de sience religieuse* 51 (1963): 96–111.

MANTESE 1952: Giovanni Mantese, *Memorie storiche della chiesa vicentina*, 5 vols. (Vicenza: Istituto S. Gaetano, 1952).

MANTESE 1964: Giovanni Mantese, "Aurelio dell'Acqua cavaliere vicentino e la sua 'Catena Evangelica,'" in *Studi in onore di A. Bardelle*, ed. Marcello Bardella (Vicenza: Istituto San Gaetano, 1964), 85–107.

MARCATTILI 2011: Francesco Marcattili, "Primo Stile e cultura della luxuria," in *Pittura ellenistica in Italia e in Sicilia, linguaggi e tradizioni*, ed. Gioacchino Francesco La Torre and Mario Torelli (Rome: Bretschneider, 2011), 415–24.

MARCHETTI 2009: N. Marchetti, "Texts Quoting Artworks: The Banquet Stele and the Palace Reliefs of Assurnasirpal II," *Revue d'Assyriologie et d'archéologie orientale* 103 (2009): 85–90.

MARCHINI 1939: Giuseppe Marchini, "L'incrostazione marmorea della Cappella Gondi in S. Maria Novella," *Palladio* 3 (1939): 205–11.

MARCHINI 2014: Maurizio Marchini, *Un tesoro della ecclesia di Brescia. La confessione di fede di Tommaso sul sarcofago da S. Afra nel Museo di S. Giulia* (Brescia: Grafo, 2014).

MARCOLINI 2001: Giuliana Marcolini, "Da Renata ad Alfonso: Scambio di 'testimone' per la 'cappella ducale' nel Castello Estense di Ferrara," in *Per l'arte: Da Venezia all'Europa*, ed. Mario Piantoni and Laura De Rossi (Monfalcone: Laguna, 2001), 229–34.

MARCONI 1994: Clemente Marconi, *Selinunte: Le metope dell'Heraion* (Modena: Panini, 1994).

MARCONI 2011: Clemente Marconi, "L'identificazione della 'Dea' di Morgantina," *Prospettiva* 141/42 (2011): 2–31.

MARCUCCI AND TORRESI 1982: Laura Marcucci and Bruno Torresi, "Palazzo Medici-Lante: Un progetto mediceo in Roma e il 'raggiustamento' di Onorio Longi (I)," *Storia Architettura* 5, no. 2 (1982): 39–62.

MARCUCCI AND TORRESI 1983: Laura Marcucci and Bruno Torresi, "Palazzo Medici-Lante: Un progetto mediceo in Roma e il 'raggiustamento' di Onorio Longi' (II)," *Storia Architettura* 6, no. 1 (1983): 21–44.

MARI 2004: Zaccaria Mari, "La Decorazione del Ninfeo Della Villa di Nerone a Subiaco (Roma)," in *Atti del IX Colloquio dell'AISCOM*, ed. Claudia Angelelli (Ravenna: Girasole, 2004), 287–99.

MARI AND FIORE CAVALIERE 2001: Zaccaria Mari and Maria Grazia Fiore Cavaliere, "Rivestimenti marmorei da una villa tiburtina e da quelle imperiali di Arcinazzo e Subiaco: Esempi tipologici ed episodi di spoliazione," in *Atti del VII Colloquio dell'AISCOM*, ed. Andrea Paribeni (Ravenna: Girasole, 2001), 427–48.

MARICHAL 1960: René Marichal, "La construction de Sainte-Sophia de Constantinople dans l'anonyme grec (xe siècle?) et les versions vieux-russes," *Byzantinoslavica* 21 (1960): 238–59.

MARKETOU 2013: Toula Marketou, "The Art of Painting at Ialysos from the Early 2nd Millennium BC to the Eruption of the Thera Volcano," in *Paintbrushes: Wall Painting and Vase-Painting of the 2nd Millennium BC in Dialogue*, ed. Andreas Vlachopoulos (Athens: Society for the Promotion of Theran Studies, 2013), 104–9.

MARKHAM SCHULZ 1977: Anne Markham Schulz, *The Sculpture of Bernardo Rossellino and His Workshop* (Princeton: Princeton University Press, 1977).

MARKHAM SCHULZ 1991: Anne Markham Schulz, *Giambattista and Lorenzo Bregno: Venetian Sculpture in the High Renaissance* (Cambridge: Cambridge University Press, 1991).

MARKHAM SCHULZ ET AL. 2003: Anne Markham Schulz, Manuela Morresi, and Toto Bergamo Rossi, *La cappella Badoer-Giustinian in San Francesco della Vigna a Venezia* (Florence: Centro Di, 2003).

MAROSI 1971: Ernö Marosi, "Einige stilistische Probleme der Inkrustationen von Gran (Esztergom)," *Acta Historiae Artium* 17 (1971): 171–229.

MAROSI 1984: Ernö Marosi, *Die Anfänge der Gotik in Ungarn. Esztergom in der Kunst des 12.–13. Jahrhunderts* (Budapest: Akademiai Kiado, 1984)

MARSHALL 1951: John H. Marshall, *Taxila: An Illustrated Account of Archaeological Excavations*, 3 vols. (Cambridge: Cambridge University Press, 1951).

MARTELLACCI 1995: Rosamaria Martellacci, "Cappelle gentilizie fiorentine in epoca di controriforma," in *Architettura di altari e spazio ecclesiale. Episodi a Firenze, Prato e Ferrara nell'eta della Controriforma*, ed. Carlo Cresti (Florence: Angelo Pontecorboli, 1995), 75–113.

MARTIN 1965: Roland Martin, *Manuel d'architecture grecque*, vol. 1, *Matériaux et techniques* (Paris: Picard, 1965).

MARTIN 1995: Janet L. Martin, *Medieval Russia, 980–1584* (Cambridge: Cambridge University Press, 1995).

MARTINEAU 1992: Jane Martineau, ed., *Andrea Mantegna* (London: Royal Academy of Arts, 1992).

MARTINIS 2014: Roberta Martinis, "Il palazzo di Eliseo Raimondi a Cremona: Abitare all'antica tra Milano e Venezia alla fine del Quattrocento," in *Modernamente antichi. Modelli, identità, tradizione nella Lombardia del Tre e Quattrocento*, ed. Pier Nicola Pagliara and Serena Romano (Rome: Viella, 2014), 257–85.

MARUCCHI 1891: Orazio Marucchi, "I lavori ad intarsio della Basilica di Giunio Basso sull'Esquilino," *Bullettino della Commissione Archeologica Comunale di Roma* 21 (1891): 81–101.

MARZAHN 1995: Joachim Marzahn, *The Ishtar Gate, the Processional Way, the New Year Festival of Babylon* (Mainz am Rhein: Philipp von Zaubern, 1995).

MASON 2007: David J. Mason, "The Location of the Treasury of Atreus," *Oxford Journal of Archaeology* 26, no. 1 (2007): 35–52.

MASON 2013: David J. Mason, "The Date of the Tomb of Clytemnestra," *Annual of the British School at Athens* 108 (2013): 97–119.

MATHEWS 1971: Thomas F. Mathews, *The Early Churches of Constantinople: Architecture and Liturgy* (University Park: Pennsylvania State University Press, 1971).

MATHEWS 1976: Thomas F. Mathews, *The Byzantine Churches of Istanbul: A Photographic Survey* (University Park: Pennsylvania State University Press, 1976).

MATHEWS 1990: Thomas F. Mathews, "Transformation Symbolism in Byzantine Architecture and the Meaning of the Pantokrator in the Dome," in *Church and People in Byzantium*, ed. Rosemary Morris (Chester: Bemrose, 1990), 191–214.

MAU 1882: August Mau, *Geschichte der decorativen Wandmalerei in Pompeji* (Berlin: G. Reimer, 1882).

MAUL 1997: Stefan Maul, "Die altorientalische Hauptstadt—Abbild und Nabel der Welt," in *Die Orientalische Stadt: Kontinuitat. Wandel. Bruch*, ed. Wilhelm Gernot (Saarbrücken: Saarbrücker, 1997), 109–24.

MAZARAKIS AINIAN 1997: Alexander Mazarakis Ainian, *From Rulers' Dwellings to Temples: Architecture, Religion and Society in Early Iron Age Greece (1100–700 B.C.)* (Jonsered: Paul Åströms, 1997).

MAZARAKIS AINIAN 2016: Alexander Mazarakis Ainian, "Early Greek Temples," in *A Companion to Greek Architecture*, ed. Margaret M. Miles (Chichester: John Wiley, 2016), 15–30.

MAZOIS 1824: François Mazois, *Les ruines de Pompéi* (Paris: F. Didot, 1824).

MAZZEI 1995: Marina Mazzei, ed., *Arpi: L'ipogeo della Medusa e la necropoli* (Bari: Edipuglia, 1995).

MAZZEI 2002: Marina Mazzei, "La Daunia e la Grecia settentrionale: Riflessioni sulle esperienze pittoriche del primo ellenismo," in *La pittura parietale in Macedonia e Magna Grecia*, ed. Angela Pontrandolfo (Paestum: Pandemos, 2002), 67–77.

MCANDREW 1980: John McAndrew, *Venetian Architecture of the Early Renaissance* (Cambridge, MA: MIT, 1980).

MCCRORY 1997: Martha McCrory, "The Symbolism of Stones: Engraved Gems at the Medici Grand-Ducal Court (1537–1609)," in *Engraved Gems: Survivals and Revivals*, ed. Clifford M. Brown (Hanover: National Gallery of Art, 1997), 158–79.

MCEWEN 2003: Indra Kagis McEwen, *Vitruvius: Writing the Body of Architecture* (Cambridge, MA: MIT, 2003).

MCKENZIE 2007: Judith McKenzie, *The Architecture of Alexandria and Egypt 300 BC–AD 700* (New Haven: Yale University Press, 2007).

MCPHAIL 2003: Eric McPhail, "The Mosaic of Speech: A Classical Topos in Renaissance Aesthetics," *Journal of the Warburg and Courtauld Institutes* 66 (2003): 249–64.

MCVEY 1983: Kathleen E. McVey, "The Domed Church as Microcosm: Literary Roots of an Architectural Symbol," *Dumbarton Oaks Papers* 37 (1983): 91–121.

MCVEY 1993: Kathleen E. McVey, "The Sogitha on the Church of Edessa in the Context of Other Early Greek and Syriac Hymns for the Consecration of Church Buildings," *Aram* 5 (1993): 329–70.

MEADOWS 1945: J. W. Meadows, "Pliny on the Smaragdus," *Classical Review* 59, no. 2 (1945): 50–51.

MEEK 1952: Harold A. Meek, "The Architect and His Profession in Byzantium," *Journal of the Royal Institute of British Architects* 59 (1952): 216–20.

MEGAW 1931–32: Arthur H. S. Megaw, "The Chronology of Some Middle-Byzantine Churches," *Annual of the British School at Athens* 32 (1931–32): 90–130.

MEGAW 1965: Arthur H. S. Megaw, "Notes on Recent Work of the Byzantine Institute in Istanbul," *Dumbarton Oaks Papers* 17 (1965): 335–64.

MÉHU 2013: Didier Méhu, "L'évidement de l'image ou la figuration de l'invisible corps du Christ (Ixe–Xie siècle)," *Images re-vues* 11 (2013): 2–37.
MEIER 1977: Christel Meier, *Gemma spiritualis: Methode und Gebrauch der Edelsteinallegorese vom frühen Christentum bis ins 18. Jahrhundert* (Munich: Fink, 1977).
MELLINK 1969: Machteld J. Mellink, "Archaeology in Asia Minor," *American Journal of Archaeology* 73, no. 2 (1969): 203–27.
MELOGRANI 1990: Anna Melograni, "Le pitture del VI e VIII secolo nella basilica inferiore di S. Crisogono in Trastevere," *Rivista dell'Istituto Nazionale di Archeologia e Storia dell'Arte* 13 (1990): 139–78.
MENNA 2005: Maria R. Menna, "Il programma imperiale dei Grandi Comneni e la chiesa della Theotokos Chrysokephalos a Trebisonda," in *Medioevo: Immagini e ideologie*, ed. Arturo C. Quintavalle (Milan: Electa, 2005), 138–44.
MERCANTINI 1981: Maria Mercantini, *Due disegni inediti Berniniani per Piazza S. Pietro* (Città di Castello: Società Tipografica Editoriale, 1981).
MERCURELLI SALARI 1996: P. Mercurelli Salari, "I Francescani e l'oratorio di San Bernardino," in *Un pittore e la sua città: Benedetto Bonfigli e Perugia*, ed. Vittoria Garibaldi (Milan: Electa, 1996), 138–47.
MEREDITH 1980: Jill Meredith, "The Impact of Italy on the Romanesque Sculpture of England" (PhD diss., Yale University, 1980).
MERITT 1970: Lucy Shoe Meritt, "The Stoa Poikile," *Hesperia* 39, no. 4 (1970): 233–64.
MERRILLEES 2005: Parvine H. Merrillees, *Catalogue of the Western Asiatic Seals in the British Museum*, vol. 6, *Cylinder Seals, Pre-Achaemenid and Achaemenid Periods* (London: British Museum, 2005).
MESNARD 1935: Maurice Mesnard, *La basilique de saint Chrysogoné à Rome* (Vatican City: Pontificio Istituto di Archeologia Cristiana, 1935).
MESSINEO 1998: Gaetano Messineo, "Fenestrae Clathratae," *Bollettino d'Arte* 105–6 (1998): 1–8.
METEVELIS 2000: Peter Metevelis, "The Lapidary Sky over Japan," *Asian Folklore Studies* 59, no. 1 (2000): 79–88.
MEYENDORFF 1959: John Meyendorff, "L'iconographie de la Sagesse Divine dans la tradition byzantine," *Cahiers archeologiques* 10 (1959): 259–77.
MEYENDORFF 1987: John Meyendorff, "Wisdom-Sophia: Contrasting Approaches to a Complex Theme," *Dumbarton Oaks Papers* 41 (1987): 391–401.
MICHAELIDES 1985: Demetrios Michaelides, "Some Aspects of Marble Imitation in Mosaic," in *Marmi antichi. Problemi d'impiego, di restauro e d'identificazione*, ed. Patrizio Pensabene (Rome: Bretschneider, 1985), 155–70.
MICHEL 1999: Cécile Michel, "Les joyaux des rois de Mari," in *Cornaline et pierres précieuses: La Méditerranée, de l'Antiquité à l'Islam*, ed. Annie Caubet (Paris: La documentation française, 1999), 403–32.
MICHEL 2004: Simone Michel, *Die Magischen Gemmen: Zu Bildern und Zauberformeln auf geschnittenen Steinen der Antike und Neuzeit* (Berlin: Akademie, 2004).
MICHIEL 1990: Dorothea Michiel, *Casa dei Cei (I 6, 15)* (Munich: Hirmer, 1990).
MICHLER 1977: Jürgen Michler, "Über die Farbfassung hochgotischer Sakralräume," *Wallraf-Richartz-Jahrbuch* 39 (1977): 29–64.
MIDDLETON 1886A: J. Henry Middleton, "The Rostra and the Graecostasis, with the Umbilicus Romae and the Milliarium Aureum," *Archaeologia* 49 (1886): 424–33.
MIDDLETON 1886B: J. Henry Middleton, "The Temple and Atrium of Vesta and the Regia," *Archaeologia* 49 (1886): 391–423.
MIDDLETON 1892: J. Henry Middleton, *The Remains of Ancient Rome*, 2 vols. (London: Adam & Charles Black, 1892).
MIELSCH 2001: Harald Mielsch, *Römische Wandmalerei* (Darmstadt: Theiss, 2001).
MILLER 1993: Stella G. Miller, *The Tomb of Lyson and Kallikles: A Painted Macedonian Tomb* (Mainz am Rhein: Zabern, 1993).
MILLET 1899: Gabriel Millet, *Le Monastère de Daphni. Histoire, Architecture, Mosaïques* (Paris: Leroux, 1899).
MILLET 1916: Gabriel Millet, *L'École grecque dans l'architecture byzantine* (Paris: E. Leroux, 1916).
MILWRIGHT 2005: Marcus Milwright, "'Waves of the Sea:' Responses to Marble in Written Sources (9th-15th Century)," in *The Iconography of Islamic Art*, ed. Bernard O'Kane (Edinburgh: Edinburgh University Press, 2005), 211–21.
MIRABELLA ROBERTI 1975: Mario Mirabella Roberti, "Motivi aquileiesi nei mosaici della val Padana," in *La mosaïque gréco-romaine II*, ed. Henri Stern and Marcel Le Glay (Paris: A. & J. Picard, 1975), 193–204.
MITCHELL 2012A: John Mitchell, "Believing Is Seeing: The Natural Image in Late Antiquity," in *Architecture and Interpretation*, ed. Jill A. Franklin, T. A. Heslop, and Christine Stevenson (Woodbridge: Boydell & Brewer, 2012), 16–41.
MITCHELL 2012B: John Mitchell, "Of Eyes and Dogs: A Late Roman Amulet from Butrint," in *Le plaisir de l'art du Moyen Âge: Commande, production et reception de l'oeuvre d'art* (Paris: Picard, 2012), 61–68.
MITCHELL AND MIDDLETON 2002: T. C. Mitchell and A. P. Middleton, "The Stones Used in the Assyrian Sculptures," *Journal of Cuneiform Studies* 54 (2002): 93–98.
MITCHELL ET AL. 2001: John Mitchell, Inge Lyse Hansen, and Catherine M. Coutts, *San Vincenzo al Volturno 3*, 2 vols. (Spoleto: Centro Italiano di Studi sull'Alto Medioevo, 2001).
MOLINARI AND PRIESTER 1999: Maria C. Molinari and Sascha Priester, "L'Excubitorium della VII Coorte a Trastevere (Roma): Nuova analisi delle fonti manoscritte e di quelle edite," *Rendiconti dell'Accademia Nazionale dei Lincei* 10, no. 4 (1999): 523–73.
MÖLLER 2000: Astrid Möller, *Naukratis: Trade in Archaic Greece* (Oxford: Oxford University Press, 2000).
MOLS 1999: Stephan T. A. M. Mols, *Wooden Furniture in Herculaneum: Form, Technique, and Function* (Amsterdam: Gieben, 1999).
MOLS 2005: Stephan T. A. M. Mols, "Il Primo Stile 'retró': Dai Propilei di Mnesicle a Pompei?," in *Omni Pede Stare. Saggi architettonici e circumvesuviani in memoriam Jos De Waele*, ed. Stephan T. A. M. Mols and Eric M. Moormann (Naples: Electa, 2005), 243–46.
MONTANARI 2004: Tommaso Montanari, "'Theatralia,' di Giovan Battista Doni: Una nuova fonte per il teatro di Bernini," in *Estetica Barocca*, ed. Sebastian Schütze (Rome: Campisano, 2004), 301–20.
MONTELLI 1998: Emanuela Montelli, "Note su alcune tecniche costruttive impiegate per l'esecuzione di accurati paramenti laterizi nel cantiere romano cinquecentesco," *Quaderni dell'Istituto di Storia dell'Architettura* 32 (1998): 77–96.
MOOREY 1994: Peter R. S. Moorey, *Ancient Mesopotamian Materials and Industries: The Archaeological Evidence* (Oxford: Clarendon, 1994).
MOORMANN AND SWINKELS 1983: Eric M. Moormann and Louis J. F. Swinkels, "Lozenges in Perspective," in *La peinture murale romaine dans le provinces de l'Empire*, ed. Alix Barbet (Oxford: Centre d'études des peintures murales romaines, 1983), 239–62.
MORA ET AL. 1977: Paolo Mora, Laura Mora, and Paul Philippot, *La conservation des peintures murales* (Bologna: Compositori, 1977).
MOREL 1984: Philippe Morel, "Morfologia Delle Cupole Dipinte Da Correggio a Lanfranco," *Bollettino d'Arte* 69, no. 23 (1984): 1–34.
MOREL 1998: Philippe Morel, *Les grottes maniéristes en Italie au XVIe siècle: Théâtre et alchimie de la nature* (Paris: Macula, 1998).
MORGAN 1952: Charles H. Morgan, "Pheidias and Olympia," *Hesperia* 21 (1952): 295–339.
MORGAN 1990: Catherine Morgan, *Athletes and Oracles: The Transformation of Olympia and Delphi in the Eighth Century B.C.* (Cambridge: Cambridge University Press, 1990).
MORGAN 2005: Lyvia Morgan, ed., *Aegean Wall Painting: A Tribute to Mark Cameron* (London: British School at Athens, 2005).
MORRESI 1996: Manuela Morresi, "Cooperation and Collaboration in Vicenza before Palladio: Jacopo Sansovino and the Pedemuro Masters at the High Altar of the Cathedral of Vicenza," *Journal of the Society of Architectural Historians* 55, no. 2 (1996): 158–77.
MORRESI 1998: Manuela Morresi, "Venezia e le città del Dominio," in

Storia dell'architettura italiana. Il Quattrocento, ed. Francesco Paolo Fiore (Milan: Electa, 1998), 200–243.

MORRESI 2000: Manuela Morresi, *Jacopo Sansovino* (Milan: Electa, 2000).

MORRIS 2007: Sarah P. Morris, "Apollo, Dionysus and Zeus: On the Sacred Landscapes of Ancient Naxos," in *ἀμύμονα ἔργα. Τιμητικός τόμος για τον καθηγητή Βασίλη Κ. Λαμπρινουδάκη*, ed. E. Simantoni-Bournia, N. Kourou, and L. Mendoni (Athens: University of Athens, 2007), 96–108.

MORROGH 1985A: Andrew Morrogh, *Disegni di architetti fiorentini, 1540–1640* (Florence: Olschki, 1985).

MORROGH 1985B: Andrew Morrogh, "Vasari and Coloured Stones," in *Giorgio Vasari: Tra decorazione ambientale e storiografia artistica*, ed. Gian Carlo Garfagnini (Florence: Olschki, 1985), 309–20.

MORROGH 2017: Andrew Morrogh, "The Cappella Dei Principi under Ferdinando I De' Medici," in *San Lorenzo: A Florentine Church*, ed. Robert W. Gaston and Louis A. Waldman (Florence: Villa i Tatti, 2017), 567–610.

MORTON 2003: Thomas J. Morton, "The Impact of Luxury: The Forum of Meninx: An Architectural Investigation" (PhD diss., University of Pennsylvania, 2003).

MÖSENEDER 1981: Karl Möseneder, "Lapides Vivi: Über die Kreuzkapelle der Burg Karlstein," *Wiener Jahrbuch für Kunstgeschichte* 34, no. 1 (1981): 39–69.

MOSER VON FILSECK 1986: Karin M. Moser von Filseck, "Der Alabasterfries von Tiryns," *Archäologischer Anzeiger* 1 (1986): 1–32.

MULLER 1995: Béatrice Muller, "Aspects de la peinture murale proche-orientale au IIe millénaire av. J.-C.," *Revue archéologique de Picardie* 10, no. 1 (1995): 131–40.

MULLIEZ-TRAMOND 2010: Maud Mulliez-Tramond, "La représentation de marbres colorés dans la peinture pariétale de la fin de la République Romaine: L'exemple de l'onyx," in *Atti del X congresso dell'AIPMA*, ed. Irene Brigantini (Naples: Università degli Studi di Napoli "L'Orientale," 2010), 809–14.

MUNCEY 1930: Raymond W. Muncey, *A History of the Consecration of Churches and Churchyards* (Cambridge: Heffer, 1930).

MUNDELL 1977: Marlia Mundell, "Monophysite Church Decoration," in *Iconoclasm*, ed. Anthony Bryer and Judith Herrin (Birmingham: Centre for Byzantine Studies, 1977), 59–74.

MUNDY 1977: E. James Mundy, "Franciscus alter Christus: The Intercessory Function of a Late Quattrocento Panel," *Record of the Art Museum, Princeton University* 36, no. 2 (1977): 4–15.

MUÑOZ 1924: Antonio Muñoz, *L'Église de Sainte Sabine à Rome* (Rome: Cuggiani, 1924).

MUÑOZ 1926: Antonio Muñoz, *Il restauro della basilica di S. Giorgio al Velabro in Roma* (Rome: Società Editrice d'Arte Illustrata, 1926).

MURARO 1961: Michelangelo Muraro, "The Statutes of the Venetian Arti and the Mosaics of the Mascoli Chapel," *Art Bulletin* 43 (1961): 263–75.

MYLONAS 1966: George E. Mylonas, *Mycenae and the Mycenaean Age* (Princeton: Princeton University Press, 1966).

MYLONAS 1990: Paulos M. Mylonas, "Δομική έρευνα στο εκκλησιαστικό συγκρότημα του Οσίου Λουκά Φωκίδος," *Arcaiologia* 36 (1990): 6–30.

NAGEL 2005: Alexander Nagel, "Experiments in Art and Reform in Italy in the Early Sixteenth Century," in *The Pontificate of Clement VII: History, Politics, Culture*, ed. Kenneth Gouwens and Sheryl E. Reiss (Aldershot: Ashgate, 2005), 385–409.

NAGEL 2011: Alexander Nagel, *The Controversy of Renaissance Art* (Chicago: University of Chicago Press, 2011).

NAPOLEONE 1996: Caterina Napoleone, "La transverberazione di Teresa. La Cappella Cornaro del Bernini," *Franco Maria Ricci* 119 (1996): 38–44.

NAPOLEONE 1998: Caterina Napoleone, "Bernini e il cantiere della Cappella Cornaro," *Antologia di Belle Arti* 55–58 (1998): 180–81.

NAPOLEONE 2001: Caterina Napoleone, "L'impiego dei diaspri e delle agate di Sicilia dal XVI al XVII secolo," in *Splendori di Sicilia: Arti decorative dal Rinascimento al Barocco*, ed. Concetta Di Natale Maria (Milan: Charta, 2001), 193–203.

NAUMANN AND BELTING 1966: Rudolf Naumann and Hans Belting, *Die Euphemia-Kirche am Hippodrom zu Istanbul und ihre Fresken* (Berlin: Gebrüder Mann, 1966).

NECIPOGLU 1992: Gülru Necipoglu, ed., *The Life of an Imperial Monument: Hagia Sophia after Byzantium* (Cambridge: Cambridge University Press, 1992).

NECIPOGLU-KAFADAR 1985: Gulru Necipoglu-Kafadar, "The Süleymaniye Complex in Istanbul: An Interpretation," *Muqarnas* 3 (1985): 92–117.

NEER 2004: Richard Neer, "The Athenian Treasury at Delphi and the Material of Politics," *Classical Antiquity* 23, no. 1 (2004): 63–93.

NEER 2010: Richard Neer, *The Emergence of the Classical Style in Greek Sculpture* (Chicago: University of Chicago Press, 2010).

NEGRO 2002: Angela Negro, *Bernini e il bel composto: La cappella de Sylva in Sant'Isidoro* (Rome: Campisano, 2002).

NEIMAN 1969: David Neiman, "The Supercaelian Sea," *Journal of Near Eastern Studies* 28, no. 4 (1969): 243–49.

NELSON 1995: Robert S. Nelson, "The Italian Appreciation and Appropriation of Illuminated Byzantine Manuscripts, ca. 1200–1450," *Dumbarton Oaks Papers* 49 (1995): 209–35.

NELSON 2003: Michael C. Nelson, "Leveling Ashlar Masonry," in *Metron: Measuring the Aegean Bronze Age: Proceedings of the 9th International Aegean Conference New Haven, Yale University, 18–21 April 2002*, ed. Karen Polinger Foster and Robert Laffineur (Liège: Université de Liège, 2003), 269–74.

NELSON 2004: Robert S. Nelson, *Hagia Sophia, 1850–1950: Holy Wisdom, Modern Monument* (Chicago: University of Chicago Press, 2004).

NELSON 2006: Michael C. Nelson, "Pylos, Block Masonry and Monumental Architecture in the Late Bronze Age Peloponnese," in *Power and Architecture: Monumental Architecture in the Bronze Age Near East and the Aegean*, ed. Joachim Bretschneider, Jan Driessen, and Karel van Lerberghe (Leuven: Peeters, 2006), 143–59.

NELSON 2007: Michael C. Nelson, "Minoan and Mycenaean Stone Revetment," in *Κριναι και Λιμενοι: Festschrift in Honor of Joseph and Maria Shaw*, ed. Philip P. Betancourt, Michael C. Nelson, and Hector Williams (Philadelphia: INSTAP, 2007), 17–22.

NESBITT 1880: Alexander Nesbitt, "On Wall Decorations in Sectile Work as Used by the Romans, with Special Reference to the Decorations of the Palace of the Bassi at Rome," *Archaeologia* 45, no. 2 (1880): 267–96.

NESSELRATH 2003: Arnold Nesselrath, "Von Volpaia bis Volpi: Die farbige Marmorverkleidung der Vorhalle des Pantheon," *Pegasus. Berliner Beiträge zum Nachleben der Antike* 4 (2003): 19–36.

NEVEROV 1986: Oleg Neverov, "Nero-Helios," in *Pagan Gods and Shrines of the Roman Empire*, ed. Martin Henig and Martin King (Oxford: Oxford University Committee for Archaeology, 1986), 189–94.

NEWBY 1987: Martine Newby, "Medieval Glass from Farfa," in *Annales du 10e Congrès de l'Association Internationale pour l'Histoire du Verre* (Amsterdam: Association Internationale pour l'histoire du verre, 1987), 255–70.

NEWBY 1991: Martine Newby, "The Glass from Farfa Abbey: An Interim Report," *Journal of Glass Studies* 33 (1991): 32–41.

NEWLANDS 2002: Carole E. Newlands, *Statius' Silvae and the Poetics of Empire* (Cambridge: Cambridge University Press, 2002).

NICHOLSON 2012: Paul T. Nicholson, "'Stone . . . That Flows': Faience and Glass as Man-Made Stones in Egypt," *Journal of Glass Studies* 54 (2012): 11–23.

NIEMEIER 1991: Wolf-Dietrich Niemeier, "Minoan Artisans Travelling Overseas: The Alalakh Frescoes and the Painted Plaster Floor at Tel Kabri (Western Galilee)," in *Thalassa. L'Égée préhistorique et la mer*, ed. Robert Laffineur and Lucien Basch (Leuven: Université de Liège, 1991), 189–200.

NIEMEIER AND NIEMEIER 2000: Barbara Niemeier and Wolf-Dietrich Niemeier, "Aegean Frescoes in Syria-Palestine: Alalakh and Tel Kabri," in Sherratt 2000, 763–802.

NOLL ET AL. 1975: R. Noll, L. Born, and R. Holm, "Keramiken und Wandmalerein der Ausgrabungen von Thera," *Die Naturwissenschaften* 62 (1975): 87–94.

NORDENFALK 1937: Carl A. J. Nordenfalk, *Vier Kanonestafeln eines spätantiken Evangelienbuches* (Berlin: Elanders boktryckeri aktiebolag, 1937).

NORDHAGEN 1962: Per J. Nordhagen, "The Earliest Decorations in Santa Maria Antiqua and Their Date," *Acta ad archaeologiam et artium historiam pertinentia* 1 (1962): 53–72.

NORRE 1966: Anastasia Demetriades Norre, *Studies in the History of the Parthenon* (Ann Arbor: UMI, 1966).

NOVARA 1998A: Paola Novara, "Il rivestimento parietale nella chiesa di San Vitale a Ravenna: Archeologia e antiquaria," *Ravenna, Studi e Ricerche* 2 (1998): 61–117.

NOVARA 1998B: Paola Novara, "Sectilia parietali dalle Catacombe di S. Gennaro a Capodimonte (Napoli)," *Rivista di Archeologia Cristiana* 74, no. 1 (1998): 149–62.

NOVARA 1998C: Paola Novara, "Sectilia parietali di ignota provenienza conservati presso il Museo Nazionale di Ravenna," *Archeologia dell'Emilia-Romagna* 2, no. 1 (1998): 153–64.

NOVARA 2000: Paola Novara, "Il 'sectile' parietale del battistero neoniano di Ravenna: Un nuovo tentativo di lettura," in *Atti del VI Colloquio dell'AISCOM*, ed. Federico Guidobaldi and Andrea Paribeni (Ravenna: Girasole, 2000), 361–72.

NOVARA PIOLANTI 1998: Paola Novara Piolanti, "Sectilia parietali dal Palatium Ravennate," in *Ricerche di archeologia e topografia*, ed. Raffaella Farioli Campanati (Ravenna: Girasole, 1998), 533–64.

NOWICKA 1969: Maria Nowicka, *La maison privée dans l'Egypte ptolémaïque* (Wroclaw: Zaklad Narodowy im. Ossolinskich-Wydawnictwo Polskiej Akademii Nauk, 1969).

NOWICKA 1984: Maria Nowicka, "Théophilos, peintre alexandrian, et son activité," in *Alessandria e il mondo ellenistico-romano*, ed. Nicola Bonacasa and Antonino di Vita (Rome: Bretschneider, 1984), 256–59.

NUNES PEDROSO 2001: Rui Nunes Pedroso, "La basilique de Tróia, un décor Luso-Romain du Ive S. ap. J.-C," in *La peinture funéraire antique: Ive siècle av. J.-C.-Ive siècle ap. J.-C.*, ed. Alix Barbet (Paris: Errance, 2001), 305–8.

NUNN 1988: Astrid Nunn, *Die Wandmalerei und der glasierte Wandschmuck im Alten Orient* (Leiden: Brill, 1988).

NYLANDER 1970: Carl Nylander, *Ionians in Pasargadae: Studies in Old Persian Architecture* (Uppsala: Almqvist & Wiksell Boktryckeri Aktiebolag, 1970).

O'BRYHIM 1996: Shawn O'Bryhim, "A New Interpretation of Hesiod, 'Theogony' 35," *Hermes* 124 (1996): 131–39.

O'GORMAN 1971: James F. O'Gorman, "The Villa Lante in Rome: Some Drawings and Some Observations," *Burlington Magazine* 113, no. 816 (1971): 133–38.

O'MEARA 2004: Dominic O'Meara, "Geometry and the Divine in Proclus," in *Mathematics and the Divine: A Historical Study*, ed. Teun Koetsier and Luc Bergmans (Amsterdam: Elsevier, 2004), 135–45.

O'SHAUGNESSY 1973: Thomas J. O'Shaugnessy, S. J., "God's Throne and the Biblical Symbolism of the Qur'an," *Numen* 20 (1973): 202–21.

OERTEL 1961: Robert Oertel, *Frühe italienische Malerei in Altenburg; beschreibender Katalog der Gemälde des 13. bis 16. Jahrhunderts im Staatlichen Lindenau-Museum* (Berlin: Henschelverlag, 1961).

OHNESORG 1993: Aenne Ohnesorg, *Inselionische Marmordächer*, vol. 2, *Architektur auf Naxos und Paros* (Berlin: Walter de Gruyter, 1993).

OHNESORG 2005: Aenne Ohnesorg, "Naxian and Parian Architecture: General Features and New Discoveries," in *Architecture and Archaeology in the Cyclades: Papers in Honour of J. J. Coulton*, ed. Marina Yeroulanou and Maria Stamatopoulou (Oxford: Archaeopress, 2005), 135–52.

OHNESORG 2011: Aenne Ohnesorg, "Der naxische Lichtdom: Das Phänomen lichtdurchlässiger inselionischer Marmordächer," in *Licht-Konzepte in der vormodernen Architektur*, ed. Ulrike Wulf-Rheidt and Peter I. Schneider (Regensburg: Schnell & Steiner, 2011), 92–100.

OHR 1991: Karlfriedrich Ohr, *Die Basilika in Pompeji* (Berlin: De Gruyter, 1991).

OLIVATO AND PUPPI 1977: Loredana Olivato and Lionello Puppi, *Mauro Codussi* (Milan: Electa, 1977).

OLIVER 2001: Andrew Oliver, "A Glass Opus Sectile Panel from Corinth," *Hesperia* 70 (2001): 349–63.

ONIANS 1971: John Onians, "Alberti and FILARETH: A Study in Their Sources," *Journal of the Warburg and Courtauld Institutes* 34 (1971): 96–114.

ONIANS 1980: John Onians, "Abstraction and Imagination in Late Antiquity," *Art History* 3 (1980): 1–23.

OPPENHEIM 1966: A. Leo Oppenheim, "Mesopotamia in the Early History of Alchemy," *Revue d'Assyriologie et d'archéologie orientale* 60, no. 1 (1966): 29–45.

OPPENHEIM ET AL. 1970: A. Leo Oppenheim, Robert H. Brill, Dan Barag, and Axel von Saldern, *Glass and Glassmaking in Ancient Mesopotamia: An Edition of the Cuneiform Texts Which Contain Instructions for Glassmakers, with a Catalogue of Surviving Objects* (Corning, NY: Corning Museum of Glass, 1970).

ORESKO AND PARROTT 1997: Robert Oresko and David Parrott, "The Sovereignty of Monferrato and the Citadel of Casale as European Problems in the Early Modern Period," in *Stefano Guazzo e Casale tra Cinque e Seicento*, ed. Daniela Ferrari (Rome: Bulzoni, 1997), 11–86.

ORLANDOS 1966–68: Anastasios K. Orlandos, *Les matériaux de construction et la technique architecturale des anciens Grecs*, trans. Vanna Hadjimichali and Krista Laumonier, 2 vols. (Paris: E. de Boccard, 1966–68).

ORTALLI 2000: Jacopo Ortalli, "Rimini: La Domus 'del Chirurgo,'" in *Aemilia: La cultura romana in Emilia Romagna dal III secolo a. c. all'età Costantiniana*, ed. Mirella Marini Calvani (Venice: Marsilio, 2000), 513–26.

OSING 1972–92: Jürgen Osing, "Pharaoh," in *Lexikon der Ägyptologie*, ed. Wolfgang Helck and Eberhard Otto (Wiesbaden: Harrassowitz, 1972–92), 4:1021.

OSTROW 1990: Steven F. Ostrow, "Marble Revetment in Late Sixteenth Century Roman Chapels," in *IL 60: Essays Honoring Irving Lavin on his Sixtieth Birthday*, ed. Marilyn Aronberg Lavin and Irving Lavin (New York: Italica, 1990), 253–76.

OSTROW 1996: Steven F. Ostrow, *Art and Spirituality in Counter-Reformation Rome: The Sistine and Pauline Chapels in S. Maria Maggiore* (Cambridge: Cambridge University Press, 1996).

OSTROW 2004: Steven F. Ostrow, "Playing with the Paragone: The Reliefs of Pietro Bernini," *Zeitschrift für Kunstgeschichte* 67 (2004): 329–64.

OSTROW 2007: Steven F. Ostrow, "Bernini e il Paragone," in *Bernini Pittore*, ed. Tommaso Montanari (Cinisello Balsamo: Silvana, 2007), 223–33.

OSTROW 2009: Steven F. Ostrow, "The 'Confessio' in Post-Tridentine Rome," in *Arte e committenza nel Lazio nell'età di Cesare Baronio*, ed. Patrizia Tosini (Rome: Gangemi, 2009), 19–32.

OSTROW 2014: Steven F. Ostrow, "'Appearing to Be What They Are Not': Bernini's Reliefs in Theory and Practice," in *Critical Perspectives on Roman Baroque Sculpture*, ed. Anthony Colantuono and Steven F. Ostrow (University Park: Pennsylvania State University Press, 2014), 165–84.

OTTO 1994: Randall E. Otto, *Coming in the Clouds: An Evangelical Case for the Invisibility of Christ at His Second Coming* (Lanham, MD: University Press of America, 1994).

OURSEL 1989: Raymond Oursel, *France romane*, 2nd ed., 2 vols. (La Pierre-qui-Vire: Zodiaque, 1989).

OUSTERHOUT 1987: Robert G. Ousterhout, *The Architecture of the Kariye Camii in Istanbul* (Washington, DC: Dumbarton Oaks, 1987).

OUSTERHOUT 1992: Robert G. Ousterhout, "Originality in Byzantine Architecture: The Case of Nea Moni," *Journal of the Society of Architectural Historians* 51, no. 1 (1992): 48–60.

OUSTERHOUT 1999: Robert G. Ousterhout, *Master Builders of Byzantium* (Princeton: Princeton University Press, 1999).

OUSTERHOUT 2001: Robert G. Ousterhout, "Art, Architecture and Komnenian Ideology at the Pantokrator Monastery," in *Byzantine Constantinople: Monuments, Topography, and Everyday Life*, ed. Nevra Necipoglu (Leiden: Brill, 2001), 133–50.

OUSTERHOUT 2010: Robert Ousterhout, "New Temples and New Solomons: The Rhetoric of Byzantine Architecture," in *The Old Testament in Byzantium*, ed. Paul Magdalino and Robert Nelson (Washington, DC: Dumbarton Oaks, 2010), 223–54.

PAATZ 1940: Walter Paatz, "Die Hauptsrömungen in der florentiner Baukunst der frühen und hohen Mittelalters und ihr geschichtlicher Hintergrund," *Mitteilungen des Kunsthistorischen Institutes in Florenz* 6 (1940): 33–72.

PÄCHT 1951: Otto Pächt, "Giovanni da Fano's Illustrations for Basinio's Epos Hesperis, with Two Appendices by A. Campana," *Studi romagnoli* 2 (1951): 91–112.

PACINI 2003: Piero Pacini, "Fasto barocco e rigore monastico per Santa Maria Maddalena de' Pazzi. La costruzione della cappella-reliquiario di Ciro Ferri," *Mitteilungen des Kunsthistorischen Institutes in Florenz* 47, nos. 2/3 (2003): 375–439.

PAGENSTECHER 1919: Rudolf Pagenstecher, *Nekropolis. Untersuchungen über Gestalt und Entwicklung der alexandrinischen Grabenlagen und ihrer Malerien* (Leipzig: Giesecke & Devrient, 1919).

PAGLIAI 1989: Daniela Pagliai, "Natura e mitologia nella villa di Blosio Palladio," in Danesi Squarzina 1989, 347–55.

PALAGIA 2000: Olga Palagia, "Meaning and Narrative Techniques in Statue-Bases of the Pheidian Circle," in *Word and Image in Ancient Greece*, ed. N. Keith Rutter and Brian A. Sparkes (Edinburgh: Edinburgh University Press, 2000), 53–78.

PALAGIA AND HERZ 2002: Olga Palagia and Norman Herz, "Investigation of Marbles at Delphi," in Hermann et al. 2002, 240–49.

PALLAS 1952: Demetrios I. Pallas, *La "Thalassa" dans l'église chrétienne. Contribution a l'histoire de l'autel chrétien et a la morphologie de la liturgie* (Athens: Institut français d'Athènes, 1952).

PALMA AND DE LACHENAL 1983: Beatrice Palma and Lucilla De Lachenal, *Museo Nazionale Romano. Le Sculture*, vol. 1.5, *I Marmi Ludovisi nel Museo Nazionale Romano* (Rome: De Luca, 1983).

PALMER 1990: Andrew Palmer, "The Inauguration Anthem of Hagia Sophia Again," *Byzantine and Modern Greek Studies* 14 (1990): 247–48.

PALMERIO AND VILLETTI 1989: Giancarlo Palmerio and Gabriella Villetti, *Storia edilizia di S. Maria sopra Minerva in Roma, 1275–1870* (Rome: Viella, 1989).

PANE 1975: Roberto Pane, "Le tre capelle toscane della chiesa di Monteoliveto," in *Il Rinascimento nell'Italia meridionale*, ed. Roberto Pane (Milan: Comunità, 1975), 1.229–34.

PANI ERMINI 1998: Letizia Pani Ermini, "Del sepolcro di San Giovenale alla cattedrale di Narni," in *San Giovenale, La Cattedrale di Narni nella storia e nell'arte*, ed. Cinzia Perissinotto (Narni: Centro Studi Storici, 1998), 85–92.

PAOLETTI 1929: Pietro Paoletti, *La Scuola Grande di San Marco* (Venice: Comune di Venezia, 1929).

PAPALEXANDROU 2001: Amy Papalexandrou, "Text in Context: Eloquent Monuments and the Byzantine Beholder," *Word and Image* 17, no. 3 (2001): 259–83.

PAPALEXANDROU 2007: Amy Papalexandrou, "Echoes of Orality in the Monumental Inscriptions of Byzantium," in *Art and Text in Byzantine Culture*, ed. Liz James (Cambridge: Cambridge University Press, 2007), 161–87.

PAPARAZZO 2011: Ernesto Paparazzo, "Philosophy and Science in the Elder Pliny's Naturalis Historia," in *Pliny the Elder: Themes and Contexts*, ed. Roy Gibson and Ruth Morello (Leiden: Brill, 2011), 89–111.

PARIBENI 1989: Andrea Paribeni, "L'uso e il gusto del marmo in età bizantina attraverso le descrizioni e le rappresentazioni antiche," in *Il marmo nella civiltà romana: La produzione e il commercio*, ed. Enrico Dolci (Carrara: Internazionale marmi e macchine, 1989), 163–83.

PARIS 1990: Rita Paris, "Il pannello di Hylas e le ninfe dalla basilica di Giunio Basso," *Bollettino di Archeologia* 1–2 (1990): 194–202.

PARIS 1991: Rita Paris, "I pannelli in opus sectile dalla Basilica di Giunio Basso: Osservazioni sul pannello con biga e fazioni del circo," *Bollettino di Archeologia* 7 (1991): 91–95.

PARIS 2000: Rita Paris, ed., *Via Appia: La Villa dei Quintili* (Milan: Electa, 2000).

PARISI 2014: Antonella Parisi, "Il Sepulcrum Scipionis e le altre piramidi di Roma," in *Apoteosi da uomini a dei*, ed. Letizia Abbondanza, Filippo Coarelli, and Eugenio Lo Sardo (Rome: Palombi, 2014), 98–109.

PARLAND 1913: Alfred A. Parland, *K' Istorii Arkhitekturnoi Dekoratsii v Italii* (St. Petersburg: P. P. Soikina, 1913).

PARLATO 2009: Enrico Parlato, "Enrico Caetani a S. Pudenziana: Antichità Cristiane, magnificenza decorativa e prestigio del casato nella Roma di fine Cinquecento," in *Arte e committenza nel Lazio nell'età di Cesare Baronio*, ed. Patrizia Tosini (Rome: Gangemi, 2009), 143–64.

PARLATO 2014: Enrico Parlato, "Esequie e sepolture romane dei cardinali veneziani," in *Arte e committenza tra Venezia e Roma (1523–1605)*, ed. Caterina Furlan and Patrizia Tosini (Cinisello Balsamo: Silvana, 2014), 329–49.

PARLATO 2016: Enrico Parlato, "Roma 1570: L'eredità di Michelangelo nelle memorie funebri Castiglioni, Savelli e Pisani," in *Dopo Il 1564 L'eredità di Michelangelo a Roma nel tardo cinquecento*, ed. Marco Simone Bolzoni, Furio Rinaldi, and Patrizia Tosini (Rome: De Luca, 2016), 142–61.

PARRISH 1995: David Parrish, "A Mythological Theme in the Decoration of Late Roman Dining Rooms," *Révue archéologique*, no. 2 (1995): 307–32.

PARROT 1958A: André Parrot, *Mission archéologique de Mari*, vol. 2, *Le Palais. Architecture* (Paris: Librairie orientaliste Paul Geuthner, 1958).

PARROT 1958B: André Parrot, *Mission archéologique de Mari*, vol. 2, *Le Palais. Peintures murales* (Paris: Librairie orientaliste Paul Geuthner, 1958).

PARTIDA 2000: Elena K. Partida, "Architecture in Parian Marble at Delphi," in Schilardi 2000, 355–64.

PARTINGTON 1934: J. R. Partington, "Chemical Arts in the Mount Athos Manual of Christian Iconography," *Isis* 22, no. 1 (1934): 136–49.

PASADAIOS 1973: Aristeides A. Pasadaios, *Ό κεραμοπλαστικος διακοσμος των βυζαντινων κτηριων της Κωνσταντινουπολεως* (Athens: Archaeological Society, 1973).

PASQUALI 1996: Susanna Pasquali, *Il Pantheon: Architettura e antiquaria nel Settecento a Roma* (Modena: Franco Cosimo Panini, 1996).

PASQUINI 2002: Laura Pasquini, *La decorazione a stucco in Italia fra Tardo Antico e Alto Medioevo* (Ravenna: Longo, 2002), 80–81.

PATON AND SCHNEIDER 1999: Sara Paton and Rolf M. Schneider, "Imperial Splendour in the Province: Imported Marble on Roman Crete," in *From Minoan Farmers to Roman Traders: Sidelights on the Economy of Ancient Crete*, ed. Angelos Chaniotis (Stuttgart: Franz Steiner, 1999), 292–304.

PAVLOVSKIS 1965: Zoja Pavlovskis, "Statius and the Late Latin Epithalamia," *Classical Philology* 60, no. 3 (1965): 164–77.

PAVOLINI 1986: Carlo Pavolini, "L'edilizia commerciale e l'edilizia abitativa nel contesto di Ostia tardoantica," in *Società romana e impero tardoantico*, ed. Andrea Giardina (Rome: Laterza, 1986), 239–98.

PAYNE 2007: Alina Payne, "Alberti and the Origins of the Paragone between Architecture and the Figurative Arts," in *Leon Battista Alberti: Teorico delle arti e gli impegni civili del "De Re Aedificatoria,"* ed. Arturo Calzona, Francesco Paolo Fiore, Alberto Tenenti, and Cesare Vasoli (Florence: Olschki, 2007), 347–68.

PAYNTER AND TITE 2001: Sarah Paynter and Michael Tite, "The Evolution of Glazing Technology in the Ancient Near East and Egypt," in *The Social Context of Technological Change: Egypt and the Near East, 1650–1550 BC*, ed. Andrew J. Shortland (Oxford: Oxbow, 2001), 239–54.

PEACOCK AND WILLIAMS 1999: David P. S. Peacock and D. F. Williams, "Ornamental Coloured Marble in Roman Britain: An Interim Report," in *Archéomatériaux: Marbres et autres roches*, ed. Max Schvoerer (Talence: CRPAA, 1999), 353–57.

PEARCE 2014: Mark Pearce, "Into the Realm of 'Obscurity and Thick Darkness': Can We Reconstruct the Belief Systems of Past Miners?," *Archeologia Postmedievale* 17 (2014): 271–75.

PEARCY 1977: Lee T. Pearcy, "Horace's Architectural Imagery," *Latomus* 36 (1977): 772–81.

PEDONE 2012: Silvia Pedone, "The Marble Omphalos of Saint Sophia in Constantinople: An Analysis of an Opus Sectile Pavement of Middle Byzantine Age," in *11th International Colloquium on Ancient Mosaics*, ed. Mustafa Şahin (Istanbul: Phoibos, 2012), 749–68.

PEJRANI BARICCO 1990: Luisella Pejrani Baricco, "Decorazione in opus sectile," in *Milano capitale dell'impero romano 286–402 D.C.*, ed. Gemma Sena Chiesa (Milan: Silvana, 1990), 298.

PEJRANI BARICCO 2000: Luisella Pejrani Baricco, "Fonti archeologiche per la storia dell'isola," in *San Giulio e la sua isola nel XVI centenario di San Giulio*, ed. Giancarlo Andenna (Novara: Interlinea, 2000), 85–111.

PELLECCHIA 1992: Linda Pellecchia, "Architects Read Vitruvius: Renaissance Interpretations of the Atrium of the Ancient House," *Journal of the Society of Architectural Historians* 51, no. 4 (1992): 377–416.

PENNI IACCO 2004: Emanuela Penni Iacco, *La Basilica di S. Apollinare Nuovo di Ravenna attraverso i secoli* (Bologna: Ante Quem, 2004).

PENSABENE 1994: Patrizio Pensabene, *Le vie del marmo: I blocchi di cava di Roma e di Ostia: Il fenomeno del marmo nella Roma antica* (Rome: Ministero per i Beni Culturali e Ambientali, 1994).

PENSABENE 1997: Patrizio Pensabene, "Elementi architettonici dalla Casa di Augusto sul Palatino," *Mitteilungen des deutschen archäologischen Instituts, Römische Abteilung* 104 (1997): 151–92.

PENSABENE AND PANELLA 1999: Patrizio Pensabene and Clementina Panella, *Arco di Costantino tra archeologia e archeometria* (Rome: Bretschneider, 1999).

PENTCHEVA 2010: Bissera V. Pentcheva, *The Sensual Icon: Space, Ritual, and the Senses in Byzantium* (University Park: Pennsylvania State University Press, 2010).

PENTCHEVA 2017: Bissera V. Pentcheva, *Hagia Sophia: Sound, Space, and Spirit in Byzantium* (University Park: Pennsylvania State University Press, 2017).

PEPE 1969: Mario Pepe, "Il 'paragone' tra pittura e scoltura nella letteratura artistica rinascimentale," *Cultura e scuola* 8 (1969): 120–31.

PERNOT 1993: Laurent Pernot, *La rhétorique de l'éloge dans le monde gréco-romain* (Paris: Études Augustiniennes, 1993).

PERONI 1977: Adriano Peroni, "Il mosaico pavimentale di San Michele Maggiore a Pavia: Materiali per un'edizione," *Studi Medievali* 18, no. 2 (1977): 705–43.

PEROSA 1960: Alessandro Perosa, *Giovanni Rucellai ed il suo Zibaldone* (London: Warburg Institute, 1960).

PERRIER 1996: R. Perrier, "Les roches ornementales du Languedoc-Roussillon," *Mines et carrières* 78 (1996): 65–75.

PERROT AND CHIPIEZ 1894: Georges Perrot and Charles Chipiez, *Histoire de l'art dans l'antiquité*, vol. 6, *La Grèce primitive, L'art Mycenien* (Paris: Hachette, 1894), 608–41.

PESCHLOW 2001: Urs Peschlow, "Die befestigte Rezidenz von Mermerkule. Beobachtungen an einem spätbyzantinischen Bau im verteidigungssystem Konstantinopels," *Jahrbuch der österreichischen Byzanzinistik* 51 (2001): 385–401.

PESCHLOW-BINDOKAT 1990: Anneliese Peschlow-Bindokat, *Die Steinbrüche von Selinunt. Die Cave di Cusa und die Cave di Barone* (Mainz am Rhein: Philipp Von Zabern, 1990).

PETERSSON 1970: Robert T. Petersson, *The Art of Ecstasy: Teresa, Bernini, and Crashaw* (New York: Atheneum, 1970).

PETILÄ-CASTRÉN 1987: Leena Petilä-Castrén, *Magnificentia publica: The Victory Monuments of the Roman Generals in the Era of the Punic Wars* (Helsinki: Societas Scientiarum Fennica, 1987).

PETRAIN 2005: David Petrain, "Gems, Metapoetics, and Value: Greek and Roman Responses to a Third-Century Discourse on Precious Stones," *Transactions of the American Philological Association* 135 (2005): 329–57.

PETTAZONI 1954: Raffaele Pettazoni, *Essays on the History of Religions*, trans. H. J. Rose (Leiden: Brill, 1954).

PFÄLZNER 2013: Peter Pfälzner, "The Qatna Wall Paintings and the Formation of Aegeo-Syrian Art," in Aruz et al. 2013, 200–215.

PIANA 2000: Mario Piana, "Il Convento della Carità: Materiali, tecniche, strutture," *Annali di Architettura* 10–11 (1998–99 [2000]): 310–21.

PIANA AND WOLTERS 2003: Mario Piana and Wolfgang Wolters, eds., *Santa Maria dei Miracoli a Venezia: La storia, la fabbrica, i restauri* (Venice: Istituto Veneto, 2003).

PICCIRILLO AND ALLIATA 1989: Michele Piccirillo and Eugenio Alliata, *Madaba: Le chiese e i mosaici* (Cinisello Balsamo: Edizioni Paoline, 1989).

PICCIRILLO ET AL. 1993: Michele Piccirillo, Patricia Maynor Bikai, and Thomas A. Dailey, *The Mosaics of Jordan* (Amman: American Center of Oriental Research, 1993).

PIETRAMELLARA 1973: Carla Pietramellara, *Battistero di S. Giovanni a Firenze: Rilievo e studio critico* (Florence: Polistampa, 1973).

PILLIOD 1989: Elizabeth Pilliod, "Studies on the Early Career of Alessandro Allori" (PhD diss., University of Michigan, 1989).

PILLIOD 1991: Elizabeth Pilliod, "Le noli me tangere de Bronzino et la décoration de la chapelle Cavalcanti de l'église Santo Spirito à Florence," *Revue du Louvre. La revue des musées de France* 41, nos. 5/6 (1991): 50–61.

PINCUS 1990: Debra Pincus, "Andrea Dandolo (1343–1354) and Visible History: The San Marco Projects," in *Art and Politics in Late Medieval and Early Renaissance Italy: 1250–1500*, ed. Charles M. Rosenberg (Notre Dame, IN: University of Notre Dame Press, 1990), 191–206.

PINCUS 1992: Debra Pincus, "Venice and the Two Romes: Byzantium and Rome as a Double Heritage in Venetian Cultural Politics," *Artibus et Historiae* 26 (1992): 101–14.

PINCUS 1996: Debra Pincus, "The Stones of Venice in the Baptistery of San Marco: Eastern Marbles in Western Mosaics," in *Architectural Studies in Memory of Richard Krautheimer*, ed. Cecil L. Striker (Mainz am Rhein: Philipp von Zabern, 1996), 137–42.

PIPER 1873: Ferdinand Piper, "Maria als Thron Salomons und ihre Tugenden bei der Verkündigung," *Jahrbücher für Kunstwissenschaft* 5 (1873): 97–137.

PLACE 1867–70: Victor Place, *Ninive et l'Assyrie*, 3 vols. (Paris: Imprimierie impériale, 1867–70).

PLANTZOS 1996A: Dimitris Plantzos, "Hellenistic Cameos: Problems of Classification and Chronology," *Bulletin of the Institute of Classical Studies* 41 (1996): 115–31.

PLANTZOS 1996B: Dimitris Plantzos, "Ptolemaic Cameos of the Second and First Centuries BC," *Oxford Journal of Archaeology* 15 (1996): 39–61.

PLANTZOS 1997: Dimitris Plantzos, "Crystals and Lenses in the Graeco-Roman World," *American Journal of Archaeology* 101, no. 3 (1997): 451–64.

PLOMMER 1965: Hugh Plommer, "A Carved Block from the Megaron of Mycenae," *Annual of the British School at Athens* 60 (1965): 207–11.

PLUMPE 1943: Joseph C. Plumpe, "Vivum Saxum, Vivi Lapides. The Concept of 'Living Stone' in Classical and Christian Antiquity," *Traditio* 1 (1943): 1–14.

POLLITT 1974: Jerome J. Pollitt, *The Ancient View of Greek Art: Criticism, History, and Terminology* (New Haven: Yale University Press, 1974).

POMPEI 1981: *Pompei, 1748–1980: I tempi della documentazione* (Rome: Multigrafica 1981).

PONGRATZ-LEISTEN 2013: Beate Pongratz-Leisten, "From Pictograph to Pictogram: The Solarization of Kingship in Syro-Anatolia and Assyria," in Aruz and Wallenfels 2013, 298–311.

POPE AND SCHULTZ 2014: Spencer Pope and Peter Schultz, "The Chryselephantine Doors of the Parthenon," *American Journal of Archaeology* 118, no. 1 (2014): 19–31.

POSTGATE 1997: Nicholas Postgate, "Mesopotamian Petrology: Stages in the Classification of the Material World," *Cambridge Archaeological Journal* 7, no. 2 (1997): 205–24.

PRANDI 1953: Adriano Prandi, *Il complesso monumentale della basilica celimontana dei SS. Giovanni e Paolo* (Vatican City: Tipografia Poliglotta Vaticana, 1953).

PRANDI 2012: Luisa Prandi, "New Evidence for the Dating of Cleitarchus (Poxy Lxxi.4808)?," *Histos* 6 (2012): 15–26.

PRATER 1985: Andreas Prater, "Mantegnas Cristo in Scurto," *Zeitschrift für Kunstgeschichte* 48 (1985): 279–99.

PREIMESBERGER 1985: Rudolf Preimesberger, "Themes from Art Theory in the Early Works of Bernini," in *Gianlorenzo Bernini: New Aspects of His Art and Thought*, ed. Irving Lavin (University Park: Pennsylvania State University Press, 1985), 1–24.

PREIMESBERGER 1986: Rudolf Preimesberger, "Berninis Cappella Cornaro. Eine Bild-Wort-Synthese des siebzehnten Jahrhunderts?," *Zeitschrift für Kunstgeschichte* 49, no. 2 (1986): 190–219.

PRESSOUYRE 1984: Sylvia Pressouyre, *Nicolas Cordier, recherches sur la sculpture à Rome autour de 1600* (Rome: École française de Rome, 1984).

PRICE 1993: Jennifer Price, "Window Glass," in *The Uley Shrines*, ed. Ann Woodward and Peter Leach (London: English Heritage, 1993), 189–92, 210–15.

PRIESTER 1990: Ann E. Priester, "The Belltowers of Medieval Rome and the Architecture of Renovatio" (PhD diss., Princeton University, 1990).

PRZYBOROWSKI 1982: Claudia Przyborowski, "Die Ausstattung der Fürstenkapelle an der Basilica von San Lorenzo in Florenz: Versuch einer Rekonstruktion" (PhD diss., Technische Universität Berlin, 1982).

PUGNO 1961: Giuseppe M. Pugno, *La Santa Sindone che si venera a Torino: Disegno storico e collegamenti* (Turin: Società Editrice Internazionale, 1961).

PUPPI 1973: Lionello Puppi, *Andrea Palladio*, 2 vols. (Milan: Electa, 1973).

PURCELL 1967: Donovan Purcell, *Cambridge Stone* (London: Faber, 1967), 88–93.

PURKART 1972: Josef Purkart, "Franciscus de Retza: 'Defensorium inviolatae virginitatis,' eine unbekannte Hs. aus dem Stift Lambach (OÖ)," *Amsterdamer Beiträge zur älteren Germanistik* 3 (1972): 181–99.

QUARANTA AND BRUTO 2006: Paola Quaranta and Maria Luisa Bruto, "L'opus Interrasile a Roma e nell'Italia centrale: Rilettura dei materiali editi e nuove acquisizioni," in *Atti dell'XI Colloquio dell'Associazione Italiana per oo Studio e la Conservazione del Mosaico, Ancona, 16–19 febbraio 2005*, ed. Claudia Angelelli (Tivoli: Edizioni Scripta Manent, 2006), 1–10.

QUEYREL 2014: François Queyrel, "Apollon et le colosse ds Naxiens," *Revue archéologique* 58, no. 2 (2014): 245–48.

RABY 1980: Julian Raby, "El Gran Turco: Mehmed the Conqueror as a Patron of the Arts of Christendom" (PhD diss., University of Oxford, 1980).

RABY 1987: Julian Raby, "Pride and Prejudice: Mehmed the Conqueror and the Italian Portrait Medal," in *Italian Medals*, ed. J. Graham Pollard (Washington, DC: National Gallery of Art, 1987), 171–96.

RACHMAN-SCHRIRE 2017: Yamit Rachman-Schrire, "Christ's Unction and the Material Realisation of a Stone in Jerusalem," in *Natural Materials of the Holy Land and the Visual Translation of Place, 500–1500*, ed. Bianca Kühnel, Renana Bartal, and Neta Bodner (Farnham: Routledge, 2017), 216–29.

RAGGIO 1960: Olga Raggio, "The Farnese Table: A Rediscovered Work by Vignola," *Metropolitan Museum of Art Bulletin* 18, no. 7 (1960): 213–31.

RAKOB 1979: Friedrich Rakob, "Numidische Königsarchitektur in Nordafrika," in *Die Numider. Reiter Und Könige Nördlich Der Sahara*, ed. Heinz Günter Horn and Christoph B. Rüger (Köln/Bonn: Rheinland and Habelt, 1979), 120–32.

RAPPOPORT 1995: Pavel A. Rappoport, *Building the Churches of Kievan Russia* (Aldershot: Variorum, 1995).

RASMUSSEN ET AL. 2004: Svend E. Rasmussen, Sidsel Grundvig, and Walter L. Friedrich, "Composition of Pigments of Santorini Frescoes: The Rietveld Method as an Aid in Qualitative Phase Analysis," *Powder Diffraction* 19 (2004): 145–48.

RAZMJOU 2004: Shahrohk Razmjou, "Glazed Bricks in the Achaemenid Period," in *Persiens Antike Pracht: Bergbau, Handwerk, Archäologie*, ed. Thomas Stöllner, Rainer Slotta, and Abdolrasool Vatandoust (Bochum: Deutsches Bergbau-Museum, 2004), 382–93.

READE 1979/80: Julian E. Reade, "Assyrian Architectural Decoration," *Baghdader Mitteilungen* 10/11 (1979/80): 71–87.

READER 1981: William W. Reader, "The Twelve Jewels of Revelation 21:19–20: Tradition, History, and Modern Interpretations," *Journal of Biblical Literature* 100, no. 3 (1981): 433–57.

REARICK 2003: William R. Rearick, "La dispersione dei dipinti già a Santa Maria dei Miracoli," in Piana and Wolters 2003, 179–92.

REBECCHI 1980: Fernando Rebecchi, "Esempi di scultura romana a Grado. Clipei ornamentali di porte urbiche: Aquileia, Parma, Ravenna," in *Grado nella storia e nell'arte* (Udine: Arti Grafiche Friulane, 1980), 85–110.

REEKMANS 1988: Louis Reekmans, *Le complexe cémétérial du pape Gaius dans la catacombe de Callixte* (Vatican City: Pontificio Istituto di Archeologia Cristiana, 1988).

REINACH 1909–12: Salomon Reinach, *Répertoire de reliefs grecs et romains*, 3 vols. (Paris: E. Leroux, 1909–12).

REUSCHE 1971: Ehrhard Reusche, "Polychromes Sichtmauerwerk byzantinischer und von Byzanz beeinflusster Bauten Südosteuropas. Überlieferung und Entwicklung einer handwerklichen Technik" (PhD diss., Universität zu Köln, 1971).

REUTERSWARD 1982: Patrik Reutersward, "The Forgotten Symbols of God," *Kunstchronik* 51, no. 3 (1982): 103–25.

REUTERSWARD 1984: Patrik Reutersward, "Windows of Divine Light," in *Interpretazioni Veneziane. Studi di storia dell'arte in onore di Michelangelo Muraro*, ed. David Rosand (Venice: Arsenale, 1984), 77–84.

REUTERSWARD 1991: Patrik Reutersward, "What Colour Is Divine Light?," in *The Visible and Invisible in Art: Essays in the History of Art* (Vienna: Irsa, 1991), 143–56.

RICCI 1905: Corrado Ricci, "Vecchie e inedite descrizioni di Ravenna," *Diario Ravennate* 88 (1905).

RICCI 1914: Corrado Ricci, *Il Mausoleo do Galla Placidia in Ravenna* (Rome: E. Calzone, 1914).

RICCI 1925: Corrado Ricci, *Il Tempio Malatestiano* (Milan: Bestetti & Tumminelli, 1925).

RICE 1997: Louise Rice, *The Altars and Altarpieces of New St. Peter's: Outfitting the Basilica, 1621–1666* (Cambridge: Cambridge University Press, 1997).

RICHARDSON 1955: Lawrence Richardson, *Pompeii: The Casa dei Dioscuri and Its Painters* (Rome: American Academy in Rome, 1955).

RICHMOND 1924: Ernest T. Richmond, *The Dome of the Rock in Jerusalem: A Description of Its Structure and Decoration* (Oxford: Clarendon, 1924).

RICHTER 1970: Gisela M. A. Richter, *Kouroi: Archaic Greek Youths*, 3rd ed. (London: Phaidon, 1970).

RICKERT 1998: Franz Rickert, "Zum Inkrustationsschmuck von S. Sabina in Rom," in *Chartulae: Festschrift für Wolfgang Speyer*, ed. Ernst Dassmann, Klaus Thraede, and Josef Engemann (Münster: Aschendorff, 1998), 263–70.

RIEBESELL 2003: Christina Riebesell, "L'arredo architettonico del Palazzo Farnese a Roma: Vignola e Guglielmo della Porta," in *Vignola e i Farnese*, ed. Christoph L. Frommel, Maurizio Ricci, and Richard J. Tuttle (Milan: Electa, 2003), 35–59.

RIGHI 1980: Roberto Righi, "La villa di Domiziano in località Palazzo sul lago di Sabaudia: Pavimenti in opus sectile dell'edificio balneare ad esedre," *Archeologia Laziale* 3 (1980): 97–110.

RINALDI 1999: Alessandro Rinaldi, "Saxum vivum e non-finito nelle grotte fiorentine del Cinquecento," in *Artifici d'acque e giardini: La cultura delle grotte e dei ninfei in Italia e in Europa*, ed. Isabella Lapi Ballerini and Litta Medri (Florence: Centro Di, 1999), 299–307.

RINGBOM 1965: Sixten Ringbom, *Icon to Narrative: The Rise of the Dramatic Close-Up in Fifteenth-Century Devotional Painting* (Abo: Abo akademi, 1965).

ROBB 1936: David M. Robb, "The Iconography of the Annunciation in the Fourteenth and Fifteenth Centuries," *Art Bulletin* 18, no. 4 (1936): 480–526.

ROBERT 1940: Louis Robert, *Les gladiateurs dans l'orient grec* (Paris: E. Champion, 1940).

ROBERT 1962: Louis Robert, "Les Kordakia de Nicée, le combustible de Synnada et les poissons-scies. Sur des lettres d'un Métropolite de Phrygie au Xe siècle. Philologie et réalités (2)," *Journal des savants* 3–4 (1962): 78–121.

ROBERTS 1989A: Michael Roberts, *The Jeweled Style: Poetry and Poetics in Late Antiquity* (Ithaca, NY: Cornell University Press, 1989).

ROBERTS 1989B: Michael Roberts, "The Use of Myth in Latin Epithalamia from Statius to Venantius Fortunatus," *Transactions of the American Philological Association* 119 (1989): 321–48.

ROBERTSON 1941: D. S. Robertson, "New Light on the Façade of the Treasury of Atreus," *Journal of Hellenic Studies* 61 (1941): 14–16.

ROBINSON AND MYLONAS 1946: David M. Robinson and George E. Mylonas, *Excavations at Olynthus 12: Domestic and Public Architecture* (Baltimore: Johns Hopkins University, 1946).

ROCA DE AMICIS 1987: Augusto Roca De Amicis, "Francesco Borromini e la tomba Merlini in Santa Maria Maggiore a Roma," *Architettura, storia e documenti* (1987): 125–34.

ROCCO 1988: Giorgio Rocco, "Alcune osservazioni sul valore architettonico della antica decorazione parietale: La Domus Aurea di Nerone," *Palladio* 1, no. 1 (1988): 121–34.

ROCKWELL 1991: Peter Rockwell, "Unfinished Statuary Associated with a Sculptor's Studio," in *Aphrodisias Papers 2: The Theatre, a Sculptor's Workshop, Philosophers, and Coin-Types*, ed. R. R. R. Smith and Kenan T. Erim (Ann Arbor: Journal of Roman Archaeology, 1991), 127–43.

RODENWALDT 1912: Gerhart Rodenwaldt, *Die Fresken des Palastes* (Athens: Eleutheroudakis & Barth, 1912).

RODENWALDT 1919: Gerhart Rodenwaldt, "Mykenische Studien I. Die Füßboden des Megarons von Mykenai," *Jahrbuch des Deutschen Archäologischen Instituts* 34 (1919): 87–106.

RODOLICO 1953: Francesco Rodolico, *Le pietre delle città d'Italia* (Florence: F. Le Monnier, 1953).

ROGERS 2008: Patrick Rogers, *The Beauty of Stone: The Westminster Cathedral Marbles* (London: Oremus, 2008).

ROHAULT DE FLEURY 1878: Charles Rohault de Fleury, *La Sainte Vierge: Études archéologiques et iconographiques*, 2 vols. (Paris: Librairie Poussielgue, 1878).

ROMANO 1991: Marco Romano, "Materiale di spoglio nel battistero di San Giovanni in Laterano: Un riesame e nuove considerazioni," *Bollettino d'Arte* 70 (1991): 31–70.

ROMANO 1996: Marco Romano, "L'Oratorio della S. Croce al Laterano. Preliminari di un'indagine archeologica-topografica," *Zeitschrift für Kunstgeschichte* 59 (1996): 341–49.

ROSEN-AYALON 1989: Myriam Rosen-Ayalon, *The Early Islamic Monuments of al-Haram al-Sharif: An Iconographic Study* (Jerusalem: Hebrew University, 1989).

ROSENBAUM 1955: Elisabeth Rosenbaum, "The Vine Columns of Old St. Peter's in Carolingian Canon Tables," *Journal of the Warburg and Courtauld Institutes* 18, nos. 1–2 (1955): 1–15.

ROSENBERG 2010: Angele Rosenberg, "Taking the First Style Seriously," *Chicago Art Journal* 20 (2010): 103–17.

ROSENMEYER 2000: Thomas G. Rosenmeyer, "Seneca and Nature," *Arethusa* 33, no. 1 (2000): 99–119.

ROSS 1943: Marvin C. Ross, "The Rubens Vase, Its History and Date," *Journal of the Walters Art Gallery* 6 (1943): 9–35.

ROSSI 2002: F. Rossi, *La pittura di pietra: Dall'arte del mosaico allo splendore delle pietre dure* (Florence: Giunti, 2002).

ROSSI 2009: Tosca Rossi, *Bergamo urbs picta: Le facciate dipinte di Bergamo tra XV e XVII secolo. Censimento e schedatura di tutti i dipinti murali degli edifici di Bergamo Alta, Bassa e dei colli esistenti o citati dalle fonti* (Treviolo: Ikonos, 2009).

ROSSI 2014: Filli Rossi, ed., *Un luogo per gli dei: L'area del Capitolium a Brescia* (Florence: All'insegna del giglio, 2014).

ROTONDI 1948: Pasquale Rotondi, *Manifestazioni di Paganesimo umanistico nella civiltà urbinate del Rinascimento. Il tempio delle Muse e la cappella del Perdono nel Palazzo ducale di Urbino* (Urbino: Istituto d'Arte, 1948).

ROUVERET 1985: Agnès Rouveret, ed., *Textes grecs et latins relatifs à l'histoire de la peinture ancienne* (Paris: Macula, 1985).

ROUVERET 2002: Agnès Rouveret, "Vitruve et les faux-semblants," in *La pittura parietale in Macedonia e Magna Grecia*, ed. Angela Pontrandolfo Greco (Salerno: Pandemos, 2002), 105–25.

ROUX 1952: Georges Roux, "La terrasse d'Attale I à Delphes," *Bulletin de correspondance hellénique* 76 (1952): 141–96.

ROUX 1979: Georges Roux, "Le vrai temple d'apollon à Délos," *Bulletin de correspondance hellénique* 103, no. 1 (1979): 109–35.

ROVETTA 2002: Alessandro Rovetta, "Cesariano, Bramante e gli studi vitruviani nell'età di Ludovico il Moro," in Frommel et al. 2002, 83–98.

ROWLAND 1986: Ingrid Rowland, "Render unto Caesar the Things Which Are Caesar's: Humanism and the Arts in the Patronage of Agostino Chigi," *Renaissance Quarterly* 39 (1986): 673–730.

ROWLEY 2016: Trevor Rowley, *An Archaeological Study of the Bayeux Tapestry: The Landscapes, Buildings, and Places* (Barnsley: Pen & Sword Archaeology, 2016).

RUGA 1996: Alfredo Ruga, "La copertura dell'Edifico A," in Spadea 1996, 99–105.

RUPP 1912: Fritz Rupp, *Inkrustationstil der romanischen Baukunst zu Florenz* (Strasbourg: Heitz & Mündel, 1912).

RUSSELL 1991: John M. Russell, *Sennacherib's Palace without Rival at Nineveh* (Chicago: University of Chicago Press, 1991).

RUSSELL AND DAHLIN 2007: Bradley W. Russell and Bruce H. Dahlin, "Traditional Burnt-Lime Production at Mayapán, Mexico," *Journal of Field Archaeology* 32, no. 4 (2007): 407–23.

RUSSO 2001: Eugenio Russo, ed., *L'edificio battesimale in Italia, aspetti e problemi. Atti dell'VIII Congresso nazionale di archeologia cristiana*, 2 vols. (Bordighera: Istituto Internazionale di Studi Liguri, 2001).

RUYSSCHAERT 1986: José Ruysschaert, "Les trois étapes de l'aménagement de la Bibliothèque Vaticane de 1471 à 1481," in *Un pontificato ed una città: Sisto IV (1471–1484)*, ed. Massimo Miglio (Vatican City: Scuola Vaticana di Paleografia Diplomatica e Archivistica, 1986), 103–14.

RYSTEDT ET AL. 1993: Eva Rystedt, Charlotte Wikander, and Örjan Wikander, *Deliciae fictiles* (Stockholm: Svenska institutet, 1993).

SACCHI 2000: Giulia Sacchi, *Mantegna Archipictor. La dimora dell'artista* (Mantua: Casa Del Mantegna, 2000).

SAFAR ET AL. 1981: Mustafa Fuad Safar, Mohammed Ali Mustafa Sayyid, and Seton Lloyd, *Eridu* (Baghdad: Iraqi Ministry of Culture and Information, 1981).

SALIBY 1954: Nassib Saliby, "Rapport préliminaire sur la deuxième campagne de fouilles a Raqqua (Automne 1952)," *Les annales archéologiques de Syrie* 4 (1954): 205–12.

SALVADORI 1998: Monica Salvadori, "Ante omnes est purpurissimum. La porpora nella pittura parietale romana," in *La porpora. Realtà e immaginario di un colore simbolico*, ed. Oddone Longo (Venice: Istituto Veneto di Scienze Lettere ed Arti, 1998), 203–25.

SALZENBERG 1854: Wilhelm Salzenberg, *Alt-christliche Baudenkmale von Constantinopel vom V. bis XII. Jahrhundert* (Leipzig: Karl W. Hirsemann, 1854).

SANTAMARIA ET AL. 2004: Ugo Santamaria, Fabio Morresi, and Maurizio Delle Rose, "Indagine scientifiche dei pigmenti e leganti delle lastre marmoree dipinte dell'Aula del Colosso del Foro di Augusto," in Bankel and Liverani 2004, 281–89.

SANTERRE 1959: Hubert Gallet de Santerre, *La terrasse des lions, le Létoon et le monument de granit à Délos: Publication topographique et architecturale* (Paris: E. de Boccard, 1959).

Santerre 1984: Hubert Gallet de Santerre, “L’oikos des Naxiens à Délos était-il un temple?,” *Bulletin de correspondance hellénique* 108 (1984): 671–93.

Santillo Frizell 1998: Barbro Santillo Frizell, “Giants or Geniuses? Monumental Buildings at Mycenae,” *Current Swedish Archaeology* 6 (1998): 167–84.

Santoro Bianchi 2001–2: Sara Santoro Bianchi, “I colori della città romana fra Campania, Roma e Italia settentrionale,” in *Amoenitas vrbivm: Les agréments de la vie urbaine en Gaule romaine et dans les régions voisines*, ed. Robert Bedon and Nicole Dupré (Limoges: Presses universitaires de Limoges, 2001–2), 171–85.

Sapirstein 2016: Philip Sapirstein, “The Columns of the Heraion at Olympia: Dörpfeld and Early Doric Architecture,” *American Journal of Archaeology* 120, no. 4 (2016): 565–601.

Saradi 1995: Helen Saradi, “The Kallos of the Byzantine City: The Development of a Rhetorical Topos and Historical Reality,” *Gesta* 34, no. 1 (1995): 37–56.

Sauter 1934: Franz Sauter, *Der Römische Kaiserkult bei Martial und Statius* (Stuttgart: W. Kohlhammer, 1934).

Savage 1989: Robert J. G. Savage, “Natural History of the Goldney Grotto, Clifton, Bristol,” *Garden History: The Journal of the Garden History Society* 17, no. 1 (1989): 1–40.

Savio 1957: Pietro Savio, *Ricerche storiche sulla santa Sindone* (Turin: Società Editrice Internazionale, 1957).

Scarisbrick 1982: Diana Scarisbrick, “Forever Adamant: A Renaissance Diamond Ring,” *Journal of the Walters Art Gallery* 40 (1982): 57–64.

Scheja 1962: Georg Scheja, “Hagia Sophia und Templum Salamonis,” *Istanbuler Mitteilungen* 12 (1962): 44–58.

Schelkle 1954: Karl H. Schelkle, “Cento,” in *Reallexikon für Antike und Christentum: Sachwörterbuch zur Auseinandersetzung des Christentums mit der antiken Welt*, ed. Theodor Klauser (Stuttgart: Hiersemann, 1954), 2:972–73.

Schibille 2009: Nadine Schibille, “The Profession of the Architect in Late Antique Byzantium,” *Byzantion* 79 (2009): 360–79.

Schibille 2014: Nadine Schibille, *Hagia Sophia and the Byzantine Aesthetic Experience* (Farnham: Ashgate, 2014).

Schiefsky 2007: Mark J. Schiefsky, “Art and Nature in Ancient Mechanics,” in *The Artificial and the Natural: An Evolving Polarity*, ed. Bernadette Bensaude-Vincent and William R. Newman (Cambridge, MA: MIT Press, 2007), 67–108.

Schiering 1960: Wolfgang Schiering, “Steine und Malerei in Der Minoischen Kunst,” *Jahrbuch des Deutschen Archäologischen Instituts* 75 (1960): 17–36.

Schiering 1965: Wolfgang Schiering, “Die Naturanschauung in der alt-kretischen Kunst,” *Antike Kunst* 8 (1965): 3–12.

Schiering 1987: Wolfgang Schiering, “Stein und Gelandemotive in der Minoische Wandmalerei auf Kreta und Thera,” in *Ägäische Bronzezeit*, ed. Hans-Günter Buchholz (Darmstadt: Wissenschaftliche Buchgesellschaft, 1987), 314–28.

Schiessl 1979: Ulrich Schiessl, *Rokokofassung und Materialillusion: Untersuchungen zur Polychromie sakraler Bildwerke im suddeutschen Rokoko* (Mittenwald: Maander, 1979).

Schilardi 2000: Demetrius U. Schilardi, ed., *Paria Lithos: Parian Quarries, Marble and Workshops of Sculpture* (Athens: Demetrius U. Schilardi, 2000).

Schlauch 1932: Margaret Schlauch, “The Palace of Hugon de Constantinople,” *Speculum* 7, no. 4 (1932): 500–514.

Schliemann 1886: Heinrich Schliemann, *Tiryns: The Prehistoric Palace of the Kings of Tiryns: The Results of the Latest Excavations* (London: John Murray, 1886).

Schmidt Arcangeli 1996: Catarina Schmidt Arcangeli, “‘Un Tempio Aperto:’ I Bellini e il loro ruolo nel progetto della facciata della Scuola Grande di San Marco,” *Studi di Storia dell’Arte* 7 (1996): 65–98.

Schneider 1941a: Alfons M. Schneider, *Die Grabung im Westhof der Sophienkirche zu Istanbul* (Berlin: Deutsches Archäologisches Institut, 1941).

Schneider 1941b: Alfons M. Schneider, “Die Kathedrale von Edessa,” *Oriens Christianus* 36 (1941): 161–67.

Schneider 1986: Rolf M. Schneider, *Bunte Barbaren: Orientalenstatuen Aus Farbigem Marmor in Der Römischen Repräsentationskunst* (Worms: Wernersche Verlagsgesellschaft, 1986).

Schneider 1998: Rolf M. Schneider, “Die Faszination des Feindes: Bilder der Parther und des Orients in Rom,” in *Das Partherreich und seine Zeugnisse*, ed. Josef Wiesehöfer (Stuttgart: Franz Steiner, 1998), 95–146.

Schneider 2001: Rolf M. Schneider, “Coloured Marble. The Splendour and Power of Imperial Rome,” *Apollo* 154, no. 473 (2001): 3–10.

Schneider 2002: Rolf M. Schneider, “Nuove immagini del potere romano. Sculture in marmo colorato nell’impero romano,” in *I marmi colorati della Roma imperiale*, ed. Matilda De Nuccio and Lucrezia Ungaro (Venice: Marsilio, 2002), 83–105.

Schofield 1992: Richard V. Schofield, “Avoiding Rome: An Introduction to Lombard Sculptors and the Antique,” *Arte Lombarda* 100, no. 1 (1992): 29–44.

Schofield 1999: Richard V. Schofield, “The Certosa Medallions: An Addendum,” *Arte Lombarda* 127 (1999): 74–85.

Schofield 2002: Richard V. Schofield, “The Colleoni Chapel and the Creation of a Local all’antica Architectural Style,” in Frommel et al. 2002, 167–92.

Schofield 2006: Richard V. Schofield, “La facciata della Scuola Grande di San Marco: Osservazioni preliminari,” in *I Lombardo: Wrchitettura e scultura a Venezia tra ’400 e ’500*, ed. Andrea Guerra, Manuela Morresi, and Richard V. Schofield (Venice: Marsilio, 2006), 161–93.

Scholem 1941: Gershom G. Scholem, *Major Trends in Jewish Mysticism* (Jerusalem: Schocken, 1941).

Scholten 1929: François Scholten, *La Palestine illustrée*, 2 vols. (Paris: Budry, 1929).

Schuler 1999: Stefan Schuler, *Vitruv im Mittelalter. Die Rezeption von “De architectura” von der Antike bis in die frühe Neuzeit* (Cologne: Hermann Böhlaus, 1999).

Schuller 2000: Manfred Schuller, “Le facciate dei palazzi medioevali di Venezia: Ricerche su singoli esempi architettonici,” in *L’architettura gotica veneziana*, ed. Wolfgang Wolters and Francesco Valcanover (Venice: Istituto Veneto di Scienze, Lettere e Arti, 2000), 290–300.

Schultz and Barnsley 1901: Robert Weir Schultz and Sidney H. Barnsley, *The Monastery of Saint Luke of Stiris, in Phocis, and the Dependent Monastery of Saint Nicolas in the Fields, near Skripou, in Boeotia* (London: Macmillan, 1901).

Schulz 1986: Hans-Joachim Schulz, *The Byzantine Liturgy: Symbolic Structure and Faith Expression*, trans. M. J. O’Connell (Collegeville, MN: Liturgical, 1986).

Schulz 2005: Jürgen Schulz, *The New Palaces of Medieval Venice* (University Park: Pennsylvania State Press, 2005).

Schulze 1985: Franz Schulze, *Mies van der Rohe: A Critical Biography* (Chicago: University of Chicago Press, 1985).

Schur 2004: David Schur, “A Garland of Stones: Hellenistic Lithika as Reflections on Poetic Transformation,” in *Labored in Papyrus Leaves: Perspectives on an Epigram Collection Attributed to Posidippus*, ed. Benjamin Acosta-Hughes, Elisabeth Kosmetatou, and Manuel Bambauch (Washington, DC: Center for Hellenic Studies, 2004), 118–22.

Schuster-Brandis 2008: Anais Schuster-Brandis, *Steine als Schutz- und Heilmittel. Untersuchung zu ihrer Verwendung in der Beschwörungskunst Mesopotamiens im 1. Jt. v. Chr.* (Münster: Ugarit, 2008).

Schwarzenberg 2000: Erkinger Schwarzenberg, “Colour, Light and Transparency in the Greek World,” in *Medieval Mosaics: Light, Color, Materials*, ed. Eve Borsook, Fiorella Gioffredi Superbi, and Giovanni Pagliarulo (Florence: Silvana, 2000), 15–34.

Schwarzenberg 2004: Erkinger Schwarzenberg, “Cristallo,” in *Vitrum: Vetro fra arte e scienza nel mondo romano*, ed. Marco Beretta and Giovanni Di Pasquale (Florence: Giunti, 2004), 61–70.

SCHWEIKHART 1973: Gunter Schweikhart, *Fassadenmalerei in Verona vom 14. bis zum 20. Jahrhundert* (Munich: F. Bruckmann, 1973).

SCHWEIKHART 1977: Gunter Schweikhart, *Le case affrescate di Verona* (Verona: Banca Popolare di Verona, 1977).

SCHWEMER AND ABUSCH 2008: Daniel Schwemer and Tzvi Abusch, "Das Abwehrzauber-Ritual Maqlû ('Verbrennung')," in *Texte aus der Umwelt des Alten Testaments, Neue Folge*, ed. Tzvi Abusch (Gütersloh: Gütersloher, 2008).

SCOPPOLA 1987: Francesco Scoppola, ed., *Palazzo Altemps. Indagini per il restauro della fabbrica Riario, Soderini, Altemps* (Rome: De Luca, 1987).

SCOTT-MEISAMI 2001: Julie Scott-Meisami, "The Palace-Complex as Emblem: Some Samarran Qasidas," in *A Medieval Islamic City Reconsidered: An Interdisciplinary Approach to Samarra*, ed. Chase F. Robinson (Oxford: Oxford University Press, 2001), 69–78.

SCULLY 1962: Vincent J. Scully, *The Earth, the Temple and the Gods: Greek Sacred Architecture* (New Haven: Yale University Press, 1962).

SEAR 1975: Frank Sear, "The Earliest Wall Mosaics in Italy," *Papers of the British School at Rome* 43 (1975): 83–97.

SEDINOVÁ 1999: Hana Sedinová, "The Symbolism of the Precious Stones in St. Wenceslas Chapel," *De Artibus et Historiae* 39 (1999): 75–94.

SEDINOVÁ 2000: Hana Sedinová, "The Precious Stones of the Heavenly Jerusalem in the Medieval Book Illustration and Their Comparison with the Wall Incrustation in St. Wenceslas Chapel," *De Artibus et Historiae* 41 (2000): 31–47.

SEIFERTOVÁ 2007: Hana Seifertová, *Malby na kameni: Umělecký experiment v 16. a na začátku 17. století (Painting on Stone: An Artistic Experiment in the 16th and Early 17th Centuries)* (Prague: National Gallery in Prague, 2007).

SERJEANT AND LEWCOCK 1983: R. B. Serjeant and Ronald Lewcock, *Sana'a': An Arabian Islamic City* (London: World of Islam Festival Trust, 1983).

SETTIS 2002: Salvatore Settis, *Le pareti ingannevoli. La Villa di Livia e la pittura di giardino* (Milan: Electa, 2002).

SETTIS ET AL. 1999: Salvatore Settis, Sonia Maffei, and Ludovico Rebaudo, *Laocoonte, fama e stile* (Rome: Donzelli, 1999).

SEVILLANO-LÓPEZ 2015: David Sevillano-López, "Mitos y realidad en las descripciones del Imperio Romano en las fuentes Chinas," in *Roma y el Mundo Mediterráneo*, ed. Noelia Vicent Ramírez and Jaime de Miguel López (Alcalá de Henares: Universidad de Alcalá, 2015), 15–42.

SHALEM 1994: Avinoam Shalem, "Fountains of Light: The Meaning of Medieval Islamic Rock Crystal Lamps," *Muqarnas* 11 (1994): 1–11.

SHALEM 1997: Avinoam Shalem, "Jewels and Journeys: The Case of the Medieval Gemstone Called al-Yatimah," *Muqarnas* 14 (1997): 42–56.

SHAW 1983: Joseph W. Shaw, "The Development of Minoan Orthostates," *American Journal of Archaeology* 87 (1983): 213–16.

SHAW 2006: Maria C. Shaw, "Plasters from the Monumental Minoan Buildings: Evidence for Painted Decoration, Architectural Appearance, and Archaeological Event," in *Kommos V: The Monumental Minoan Buildings at Kommos (Kommos: An Excavation of the South Coast of Crete)*, ed. Joseph W. Shaw and Maria C. Shaw (Princeton: Princeton University Press, 2006).

SHAW 2015: Joseph W. Shaw, *Elite Minoan Architecture: Its Development at Knossos, Phaistos, and Malia* (Philadelphia: INSTAP, 2015).

SHAW AND SHAW 2006: Joseph W. Shaw and Maria C. Shaw, eds., *Kommos V: The Monumental Minoan Buildings at Kommos (Kommos: An Excavation of the South Coast of Crete)* (Princeton: Princeton University Press, 2006).

SHEARMAN 1961: John Shearman, "The Chigi Chapel in S. Maria del Popolo," *Journal of the Warburg and Courtauld Institutes* 24 (1961): 129–60.

SHEARMAN 1972: John Shearman, *The Vatican Stanze: Functions and Decorations* (London: Oxford University Press, 1972).

SHEARMAN 1977: John Shearman, "Raphael, Rome and the Codex Escurialensis," *Master Drawings* 15, no. 2 (1977): 107–46.

SHEARMAN 1981: John Shearman, "Pentimenti in the Chigi Chapel," in *Art the Ape of Nature*, ed. Luc Freeman Sandler and Moshe Barasch (New York: Prentice-Hall, 1981), 219–22.

SHEARMAN 1987: John Shearman, "Raphael's Clouds and Correggio's," in *Studi su Raffaello*, ed. Micaela Sambucco Hamoud and Maria L. Strocchi (Urbino: QuattroVenti, 1987), 657–68.

SHEARMAN 1992: John Shearman, *Only Connect . . . Art and the Spectator in the Italian Renaissance* (Princeton: Princeton University Press, 1992).

SHEEDY AND PIKE 2008: Kenneth A. Sheedy and Scott Pike, "The Delos Colossus Project," in *La sculpture des Cyclades à l'époque archaïque, actes du colloque international organisé par l'éphorie des antiquités préhistoriques et classiques des Cyclades et l'École française d'Athènes (7–9 Septembre 1998)*, ed. Yannos Kourayos and Francis Prost (Athens: École française d'Athènes, 2008), 493–509.

SHELL 1993: Janice Shell, "Amadeo, the Mantegazza, and the Façade of the Certosa di Pavia," in *Giovanni Antonio Amadeo: Scultura e architettura del suo tempo*, ed. Janice Shell and Liana Castelfranchi Vegas (Milan: Cisalpino, 1993), 188–222.

SHERRAT 2000: Susan Sherrat, ed., *The Wall Paintings of Thera: Proceedings of the First International Symposium: Petros M. Nomikos Conference Centre, Thera, Hellas, 30 August–4 September 1997* (Piraeus: Petros M. Nomikos and Thera Foundation, 2000).

SHOE 1949: Lucy T. Shoe, "Dark Stone in Greek Architecture," *Hesperia*, supp. 8, *Commemorative Studies in Honor of Theodore Leslie Shear* (1949): 341–52.

SHORR 1946: Dorothy C. Shorr, "The Iconographical Development of the Presentation in the Temple," *Art Bulletin* 28, no. 1 (1946): 17–32.

SHORTLAND 2000: Andrew Shortland, "Depictions of Glass Vessels in Two Theban Tombs and Their Role in the Dating of Early Glass," *Journal of Egyptian Archaeology* 86 (2000): 159–61.

SICCA 2002: Cinzia M. Sicca, "Consumption and Trade of Art between Italy and England in the First Half of the Sixteenth Century: The London House of the Bardi and Cavalcanti Company," *Renaissance Studies* 16, no. 2 (2002): 163–201.

SIEBERT 1990: Gérard Siebert, "Imaginaire de la grotte dans la Grèce archaïque et classique," *KTÈMA Civilisations de l'Orient, de la Grèce et de Rome antiques* 15 (1990): 151–61.

SIEGFRIED 2002: Robert Siegfried, "From Elements to Atoms: A History of Chemical Composition," *Transactions of the American Philosophical Society* 92, no. 4 (2002): 1–278.

SIGANIDOU 1982: Maria Siganidou, "Η ἰδιοτίκι κατικία στην αρχαία Πελλας," *Archaiologia* 2 (February 1982): 31–36.

SIGNORINI 1991: Rodolfo Signorini, "'Una porta gemmea.' Il portale della Grotta di Isabella d'Este in Corte Vecchia," in *Per Mantova una vita. Studi in memoria di Rita Castagna*, ed. Maria Lorenzoni Anna and Roberto Navarrini (Mantua: Publi-Paolini, 1991), 25–52.

SIGNORINI 2006: Rodolfo Signorini, ed., *A casa di Andrea Mantegna. Cultura artistica a Mantova nel Quattrocento* (Cinisello Balsamo: Silvana, 2006).

SIGNORINI AND RADCLIFFE 1997: Rodolfo Signorini and Anthony Radcliffe, "'Una figura nuda legata a un tronco,' A Gilt Bronze Statuette Here Attributed to Andrea Mantegna," *Atti e memorie dell'Accademia Nazionale Virgiliana di scienze e lettere ed arti* 65 (1997): 49–104.

SILBERMAN 1974: Lou H. Silberman, "The Queen of Sheba in Judaic Tradition," in *Solomon and Sheba*, ed. James B. Pritchard (New York: Phaidon, 1974), 65–84.

SIMKÓ 2015: Krisztián Simkó, "On the Meaning and Cultural Setting of the Phrase *lā kišitti* šamme," *Kaskal* 12 (2015): 337–48.

SINDING-LARSEN 1975: Staale Sinding-Larsen, "A Tale of Two Cities: Florentine and Roman Visual Context for Fifteenth-Century Palaces," *Acta ad archaeologiam et artium historiam pertinentia* 6 (1975): 163–212.

SKON-JEDELE 1994: Nancy J. Skon-Jedele, "Aigyptaka: A Catalogue of Egyptian and Egyptianizing Objects Excavated from Greek Archaeological Sites, ca. 1100–525 B.C., with Historical Commentary" (PhD diss., University of Pennsylvania, 1994).

SLINGS 1978: S. R. Slings, "Anacreon's Two Meadows," *Zeitschrift für Papyrologie und Epigraphik* 30 (1978): 38.

SMITH 1987: Christine Smith, "Cyriacus of Ancona's Seven Drawings of Hagia Sophia," *Art Bulletin* 69, no. 1 (1987): 16–32.

SMITH 1988: Warren Smith, "The Disguises of the Gods in the 'Iliad,'" *Numen* 35, no. 2 (1988): 161–78.

SMITH 1995: Christine Smith, "Piero's Painted Architecture: Analysis of His Vocabulary," in *Piero della Francesca and His Legacy*, ed. Marilyn Aronberg Lavin (Washington, DC: National Gallery of Art, 1995), 223–54.

SODINI 1989: Jean-Pierre Sodini, "Le commerce des marbres à l'époque protobyzantine," in *Hommes et richesses dans l'Empire byzantin*, ed. Catherine Abadie-Reynal (Paris: Lethielleux, 1989), 163–86.

SODINI 1994: Jean-Pierre Sodini, "Le goût du marbre à Byzance: Sa signification pour les Byzantins et les non-Byzantins," in *Modes de vie et modes de pensée à Byzance* (Paris: De Boccard, 1994), 177–201.

SOERGEL PANOFSKY 1996: Gerda Soergel Panofsky, "An Artist's Library in Rome around 1600," in *Ars naturam adiuvans. Festschrift für Matthias Winner*, ed. Sebastian Schütze and Victoria von Flemming (Mainz am Rhein: Philipp von Zabern, 1996), 367–80.

SOGGIA AND ZUCCOLI 1994: Roberto Soggia and Noris Zuccoli, "Finiture di facciata nei costrutti albertiani. San Sebastiano e Sant'Andrea a Mantova," in *Leon Battista Alberti*, ed. Joseph Rykwert and Anne Engel (Ivrea: Olivetti/Electa, 1994), 392–401.

SOHM 1982: Philip L. Sohm, *The Scuola Grande di San Marco, 1437–1550: The Architecture of a Venetian Lay Confraternity* (New York: Garland, 1982).

SOHM 2000: Philip Sohm, "Baroque Piles and Other Decompositions," in *Pictorial Composition from Medieval to Modern Art*, ed. François Quiviger and Paul Taylor (London: Nuovo Aragno, 2000), 58–90.

SOKOLOWSKI 1970: Robert Sokolowski, "Matter, Elements and Substance in Aristotle," *Journal of the History of Philosophy* 8 (1970): 263–88.

SOLERO 1956: Silvio Solero, *Il Duomo di Torino e la R. Cappella della Sindone* (Pinerolo: Alzani, 1956).

SOLLBERGER 1965: Edmond Sollberger, "A New Inscription of Šilhak-Inšušinak," *Journal of Cuneiform Studies* 19, no. 1 (1965): 31–32.

SOLLBERGER 1987: Edmond Sollberger, "A Bead for Sennacherib," in *Language, Literature, and History: Philological and Historical Studies Presented to Erica Reiner*, ed. Francesca Rochberg-Halton (New Haven: American Oriental Society, 1987), 379–81.

SOTERIOU AND SOTERIOU 1952: Maria Georgiou Soteriou and Georgious Angelou Soteriou, *He basilike tou Hagiou Demetriou Thessalonikes* (Athens: He en Athenais Archaiologike Hetaireia, 1952).

SOURVINOU-INWOOD 1993: Christiane Sourvinou-Inwood, "Early Sanctuaries, the Eighth Century and Ritual Space," in *Greek Sanctuaries: New Approaches*, ed. Nanno Marinatos and Robin Hägg (London: Routledge, 1993), 1–17.

SPADEA 1996: Roberto Spadea, ed., *Il tesoro di Hera: Scoperte nel santuario di Hera Lacinia a Capo Colonna di Crotone* (Milan: ET, 1996).

SPARGO 1934: John W. Spargo, *Virgil the Necromancer: Studies in Virgilian Legends* (Cambridge, MA: Harvard University Press, 1934).

SPENCER 1955: John R. Spencer, "Spatial Imagery of the Annunciation in Fifteenth Century Florence," *Art Bulletin* 37, no. 4 (1955): 273–80.

SPERA 1992: Lucrezia Spera, "Un cubiculo monumentale nella catacomba di Pretestato," *Rivista di Archeologia Cristiana* 68 (1992): 271–307.

SPERA 1995: Lucrezia Spera, "Decorazioni in marmo delle catacombe romane: Osservazioni preliminari," in *Atti del II colloquio dell'ASMOM*, ed. Irene Bragantini and Federico Guidobaldi (Bordighera: Istituto Internazionale di Studi Liguri, 1995), 433–46.

SPERTI 1996: Luigi Sperti, "Sul reimpiego di scultura antica a Venezia: L'altare di Palazzo Mastelli," *Rivista di Archeologia* 20 (1996): 119–39.

SPIAZZI 2005: Anna Maria Spiazzi, "Il Castello Carrarese. Per la storia delle decorazioni d'interni a Padova nella seconda metà del Trecento," in *Dipinti e sculture del Trecento e Quattrocento restaurati in Veneto*, ed. Anna Maria Spiazzi and Fabrizio Magani (Treviso: Canova, 2005), 9–20.

SPIESER 1984: Jean-Michel Spieser, *Thessalonique et ses monuments du IVe au VIe siècle. Contribution a l'étude d'une ville paléochrétienne* (Athens: École française d'Athènes, 1984).

SPINELLI 1992: Riccardo Spinelli, "Giovanni Antonio Dosio e il progetto della cappella Niccolini in Santa Croce a Firenze: Quarantatre lettere inedite," *Rivista d'arte* 44 (1992): 199–296.

SPORN 2010: Katja Sporn, "Espace naturel et paysages religieux: Les grottes dans le monde grec," *Revue de l'histoire des religions* 227, no. 4 (2010): 553–71.

SPORN 2013: Katja Sporn, "Mapping Greek Sacred Caves: Sources, Features, Cults," in *Stable Places and Changing Perceptions: Cave Archaeology in Greece*, ed. Fanis Mavridis and Jesper Tae Jensen (Oxford: Archaeopress, 2013), 202–16.

STAMBAUGH 1978: John E. Stambaugh, "The Functions of Roman Temples," in *Aufstieg und Niedergang der Römischen Welt*, ed. Hildegard Temporini and Wolfgang Haase (Berlin: De Gruyter, 1978), 554–608.

STANIER 1953: Robert S. Stanier, "The Cost of the Parthenon," *Journal of Hellenic Studies* 73 (1953): 68–76.

STEINBY 1993: Eva M. Steinby, ed., *Lexicon topographicum urbis Romae*, 6 vols. (Rome: Quasar, 1993).

STEINGRÄBER 1988: Stephan Steingräber, "Die Tomba dei Festoni in Tarquinia und die Deckenmalereien der jüngeren etruskischen Kammergräber," *Jahrbuch des Deutschen Archäologischen Instituts* 103 (1988): 217–45.

STEINGRÄBER 2000: Stephan Steingräber, *Arpi–Apulien–Makedonien. Studien zum Unteritalischen Grabwesen in Hellenistischer Zeit* (Mainz am Rhein: Philipp von Zabern, 2000).

STELLA ARENA AND CARRUBA 2005: Maria Stella Arena and Anna Maria Carruba, *Opus sectile di Porta Marina* (Rome: Direzione Generale per i Beni Archeologici, 2005).

STERN 1958: Henri Stern, "Les mosaïques de l'église de Sainte-Constance à Rome," *Dumbarton Oaks Papers* 12 (1958): 159–218.

STERN 1985: Eva M. Stern, "Die Kapitelle der Nordhalle des Erechtheions," *Mitteilungen des Deutschen Archäologischen Instituts. Athenische Abteilung* 100 (1985): 405–26.

STERNBACH 1902: Leo Sternbach, "Beiträge zur Kunstgeschichte," *Jahreshefte des Österreischischen Archäologische Instituts in Wien* 5 (1902): cols. 65–94.

STEVENS 1955: Gorham P. Stevens, "Remarks upon the Colossal Chryselephantine Statue of Athena in the Parthenon," *Hesperia* 24, no. 3 (1955): 240–76.

STOKES 1934: Adrian Stokes, *The Stones of Rimini* (New York: Putnam, 1934).

STONEMAN 1992: Richard Stoneman, "Oriental Motifs in the Alexander Romance," *Antichthon* 26 (1992): 95–113.

STONEMAN 2003: Richard Stoneman, "The Metamorphosis of the Alexander Romance," in *The Novel in the Ancient World*, ed. Gareth Schmeling (Leiden: Brill, 2003), 601–12.

STOOKEY 1969: Laurence H. Stookey, "The Gothic Cathedral as the Heavenly Jerusalem: Liturgical and Theological Sources," *Gesta* 8, no. 1 (1969): 35–41.

STROCKA 1977: Volker M. Strocka, *Die Wandmalerei Der Hangehäuser in Ephesos* (Vienna: Österreichischen Akademie der Wissenschaften, 1977).

STRONG AND WARD-PERKINS 1960: Donald E., Strong and John B. Ward-Perkins, "The Round Temple in the Forum Boarium," *Papers of the British School at Rome* 28 (1960): 7–30.

STRUBE 1973: Christine Strube, *Die westliche Eingangsseite der Kirchen von Konstantinopel in justinianischer Zeit: Architektonische und quellenkritische Untersuchungen* (Wiesbaden: Harrassowitz, 1973).

STRUNCK 2003: Christina Strunck, "Bellori und Bernini rezipieren Raffael. Unbekannte Dokumente zur Cappella Chigi in Santa Maria del Popolo," *Marburger Jahrbuch für Kunstwissenschaft* 30 (2003): 131–82.

STRUPP 1993: Joachim Strupp, "The Colour of Money: Use, Cost and Aesthetic Appreciation of Marble in Venice ca. 1500," *Venezia '500* (1993): 7–32.

STURGEON 2006: Mary C. Sturgeon, "Archaic Athens and the Cyclades," in *Greek Sculpture: Function, Materials, and Techniques in the Archaic and Classical Periods,* ed. Olga Palagia (New York: Cambridge University Press, 2006), 32–76.

SUMMITT 2000: James B. Summitt, "Greek Architectural Polychromy from the Seventh to Second Centuries B.C.: History and Significance" (PhD diss., University of Michigan, 2000).

SWIFT 1940: Emerson H. Swift, *Hagia Sophia* (New York: Columbia University Press, 1940).

SZAFRANSKA 1989: Malgorzata Szafranska, "The Philosophy of Nature and the Grotto in the Renaissance Garden," *Journal of Garden History* 9, no. 2 (1989): 76–85.

TAFT 1975: Robert F. Taft, *The Great Entrance: A History of the Transfer of Gifts and Other Preanaphoral Rites of the Liturgy of St. John Chrysostom* (Rome: Pontifium Institutum Studiorum Orientalium, 1975).

TAFURI 1969: Manfredo Tafuri, *Jacopo Sansovino e l'architettura del '500 a Venezia* (Padua: Marsilio, 1969).

TAFURI 1985: Manfredo Tafuri, *Venezia e il rinascimento: Religione, scienza, architettura* (Turin: Giulio Einaudi, 1985).

TALBOT 1999: Alice-Mary Talbot, "Epigrams in Context: Metrical Inscriptions on Art and Architecture of the Palaiologan Era," *Dumbarton Oaks Papers* 53 (1999): 75–90.

TALBOT RICE 1930: David Talbot Rice, "Notes on Some Religious Buildings in the City and Vilayet of Trebizond," *Byzantion* 5 (1930): 47–81.

TAMBAKOPOULOS AND MANIATIS 2017: Dimitris Tambakopoulos and Yannis Maniatis, "The Search for the Prehistoric Marble Sources in the Cyclades," in *Early Cycladic Sculpture in Context,* ed. Marissa Marthari, Colin Renfrew, and Michael J. Boyd (Oxford: Oxbow, 2017), 468–82.

TASSINI 1993: Giuseppe Tassini, *Alcuni palazzi ed antichi edificii di Venezia* (Venice: Filippi, 1993).

TAUT 1936: Bruno Taut, *Fundamentals of Japanese Architecture* (Tokyo: Kokusai bunka shinkokai, 1936).

TAVANI 1999: Sergio Tavani, *Aquileia e Grado. Storia-Arte-Cultura,* 3rd ed. (Trieste: Lint, 1999).

TAYLOR 1996: Rabun Taylor, "A Literary and Structural Analysis of the First Dome on Justinian's Hagia Sophia, Constantinople," *Journal of the Society of Architectural Historians* 55, no. 1 (1996): 66–78.

TELEVANTOU 1990: Christina A. Televantou, "New Light on the West House Wall-Paintings," in *Thera and the Aegean World III: Proceedings of the Third International Congress, Santorini, Greece, 3–9 September 1989,* ed. D. A. Hardy and Colin Renfrew (London: Thera Foundation, 1990), 309–26.

TELEVANTOU 1994: Christina A. Televantou, *Ἀκρωτήρι Θήρι: οἱ τοιχωγραφίες τῆς Δυτικῆς Οἰκιας* (Athens: Hē en Athēnais Archaiologikē Hetaireia, 1994).

TERHALLE 1996: Johannes Terhalle, "Quantity as Quality: The Marble of S. Andrea al Quirinale," *Daidalos* 61 (1996): 44–49.

TERRY 1986: Ann Terry, "The Opus Sectile in the Eufrasius Cathedral," *Dumbarton Oaks Papers* 40 (1986): 147–64.

TESSARI 1995: Cristiano Tessari, *Baldassarre Peruzzi: Il progetto dell'antico* (Milan: Electa, 1995).

THEIS 1999: Lioba Theis, "Zur Geschichte der wissenschaftlichen Erforschung der Hagia Sophia," in *Die Hagia Sophia in Istanbul. Bilder aus sechs Jahrhunderten und Gaspare Fossatis Restaurierung der Jahre 1847 bis 1849,* ed. Volker Hoffmann (Bern: Peter Lang, 1999), 55–80.

THIEME AND BECKER 1907–50: Ulrich Thieme and Felix Becker, *Allgemeines lexikon der bildenden künstler von der antike bis zur gegenwart,* 37 vols. (Leipzig: Engelmann, 1907–50).

THIERSCH 1879: Friedrich Thiersch, "Die Tholos des Atreus zu Mykene," *Mitteilungen des Deutschen Archäologischen Instituts Athenische Abteilung* 4 (1879): 177–82 and pls. XI–XIII.

THOMAS 1982: John Thomas, "Sir Thomas Graham Jackson and the Church of Hagia Sophia," *Architectural History* 25 (1982): 98–101.

THOMAS 1991: Christopher A. Thomas, *The Lincoln Memorial and Its Architect, Henry Bacon (1866–1924),* 2 vols. (Ann Arbor: UMI, 1991).

THOMAS AND WITSCHEL 1992: Edmund Thomas and Christian Witschel, "Constructing Reconstruction: Claim and Reality of Roman Building Inscriptions from the Latin West," *Papers of the British School at Rome* 60 (1992): 135–78.

THOMPSON 1980: Homer A. Thompson, "Stone, Tile and Timber: Commerce in Building Materials in Classical Athens," *Expedition* 22, no. 3 (1980): 12–26.

TINÈ BERTOCCHI 1964: Fernanda Tinè Bertocchi, *La pittura funeraria apula* (Naples: G. Macchiaroli, 1964).

TIONE 1999: Raffaella Tione, "Le domus tardoantiche: Nuovi elementi per l'interpretazione dell'edilizia abitativa attraverso la lettura stratigrafica degli elevati," *Mededelingen van het Nederlands Institut te Rome* 58 (1999): 191–207.

TITE AND BIMSON 1989: Michael S. Tite and Mavis Bimson, "Glazed Steatite: An Investigation of the Methods of Glazing Used in Ancient Egypt," *World Archaeology* 21, no. 1 (1989): 87–100.

TKACZOW 1995: Barbara Tkaczow, "An Imitation of Opus Alexandrinum in Wall Painting? Two Wall Paintings from Kom El-Dikka," *Études et travaux/Studia i prace* 17 (1995): 319–26.

TKACZOW 2014: Barbara Tkaczow, "Fragments of Wall Paintings from Polish Excavations at Kom El-Dikka, Alexandria," *Études et travaux/ Studia i prace* 27 (2014): 422–39.

TOBER 2011: Barbara Tober, "Late Hellenistic 'Masonry Style' at Iasos," *Bollettino dell'Associazione Iasos di Caria* 17 (2011): 17–24.

TOMLINSON 2000: Richard Tomlinson, "Towards a Distribution Pattern for Parian Marble in the Architecture of the 6th C. B.C.," in Schilardi 2000, 139–42.

TÖNNESMANN 1984: Andreas Tönnesmann, "Palatium Nervae. Ein antikes Vorbild für Florentiner Rustikafasaden," *Römisches Jahrbuch für Kunstgeschichte* 21 (1984): 61–69.

TORBOLI 2010: Micaela Torboli, *Diamante! Curiosità araldiche nell'arte Estense del Quattrocento* (Ferrara: Cartografica, 2010).

TORRICELLI ET AL. 1993: Maria Pia Torricelli, Sandra Tugnoli Pattaro, and Cristiana Scappini, *Lo Studio Aldrovandi in Palazzo pubblico, 1617–1742* (Bologna: Clueb, 1993).

TRATZ 1991–92: Helga Tratz, "Die Ausstattung des Langhauses von St. Peter unter Innozenz X," *Römisches Jahrbuch der Bibliotheca Hertziana* 27 (1991–92): 339–74.

TRILLING 1987: James Trilling, "Late Antique and Sub-Antique, or the 'Decline of Form' Reconsidered," *Dumbarton Oaks Papers* 41 (1987): 469–76.

TRILLING 1994: James Trilling, "Modernism and the Rejection of Ornament: The Revolution That Never Happened," *Common Knowledge* 3, no. 2 (1994): 79–110.

TRILLING 1997: James Trilling, "Daedalus and the Nightingale: Art and Technology in the Myth of the Byzantine Court," in *Byzantine Court Culture from 829 to 1204,* ed. Henry Maguire (Washington, DC: Dumbarton Oaks, 1997), 217–30.

TRILLING 1998: James Trilling, "The Image Not Made by Hands and the Byzantine Way of Seeing," in *The Holy Face and the Paradox of Representation,* ed. Herbert L. Kessler and Gerhardt Wolf (Bologna: Nuova Alfa, 1998), 109–28.

TRINQUIER 2002: Jean Trinquier, "'Confusis oculis prosunt uirentia' (Sénèque, De ira, 3, 9, 2): Les vertus magiques et hygiéniques du vert dans l'antiquité," in *Couleurs et vision dans l'antiquité classique,* ed. Laurence Villard (Mont-Saint-Aignan: L'Université de Rouen, 2002), 97–128.

TRIONFI HONORATI 1983: Maddalena Trionfi Honorati, "Prospettive architettoniche a tarsia: Le porte del palazzo ducale di Urbino," *Notizie da Palazzo Albani* 1–2 (1983): 38–50.

TRONZO 1997: William Tronzo, *The Cultures of His Kingdom: Roger II and the Cappella Palatina in Palermo* (Princeton: Princeton University Press, 1997).

TROVABENE 1996: Giordana Trovabene, "Il mosaico pavimentale della chiesa di San Tomaso a Reggio Emilia," in *Atti del III colloquio dell'AISCOM*, ed. Federico Guidobaldi and Alessandra Guiglia Guidobaldi (Bordighera: Istituto Internazionale di Studi Liguri, 1996), 383–400.

TROVABENE 2001: Giordana Trovabene, "Poesia musiva medievale: Epigrafi didascaliche in versi nei pavimenti a mosaico," in *Atti del VII colloquio dell'AISCOM*, ed. Andrea Paribeni (Ravenna: Girasole, 2001), 353–66.

TSAKIRGIS 2016: Barbara Tsakirgis, "The Architecture of Greek Houses," in *A Companion to Greek Architecture*, ed. Margaret M. Miles (Chichester: John Wiley, 2016), 273–87.

TUENA 1989: Filippo Tuena, "Un bando del 1572 per la regolamentazione degli scavi in Roma," *Xenia* 17 (1989): 91–94.

TUENA 1990: Filippo Tuena, "I marmi commessi nel tardo rinascimento romano," in Borghini 1990, 80–97.

TUMMERS 1980: H. A. Tummers, *Early Secular Effigies in England: The Thirteenth Century* (Leiden: Brill, 1980).

TUOHY 1996: Thomas Tuohy, *Herculean Ferrara: Ercole d'Este, 1471–1505, and the Invention of a Ducal Capital* (Cambridge: Cambridge University Press, 1996).

TURCO 1997: Maria Grazia Turco, *Il titulus dei Santi Nereo ed Achilleo emblema della riforma cattolica* (Rome: Dedalo, 1997).

TUTTLE 2002: Richard J. Tuttle, ed., *Jacopo Barozzi da Vignola* (Milan: Electa, 2002).

UEBLACKER AND CAPRINO 1985: Mathias Ueblacker and Catia Caprino, *Das Teatro Marittimo in der Villa Hadriana* (Mainz am Rhein: Philipp von Zabern, 1985).

UFFLER 1971: A.-M. Uffler, "Fresquistes gallo-romains: Le bas-relief du musée de Sens," *Revue archéologique de l'Est. Du paléolithique au moyen âge* 22 (1971): 393–401.

UGUCCIONI 1989: Alessandra Uguccioni, "Decorazione e collezionismo antiquario nella sala grande di palazzo Della Valle," in Danesi Squarzina 1989, 356–70.

UNDERWOOD 1960: Paul A. Underwood, "Notes on the Work of the Byzantine Institute in Istanbul: 1957–1959," *Dumbarton Oaks Papers* 14 (1960): 205–22.

UNDERWOOD 1967–75: Paul A. Underwood, *The Kariye Djami*, 4 vols. (London: Pantheon, 1967–75).

UNGARO 2004: Lucrezia Ungaro, "Il rivestimento dipinto dell' 'Aula del Colosso' nel Foro di Augusto," in Bankel and Liverani 2004, 275–80.

VACCARO AND VETRI 2018: Vincenzo Vaccaro and Franco Vetri, *La Cappella dei Principi: Il restauro delle pietre dure* (Livorno: Sillabe, 2018).

VAI AND CAVAZZA 2006: Gian Battista Vai and William Cavazza, "Ulisse Aldrovandi and the Origin of Geology and Science," *Geological Society of America*, special paper 411 (2006): 43–63.

VALAZZI 1988: Maria Rosaria Valazzi, *La Pala ricostituita: L'Incoronazione della Vergine e la cimasa vaticana di Giovanni Bellini. Indagini e restauri* (Venice: Cataloghi Marsilio, 1988).

VALLA 1992: Francesca L. Valla, "Per la cronologia dei mosaici di San Savino a Piacenza," *Bollettino Storico Piacentino* 87, no. 1 (1992): 77–98.

VALLOIS 1944–78: René Vallois, *L'architecture Hellénique et Hellénistique a Délos jusqu'à l'éviction des Déliens (166 Av. J.-C.)*, 2 vols. (Paris: E. de Boccard, 1944–78).

VALVO 2008: Alfredo Valvo, "Le iscrizioni latine di Brescia di età repubblicana e triumvirale," in *Epigrafia 2006, atti della XIVe rencontre sur l'épigraphie in onore di Silvio Panciera con altri contributi di colleghi, allievi e collaboratori (Roma, 18–21 Ottobre 2006)*, ed. Maria Letizia Caldelli, Gian Luca Gregori, Silvia Orlandi, and Silvio Panciera (Rome: Quasar, 2008), 137–54.

VAN BUREN 1946: E. Douglas Van Buren, "Archaic Mosaic Wall Decoration," *Artibus Asiae* 9, no. 4 (1946): 323–45.

VAN BUREN 1956: E. Douglas Van Buren, "The Sceptre, Its Origin and Significance," *Revue d'Assyriologie et d'archéologie orientale* 50, no. 2 (1956): 101–3.

VAN DALEN 1991: Jan H. Van Dalen, "The Late Use of Opus Reticulatum in Ostia," *Mededelingen van het Nederlands Instituut te Rome* 50 (1991): 236–80.

VAN DE MIEROOP 2003: Marc van de Mieroop, "Reading Babylon," *American Journal of Archaeology* 107, no. 2 (2003): 257–75.

VAN ESS 2013: Margarete van Ess, "Neue Radiocarbondatierungen aus Uruk," in *Uruk: 5000 Jahre Megacity*, ed. Nicola Crüsemann Margarete van Ess, Friedhelm Pedde, and Judith Ramadan (Petersberg: Michael Imhof, 2013), 362–63.

VAN HAARLEM 1997: Willem M. Van Haarlem, "Imitations in Pottery of Stone Vessels in a Protodynastic Tomb from Tell Ibrahim Awad," *Archéo-Nil* 7 (1997): 145–50.

VAN OPSTALL 2013: Emilie van Opstall, "Cave and Cosmos: Sacred Caves in Greek Epic Poetry from Homer (Eighth Century BCE) to Nonnus (Fifth Century CE)," in *The Ideologies of Lived Space in Literary Texts, Ancient and Modern*, ed. Jacqueline Klooster and Jo Heirman (Gent: Academia, 2013), 15–33.

VAN STEKELENBURG 1992/93: Albert V. Van Stekelenburg, "Some Statistics Regarding the Spolia Columns in the Christian Basilicas of Rome," *Mededelingen van het Nederlands Institut te Rome* 51/52 (1992/93): 114–21.

VANDERPOOL 1953–54: Eugene Vanderpool, "The Museum and Garden of the Peripatetics," *Archaiologikē ephēmeris: Ekdidomenē tēs Archaiologikēs Hetaireias*, pt. 2, nos. 92/93 (1953–54): 126–28.

VANHOVE ET AL. 1995: Doris Vanhove, Alain de Wulf, Paul de Paepe, and Luc Moens, *Roman Marble Quarries in Southern Euboea and the Associated Road Systems* (Leiden: Brill, 1995).

VANNUTELLI 1896: Vincenzo Vannutelli, *La Santa Colonna che si venera in Roma a S. Prassede: Conferenza archeologica, tenuta domenica 16 febbraio 1896* (Rome: Tipografia Editrice Romana, 1896).

VANORIO AND KANITPANYACHAROEN 2015: Tiziana Vanorio and Waruntorn Kanitpanyacharoen, "Rock Physics of Fibrous Rocks Akin to Roman Concrete Explains Uplifts at Campi Flegrei Caldera," *Science* 349, no. 6248 (2015): 617–21.

VARESE 1978: Ranieri Varese, "La vera facciata di Schifanoia," *Critica d'Arte* 43, nos. 160–62 (1978): 44–66.

VARESE 1989: Ranieri Varese, *Atlante di Schifanoia* (Modena: Panini, 1989).

VASILIEV 1948: Alexander A. Vasiliev, "Imperial Porphyry Sarcophagi in Constantinople," *Dumbarton Oaks Papers* 4 (1948): 1–26.

VELLACOTT 1908: C. H. Vellacott, "Quarrying," *Victoria History of Dorset* 2 (1908): 331–44.

VENIT 1988: Marjorie S. Venit, "The Painted Tomb from Wardian and the Decoration of Alexandrian Tombs," *Journal of the American Research Center in Egypt* 25 (1988): 71–91.

VENIT 2002: Marjorie S. Venit, *The Monumental Tombs of Ancient Alexandria: The Theater of the Dead* (Cambridge: Cambridge University Press, 2002).

VENIT 2015: Marjorie S. Venit, *Visualizing the Afterlife in the Tombs of Graeco-Roman Egypt* (Cambridge: Cambridge University Press, 2015).

VENTURA 1991: Leandro Ventura, "Un bozzetto barocco riscoperto: Una nuova attribuzione per Melchiorre Caffà," *Ricerche di Storia dell'arte* 45 (1991): 77–84.

VENTURA FOLLI 1993: Irene Ventura Folli, "La Natura 'Scritta': La 'Libraria' di Ulisse Aldrovandi (1522–1605)," in *Bibliothecae selectae: Da Cusano a Leopardi*, ed. Eugenio Canone (Florence: Olschki, 1993), 495–506.

VERITÀ ET AL. 2008: Marco Verità, Maria Stella Arena, Anna Maria Carruba, and Paola Santopadre, "Roman Glass: Art and Technology in a 4th Century A.D. Opus Sectile in Ostia (Rome)," *Journal of Cultural Heritage* 9 (2008): 16–20.

VERNANT 1991: Jean Pierre Vernant, "Mortals and Immortals: The Body of the Divine," in *Mortals and Immortals: Collected Essays*, ed. Froma I. Zeitlin (Princeton: Princeton University Press, 1991), 27–49.

VERNIA 2005: Barbara Vernia, "L'arredo liturgico della basilica di Sant'Apollinare Nuovo a Ravenna," in *Venezia e Bisanzio. Aspetti della cultura artistica bizantina da Ravenna a Venezia (V–XIV secolo)*, ed. Clementina Rizzardi (Venice: Istituto Veneto di Scienze, Lettere ed Arti, 2005), 363–89.

VERSLUYS 2002: Miguel J. Versluys, *Aegyptiaca Romana: Nilotic Scenes and the Roman Views of Egypt* (Leiden: Brill, 2002).

VESSEY 1983: David W. T. Vessey, "Mediis discumbere in astris: Silvae IV.2," *Acta Classica* 52 (1983): 206–20.

VIAUD 1910: Prosper Viaud, *Nazareth et ses deux églises de l'Annonciation et de Saint-Joseph, d'après les fouilles récentes pratiquées* (Paris: A. Picard, 1910).

VILLER ET AL. 1937–95: Marcel Viller, F. Cavallera, and J. de Guibert, eds., *Dictionnaire de spiritualité ascétique et mystique, doctrine et histoire*, 17 vols. (Paris: Beauchesne, 1937–95).

VINCENT AND ABEL 1912–26: Louis H. Vincent and Félix-Marie Abel, *Jérusalem: Recherches de topographie, d'archéologie, et d'histoire*, 4 vols. (Paris: J. Gabalda, 1912–26).

VISCOGLIOSI 1996: Alessandro Viscogliosi, *Il tempio di Apollo in Circo e la formazione del linguaggio architettonico augusteo* (Rome: Bretschneider, 1996).

VISCOGLIOSI 2000: Alessandro Viscogliosi, *I Fori Imperiali nei disegni d'architettura del primo cinquecento. Ricerche sull'architettura e l'urbanistica di Roma* (Rome: Gangemi, 2000).

VISIOLI 1993: Monica Visioli, "Documenti per Palazzo Raimondi," *Artes* 1 (1993): 84–87.

VISIOLI 2001: Monica Visioli, *Palazzo Raimondi. Nuove ricerche in occasione dei restauri alla facciata* (Viareggio: Marco Baroni, 2001).

VOISIN 1987: Jean-Louis Voisin, "Exoriente sole (SUÉTONE, Ner. 6). D'Alexandrie à la Domus Aurea," in *L'Urbs. Espace urbain et histoire (Ier siècle av. J.-C.–IIIe siècle ap. J.-C.)* (Paris: CNRS/École française de Rome, 1987), 509–43.

VOLPE ET AL. 2004: Giuliano Volpe, Giuliano De Felice, and Maria Turchiano, "Musiva e sectilia in una lussuosa residenza rurale dell'Apulia tardoantica: La villa di Faragola (Ascoli Satriano, Foggia)," *Musiva e Sectilia* 1 (2004): 127–58.

VON BECKERATH 1999: Jürgen von Beckerath, *Handbuch der ägyptischen Königsnamen* (Mainz: Philipp von Zabern, 1999).

VON RÜDEN 2011: Constance von Rüden, *Die Wandmalerein aus Tall Mišrife/Qaṭna im Kontext überregionaler Kommunikation. Mit Beiträgen von Ann Brysbaert und Ilka Weisser* (Wiesbaden: Harrassowitz, 2011).

WACE 1949: Alan J. B. Wace, *Mycenae: An Archaeological History and Guide* (Princeton: Princeton University Press, 1949).

WACE ET AL. 1921–23: Alan J. B. Wace, W. A. Heurtley, Winifred Lamb, Leicester B. Holland, and C. A. Boethius, "The Report of the School Excavations at Mycenae, 1921–1923," *Annual of the British School at Athens* 25 (1921/22–1922/23): 1–434.

WAELKENS 1985: Marc Waelkens, "From a Phrygian Quarry: The Provenance of the Statues of the Dacian Prisoners in Trajan's Forum at Rome," *American Journal of Archaeology* 89 (1985): 641–53.

WALKER AND MATTHEWS 1997: Susan Walker and K. J. Matthews, "The Marbles of the Mausoleum," in *Sculptors and Sculpture of Caria and the Dodecanese*, ed. Ian Jenkins and Geoffrey B. Waywell (London: British Museum, 1997), 49–59.

WALLACE-HADRILL 1988: Andrew Wallace-Hadrill, "The Social Structure of the Roman House," *Papers of the British School at Rome* 56 (1988): 43–97.

WALLACE-HADRILL 1990: Andrew Wallace-Hadrill, "Pliny the Elder and Man's Unnatural History," *Greece and Rome* 37, no. 1 (1990): 80–96.

WALLACE-HADRILL 1994: Andrew Wallace-Hadrill, *Houses and Society in Pompeii and Herculaneum* (Princeton: Princeton University Press, 1994).

WALLACE MAZE 2013: Daniel Wallace Maze, "Giovanni Bellini: Birth, Parentage, and Independence," *Renaissance Quarterly* 66, no. 3 (2013): 783–823.

WALTER-KARYDI 1998: Elena Walter-Karydi, *The Greek House: The Rise of Noble Houses in Late Classical Times* (Athens: Archaeological Society at Athens, 1998).

WALTERS 2000: E. J. Walters, "The Black and the Red: Isis and the Egyptian Tradition for Contrasts in Stone," in *Asmosia VI: Proceedings of the Sixth International Conference of the "Association for the Study of Marble and Other Stones in Antiquity," Venice, June 15–18 2000*, ed. Lorenzo Lazzarini (Venice: Laboratorio di Analisi dei Materiali Antichi, 2000), 127–34.

WARBURG 1912: Aby Warburg, "Italienische Kunst und internationale Astrologie im Palazzo Schifanoja zu Ferrara," in *L'Italia e l'arte straniera*, ed. Adolfo Venturi (Rome: Maglione & Strini, 1912), 179–93.

WARD-PERKINS 1977: John B. Ward-Perkins, *Roman Architecture* (New York: Abrams, 1977).

WARLAND 2003: Rainer Warland, "The Concept of Rome in Late Antiquity Reflected in the Mosaics of the Triumphal Arch of Santa Maria Maggiore in Rome," *Acta ad Archaeologiam et Artium Historiam Pertinentia* 17 (2003): 127–41.

WARNER 2008: Marina Warner, "The Writing of Stones," *Cabinet* 29 (2008), http://cabinetmagazine.org/issues/29/warner.php.

WARREN 1969: Peter M. Warren, *Minoan Stone Vases* (London: Cambridge University Press, 1969).

WARREN 1989: Peter M. Warren, "Egyptian Stone Vessels from the City of Knossos: Contributions Towards Minoan Economic and Social Structure," *Αριάδνη* 5 (1989): 1–9.

WARREN 1996: Peter M. Warren, "A New Minoan Decoration–the Jackson Pollock Style–and Its Place in Minoan Art," in *Minotaur and Centaur: Studies in the Archaeology of Crete and Euboea Presented to Mervyn Popham*, ed. Doniert Evely, Irene S. Lemos, and Susan Sherratt (Oxford: Tempus Reparatum, 1996), 46–50.

WARREN 2012: Peter Warren, "The Rediscovery of Greek Rosso Antico Marble and Its Use in Britain in the Nineteenth and Twentieth Centuries," *Annual of the British School at Athens* 107 (2012): 341–86, 389–93.

WEBB 1999: Ruth Webb, "Ekphrasis Ancient and Modern: The Invention of the Genre," *Word and Image* 15, no. 1 (1999): 7–18.

WEDDIGEN 2001: Tristan Weddigen, "Raffaels Papageizimmer. Raumfunktion und Dekoration im Vatikanpalast der Renaissance" (PhD diss., Technische Universität, Berlin, 2001).

WEDEPOHL 2004: Claudia Wedepohl, "Die Cappella del Perdono und der Tempietto delle Muse im Herzogspalast von Urbino. Kontextualisierung eines humanistischen Raumprogramms" (PhD diss., Universität Hamburg, 2004).

WEIL-GARRIS BRANDT 1986: Kathleen Weil-Garris Brandt, "Cosmological Patterns in Raphael's Chigi Chapel in S. Maria del Popolo," in Frommel and Winner 1986, 125–57.

WEINBERGER 1940/41: Martin Weinberger, "The First Facade of the Cathedral of Florence," *Journal of the Warburg and Courtauld Institutes* 4, nos. 1/2 (1940/41): 67–79.

WEIS 1992: Anne Weis, *The Hanging Marsyas and Its Copies: Roman Innovations in a Hellenistic Sculptural Tradition* (Rome: Bretschneider, 1992).

WEISS 1985: Harvey Weiss, ed., *Ebla to Damascus: Art and Archaeology of Ancient Syria* (Washington, DC: Smithsonian, 1985).

WEISS 1995: Daniel H. Weiss, "Architectural Symbolism and the Decoration of the Ste.-Chapelle," *Art Bulletin* 77, no. 2 (1995): 308–20.

WELLMANN 1935: Max Wellmann, "Die Stein- und Gemmenbücher der Antike," *Quellen und Studien zur Geschichte der Naturwissenschaften und der Medizin* 4 (1935): 86–149.

WESCOAT 2018: Bonna D. Wescoat, *Samothrace: Excavations Conducted by the Institute of Fine Arts of New York University*, vol. 9, *The Monuments of the Eastern Hill* (Princeton: American School of Classical Studies at Athens, 2018).

WESENBERG 1975–76: Burkhardt Wesenberg, "Certae rationes picturarum," *Marburger Wincklemann-program* (1975–76): 23–43.

WESENBERG 2000: Burkhardt Wesenberg, "Ein 'delisches Problem': Der Koloss der Naxier," in *Munus. Festschrift für Hans Wiegartz*, ed. Torsten Mattern and Dieter Korol (Münster: Scriptorium, 2000), 313–24.

WESTGATE 2000A: Ruth Westgate, "Pavimenta atque emblemata vermiculata: Regional Styles in Hellenistic Mosaic and the First Mosaics at Pompei," *American Journal of Archaeology* 104 (2000): 255–75.

WESTGATE 2000B: Ruth C. Westgate, "Space and Decoration in Hellenistic Houses," *Annual of the British School at Athens* 95 (2000): 391–426.

WHITBY 1985: Mary Whitby, "The Occasion of Paul the Silentiary's Ekphrasis of S. Sophia," *Classical Quarterly* 35, no. 1 (1985): 215–28.

WHITBY 2006: Mary Whitby, "The St. Polyeuktos Epigram (Ap 1.10): A Literary Perspective," in *Greek Literature in Late Antiquity: Dynamism, Didacticism, Classicism,* ed. Scott Fitzgerald Johnson (Aldershot: Ashgate, 2006), 159–87.

WHITE 1969A: John White, "Donatello's High Altar in the Santo at Padua. Part One: The Documents and Their Implications," *Art Bulletin* 51, no. 1 (1969): 1–14.

WHITE 1969B: John White, "Donatello's High Altar in the Santo at Padua. Part Two: The Reconstruction," *Art Bulletin* 51, no. 2 (1969): 119–41.

WHITEHOUSE 2001: Helen Whitehouse, *Ancient Mosaics and Wallpaintings* (London: Harvey Miller, 2001): 138–65.

WICKENS 1986: Jere M. Wickens, "The Archaeology and History of Cave Use in Attica, Greece, from Prehistoric Through Late Roman Times" (PhD diss., Indiana University, 1986).

WILKINSON 1982: John Wilkinson, "Paulinus' Temple at Tyre," in *XVI. Internationaler Byzantinistenkongress,* ed. Herbert Hunger (Vienna: Österreichische Akademie der Wissenschaften, 1982), 553–61.

WILMERING AND RAGGIO 1999: Antoine M. Wilmering and Olga Raggio, *The Gubbio Studiolo and Its Conservation,* 2 vols. (New York: Metropolitan Museum of Art, 1999).

WILPERT 1916: Josef Wilpert, *Die römischen mosaiken und malereien der kirchlichen bauten vom IV. bis XIII. jahrundret,* 2 vols. (Freiburg im Breisgau: Herder, 1916).

WILSON 1937: Benjamin F. Wilson, *The Parthenon of Pericles and Its Reproduction in America* (Nashville: Parthenon, 1937).

WILSON 1976: Carolyn C. Wilson, "Giovanni Bellini's Pesaro Altarpiece: Studies in Its Context and Meaning" (PhD diss., Institute of Fine Arts, 1976).

WILSON 1978: Ian Wilson, *The Turin Shroud* (London: V. Gollancz, 1978).

WILSON 2013: Malcolm Wilson, *Structure and Method in Aristotle's Meteorologica: A More Disorderly Nature* (Cambridge: Cambridge University Press, 2013).

WILSON JONES 2014: Mark Wilson Jones, *Origins of Classical Architecture: Temples, Orders and Gifts to the Gods in Ancient Greece* (New Haven: Yale University Press, 2014).

WINTER 1994: Irene J. Winter, "Radiance as an Aesthetic Value in the Art of Mesopotamia," in *Art: The Integral Vision: A Volume of Essays in Felicitation of Kapila Vatsyayan,* ed. Baidyanath Saraswati, Subhash C. Malik, and Madhu Khanna (New Delhi: D. K. Printworld, 1994), 123–32.

WINTER 2000: Irene J. Winter, "Thera Paintings and the Ancient Near East: The Private and Public Domains of Wall Decoration," in Sherrat 2000, 745–62.

WIRTH 1931: Fritz Wirth, "Wanddekoration ersten Stils in Athen," *Mitteilungen des Deutschen Archäologischen Instituts, Athenische Abteilung* 56 (1931): 33–58.

WISCH 1991: Barbara Wisch, "The Passion of Christ in the Art, Theater, and Penitential Rituals of the Roman Confraternity of the Gonfalone," in *Crossing the Boundaries: Christian Piety and the Arts in Italian Medieval and Renaissance Confraternities,* ed. Konrad Eisenbichler (Kalamazoo: Medieval Institute Publications, Western Michigan University, 1991), 246–52.

WISEMAN 1987: Timothy P. Wiseman, "Conspicui postes tectaque digna deo: The Public Image of Aristocratic and Imperial Houses in the Late Republic and Early Empire," in *L'Urbs. Espace urbain et histoire (Ier siècle av. J.-C.–IIIe siècle ap. J.-C.)* (Paris/Rome: CNRS and École française de Rome, 1987), 393–413.

WITTKOWER 1940/41: Rudolf Wittkower, "Alberti's Approach to Antiquity in Architecture," *Journal of the Warburg and Courtauld Institutes* 4, nos. 1/2 (1940/41): 1–18.

WITTKOWER 1981: Rudolf Wittkower, *Bernini: The Sculptor of the Roman Baroque,* 4th ed. (London: Phaidon, 1981).

WOLTERS 1976: Wolfgang Wolters, *La scultura veneziana gotica (1300–1460)* (Venice: Alfieri, 1976).

WOLTERS 1986: Wolfgang Wolters, "Die ursprüngliche Gestalt der Fassade von S. Andrea in Mantua," *Mitteilungen des Kunsthistorischen Institutes in Florenz* 30, no. 3 (1986): 425–32.

WOLTERS 2000: Wolfgang Wolters, *Architektur und Ornament: Venezianischer Bauschmuck der Renaissance* (Munich: C.H. Beck, 2000).

WOOD 1986: Ian Wood, "The Audience of Architecture in Post-Roman Gaul," in *The Anglo-Saxon Church,* ed. L. A. S. Butler and R. K. Morris (London: Council for British Archaeology, 1986), 74–79.

WOOLLEY 1955: Leonard Woolley, *Alalakh: An Account of the Excavations at Tell Atchana in the Hatay, 1937–1949* (Oxford: Society of Antiquaries, 1955).

WOOLLEY AND MOOREY 1982: Leonard Woolley and Peter R. S. Moorey, *Ur "of the Chaldees": A Revised and Updated Edition of Sir Leonard Woolley's Excavations at Ur* (Ithaca, NY: Cornell University Press, 1982).

WORMALD 1961: Frances Wormald, "The Throne of Solomon and St. Edward's Chair," in *De artibus opuscula XL: Essays in Honor of Erwin Panofsky,* ed. Millard Meiss, 2 vols. (New York: New York University Press, 1961), 1:532–39.

WORRELL 1948: W. H. Worrell, "Note on Mishkāt," in *Ignace Goldziher Memorial Volume,* ed. Sandor Scheiber, D. S. Loewinger, and Joseph De Somogyi (Budapest: Globus, 1948), 46–47.

WRIGHT 1978: James C. Wright, "Mycenaean Masonry Practices and Elements of Construction" (PhD diss., Bryn Mawr College, 1978).

WRIGHT 1987: James C. Wright, "Death and Power at Mycenae: Changing Symbols in Mortuary Practice," in *Thanatos: Les coutumes funéraires en Egée a l'âge du bronze. Actes du colloque de Liège, 21–23 Avril 1986,* ed. Robert Laffineur (Liège: Université de Liège, 1987), 177–83.

WRIGHT 1994: James C. Wright, "The Spatial Configuration of Belief: The Archaeology of Mycenaean Religion," in *Placing the Gods: Sanctuaries and Sacred Space in Classical Greece,* ed. Susan E. Alcock and Robin Osborne (Oxford/New York: Clarendon and Oxford University Press, 1994), 37–78.

WURM 1984: Heinrich Wurm, *Baldassarre Peruzzi. Architekturzeichnungen* (Tübingen: Wasmuth, 1984).

WYCHERLEY 1974: Richard E. Wycherley, "Poros: Notes on Greek Building Stones," in *Phoros, Tribute to Benjamin Dean Meritt,* ed. Donald W. Bradeen and Malcolm F. McGregor (Locust Valley, NY: J. J. Augustin, 1974), 179–87.

YANNI 1999: Carla Yanni, *Nature's Museums: Victorian Science and the Architecture of Display* (Baltimore: Johns Hopkins University Press, 1999).

YANNOPOULOS 1991: Panayotis Yannopoulos, "Le couronnement de l'empereur à Byzance: Rituel et fond institutionnel," *Byzantion* 61 (1991): 71–92.

YENISEHIRLIOGLU 1982: Filiz Yenisehirlioglu, "L'emploi de la brique sur les façades des édifices byzantins et ottomans au XIVe et XVe siècles," in *XVI. Internationaler Byzantinistenkongress,* ed. Herbert Hunger (Vienna: Österreichische Akademie der Wissenschaften, 1982), 327–37.

YERASIMOS 1990: Stefanos Yerasimos, *La fondation de Constantinople et de Sainte-Sophie dans les traditions turques: Légendes d'Empire* (Istanbul: Institut français d'Études Anatoliennes d'Istanbul, 1990).

YOUNG 1975: Eric Young, *Bartolomé Bermejo, the Great Hispano-Flemish Master* (London: Elek, 1975).

YOUNGER 1987: John C. Younger, "The Elgin Plaques from the Treasury of Atreus: Evidence for a New Reconstruction of the Façade," in *Kolloquium zur ägäischen Vorgeschichte, Mannhein 20–22.2.1986,* ed. Wolfgang Schiering (Mannheim: Deutschen Archäologen-Verband, 1987), 138–50.

YUEN 1970: Toby Yuen, "The 'Bibliotheca Graeca.' Castagno, Alberti and Ancient Sources," *Burlington Magazine* 112, no. 812 (1970): 725–36.

ZAFFAGNINI 1965: Giulia Maria Zaffagnini, "La Basilica Ursiana di Ravenna," *Felix Ravenna* 91 (1965): 5–76.

ZANKER 1995: Paul Zanker, *The Mask of Socrates: The Image of the Intellectual in Antiquity*, trans. Alan Shapiro (Berkeley: University of California Press, 1995).

ZECCHIN 1987–90: Luigi Zecchin, *Vetro e vetrai di Murano: Studi sulla storia del vetro*, 3 vols. (Venice: Arsenale, 1987–90).

ZEVI 1960: Bruno Zevi, *Biagio Rossetti: Architetto ferrarese, il primo urbanista moderno europeo* (Turin: Einaudi, 1960).

ZEVI 1976: Fausto Zevi, "L'identificazione del tempio di Marte 'in circo' e altre osservazioni," in *L'Italie préromaine et la Rome républicaine: Mélanges offerts à Jacques Heurgon* (Rome: École française de Rome, 1976), 1047–60.

ZEVI 1991: Fausto Zevi, "Il terremotò del 62 e l'edilizia privata," in *Pompei*, ed. Fausto Zevi, 2 vols. (Naples: Banco di Napoli, 1991), 2:39–58.

ZEZZA 2005: Fulvio Zezza, *Castel del Monte. La pietra e i marmi* (Bari: Adda, 2005).

ZIMMER 2010: Torsten Zimmer, "Repräsentatives Wohnen am Beispiel der Palastanlagen von Pergamon," in *Städtisches Wohnen im Östlichen Mittelmeerraum 4. Jh. v. Chr.–1. Jh. n. Chr.*, ed. Sabine Ladstätter and Veronika Scheibelreiter (Vienna: Verlag der Österreichischen Akademie der Wissenschaften, 2010), 155–66.

ZINK 2008: Stephan Zink, "Reconstructing the Palatine Temple of Apollo: A Case Study in Early Augustan Temple Design," *Journal of Roman Archaeology* 21, no. 1 (2008): 47–63.

ZIOLKOWSKI 1986: Adam Ziolkowski, "Les temples A et C du Largo Argentina: Quelques considérations," *Mélanges de l'École française de Rome. Antiquité* 98, no. 2 (1986): 623–47.

ZIOLKOWSKI 1988: Adam Ziolkowski, "Mummius' Temple of Hercules Victor and the Round Temple on the Tiber," *Phoenix* 42, no. 4 (1988): 309–33.

ZOLLIKOFER 2017: Kaspar Zollikofer, *Die Cappella Gregoriana. Der erste Innenraum von Neu-Sankt-Peter in Rom und seine Genese* (Basel: Schwabe, 2017).

ZORZI 1926: Giangiorgio Zorzi, "Contributo alla Storia dell'Arte Vicentina nei secoli XV e XVI," *Miscellanea di Storia veneto-tridentina della Regia Deputazione veneta-tridentina di Storia Patria* 2 (1926): 1–330.

ZORZI 1958: Giangiorgio Zorzi, "L'abside della cattedrale di Vicenza e il contributo di Andrea Palladio al suo compimento," in *Studi in onore di Federico M. Mistrorigo*, ed. Aristide Dani (Vicenza: Comune di Vicenza, 1958), 271–310.

ZUCCOLI 1986: Noris Zuccoli, *Mantova. Intonaci e coloriture architettoniche* (Mantua: Alinea, 1986): 33.

Index

Page numbers in *italics* indicate illustrations.

Q

S

X

Z

Z

Illustration Credits

The photographers and the sources of visual material other than the owners indicated in the captions are as follows. Every effort has been made to supply complete and correct credits; if there are errors or omissions, please contact Yale University Press so that corrections can be made in any subsequent edition.

Académie d'Architecture, Paris. III6 HP.DES.10/9.I.: 7.3
Adriani et al. 2000: 5.9
Alamy: 9.5
Alexander 1994: 11.11
Alinari: 10.16, 12.11
Amelli 1896: 10.18
Artefacts, Berlin: 1.11
Artists Rights Society: 2.28
Ashby 1904: 10.15
Aurigemma and Cavallaro 1999: 11.22
Author: 1.7, 1.10, 1.16, 2.1, 2.2, 2.4, 2.11, 2.12, 2.14, 2.17–2.19, 2.21, 2.22, 2.24, 2.26, 3.3–3.7, 3.12, 3.13, 3.15–3.17a, 3.18b, 3.18c, 3.19, 4.2, 4.4a–c, 4.14–4.17, 4.22, 4.24, 4.26, 4.27, 4.29, 4.32, 5.2, 5.4, 5.6, 5.11–5.14, 5.18, 5.20, 5.21, 6.1–6.3, 6.7, 6.8, 6.12, 6.20, 6.22, 6.27, 6.28, 7.4–7.8, 7.10, 7.12–7.14, 7.18, 8.1–8.3, 8.6, 8.8, 8.10, 8.11, 9.2, 9.3, 9.8, 9.32–9.34, 10.8, 10.9, 10.12, 10.14, 11.2, 11.3, 11.5–11.8, 11.15–11.18, 11.20, 11.23–11.25, 11.27–11.29, 11.31, 12.3, 12.10, 12.12, 12.14, 12.17, 12.22–12.24, 12.27, 12.30, 12.32–12.34, 12.37–12.39, 13.2
Bazzotti et al. 2009: 11.10
Becatti 1969: 5.17
Benaki Museum, Athens: 8.12
Bernia et al. 1648, pl. 758: 7.15
BNF, Paris: 9.25
Bombi et al. 1988: 11.12
Bona Castellotti 2000: 12.13
Borsi 1985: 5.19
Bruce Boucher, *Andrea Palladio: The Architect in His Time* (New York: Abbeville, 1994): 10.17
Brera, Milan: 6.17, 9.16
British Museum, London: 2.30
Brooklyn Museum Archives (S10|08 Sakkara, image 9645). Wikimedia: 1.8
Bruneau 1970: 4.23
Buckton 1984: 7.19, 7.20
Cameraphoto, Venice: 8.4, 8.5
Marcello Castrichini: 2.29
Caubet and Pierrat-Bonnefois 2005, fig. 292: 5.5
Cavari and Donati 2015: 4.19
Robert Coates-Stevens: 3.9, 6.6
Mario Coen: 12.2
Comune di Ferrara: 11.9
Conze 1912: 5.15
Corsi Collection, Oxford University: 4.30
Courby 1927: 2.20a
Stephen Cox: 13.6
Curtius et al. 1890, vol. 1, pls. 11, 9: 2.9a, 2.9b
Dean Dixon: 2.8
de Garis Davies 1935 (pl. ix): 1.4
Xavier de Jauréguiberry: 5.16
De Nuccio and Ungaro 2002: 4.33, 5.10
Department of Art History and Archaeology, Columbia University: 4.21
Department of Classics, University of Cincinnati: 4.12, 4.13
Dimitrios Dimos: 3.1
Doumas 1992: 4.10
du Buisson 1935: 4.11
Dunbabin 1999: 4.1
Hélène Eristov: 4.25
Eski İstanbul Fotoğrafları Arşivi: 11.1
Evans and Evans 1921: 4.3, 4.6
Fabbrica di S. Pietro: 12.5, 12.20
Factum Arte, Madrid: 13.7a–c
Flickr.com: 6.17, 9.21, 9.22, 11.4, 12.4
Giulio Romano 1989: 11.26
González-Palacios 1984: 12.9
Granger Historical Picture Archive: 6.23
Gruben 1997: 2.5
Hall and Woolley 1927, XXXV.7: 1.13b
Gardiner Hallock: 11.30
Higgins et al. 1968: 2.3a
Image Resource: 8.9
Istock: 6.5, 10.11, 12.40
istockphoto.com: 1.3
J. Paul Getty Museum, Los Angeles. 87.MN.141: 9.7
J. Paul Getty Museum, Malibu: 12.15
Nathaniel Jones: 4.35
Kircher 1650: 12.31a, 12.31b
Kircher 1665: 12.7
Manolis Korres: 3.2b, 3.2c
Didier Laroche: 2.20b
Dario Lasagni: 9.11
Lauer 1936: 1.9
Layard 1849–53, vol. 2, pl. 13: 5.3
Lazzari 1753: 12.8
Brian Lee: 6.16, 10.5

Levi 1942: 3.10
Library, Hanover: 9.6
Lindenau Museum, Altenburg: 9.12a, 9.12b
Machteld J. Mellink Archaeological Site Photography, Bryn Mawr Collections: 2.6
Manar Al-Athar Archive: 5.7, 8.13
Matthew Mason: 2.3b–d
McCullagh Photo Archive, McGill University: 3.18c
Michele Mercati, *Metallotheca Vaticana Michaelis Mercati* (1719): 12.6
Metropolitan Museum of Art, New York: 1.5, 1.6, 3.18a, 5.1, 12.1
John Mitchell: 6.14
Rebecca Molholt: 4.31
National Gallery of Art, Washington, DC: 6.15, 9.24, 10.3
Negro 2002: 12.36
G. Oikonomake-Papadopoulou, B. Pitarakis, and K. Loverdou-Tsigarida, Ἱερά Μεγίστη Μονή Βατοπαιδίου: Ἐγκόλπια (Mount Athos, 2000), 167: 9.15
Giorgio Ortolani: 4.34
Palladio 1570: 11.32
Parrot 1958b: 1.2 (color pl. E), 4.8
Place 1867–70, 3: pl. 37: 1.15
Princeton Art Gallery: 9.13a, 9.13b
Princeton University Art Museum: 3.8
Reinach 1909–12: 3.2a
Egisto Sani: 3.11
Scholten 1929, 1: fig. 93: 7.11
Ben Schultz: 13.5
Serlio 1584: 12.28
Shutterstock: 7.1
Peter J. Sieger: 13.4
Sovrintendenza: 4.18
Sovrintendenza, Brescia: 2.23
Sovrintendenza, Pompei: 4.28, 6.9
Staatliche Kunstsammlungen, Dresden: 9.14
Staatliches Museum Schwerin: 12.26
Tate Modern, London: 7.2
Televantou 1994: 4.9
Trustees of the British Museum: 3.17b
E. N. Tsigaridas, "The Mosaics and the Byzantine Wall-Paintings," in *The Holy and Great Monastery of Vatopaidi: Tradition, History, Art* (Mount Athos, 1998), I, fig. 224: 9.17
University of Pennsylvania Museum of Archaeology and Anthropology, Philadelphia: 1.13a
Alessandro Vasari: 11.21
Michael Waters: 3.18a, 9.1
Web Gallery of Art: 10.4
Westminster Abbey: 6.26
Westminster Cathedral: 13.1
Wikimedia: 1.1, 1.12 (assembled by author), 1.14, 2.7, 2.8, 2.10, 2.13, 2.15, 2.25, 2.27, 3.14, 4.5, 4.7, 4.20, 5.8, 6.4, 6.13, 6.14, 6.21, 6.24, 7.9, 7.17, 8.7, 9.4 (assembled by author), 9.9, 9.10, 9.18–9.23, 9.27–9.31, 9.35, 10.1, 10.2, 10.6, 10.7, 10.10, 10.13, 10.19, 11.4, 11.13, 11.19, 12.16, 12.18, 12.19, 12.25, 12.35, 13.3
Richard Wilson: 2.16
Richard Wittman: 6.25, 12.21, 12.29